AF173453

During the past few decades, the gradual merger of Discrete Geometry and the newer discipline of Computational Geometry has provided enormous impetus to mathematicians and computer scientists interested in geometric problems. This volume, which contains 32 papers on a broad range of topics of current interest in the field, is an outgrowth of that synergism. It includes surveys and research articles exploring geometric arrangements, polytopes, packing, covering, discrete convexity, geometric algorithms and their complexity, and the combinatorial complexity of geometric objects, particularly in low dimension. There are points of contact with many applied areas such as mathematical programming, visibility problems, kinetic data structures, and biochemistry, as well as with algebraic topology, geometric probability, real algebraic geometry, and combinatorics.

Mathematical Sciences Research Institute
Publications

52

Combinatorial and Computational Geometry

Mathematical Sciences Research Institute Publications

1   Freed/Uhlenbeck: *Instantons and Four-Manifolds*, second edition
2   Chern (ed.): *Seminar on Nonlinear Partial Differential Equations*
3   Lepowsky/Mandelstam/Singer (eds.): *Vertex Operators in Mathematics and Physics*
4   Kac (ed.): *Infinite Dimensional Groups with Applications*
5   Blackadar: *K-Theory for Operator Algebras*, second edition
6   Moore (ed.): *Group Representations, Ergodic Theory, Operator Algebras, and Mathematical Physics*
7   Chorin/Majda (eds.): *Wave Motion: Theory, Modelling, and Computation*
8   Gersten (ed.): *Essays in Group Theory*
9   Moore/Schochet: *Global Analysis on Foliated Spaces*
10–11   Drasin/Earle/Gehring/Kra/Marden (eds.): *Holomorphic Functions and Moduli*
12–13   Ni/Peletier/Serrin (eds.): *Nonlinear Diffusion Equations and Their Equilibrium States*
14   Goodman/de la Harpe/Jones: *Coxeter Graphs and Towers of Algebras*
15   Hochster/Huneke/Sally (eds.): *Commutative Algebra*
16   Ihara/Ribet/Serre (eds.): *Galois Groups over* $\mathbb{Q}$
17   Concus/Finn/Hoffman (eds.): *Geometric Analysis and Computer Graphics*
18   Bryant/Chern/Gardner/Goldschmidt/Griffiths: *Exterior Differential Systems*
19   Alperin (ed.): *Arboreal Group Theory*
20   Dazord/Weinstein (eds.): *Symplectic Geometry, Groupoids, and Integrable Systems*
21   Moschovakis (ed.): *Logic from Computer Science*
22   Ratiu (ed.): *The Geometry of Hamiltonian Systems*
23   Baumslag/Miller (eds.): *Algorithms and Classification in Combinatorial Group Theory*
24   Montgomery/Small (eds.): *Noncommutative Rings*
25   Akbulut/King: *Topology of Real Algebraic Sets*
26   Judah/Just/Woodin (eds.): *Set Theory of the Continuum*
27   Carlsson/Cohen/Hsiang/Jones (eds.): *Algebraic Topology and Its Applications*
28   Clemens/Kollár (eds.): *Current Topics in Complex Algebraic Geometry*
29   Nowakowski (ed.): *Games of No Chance*
30   Grove/Petersen (eds.): *Comparison Geometry*
31   Levy (ed.): *Flavors of Geometry*
32   Cecil/Chern (eds.): *Tight and Taut Submanifolds*
33   Axler/McCarthy/Sarason (eds.): *Holomorphic Spaces*
34   Ball/Milman (eds.): *Convex Geometric Analysis*
35   Levy (ed.): *The Eightfold Way*
36   Gavosto/Krantz/McCallum (eds.): *Contemporary Issues in Mathematics Education*
37   Schneider/Siu (eds.): *Several Complex Variables*
38   Billera/Björner/Green/Simion/Stanley (eds.): *New Perspectives in Geometric Combinatorics*
39   Haskell/Pillay/Steinhorn (eds.): *Model Theory, Algebra, and Geometry*
40   Bleher/Its (eds.): *Random Matrix Models and Their Applications*
41   Schneps (ed.): *Galois Groups and Fundamental Groups*
42   Nowakowski (ed.): *More Games of No Chance*
43   Montgomery/Schneider (eds.): *New Directions in Hopf Algebras*
44   Buhler/Stevenhagen (eds.): *Algorithmic Number Theory*
45   Jensen/Ledet/Yui: *Generic Polynomials: Constructive Aspects of the Inverse Galois Problem*
46   Rockmore/Healy (eds.): *Modern Signal Processing*
47   Uhlmann (ed.): *Inside Out: Inverse Problems and Applications*
48   Gross/Kotiuga: *Electromagnetic Theory and Computation: A Topological Approach*
49   Darmon/Zhang (eds.): *Heegner Points and Rankin L-Series*
50   Bao/Bryant/Chern/Shen (eds.): *A Sampler of Riemann–Finsler Geometry*
51   Avramov/Green/Huneke/Smith/Sturmfels (eds.): *Trends in Commutative Algebra*
52   Goodman/Pach/Welzl (eds.): *Combinatorial and Computational Geometry*

Volumes 1–4 and 6–27 are published by Springer-Verlag

Combinatorial and Computational Geometry

Edited by

Jacob E. Goodman
City College, CUNY

János Pach
City College, CUNY and
Courant Institute, NYU

Emo Welzl
ETH Zürich

CAMBRIDGE
UNIVERSITY PRESS

Jacob E. Goodman: Department of Mathematics, City College, CUNY,
New York, NY 10031   goodman@sci.ccny.cuny.edu

János Pach: Courant Institute, NYU, 251 Mercer Street, New York, NY 10012
and   City College, CUNY, New York, NY 10031   pach@courant.nyu.edu

Emo Welzl: Informatik, Eidgenössische Technische Hochschule,
Rämistrasse 101, CH-8092 Zürich, Switzerland   emo@inf.ethz.ch

Silvio Levy (*Series Editor*):   Mathematical Sciences Research Institute,
17 Gauss Way, Berkeley, CA 94720, United States   levy@msri.org

The Mathematical Sciences Research Institute wishes to acknowledge support by
the National Science Foundation. This material is based upon work supported by
NSF Grant 9810361.

CAMBRIDGE UNIVERSITY PRESS
Cambridge, New York, Melbourne, Madrid, Cape Town,
Singapore, São Paulo, Delhi, Tokyo, Mexico City

Cambridge University Press
The Edinburgh Building, Cambridge CB2 8RU, UK

Published in the United States of America by Cambridge University Press, New York

www.cambridge.org
Information on this title: www.cambridge.org/9780521178396

© Mathematical Sciences Research Institute 2005

This publication is in copyright. Subject to statutory exception
and to the provisions of relevant collective licensing agreements,
no reproduction of any part may take place without the written
permission of Cambridge University Press.

First published 2005
First paperback edition 2011

A catalogue record for this publication is available from the British Library

Library of Congress Cataloguing in Publication data

Combinatorial and computational geometry / edited by Jacob E. Goodman, János
Pach, Emo Welzl.
    p.   cm. – (Mathematical Sciences Research Institute publications ; 52)
Includes bibliographical references and index.
ISBN 0-521-84862-8 (hb)
    1. Discrete geometry. 2. Combinatorial geometry. 3. Geometry–Data processing.
I. Goodman, Jacob E. II. Pach, János. III. Welzl, Emo. IV. Series.

QA640.7.D54   2005
516′.11–dc22

                                                           2005042199

ISBN 978-0-521-84862-6 Hardback
ISBN 978-0-521-17839-6 Paperback

Cambridge University Press has no responsibility for the persistence or
accuracy of URLs for external or third-party internet websites referred to in
this publication, and does not guarantee that any content on such websites is,
or will remain, accurate or appropriate.

Combinatorial and Computational Geometry
MSRI Publications
Volume **52**, 2005

Contents

Preface   xi

Geometric Approximation via Coresets   1
    P. K. AGARWAL, S. HAR-PELED AND K. R. VARADARAJAN

Applications of Graph and Hypergraph Theory in Geometry   31
    IMRE BÁRÁNY

Convex Geometry of Orbits   51
    ALEXANDER BARVINOK AND GRIGORIY BLEKHERMAN

The Hadwiger Transversal Theorem for Pseudolines   79
    SAUGATA BASU, JACOB E. GOODMAN, ANDREAS HOLMSEN, AND
    RICHARD POLLACK

Betti Number Bounds, Applications and Algorithms   87
    SAUGATA BASU, RICHARD POLLACK, AND MARIE-FRANÇOISE ROY

Shelling and the h-Vector of the (Extra)ordinary Polytope   97
    MARGARET M. BAYER

On the Number of Mutually Touching Cylinders   121
    ANDRÁS BEZDEK

Edge-Antipodal 3-Polytopes   129
    KÁROLY BEZDEK, TIBOR BISZTRICZKY, AND KÁROLY BÖRÖCZKY

A Conformal Energy for Simplicial Surfaces   135
    ALEXANDER BOBENKO

On the Size of Higher-Dimensional Triangulations   147
    PETER BRASS

The Carpenter's Ruler Folding Problem   155
    GRUIA CĂLINESCU AND ADRIAN DUMITRESCU

A Survey of Folding and Unfolding in Computational Geometry   167
    ERIK D. DEMAINE AND JOSEPH O'ROURKE

On the Rank of a Tropical Matrix 213
  MIKE DEVELIN, FRANCISCO SANTOS, AND BERND STURMFELS

The Geometry of Biomolecular Solvation 243
  HERBERT EDELSBRUNNER AND PATRICE KOEHL

Inequalities for Zonotopes 277
  RICHARD EHRENBORG

Quasiconvex Programming 287
  DAVID EPPSTEIN

De Concini–Procesi Wonderful Arrangement Models: A Discrete
Geometer's Point of View 333
  EVA MARIA FEICHTNER

Thinnest Covering of a Circle by Eight, Nine, or Ten Congruent Circles 361
  GÁBOR FEJES TÓTH

On the Complexity of Visibility Problems with Moving Viewpoints 377
  PETER GRITZMANN AND THORSTEN THEOBALD

Cylindrical Partitions of Convex Bodies 399
  ALADÁR HEPPES AND WŁODZIMIERZ KUPERBERG

Tropical Halfspaces 409
  MICHAEL JOSWIG

Two Proofs for Sylvester's Problem Using an Allowable Sequence of
Permutations 433
  HAGIT LAST

A Comparison of Five Implementations of 3D Delaunay Tessellation 439
  YUANXIN LIU AND JACK SNOEYINK

The Bernstein Basis and Real Root Isolation 459
  BERNARD MOURRAIN, FABRICE ROUILLIER, AND
  MARIE-FRANÇOISE ROY

Extremal Problems Related to the Sylvester–Gallai Theorem 479
  NIRANJAN NILAKANTAN

A Long Noncrossing Path Among Disjoint Segments in the Plane 495
  JÁNOS PACH AND ROM PINCHASI

On a Generalization of Schönhardt's Polyhedron 501
  JÖRG RAMBAU

On Hadwiger Numbers of Direct Products of Convex Bodies 517
  ISTVÁN TALATA

Binary Space Partitions: Recent Developments   529
    CSABA D. TOTH

The Erdős–Szekeres Theorem: Upper Bounds and Related Results   557
    GÉZA TÓTH AND PAVEL VALTR

On the Pair-Crossing Number   569
    PAVEL VALTR

Geometric Random Walks: A Survey   577
    SANTOSH VEMPALA

Combinatorial and Computational Geometry
MSRI Publications
Volume **52**, 2005

Preface

> The Great Bear is looking so geometrical,
> One would think that something or other could be proved.
> — Christopher Fry, "The Lady's Not for Burning"

During the past several decades, the gradual merger of the field of discrete geometry and the newer discipline of computational geometry has provided a significant impetus to mathematicians and computer scientists interested in geometric problems. The resulting field of discrete and computational geometry has now grown to the point where not even a semester program, such as the one held at the Mathematical Sciences Research Institute in the fall of 2003, with its three workshops and nearly 200 participants, could include everyone involved in making important contributions to the area. The same holds true for the present volume, which presents just a sampling of the work generated during the MSRI program; we have tried to assemble a sample that is representative of the program.

The volume includes 32 papers on topics ranging from polytopes to complexity questions on geometric arrangements, from geometric algorithms to packing and covering, from visibility problems to geometric graph theory. There are points of contact with both mathematical and applied areas such as algebraic topology, geometric probability, algebraic geometry, combinatorics, differential geometry, mathematical programming, data structures, and biochemistry.

We hope the articles in this volume—surveys as well as research papers—will serve to give the interested reader a glimpse of the current state of discrete, combinatorial and computational geometry as we stand poised at the beginning of a new century.

Jacob E. Goodman
János Pach
Emo Welzl

Combinatorial and Computational Geometry
MSRI Publications
Volume **52**, 2005

Geometric Approximation via Coresets

PANKAJ K. AGARWAL, SARIEL HAR-PELED, AND KASTURI R. VARADARAJAN

ABSTRACT. The paradigm of coresets has recently emerged as a powerful tool for efficiently approximating various extent measures of a point set P. Using this paradigm, one quickly computes a small subset Q of P, called a *coreset*, that approximates the original set P and and then solves the problem on Q using a relatively inefficient algorithm. The solution for Q is then translated to an approximate solution to the original point set P. This paper describes the ways in which this paradigm has been successfully applied to various optimization and extent measure problems.

1. Introduction

One of the classical techniques in developing approximation algorithms is the extraction of "small" amount of "most relevant" information from the given data, and performing the computation on this extracted data. Examples of the use of this technique in a geometric context include random sampling [Chazelle 2000; Mulmuley 1993], convex approximation [Dudley 1974; Bronshteyn and Ivanov 1976], surface simplification [Heckbert and Garland 1997], feature extraction and shape descriptors [Dryden and Mardia 1998; Costa and César 2001]. For geometric problems where the input is a set of points, the question reduces to finding a small subset (a *coreset*) of the points, such that one can perform the desired computation on the coreset.

As a concrete example, consider the problem of computing the diameter of a point set. Here it is clear that, in the worst case, classical sampling techniques like ε-approximation and ε-net would fail to compute a subset of points that contain a good approximation to the diameter [Vapnik and Chervonenkis 1971; Haussler and Welzl 1987]. While in this problem it is clear that convex approximation

Research by the first author is supported by NSF under grants CCR-00-86013, EIA-98-70724, EIA-01-31905, and CCR-02-04118, and by a grant from the U.S.-Israel Binational Science Foundation. Research by the second author is supported by NSF CAREER award CCR-0132901. Research by the third author is supported by NSF CAREER award CCR-0237431.

(i.e., an approximation of the convex hull of the point set) is helpful and provides us with the desired coreset, convex approximation of the point set is not useful for computing the narrowest annulus containing a point set in the plane.

In this paper, we describe several recent results which employ the idea of coresets to develop efficient approximation algorithms for various geometric problems. In particular, motivated by a variety of applications, considerable work has been done on measuring various descriptors of the extent of a set P of n points in $\mathbb{R}^d$. We refer to such measures as *extent measures* of P. Roughly speaking, an extent measure of P either computes certain statistics of P itself or of a (possibly nonconvex) geometric shape (e.g. sphere, box, cylinder, etc.) enclosing P. Examples of the former include computing the k-th largest distance between pairs of points in P, and the examples of the latter include computing the smallest radius of a sphere (or cylinder), the minimum volume (or surface area) of a box, and the smallest width of a slab (or a spherical or cylindrical shell) that contain P. There has also been some recent work on maintaining extent measures of a set of moving points [Agarwal et al. 2001b].

Shape fitting, a fundamental problem in computational geometry, computer vision, machine learning, data mining, and many other areas, is closely related to computing extent measures. The shape fitting problem asks for finding a shape that best fits P under some "fitting" criterion. A typical criterion for measuring how well a shape γ fits P, denoted as $\mu(P, \gamma)$, is the maximum distance between a point of P and its nearest point on γ, i.e., $\mu(P, \gamma) = \max_{p \in P} \min_{q \in \gamma} \|p - q\|$. Then one can define the extent measure of P to be $\mu(P) = \min_\gamma \mu(P, \gamma)$, where the minimum is taken over a family of shapes (such as points, lines, hyperplanes, spheres, etc.). For example, the problem of finding the minimum radius sphere (resp. cylinder) enclosing P is the same as finding the point (resp. line) that fits P best, and the problem of finding the smallest width slab (resp. spherical shell, cylindrical shell)[1] is the same as finding the hyperplane (resp. sphere, cylinder) that fits P best.

The exact algorithms for computing extent measures are generally expensive, e.g., the best known algorithms for computing the smallest volume bounding box containing P in $\mathbb{R}^3$ run in $O(n^3)$ time. Consequently, attention has shifted to developing approximation algorithms [Barequet and Har-Peled 2001]. The goal is to compute an $(1+\varepsilon)$-approximation, for some $0 < \varepsilon < 1$, of the extent measure in roughly $O(nf(\varepsilon))$ or even $O(n+f(\varepsilon))$ time, that is, in time near-linear or linear in n. The framework of coresets has recently emerged as a general approach to achieve this goal. For any extent measure μ and an input point set P for which we wish to compute the extent measure, the general idea is to argue that there exists an easily computable subset $Q \subseteq P$, called a *coreset*, of size $1/\varepsilon^{O(1)}$, so

---

[1] A *slab* is a region lying between two parallel hyperplanes; a *spherical shell* is the region lying between two concentric spheres; a *cylindrical shell* is the region lying between two coaxial cylinders.

that solving the underlying problem on Q gives an approximate solution to the original problem. For example, if $\mu(Q) \geq (1 - \varepsilon)\mu(P)$, then this approach gives an approximation to the extent measure of P. In the context of shape fitting, an appropriate property for Q is that for any shape γ from the underlying family, $\mu(Q, \gamma) \geq (1 - \varepsilon)\mu(P, \gamma)$. With this property, the approach returns a shape γ^* that is an approximate best fit to P.

Following earlier work [Barequet and Har-Peled 2001; Chan 2002; Zhou and Suri 2002] that hinted at the generality of this approach, [Agarwal et al. 2004] provided a formal framework by introducing the notion of ε-kernel and showing that it yields a coreset for many optimization problems. They also showed that this technique yields approximation algorithms for a wide range of problems. Since the appearance of preliminary versions of their work, many subsequent papers have used a coreset based approach for other geometric optimization problems, including clustering and other extent-measure problems [Agarwal et al. 2002; Bădoiu and Clarkson 2003b; Bădoiu et al. 2002; Har-Peled and Wang 2004; Kumar et al. 2003; Kumar and Yildirim ≥ 2005].

In this paper, we have attempted to review coreset based algorithms for approximating extent measure and other optimization problems. Our aim is to communicate the flavor of the techniques involved and a sense of the power of this paradigm by discussing a number of its applications. We begin in Section 2 by describing ε-kernels of point sets and algorithms for constructing them. Section 3 defines the notion of ε-kernel for functions and describes a few of its applications. We then describe in Section 4 a simple incremental algorithm for shape fitting. Section 5 discusses the computation of ε-kernels in the streaming model. Although ε-kernels provide coresets for a variety of extent measures, they do not give coresets for many other problems, including clustering. Section 6 surveys the known results on coresets for clustering. The size of the coresets discussed in these sections increases exponentially with the dimension, so we conclude in Section 7 by discussing coresets for points in very high dimensions whose size depends polynomially on the dimension, or is independent of the dimension altogether.

2. Kernels for Point Sets

Let μ be a measure function (e.g., the width of a point set) from subsets of $\mathbb{R}^d$ to the nonnegative reals $\mathbb{R}^+ \cup \{0\}$ that is monotone, i.e., for $P_1 \subseteq P_2$, $\mu(P_1) \leq \mu(P_2)$. Given a parameter $\varepsilon > 0$, we call a subset $Q \subseteq P$ an ε-coreset of P (with respect to μ) if

$$(1 - \varepsilon)\mu(P) \leq \mu(Q).$$

Agarwal et al. [2004] introduced the notion of ε-kernels and showed that it is an $f(\varepsilon)$-coreset for numerous minimization problems. We begin by defining ε-kernels and related concepts.

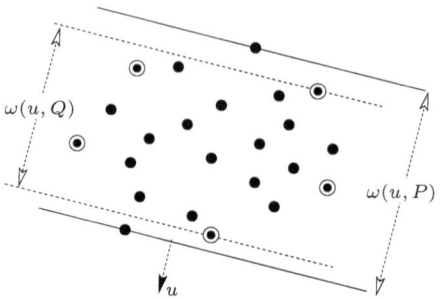

Figure 1. Directional width and ε-kernel.

ε-**kernel.** Let $\mathbb{S}^{d-1}$ denote the unit sphere centered at the origin in $\mathbb{R}^d$. For any set P of points in $\mathbb{R}^d$ and any direction $u \in \mathbb{S}^{d-1}$, we define the *directional width* of P in direction u, denoted by $\omega(u, P)$, to be

$$\omega(u, P) = \max_{p \in P} \langle u, p \rangle - \min_{p \in P} \langle u, p \rangle,$$

where $\langle \cdot, \cdot \rangle$ is the standard inner product. Let $\varepsilon > 0$ be a parameter. A subset $Q \subseteq P$ is called an ε-*kernel* of P if for each $u \in \mathbb{S}^{d-1}$,

$$(1 - \varepsilon)\omega(u, P) \leq \omega(u, Q).$$

Clearly, $\omega(u, Q) \leq \omega(u, P)$. Agarwal et al. [2004] call a measure function μ *faithful* if there exists a constant c, depending on μ, so that for any $P \subseteq \mathbb{R}^d$ and for any ε, an ε-kernel of P is a $c\varepsilon$-coreset for P with respect to μ. Examples of faithful measures considered in that reference include diameter, width, radius of the smallest enclosing ball, and volume of the smallest enclosing box. A common property of these measures is that $\mu(P) = \mu(\mathrm{conv}(P))$. We can thus compute an ε-coreset of P with respect to several measures by simply computing an (ε/c)-kernel of P.

Algorithms for computing kernels. An ε-kernel of P is a subset whose convex hull approximates, in a certain sense, the convex hull of P. Other notions of convex hull approximation have been studied and methods have been developed to compute them; see [Bentley et al. 1982; Bronshteyn and Ivanov 1976; Dudley 1974] for a sample. For example, in the first of these articles Bentley, Faust, and Preparata show that for any point set $P \subseteq \mathbb{R}^2$ and $\varepsilon > 0$, a subset Q of P whose size is $O(1/\varepsilon)$ can be computed in $O(|P| + 1/\varepsilon)$ time such that for any $p \in P$, the distance of p to $\mathrm{conv}(Q)$ is at most $\varepsilon\mathrm{diam}(Q)$. Note however that such a guarantee is not enough if we want Q to be a coreset of P with respect to faithful measures. For instance, the width of Q could be arbitrarily small compared to the width of P. The width of an ε-kernel of P, on the other hand, is easily seen to be a good approximation to the width of P. To the best of our knowledge, the first efficient method for computing a small ε-kernel of an arbitrary point set is implicit in [Barequet and Har-Peled 2001].

We call P α-fat, for $\alpha \leq 1$, if there exists a point $p \in \mathbb{R}^d$ and a hypercube $\overline{\mathbb{C}}$ centered at the origin so that

$$p + \alpha\overline{\mathbb{C}} \subset \text{conv}(P) \subset p + \overline{\mathbb{C}}.$$

A stronger version of the following lemma, which is very useful for constructing an ε-kernel, was proved in [Agarwal et al. 2004] by adapting a scheme from [Barequet and Har-Peled 2001]. Their scheme can be thought of as one that quickly computes an approximation to the Löwner–John Ellipsoid [John 1948].

LEMMA 2.1. *Let P be a set of n points in $\mathbb{R}^d$ such that the volume of $\text{conv}(P)$ is nonzero, and let $\mathbb{C} = [-1,1]^d$. One can compute in $O(n)$ time an affine transform τ so that $\tau(P)$ is an α-fat point set satisfying $\alpha\mathbb{C} \subset \text{conv}(\tau(P)) \subset \mathbb{C}$, where α is a positive constant depending on d, and so that a subset $Q \subseteq P$ is an ε-kernel of P if and only if $\tau(Q)$ is an ε-kernel of $\tau(P)$.*

The importance of Lemma 2.1 is that it allows us to adapt some classical approaches for convex hull approximation [Bentley et al. 1982; Bronshteyn and Ivanov 1976; Dudley 1974] which in fact do compute an ε-kernel when applied to fat point sets.

We now describe algorithms for computing ε-kernels. By Lemma 2.1, we can assume that $P \subseteq [-1,+1]^d$ is α-fat. We begin with a very simple algorithm.

Let δ be the largest value such that $\delta \leq (\varepsilon/\sqrt{d})\alpha$ and $1/\delta$ is an integer. We consider the d-dimensional grid $\mathbb{Z}$ of size δ. That is,

$$\mathbb{Z} = \{(\delta i_1, \ldots, \delta i_d) \mid i_1, \ldots, i_d \in \mathbb{Z}\}.$$

For each column along the x_d-axis in $\mathbb{Z}$, we choose one point from the highest nonempty cell of the column and one point from the lowest nonempty cell of the column; see Figure 2, top left. Let Q be the set of chosen points. Since $P \subseteq [-1,+1]^d$, $|Q| = O(1/(\alpha\varepsilon)^{d-1})$. Moreover Q can be constructed in time $O(n + 1/(\alpha\varepsilon)^{d-1})$ provided that the ceiling operation can be performed in constant time. Agarwal et al. [2004] showed that Q is an ε-kernel of P. Hence, we can compute an ε-kernel of P of size $O(1/\varepsilon^{d-1})$ in time $O(n+1/\varepsilon^{d-1})$. This approach resembles the algorithm of [Bentley et al. 1982].

Next we describe an improved construction, observed independently in [Chan 2004] and [Yu et al. 2004], which is a simplification of an algorithm of [Agarwal et al. 2004], which in turn is an adaptation of a method of Dudley [1974]. Let $\mathcal{S}$ be the sphere of radius $\sqrt{d}+1$ centered at the origin. Set $\delta = \sqrt{\varepsilon\alpha} \leq 1/2$. One can construct a set $\mathcal{I}$ of $O(1/\delta^{d-1}) = O(1/\varepsilon^{(d-1)/2})$ points on the sphere $\mathcal{S}$ so that for any point x on $\mathcal{S}$, there exists a point $y \in \mathcal{I}$ such that $\|x - y\| \leq \delta$. We process P into a data structure that can answer ε-approximate nearest-neighbor queries [Arya et al. 1998]. For a query point q, let $\varphi(q)$ be the point of P returned by this data structure. For each point $y \in \mathcal{I}$, we compute $\varphi(y)$ using this data structure. We return the set $Q = \{\varphi(y) \mid y \in \mathcal{I}\}$; see Figure 2, top right.

We now briefly sketch, following the argument in [Yu et al. 2004], why Q is is an ε-kernel of P. For simplicity, we prove the claim under the assumption that $\varphi(y)$ is the *exact* nearest-neighbor of y in P. Fix a direction $u \in \mathbb{S}^{d-1}$. Let $\sigma \in P$ be the point that maximizes $\langle u, p \rangle$ over all $p \in P$. Suppose the ray emanating from σ in direction u hits $\mathcal{S}$ at a point x. We know that there exists a point $y \in \mathcal{I}$ such that $\|x - y\| \leq \delta$. If $\varphi(y) = \sigma$, then $\sigma \in Q$ and

$$\max_{p \in P} \langle u, p \rangle - \max_{q \in Q} \langle u, q \rangle = 0.$$

Now suppose $\varphi(y) \neq \sigma$. Let B be the d-dimensional ball of radius $\|y - \sigma\|$ centered at y. Since $\|y - \varphi(y)\| \leq \|y - \sigma\|$, $\varphi(y) \in B$. Let us denote by z the point on the sphere ∂B that is hit by the ray emanating from y in direction $-u$. Let w be the point on zy such that $zy \perp \sigma w$ and h the point on σx such that $yh \perp \sigma x$; see Figure 2, bottom.

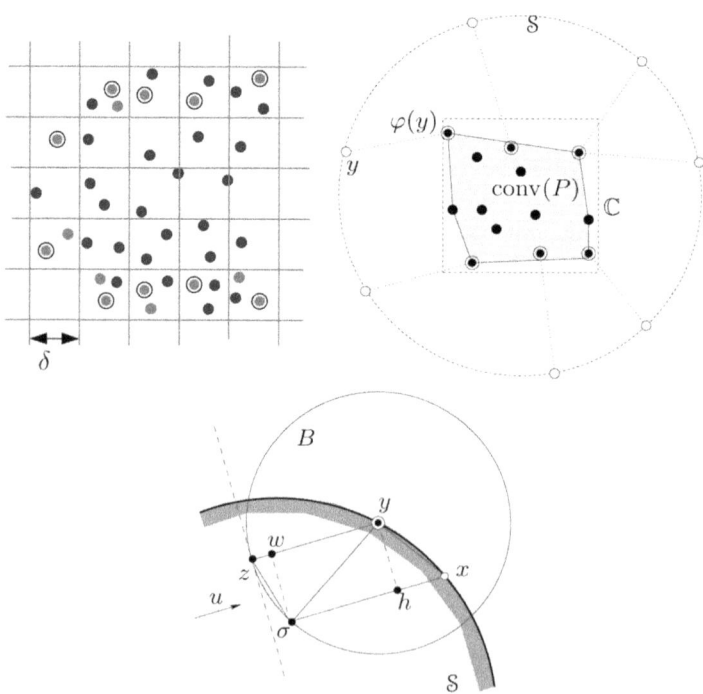

Figure 2. Top left: A grid based algorithm for constructing an ε-kernel. Top right: An improved algorithm. Bottom: Correctness of the improved algorithm.

The hyperplane normal to u and passing through z is tangent to B. Since $\varphi(y)$ lies inside B, $\langle u, \varphi(y) \rangle \geq \langle u, z \rangle$. Moreover, it can be shown that $\langle u, \sigma \rangle - \langle u, \varphi(y) \rangle \leq \alpha \varepsilon$. Thus, we can write

$$\max_{p \in P} \langle u, p \rangle - \max_{q \in Q} \langle u, q \rangle \leq \langle u, \sigma \rangle - \langle u, \varphi(y) \rangle \leq \alpha \varepsilon.$$

Similarly, we have $\min_{p \in P} \langle u, p \rangle - \min_{q \in Q} \langle u, q \rangle \geq -\alpha\varepsilon$.

The above two inequalities together imply that $w(u, Q) \geq w(u, P) - 2\alpha\varepsilon$. Since $\alpha\mathbb{C} \subset \mathrm{conv}(P)$, $w(u, P) \geq 2\alpha$. Hence $w(u, Q) \geq (1 - \varepsilon)w(u, P)$, for any $u \in \mathbb{S}^{d-1}$, thereby implying that Q is an ε-kernel of P.

A straightforward implementation of the above algorithm, i.e., the one that answers a nearest-neighbor query by comparing the distances to all the points, runs in $O(n/\varepsilon^{(d-1)/2})$ time. However, we can first compute an $(\varepsilon/2)$-kernel Q' of P of size $O(1/\varepsilon^{d-1})$ using the simple algorithm and then compute an $(\varepsilon/4)$-kernel using the improved algorithm. Chan [2004] introduced the notion of discrete Voronoi diagrams, which can be used for computing the nearest neighbors of a set of grid points among the sites that are also a subset of a grid. Using this structure Chan showed that $\varphi(y)$, for all $y \in \mathfrak{I}$, can be computed in a total time of $O(n + 1/\varepsilon^{d-1})$ time. Putting everything together, one obtains an algorithm that runs in $O(n + 1/\varepsilon^{d-1})$ time. Chan in fact gives a slightly improved result:

THEOREM 2.2 [Chan 2004]. *Given a set P of n points in $\mathbb{R}^d$ and a parameter $\varepsilon > 0$, one can compute an ε-kernel of P of size $O(1/\varepsilon^{(d-1)/2})$ in time $O(n + 1/\varepsilon^{d-(3/2)})$.*

Experimental results. Yu et al. [2004] implemented their ε-kernel algorithm and tested its performance on a variety of inputs. They measure the quality of an ε-kernel Q of P as the maximum relative error in the directional width of P and Q. Since it is hard to compute the maximum error over all directions, they sampled a set Δ of 1000 directions in $\mathbb{S}^{d-1}$ and computed the maximum relative error with respect to these directions, i.e.,

$$\mathrm{err}(Q, P) = \max_{u \in \Delta} \frac{w(u, P) - w(u, Q)}{w(u, P)}. \qquad (2\text{-}1)$$

They implemented the constant-factor approximation algorithm of [Barequet and Har-Peled 2001] for computing the minimum-volume bounding box to convert P into an α-fat set, and they used the ANN library [Arya and Mount 1998] for answering approximate nearest-neighbor queries. Table 1 shows the running time of their algorithm for a variety of synthetic inputs: (i) points uniformly distributed on a sphere, (ii) points distributed on a cylinder, and (iii) clustered point sets, consisting of 20 equal sized clusters. The running time is decomposed into two components: (i) preprocessing time that includes the time spent in converting P into a fat set and in preprocessing P for approximate nearest-neighbor queries, and (ii) query time that includes the time spent in computing $\varphi(x)$ for $x \in \mathfrak{I}$. Figure 3 shows how the error $\mathrm{err}(Q, P)$ changes as the function of kernel. These experiments show that their algorithm works extremely well in low dimensions (≤ 4) both in terms of size and running time. See [Yu et al. 2004] for more detailed experiments.

| Input | Input | $d = 2$ | | $d = 4$ | | $d = 6$ | | $d = 8$ | |
Type	Size	Pre	Que	Pre	Que	Pre	Que	Pre	Que
	10^4	0.03	0.01	0.06	0.05	0.10	9.40	0.15	52.80
sphere	10^5	0.54	0.01	0.90	0.50	1.38	67.22	1.97	1393.88
	10^6	9.25	0.01	13.08	1.35	19.26	227.20	26.77	5944.89
	10^4	0.03	0.01	0.06	0.03	0.10	2.46	0.16	17.29
cylinder	10^5	0.60	0.01	0.91	0.34	1.39	30.03	1.94	1383.27
	10^6	9.93	0.01	13.09	0.31	18.94	87.29	26.12	5221.13
	10^4	0.03	0.01	0.06	0.01	0.10	0.08	0.15	2.99
clustered	10^5	0.31	0.01	0.63	0.02	1.07	1.34	1.64	18.39
	10^6	5.41	0.01	8.76	0.02	14.75	1.08	22.51	54.12

Table 1. Running time for computing ε-kernels of various synthetic data sets, $\varepsilon < 0.05$. *Prepr* denotes the preprocessing time, including converting P into a fat set and building ANN data structures. *Query* denotes the time for performing approximate nearest-neighbor queries. Running time is measured in seconds. The experiments were conducted on a Dell PowerEdge 650 server with a 3.06GHz Pentium IV processor and 3GB memory, running Linux 2.4.20.

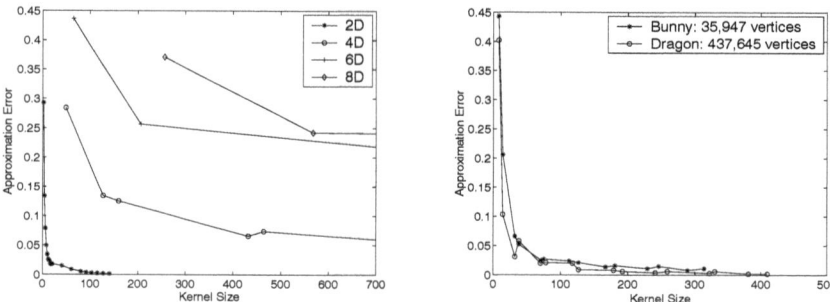

Figure 3. Approximation errors under different sizes of computed ε-kernels. Left: *sphere*. Right: various geometric models. All synthetic inputs had 100,000 points.

Applications. Theorem 2.2 can be used to compute coresets for faithful measures, defined in Section 2. In particular, if we have a faithful measure μ that can be computed in $O(n^\alpha)$ time, then by Theorem 2.2, we can compute a value $\overline{\mu}$, $(1-\varepsilon)\mu(P) \leq \overline{\mu} \leq \mu(P)$ by first computing an (ε/c)-kernel Q of P and then using an exact algorithm for computing $\mu(Q)$. The total running time of the algorithm is $O(n + 1/\varepsilon^{d-(3/2)} + 1/\varepsilon^{\alpha(d-1)/2})$. For example, a $(1 + \varepsilon)$-approximation of the diameter of a point set can be computed in time $O(n + 1/\varepsilon^{d-1})$ since the exact diameter can be computed in quadratic time. By being a little more careful, the running time of the diameter algorithm can be improved to $O(n + 1/\varepsilon^{d-(3/2)})$ [Chan 2004]. Table 2 gives running times for computing an $(1+\varepsilon)$-approximation of a few faithful measures.

We note that ε-kernels in fact guarantee a stronger property for several faithful measures. For instance, if Q is an ε-kernel of P, and C is some cylinder containing

Extent	Time complexity
Diameter	$n + 1/\varepsilon^{d-(3/2)}$
Width	$(n + 1/\varepsilon^{d-2}) \log(1/\varepsilon)$
Minimum enclosing cylinder	$n + 1/\varepsilon^{d-1}$
Minimum enclosing box(3D)	$n + 1/\varepsilon^3$

Table 2. Time complexity of computing $(1 + \varepsilon)$-approximations for certain faithful measures.

Q, then a "concentric" scaling of C by a factor of $(1 + c\varepsilon)$, for some constant c, contains P. Thus we can compute not only an approximation to the minimum radius r^* of a cylinder containing P, but also a cylinder of radius at most $(1+\varepsilon)r^*$ that contains P.

The approach described in this section for approximating faithful measures had been used for geometric approximation algorithms before the framework of ε-kernels was introduced; see [Agarwal and Procopiuc 2002; Barequet and Har-Peled 2001; Chan 2002; Zhou and Suri 2002], for example. The framework of ε-kernels, however, provides a unified approach and turns out to be crucial for the approach developed in the next section for approximating measures that are not faithful.

3. Kernels for Sets of Functions

The crucial notion used to derive coresets and efficient approximation algorithms for measures that are not faithful is that of a kernel of a set of functions.

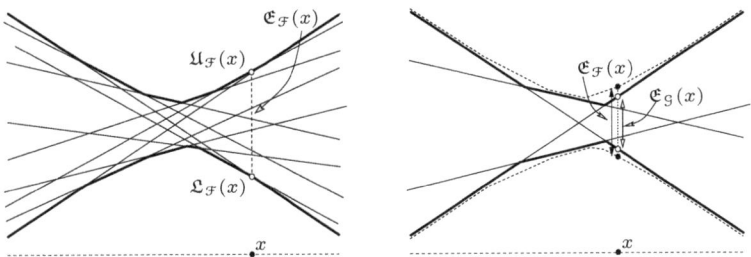

Figure 4. Envelopes, extent, and ε-kernel.

Envelopes and extent. Let $\mathcal{F} = \{f_1, \ldots, f_n\}$ be a set of n d-variate real-valued functions defined over $x = (x_1, \ldots, x_{d-1}, x_d) \in \mathbb{R}^d$. The *lower envelope* of $\mathcal{F}$ is the graph of the function $\mathcal{L}_{\mathcal{F}} : \mathbb{R}^d \to \mathbb{R}$ defined as $\mathcal{L}_{\mathcal{F}}(x) = \min_{f \in \mathcal{F}} f(x)$. Similarly, the *upper envelope* of $\mathcal{F}$ is the graph of the function $\mathfrak{U}_{\mathcal{F}} : \mathbb{R}^d \to \mathbb{R}$ defined as $\mathfrak{U}_{\mathcal{F}}(x) = \max_{f \in \mathcal{F}} f(x)$. The *extent* $\mathfrak{E}_{\mathcal{F}} : \mathbb{R}^d \to \mathbb{R}$ of $\mathcal{F}$ is defined as

$$\mathfrak{E}_{\mathcal{F}}(x) = \mathfrak{U}_{\mathcal{F}}(x) - \mathcal{L}_{\mathcal{F}}(x).$$

Let $\varepsilon > 0$ be a parameter. We say that a subset $\mathcal{G} \subseteq \mathcal{F}$ is an ε-*kernel* of $\mathcal{F}$ if

$$(1 - \varepsilon)\mathfrak{E}_{\mathcal{F}}(x) \le \mathfrak{E}_{\mathcal{G}}(x) \qquad \forall x \in \mathbb{R}^d.$$

Obviously, $\mathfrak{E}_{\mathcal{G}}(x) \le \mathfrak{E}_{\mathcal{F}}(x)$, as $\mathcal{G} \subseteq \mathcal{F}$.

Let $\mathcal{H} = \{h_1, \ldots, h_n\}$ be a family of d-variate linear functions and $\varepsilon > 0$ a parameter. We define a *duality* transformation that maps the d-variate function (or a hyperplane in $\mathbb{R}^{d+1}$) $h : x_{d+1} = a_1 x_1 + a_2 x_2 + \cdots + a_d x_d + a_{d+1}$ to the point $h^{\star} = (a_1, a_2, \ldots, a_d, a_{d+1})$ in $\mathbb{R}^{d+1}$. Let $\mathcal{H}^{\star} = \{h^{\star} \mid h \in \mathcal{H}\}$. It can be proved [Agarwal et al. 2004] that $\mathcal{K} \subseteq \mathcal{H}$ is an ε-kernel of $\mathcal{H}$ if and only if $\mathcal{K}^*$ is an ε-kernel of $\mathcal{H}^*$. Hence, by computing an ε-kernel of $\mathcal{H}^*$ we can also compute an ε-kernel of $\mathcal{H}$. The following is therefore a corollary of Theorem 2.2.

COROLLARY 3.1 [Agarwal et al. 2004; Chan 2004]. *Given a set $\mathcal{F}$ of n d-variate linear functions and a parameter $\varepsilon > 0$, one can compute an ε-kernel of $\mathcal{F}$ of size $O(1/\varepsilon^{d/2})$ in time $O(n + 1/\varepsilon^{d-(1/2)})$.*

We can compute ε-kernels of a set of polynomial functions by using the notion of linearization.

Linearization. Let $f(x, a)$ be a $(d+p)$-variate polynomial, $x \in \mathbb{R}^d$ and $a \in \mathbb{R}^p$. Let $a^1, \ldots, a^n \in \mathbb{R}^p$, and set $\mathcal{F} = \{f_i(x) \equiv f(x, a^i) \mid 1 \le i \le n\}$. Suppose we can express $f(x, a)$ in the form

$$f(x, a) = \psi_0(a) + \psi_1(a)\varphi_1(x) + \cdots + \psi_k(a)\varphi_k(x), \qquad (3\text{-}1)$$

where $\psi_0, \ldots, \psi_k$ are p-variate polynomials and $\varphi_1, \ldots, \varphi_k$ are d-variate polynomials. We define the map $\varphi : \mathbb{R}^d \to \mathbb{R}^k$

$$\varphi(x) = (\varphi_1(x), \ldots, \varphi_k(x)).$$

Then the image $\Gamma = \{\varphi(x) \mid x \in \mathbb{R}^d\}$ of $\mathbb{R}^d$ is a d-dimensional surface in $\mathbb{R}^k$ (if $k \ge d$), and for any $a \in \mathbb{R}^p$, $f(x, a)$ maps to a k-variate linear function

$$h_a(y_1, \ldots, y_k) = \psi_0(a) + \psi_1(a)y_1 + \cdots + \psi_k(a)y_k$$

in the sense that for any $x \in \mathbb{R}^d$, $f(x, a) = h_a(\varphi(x))$. We refer to k as the *dimension* of the *linearization* φ, and say that $\mathcal{F}$ admits a linearization of dimension k. The most popular example of linearization is perhaps the so-called lifting transform that maps $\mathbb{R}^d$ to a unit paraboloid in $\mathbb{R}^{d+1}$. For example, let $f(x_1, x_2, a_1, a_2, a_3)$ be the function whose absolute value is some measure of the "distance" between a point $(x_1, x_2) \in \mathbb{R}^2$ and a circle with center (a_1, a_2) and radius a_3, which is the 5-variate polynomial

$$f(x_1, x_2, a_1, a_2, a_3) = a_3^2 - (x_1 - a_1)^2 - (x_2 - a_2)^2.$$

We can rewrite f in the form

$$f(x_1, x_2, a_1, a_2, a_3) = [a_3^2 - a_1^2 - a_2^2] + [2a_1 x_1] + [2a_2 x_2] - [x_1^2 + x_2^2], \qquad (3\text{-}2)$$

thus, setting

$$\psi_0(a) = a_3^2 - a_1^2 - a_2^2, \psi_1(a) = 2a_1, \psi_2(a) = 2a_2, \psi_3(a) = -1,$$
$$\varphi_1(x) = x_1, \; \varphi_2(x) = x_2, \; \varphi_3(x) = x_1^2 + x_2^2,$$

we get a linearization of dimension 3. Agarwal and Matoušek [1994] describe an algorithm that computes a linearization of the smallest dimension under certain mild assumptions.

Returning to the set $\mathcal{F}$, let $\mathcal{H} = \{h_{a^i} \mid 1 \leq i \leq n\}$. It can be verified [Agarwal et al. 2004] that a subset $\mathcal{K} \subseteq \mathcal{H}$ is an ε-kernel if and only if the set $\mathcal{G} = \{f_i \mid h_{a^i} \in \mathcal{K}\}$ is an ε-kernel of $\mathcal{F}$.

Combining the linearization technique with Corollary 3.1, one obtains:

THEOREM 3.2 [Agarwal et al. 2004]. *Let $\mathcal{F} = \{f_1(x), \ldots, f_n(x)\}$ be a family of d-variate polynomials, where $f_i(x) \equiv f(x, a^i)$ and $a^i \in \mathbb{R}^p$ for each $1 \leq i \leq n$, and $f(x, a)$ is a $(d+p)$-variate polynomial. Suppose that $\mathcal{F}$ admits a linearization of dimension k, and let $\varepsilon > 0$ be a parameter. We can compute an ε-kernel of $\mathcal{F}$ of size $O(1/\varepsilon^\sigma)$ in time $O(n + 1/\varepsilon^{k-1/2})$, where $\sigma = \min\{d, k/2\}$.*

Let $\mathcal{F} = \{(f_1)^{1/r}, \ldots, (f_n)^{1/r}\}$, where $r \geq 1$ is an integer and each f_i is a polynomial of some bounded degree. Agarwal et al. [2004] showed that if $\mathcal{G}$ is an $(\varepsilon/2(r-1))^r$-kernel of $\{f_1, \ldots, f_n\}$, then $\{(f_i)^{1/r} \mid f_i \in \mathcal{G}\}$ is an ε-kernel of $\mathcal{F}$. Hence, we obtain the following.

THEOREM 3.3. *Let $\mathcal{F} = \{(f_1)^{1/r}, \ldots, (f_n)^{1/r}\}$ be a family of d-variate functions as in Theorem 3.2, each f_i is a polynomial that is nonnegative for every $x \in \mathbb{R}^d$, and $r \geq 2$ is an integer constant. Let $\varepsilon > 0$ be a parameter. Suppose that $\mathcal{F}$ admits a linearization of dimension k. We can compute in $O(n + 1/\varepsilon^{r(k-1/2)})$ time an ε-kernel of size $O(1/\varepsilon^{r\sigma})$ where $\sigma = \min\{d, k/2\}$.*

Applications to shape fitting problems. Agarwal et al. [2004] showed that Theorem 3.3 can be used to compute coresets for a number of unfaithful measures as well. We illustrate the idea by sketching their $(1+\varepsilon)$-approximation algorithm for computing a minimum-width spherical shell that contains $P = \{p_1, \ldots, p_n\}$. A spherical shell is (the closure of) the region bounded by two concentric spheres: the width of the shell is the difference of their radii. Let $f_i(x) = \|x - p_i\|$. Set $\mathcal{F} = \{f_1, \ldots, f_n\}$. Let $w(x, S)$ denote the width of the thinnest spherical shell centered at x that contains a point set S, and let $w^* = w^*(S) = \min_{x \in \mathbb{R}^d} w(x, S)$ be the width of the thinnest spherical shell containing S. Then

$$w(x, P) = \max_{p \in P} \|x - p\| - \min_{p \in P} \|x - p\| = \max_{f_p \in \mathcal{F}} f_p(x) - \min_{f_p \in \mathcal{F}} f_p(x) = \mathfrak{E}_{\mathcal{F}}(x).$$

Let $\mathcal{G}$ be an ε-kernel of $\mathcal{F}$, and suppose $Q \subseteq P$ is the set of points corresponding to $\mathcal{G}$. Then for any $x \in \mathbb{R}^d$, we have $w(x, Q) \geq (1-\varepsilon)w(x, P)$. So if we first compute $\mathcal{G}$ (and therefore Q) using Theorem 3.3, compute the minimum-width spherical shell A^* containing Q, and take the smallest spherical shell containing P centered at the center of A^*, we get a $(1 + O(\varepsilon))$-approximation to the minimum-width spherical shell containing P. The running time of such an approach is $O(n+f(\varepsilon))$. It is a simple and instructive exercise to translate this approach to the problem of computing a $(1 + \varepsilon)$-approximation of the minimum-width cylindrical shell enclosing a set of points.

Using the kernel framework, Har-Peled and Wang [2004] have shown that shape fitting problems can be approximated efficiently even in the presence of a few outliers. Let us consider the following problem: Given a set P of n points in $\mathbb{R}^d$, and an integer $1 \le k \le n$, find the minimum-width slab that contains $n - k$ points from P. They present an ε-approximation algorithm for this problem whose running time is near-linear in n. They obtain similar results for problems like minimum-width spherical/cylindrical shell and indeed all the shape fitting problems to which the kernel framework applies. Their algorithm works well if the number of outliers k is small. Erickson et al. [2004] show that for large values of k, say roughly $n/2$, the problem is as hard as the $(d - 1)$-dimensional affine degeneracy problem: Given a set of n points (with integer co-ordinates) in $\mathbb{R}^{d-1}$, do any d of them lie on a common hyperplane? It is widely believed that the affine degeneracy problem requires $\Omega(n^{d-1})$ time.

Points in motion. Theorems 3.2 and 3.3 can be used to maintain various extent measures of a set of moving points. Let $P = \{p_1, \ldots, p_n\}$ be a set of n points in $\mathbb{R}^d$, each moving independently. Let $p_i(t) = (p_{i1}(t), \ldots, p_{id}(t))$ denote the position of point p_i at time t. Set $P(t) = \{p_i(t) \mid 1 \le i \le n\}$. If each p_{ij} is a polynomial of degree at most r, we say that the motion of P has *degree* r. We call the motion of P *linear* if $r = 1$ and *algebraic* if r is bounded by a constant.

Given a parameter $\varepsilon > 0$, we call a subset $Q \subseteq P$ an ε-*kernel* of P if for any direction $u \in \mathbb{S}^{d-1}$ and for all $t \in \mathbb{R}$,

$$(1 - \varepsilon)\omega(u, P(t)) \le \omega(u, Q(t)),$$

where $\omega()$ is the directional width. Assume that the motion of P is linear, i.e., $p_i(t) = a_i + b_i t$, for $1 \le i \le n$, where $a_i, b_i \in \mathbb{R}^d$. For a direction $u = (u_1, \ldots, u_d) \in \mathbb{S}^{d-1}$, we define a polynomial

$$f_i(u, t) = \langle p_i(t), u \rangle = \langle a_i + b_i t, u \rangle = \sum_{j=1}^{d} a_{ij} u_j + \sum_{j=1}^{d} b_{ij} \cdot (t u_j).$$

Set $\mathcal{F} = \{f_1, \ldots, f_n\}$. Then

$$\omega(u, P(t)) = \max_i \langle p_i(t), u \rangle - \min_i \langle p_i(t), u \rangle = \max_i f_i(u, t) - \min_i f_i(u, t) = \mathfrak{E}_{\mathcal{F}}(u, t).$$

Evidently, $\mathcal{F}$ is a family of $(d+1)$-variate polynomials that admits a linearization of dimension $2d$ (there are $2d$ monomials). Exploiting the fact that $u \in \mathbb{S}^{d-1}$, Agarwal et al. [2004] show that $\mathcal{F}$ is actually a family of d-variate polynomials that admits a linearization of dimension $2d-1$. Using Theorem 3.2, we can therefore compute an ε-kernel of P of size $O(1/\varepsilon^{d-(1/2)})$ in time $O(n + 1/\varepsilon^{2d-(3/2)})$.

The above argument can be extended to higher degree motions in a straightforward manner. The following theorem summarizes the main result.

THEOREM 3.4. *Given a set P of n moving points in $\mathbb{R}^d$ whose motion has degree $r > 1$ and a parameter $\varepsilon > 0$, we can compute an ε-kernel Q of P of size $O(1/\varepsilon^d)$ in $O(n + 1/\varepsilon^{(r+1)d-(3/2)})$ time.*

The theorem implies that at any time t, $Q(t)$ is a coreset for $P(t)$ with respect to all faithful measures. Using the same technique, a similar result can be obtained for unfaithful measures such as the minimum-width spherical shell.

Yu et al. [2004] have performed experiments with kinetic data structures that maintain the axes-parallel bounding box and convex hull of a set of points P with algebraic motion. They compare the performance of the kinetic data structure for the entire point set P with that of the data structure for a kernel Q computed by methods similar to Theorem 3.4. The experiments indicate that the number of events that the data structure for Q needs to process is significantly lower than for P even when Q is a very good approximation to P.

4. An Incremental Algorithm for Shape Fitting

Let P be a set of n points in $\mathbb{R}^d$. In [Bădoiu et al. 2002] a simple incremental algorithm is given for computing an ε-approximation to the minimum-enclosing ball of P. They showed, rather surprisingly, that the number of iterations of their algorithm depends only on ε and is independent of both d and n. The bound was improved by Bădoiu and Clarkson [2003b; 2003a] and by Kumar et al. [2003]. Kumar and Yıldırım [≥ 2005] analyzed a similar algorithm for the minimum-volume enclosing ellipsoid and gave a bound on the number of iterations that is independent of d. The minimum-enclosing ball and minimum-enclosing ellipsoid are convex optimization problems, and it is somewhat surprising that a variant of this iterative algorithm works for nonconvex optimization problems, e.g., the minimum-width cylinder, slab, spherical shell, and cylindrical shell containing P. As shown in [Yu et al. 2004], the number of iterations of the incremental algorithm is independent of the number n of points in P for all of these problems.

We describe here the version of the algorithm for computing the minimum-width slab containing P. The algorithm and its proof of convergence are readily translated to the other problems mentioned. Let Q be any affinely independent subset of $d + 1$ points in P.

(i) Let S be the minimum-width slab containing Q, computed by some brute-force method. If a $(1 + \varepsilon)$-expansion of S contains P, we return this $(1 + \varepsilon)$-expansion.

(ii) Otherwise, let $p \in P$ be the point farthest from S.

(iii) Set $Q = Q \cup \{p\}$ and go to Step 1.

It is clear that when the algorithm terminates, it does so with an ε-approximation to the minimum-width slab containing P. Also, the running time of the algorithm is $O(k(n + f(O(k))))$, where k is the number of iterations of the algorithm, and $f(t)$ is the running time of the brute-force algorithm for computing a minimum-enclosing slab of t points. Following an argument similar to the one used for proving the correctness of the algorithm for constructing ε-kernels, Yu et al. [2004] proved that the above algorithm converges within $O(1/\varepsilon^{(d-1)/2})$ iterations. They also do an experimental analysis of this algorithm and conclude that its typical performance is quite good in comparison with even the coreset based algorithms. This is because the number of iterations for typical point sets is quite small, as might be expected. See the original paper for details.

We conclude this section with an interesting open problem: Does the incremental algorithm for the minimum-enclosing cylinder problem terminate in $O(f(d) \cdot g(d, \varepsilon))$ iterations, where $f(d)$ is a function of d only, and $g(d, \varepsilon)$ is a function that depends only polynomially on d? Note that the algorithm for the minimum-enclosing ball terminates in $O(1/\varepsilon)$ iterations, while the algorithm for the minimum-enclosing slab can be shown to require $\Omega(1/\varepsilon^{(d-1)/2})$ iterations.

5. Coresets in a Streaming Setting

Algorithms for computing an ε-kernel for a given set of points in $\mathbb{R}^d$ can be adapted for efficiently maintaining an ε-kernel of a set of points under insertions and deletions. Here we describe the algorithm from [Agarwal et al. 2004] for maintaining ε-kernels in the streaming setting. Suppose we are receiving a stream of points $p_1, p_2, \ldots$ in $\mathbb{R}^d$. Given a parameter $\varepsilon > 0$, we wish to maintain an ε-kernel of the n points received so far. The resource that we are interested in minimizing is the space used by the data structure. Note that our analysis is in terms of n, the number of points inserted into the data structure. However, n does not need to be specified in advance. We assume the existence of an algorithm $\mathbb{A}$ that can compute a δ-kernel of a subset $S \subseteq P$ of size $O(1/\delta^k)$ in time $O(|S| + T_{\mathbb{A}}(\delta))$; obviously $T_{\mathbb{A}}(\delta) = \Omega(1/\delta^k)$. We will use $\mathbb{A}$ to maintain an ε-kernel dynamically. Besides such an algorithm, our scheme only uses abstract properties of kernels such as the following:

(1) If P_2 is an ε-kernel of P_1, and P_3 is a δ-kernel of P_2, then P_3 is a $(\delta+\varepsilon)$-kernel of P_1;
(2) If P_2 is an ε-kernel of P_1, and Q_2 is an ε-kernel of Q_1, then $P_2 \cup Q_2$ is an ε-kernel of $P_1 \cup Q_1$.[2]

Thus the scheme applies more generally, for instance, to some notions of coresets defined in the clustering context.

---

[2]This property is, strictly speaking, not true for kernels. However, if we slightly modify the definition to say that $Q \subseteq P$ is an ε-kernel of P if the $1/(1 - \varepsilon)$-expansion of any slab that contains Q also contains P, both properties are seen to hold. Since the modified definition is intimately connected with the definition we use, we feel justified in pretending that the second property also holds for kernels.

We assume without loss of generality that $1/\varepsilon$ is an integer. We use the dynamization technique of [Bentley and Saxe 1980], as follows: Let $P = \langle p_1, \ldots, p_n \rangle$ be the sequence of points that we have received so far. For integers $i \geq 1$, let $\rho_i = \varepsilon/ci^2$, where $c > 0$ is a constant, and set $\delta_i = \prod_{l=1}^{i}(1 + \rho_l) - 1$. We partition P into subsets $P_0, P_1, \ldots, P_u$, where $u = \lfloor \log_2 \varepsilon^k n \rfloor + 1$, as follows. $|P_0| = n \mod 1/\varepsilon^k$, and for $1 \leq i \leq u$, $|P_i| = 2^{i-1}/\varepsilon^k$ if the i-th rightmost bit in the binary expansion of $\lfloor \varepsilon^k n \rfloor$ is 1, otherwise $|P_i| = 0$. Furthermore, if $0 \leq i < j \leq u$, the points in P_j arrived before any point in P_i. These conditions uniquely specify $P_0, \ldots, P_u$. We refer to i as the *rank* of P_i. Note that for $i \geq 1$, there is at most one nonempty subset of rank i.

Unlike the standard Bentley–Saxe technique, we do not maintain each P_i explicitly. Instead, for each nonempty subset P_i, we maintain a δ_i-kernel Q_i of P_i; if $P_i = \varnothing$, we set $Q_i = \varnothing$ as well. We also let $Q_0 = P_0$. Since

$$1 + \delta_i = \prod_{l=1}^{i}\left(1 + \frac{\varepsilon}{cl^2}\right) \leq \exp\left(\sum_{l=1}^{i} \frac{\varepsilon}{cl^2}\right)$$

$$= \exp\left(\frac{\varepsilon}{c}\sum_{l=1}^{i}\frac{1}{l^2}\right) \leq \exp\left(\frac{\pi^2\varepsilon}{6c}\right) \leq 1 + \frac{\varepsilon}{3}, \tag{5-1}$$

provided c is chosen sufficiently large, Q_i is an $(\varepsilon/3)$-kernel of P_i. Therefore, $\bigcup_{i=0}^{u} Q_i$ is an $(\varepsilon/3)$-kernel of P. We define the *rank* of a set Q_i to be i. For $i \geq 1$, if P_i is nonempty, $|Q_i|$ will be $O(1/\rho_i^k)$ because $\rho_i \leq \delta_i$; note that $|Q_0| = |P_0| < 1/\varepsilon^k$.

For each $i \geq 0$, we also maintain an $\varepsilon/3$-kernel K_i of $\bigcup_{j\geq i} Q_j$, as follows. Let $u = \lfloor \log_2(\varepsilon^k n) \rfloor + 1$ be the largest value of i for which P_i is nonempty. We have $K_u = Q_u$, and for $1 \leq i < u$, K_i is a ρ_i-kernel of $K_{i+1} \cup Q_i$. Finally, $K_0 = Q_0 \cup K_1$. The argument in (5-1), by the coreset properties (1) and (2), implies that K_i is an $(\varepsilon/3)$-kernel of $\bigcup_{j\geq i} Q_j$, and thus K_0 is the required ε-kernel of P. The size of the entire data structure is

$$\sum_{i=0}^{u}(|Q_i| + |K_i|) \leq |Q_0| + |K_0| + \sum_{i=1}^{u} O(1/\rho_i^k)$$

$$= O(1/\varepsilon^k) + \sum_{i=1}^{\lfloor \log_2 \varepsilon^k n \rfloor + 1} O\left(\frac{i^{2k}}{\varepsilon^k}\right) = O\left(\frac{\log^{2k+1} n}{\varepsilon^k}\right).$$

At the arrival of the next point p_{n+1}, the data structure is updated as follows. We add p_{n+1} to Q_0 (and conceptually to P_0). If $|Q_0| < 1/\varepsilon^k$ then we are done. Otherwise, we promote Q_0 to have rank 1. Next, if there are two δ_j-kernels Q_x, Q_y of rank j, for some $j \leq \lfloor \log_2 \varepsilon^k(n+1) \rfloor + 1$, we compute a ρ_{j+1}-kernel Q_z of $Q_x \cup Q_y$ using algorithm $\mathbb{A}$, set the rank of Q_z to $j + 1$, and discard the sets Q_x and Q_y. By construction, Q_z is a δ_{j+1}-kernel of $P_z = P_x \cup P_y$ of size $O(1/\rho_{j+1}^k)$ and $|P_z| = 2^j/\varepsilon^k$. We repeat this step until the ranks of all Q_i's are distinct. Suppose ξ is the maximum rank of a Q_i that was reconstructed, then we recompute $K_\xi, \ldots, K_0$ in that order. That is, for $\xi \geq i \geq 1$, we compute a ρ_i-kernel of $K_{i+1} \cup Q_i$ and set this to be K_i; finally, we set $K_0 = K_1 \cup Q_0$.

For any fixed $i \geq 1$, Q_i and K_i are constructed after every $2^{i-1}/\varepsilon^k$ insertions, therefore the amortized time spent in updating Q after inserting a point is

$$\sum_{i=1}^{\lfloor \log_2 \varepsilon^k n \rfloor + 1} \frac{\varepsilon^k}{2^{i-1}} O\left(\frac{i^{2k}}{\varepsilon^k} + T_\mathbb{A}\left(\frac{\varepsilon}{ci^2} \right) \right) = O\left(\sum_{i=1}^{\lfloor \log_2 \varepsilon^k n \rfloor + 1} \frac{\varepsilon^k}{2^{i-1}} T_\mathbb{A}\left(\frac{\varepsilon}{ci^2} \right) \right).$$

If $T_\mathbb{A}(x)$ is bounded by a polynomial in $1/x$, then the above expression is bounded by $O(\varepsilon^k T_\mathbb{A}(\varepsilon))$.

THEOREM 5.1 [Agarwal et al. 2004]. *Let P be a stream of points in $\mathbb{R}^d$, and let $\varepsilon > 0$ be a parameter. Suppose that for any subset $S \subseteq P$, we can compute an ε-kernel of S of size $O(1/\varepsilon^k)$ in $O(|S| + T_\mathbb{A}(\varepsilon))$ time, where $T_\mathbb{A}(\varepsilon) \geq 1/\varepsilon^k$ is bounded by a polynomial in $1/\varepsilon$. Then we can maintain an ε-kernel of P of size $O(1/\varepsilon^k)$ using a data structure of size $O(\log^{2k+1}(n)/\varepsilon^k)$. The amortized time to insert a point is $O(\varepsilon^k T_\mathbb{A}(\varepsilon))$, and the running time in the worst case is $O\left((\log^{2k+1} n)/\varepsilon^k + T_\mathbb{A}(\varepsilon/\log^2 n) \log n \right)$.*

Combined with Theorem 2.2, we get a data-structure using $(\log n/\varepsilon)^{O(d)}$ space to maintain an ε-kernel of size $O(1/\varepsilon^{(d-1)/2})$ using $(1/\varepsilon)^{O(d)}$ amortized time for each insertion.

Improvements. The previous scheme raises the question of whether there is a data structure that uses space independent of the size of the point set to maintain an ε-kernel. Chan [2004] shows that the answer is "yes" by presenting a scheme that uses only $(1/\varepsilon)^{O(d)}$ storage. This result implies a similar result for maintaining coresets for all the extent measures that can be handled by the framework of kernels. His scheme is somewhat involved, but the main ideas and difficulties are illustrated by a simple scheme, reproduced below, that he describes that uses constant storage for maintaining a constant-factor approximation to the radius of the smallest enclosing cylinder containing the point set. We emphasize that the question is that of maintaining an approximation to the radius: it is not hard to maintain the axis of an approximately optimal cylinder.

A simple constant-factor offline algorithm for approximating the minimum-width cylinder enclosing a set P of points was proposed in [Agarwal et al. 2001a]. The algorithm picks an arbitrary input point, say o, finds the farthest point v from o, and returns the farthest point from the line $\overline{ov}$.

Let $\mathrm{rad}(P)$ denote the minimum radius of all cylinders enclosing P, and let $d(p, \ell)$ denote the distance between point p and line ℓ. The following observation immediately implies an upper bound of 4 on the approximation factor of the above algorithm.

OBSERVATION 5.2. $d(p, \overline{ov}) \leq 2\left(\frac{\|o - p\|}{\|o - v\|} + 1 \right) \mathrm{rad}(\{o, v, p\})$.

Unfortunately, the above algorithm requires two passes, one to find v and one to find the radius, and thus does not fit in the streaming framework. Nevertheless, a simple variant of the algorithm, which maintains an approximate candidate for v on-line, works, albeit with a larger constant:

THEOREM 5.3 [Chan 2004]. *Given a stream of points in $\mathbb{R}^d$ (where d is not necessarily constant), we can maintain a factor-18 approximation of the minimum radius over all enclosing cylinders with $O(d)$ space and update time.*

PROOF. Initially, say o and v are the first two points, and set $w = 0$. We may assume that o is the origin. A new point is inserted as follows:

> insert(p):
> 1. $w := \max\{w, \mathrm{rad}(\{o, v, p\})\}$.
> 2. if $\|p\| > 2\|v\|$ then $v := p$.
> 3. Return w.

After each point is inserted, the algorithm returns a quantity that is shown below to be an approximation to the radius of the smallest enclosing cylinder of all the points inserted thus far.

In the following analysis, w_f and v_f refer to the final values of w and v, and v_i refers to the value of v after its i-th change. Note that $\|v_i\| > 2\|v_{i-1}\|$ for all i. Also, we have $w_f \geq \mathrm{rad}(\{o, v_{i-1}, v_i\})$ since $\mathrm{rad}(\{o, v_{i-1}, v_i\})$ was one of the "candidates" for w. From Observation 5.2, it follows that

$$d(v_{i-1}, \overline{ov_i}) \leq 2\left(\frac{\|v_{i-1}\|}{\|v_i\|} + 1\right)\mathrm{rad}(\{o, v_{i-1}, v_i\}) \leq 3\mathrm{rad}(\{o, v_{i-1}, v_i\}) \leq 3w_f.$$

Fix a point $q \in P$, where P denotes the entire input point set. Suppose that $v = v_j$ just after q is inserted. Since $\|q\| \leq 2\|v_j\|$, Observation 5.2 implies that $d(q, \overline{ov_j}) \leq 6w_f$.

For $i > j$, we have $d(q, \overline{ov_i}) \leq d(q, \overline{ov_{i-1}}) + d(\hat{q}, \overline{ov_i})$, where $\hat{q}$ is the orthogonal projection of q to $\overline{ov_{i-1}}$. By similarity of triangles,

$$d(\hat{q}, \overline{ov_i}) = (\|\hat{q}\| / \|v_{i-1}\|)d(v_{i-1}, \overline{ov_i}) \leq (\|q\| / \|v_{i-1}\|)3w_f.$$

Therefore,

$$d(q, \overline{ov_i}) \leq \begin{cases} 6w_f & \text{if } i = j, \\ d(q, \overline{ov_{i-1}}) + \dfrac{\|q\|}{\|v_{i-1}\|}3w_f & \text{if } i > j. \end{cases}$$

Expanding the recurrence, one can obtain that $d(q, \overline{ov_f}) \leq 18w_f$. So, $w_f \leq \mathrm{rad}(P) \leq 18w_f$. $\qquad\square$

6. Coresets for Clustering

Given a set P of n points in $\mathbb{R}^d$ and an integer $k > 0$, a typical clustering problem asks for partitioning P into k subsets (called *clusters*), $P_1, \ldots, P_k$, so that certain *objective* function is minimized. Given a function μ that measures the extent of a cluster, we consider two types of clustering objective functions: *centered* clustering in which the objective function is $\max_{1\leq i\leq k}\mu(P_i)$, and the *summed* clustering in which the objective function is $\sum_{i=1}^{k}\mu(P_i)$; k-center and k-line-center are two examples of the first type, and k-median and k-means are two examples of the second type.

It is natural to ask whether coresets can be used to compute clusterings efficiently. In the previous sections we showed that an ε-kernel of a point set provides a coreset for several extent measures of P. However, the notion of ε-kernel is too weak to provide a coreset for a clustering problem because it approximates the extent of the entire P while for clustering problems we need a subset that approximates the extent of "relevant" subsets of P as well. Nevertheless, coresets exist for many clustering problems, though the precise definition of coreset depends on the type of clustering problem we are considering. We review some of these results in this section.

6.1. k-center and its variants. We begin by defining generalized k-clustering: we define a *cluster* to be a pair (f, S), where f is a q-dimensional subspace for some $q \leq d$ and $S \subseteq P$. Define $\mu(f, S) = \max_{p \in S} d(p, f)$. We define $\mathcal{B}(f, r)$ to be the Minkowski sum of f and the ball of radius r centered at the origin; $\mathcal{B}(f, r)$ is a ball (resp. cylinder) of radius r if f is a point (resp. line), and a slab of width $2r$ if f is a hyperplane. Obviously, $S \subseteq \mathcal{B}(f, \mu(f, S))$. We call $\mathbb{C} = \{(f_1, P_1), \ldots, (f_k, P_k)\}$ a k-*clustering* (of dimension q) if each f_i is a q-dimensional subspace and $P = \bigcup_{i=1}^{k} P_i$. We define $\mu(\mathbb{C}) = \max_{1\leq i\leq k}\mu(f_i, P_i)$, and set $r_{\text{opt}}(P, k, q) = \min_{\mathbb{C}} \mu(\mathbb{C})$, where the minimum is taken over all k-clusterings (of dimension q) of P. We use $C_{\text{opt}}(P, k, q)$ to denote an optimal k-clustering (of dimension q) of P. For $q = 0, 1, d - 1$, the above clustering problems are called k-center, k-line-center, and k-hyperplane-center problems, respectively; they are equivalent to covering P by k balls, cylinders, and slabs of minimum radius, respectively.

We call $Q \subseteq P$ an *additive ε-coreset* of P if for every k-clustering $\mathbb{C} = \{(f_1, Q_1), \ldots, (f_k, Q_k)\}$ of Q, with $r_i = \mu(f_i, Q_i)$,

$$P \subseteq \bigcup_{i=1}^{k} \mathcal{B}(f_i, r_i + \varepsilon\mu(\mathbb{C})),$$

i.e., union of the expansion of each $\mathcal{B}(f_i, r_i)$ by $\varepsilon\mu(\mathbb{C})$ covers P. If for every k-clustering $\mathbb{C}$ of Q, with $r_i = \mu(f_i, Q_i)$, we have the stronger property

$$P \subseteq \bigcup_{i=1}^{k} \mathcal{B}(f_i, (1 + \varepsilon)r_i),$$

then we call Q a *multiplicative ε-coreset*.

We review the known results on additive and multiplicative coreset for k-center, k-line-center, and k-hyperplane-center.

k-center. The existence of an additive coreset for k-center follows from the following simple observation. Let $r^* = r_{\text{opt}}(P, k, 0)$, and let $\mathcal{B} = \{B_1, \ldots, B_k\}$ be a family of k balls of radius r^* that cover P. Draw a d-dimensional Cartesian grid of side length $\varepsilon r^*/2d$; $O(k/\varepsilon^d)$ of these grid cells intersect the balls in $\mathcal{B}$. For each such cell τ that also contains a point of P, we arbitrarily choose a point from $P \cap \tau$. The resulting set $\mathcal{S}$ of $O(k/\varepsilon^d)$ points is an additive ε-coreset of P, as proved by Agarwal and Procopiuc [2002]. In order to construct $\mathcal{S}$ efficiently, we use Gonzalez's greedy algorithm [1985] to compute a factor-2 approximation of k-center, which returns a value $\tilde{r} \leq 2r^*$. We then draw the grid of side length $\varepsilon\tilde{r}/4d$ and proceed as above. Using a fast implementation of Gonzalez's algorithm as proposed in [Feder and Greene 1988; Har-Peled 2004a], one can compute an additive ε-coreset of size $O(k/\varepsilon^d)$ in time $O(n + k/\varepsilon^d)$.

Agarwal et al. [2002] proved the existence of a small multiplicative ε-coreset for k-center in $\mathbb{R}^1$. It was subsequently extended to higher dimensions by Har-Peled [2004b]. We sketch their construction.

THEOREM 6.1 [Agarwal et al. 2002; Har-Peled 2004b]. *Let P be a set of n points in $\mathbb{R}^d$, and $0 < \varepsilon < 1/2$ a parameter. There exists a multiplicative ε-coreset of size $O(k!/\varepsilon^{dk})$ of P for k-center.*

PROOF. For $k = 1$, by definition, an additive ε-coreset of P is also a multiplicative ε-coreset of P. For $k > 1$, let $r^* = r_{\text{opt}}(P, k, 0)$ denote the smallest r for which k balls of radius r cover P. We draw a d-dimensional grid of side length $\varepsilon r_{\text{opt}}/(5d)$, and let $\mathbb{C}$ be the set of (hyper-)cubes of this grid that contain points of P. Clearly, $|\mathbb{C}| = O(k/\varepsilon^d)$. Let Q' be an additive $(\varepsilon/2)$-coreset of P. For every cell Δ in $\mathbb{C}$, we inductively compute an ε-multiplicative coreset of $P \cap \Delta$ with respect to $(k-1)$-center. Let Q_Δ be this set, and let $\mathbb{Q} = \bigcup_{\Delta \in \mathbb{C}} Q_\Delta \cup Q'$. We argue below that the set $\mathbb{Q}$ is the required multiplicative coreset. The bound on its size follows by a simple calculation.

Let $\mathcal{B}$ be any family of k balls that covers $\mathbb{Q}$. Consider any hypercube Δ of $\mathbb{C}$. Suppose Δ intersects all the k balls of $\mathcal{B}$. Since Q' is an additive $(\varepsilon/2)$-coreset of P, one of the balls in $\mathcal{B}$ must be of radius at least $r^*/(1 + \varepsilon/2) \geq r^*(1 - \varepsilon/2)$. Clearly, if we expand such a ball by a factor of $(1 + \varepsilon)$, it completely covers Δ, and therefore also covers all the points of $\Delta \cap P$.

We now consider the case when Δ intersects at most $k - 1$ balls of $\mathcal{B}$. By induction, $Q_\Delta \subseteq \mathbb{Q}$ is an ε-multiplicative coreset of $P \cap \Delta$ for $(k-1)$-center. Therefore, if we expand each ball in $\mathcal{B}$ that intersects Δ by a factor of $(1 + \varepsilon)$, the resulting set of balls will cover $P \cap \Delta$. □

Surprisingly, additive coresets for k-center exist even for a set of moving points in $\mathbb{R}^d$. More precisely, let P be a set of n points in $\mathbb{R}^d$ with algebraic motion of degree at most Δ, and let $0 < \varepsilon \leq 1/2$ be a parameter. Har-Peled [2004a] showed that there exists a subset $Q \subseteq P$ of size $O((k/\varepsilon^d)^{\Delta+1})$ so that for all $t \in \mathbb{R}$, $Q(t)$ is an additive ε-coreset of $P(t)$. For $k = O(n^{1/4}\varepsilon^d)$, Q can be computed in time $O(nk/\varepsilon^d)$.

k-line-center. The existence of an additive coreset for k-line-center, i.e., for the problem of covering P by k congruent cylinders of the minimum radius, was first proved in [Agarwal et al. 2002].

THEOREM 6.2 [Agarwal et al. 2002]. *Given a set P of finite points in $\mathbb{R}^d$ and a parameter $0 < \varepsilon < 1/2$, there exists an additive ε-coreset of size*

$$O((k+1)!/\varepsilon^{d-1+k})$$

of P for the k-line-center problem.

PROOF. Let $C_{\mathrm{opt}} = \{(\ell_1, P_1), \ldots, (\ell_k, P_k)\}$ be an optimal k-clustering (of dimension 1) of P, and let $r^* = \mu(P, k, 1)$, i.e., the cylinders of radius r^* with axes $\ell_1, \ldots, \ell_k$ cover P and $P_i \subset \mathcal{B}(\ell_i, r^*)$. For each $1 \leq i \leq k$, draw a family L_i of $O(1/\varepsilon^{d-1})$ lines parallel to ℓ_i so that for any point in P_i there is a line in L_i within distance $\varepsilon r^*/2$. Set $L = \bigcup_i L_i$. We project each point $p \in P_i$ to the line in L_i that is nearest to p. Let $\bar{p}$ be the resulting projection of p, and let $\bar{P}_\ell$ be the set of points that project onto $\ell \in L$. Set $\bar{P} = \bigcup_{\ell \in L} \bar{P}_\ell$. It can be argued that a multiplicative $(\varepsilon/3)$-coreset of $\bar{P}$ is an additive ε-coreset of P. Since the points in $\bar{P}_\ell$ lie on a line, by Theorem 6.1, a multiplicative $(\varepsilon/3)$-coreset $\bar{Q}_\ell$ of $\bar{P}_\ell$ of size $O(k!/\varepsilon^k)$ exists. Observing that $\bar{Q} = \bigcup_{\ell \in L} \bar{Q}_\ell$ is a multiplicative $(\varepsilon/3)$-coreset of $\bar{P}$, and thus $Q = \{p \mid \bar{p} \in \bar{Q}\}$ is an additive ε-coreset of P of size $O((k+1)!/\varepsilon^{d-1+k})$.   □

Although Theorem 6.2 proves the existence of an additive coreset for k-line-center, the proof is nonconstructive. However, Agarwal et al. [2002] have shown that the iterated reweighting technique of Clarkson [1993] can be used in conjunction with Theorem 6.2 to compute an ε-approximate solution to the k-line-center problem in $O(n \log n)$ expected time, with constants depending on k, ε, and d.

When coresets do not exist. We now present two negative results on coresets for centered clustering problems. Surprisingly, there are no multiplicative coresets for k-line-center even in $\mathbb{R}^2$.

THEOREM 6.3 [Har-Peled 2004b]. *For any $n \geq 3$, there exists a point set $P = \{p_1, \ldots, p_n\}$ in $\mathbb{R}^2$, such that the size of any multiplicative $(1/2)$-coreset of P with for 2-line-center is at least $|P| - 2$.*

PROOF. Let $p_i = (1/2^i, 2^i)$ and $P(i) = \{p_1, \ldots, p_i\}$. Let Q be a $(1/2)$-coreset of $P = P(n)$. Let $Q_i^- = Q \cap P(i)$ and $Q_i^+ = Q \setminus Q_i^-$.

If the set Q does not contain the point $p_i = (1/2^i, 2^i)$, for some $2 \leq i \leq n-1$, then Q_i^- can be covered by a horizontal strip h^- of width $\leq 2^{i-1}$ that has the x-axis as its lower boundary. Clearly, if we expand h^- by a factor of $3/2$, it still will not cover p_i. Similarly, we can cover Q_i^+ by a vertical strip h^+ of width $1/2^{i+1}$ that has the y-axis as its left boundary. Again, if we expand h^+ by a factor of $3/2$, it will still not cover p_i. We conclude, that any multiplicative $(1/2)$-coreset for P must include all the points $p_2, p_3, \ldots, p_{n-1}$.   □

This construction can be embedded in $\mathbb{R}^3$, as described in [Har-Peled 2004b], to show that even an additive coreset does not exist for 2-plane-clustering in $\mathbb{R}^3$, i.e., the problem of covering the input point set of two slabs of the minimum width.

For the special case of 2-plane-center in $\mathbb{R}^3$, a near-linear-time approximation algorithm is known [Har-Peled 2004b]. The problem of approximating the best k-hyperplane-clustering for $k \geq 3$ in $\mathbb{R}^3$ and $k \geq 2$ in higher dimensions in near-linear time is still open.

6.2. k-median and k-means clustering. Next we focus our attention to coresets for the summed clustering problem. For simplicity, we consider the k-median clustering problem, which calls for computing k "facility" points so that the average distance between the points of C and their nearest facility is minimized. Since the objective function involves sum of distances, we need to assign weights to points in coresets to approximate the objective function of the clustering for the entire point set. We therefore define k-median clustering for a weighted point set.

Let P be a set of n points in $\mathbb{R}^d$, and let $w : P \to \mathbb{Z}^+$ be a weight function. For a point set $C \subseteq \mathbb{R}^d$, let $\mu(P, w, C) = \sum_{p \in P} w(p)d(p, C)$, where $d(p, C) = \min_{q \in C} d(p, q)$. Given C, we partition P into k clusters by assigning each point in P to its nearest neighbor in C. Define

$$\mu(P, w, k) = \min_{\substack{C \subset \mathbb{R}^d \\ |C| = k}} \mu(P, w, C).$$

For $k = 1$, this is the so-called Fermat–Weber problem [Wesolowsky 1993]. A subset $Q \subseteq P$ with a weight function $\chi : P \to \mathbb{Z}^+$ is called an ε-coreset for k-median if for any set C of k points in $\mathbb{R}^d$,

$$(1 - \varepsilon)\mu(P, w, C) \leq \mu(Q, \chi, C) \leq (1 + \varepsilon)\mu(P, w, C).$$

Here we sketch the proof from [Har-Peled and Mazumdar 2004] for the existence of a small coreset for the k-median problem. There are two main ingredients in their construction. First suppose we have at our disposal a set $A = \{a_1, \ldots, a_m\}$ of "support" points in $\mathbb{R}^d$ so that $\mu(P, w, A) \leq c\mu(P, w, k)$ for a constant $c \geq 1$, i.e., A is a good approximation of the "centers" of an optimal k-median clustering. We construct an ε-coreset S of size $O((|A| \log n)/\varepsilon^d)$ using A, as follows.

Let $P_i \subseteq P$, for $1 \leq i \leq m$, be the set of points for which a_i is the nearest neighbor in A. We draw an exponential grid around a_i and choose a subset of $O((\log n)/\varepsilon^d)$ points of P_i, with appropriate weights, for S. Set $\rho = \mu(P, w, A)/cn$, which is a lower bound on the average radius $\mu(P, w, k)/n$ of the optimal k-median clustering. Let $\mathbb{C}_j$ be the axis-parallel hypercube with side length $\rho 2^j$ centered at a_i, for $0 \leq i \leq \lceil 2 \log(cn) \rceil$. Set $V_0 = \mathbb{C}_0$ and $V_i = \mathbb{C}_i \setminus \mathbb{C}_{i-1}$ for $i \geq 1$. We partition each V_i into a grid of side length $\varepsilon \rho 2^j / \alpha$, where $\alpha \geq 1$ is a constant. For each grid cell τ in the resulting exponential grid that contains at least one point of P_i, we choose an arbitrary point in $P_i \cap \tau$ and set its weight to $\sum_{p \in P_i \cap \tau} w(p)$. Let $\mathcal{S}_i$ be the resulting set of weighted points. We repeat this step for all points in A, and set $\mathcal{S} = \bigcup_{i=1}^{m} \mathcal{S}_i$. Har-Peled and Mazumdar showed that $\mathcal{S}$ is indeed an ε-coreset of P for the k-median problem, provided α is chosen appropriately.

The second ingredient of their construction is the existence of a small "support" set A. Initially, a random sample of P of $O(k \log n)$ points is chosen and the points of P that are "well-served" by this set of random centers are filtered out. The process is repeated for the remaining points of P until we get a set A' of $O(k \log^2 n)$ support points. Using the above procedure, we can construct an $(1/2)$-coreset $\mathcal{S}$ of size $O(k \log^3 n)$. Next, a simple polynomial-time local-search algorithm, described in [Har-Peled and Mazumdar 2004], can be applied to this coreset and a support set A of size k can be constructed, which is a constant-factor approximation to the optimal k-median/means clustering. Plugging this A back into the above coreset construction yields an ε-coreset of size $O((k/\varepsilon^d) \log n)$.

THEOREM 6.4 [Har-Peled and Mazumdar 2004]. *Given a set P of n points in $\mathbb{R}^d$, and parameters $\varepsilon > 0$ and k, one can compute a coreset of P for k-means and k-median clustering of size $O((k/\varepsilon^d) \log n)$. The running time of this algorithm is $O(n + \text{poly}(k, \log n, 1/\varepsilon))$, where $\text{poly}(\cdot)$ is a polynomial.*

Using a more involved construction, Har-Peled and Kushal [2004] showed that for both k-median and k-means clustering, one can construct a coreset whose size is independent of the size of the input point set. In particular, they show that there is a coreset of size $O(k^2/\varepsilon^d)$ for k-median and $O(k^3/\varepsilon^{d+1})$ for k-means. Chen [2004] recently showed that for both k-median and k-means clustering, there are coresets whose size is $O(dk\varepsilon^{-2} \log n)$, which has linear dependence on d. In particular, this implies a streaming algorithm for k-means and k-median clustering using (roughly) $O(dk\varepsilon^{-2} \log^3 n)$ space. The question of whether the dependence on n can be removed altogether is still open.

7. Coresets in High Dimensions

Most of the coreset constructions have exponential dependence on the dimensions. In this section, we do not consider d to be a fixed constant but assume that it can be as large as the number of input points. It is natural to ask whether the dependence on the dimension can be reduced or removed altogether. For example, consider a set P of n points in $\mathbb{R}^d$. A 2-approximate coreset for the minimum enclosing ball of P has size 2 (just pick a point in P, and its furthest neighbor in P). Thus, dimension-independent coresets do exist.

As another example, consider the question of whether a small coreset exists for the width measure of P (i.e., the width of the thinnest slab containing P). It is easy to verify that any ε-approximate coreset for the width needs to be of size at least $1/\varepsilon^{\Omega((d-1)/2)}$. Indeed, consider spherical cap on the unit hypersphere, with angular radius $c\sqrt{\varepsilon}$, for appropriate constant c. The height of this cap is $1 - \cos(c\sqrt{\varepsilon}) \leq 2\varepsilon$. Thus, a coreset of the hypersphere, for the measure of width, in high dimension, would require any such cap to contain at least one point of the coreset. As such, its size must be exponential, and we conclude that high-dimensional coresets (with size polynomial in the dimension) do not always exist.

7.1. Minimum enclosing ball.

Given a set of points P, an approximation of the minimum radius ball enclosing P can be computed in polynomial time using the ellipsoid method since this is a quadratic convex programming problem [Gärtner 1995; Grötschel et al. 1988]. However, the natural question is whether one can compute a small coreset, $Q \subseteq P$, such that the minimum enclosing ball for Q is a good approximation to the real minimum enclosing ball.

Bădoiu et al. [2002] presented an algorithm, which we have already mentioned in Section 4, that generates a coreset of size $O(1/\varepsilon^2)$. The algorithms starts with a set C_0 that contains a single (arbitrary) point of P. Next, in the i-th iteration, the algorithm computes the smallest enclosing ball for C_{i-1}. If the $(1 + \varepsilon)$-expansion of the ball contains P, then we are done, as we have computed the required coreset. Otherwise, take the point from P furthest from the center of the ball and add it to the coreset. The authors show that this algorithm terminates within $O(1/\varepsilon^2)$ iterations. The bound was later improved to $O(1/\varepsilon)$ in [Kumar et al. 2003; Bădoiu and Clarkson 2003b]. Bădoiu and Clarkson showed a matching lower bound and gave an elementary algorithm that uses the "hill climbing" technique. Using this algorithm instead of the ellipsoid method, we obtain a simple algorithm with running time $O(dn/\varepsilon + 1/\varepsilon^{O(1)})$ [Bădoiu and Clarkson 2003a].

It is important to note that this coreset Q is *weaker* than its low dimensional counterpart: it is not necessarily true that the $(1 + \varepsilon)$-expansion of *any* ball containing Q contains P. What is true is that the smallest ball containing Q, when $(1 + \varepsilon)$-expanded, contains P. In fact, it is easy to verify that the size of a coreset guaranteeing the stronger property is exponential in the dimension in the worst case.

Smallest enclosing ball with outliers.

As an application of this coreset, one can compute approximately the smallest ball containing all but k of the points. Indeed, consider the smallest such ball b_{opt}, and consider $P' = P \cap b_{\mathrm{opt}}$. There is a coreset $Q \subseteq P'$ such that

(1) $|Q| = O(1/\varepsilon)$, and

(2) the smallest enclosing ball for Q, if ε-expanded, contains at least $n - k$ points of P.

Thus, one can just enumerate all possible subsets of size $O(1/\varepsilon)$ as "candidates" for Q, and for each such subset, compute its smallest enclosing ball, expand the ball, and check how many points of P it contains. Finally, the smallest candidate ball that contains at least $n - k$ points of P is the required approximation. The running time of this algorithm is $dn^{O(1/\varepsilon)}$.

k-center. We execute simultaneously k copies of the incremental algorithm for the min-enclosing ball. Whenever getting a new point, we need to determine to which of the k clusters it belongs to. To this end, we ask an oracle to identify the cluster it belongs to. It is easy to verify that this algorithm generates an ε-approximate k-center clustering in k/ε iterations. The running time is $O(dkn/\varepsilon + dk/\varepsilon^{O(1)})$.

To remove the oracle, which generates $O(k/\varepsilon)$ integer numbers between 1 and k, we just generate all possible sequence answers that the oracle might give. Since there are $O(k^{O(k/\varepsilon)})$ sequences, we get that the running time of the new algorithm (which is oracle free) is $O(dnk^{O(k/\varepsilon)})$. One can even handle outliers; see [Bădoiu et al. 2002] for details.

7.2. Minimum enclosing cylinder. One natural problem is the computation of a cylinder of minimum radius containing the points of P. We saw in Section 5 that the line through any point in P and its furthest neighbor is the axis for a constant-factor approximation. Har-Peled and Varadarajan [2002] showed that there is a subset $Q \subseteq P$ of $(1/\varepsilon)^{O(1)}$ points such that the axis of an ε-approximate cylinder lies in the subspace spanned by Q. By enumerating all possible candidates for Q, and solving a "low-dimensional" problem for each of the resulting candidate subspaces, they obtain an algorithm that runs in $dn^{(1/\varepsilon)^{O(1)}}$ time. A slightly faster, but more involved algorithm, was described earlier in [Bădoiu et al. 2002].

The algorithm of Har-Peled and Varadarajan extends immediately to the problem of computing a k-flat (i.e., an affine subspace of dimension k) that minimizes the maximum distance to a point in P. The resulting running time is $dn^{(k/\varepsilon)^{O(1)}}$. The approach also handles outliers and multiple (but constant number of) flats.

Linear-time algorithm. A natural approach for improving the running time of the minimum enclosing cylinder, is to adapt the general approach underlying the algorithm of [Bădoiu and Clarkson 2003a] to the cylinder case. Here, the idea is that we start from a center line ℓ_0. At each iteration, we find the furthest point $p_i \in P$ from ℓ_{i-1}. We then generate a line ℓ_i which is "closer" to the optimal center line. This can be done by consulting with an oracle, that provides us with information about how to move the line. By careful implementation, and removing the oracle, the resulting algorithm takes $O(ndC_\varepsilon)$ time, where $C_\varepsilon = \exp\left(\frac{1}{\varepsilon^3}\log^2\frac{1}{\varepsilon}\right)$. See [Har-Peled and Varadarajan 2004] for more details.

This also implies a linear-time algorithm for computing the minimum radius k-flat. The exact running time is

$$n \cdot d \cdot \exp\left(\frac{e^{O(k^2)}}{\varepsilon^{2k+1}} \log^2 \frac{1}{\varepsilon} \right).$$

The constants involved were recently improved by Panigrahy [2004], who also simplified the analysis.

Handling multiple slabs in linear time is an open problem for further research. Furthermore, computing the best k-flat in the presence of outliers in near-linear time is also an open problem.

The L_2 measure. A natural problem is to compute the k-flat minimizing not the maximum distance, but rather the sum of squared distances; this is known as the L_2 measure, and it can be solved in $O(\min(dn^2, nd^2))$ time, using singular value decomposition [Golub and Van Loan 1996]. Recently, Rademacher et al. [2004] showed that there exists a coreset for this problem. Namely, there are $O(k^2/\varepsilon)$ points in P, such that their span contains a k-flat which is a $(1+\varepsilon)$-approximation to the best k-flat approximating the point set under the L_2 measure. Their proof also yields a polynomial time algorithm to construct such a coreset. An interesting question is whether there is a significantly more efficient algorithm for computing a coreset. Rademacher et al. also show that their approach leads to a polynomial time approximation scheme for fitting multiple k-flats, when k and the number of flats are constants.

7.3. k-means and k-median clustering. Bădoiu et al. [2002] consider the problem of computing a k-median clustering of a set P of n points in $\mathbb{R}^d$. They show that for a random sample X from P of size $O(1/\varepsilon^3 \log 1/\varepsilon)$, the following two events happen with probability bounded below by a positive constant: (i) The flat span(X) contains a $(1 + \varepsilon)$-approximate 1-median for P, and (ii) X contains a point close to the center of a 1-median of P. Thus, one can generate a small number of candidate points on span(X), such that one of those points is a median which is an $(1 + \varepsilon)$-approximate 1-median for P.

To get k-median clustering, one needs to do this random sampling in each of the k clusters. It is unclear how to do this if those clusters are of completely different cardinality. Bădoiu et al. [2002] suggest an elaborate procedure to do so, by guessing the average radius and cardinality of the heaviest cluster, generating a candidate set for centers for this cluster using random sampling, and then recursing on the remaining points. The resulting running time is

$$2^{(k/\varepsilon)^{O(1)}} d^{O(1)} n \log^{O(k)} n,$$

and the results are correct with high-probability.

A similar procedure works for k-means; see [de la Vega et al. 2003]. Those algorithms were recently improved to have running time with linear dependency on n, both for the case of k-median and k-means [Kumar et al. 2004].

7.4. Maximum margin classifier. Let P^+ and P^- be two sets of points, labeled as positive and negative, respectively. In support vector machines, one is looking for a hyperplane h such that P^+ and P^- are on different sides of h, and the minimum distance between h and the points of $P = P^+ \cup P^-$ is maximized. The distance between h and the closest point of P is known as the *margin* of h. In particular, the larger the margin is, the better generalization bounds one can prove on h. See [Cristianini and Shaw-Taylor 2000] for more information about learning and support vector machines.

In the following, let $\Delta = \Delta(P)$ denote the diameter of P, and let ρ denote the width of the maximum width margin for P. Har-Peled and Zimak [2004] showed an iterative algorithm for computing a coreset for this problem. Specifically, by iteratively picking the point that has maximum violation of the current classifier to be in the coreset, they show that the algorithm terminates after $O((\Delta/\rho)^2/\varepsilon)$ iterations. Thus, there exist subsets $Q^- \subseteq P^-$ and $Q^+ \subseteq P^+$, such that the maximum margin linear classifier h for Q^+ and Q^- has a $\geq (1-\varepsilon)\rho$ margin for P. As in the case of computing the minimum enclosing ball, one calls a procedure for computing the best linear separator only on the growing coresets, which are small. Kowalczyk [2000] presented a similar iterative algorithm, but the size of the resulting coreset seems to be larger.

8. Conclusions

In this paper, we have surveyed several approximation algorithms for geometric problems that use the coreset paradigm. We have certainly not attempted to be comprehensive and our paper does not reflect all the research work that can be viewed as employing this paradigm. For example, we do not touch upon the body of work on sublinear algorithms [Chazelle et al. 2003] or on property testing in the geometric context [Czumaj and Sohler 2001]. Even among the results that we do cover, the choice of topics for detailed exposition is (necessarily) somewhat subjective.

Acknowledgements.

We are grateful to the referees for their detailed, helpful comments.

References

[Agarwal and Matoušek 1994] P. K. Agarwal and J. Matoušek, "On range searching with semialgebraic sets", *Discrete Comput. Geom.* **11**:4 (1994), 393–418.

[Agarwal and Procopiuc 2002] P. K. Agarwal and C. M. Procopiuc, "Exact and approximation algorithms for clustering", *Algorithmica* **33**:2 (2002), 201–226.

[Agarwal et al. 2001a] P. K. Agarwal, B. Aronov, and M. Sharir, "Exact and approximation algorithms for minimum-width cylindrical shells", *Discrete Comput. Geom.* **26**:3 (2001), 307–320.

[Agarwal et al. 2001b] P. K. Agarwal, L. J. Guibas, J. Hershberger, and E. Veach, "Maintaining the extent of a moving point set", *Discrete Comput. Geom.* **26**:3 (2001), 353–374.

[Agarwal et al. 2002] P. K. Agarwal, C. M. Procopiuc, and K. R. Varadarajan, "Approximation algorithms for k-line center", pp. 54–63 in *Algorithms—ESA 2002*, Lecture Notes in Comput. Sci. **2461**, Springer, Berlin, 2002.

[Agarwal et al. 2004] P. K. Agarwal, S. Har-Peled, and K. R. Varadarajan, "Approximating extent measures of points", *J. Assoc. Comput. Mach.* **51** (2004), 606–635.

[Arya and Mount 1998] S. Arya and D. Mount, "ANN: Library for approximate nearest neighbor searching", 1998. Available at http://www.cs.umd.edu/~mount/ANN/.

[Arya et al. 1998] S. Arya, D. M. Mount, N. S. Netanyahu, R. Silverman, and A. Y. Wu, "An optimal algorithm for approximate nearest neighbor searching in fixed dimensions", *J. ACM* **45**:6 (1998), 891–923.

[Bădoiu and Clarkson 2003a] M. Bădoiu and K. L. Clarkson, "Optimal core-sets for balls", 2003. Available at http://cm.bell-labs.com/who/clarkson/coresets2.pdf.

[Bădoiu and Clarkson 2003b] M. Bădoiu and K. L. Clarkson, "Smaller core-sets for balls", pp. 801–802 in *Proceedings of the Fourteenth Annual ACM-SIAM Symposium on Discrete Algorithms*, ACM, New York, 2003.

[Bădoiu et al. 2002] M. Bădoiu, S. Har-Peled, and P. Indyk, "Approximate clustering via core-sets", pp. 250–257 in *Proc. 34th Annu. ACM Sympos. Theory Comput.*, 2002. Available at http://www.uiuc.edu/~sariel/research/papers/02/coreset/.

[Barequet and Har-Peled 2001] G. Barequet and S. Har-Peled, "Efficiently approximating the minimum-volume bounding box of a point set in three dimensions", *J. Algorithms* **38**:1 (2001), 91–109.

[Bentley and Saxe 1980] J. L. Bentley and J. B. Saxe, "Decomposable searching problems. I. Static-to-dynamic transformation", *J. Algorithms* **1**:4 (1980), 301–358.

[Bentley et al. 1982] J. L. Bentley, M. G. Faust, and F. P. Preparata, "Approximation algorithms for convex hulls", *Comm. ACM* **25**:1 (1982), 64–68.

[Bronshteyn and Ivanov 1976] E. M. Bronshteyn and L. D. Ivanov, "The approximation of convex sets by polyhedra", *Siberian Math. J.* **16** (1976), 852–853.

[Chan 2002] T. M. Chan, "Approximating the diameter, width, smallest enclosing cylinder, and minimum-width annulus", *Internat. J. Comput. Geom. Appl.* **12** (2002), 67–85.

[Chan 2004] T. M. Chan, "Faster core-set constructions and data stream algorithms in fixed dimensions", pp. 152–159 in *Proc. 20th Annu. ACM Sympos. Comput. Geom.*, 2004.

[Chazelle 2000] B. Chazelle, *The discrepancy method*, Cambridge University Press, Cambridge, 2000.

[Chazelle et al. 2003] B. Chazelle, D. Liu, and A. Magen, "Sublinear geometric algorithms", pp. 531–540 in *Proc. 35th ACM Symp. Theory of Comput.*, 2003.

[Chen 2004] K. Chen, "Clustering algorithms using adaptive sampling", 2004. Manuscript.

[Clarkson 1993] K. L. Clarkson, "Algorithms for polytope covering and approximation", pp. 246–252 in *Algorithms and data structures* (Montreal, PQ, 1993), Lecture Notes in Comput. Sci. **709**, Springer, Berlin, 1993.

[Costa and César 2001] L. Costa and R. M. César, Jr., *Shape analysis and classification*, CRC Press, Boca Raton (FL), 2001.

[Cristianini and Shaw-Taylor 2000] N. Cristianini and J. Shaw-Taylor, *Support vector machines*, Cambridge Univ. Press, New York, 2000.

[Czumaj and Sohler 2001] A. Czumaj and C. Sohler, "Property testing with geometric queries (extended abstract)", pp. 266–277 in *Algorithms—ESA* (Århus, 2001), Lecture Notes in Comput. Sci. **2161**, Springer, Berlin, 2001.

[Dryden and Mardia 1998] I. L. Dryden and K. V. Mardia, *Statistical shape analysis*, Wiley, Chichester, 1998.

[Dudley 1974] R. M. Dudley, "Metric entropy of some classes of sets with differentiable boundaries", *J. Approximation Theory* **10** (1974), 227–236.

[Erickson and Har-Peled 2004] J. Erickson and S. Har-Peled, "Optimally cutting a surface into a disk", *Discrete Comput. Geom.* **31**:1 (2004), 37–59.

[Feder and Greene 1988] T. Feder and D. H. Greene, "Optimal algorithms for approximate clustering", pp. 434–444 in *Proc. 20th Annu. ACM Sympos. Theory Comput.*, 1988.

[Gärtner 1995] B. Gärtner, "A subexponential algorithm for abstract optimization problems", *SIAM J. Comput.* **24**:5 (1995), 1018–1035.

[Golub and Van Loan 1996] G. H. Golub and C. F. Van Loan, *Matrix computations*, 3rd ed., Johns Hopkins University Press, Baltimore, MD, 1996.

[Gonzalez 1985] T. F. Gonzalez, "Clustering to minimize the maximum intercluster distance", *Theoret. Comput. Sci.* **38**:2-3 (1985), 293–306.

[Grötschel et al. 1988] M. Grötschel, L. Lovász, and A. Schrijver, *Geometric algorithms and combinatorial optimization*, Algorithms and Combinatorics **2**, Springer, Berlin, 1988. Second edition, 1994.

[Har-Peled 2004a] S. Har-Peled, "Clustering motion", *Discrete Comput. Geom.* **31**:4 (2004), 545–565.

[Har-Peled 2004b] S. Har-Peled, "No Coreset, No Cry", in *Proc. 24th Conf. Found. Soft. Tech. Theoret. Comput. Sci.*, 2004. Available at http://www.uiuc.edu/~sariel/papers/02/2slab/. To appear.

[Har-Peled and Kushal 2004] S. Har-Peled and A. Kushal, "Smaller coresets for *k*-median and *k*-means clustering", 2004. Manuscript.

[Har-Peled and Mazumdar 2004] S. Har-Peled and S. Mazumdar, "Coresets for *k*-means and *k*-median clustering and their applications", pp. 291–300 in *Proc. 36th Annu. ACM Sympos. Theory Comput.*, 2004. Available at http://www.uiuc.edu/~sariel/research/papers/03/kcoreset/.

[Har-Peled and Varadarajan 2002] S. Har-Peled and K. R. Varadarajan, "Projective clustering in high dimensions using core-sets", pp. 312–318 in *Proc. 18th Annu. ACM Sympos. Comput. Geom.*, 2002. Available at http://www.uiuc.edu/~sariel/research/papers/01/kflat/.

[Har-Peled and Varadarajan 2004] S. Har-Peled and K. R. Varadarajan, "High-dimensional shape fitting in linear time", *Discrete Comput. Geom.* **32**:2 (2004), 269–288.

[Har-Peled and Wang 2004] S. Har-Peled and Y. Wang, "Shape fitting with outliers", *SIAM J. Comput.* **33**:2 (2004), 269–285.

[Har-Peled and Zimak 2004] S. Har-Peled and D. Zimak, "Coresets for SVM", 2004. Manuscript.

[Haussler and Welzl 1987] D. Haussler and E. Welzl, "ε-nets and simplex range queries", *Discrete Comput. Geom.* **2**:2 (1987), 127–151.

[Heckbert and Garland 1997] P. S. Heckbert and M. Garland, "Survey of polygonal surface simplification algorithms", Technical report, CMU-CS, 1997. Available at http://www.uiuc.edu/~garland/papers.html.

[John 1948] F. John, "Extremum problems with inequalities as subsidiary conditions", pp. 187–204 in *Studies and essays presented to R. Courant on his 60th birthday, January 8, 1948*, Interscience, 1948.

[Kowalczyk 2000] A. Kowalczyk, *Maximal margin perceptron*, edited by A. Smola et al., MIT Press, Cambridge (MA), 2000.

[Kumar and Yildirim ≥ 2005] P. Kumar and E. Yildirim, "Approximating minimum volume enclosing ellipsoids using core sets", *J. Opt. Theo. Appl.*. To appear.

[Kumar et al. 2003] P. Kumar, J. S. B. Mitchell, and E. A. Yildirim, "Approximate minimum enclosing balls in high dimensions using core-sets", *J. Exp. Algorithmics* **8** (2003), 1.1. Available at http://www.compgeom.com/~piyush/meb/journal.pdf.

[Kumar et al. 2004] A. Kumar, Y. Sabharwal, and S. Sen, "A simple linear time $(1+\varepsilon)$-approximation algorithm for k-means clustering in any dimensions", in *Proc. 45th Annu. IEEE Sympos. Found. Comput. Sci.*, 2004.

[Mulmuley 1993] K. Mulmuley, *Computational geometry: an introduction through randomized algorithms*, Prentice Hall, Englewood Cliffs, NJ, 1993.

[Panigrahy 2004] R. Panigrahy, "Minimum enclosing polytope in high dimensions", 2004. Manuscript.

[Rademacher et al. 2004] L. Rademacher, S. Vempala, and G. Wang, "Matrix approximation and projective clustering via iterative sampling", 2004. Manuscript.

[Vapnik and Chervonenkis 1971] V. N. Vapnik and A. Y. Chervonenkis, "On the uniform convergence of relative frequencies of events to their probabilities", *Theory Probab. Appl.* **16** (1971), 264–280.

[de la Vega et al. 2003] W. F. de la Vega, M. Karpinski, C. Kenyon, and Y. Rabani, "Approximation schemes for clustering problems", pp. 50–58 in *Proc. 35th Annu. ACM Sympos. Theory Comput.*, 2003.

[Wesolowsky 1993] G. Wesolowsky, "The Weber problem: History and perspective", *Location Science* **1** (1993), 5–23.

[Yu et al. 2004] H. Yu, P. K. Agarwal, R. Poreddy, and K. R. Varadarajan, "Practical methods for shape fitting and kinetic data structures using core sets", pp. 263–272 in *Proc. 20th Annu. ACM Sympos. Comput. Geom.*, 2004.

[Zhou and Suri 2002] Y. Zhou and S. Suri, "Algorithms for a minimum volume enclosing simplex in three dimensions", *SIAM J. Comput.* **31**:5 (2002), 1339–1357.

PANKAJ K. AGARWAL
DEPARTMENT OF COMPUTER SCIENCE
BOX 90129
DUKE UNIVERSITY
DURHAM NC 27708-0129
pankaj@cs.duke.edu

SARIEL HAR-PELED
DEPARTMENT OF COMPUTER SCIENCE
DCL 2111
UNIVERSITY OF ILLINOIS
1304 WEST SPRINGFIELD AVE.
URBANA, IL 61801
sariel@uiuc.edu

KASTURI R. VARADARAJAN
DEPARTMENT OF COMPUTER SCIENCE
THE UNIVERSITY OF IOWA
IOWA CITY, IA 52242-1419
kvaradar@cs.uiowa.edu

Combinatorial and Computational Geometry
MSRI Publications
Volume **52**, 2005

Applications of Graph and Hypergraph Theory in Geometry

IMRE BÁRÁNY

ABSTRACT. The aim of this survey is to collect and explain some geometric results whose proof uses graph or hypergraph theory. No attempt has been made to give a complete list of such results. We rather focus on typical and recent examples showing the power and limitations of the method. The topics covered include forbidden configurations, geometric constructions, saturated hypergraphs in geometry, independent sets in graphs, the regularity lemma, and VC-dimension.

1. Introduction

Among n distinct points in the plane the unit distance occurs at most $O(n^{3/2})$ times. The proof of this fact uses two things. The first is a theorem from graph theory saying that a graph on n vertices containing no $K_{2,3}$ can have at most $O(n^{3/2})$ edges. The second is a simple fact from plane geometry: the unit distance graph contains no $K_{2,3}$.

This is the first application of graph theory in geometry, and is contained in a short and extremely influential paper of Paul Erdős [1946]. The first application of hypergraph theory in geometry is even earlier: it is the use of Ramsey's theorem in the famous Erdős and Szekeres result from 1935 (see below in the next section). Actually, Erdős and Szekeres proved Ramsey's theorem (without knowing it had been proved earlier) since they needed it for the geometric result.

The aim of this survey is to collect and explain some geometric results whose proof uses graph or hypergraph theory. Such applications vary in depth and difficulty. Often a very simple geometric statement adds an extra condition to the combinatorial structure at hand, which helps in the proof. At other times, the geometry is not so simple but is dictated by the combinatorics of the objects in question.

I do not attempt to give a complete list of such results, but rather concentrate on typical or recent examples showing the power and limitations of such methods. Instead of presenting complete proofs I have tried to give a sketch emphasizing the interaction between geometry and (hyper)graph theory. To fill in the details the reader is advised to consult the original papers and the excellent books [Matoušek 2002] and [Pach and Agarwal 1995]. Although I've tried to incorporate every important result, the choice of material, of course, reflects my personal preferences. Also, several further examples could have been included: the Lovász Local Lemma, discrepancy results, planar graphs and geometric graphs, etc. But in these cases I felt that either the method is more probabilistic than combinatorial, or the question is not so much geometric.

Some remarks on notation are in place here: b, c, c_i, C denote different constants. The $O(\ )$ and $o(\ )$ notation is often used. $K_{n,m}$ denotes the complete bipartite graph with classes of size n and m. $K^k(t)$ stands for the complete k-partite k-uniform hypergraph with t vertices in each class. The set $\{1, 2, \ldots, n\}$ will be denoted simply by $[n]$. A graph is denoted by $G = (V, E)$ where V is the set of vertices, and E the set of edges. The independence number $\alpha(G)$ of a graph G is the maximum size independent set in G, and a subset $W \subset V$ is independent if there are no edges between vertices of W. A hypergraph, or set system, is usually denoted by H, its ground set (or vertex set) by V, its (hyper)edges are $e \in H$, or sometimes $E \in H$. A transversal of H is a set $T \subset V$ intersecting every edge in H.

2. Forbidden Configurations

This method is typically used for counting geometric objects. It is usually based on a simple geometric fact (showing that some configuration cannot occur) combined with a graph or hypergraph theorem saying that, if certain configuration is forbidden, then the number of edges is bounded. The case of the unit distance graph in the introduction illustrates the method quite clearly; this section gives a few more examples. We mention in passing that the unit distance problem is still wide open: the maximal number of unit distances among n points is somewhere between $n^{1+(c/\ln\ln n)}$ and $cn^{4/3}$.

The first example is counting point-line incidences: Given a set of lines, L, and a set of points, P, both of them finite, how many incidences can there be? We only assume that two lines have at most one point in common and there is at most one line passing through two points. (So we are not working in the Euclidean plane.) The setting immediately defines a bipartite graph with bipartition classes L and P, with $(\ell, p) \in L \times P$ forming an edge if they are incident. This is a bipartite graph containing no $K_{2,2}$. Then a theorem of Kővári, T. Sós, and Turán [Kővári et al. 1954] applies. We state the result for the case when $|L| = |P| = n$: such a graph has at most

$$\frac{n}{2}(1 + \sqrt{4n - 3})$$

edges. This bound is asymptotically tight: the example of the projective plane of order q (where q is a prime power) shows $n = q^2 + q + 1$ points and the same number of lines while the number of incidences is exactly

$$(q^2 + q + 1)(q + 1) = \frac{n}{2}(1 + \sqrt{4n - 3}).$$

A miracle has happened: of the whole point-line structure, only the bipartiteness and the forbidden subgraph $K_{2,2}$ are needed to obtain the exact bound. It is worth mentioning that while this exact bound follows from the forbidden subgraph theorem [Kővári et al. 1954], the sharpness of the forbidden subgraph theorem is implied by the example of the projective plane. So geometry pays back its due to combinatorics.

Remark. The situation is different when the points and lines belong to the Euclidean plane (cf. the Szemerédi–Trotter theorem [1983]) but there, the structure is richer. The actual bound is $O(|P|^{2/3}|L|^{2/3}+|P|+|L|)$ which is tight apart from the implied constant. There are several proofs available now: the simplest is by L. Székely [1997] based on the crossing lemma. The above forbidden subgraph argument, combined with the so-called cutting lemma, also provides a nice proof, for details see [Matoušek 2002].

Remark. The original motivation for bounding the number of edges in a (bipartite) graph with no $K_{2,2}$ comes from number theory, see [Erdős 1938]. Erdős proves the weaker bound $3n^{3/2}$ on the number of edges but gives the example of the finite projective plane (in disguise) to show that the bound is quite good.

Examples of this type abound. Here is a less well known one due to Turán [1970].

THEOREM 1. *If $X \subset \mathbb{R}^2$ has n elements and is of diameter one, then there are at least $n^2/6 - O(n)$ pairs $x, y \in X$ whose distance is at most $1/\sqrt{2}$.*

The proof is simple. First a little geometry: Among any four points of X there are two that are at distance $1/\sqrt{2}$ or closer. (One cannot give a bound smaller than $1/\sqrt{2}$: see the square of diameter one.) So the graph $G(X, E)$, whose edges are the pairs with distance larger than $1/\sqrt{2}$, contains no K_4. By Turán's theorem [1941] the complementary graph has at least $n^2/6 - O(n)$ edges. This proof also indicates which set of n points shows that the bound $n^2/6 - O(n)$ is tight.

The classical Erdős–Szekeres theorem [1935] uses, in its proof, a certain forbidden configuration. We say that n points in the plane are *in convex position* if they form the vertices of a convex n-gon. We now state the Erdős–Szekeres theorem:

THEOREM 2. *For every $n \geq 3$ there is $N = N(n)$ such that every point set $X \subset R^2$ in general position with $|X| \geq N$ contains a subset of size n that is in convex position.*

For the proof one checks that $N(4) = 5$, that is, among 5 points in the plane there are 4 in convex position. Now set $N(n) = R_4(5, n)$, the Ramsey number, which means that in every red-blue colouring of all quadruples of an $R_4(5, n)$-set either there are 5 points whose all quadruples are red or there are n points whose all quadruples are blue. This number is finite (by the Ramsey theorem, [Ramsey 1930]). Now let $X \subset R^2$ contain N or more elements. Colour its quadruples in convex position Blue, and colour the rest Red. There are no 5 points whose all quadruples are Red (since $N(4) = 5$), so there are n points in X with all of their quadruples in convex position. It is very simple to see now that these n points are also in convex position. Here the forbidden configuration was 5 points with all of its quadruples nonconvex.

Our examples so far have shown forbidden subgraphs. Often other structures are forbidden. Here comes the beautiful case of *lower envelope of segments* in R^2. The setting is this: given n line segments in the plane, none of them vertical, what is the complexity of their lower envelope? That is, consider the segments as linear functions, each defined on some interval, take the pointwise minimum, f, of these functions. How many segments make up the graph of this minimum? The answer is $cn\alpha(n)$, where $\alpha(n)$ is a very slowly increasing function, the inverse of the Ackerman function. Without going into the details (which can be found in [Hart and Sharir 1986] and [Matoušek 2002]), I explain what kind of forbidden structure appears here.

Index the segments by $1, \ldots, n$. The function $f(x)$ is piecewise linear. Assume $I_1, I_2, \ldots, I_t$ are the intervals (in this order on the horizontal axis) where f is linear. (So we want to estimate t, the number segments on the graph of f.) Attach index i to the interval I_k if the graph of f coincides with the ith segment on I_k. Writing the various indexes, as they appear on the horizontal axis from left to right, we get a sequence $a_1, a_2, \ldots, a_t$ of numbers from $[n]$ that has the following properties:

- $a_i \neq a_{i+1}$,
- there are no indices $i_1 < i_2 < i_3 < i_4 < i_5$ such that $a_{i_1} = a_{i_3} = a_{i_5} \neq a_{i_2} = a_{i_4}$.

Only the second property (saying that a, b, a, b, a cannot be a subsequence of our sequence) needs a proof, and we leave it to the reader. This is a *forbidden subsequence* condition. Sequences with these properties are called Davenport–Schinzel sequences of order 3. Determining the maximal length of such a sequence on $[n]$ had been an open problem from 1965 until Hart and Sharir [1986] proved, by combinatorial methods, that the maximal length is $O(n\alpha(n))$. That this bound is sharp was shown later (by Peter Shor; see [Matoušek 2002]). The ingenious construction gives n segments whose lower envelope has $cn\alpha(n)$ segments. Once again, combinatorics gives the upper bound in a geometric problem, and a geometric construction shows that this bound is precise.

Further examples of forbidden configurations can be found in the books [Pach and Agarwal 1995] and [Matoušek 2002].

3. Constructions

Any hypergraph H on n vertices gives rise, in a natural way, to a point set $X(H)$ in R^n. Simply represent each $S \in H$ by its characteristic vector $x(S)$ whose ith component is one if the ith element of the ground set is in S and is zero otherwise. This set $X(H)$ is, in fact, a subset of the vertices of the unit cube. The properties of the hypergraph are reflected in the properties of $X(H)$ and vice versa. This simple connection, combined with powerful results from extremal set theory, can have amazing results, like the counterexample to Borsuk's conjecture.

In 1933 Borsuk asked whether every set of diameter one in R^d can be partitioned into $d+1$ sets of diameter smaller than one. One may immediately assume that the sets in question are convex since taking convex hull does not increase the diameter. Among convex sets, the regular simplex and the unit ball can indeed be partitioned into $d+1$ sets of smaller diameter (but not into fewer sets). This had been known for smooth convex bodies as well (with a fairly simple proof), but for polytopes, despite many efforts, there had been no proof in sight. Then, in 1992, an ingenious construction was found by Kahn and Kalai [1993] showing that the conjecture is far from being true: the smallest number of sets in a suitable partition must be at least $2^{c\sqrt{d}}$ for some small positive c. Their construction is based on the following, equally beautiful, result of Frankl and Wilson [1981]:

THEOREM 3. *Let q be a prime power. Let F be a family of $2q$-subsets of $[4q]$ so that no two sets in F have intersection of size q. Then*

$$|F| \leq 2\binom{4q-1}{q-1}.$$

How does one use this result to produce a counterexample? Consider the edges of the complete graph $K(V,E)$ whose vertex set is $V = [4q]$. For every partition $P = \{A,B\}$ of V let $S(A,B)$ be the set of edges connecting a vertex in A to one in B. Now define H to be the family of sets $S(A,B)$ where $|A| = |B| = 2q$. So H is a $4q^2$-uniform hypergraph on the set E, $|E| = 2q(4q-1)$, which gives rise to a point set $X(H)$ in $R^{|E|}$. As is easy to see, the smallest intersection between $S_1 = S(A_1,B_1) \in H$ and $S_2 = S(A_2,B_2) \in H$ occurs when $|A_1 \cap A_2| = q$. It follows that the Euclidean distance between $x(S_1)$ and $x(S_2)$ is the largest when $|A_1 \cap A_2| = q$. By the Frankl–Wilson theorem every subfamily of H with more than $2\binom{4q-1}{q-1}$ sets contains two sets, S_1 and S_2 with $|A_1 \cap A_2| = q$. That is, when partitioning H into fewer than

$$h(q) = \frac{\frac{1}{2}\binom{4q}{2q}}{2\binom{4q-1}{q-1}}$$

subfamilies, one of them contains a pair S_1 and S_2 with $|A_1 \cap A_2| = q$. The same applies to $X(H)$ which sits in $d = 2q(4q-1)$-dimensional space: in any partition of $X(H)$ into fewer than $h(q)$ sets one of the sets has the same diameter as $X(H)$. It is easy to see that $h(q)$ grows faster than $1.2^{\sqrt{d}} > d + 1$ if d is large enough. This is the first counterexample to Borsuk's conjecture. Several others with direct proofs and better estimates are available now. For a comprehensive survey, see [Raĭgorodskiĭ 2001].

The morale is that geometric intuition can be misleading in higher dimension. Taking convex hulls may not help at all and the discrete structure of the point set can be more important.

The Frankl–Wilson theorem has further geometric applications, many of them given in the original paper [Frankl and Wilson 1981]. They show for instance that the chromatic number, $g(d)$, of R^d is exponential: $g(d) > (1 + o(1))1.2^d$. Here $g(d)$ is defined as the smallest number n such that R^d can be coloured by n colours so that no two points of the same colour are distance one apart. The question of estimating $g(2)$ and more generally $g(d)$ goes back to E. Nelson, J. Isbell, and P. Erdős; see [Hadwiger 1961]. Determining $g(d)$ has turned out to be hard. For instance, the value of $g(2)$ is known to be either 4,5,6, or 7, but which of these numbers it is remains a mystery, after 60 years. Larman and Rogers [1972] proved that $g(d) \leq 3^d$. This, together with the Frankl–Wilson theorem shows that the chromatic number of R^d is exponential in d.

Geometric intuition did not help in the following construction, which is based on extremal hypergraph theory. Danzer and Grünbaum [1962] showed that among $2^d + 1$ points in R^d there are three that form an acute triangle. (The proof is beautiful!) This raised the question to determine the smallest N such that among any set of N points in R^d, there are three that form an angle $\geq \pi/2$. It was conjectured that the smallest such N is $2d - 1$. But this was soundly refuted by Erdős and Füredi [1983] with the following example, which is quite natural once you have seen it. Consider the vertices of the unit cube. Clearly, no angle is larger than $\pi/2$. Three vertices a, b, c give angle $\pi/2$ at b if and only if the vectors $a - b$ and $c - b$ are orthogonal. As a, b, c are 0-1 vectors, they are characteristic vectors of sets $A, B, C \subset [d]$. The condition $(a - b)(c - b) = 0$ translates directly to $A \cap C \subset B \subset A \cup C$. Thus the target is to construct a *large* family H of sets on the ground set $[d]$ with no three sets $A, B, C \in H$ satisfying $B \subset A \cup C$ (a slightly weaker yet sufficient condition). A quite natural random hypergraph with 1.13^d edges has this property. In the corresponding set in R^d, with 1.13^d points, all angles are smaller than $\pi/2$. For details see [Erdős and Füredi 1983], where the authors also prove, with similar methods, the existence of a set in R^d of size exponential in d such that all distances between two points of the set are between .99 and 1.01.

4. Saturated Hypergraphs

The saturated hypergraph theorem of Erdős and Simonovits [1983] says the following:

THEOREM 4. *For every positive integer k and t and every $p > 0$ there exists $\delta > 0$ with the following property. Let H be a k-uniform hypergraph on n vertices and with at least $p\binom{n}{k}$ edges. Then H contains at least*

$$\lfloor \delta n^{kt} \rfloor$$

copies (not necessarily induced) of $K^k(t)$.

One way to remember the statement is to assume that H is a random k-uniform hypergraph with edge-probability p. Then the expected number of copies of $K^k(t)$ is $p^{t^k}\binom{n}{t,\ldots,t} \geq \text{const } n^{kt}$. The saturated hypergraph theorem says that a hypergraph with positive edge density behaves like a "random hypergraph" of the same edge density. It is not surprising then that the proof of Theorem 4 goes by averaging.

This theorem is very useful when one has a family F of geometric objects and happens to know that a positive fraction of the k element subfamilies of F have a certain property, and one wants to show that, say, F has a large subfamily with some other property. Our example is the following point-selection theorem of Alon et al. [1992], a similar and earlier example is in [Bárány et al. 1990].

THEOREM 5. *Let $X \subset \mathbb{R}^d$ be an n-point set and let F be a family of some $(d+1)$-tuples of X with $|F| = \alpha\binom{n}{d+1}$, where $\alpha \in (0,1]$. Then F contains a subfamily F' of size*

$$c_d \alpha^{s_d}\binom{n}{d+1}$$

(where $c_d > 0$ and s_d are constants) such that $\bigcap_{S \in F'} \text{conv } S$ is nonempty.

In this theorem α may even depend on n, a case which is needed when bounding the number of halving hyperplanes of a given n-set in R^d (see [Bárány et al. 1990] and [Alon et al. 1992]).

What is the way of proving such a result? The first (geometric) idea is to use the fractional Helly theorem of [Katchalski and Liu 1979]. It says that if in a family of N convex sets (in R^d) a positive fraction of the $(d+1)$-tuples are intersecting, then the family has a large, cN size intersecting subfamily. So we call the convex hull of an edge in F a *simplex* of F, and try to show that a positive fraction of the $(d+1)$-tuples of the simplices of F are intersecting. Then comes the second (combinatorial) idea: F is a $(d+1)$-uniform hypergraph with positive edge density, thus the saturated hypergraph theorem stated above ensures that there are many copies of $K^{d+1}(t)$ for any fixed number t. So the next target is to prove that such a $K^{d+1}(t)$ contains $(d+1)$ vertex-disjoint simplices that intersect, provided t is large enough. Actually one has the freedom of choosing t as large as needed provided it depends only on d. Once this is proved, a routine double-counting argument shows that a positive fraction of the $(d+1)$-tuples of simplices of F are intersecting. So what remains to be shown is a geometric statement, called the Coloured Tverberg Theorem:

THEOREM 6. *Given pairwise disjoint sets $C_1, \ldots, C_{d+1} \subset R^d$ each with $|C_i| = 4d+3$, there are pairwise disjoint sets $S_1, \ldots, S_{d+1} \subset R^d$, each with $|S_j| = d+1$, such that $|C_i \cap S_j| = 1$ for all i, j and*

$$\bigcap_i^{d+1} \text{conv } S_j \neq \varnothing.$$

Here the C_i are the classes (called colours) of $K^{d+1}(t)$, the convex hull of each edge of $K^{d+1}(t)$ is a simplex of F, and the S_j are what we are after: an inter-secting $(d+1)$-tuple of pairwise vertex-disjoint simplices of F. The proof of this theorem, which is due to Živaljević and Vrećica [1992], is difficult and unusual since it is based on equivariant algebraic topology, although the statement is from convex geometry, or linear algebra, if you wish. In fact, all proofs for $d > 2$ use algebraic topology.

Another example of this kind is a lattice-point version of the fractional Helly theorem, due to Bárány and Matoušek [2003]. Assume that in a finite family F of convex sets in R^d the intersection of every $(d+1)$ sets contains a lattice point, i.e., a point all of whose coordinates are integral. Helly's theorem says that all the sets have a common point. But this may not be a lattice point: take, for instance, the convex hull of all but one vertices of the unit cube in R^d, this is one convex set for each (missing) vertex of the cube. They form a family F where every $2^d - 1$ sets share a lattice point, but $\bigcap F$ contains no lattice point whatsoever. However, it is known (see [Doignon 1973] or [Scarf 1977]) that the Helly number of lattice convex sets in R^d is 2^d, that is, if in a finite family F of convex sets in R^d every 2^d or fewer sets have a lattice point in common, then $\bigcap F$ contains a lattice point. In the given case this implies that the fractional Helly number of lattice convex sets in R^d is (at most) 2^d. (This fact is proved in [Alon et al. 2002].) So what is the precise value of this number? The answer is $d + 1$:

THEOREM 7. *For every $d \geq 1$ and every $\alpha \in (0, 1]$ there is a $\beta > 0$ with the following property. Let $K_1, \ldots, K_N$ be convex sets in R^d such that $\bigcap_{i \in I} K_i$ contains a lattice point for at least $\alpha \binom{N}{d+1}$ index sets $I \subset [N]$ of size $(d + 1)$. Then there is a lattice point common to at least βN sets among the K_i.*

In the proof the application of the saturated hypergraph theorem leads to what we call the coloured Helly theorem for convex lattice sets:

THEOREM 8. *For every integer d and r, there is an integer t such that the following holds. Assume that for each vertex v of $K^{d+1}(t)$ there is a convex set $K_v \in R^d$, such that for each edge e of $K^{d+1}(t)$, the intersection $\bigcap_{v \in e} K_v$ contains a lattice point. Then there is a set R, of size r, in one of the classes of $K^{d+1}(t)$ such that the intersection $\bigcap_{v \in R} K_v$ contains a lattice point.*

This is only needed for $r = 2^d$, but that does not seem to make any difference in the proof, which, besides using two distinct pieces of geometry, is technical, difficult, and combinatorial in nature. The method can be developed further and, when combined with the Alon–Kleitman technique [1992], it shows what can be saved from Helly's theorem when every $(d+1)$ of the sets have a lattice point in common:

THEOREM 9. *For every integer $d \geq 2$ there is an integer $H(d)$ such that the following holds. Let F be a finite family of convex sets in R^d. Assume that the intersection of every $(d+1)$ sets from F contains a lattice point. Then there is a set S of lattice points with $|S| \leq H(d)$ such that S intersects every set in F.*

For $d = 2$ this was proved by T. Hausel [1995] with $H(2) = 2$.

The applications of the saturated hypergraph theorem always lead to new, and often difficult, problems in geometry. In such problems the vertices of a $K^{d+1}(t)$ are some geometric objects, the objects in each edge satisfy a certain property, and one wants to find a special subfamily of these objects, like in Theorem 9 or in the Coloured Tverberg Theorem.

5. Independent Sets in Graphs

Given a graph $G(V, E)$ on n vertices and maximum degree d, the simplest possible greedy algorithm produces an independent set W of size $n/(d+1)$. (An equally simple random choice gives an independent set of size $n/4d$.) In a seminal paper [Ajtai et al. 1981], Ajtai, Komlós, and Szemerédi showed that this can be improved for triangle-free graphs: if G is triangle free, then

$$\alpha(G) \geq \frac{cn \log d}{d}$$

with some universal constant $c > 0$. Subsequently $c = 1 + o(1)$ was shown by Shearer [1983]. Here d is fixed and n goes to infinity. The original proof goes via sequential random choices, and the difficulty is to ensure that after each iteration, the remaining structure is still random, or behaves as if it were random. According to his coauthors, Szemerédi's philosophy, that random subgraphs of a graph behave very regularly, and his vision that such a proof should work, proved decisive. Since then, the method has been applied several times and with great success.

This lower bound on $\alpha(G)$ has the immediate corollary (see [Ajtai et al. 1980]) that the Ramsey number $R_2(n, 3)$ is $O(n^2/\ln n)$ which turned out to be the right order of magnitude (see [Kim 1995]). The result on $\alpha(G)$ has been generalized from triangle-free graphs to "locally sparse" graphs and hypergraphs in various ways. Locally sparse here means, for instance, that there are few edges connecting the neighbours of every vertex, or that two vertices don't have too many common neighbours. We are going to explain two such cases: the problems come from geometry and the solution, or the crucial step of the solution, from hypergraph theory.

The first concerns Heilbronn's conjecture which says that every set of N points in the unit disk B contains three points such that the triangle spanned by them has area less then $const/N^2$. In 1982 Komlós, Pintz, and Szemerédi [Komlós et al. 1982] constructed a counterexample to this conjecture. In the next few paragraphs I describe their construction, starting with the geometric part which is simpler and perhaps more probabilistic than geometric.

Choose first n points randomly, independently, and uniformly from B, set $t = n^{0.1}$ and $\triangle = \frac{t^2}{100n^2}$. ($N$ is going to be smaller than n.) Write V for the set of these points and call a triangle with vertices from V *small* if its area less than $\triangle$. The small triangles define a hypergraph H on V. The target is to show that H contains a large independent set $W \subset V$. The probability that three random points span a small triangle is less than

$$\int_0^2 \frac{8\triangle}{r} 2r\pi dr = 32\pi\triangle < \frac{t^2}{n^2}.$$

This can be seen by fixing two points at distance r, and then averaging over r. The expected size of H is less than $nt^2/6$. Hence by Markov's inequality,

$$|H| < nt^2/3$$

with probability at least $1/2$.

A 2-cycle in H is $e_1, e_2 \in H$ with $|e_1 \cap e_2| = 2$, a 3-cycle is $e_1, e_2, e_3 \in H$ with $|e_i \cap e_j| = 1$ for all distinct i, j, and a 4-cycle is $e_1, e_2, e_3, e_4 \in H$ with $|e_i \cap e_j| = 1$ if $j = i + 1 \mod 4$ and 0 if $j = i + 2 \mod 4$. The following facts are checked easily: with high probability

- the number of 2-cycles is less than $n^{0.1}$,
- the number of 3-cycles is less than $n^{0.7}$,
- the number of 4-cycles is less than $n^{0.7}$.

Thus deleting all vertices in 2-,3-, or 4-cycles you get, with positive probability, a new 3-uniform hypergraph H^* on ground set V^* where $|V^*| = n(1 - o(1))$. The next, and crucial, step is plain hypergraph theory.

LEMMA 10. *Assume H is a 3-uniform hypergraph on $[n]$ with at most $nt^2/3$ edges, without cycles of length $2, 3, 4$, and let $t \leq n^{0.1}$. Then H contains an independent set W with*

$$|W| > const\frac{n}{t}\sqrt{\ln t}.$$

Setting $N = const\frac{n}{t}\sqrt{\ln t}$ we have a point set W in the unit disk, of N points, without small triangles; moreover $\triangle = ct^2/n^2 = c(\ln N)/N^2$. This is the counterexample to Heilbronn's conjecture.

The crucial Lemma 10 is an improvement over the simple estimate $\alpha(H) > n/(3t)$ which is true even if short cycles are not excluded. The proof is by sequential random choices, validating, once more, Szemerédi's philosophy. The short cycle condition guarantees that the hypergraph is locally sparse, in the sense that the neighbourhoods of two distinct vertices are "independent".

Although Lemma 10 is often very useful, it typically improves an existing estimate by a log-factor. In the Heilbronn case, for instance, it is not at all clear where the truth lies. To decide where it lies, most probably, quite different methods will be needed.

In contrast with Heilbronn's problem, the next application of the improved independence number method gives an almost precise answer to a geometric problem. It is a recent result of Kim and Vu [2004]. We need to introduce some terminology.

A graph $G(V, E)$ is (d, ε)-regular if its degrees are between $d(1 - \varepsilon)$ and d. The *codegree* of a the graph, $D = D(G)$ is the maximum number of common neighbours of $x, y \in V$, $x \neq y$. An independent set $W \subset V$ is called *maximal* if it is not contained in a larger independent set. In the following theorem, which is from [Kim and Vu 2004], the asymptotics is understood with $d \to \infty$ and $w(d)$ denotes a function that tends to infinity as $d \to \infty$.

THEOREM 11. *Let G be a (d, ε)-regular graph on n vertices, where*

$$\varepsilon = (w(d) \ln d)^{-1}.$$

If

$$D(G) \leq \frac{d}{w(d) \ln^2 d},$$

then G contains a maximal independent set W with

$$(1 + o(1)) \frac{n}{d} \ln \frac{d}{D} \leq |W| \leq (1 + o(1)) \frac{n}{d} \ln \frac{d}{D} + w(d) \frac{n}{d} D \ln^2 D.$$

The error term $w(d) \frac{n}{d} D \ln^2 D$ is dominating if $w(d) D \ln^2 D$ is larger than $\ln \frac{d}{D}$. Otherwise, that is, when $w(d) D \ln^2 D = o(\ln \frac{d}{D})$, G contains a maximal independent set of size $(1 + o(1)) \frac{n}{d} \ln \frac{d}{D}$. The method is, again, a sequential random choice of vertices but the remainder term has to be estimated precisely which makes the proof hard.

This result is used in [Kim and Vu 2004] to answer a question of Segre from 1959 (see [Szőnyi 1997]) on arcs in projective planes. An *arc* in a projective plane P of order q is a set $A \subset P$ containing no three points on a line. An arc is *complete* if it is not contained in a larger arc. Segre's question is this: What are the possible sizes of complete arcs in P? Simple counting arguments, using properties of the projective plane, show that the size of a complete arc is always between $\sqrt{2q}$ and $q + 2$. Szőnyi [1997] showed that almost all values in the interval $[cq^{3/4}, q]$ can be the size of a complete arc. Kim and Vu [2003] showed the existence of complete arcs whose size is $\sqrt{q}(\ln q)^b$ with some universal constant b. This is close to the lower bound $\sqrt{2q}$. Further, it is proved in [Kim and Vu 2004] that sizes of complete arcs in P are almost dense in the interval $[\sqrt{2q}, q]$.

THEOREM 12. *There are positive constants b, c and Q such that the following holds. For every plane P of order $q \geq Q$ and every $q* \in [\sqrt{q} \ln^4 q, q]$, P contains a complete arc A with*

$$cq^* \leq |A| \leq q^* \ln^b q.$$

The proof uses the fact that the conic $C = \{(x, x^2) : x \in GF(q)\}$ is an arc in P whose secants cover every point of $P \setminus C$ (except the one at infinity) $q/2 - O(1)$ times. Set $D = P \setminus C$ and $\varepsilon = (\sqrt{q} \ln q)/q^*$, so ε is small when q is large. Given an arc $A \subset D$ one defines a graph $G_A(V, E)$ as follows: V is the set of points $v \in C$ not covered by secants from A, and $u, v \in V$ form an edge in E if there is $a \in A$ with a, u, v collinear. One has to show next (the proof is hard and probabilistic) that there is an arc $A \subset D$, of size at most $2\varepsilon\sqrt{q}$, such that G_A satisfies the conditions of Theorem 11. Then one applies Theorem 11 and an additional argument to show that G_A contains a maximal independent set of the desired size such that its secants cover $D \setminus A$. Further details of the proof (that are even less geometric) can be found in the forthcoming [Kim and Vu 2004].

Results like Lemma 10 and Theorem 11 have been used to find a large matching in a hypergraph: Given a hypergraph H, a matching M is a collection of pairwise disjoint edges. Define the *intersection graph*, $G(H)$ of H as follows: its vertex set is H, and two vertices, $e, f \in H$ form an edge in $G(H)$ if $e \cap f = \varnothing$. So a matching in H corresponds to an independent set in $G(H)$, and a large independent set corresponds to many pairwise disjoint edges. Further, using such a matching one can find an economic cover of the ground set by edges. This happens if the set of vertices left uncovered by the matching is small. In other words, if the estimate of error term is precise. This is a very promising area with plenty of results and conjectures. Their geometric applications are waiting to be discovered.

6. The Regularity Lemma

Szemerédi's famous regularity lemma is one of the most important and useful results in combinatorics, it has millions of applications in discrete mathematics, but surprisingly few in geometry. Here is a remarkably elegant one, due to János Pach [1998].

THEOREM 13. *For every $d \geq 2$ there is a positive constant c_d with the following property. Given sets $X_1, \ldots, X_{d+1} \subset R^d$, each of size n, there are subsets $Z_i \subset X_i$, ($i \in [d+1]$), each of size at least $c_d n$ such that*

$$\bigcap \text{conv}\{z_1, \ldots, z_{d+1}\} \neq \varnothing,$$

where the intersection is taken over all transversals $z_i \in Z_i$, $i \in [d+1]$.

The proof uses several ingredients: the point selection theorem (Theorem 5), a weak form of the regularity lemma for hypergraphs, and the so-called same-type lemma from [Bárány and Valtr 1998]. To state the last one we say that the sets $Z_1, \ldots, Z_k$ in R^d have *same type transversals* if there is no hyperplane intersecting the convex hull of any $d+1$ of them. (For various equivalent definitions see [Bárány and Valtr 1998] or [Matoušek 2002].) What we will need is the following fact. If $Z_1, \ldots, Z_{d+2}$ have same type transversals, and if some $z_1 \in Z_1, \ldots, z_{d+2} \in Z_{d+2}$ satisfies $z_{d+2} \in \operatorname{conv}\{z_1, \ldots, z_{d+1}\}$, then $w_{d+2} \in \operatorname{conv}\{w_1, \ldots, w_{d+1}\}$ holds for all $w_1 \in Z_1, \ldots, w_{d+2} \in Z_{d+2}$. (Hopefully, this also explains the meaning of "same type".)

LEMMA 14. *For every $d \geq 2$ and every $k \geq d+1$ there is a positive constant $b(d,k)$ with the following property. Given nonempty sets $X_1, \ldots, X_k \subset R^d$ in general position, there are subsets $Z_i \subset X_i$, $(i \in [k])$, each with $|Z_i| \geq b(d,k)|X_i|$ such that $Z_1, \ldots, Z_k$ have the same type transversals.*

Remark. Ramsey's theorem guarantees the existence of sets Z_i with this property but their size is much smaller than cn. Here geometry is needed to guarantee linear size.

The proof of Theorem 13 begins by forming the $(d+1)$-uniform hypergraph H whose edges are the sets $\{x_1, \ldots, x_{d+1}\}$ with $x_i \in X_i$. H has $(d+1)n$ vertices and n^{d+1} edges, so Theorem 5 gives a subhypergraph $H^* \subset H$ and a point $z \in R^d$ such that $|H^*| \geq \beta n^{d+1}$ and $z \in \operatorname{conv} e$ for each edge $e \in H^*$, where $\beta > 0$ depends only on d.

Next, a weak form of the regularity lemma for hypergraph (see [Pach 1998]) is needed. Without stating it we just claim that it ensures the existence of $Y_i \subset X_i$, $|Y_i| \geq \gamma |X_i|$ such that for every subset $Z_1 \subset Y_1, \ldots, Z_{d+1} \subset Y_{d+1}$ with $|Z_i| \geq b(d, d+2)|Y_i|$ there are vertices $z_i \in Z_i$ $i \in [d+1]$ such that $\{z_1, \ldots, z_{d+1}\}$ is an edge of H^*. Here $\gamma > 0$ depends only on d.

Finally, one applies the same type lemma for the sets $Y_1, \ldots, Y_{d+1}$ and $Y_{d+2} = \{z\}$. This gives sets $Z_i \subset Y_i$ ($i \in [d+1]$), each of size at least $b(d, d+2)|Y_I|$, and $Z_{d+2} = \{z\}$ with same type transversals. By the weak regularity lemma, there is at least one simplex with vertices $z_i \in Z_i$, $i \in [d+1]$ that contains z. Then, by the same type lemma, all such simplices contain z. This finishes the proof.

It is high time to state the original regularity lemma now. We need some terminology: Given a graph $G(V, E)$, and disjoint sets $X, Y \subset V$, their *density* is defined as

$$d(X, Y) = \frac{|E(X, Y)|}{|X| \cdot |Y|},$$

where $E(X, Y)$ is the set of edges between X and Y. Given some $\delta > 0$, and disjoint $A, B \subset V$, the pair (A, B) is called δ-*regular* if, for every $X \subset A$ and $Y \subset B$ satisfying $|X| > \delta|A|$ and $|Y| > \delta|B|$ we have

$$|d(X, Y) - d(A, B)| < \delta.$$

Now we state the regularity lemma of Szemerédi [1978] in the hope that it will find further geometric applications.

THEOREM 15. *Given $\delta > 0$ and an integer m, there is an $M = M(\delta, m)$ such that the vertex set of every graph $G(V, E)$ with $|V| > m$ can be partitioned into classes $V_0, V_1, \ldots, V_k$, where $m \leq k \leq M$, such that $|V_0| \leq |V_1| = \ldots = |V_k|$ and all but at most δk^2 of the pairs (V_i, V_j), $i, j \in [k]$ are δ-regular.*

A proper illustration of the use of this lemma is a very recent result of Pach, Pinchasi, and Vondrák (manuscript, 2004). This result answers a question of Erdős in the following form: Assume $\varepsilon > 0$, X is a set of n points in R^3, and every two points in X are at distance one at least. If there are εn^2 pairs in X whose distance is between t and $t + 1$ for some $t > 0$, then the diameter of X is at least cn where c only depends on ε.

The conditions immediately cry out for the regularity lemma. In the graph $G(X, E)$, $x, y \in X$ form an edge if $\|x - y\| \in [t, t + 1]$. One obtains two disjoint sets $A, B \subset X$ of size $c_1 n$ with (A, B) ε-regular. This is a very strong condition on the point sets A, B. Using geometry one can find subsets $X \subset A$ and $Y \subset B$, each of size $c_2 n$ and such that $\|x - y\| \in [t, t+1]$ for *every* $x \in X, y \in Y$. Here $c_2 > 0$ depends on ε only. The rest of the proof is 3-dimensional geometry.

Szemerédi's regularity lemma has recently been generalized for hypergraphs by Gowers and by Rödl et al. (unpublished yet) with the potential of having further geometric applications. The regularity lemma is extremely useful in discrete mathematics, but, so far, it has not been applied in geometry very often.

7. VC-Dimension and ε-Nets

Given a hypergraph H with vertex set V, and ε-net (where $\varepsilon \in (0, 1]$) is a subset $N \subset V$ that intersects each edge $E \in H$ with $|E| \geq \varepsilon|V|$. In other words, N is an ε-net for H if it is a transversal for the edges with at least $\varepsilon|V|$ elements. This definition can be extended to "infinite hypergraphs": Assume V is a set, μ is a probability measure on V, and H is a system of μ-measurable sets. Then $N \subset V$ is called an ε-net for H with respect to μ if it intersects every set $E \in H$ whose measure is at least ε.

There is a special condition, of combinatorial nature, that ensures the existence of "very finite" ε-nets. Given a set system H on a finite or infinite ground set V, a set $A \subset V$ is *shattered by H* if each subset of A can be produced as $A \cap E$ for a suitable $E \in H$. The VC-dimension of the set system H, denoted by $\dim H$, is the maximum of the sizes of all finite shattered subsets of V, or ∞ if there are arbitrarily large shattered subsets. The VC-dimension, introduced by Vapnik and Chervonenkis in [1971] has turned out to be a very powerful tool everywhere: in statistics (the original motivation for the VC-dimension), discrete geometry, computational geometry, combinatorics of hypergraphs, and discrepancy theory. The terminology is sometimes different, for instance in computational geometry, the set system H is called *range space* and its edges *ranges*.

A simple example is a set of points V in R^d for which H is formed by the sets of type $V \cap h$ where h is a half-space. The VC-dimension of H is then $d+1$ since, by Radon's theorem, no $(d+2)$-set is shattered by half-spaces in R^d. Another example with finite VC-dimension, on the same ground set V, is the collection of all Euclidean balls.

The reason for the wide range of applications of VC-dimension lies in the very general setting and in the so called ε-net theorem (see [Haussler and Welzl 1987]) and the ε-approximation theorem (introduced in to [Vapnik and Chervonenkis 1971]).

THEOREM 16. *Let V be a set, and μ be a probability measure on V, H a system of μ-measurable subsets of V, and $\varepsilon \in (0,1]$. If $\dim H \leq d$ where $d \geq 2$, then there exists an ε-net for H of size at most $\frac{4d}{\varepsilon} \ln \frac{1}{\varepsilon}$.*

While an ε-net intersects each (large enough) set in H in at least one point, an ε-approximation $M \subset V$ provides a "proportional representation" of each set in H: for each $E \in H$

$$\left| \mu(E) - \frac{|M \cap E|}{|M|} \right| < \varepsilon.$$

THEOREM 17. *Let V be a set, and μ be a probability measure on V, H a system of μ-measurable subsets of V, and $\varepsilon \in (0,1]$. If $\dim H \leq d$ where $d \geq 2$, then there exists an ε-approximation for H, of size at most*

$$\frac{Cd}{\varepsilon^2} \ln \frac{1}{\varepsilon}.$$

The ε-net theorem is more often used in geometry. The following application to an *art gallery* problem is due to Kalai and Matoušek [1997]. An art gallery is a simply connected compact set T in the plane, and the set of points visible from $x \in T$ is, by definition,

$$V(x) = \{y \in T : [x, y] \subset T\}.$$

In other words, x sees or guards the points in $V(x)$.

THEOREM 18. *Let $T \subset R^2$ be a simply connected art gallery of Lebesgue measure one. Assume that for some $r \geq 2$ the Lebesgue measure of each $V(x)$ is at least $1/r$. Then T can be guarded by at most $Cr \ln r$ points, that is, there is a set $N \subset T$, having at most $Cr \ln r$ points, with $T = \cup_{x \in N} V(x)$.*

The proof begins by introducing the set system $H = \{V(x) : x \in T\}$ and noting that a set $N \subset T$ guards T iff it intersects each set in H. So we are done if H admits an $(1/r)$-net of the required size. This is guaranteed by the ε-net theorem provided the VC-dimension of H is bounded by some constant independent of T. This can be shown by a geometric argument using the fact that T is simply connected. The details can be found in [Kalai and Matoušek 1997], or in [Valtr 1998] where $\dim H \leq 23$ is shown.

There are several geometric applications of VC-dimension and the ε-net theorem, see for instance the books [Chazelle 2000], [Matoušek 2002], and [Pach and Agarwal 1995]. Since most of them require new concepts and further preparations that go beyond the limits of this survey, I only explain one more case, that of a spanning tree with low crossing number. The setting is this. Given a set X of n points in R^2 in general position, we want to build a spanning tree (with vertex set X and edge set segments connecting certain pairs of X) such that no line meets too many of the edges. The following beautiful theorem is due to Welzl [1988] (the $\ln n$ factor has been since then removed).

THEOREM 19. *Given a set X of n points in R^2 in general position, there is a spanning tree with vertex set X such that no line meets more than $O(\sqrt{n}\ln n)$ edges of the tree.*

For the proof one checks first that the following set system H has finite VC-dimension: The ground set is the collection of all lines in R^2 and H consists of sets of lines L_s that intersect a fixed segment s. To see that $\dim H$ is finite assume an n element set of lines A is shattered by H. These lines divide the plane into $m \leq \binom{n}{2}+n+1$ cells, and if s and t are two segments whose endpoints (in pairs) belong to the same cell, then L_s and L_t have the same intersection with A. Consequently there are at most $\binom{m}{2}$ segments s for which $L_s \cap A$ are pairwise distinct, so $2^n \leq \binom{m}{2}$ implying that $\dim H \leq n \leq 12$.

LEMMA 20. *Given a set S of k points in general position, and a set L of m lines in R^2 with no point incident to any of the lines, there exist $x, y \in S$ such that the line segment $[x, y]$ intersects at most $(cm \ln k)/\sqrt{k}$ lines from L.*

For the proof one notes that the set system H has finite VC-dimension, so the ε-net theorem applies: with $\varepsilon = c_1(\ln k)/k$ we get a collection of lines $L' \subset L$ of size $c_2\varepsilon^{-1}\ln\varepsilon^{-1} < \sqrt{k}/2$ such that every open segment crossing

$$\varepsilon m = c\frac{m \ln k}{\sqrt{k}}$$

elements of L crosses some line in L'. The lines in L' divide the plane into less than k cells. Thus one cell contains two points of S; the segment connecting them satisfies the requirements of the lemma.

To finish the proof of the spanning tree theorem one starts with constructing a set, L, of $\binom{n}{2}$ lines that represent all possible partitions of X by lines. Setting $S_0 = X$ and $L_0 = L$ one applies the lemma to S_i, L_i ($i = 0, 1, \ldots, n-2$) to obtain a segment $[x_i, y_i]$ intersecting at most $c\frac{m_i \ln n_i}{\sqrt{n_i}}$ from L_i. For the next iteration $S_{i+1} = S_i \setminus \{x_i\}$ and L_{i+1} is the set of lines consisting of L_i plus one more, slightly perturbed, copy of each line in L_i intersecting $[x_i, y_i]$. The analysis of this algorithm finishes the proof; the details can be found in [Welzl 1988] or [Pach and Agarwal 1995].

8. Epilogue

László Fejes Tóth asked in 1976 whether the densest packing of congruent circles in the plane is unique or not in the following sense: Assume that in a circle packing, $\mathcal{C}$, in the plane, every circle is touched by at least six others. Is it true then, that arbitrarily large or arbitrarily small circles occur in $\mathcal{C}$ unless it is the densest packing of congruent circles. The answer is yes and is the content of [Bárány et al. 1984]:

THEOREM 21. *Under the conditions above arbitrarily small circles occur in $\mathcal{C}$ unless $\mathcal{C}$ is the densest packing of congruent circles.*

For the proof one defines the graph $G(V, E)$ whose vertices are the circles with two of them forming an edge if the corresponding circles are touching each other. G is a planar graph. Define the function $f : V \to R$ by $f(v) = 1/r$ when r is the radius of the circle corresponding to $v \in V$. Surprisingly, this function is *subharmonic* on G, that is, $f(v)$ is less than or equal to the average of f on the neighbours of v. This is the first geometric component in the proof. Then one uses, or rather proves a theorem saying that, under suitable conditions on the underlying graph, if a subharmonic function is bounded from above, then it is necessarily constant. Finally, the "suitable" condition follows from the planarity of G. I'm sure that, in the world of geometry, there are hundreds of similar proofs waiting to be discovered.

Acknowledgment

The demand for such a survey, and the idea of writing one, emerged in the semester on Discrete and Computational Geometry at MSRI in Berkeley, 2003, autumn term. I thank the organizers and the participants for the excellent atmosphere and working conditions provided. I'm also grateful to Microsoft Research (in Redmond, WA), as this paper was written on a very pleasant and fruitful visit there. My thanks are due to Anders Björner, Jiří Matoušek, and Van Vu Ha for valuable comments on an earlier version of this paper. Partial support from Hungarian National Foundation Grants No. 046246 and 037846 is acknowledged.

References

[Ajtai et al. 1980] M. Ajtai, J. Komlós, and E. Szemerédi, "A note on Ramsey numbers", *J. Combin. Theory Ser. A* **29**:3 (1980), 354–360.

[Ajtai et al. 1981] M. Ajtai, J. Komlós, and E. Szemerédi, "A dense infinite Sidon sequence", *European J. Combin.* **2**:1 (1981), 1–11.

[Alon and Kleitman 1992] N. Alon and D. J. Kleitman, "Piercing convex sets and the Hadwiger-Debrunner (p, q)-problem", *Adv. Math.* **96**:1 (1992), 103–112.

[Alon et al. 1992] N. Alon, I. Bárány, Z. Füredi, and D. J. Kleitman, "Point selections and weak ε-nets for convex hulls", *Combin. Probab. Comput.* **1**:3 (1992), 189–200.

[Alon et al. 2002] N. Alon, G. Kalai, J. Matoušek, and R. Meshulam, "Transversal numbers for hypergraphs arising in geometry", *Adv. in Appl. Math.* **29**:1 (2002), 79–101.

[Bárány and Matoušek 2003] I. Bárány and J. Matoušek, "A fractional Helly theorem for convex lattice sets", *Adv. Math.* **174**:2 (2003), 227–235.

[Bárány and Valtr 1998] I. Bárány and P. Valtr, "A positive fraction Erdős-Szekeres theorem", *Discrete Comput. Geom.* **19**:3 (1998), 335–342.

[Bárány et al. 1984] I. Bárány, Z. Füredi, and J. Pach, "Discrete convex functions and proof of the six circle conjecture of Fejes Tóth", *Canad. J. Math.* **36**:3 (1984), 569–576.

[Bárány et al. 1990] I. Bárány, Z. Füredi, and L. Lovász, "On the number of halving planes", *Combinatorica* **10**:2 (1990), 175–183.

[Chazelle 2000] B. Chazelle, *The discrepancy method*, Cambridge University Press, Cambridge, 2000.

[Danzer and Grünbaum 1962] L. Danzer and B. Grünbaum, "Über zwei Probleme bezüglich konvexer Körper von P. Erdős und von V. L. Klee", *Math. Z.* **79** (1962), 95–99.

[Doignon 1973] J.-P. Doignon, "Convexity in cristallographical lattices", *J. Geometry* **3** (1973), 71–85.

[Erdős 1938] P. Erdős, "On sequences of integers no one of which divides the product of two others and related problems", *Mitt. Forsch. Institut. Mat. Tomsk.* **2** (1938), 38–42.

[Erdős 1946] P. Erdős, "On sets of distances of n points", *Amer. Math. Monthly* **53** (1946), 248–250.

[Erdős and Füredi 1983] P. Erdős and Z. Füredi, "The greatest angle among n points in the d-dimensional Euclidean space", pp. 275–283 in *Combinatorial mathematics* (Marseille–Luminy, 1981), edited by C. Berge et al., North-Holland Math. Stud. **75**, North-Holland, Amsterdam, 1983.

[Erdős and Simonovits 1983] P. Erdős and M. Simonovits, "Supersaturated graphs and hypergraphs", *Combinatorica* **3**:2 (1983), 181–192.

[Erdős and Szekeres 1935] P. Erdős and G. Szekeres, "A combinatorial problem in geometry", *Compositio Math.* **2** (1935), 463–470.

[Frankl and Wilson 1981] P. Frankl and R. M. Wilson, "Intersection theorems with geometric consequences", *Combinatorica* **1**:4 (1981), 357–368.

[Hadwiger 1961] H. Hadwiger, "Ungelöste Probleme N. 40", *Elem. Math.* **16** (1961), 103–104.

[Hart and Sharir 1986] S. Hart and M. Sharir, "Nonlinearity of Davenport–Schinzel sequences and of generalized path compression schemes", *Combinatorica* **6**:2 (1986), 151–177.

[Hausel 1995] T. Hausel, "On a Gallai-type problem for lattices", *Acta Math. Hungar.* **66**:1-2 (1995), 127–145.

[Haussler and Welzl 1987] D. Haussler and E. Welzl, "ε-nets and simplex range queries", *Discrete Comput. Geom.* **2**:2 (1987), 127–151.

[Kahn and Kalai 1993] J. Kahn and G. Kalai, "A counterexample to Borsuk's conjecture", *Bull. Amer. Math. Soc. (N.S.)* **29**:1 (1993), 60–62.

[Kalai and Matoušek 1997] G. Kalai and J. Matoušek, "Guarding galleries where every point sees a large area", *Israel J. Math.* **101** (1997), 125–139.

[Katchalski and Liu 1979] M. Katchalski and A. a. Liu, "A problem of geometry in $\mathbf{R}^n$", *Proc. Amer. Math. Soc.* **75**:2 (1979), 284–288.

[Kim 1995] J. H. Kim, "The Ramsey number $R(3,t)$ has order of magnitude $t^2/\log t$", *Random Structures Algorithms* **7**:3 (1995), 173–207.

[Kim and Vu 2003] J. H. Kim and V. H. Vu, "Small complete arcs in projective planes", *Combinatorica* **23**:2 (2003), 311–363.

[Kim and Vu 2004] J. H. Kim and V. H. Vu, "Maximal independent sets and Segre's problem in finite geometry", 2004. Manuscript.

[Komlós et al. 1982] J. Komlós, J. Pintz, and E. Szemerédi, "A lower bound for Heilbronn's problem", *J. London Math. Soc.* (2) **25**:1 (1982), 13–24.

[Kővári et al. 1954] T. Kővári, V. T. Sós, and P. Turán, "On a problem of K. Zarankiewicz", *Colloquium Math.* **3** (1954), 50–57.

[Larman and Rogers 1972] D. G. Larman and C. A. Rogers, "The realization of distances within sets in Euclidean space", *Mathematika* **19** (1972), 1–24.

[Matoušek 2002] J. Matoušek, *Lectures on discrete geometry*, Graduate Texts in Mathematics **212**, Springer, New York, 2002.

[Pach 1998] J. Pach, "A Tverberg-type result on multicolored simplices", *Comput. Geom.* **10**:2 (1998), 71–76.

[Pach and Agarwal 1995] J. Pach and P. K. Agarwal, *Combinatorial geometry*, Wiley, New York, 1995.

[Raĭgorodskiĭ 2001] A. M. Raĭgorodskiĭ, "The Borsuk problem and the chromatic numbers of some metric spaces", *Uspekhi Mat. Nauk* **56**:1 (2001), 107–146. In Russian. Translation in *Russian Math. Surveys* **56** (2001), 102–139.

[Ramsey 1930] F. Ramsey, "On a problem of formal logic", *Proc. London Math. Soc.* **30** (1930).

[Scarf 1977] H. E. Scarf, "An observation on the structure of production sets with indivisibilities", *Proc. Nat. Acad. Sci. U.S.A.* **74**:9 (1977), 3637–3641.

[Shearer 1983] J. B. Shearer, "A note on the independence number of triangle-free graphs", *Discrete Math.* **46**:1 (1983), 83–87.

[Székely 1997] L. A. Székely, "Crossing numbers and hard Erdős problems in discrete geometry", *Combin. Probab. Comput.* **6**:3 (1997), 353–358.

[Szemerédi 1978] E. Szemerédi, "Regular partitions of graphs", pp. 399–401 in *Problèmes combinatoires et théorie des graphes* (Orsay, 1976), Colloq. Internat. CNRS **260**, CNRS, Paris, 1978.

[Szemerédi and Trotter 1983] E. Szemerédi and W. T. Trotter, Jr., "A combinatorial distinction between the Euclidean and projective planes", *European J. Combin.* **4**:4 (1983), 385–394.

[Szőnyi 1997] T. Szőnyi, "Some applications of algebraic curves in finite geometry and combinatorics", pp. 197–236 in *Surveys in combinatorics* (London, 1997), edited by R. A. Bailey, London Math. Soc. Lecture Note Ser. **241**, Cambridge Univ. Press, Cambridge, 1997.

[Turán 1941] P. Turán, "On an extremal problem in graph theory", *Matematikai Lapok* **48** (1941), 436–452. In Hungarian.

[Turán 1970] P. a. Turán, "Applications of graph theory to geometry and potential theory", pp. 423–434 in *Combinatorial structures and their applications* (Calgary, 1969), Gordon and Breach, New York, 1970.

[Valtr 1998] P. Valtr, "Guarding galleries where no point sees a small area", *Israel J. Math.* **104** (1998), 1–16.

[Vapnik and Chervonenkis 1971] V. N. Vapnik and A. Y. Chervonenkis, "On the uniform convergence of relative frequencies of events to their probablilities", *Theory Probab. Appl.* **16** (1971), 264–280.

[Welzl 1988] E. Welzl, "Partition trees for triangle counting and other range searching problems", pp. 23–33 in *Proceedings of the Fourth Annual Symposium on Computational Geometry* (Urbana, IL, 1988), ACM, New York, 1988.

[Živaljević and Vrećica 1992] R. T. Živaljević and S. T. Vrećica, "The colored Tverberg's problem and complexes of injective functions", *J. Combin. Theory Ser. A* **61**:2 (1992), 309–318.

IMRE BÁRÁNY
RÉNYI INSTITUTE
POBOX 127,
BUDAPEST, 1364
HUNGARY

AND

DEPARTMENT OF MATHEMATICS
UNIVERSITY COLLEGE LONDON
GOWER STREET
LONDON WC1E 6BT
UNITED KINGDOM
barany@renyi.hu

Combinatorial and Computational Geometry
MSRI Publications
Volume **52**, 2005

Convex Geometry of Orbits

ALEXANDER BARVINOK AND GRIGORIY BLEKHERMAN

ABSTRACT. We study metric properties of convex bodies B and their polars B°, where B is the convex hull of an orbit under the action of a compact group G. Examples include the Traveling Salesman Polytope in polyhedral combinatorics ($G = S_n$, the symmetric group), the set of nonnegative polynomials in real algebraic geometry ($G = \mathrm{SO}(n)$, the special orthogonal group), and the convex hull of the Grassmannian and the unit comass ball in the theory of calibrated geometries ($G = \mathrm{SO}(n)$, but with a different action). We compute the radius of the largest ball contained in the symmetric Traveling Salesman Polytope, give a reasonably tight estimate for the radius of the Euclidean ball containing the unit comass ball and review (sometimes with simpler and unified proofs) recent results on the structure of the set of nonnegative polynomials (the radius of the inscribed ball, volume estimates, and relations to the sums of squares). Our main tool is a new simple description of the ellipsoid of the largest volume contained in B°.

1. Introduction and Examples

Let G be a compact group acting in a finite-dimensional real vector space V and let $v \in V$ be a point. The main object of this paper is the convex hull

$$B = B(v) = \mathrm{conv}(gv : g \in G)$$

of the orbit as well as its polar

$$B^\circ = B^\circ(v) = \{\ell \in V^* : \ell(gv) \leq 1 \text{ for all } g \in G\}.$$

Objects such as B and B° appear in many different contexts. We give three examples below.

EXAMPLE 1.1 (COMBINATORIAL OPTIMIZATION POLYTOPES). Let $G = S_n$ be the symmetric group, that is, the group of permutations of $\{1, \ldots, n\}$. Then

---

Mathematics Subject Classification: 52A20, 52A27, 52A21, 53C38, 52B12, 14P05.

Keywords: convex bodies, ellipsoids, representations of compact groups, polyhedral combinatorics, Traveling Salesman Polytope, Grassmannian, calibrations, nonnegative polynomials.

This research was partially supported by NSF Grant DMS 9734138.

$B(v)$ is a polytope and varying V and v, one can obtain various polytopes of interest in combinatorial optimization. This idea is due to A.M. Vershik (see [Barvinok and Vershik 1988]) and some polytopes of this kind were studied in [Barvinok 1992].

Here we describe perhaps the most famous polytope in this family, the Traveling Salesman Polytope (see, for example, Chapter 58 of [Schrijver 2003]), which exists in two major versions, symmetric and asymmetric. Let V be the space of $n \times n$ real matrices $A = (a_{ij})$ and let S_n act in V by simultaneous permutations of rows and columns: $(ga)_{ij} = a_{g^{-1}(i)g^{-1}(j)}$ (we assume that $n \geq 4$). Let us choose v such that $v_{ij} = 1$ provided $|i - j| = 1 \mod n$ and $v_{ij} = 0$ otherwise. Then, as g ranges over the symmetric group S_n, matrix gv ranges over the adjacency matrices of Hamiltonian cycles in a complete undirected graph with n vertices. The convex hull $B(v)$ is called the *symmetric Traveling Salesman Polytope* (we denote it by ST_n). It has $(n - 1)!/2$ vertices and its dimension is $(n^2 - 3n)/2$.

Let us choose $v \in V$ such that $v_{ij} = 1$ provided $i - j = 1 \mod n$ and $v_{ij} = 0$ otherwise. Then, as g ranges over the symmetric group S_n, matrix gv ranges over the adjacency matrices of Hamiltonian circuits in a complete directed graph with n vertices. The convex hull $B(v)$ is called the *asymmetric Traveling Salesman Polytope* (we denote it by AT_n). It has $(n - 1)!$ vertices and its dimension is $n^2 - 3n + 1$.

A lot of effort has been put into understanding of the facial structure of the symmetric and asymmetric Traveling Salesman Polytopes, in particular, what are the linear inequalities that define the facets of AT_n and ST_n, see Chapter 58 of [Schrijver 2003]. It follows from the computational complexity theory that in some sense one *cannot* describe efficiently the facets of the Traveling Salesman Polytope. More precisely, if *NP* $\neq$ *co-NP* (as is widely believed), then there is no polynomial time algorithm, which, given an inequality, decides if it determines a facet of the Traveling Salesman Polytope, symmetric or asymmetric, see, for example, Section 5.12 of [Schrijver 2003]. In a similar spirit, Billera and Sarangarajan proved that any 0-1 polytope (that is, a polytope whose vertices are 0-1 vectors), appears as a face of some AT_n (up to an affine equivalence) [Billera and Sarangarajan 1996].

EXAMPLE 1.2 (NONNEGATIVE POLYNOMIALS). Let us fix integers $n \geq 2$ and $k \geq 1$. We are interested in homogeneous polynomials $p : \mathbb{R}^n \to \mathbb{R}$ of degree $2k$ that are nonnegative for all $x = (x_1, \ldots, x_n)$. Such polynomials form a convex cone and we consider its compact base:

$$\text{Pos}_{2k,n} = \left\{ p : p(x) \geq 0 \text{ for all } x \in \mathbb{R}^n \text{ and } \int_{\mathbb{S}^{n-1}} p(x) \, dx = 1 \right\}, \qquad (1.2.1)$$

where dx is the rotation-invariant probability measure on the unit sphere $\mathbb{S}^{n-1}$.

It is not hard to see that $\dim \text{Pos}_{2k,n} = \binom{n+2k-1}{2k} - 1$.

It is convenient to consider a translation $\text{Pos}'_{2k,n}$, $p \mapsto p - (x_1^2 + \cdots + x_n^2)^k$ of $\text{Pos}_{2k,n}$:

$$\text{Pos}'_{2k,n} = \left\{ p : p(x) \geq -1 \text{ for all } x \in \mathbb{R}^n \text{ and } \int_{\mathbb{S}^{n-1}} p(x)\, dx = 0 \right\}. \qquad (1.2.2)$$

Let $U_{m,n}$ be the real vector space of all homogeneous polynomials $p : \mathbb{R}^n \to \mathbb{R}$ of degree m such that the average value of p on $\mathbb{S}^{n-1}$ is 0. Then, for $m = 2k$, the set $\text{Pos}'_{2k,n}$ is a full-dimensional convex body in $U_{2k,n}$.

One can view $\text{Pos}'_{2k,n}$ as the *negative polar* $-B^\circ(v)$ of some orbit.

We consider the m-th tensor power $(\mathbb{R}^n)^{\otimes m}$ of $\mathbb{R}^n$, which we view as the vector space of all m-dimensional arrays $\left(x_{i_1,\ldots,i_m} : 1 \leq i_1,\ldots,i_m \leq n \right)$. For $x \in \mathbb{R}^n$, let $y = x^{\otimes m}$ be the tensor with the coordinates $y_{i_1,\ldots,i_m} = x_{i_1} \cdots x_{i_m}$. The group $G = \text{SO}(n)$ of orientation preserving orthogonal transformations of $\mathbb{R}^n$ acts in $(\mathbb{R}^n)^{\otimes m}$ by the m-th tensor power of its natural action in $\mathbb{R}^n$. In particular, $gy = (gx)^{\otimes m}$ for $y = x^{\otimes m}$.

Let us choose $e \in \mathbb{S}^{n-1}$ and let $w = e^{\otimes m}$. Then the orbit $\{gw : g \in G\}$ consists of the tensors $x^{\otimes m}$, where x ranges over the unit sphere in $\mathbb{R}^n$. The orbit $\{gw : g \in G\}$ lies in the symmetric part of $(\mathbb{R}^n)^{\otimes m}$. Let $q = \int_{\mathbb{S}^{n-1}} gw\, dg$ be the center of the orbit (we have $q = 0$ if m is odd). We translate the orbit by shifting q to the origin, so in the end we consider the convex hull B of the orbit of $v = w - q$:

$$B = \text{conv}\left(gv : g \in G \right).$$

A homogeneous polynomial

$$p(x_1,\ldots,x_n) = \sum_{1 \leq i_1,\ldots,i_m \leq n} c_{i_1,\ldots,i_m} x_{i_1} \cdots x_{i_m}$$

of degree m, viewed as a function on the unit sphere in $\mathbb{R}^n$, is identified with the restriction onto the orbit $\{gw : g \in G\}$ of the linear functional $\ell : (\mathbb{R}^n)^{\otimes m} \to \mathbb{R}$ defined by the coefficients $c_{i_1,\ldots,i_m}$. Consequently, the linear functionals ℓ on B are in one-to-one correspondence with the polynomials $p \in U_{m,n}$. Moreover, for $m = 2k$, the negative polar $-B^\circ$ is identified with $\text{Pos}'_{2k,n}$. If m is odd, then $B^\circ = -B^\circ$ is the set of polynomials p such that $|p(x)| \leq 1$ for all $x \in \mathbb{S}^{n-1}$.

The facial structure of $\text{Pos}_{2k,n}$ is well-understood if $k = 1$ or if $n = 2$, see, for example, Section II.11 (for $n = 2$) and Section II.12 (for $k = 1$) of [Barvinok 2002b]. In particular, for $k = 1$, the set $\text{Pos}_{2,n}$ is the convex body of positive semidefinite n-variate quadratic forms of trace n. The faces of $\text{Pos}_{2,n}$ are parameterized by the subspaces of $\mathbb{R}^n$: if $L \subset \mathbb{R}^n$ is a subspace then the corresponding face is

$$F_L = \left\{ p \in \text{Pos}_{2,n} : p(x) = 0 \text{ for all } x \in L \right\}$$

and $\dim F_L = r(r+1)/2 - 1$, where $r = \text{codim}\, L$. Interestingly, for large n, the set $\text{Pos}_{2,n}$ is a counterexample to famous Borsuk's conjecture [Kalai 1995].

For any $k \geq 2$, the situation is much more complicated: the *membership problem* for $\mathrm{Pos}_{2k,n}$:

> *given a polynomial, decide whether it belongs to* $\mathrm{Pos}_{2k,n}$,

is NP-hard, which indicates that the facial structure of $\mathrm{Pos}_{2k,n}$ is probably hard to describe.

EXAMPLE 1.3 (CONVEX HULLS OF GRASSMANNIANS AND CALIBRATIONS). Let $G_m(\mathbb{R}^n)$ be the Grassmannian of all oriented m-dimensional subspaces of $\mathbb{R}^n$, $n > 1$. Let us consider $G_m(\mathbb{R}^n)$ as a subset of $V_{m,n} = \bigwedge^m \mathbb{R}^n$ via the Plücker embedding. Namely, let $e_1, \ldots, e_n$ be the standard basis of $\mathbb{R}^n$. We make $V_{m,n}$ a Euclidean space by choosing an orthonormal basis $e_{i_1} \wedge \cdots \wedge e_{i_m}$ for $1 \leq i_1 < \cdots < i_m \leq n$. Thus the coordinates of a subspace $x \in G_m(\mathbb{R}^n)$ are indexed by m-subsets $1 \leq i_1 < i_2 < \cdots < i_m \leq n$ of $\{1, \ldots, n\}$ and the coordinate $x_{i_1,\ldots,i_m}$ is equal to the oriented volume of the parallelepiped spanned by the orthogonal projection of $e_{1_1}, \ldots, e_{i_m}$ onto x. This identifies $G_m(\mathbb{R}^n)$ with a subset of the unit sphere in $V_{m,n}$. The convex hull $B = \mathrm{conv}\,(G_m(\mathbb{R}^n))$, called the *unit mass ball*, turns out to be of interest in the theory of calibrations and area-minimizing surfaces: a face of B gives rise to a family of m-dimensional area-minimizing surfaces whose tangent planes belong to the face, see [Harvey and Lawson 1982] and [Morgan 1988]. The *comass* of a linear functional $\ell : V_{m,n} \to \mathbb{R}$ is the maximum value of ℓ on $G_m(\mathbb{R}^n)$. A *calibration* is a linear functional $\ell : V_{m,n} \to \mathbb{R}$ of comass 1. The polar B° is called the *unit comass ball*.

One can easily view $G_m(\mathbb{R}^n)$ as an orbit. We let $G = \mathrm{SO}(n)$, the group of orientation-preserving orthogonal transformations of $\mathbb{R}^n$, and consider the action of $\mathrm{SO}(n)$ in $V_{m,n}$ by the m-th exterior power of its defining action in $\mathbb{R}^n$. Choosing $v = e_1 \wedge \cdots \wedge e_m$, we observe that $G_m(\mathbb{R}^n)$ is the orbit $\{gv : g \in G\}$. It is easy to see that $\dim \mathrm{conv}(G_m(\mathbb{R}^n)) = \binom{n}{m}$.

This example was suggested to the authors by B. Sturmfels and J. Sullivan.

The facial structure of the convex hull of $G_m(\mathbb{R}^n)$ is understood for $m \leq 2$, for $m \geq n - 2$ and for some special values of m and n, see [Harvey and Lawson 1982], [Harvey and Morgan 1986] and [Morgan 1988]. If $m = 2$, then the faces of the unit mass ball are as follows: let us choose an even-dimensional subspace $U \subset \mathbb{R}^m$ and an orthogonal complex structure on U, thus identifying $U = \mathbb{C}^{2k}$ for some k. Then the corresponding face of $\mathrm{conv}\,(G_m(\mathbb{R}^n))$ is the convex hull of all oriented planes in U identified with complex lines in $\mathbb{C}^{2k}$.

In general, it appears to be difficult to describe the facial structure of the unit mass ball. The authors do not know the complexity status of the *membership problem* for the unit mass ball:

> *given a point* $x \in \bigwedge^m \mathbb{R}^n$, *decide if it lies in* $\mathrm{conv}\,(G_m(\mathbb{R}^n))$,

but suspect that the problem is NP-hard if $m \geq 3$ is fixed and n is allowed to grow.

The examples above suggest that the boundary of B and $B°$ can get very complicated, so there is little hope in understanding the combinatorics (the facial structure) of general convex hulls of orbits and their polars. Instead, we study metric properties of convex hulls. Our approach is through approximation of a complicated convex body by a simpler one.

As is known, every convex body contains a unique ellipsoid $E_{\max}$ of the maximum volume and is contained in a unique ellipsoid $E_{\min}$ of the minimum volume, see [Ball 1997]. Thus ellipsoids $E_{\max}$ and $E_{\min}$ provide reasonable "first approximations" to a convex body.

The main result of Section 2 is Theorem 2.4 which states that the maximum volume ellipsoid of $B°$ consists of the linear functionals $\ell : V \to \mathbb{R}$ such that the average value of ℓ^2 on the orbit does not exceed $(\dim V)^{-1}$. We compute the minimum- and maximum- volume ellipsoids of the symmetric Traveling Salesman Polytope, which both turn out to be balls under the "natural" Euclidean metric and ellipsoid $E_{\min}$ of the asymmetric Traveling Salesman Polytope, which turns out to be slightly stretched in the direction of the skew-symmetric matrices. As an immediate corollary of Theorem 2.4, we obtain the description of the maximum volume ellipsoid of the set of nonnegative polynomials (Example 1.2), as a ball of radius

$$\left(\binom{n + 2k - 1}{2k} - 1 \right)^{-1/2}$$

in the L^2-metric. We also compute the minimum volume ellipsoid of the convex hull of the Grassmannian and hence the maximum volume ellipsoid of the unit comass ball (Example 1.3).

In Section 3, we obtain some inequalities which allow us to approximate the maximum value of a linear functional ℓ on the orbit by an L^p-norm of ℓ. We apply those inequalities in Section 4. We obtain a reasonably tight estimate of the radius of the Euclidean ball containing the unit comass ball and show that the classical Kähler and special Lagrangian faces of the Grassmannian, are, in fact, rather "shallow" (Example 1.3). Also, we review (with some proofs and some sketches) the recent results of [Blekherman 2003], which show that for most values of n and k the set of nonnegative n-variate polynomials of degree $2k$ is much larger than its subset consisting of the sums of squares of polynomials of degree k.

2. Approximation by Ellipsoids

Let $B \subset V$ be a convex body in a finite-dimensional real vector space. We assume that $\dim B = \dim V$. Among all ellipsoids contained in B there is a unique ellipsoid $E_{\max}$ of the maximum volume, which we call the maximum volume ellipsoid of B and which is also called the John ellipsoid of B or the Löwner-John ellipsoid of B. Similarly, among all ellipsoids containing B there is a unique ellipsoid $E_{\min}$ of the minimum volume, which we call the minimum volume ellipsoid of B and which is also called the Löwner or the Löwner-John ellipsoid. The maximum and minimum volume ellipsoids of B do not depend on the volume form chosen in V, they are intrinsic to B.

Assuming that the center of $E_{\max}$ is the origin, we have

$$E_{\max} \subset B \subset (\dim B)\, E_{\max}.$$

If B is symmetric about the origin, that is, if $B = -B$ then the bound can be strengthened:

$$E_{\max} \subset B \subset \left(\sqrt{\dim B}\right) E_{\max}.$$

More generally, let us suppose that $E_{\max}$ is centered at the origin. The *symmetry coefficient* of B with respect to the origin is the largest $\alpha > 0$ such that $-\alpha B \subset B$. Then

$$E_{\max} \subset B \subset \left(\sqrt{\frac{\dim B}{\alpha}}\right) E_{\max},$$

where α is the symmetry coefficient of B with respect to the origin.

Similarly, assuming that $E_{\min}$ is centered at the origin, we have

$$(\dim B)^{-1}\, E_{\min} \subset B \subset E_{\min}.$$

If, additionally, α is the symmetry coefficient of B with respect to the origin, then

$$\left(\sqrt{\frac{\alpha}{\dim B}}\right) E_{\min} \subset B \subset E_{\min}.$$

In particular, if B is symmetric about the origin, then

$$(\dim B)^{-1/2}\, E_{\min} \subset B \subset E_{\min}.$$

These, and other interesting properties of the minimum- and maximum- volume ellipsoids can be found in [Ball 1997], see also the original paper [John 1948], [Blekherman 2003], and Chapter V of [Barvinok 2002a]. There are many others interesting ellipsoids associated with a convex body, such as the minimum width and minimum surface area ellipsoids [Giannopoulos and Milman 2000]. The advantage of using $E_{\max}$ and $E_{\min}$ is that these ellipsoids do not depend on the Euclidean structure of the ambient space and even on the volume form in the space, which often makes calculations particularly easy.

Suppose that a compact group G acts in V by linear transformations and that B is invariant under the action: $gB = B$ for all $g \in G$. Let $\langle \cdot, \cdot \rangle$ be a G-invariant scalar product in V, so G acts in V by isometries. Since the ellipsoids $E_{\max}$ and $E_{\min}$ associated with B are unique, they also have to be invariant under the action of G. If the group of symmetries of B is sufficiently rich, we may be able to describe $E_{\max}$ or $E_{\min}$ precisely.

The following simple observation will be used throughout this section. Let us suppose that the action of G in V is irreducible: if $W \subset V$ is a G-invariant subspace, then either $W = \{0\}$ or $W = V$. Then, the ellipsoids $E_{\max}$ and $E_{\min}$ of a G-invariant convex body B are necessarily balls centered at the origin:

$$E_{\max} = \{x \in V : \langle x, x \rangle \leq r^2\} \quad \text{and} \quad E_{\min} = \{x \in V : \langle x, x \rangle \leq R^2\}$$

for some $r, R > 0$.

Indeed, since the action of G is irreducible, the origin is the only G-invariant point and hence both $E_{\max}$ and $E_{\min}$ must be centered at the origin. Furthermore, an ellipsoid $E \subset V$ centered at the origin is defined by the inequality $E = \{x : q(x) \leq 1\}$, where $q : V \to \mathbb{R}$ is a positive definite quadratic form. If E is G-invariant, then $q(gx) = q(x)$ for all $g \in G$ and hence the eigenspaces of q must be G-invariant. Since the action of G is irreducible, there is only one eigenspace which coincides with V, from which $q(x) = \lambda \langle x, x \rangle$ for some $\lambda > 0$ and all $x \in V$ and E is a ball.

This simple observation allows us to compute ellipsoids $E_{\max}$ and $E_{\min}$ of the Symmetric Traveling Salesman Polytope (Example 1.1).

EXAMPLE 2.1 (THE MINIMUM AND MAXIMUM VOLUME ELLIPSOIDS OF THE SYMMETRIC TRAVELING SALESMAN POLYTOPE). In this case, V is the space of $n \times n$ real matrices, on which the symmetric group S_n acts by simultaneous permutations of rows and columns, see Example 1.1. Introduce an S_n-invariant scalar product by

$$\langle a, b \rangle = \sum_{i,j=1}^{n} a_{ij} b_{ij} \quad \text{for } a = (a_{ij}) \text{ and } b = (b_{ij})$$

and the corresponding Euclidean norm $\|a\| = \sqrt{\langle a, a \rangle}$. It is not hard to see that the affine hull of the symmetric Traveling Salesman Polytope ST_n consists of the symmetric matrices with 0 diagonal and row and column sums equal to 2, from which one can deduce the formula $\dim ST_n = (n^2 - 3n)/2$. Let us make the affine hull of ST_n a vector space by choosing the origin at $c = (c_{ij})$ with $c_{ij} = 2/(n-1)$ for $i \neq j$ and $c_{ii} = 0$, the only fixed point of the action. One can see that the action of S_n on the affine hull of ST_n is irreducible and corresponds to the Young diagram $(n - 2, 2)$, see, for example, Chapter 4 of [Fulton and Harris 1991].

Hence the maximum- and minimum- volume ellipsoids of ST_n must be balls in the affine hull of ST_n centered at c. Moreover, since the boundary of the minimum volume ellipsoid $E_{\min}$ must contain the vertices of ST_n, we conclude that the radius of the ball representing $E_{\min}$ is equal to $\sqrt{2n(n-3)/(n-1)}$.

One can compute the symmetry coefficient of ST_n with respect to the center c. Suppose that $n \geq 5$. Let us choose a vertex v of ST_n and let us consider the functional $\ell(x) = \langle v - c, x - c \rangle$ on ST_n. The maximum value of $2n(n-3)/(n-1)$ is attained at $x = v$ while the minimum value of $-4n/(n-1)$ is attained at the face F_v of ST_n with the vertices h such that $\langle v, h \rangle = 0$ (combinatorially, h correspond to Hamiltonian cycles in the graph obtained from the complete graph on n vertices by deleting the edges of the Hamiltonian cycle encoded by

v). Moreover, one can show that for $\lambda = 2/(n-3)$, we have $-\lambda(v-c) + c \in F_v$. This implies that the coefficient of symmetry of ST_n with respect to c is equal to $2/(n-3)$. Therefore ST_n contains the ball centered at c and of the radius $\sqrt{8/((n-1)(n-3))}$ (for $n \geq 5$).

The ball centered at c and of the radius $\sqrt{8/((n-1)(n-3))}$ touches the boundary of ST_n. Indeed, let $b = (b_{ij})$ be the centroid of the set of vertices x of ST_n with $x_{12} = x_{21} = 0$. Then

$$
b_{ij} = \begin{cases}
0 & \text{if } 1 \leq i, j \leq 2, \\[2mm]
\dfrac{2}{n-2} & \text{if } i = 1, 2 \text{ and } j > 2 \quad \text{or} \quad j = 1, 2 \text{ and } i > 2, \\[3mm]
\dfrac{2(n-4)}{(n-2)(n-3)} & \text{if } i, j \geq 3,
\end{cases}
$$

and the distance from c to b is precisely $\sqrt{8/((n-1)(n-3))}$.

Hence for $n \geq 5$ the maximum volume ellipsoid $E_{\max}$ is the ball centered at c of the radius $\sqrt{8/((n-1)(n-3))}$.

Some bounds on the radius of the largest inscribed ball for a polytope from a particular family of combinatorially defined polytopes are computed in [Vyalyĭ 1995]. The family of polytopes includes the symmetric Traveling Salesman Polytope, although in its case the bound from [Vyalyĭ 1995] is not optimal.

If the action of G in the ambient space V is not irreducible, the situation is more complicated. For one thing, there is more than one (up to a scaling factor) G-invariant scalar product, hence the notion of a "ball" is not really defined. However, we are still able to describe the minimum volume ellipsoid of the convex hull of an orbit.

Without loss of generality, we assume that the orbit $\{gv : g \in G\}$ spans V affinely. Let $\langle \cdot, \cdot \rangle$ be a G-invariant scalar product in V. As is known, V can be decomposed into the direct sum of pairwise orthogonal invariant subspaces V_i, such that the action of G in each V_i is irreducible. It is important to note that the decomposition is *not* unique: nonuniqueness appears when some of V_i are isomorphic, that, is, when there exists an isomorphism $V_i \to V_j$ which commutes with G. If the decomposition is unique, we say that the action of G is *multiplicity-free*.

Since the orbit spans V affinely, the orthogonal projection v_i of v onto each V_i must be nonzero (if $v_i = 0$ then the orbit lies in $V_i^{\perp}$). Also, the origin in V must be the only invariant point of the action of G (otherwise, the orbit is contained in the hyperplane $\langle x, u \rangle = \langle v, u \rangle$, where $u \in V$ is a nonzero vector fixed by the action of G).

THEOREM 2.2. *Let B be the convex hull of the orbit of a vector $v \in V$:*

$$
B = \operatorname{conv}(gv : g \in G).
$$

Suppose that the affine hull of B is V.

Then there exists a decomposition

$$V = \bigoplus_i V_i$$

of V into the direct sum of pairwise orthogonal irreducible components such that the following holds.

The minimum volume ellipsoid $E_{\min}$ of B is defined by the inequality

$$E_{\min} = \Big\{ x : \sum_i \frac{\dim V_i}{\dim V} \cdot \frac{\langle x_i, x_i \rangle}{\langle v_i, v_i \rangle} \le 1 \Big\}, \qquad 2.2.1$$

where x_i (resp. v_i) is the orthogonal projection of x (resp. v) onto V_i.

We have

$$\int_G \langle x, gv \rangle^2 \, dg = \sum_i \frac{\langle x_i, x_i \rangle \langle v_i, v_i \rangle}{\dim V_i} \quad \text{for all } x \in V, \qquad 2.2.2$$

where dg is the Haar probability measure on G.

PROOF. Let us consider the quadratic form $q : V \to \mathbb{R}$ defined by

$$q(x) = \int_G \langle x, gv \rangle^2 \, dg.$$

We observe that q is G-invariant, that is, $q(gx) = q(x)$ for all $x \in V$ and all $g \in G$. Therefore, the eigenspaces of q are G-invariant. Writing the eigenspaces as direct sums of pairwise orthogonal invariant subspaces where the action of G is irreducible, we obtain a decomposition $V = \bigoplus_i V_i$ such that

$$q(x) = \sum_i \lambda_i \langle x_i, x_i \rangle \quad \text{for all } x \in V$$

and some $\lambda_i \ge 0$. Recall that $v_i \ne 0$ for all i since the orbit $\{gv : g \in G\}$ spans V affinely.

To compute λ_i, we substitute $x \in V_i$ and observe that the trace of

$$q_i(x) = \int_G \langle x, gv_i \rangle^2 \, dg$$

as a quadratic form $q_i : V_i \to \mathbb{R}$ is equal to $\langle v_i, v_i \rangle$. Hence we must have $\lambda_i = \langle v_i, v_i \rangle / \dim V_i$, which proves (2.2.2) [Barvinok 2002b].

We will also use the polarized form of (2.2.2):

$$\int_G \langle x, gv \rangle \langle y, gv \rangle \, dg = \sum_i \frac{\langle x_i, y_i \rangle \langle v_i, v_i \rangle}{\dim V_i}, \qquad 2.2.3$$

obtained by applying (2.2.2) to $q(x + y) - q(x) - q(y)$.

Next, we observe that the ellipsoid E defined by the inequality (2.2.1) contains the orbit $\{gv : g \in G\}$ on its boundary and hence contains B.

Our goal is to show that E is the minimum volume ellipsoid. It is convenient to introduce a new scalar product:

$$(a, b) = \sum_i \frac{\dim V_i}{\dim V} \cdot \frac{\langle a_i, b_i \rangle}{\langle v_i, v_i \rangle} \quad \text{for all } a, b \in V.$$

Obviously $(\cdot, \cdot)$ is a G-invariant scalar product. Furthermore, the ellipsoid E defined by (2.2.1) is the unit ball in the scalar product $(\cdot, \cdot)$.

Now,

$$(c, gv) = \sum_i \frac{\dim V_i}{\dim V} \cdot \frac{\langle c_i, gv \rangle}{\langle v_i, v_i \rangle}$$

and hence

$$(c, gv)^2 = \sum_{i,j} \frac{(\dim V_i)(\dim V_j)}{(\dim V)^2} \cdot \frac{\langle c_i, gv \rangle \langle c_j, gv \rangle}{\langle v_i, v_i \rangle^2}.$$

Integrating and using (2.2.3), we get

$$\int_G (c, gv)^2 \, dg = \frac{1}{\dim V} \sum_i \frac{\dim V_i}{\dim V} \cdot \frac{\langle c_i, c_i \rangle}{\langle v_i, v_i \rangle} = \frac{(c, c)}{\dim V}. \qquad 2.2.4$$

Since the origin is the only fixed point of the action of G, the minimum volume ellipsoid should be centered at the origin.

Let $e_1, \ldots, e_k$ for $k = \dim V$ be an orthonormal basis with respect to the scalar product $(\cdot, \cdot)$. Suppose that $E' \subset V$ is an ellipsoid defined by

$$E' = \left\{ x \in V : \sum_{j=1}^k \frac{(x, e_j)^2}{\alpha_j^2} \leq 1 \right\}$$

for some $\alpha_1, \ldots, \alpha_k > 0$. To show that E is the minimum volume ellipsoid, it suffices to show that as long as E' contains the orbit $\{gv : g \in G\}$, we must have $\operatorname{vol} E' \geq \operatorname{vol} E$, which is equivalent to $\alpha_1 \cdots \alpha_k \geq 1$.

Indeed, since $gv \in E'$, we must have

$$\sum_{j=1}^k \frac{(e_j, gv)^2}{\alpha_j^2} \leq 1 \quad \text{for all } g \in G.$$

Integrating, we obtain

$$\sum_{j=1}^k \frac{1}{\alpha_j^2} \int_G (e_j, gv)^2 \, dg \leq 1.$$

Applying (2.2.4), we get

$$\frac{1}{\dim V} \sum_{j=1}^k \frac{1}{\alpha_j^2} \leq 1.$$

Since $k = \dim V$, from the inequality between the arithmetic and geometric means, we get $\alpha_1 \ldots \alpha_k \geq 1$, which completes the proof. $\qquad \square$

REMARK. In the part of the proof where we compare the volumes of E' and E, we reproduce the "sufficiency" (that is, "the easy") part of John's criterion for optimality of an ellipsoid; see, for example, [Ball 1997].

Theorem 2.2 allows us to compute the minimum volume ellipsoid of the asymmetric Traveling Salesman Polytope, see Example 1.1.

EXAMPLE 2.3 (THE MINIMUM VOLUME ELLIPSOID OF THE ASYMMETRIC TRAVELING SALESMAN POLYTOPE). In this case (compare Examples 1.1 and 2.1), V is the space of $n \times n$ matrices with the scalar product and the action of the symmetric group S_n defined as in Example 2.1. On can observe that the affine hull of AT_n consists of the matrices with zero diagonal and row and column sums equal to 1, from which one can deduce the formula $\dim AT_n = n^2 - 3n + 1$.

The affine hull of AT_n is S_n-invariant. We make the affine hull of AT_n a vector space by choosing the origin at $c = (c_{ij})$ with $c_{ij} = 1/(n-1)$ for $i \neq j$ and $c_{ii} = 0$, the only fixed point of the action. The action of S_n on the affine hull of AT_n is reducible and multiplicity-free, so there is no ambiguity in choosing the irreducible components. The affine hull is the sum of two irreducible invariant subspaces V_s and V_a.

Subspace V_s consists of the matrices $x + c$, where x is a symmetric matrix with zero diagonal and zero row and column sums. One can see that the action of S_n in V_s is irreducible and corresponds to the Young diagram $(n-2, 2)$, see, for example, Chapter 4 of [Fulton and Harris 1991]. We have $\dim V_s = (n^2 - 3n)/2$.

Subspace V_a consists of the matrices $x + c$, where x is a skew-symmetric matrix with zero row and column sums. One can see that the action of S_n in V_a is irreducible and corresponds to the Young diagram $(n-2, 1, 1)$, see, for example, Chapter 4 of [Fulton and Harris 1991]. We have $\dim V_a = (n-1)(n-2)/2$.

The orthogonal projection onto V_s is defined by $x \mapsto (x + x^t)/2$, while the orthogonal projection onto V_a is defined by $x \mapsto (x - x^t)/2 + c$.

Applying Theorem 2.2, we conclude that the minimum volume ellipsoid of AT_n is defined in the affine hull of AT_n by the inequality:

$$(n-1) \sum_{1 \leq i \neq j \leq n} \left(\frac{x_{ij} + x_{ji}}{2} - \frac{1}{n-1} \right)^2$$
$$+ \frac{(n-1)(n-2)}{n} \sum_{1 \leq i \neq j \leq n} \left(\frac{x_{ij} - x_{ji}}{2} \right)^2 \leq n^2 - 3n + 1.$$

Thus one can say that the minimum volume ellipsoid of the asymmetric Traveling Salesman Polytope is slightly stretched in the direction of skew-symmetric matrices.

The dual version of Theorem 2.2 is especially simple.

THEOREM 2.4. *Let G be a compact group acting in a finite-dimensional real vector space V. Let B be the convex hull of the orbit of a vector $v \in V$:*

$$B = \mathrm{conv}\Big(gv : g \in G \Big).$$

Suppose that the affine hull of B is V.

Let V^* be the dual to V and let

$$B^\circ = \Big\{ \ell \in V^* : \ell(x) \leq 1 \text{ for all } x \in B \Big\}$$

be the polar of B. Then the maximum volume ellipsoid of B° is defined by the inequality

$$E_{\max} = \Big\{ \ell \in V^* : \int_G \ell^2(gv)\, dg \leq \frac{1}{\dim V} \Big\}.$$

PROOF. Let us introduce a G-invariant scalar product $\langle \cdot, \cdot \rangle$ in V, thus identifying V and V^*. Then

$$B^\circ = \Big\{ c \in V : \langle c, gv \rangle \leq 1 \text{ for all } g \in G \Big\}.$$

Since the origin is the only point fixed by the action of G, the maximum volume ellipsoid $E_{\max}$ of B° is centered at the origin. Therefore, $E_{\max}$ must be the polar of the minimum volume ellipsoid of B.

Let $V = \bigoplus_i V_i$ be the decomposition of Theorem 2.2. Since $E_{\max}$ is the polar of the ellipsoid $E_{\min}$ associated with B, from (2.2.1), we get

$$E_{\max} = \Big\{ c : \dim V \sum_i \frac{\langle c_i, c_i \rangle \langle v_i, v_i \rangle}{\dim V_i} \leq 1 \Big\}.$$

Applying (2.2.2), we get

$$E_{\max} = \Big\{ c : \int_G \langle c, gv \rangle^2\, dg \leq \frac{1}{\dim V} \Big\},$$

which completes the proof. $\square$

REMARK. Let G be a compact group acting in a finite-dimensional real vector space V and let $v \in V$ be a point such that the orbit $\{ gv : g \in V \}$ spans V affinely. Then the dual space V^* acquires a natural scalar product

$$\langle \ell_1, \ell_2 \rangle = \int_G \ell_1(gv)\ell_2(gv)\, dg$$

induced by the scalar product in $L^2(G)$. Theorem 2.4 states that the maximum volume ellipsoid of the polar of the orbit is the ball of radius $(\dim V)^{-1/2}$ in this scalar product.

By duality, V acquires the dual scalar product (which we denote below by $\langle \cdot, \cdot \rangle$ as well). It is a constant multiple of the product $(\cdot, \cdot)$ introduced in the proof of Theorem 2.2: $\langle u_1, u_2 \rangle = (\dim V)(u_1, u_2)$. We have $\langle v, v \rangle = \dim V$ and the minimum volume ellipsoid of the convex hull of the orbit of v is the ball of radius $\sqrt{\dim V}$.

As an immediate application of Theorem 2.4, we compute the maximum volume ellipsoid of the set of nonnegative polynomials, see Example 1.2.

EXAMPLE 2.5 (THE MAXIMUM VOLUME ELLIPSOID OF THE SET OF NONNEGA-
TIVE POLYNOMIALS). In this case, $U^*_{2k,n}$ is the space of all homogeneous polynomials $p : \mathbb{R}^n \to \mathbb{R}$ of degree $2k$ with the zero average on the unit sphere $\mathbb{S}^{n-1}$, so $\dim U^*_{2k,n} = \binom{n+2k-1}{2k} - 1$. We view such a polynomial p as a linear functional ℓ on an orbit $\{gv : g \in G\}$ in the action of the orthogonal group $G = \mathrm{SO}(n)$ in $(\mathbb{R}^n)^{\otimes 2k}$ and the shifted set $\mathrm{Pos}'_{2k,n}$ of nonnegative polynomials as the negative polar $-B^\circ$ of the orbit, see Example 1.2. In particular, under this identification $p \longleftrightarrow \ell$, we have

$$\int_{\mathbb{S}^{n-1}} p^2(x)\, dx = \int_G \ell^2(gv)\, dg,$$

where dx and dg are the Haar probability measures on $\mathbb{S}^{n-1}$ and $\mathrm{SO}(n)$ respectively.

Applying Theorem 2.4 to $-B^\circ$, we conclude that the maximum volume ellipsoid of $-B^\circ = \mathrm{Pos}'_{2k,n}$ consists of the polynomials p such that

$$\int_{\mathbb{S}^{n-1}} p(x)\, dx = 0 \quad \text{and} \quad \int_{\mathbb{S}^{n-1}} p^2(x)\, dx \le \left(\binom{n+2k-1}{2k} - 1 \right)^{-1}.$$

Consequently, the maximum volume ellipsoid of $\mathrm{Pos}_{2k,n}$ consists of the polynomials p such that

$$\int_{\mathbb{S}^{n-1}} p(x)\, dx = 1 \quad \text{and} \quad \int_{\mathbb{S}^{n-1}} (p(x)-1)^2\, dx \le \left(\binom{n+2k-1}{2k} - 1 \right)^{-1}.$$

Geometrically, the maximum volume ellipsoid of $\mathrm{Pos}_{2k,n}$ can be described as follows. Let us introduce a scalar product in the space of polynomials by

$$\langle f,\, g \rangle = \int_{\mathbb{S}^{n-1}} f(x)g(x)\, dx,$$

where dx is the rotation-invariant probability measure, as above. Then the maximum volume ellipsoid of $\mathrm{Pos}_{2k,n}$ is the ball centered at $r(x) = (x_1^2 + \cdots + x_n^2)^k$ and having radius

$$\left(\binom{n+2k-1}{2k} - 1 \right)^{-1/2}$$

(note that multiples of $r(x)$ are the only $\mathrm{SO}(n)$-invariant polynomials, see for example, p. 13 of [Barvinok 2002a]). This result was first obtained by more direct and complicated computations in [Blekherman 2004]. In the same paper,

G. Blekherman also determined the coefficient of symmetry of $\text{Pos}_{2k,n}$ (with respect to the center r), it turns out to be equal to

$$\left(\binom{n+k-1}{k} - 1\right)^{-1}.$$

It follows then that $\text{Pos}_{2k,n}$ is contained in the ball centered at r and of the radius

$$\left(\binom{n+k-1}{k} - 1\right)^{1/2}.$$

This estimate is poor if k is fixed and n is allowed to grow: as follows from results of Duoandikoetxea [1987], for any fixed k, the set $\text{Pos}_{2k,n}$ is contained in a ball of a fixed radius, as n grows. However, the estimate gives the right logarithmic order if $k \gg n$, which one can observe by inspecting a polynomial $p \in \text{Pos}_{2k,n}$ that is the $2k$-th power of a linear function.

We conclude this section by computing the minimum volume ellipsoid of the convex hull of the Grassmannian and, consequently, the maximum volume ellipsoid of the unit comass ball, see Example 1.3.

EXAMPLE 2.6 (THE MINIMUM VOLUME ELLIPSOID OF THE CONVEX HULL OF THE GRASSMANNIAN). In this case, $V_{m,n} = \bigwedge^m \mathbb{R}^n$ with the orthonormal basis $e_I = e_{i_1} \wedge \cdots \wedge e_{i_m}$, where I is an m-subset $1 \leq i_1 < i_2 < \cdots < i_m \leq n$ of the set $\{1, \ldots, n\}$ and $e_1, \ldots, e_n$ is the standard orthonormal basis of $\mathbb{R}^n$.

Let $\langle \cdot, \cdot \rangle$ be the corresponding scalar product in $V_{m,n}$, so that

$$\langle a, b \rangle = \sum_I a_I b_I,$$

where I ranges over all m-subsets of $\{1, \ldots, n\}$. The scalar product allows us to identify $V_{m,n}^*$ with $V_{m,n}$. First, we find the maximum volume ellipsoid of the unit comass ball B°, that is the polar of the convex hull $B = \text{conv}\,(G_m(\mathbb{R}^n))$ of the Grassmannian.

A linear functional $a \in V_{m,n}^* = V_{m,n}$ is defined by its coefficients a_I. To apply Theorem 2.4, we have to compute

$$\int_{SO(n)} \langle a, gv \rangle^2 \, dg = \int_{G_m(\mathbb{R}^n)} \langle a, x \rangle^2 \, dx,$$

where dx is the Haar probability measure on the Grassmannian $G_m(\mathbb{R}^n)$. We note that

$$\int_{G_m(\mathbb{R}^n)} \langle e_I, x \rangle \langle e_J, x \rangle \, dx = 0$$

for $I \neq J$, since for $i \in I \setminus J$, the reflection $e_i \mapsto -e_i$ of $\mathbb{R}^n$ induces an isometry of $V_{m,n}$, which maps $G_m(\mathbb{R}^n)$ onto itself, reverses the sign of $\langle e_I, x \rangle$ and does not change $\langle e_J, x \rangle$. Also,

$$\int_{G_m(\mathbb{R}^n)} \langle e_I, x \rangle^2 \, dx = \binom{n}{m}^{-1},$$

since the integral does not depend on I and $\sum_I \langle e_I, x \rangle^2 = 1$ for all $x \in G_m(\mathbb{R}^n)$.

By Theorem 2.4, we conclude that the maximum volume ellipsoid of the unit comass ball B° is defined by the inequality

$$E_{\max} = \left\{ a \in V_{m,n} : \sum_I a_I^2 \le 1 \right\},$$

that is, the unit ball in the Euclidean metric of $V_{m,n}$. Since B° is centrally symmetric, we conclude that B° is contained in the ball of radius $\binom{n}{m}^{1/2}$. As follows from Theorem 4.1, this estimate is optimal up to a factor of $\sqrt{m(n-1)(1+\ln m)}$.

Consequently, the convex hull B of the Grassmannian is contained in the unit ball of $V_{m,n}$, which is the minimum volume ellipsoid of B, and contains a ball of radius $\binom{n}{m}^{-1/2}$. Again, the estimate of the radius of the inner ball is optimal up to a factor of $\sqrt{m(n-1)(1+\ln m)}$.

3. Higher Order Estimates

The following construction can be used to get a better understanding of metric properties of an orbit $\{gv : g \in G\}$. Let us choose a positive integer k and let us consider the k-th tensor power

$$V^{\otimes k} = \underbrace{V \otimes \cdots \otimes V}_{k \text{ times}}.$$

The group G acts in $V^{\otimes k}$ by the k-th tensor power of its action in V: on decomposable tensors we have

$$g(v_1 \otimes \cdots \otimes v_k) = g(v_1) \otimes \cdots \otimes g(v_k).$$

Let us consider the orbit $\{gv^{\otimes k} : g \in G\}$ for

$$v^{\otimes k} = \underbrace{v \otimes \cdots \otimes v}_{k \text{ times}}.$$

Then, a linear functional on the orbit of $v^{\otimes k}$ is a polynomial of degree k on the orbit of v and hence we can extract some new "higher order" information about the orbit of v by applying already developed methods to the orbit of $v^{\otimes k}$. An important observation is that the orbit $\{gv^{\otimes k} : g \in G\}$ lies in the symmetric part of $V^{\otimes k}$, so the dimension of the affine hull of the orbit of $v^{\otimes k}$ does not exceed $\binom{\dim V + k - 1}{k}$.

THEOREM 3.1. *Let G be a compact group acting in a finite-dimensional real vector space V, let $v \in V$ be a point, and let $\ell : V \to \mathbb{R}$ be a linear functional. Let us define*

$$f : G \to \mathbb{R} \qquad by \qquad f(g) = \ell(gv).$$

For an integer $k > 0$, let d_k be the dimension of the subspace spanned by the orbit $\{gv^{\otimes k} : g \in G\}$ in $V^{\otimes k}$. In particular, $d_k \leq \binom{\dim V + k - 1}{k}$. Let

$$\|f\|_{2k} = \left(\int_G f^{2k}(g) \, dg \right)^{1/2k}.$$

(i) *Suppose that k is odd and that*

$$\int_G f^k(g) \, dg = 0.$$

   Then

$$d_k^{-1/2k} \|f\|_{2k} \leq \max_{g \in G} f(g) \leq d_k^{1/2k} \|f\|_{2k}.$$

(ii) *We have*

$$\|f\|_{2k} \leq \max_{g \in G} |f(g)| \leq d_k^{1/2k} \|f\|_{2k}.$$

PROOF. Without loss of generality, we assume that $f \not\equiv 0$.

Let

$$B_k(v) = \operatorname{conv}\left(gv^{\otimes k} : g \in G \right)$$

be the convex hull of the orbit of $v^{\otimes k}$. We have $\dim B_k(v) \leq d_k$.

Let $\ell^{\otimes k} \in (V^*)^{\otimes k}$ be the k-th tensor power of the linear functional $\ell \in V^*$. Thus $f^k(g) = \ell^{\otimes k}\left(gv^{\otimes k} \right)$.

To prove Part (1), we note that since k is odd,

$$\max_{g \in G} f^k(g) = \left(\max_{g \in G} f(g) \right)^k.$$

Let

$$u = \int_G g\left(v^{\otimes k} \right) \, dg$$

be the center of $B_k(v)$. Since the average value of $f^k(g)$ is equal to 0, we have $\ell^{\otimes k}(u) = 0$ and hence $\ell^{\otimes k}(x) = \ell^{\otimes k}(x - u)$ for all $x \in V^{\otimes k}$. Let us translate $B_k(v)' = B_k(v) - u$ to the origin and let us consider the maximum volume ellipsoid E of the polar of $B_k(v)'$ in its affine hull. By Theorem 2.4, we have

$$E = \left\{ \mathcal{L} \in \left(V^{\otimes k} \right)^* : \int_G \mathcal{L}^2 \left(gv^{\otimes k} - u \right) \, dg \leq \frac{1}{\dim B_k(v)} \right\}.$$

Since the ellipsoid E is contained in the polar of $B_k(v)'$, for any linear functional $\mathcal{L} : V^{\otimes k} \to \mathbb{R}$, the inequality

$$\int_G \mathcal{L}^2 \left(gv^{\otimes k} - u \right) \, dg \leq \frac{1}{d_k} \leq \frac{1}{\dim B_k(v)}$$

implies the inequality

$$\max_{g \in G} \mathcal{L}\left(gv^{\otimes k} - u\right) \leq 1.$$

Choosing $\mathcal{L} = \lambda \ell^{\otimes k}$ with $\lambda = d_k^{-1/2}\|f\|_{2k}^{-k}$, we then obtain the upper bound for $\max_{g \in G} f(g)$.

Since the ellipsoid $(\dim E)E$ contains the polar of $B_k(v)'$, for any linear functional $\mathcal{L} : V^{\otimes k} \to \mathbb{R}$, the inequality

$$\max_{g \in G} \mathcal{L}\left(gv^{\otimes k} - u\right) \leq 1$$

implies the inequality

$$\int_G \mathcal{L}^2\left(gv^{\otimes k} - u\right) dg \leq \dim B_k(v) \leq d_k.$$

Choosing $\mathcal{L} = \lambda \ell^{\otimes k}$ with any $\lambda > \|f\|_{2k}^{-k}d_k^{1/2}$, we obtain the lower bound for $\max_{g \in G} f(g)$.

The proof of Part (2) is similar. We modify the definition of $B_k(v)$ by letting

$$B_k(v) = \operatorname{conv}\left(gv^{\otimes k}, -gv^{\otimes k} : g \in G\right).$$

The set $B_k(v)$ so defined can be considered as the convex hull of an orbit of $G \times \mathbb{Z}_2$ and is centrally symmetric, so the ellipsoid $(\sqrt{\dim E})E$ contains the polar of $B_k(v)$.

Part (2) is also proven by a different method in [Barvinok 2002b]. □

REMARK. Since $d_k \leq \binom{\dim V + k - 1}{k}$, the upper and lower bounds in Theorem 3.1 are asymptotically equivalent as long as $k^{-1}\dim V \to 0$. In many interesting cases we have $d_k \ll \binom{\dim V + k - 1}{k}$, which results in stronger inequalities.

Polynomials on the unit sphere. As is discussed in Examples 1.2 and 2.5, the restriction of a homogeneous polynomial $f : \mathbb{R}^n \to \mathbb{R}$ of degree m onto the unit sphere $\mathbb{S}^{n-1} \subset \mathbb{R}^n$ can be viewed as the restriction of a linear functional $\ell : (\mathbb{R}^n)^{\otimes m} \to \mathbb{R}$ onto the orbit of a vector $v = e^{\otimes m}$ for some $e \in \mathbb{S}^{n-1}$ in the action of the special orthogonal group $\mathrm{SO}(n)$. In this case, the orbit of $v^{\otimes k} = e^{\otimes mk}$ spans the symmetric part of $(\mathbb{R}^n)^{mk}$, so we have $d_k = \binom{n+mk-1}{mk}$ in Theorem 3.1.

Hence Part (1) of Theorem 3.1 implies that if f is an n-variate homogeneous polynomial of degree m such that

$$\int_{\mathbb{S}^{n-1}} f^k(x)\, dx = 0,$$

where dx is the rotation-invariant probability measure on $\mathbb{S}^{n-1}$, then

$$\binom{n+mk-1}{mk}^{-1/2k}\|f\|_{2k} \leq \max_{x \in \mathbb{S}^{n-1}} f(x) \leq \binom{n+mk-1}{mk}^{1/2k}\|f\|_{2k},$$

where

$$\|f\|_{2k} = \left(\int_{\mathbb{S}^{n-1}} f^{2k}(x)\, dx \right)^{1/2k}.$$

We obtain the following corollary.

COROLLARY 3.2. *Suppose that* $k \geq (n-1)\max\{\ln(m+1),\, 1\}$. *Then*

$$\|f\|_{2k} \leq \max_{x \in \mathbb{S}^{n-1}} |f(x)| \leq \alpha \|f\|_{2k},$$

for some absolute constant $\alpha > 0$ *and all homogeneous polynomials* $f : \mathbb{R}^n \to \mathbb{R}$ *of degree* m. *One can take* $\alpha = \exp(1 + 0.5e^{-1}) \approx 3.27$.

PROOF. Applying Part(2) of Theorem 3.1 as above, we conclude that for any homogeneous polynomial $f : \mathbb{R}^n \to \mathbb{R}$ of degree m,

$$\|f\|_{2k} \leq \max_{x \in \mathbb{S}^{n-1}} |f(x)| \leq \binom{n+mk-1}{mk}^{1/2k} \|f\|_{2k}.$$

This inequality is also proved in [Barvinok 2002b]. Besides, it can be deduced from some classical estimates for spherical harmonics; see p. 14 of [Müller 1966].
    We use the estimate

$$\ln\binom{a}{b} \leq b \ln \frac{a}{b} + (a-b)\ln \frac{a}{a-b};$$

see, for example, Theorem 1.4.5 of [van Lint 1999]. Applying the inequality with $b = mk$ and $a = n + mk - 1$, we get

$$b \ln \frac{a}{b} = mk \ln\left(1 + \frac{n-1}{mk}\right) \leq n - 1$$

and

$$(a-b)\ln \frac{a}{a-b} = (n-1)\ln \frac{n+mk-1}{n-1} \leq (n-1)\left(\ln(m+1) + \ln \frac{k}{n-1}\right).$$

Summarizing,

$$\frac{1}{2k} \ln\binom{n+mk-1}{mk} \leq \frac{1}{2} + \frac{1}{2} + \frac{1}{2\rho}\ln \rho \quad \text{for} \quad \rho = \frac{k}{n-1}.$$

Since $\rho^{-1}\ln \rho \leq e^{-1}$ for all $\rho \geq 1$, the proof follows.   □

Our next application concerns calibrations; compare Examples 1.3 and 2.6.

THEOREM 3.3. *Let* $G_m(\mathbb{R}^n) \subset \bigwedge^m \mathbb{R}^n$ *be the Plücker embedding of the Grassmannian of oriented* m-*subspaces of* $\mathbb{R}^n$. *Let* $\ell : \bigwedge^m \mathbb{R}^n \to \mathbb{R}$ *be a linear functional. Let*

$$\|\ell\|_{2k} = \left(\int_{G_m(\mathbb{R}^n)} \ell^{2k}(x)\, dx \right)^{1/2k},$$

where dx is the Haar probability measure on $G_m(\mathbb{R}^n)$. Then, for any positive integer k,

$$\|\ell\|_{2k} \leq \max_{x \in G_m(\mathbb{R}^n)} |\ell(x)| \leq (d_k)^{1/2k} \|\ell\|_{2k},$$

where $d_k = \prod_{i=1}^{m} \prod_{j=1}^{k} \frac{n+j-i}{m+k-i-j+1}$.

PROOF. As we discussed in Example 1.3, the Grassmannian $G_m(\mathbb{R}^n)$ can be viewed as the orbit of $v = e_1 \wedge \cdots \wedge e_m$, where $e_1, \ldots, e_n$ is the standard basis of $\mathbb{R}^n$, under the action of the special orthogonal group $SO(n)$ by the m-th exterior power of its defining representation in $\mathbb{R}^n$. We are going to apply Part (2) of Theorem 3.1 and for that we need to estimate the dimension of the subspace spanned by the orbit of $v^{\otimes k}$. First, we identify $\bigwedge^m \mathbb{R}^n$ with the subspace of skew-symmetric tensors in $(\mathbb{R}^n)^{\otimes m}$ and v with the point

$$\sum_{\sigma \in S_m} (\operatorname{sgn} \sigma) e_{\sigma(1)} \otimes \cdots \otimes e_{\sigma(m)},$$

where S_m is the symmetric group of all permutations of $\{1, \ldots, m\}$.

Let us consider $W = (\mathbb{R}^n)^{\otimes mk}$. We introduce the right action of the symmetric group S_{mk} on W by permutations of the factors in the tensor product:

$$(u_1 \otimes \cdots \otimes u_{mk})\sigma = u_{\sigma(1)} \otimes \cdots \otimes u_{\sigma(mk)}.$$

For $i = 1, \ldots, m$, let $R_i \subset S_{mk}$ be the subgroup permuting the numbers $1 \leq a \leq mk$ such that $a \equiv i \mod m$ and leaving all other numbers intact and for $j = 1, \ldots, k$, let $C_i \subset S_{mk}$ be the subgroup permuting the numbers $m(i-1)+1 \leq a \leq mi$ and leaving all other numbers intact.

Let $w = e_1 \otimes \cdots \otimes e_m$. Then

$$v^{\otimes k} = (k!)^{-m} w^{\otimes k} \left(\sum_{\sigma \in R_1 \times \cdots \times R_m} \sigma \right) \left(\sum_{\sigma \in C_1 \times \cdots \times C_k} (\operatorname{sgn} \sigma)\sigma \right).$$

It follows then that $v^{\otimes k}$ generates the GL_n-module indexed by the rectangular $m \times k$ Young diagram, so its dimension d_k is given by the formula of the Theorem, see Chapter 6 of [Fulton and Harris 1991]. □

COROLLARY 3.4. *Under the conditions of Theorem 3.3, let*

$$k \geq m(n-1) \max\{\ln m, 1\}.$$

Then

$$\|\ell\|_{2k} \leq \text{comass of } \ell \leq \alpha \|\ell\|_{2k}$$

for some absolute constant $\alpha > 0$.

One can choose $\alpha = \exp\left(0.5 + 0.5e^{-1} + 1/\ln 3\right) \approx 4.93$.

PROOF. We have

$$d_k \leq \prod_{i=1}^{m}\prod_{j=1}^{k} \frac{n+j-i}{k-j+1} \leq \left(\prod_{j=1}^{k} \frac{n+j-1}{k-j+1}\right)^m = \binom{n+k-1}{n-1}^m.$$

Hence

$$\ln d_k \leq m \ln \binom{n+k-1}{n-1} \leq m(n-1)\ln\frac{n+k-1}{n-1} + mk\ln\frac{n+k-1}{k}$$

$$\leq m(n-1)\left(\ln\frac{n+k-1}{n-1}+1\right) = m(n-1)\left(\ln\frac{k}{n-1}+2\right);$$

compare the proof of Corollary 3.2.

If $m \geq 3$ then $\ln m \geq 1$ and $k/(n-1) \geq m \ln m$. Since the function $\rho^{-1}\ln\rho$ is decreasing for $\rho \geq e$, substituting $\rho = k/(n-1)$, we get

$$\rho^{-1}\ln\rho = \frac{n-1}{k}\ln\frac{k}{n-1} \leq \frac{\ln m + \ln\ln m}{m\ln m}.$$

Therefore, for $m \geq 3$, we have

$$\frac{1}{2k}\ln d_k \leq \frac{\ln m + \ln\ln m}{2\ln m} + \frac{1}{\ln m} \leq \frac{1}{2} + \frac{1}{2e} + \frac{1}{\ln 3}.$$

If $m \leq 2$ then

$$\frac{n-1}{k}\ln\frac{k}{n-1} \leq e^{-1},$$

since the maximum of $\rho^{-1}\ln\rho$ for is attained at $\rho = e$. Therefore,

$$\frac{1}{2k}\ln d_k \leq e^{-1} + 1 < \frac{1}{2} + \frac{1}{2e} + \frac{1}{\ln 3}$$

The proof now follows.   □

To understand the convex geometry of an orbit, we would like to compute the maximum value of a "typical" linear functional on the orbit. Theorem 3.1 allows us to replace the maximum value by an L^p norm. To estimate the average value of an L^p norm, we use the following simple computation.

LEMMA 3.5. *Let G be a compact group acting in a d-dimensional real vector space V endowed with a G-invariant scalar product $\langle\cdot,\cdot\rangle$ and let $v \in V$ be a point. Let $\mathbb{S}^{d-1} \subset V$ be the unit sphere endowed with the Haar probability measure dc. Then, for every positive integer k, we have*

$$\int_{\mathbb{S}^{d-1}}\left(\int_G \langle c, gv\rangle^{2k}\,dg\right)^{1/2k}dc \leq \sqrt{\frac{2k\langle v,v\rangle}{d}}.$$

PROOF. Applying Hölder's inequality, we get

$$\int_{\mathbb{S}^{d-1}}\left(\int_G \langle c, gv\rangle^{2k}\,dg\right)^{1/2k}dc \leq \left(\int_{\mathbb{S}^{d-1}}\int_G \langle c, gv\rangle^{2k}\,dg\,dc\right)^{1/2k}.$$

Interchanging the integrals, we get

$$\int_{\mathbb{S}^{d-1}} \int_G \langle c, gv \rangle^{2k} \, dg \, dc = \int_G \left(\int_{\mathbb{S}^{d-1}} \langle c, gv \rangle^{2k} \, dc \right) dg. \qquad 3.5.1$$

Now we observe that the integral inside has the same value for all $g \in G$. Therefore, (3.5.1) is equal to

$$\int_{\mathbb{S}^{d-1}} \langle c, v \rangle^{2k} \, dc = \langle v, v \rangle^k \frac{\Gamma(d/2)\Gamma(k+1/2)}{\sqrt{\pi}\Gamma(k+d/2)},$$

see, for example, [Barvinok 2002b].

Now we use that $\Gamma(k+1/2) \leq \Gamma(k+1) \leq k^k$ and

$$\frac{\Gamma(d/2)}{\Gamma(k+d/2)} = \frac{1}{(d/2)(d/2+1)\cdots(d/2+k-1)} \leq (d/2)^{-k}. \qquad \square$$

4. Some Geometric Corollaries

The metric structure of the unit comass ball. Let $V_{m,n} = \bigwedge^m \mathbb{R}^n$ with the orthonormal basis $e_I = e_{i_1} \wedge \cdots \wedge e_{i_m}$, where I is an m-subset $1 \leq i_1 < i_2 < \cdots < i_m \leq n$ of the set $\{1, \ldots, n\}$, and the corresponding scalar product $\langle \cdot, \cdot \rangle$. Let $G_m(\mathbb{R}^n) \subset V_{m,n}$ be the Plücker embedding of the Grassmannian of oriented m-subspaces of $\mathbb{R}^n$, let $B = \operatorname{conv}(G_m(\mathbb{R}^n))$ be the unit mass ball, and let $B^\circ \subset V_{m,n}^* = V_{m,n}$ be the unit comass ball, consisting of the linear functionals with the maximum value on $G_m(\mathbb{R}^n)$ not exceeding 1, see Examples 1.3 and 2.6.

The most well-known example of a linear functional $\ell : V_{m,n} \to \mathbb{R}$ of comass 1 is given by an exterior power of the Kähler form. Let us suppose that m and n are even, so $m = 2p$ and $n = 2q$. Let

$$\omega = e_1 \wedge e_2 + e_3 \wedge e_4 + \cdots + e_{q-1} \wedge e_q$$

and

$$f = \frac{1}{p!} \underbrace{\omega \wedge \cdots \wedge \omega}_{p \text{ times}} \in V_{m,n}.$$

Then

$$\max_{x \in G_m(\mathbb{R}^n)} \langle f, x \rangle = 1,$$

and, moreover, the subspaces $x \in G_m(\mathbb{R}^n)$ where the maximum value 1 is attained look as follows. We identify $\mathbb{R}^n$ with $\mathbb{C}^q$ by identifying

$$\mathbb{R}e_1 \oplus \mathbb{R}e_2 = \mathbb{R}e_3 \oplus \mathbb{R}e_4 = \cdots = \mathbb{R}e_{q-1} \oplus \mathbb{R}e_q = \mathbb{C}.$$

Then the subspaces $x \in G_m(\mathbb{R}^n)$ with $\langle f, x \rangle = 1$ are exactly those identified with the complex p-dimensional subspaces of $\mathbb{C}^q$, see [Harvey and Lawson 1982].

We note that the Euclidean length $\langle f, f \rangle^{1/2}$ of f is equal to $\binom{q}{p}^{1/2}$. In particular, if $m = 2p$ is fixed and $n = 2q$ grows, the length of f grows as $q^{p/2} = (n/2)^{m/4}$.

Another example is provided by the special Lagrangian calibration a. In this case, $n = 2m$ and

$$a = \operatorname{Re}\,(e_1 + ie_2) \wedge \cdots \wedge (e_{2m-1} + ie_{2m}).$$

The length $\langle a, a\rangle^{1/2}$ of a is $2^{(m-1)/2}$. The maximum value of $\langle a, x\rangle$ for $x \in G_m(\mathbb{R}^n)$ is 1 and it is attained on the "special Lagrangian subspaces", see [Harvey and Lawson 1982].

The following result shows that there exist calibrations with a much larger Euclidean length than that of the power f of the Kähler form or the special Lagrangian calibration a.

THEOREM 4.1. (i) *Let $c \in V_{m,n}$ be a vector such that*

$$\max_{x \in G_m(\mathbb{R}^n)} \langle c, x\rangle = 1.$$

Then

$$\langle c, c\rangle^{1/2} \leq \binom{n}{m}^{1/2}.$$

(ii) *There exists $c \in V_{m,n}$ such that*

$$\max_{x \in G_m(\mathbb{R}^n)} \langle c, x\rangle = 1$$

and

$$\langle c, c\rangle^{1/2} \geq \frac{\beta}{\sqrt{m(n-1)(1+\ln m)}} \binom{n}{m}^{1/2},$$

where $\beta > 0$ is an absolute constant.

One can choose $\beta = \exp\bigl(-0.5 - 0.5e^{-1} - 1/\ln 3\bigr)/\sqrt{2} \approx 0.14$.

PROOF. Part (1) follows since the convex hull of the Grassmannian contains a ball of radius $\binom{n}{m}^{-1/2}$; see Example 2.6.

To prove Part (2), let us choose $k = \lfloor m(n-1)(1+\ln m)\rfloor$ in Lemma 3.5. Then, by Corollary 3.4,

$$\alpha^{-1} \max_{x \in G_m(\mathbb{R}^n)} \langle c, x\rangle \leq \left(\int_{G_m(\mathbb{R}^n)} \langle c, x\rangle^{2k}\, dx\right)^{1/2k},$$

for some absolute constant $\alpha > 1$. We apply Lemma 3.5 with $V = V_{m,n}$, $d = \binom{n}{m}$, $G = \mathrm{SO}(n)$, and $v = e_1 \wedge \cdots \wedge e_m$. Hence $\langle v, v\rangle = 1$ and there exists $c \in V_{m,n}$ with $\langle c, c\rangle = 1$ and such that

$$\left(\int_{G_m(\mathbb{R}^n)} \langle c, x\rangle^{2k}\, dx\right)^{1/2k} \leq \sqrt{2k}\binom{n}{m}^{-1/2}.$$

Rescaling c to a comass 1 functional, we complete the proof of Part (2). □

For $m = 2$ the estimate of Part (2) is exact up to an absolute constant, as witnessed by the Kähler calibration. However, for $m \geq 3$, the calibration c of Part (2) has a larger length than the Kähler or special Lagrangian calibrations. The gap only increases when m and n grow. The distance to the origin of the supporting hyperplane $\langle c, x \rangle = 1$ of the face of the convex hull of the Grassmannian is equal to $\langle c, c \rangle^{-1/2}$ so the faces defined by longer calibrations c are closer to the origin. Thus, the faces spanned by complex subspaces or the faces spanned by special Lagrangian subspaces are much more "shallow" than the faces defined by calibrations c in Part (2) of the Theorem. We do not know if those "deep" faces are related to any interesting geometry. Intuitively, the closer the face to the origin, the larger piece of the Grassmannian it contains, so it is quite possible that some interesting classes of manifolds are associated with the "long" calibrations c [Morgan 1992].

The volume of the set of nonnegative polynomials. Let $U_{m,n}$ be the space of real homogeneous polynomials p of degree m in n variables such that the average value of p on the unit sphere $\mathbb{S}^{n-1} \subset \mathbb{R}^n$ is 0, so $\dim U_{m,n} = \binom{n+m-1}{m} - 1$ for m even and $\dim U_{m,n} = \binom{n+m-1}{m}$ for m odd. As before, we make $U_{m,n}$ a Euclidean space with the L^2 inner product

$$\langle f, g \rangle = \int_{\mathbb{S}^{n-1}} f(x)g(x)\, dx.$$

We obtain the following corollary.

COROLLARY 4.2. *Let $\Sigma_{m,n} \subset U_{m,n}$ be the unit sphere, consisting of the polynomials with L^2-norm equal to 1. For a polynomial $p \in U_{m,n}$, let*

$$\|p\|_\infty = \max_{x \in \mathbb{S}^{n-1}} |p(x)|.$$

Then

$$\int_{\Sigma_{m,n}} \|p\|_\infty\, dp \leq \beta \sqrt{(n-1)\ln(m+1) + 1}$$

for some absolute constant $\beta > 0$. One can take $\beta = \sqrt{2}\exp(1 + 0.5e^{-1}) \approx 4.63$.

PROOF. Let us choose $k = \lfloor (n-1)\ln(m+1) + 1 \rfloor$. Then, by Corollary 3.2,

$$\|p\|_\infty \leq \alpha \left(\int_{\mathbb{S}^{n-1}} p^{2k}\, dx \right)^{1/2k},$$

where we can take $\alpha = \exp(1 + 0.5e^{-1})$. Now we use Lemma 3.5. As in Examples 1.2 and 2.5, we identify the space $U_{m,n}$ with the space of linear functionals $\langle c, gv \rangle$ on the orbit $\{gv : g \in \mathrm{SO}(n)\}$ of v. By the remark after the proof of Theorem 2.4, we have $\langle v, v \rangle = \dim U_{m,n}$. The proof now follows. $\square$

Thus the L^∞-norm of a typical n-variate polynomial of degree m of the unit L^2-norm in $U_{m,n}$ is $O\left(\sqrt{(n-1)\ln(m+1) + 1}\right)$. In contrast, the L^∞ norm of a *particular* polynomial can be of the order of $n^{m/2}$, that is, substantially bigger.

Corollary 4.2 was used by the second author to obtain a bound on the volume of the set of nonnegative polynomials.

Let us consider the shifted set $\mathrm{Pos}'_{2k,n} \subset U_{2k,n}$ of nonnegative polynomials defined by (1.2.2). We measure the size of a set $X \subset U_{2k,n}$ by the quantity

$$\left(\frac{\mathrm{vol}\, X}{\mathrm{vol}\, K}\right)^{1/d},$$

where $d = \dim U_{2k,n}$ and K is the unit ball in $U_{2k,n}$, which is more "robust" than just the volume $\mathrm{vol}\, X$, as it takes into account the effect of a high dimension; see Chapter 6 of [Pisier 1989].

The following result is from [Blekherman 2003], we made some trivial improvement in the dependence on the degree $2k$.

THEOREM 4.3. *Let* $\mathrm{Pos}'_{2k,n} \subset U_{2k,n}$ *be the shifted set of nonnegative polynomials, let* $K \subset U_{2k,n}$ *be the unit ball and let* $d = \dim U_{2k,n} = \binom{n+2k-1}{2k} - 1$. *Then*

$$\left(\frac{\mathrm{vol}\, \mathrm{Pos}_{2k,n}}{\mathrm{vol}\, K}\right)^{1/d} \geq \frac{\gamma}{\sqrt{(n-1)\ln(2k+1)+1}}$$

for some absolute constant $\gamma > 0$. *One can take* $\gamma = \exp\big(-1 - 0.5e^{-1}\big)/\sqrt{2} \approx 0.21$.

PROOF. Let $\Sigma_{2k,n} \subset U_{2k,n}$ be the unit sphere. Let $p \in \Sigma_{2k,n}$ be a point. The ray $\lambda p : \lambda \geq 0$ intersects the boundary of $\mathrm{Pos}'_{2k,n}$ at a point p_1 such that $\min_{x \in \mathbb{S}^{n-1}} p_1(x) = -1$, so the length of the interval $[0, p_1]$ is $|\min_{x \in \mathbb{S}^{n-1}} p(x)| \leq \|p\|_\infty$.

Hence

$$\left(\frac{\mathrm{vol}\, \mathrm{Pos}'_{2k,n}}{\mathrm{vol}\, K}\right)^{1/d} = \left(\int_{\Sigma_{2k,n}} \Big|\min_{x \in \mathbb{S}^{n-1}} p(x)\Big|^{-d} dp\right)^{1/d} \geq \left(\int_{\Sigma_{2k,n}} \|p\|_\infty^{-d}\, dp\right)^{1/d}$$

$$\geq \int_{\Sigma_{2k,n}} \|p\|_\infty^{-1}\, dp \geq \left(\int_{\Sigma_{2k,n}} \|p\|_\infty\, dp\right)^{-1},$$

by the consecutive application of Hölder's and Jensen's inequalities, so the proof follows by Corollary 4.2. □

We defined $\mathrm{Pos}_{2k,n}$ as the set of nonnegative polynomials with the average value 1 on the unit sphere, see (1.2.1). There is an important subset $Sq_{2k,n} \subset \mathrm{Pos}_{2k,n}$, consisting of the polynomials that are sums of squares of homogeneous polynomials of degree k. It is known that $\mathrm{Pos}_{2k,n} = Sq_{2k,n}$ if $k = 1$, $n = 2$, or $k = 2$ and $n = 3$, see Chapter 6 of [Bochnak et al. 1998]. The following result from [Blekherman 2003] shows that, in general, $Sq_{2k,n}$ is a rather small subset of $\mathrm{Pos}_{2k,n}$.

Translating $p \mapsto p - (x_1^2 + \cdots + x_n^2)^k$, we identify $Sq_{2k,n}$ with a subset $Sq'_{2k,n}$ of $U_{2k,n}$.

THEOREM 4.4. *Let $Sq'_{2k,n} \subset U_{2k,n}$ be the shifted set of sums of squares, let $K \subset U_{2k,n}$ be the unit ball and let $d = \dim U_{2k,n} = \binom{n+2k-1}{2k} - 1$. Then*

$$\left(\frac{\operatorname{vol} Sq_{2k,n}}{\operatorname{vol} K}\right)^{1/d} \leq \gamma 2^{4k} \binom{n+k-1}{k}^{1/2} \binom{n+2k-1}{2k}^{-1/2}$$

for some absolute constant $\gamma > 0$. One can choose $\gamma = \exp(1 + 0.5e^{-1}) \approx 3.27$.

In particular, if k is fixed and n grows, the upper bound has the form $c(k)n^{-k/2}$ for some $c(k) > 0$.

The proof is based on bounding the right hand side of the inequality of Theorem 4.4 by the average width of $Sq_{2k,n}$; see Section 6.2 of [Schneider 1993]. The average width is represented by the integral

$$\int_{\Sigma_{2k,n}} \max_{f \in \Sigma_{k,n}} \langle g, f^2 \rangle \, dg.$$

By Corollary 3.2, we can bound the integrand by

$$\alpha \left(\int_{\Sigma_{k,n}} \langle g, f^2 \rangle^{2q} \, df\right)^{1/2q}$$

for some absolute constant α and $q = \binom{n+k-1}{k}$ and proceed as in the proof of Lemma 3.5. The factor 2^{4k} comes from an inequality of [Duoandikoetxea 1987], which allows us to bound the L^2-norm f^2 by 2^{4k} for every polynomial $f \in \Sigma_{k,n}$.

Acknowledgment

We thank B. Sturmfels for suggesting to us to consider the convex hull of the Grassmannian and J. Sullivan for pointing to connections with calibrated geometries. We thank F. Morgan and an anonymous referee for comments.

References

[Ball 1997] K. Ball, "An elementary introduction to modern convex geometry", pp. 1–58 in *Flavors of geometry*, edited by S. Levy, Math. Sci. Res. Inst. Publ. **31**, Cambridge Univ. Press, Cambridge, 1997.

[Barvinok 1992] A. I. Barvinok, "Combinatorial complexity of orbits in representations of the symmetric group", pp. 161–182 in *Representation theory and dynamical systems*, Adv. Soviet Math. **9**, Amer. Math. Soc., Providence, RI, 1992.

[Barvinok 2002a] A. Barvinok, *A course in convexity*, vol. 54, Graduate Studies in Mathematics, American Mathematical Society, Providence, RI, 2002.

[Barvinok 2002b] A. Barvinok, "Estimating L^∞ norms by L^{2k} norms for functions on orbits", *Found. Comput. Math.* **2**:4 (2002), 393–412.

[Barvinok and Vershik 1988] A. I. Barvinok and A. M. Vershik, "Convex hulls of orbits of representations of finite groups, and combinatorial optimization", *Funktsional. Anal. i Prilozhen.* **22**:3 (1988), 66–67.

[Billera and Sarangarajan 1996] L. J. Billera and A. Sarangarajan, "All 0-1 polytopes are traveling salesman polytopes", *Combinatorica* **16**:2 (1996), 175–188.

[Blekherman 2003] G. Blekherman, "There are significantly more nonnegative polynomials than sums of squares", 2003. Available at math.AG/0309130.

[Blekherman 2004] G. Blekherman, "Convexity properties of the cone of nonnegative polynomials", *Discrete Comput. Geom.* **32**:3 (2004), 345–371.

[Bochnak et al. 1998] J. Bochnak, M. Coste, and M.-F. Roy, *Real algebraic geometry*, vol. 36, Ergebnisse der Mathematik, Springer, Berlin, 1998.

[Duoandikoetxea 1987] J. Duoandikoetxea, "Reverse Hölder inequalities for spherical harmonics", *Proc. Amer. Math. Soc.* **101**:3 (1987), 487–491.

[Fulton and Harris 1991] W. Fulton and J. Harris, *Representation theory*, vol. 129, Graduate Texts in Mathematics, Springer, New York, 1991.

[Giannopoulos and Milman 2000] A. A. Giannopoulos and V. D. Milman, "Extremal problems and isotropic positions of convex bodies", *Israel J. Math.* **117** (2000), 29–60.

[Harvey and Lawson 1982] R. Harvey and H. B. Lawson, Jr., "Calibrated geometries", *Acta Math.* **148** (1982), 47–157.

[Harvey and Morgan 1986] R. Harvey and F. Morgan, "The faces of the Grassmannian of three-planes in $\mathbf{R}^7$ (calibrated geometries on $\mathbf{R}^7$)", *Invent. Math.* **83**:2 (1986), 191–228.

[John 1948] F. John, "Extremum problems with inequalities as subsidiary conditions", pp. 187–204 in *Studies and Essays Presented to R. Courant on his 60th Birthday, January 8, 1948*, Interscience, 1948.

[Kalai 1995] G. Kalai, "Combinatorics and convexity", pp. 1363–1374 in *Proceedings of the International Congress of Mathematicians* (Zürich, 1994), vol. 2, Birkhäuser, Basel, 1995.

[van Lint 1999] J. H. van Lint, *Introduction to coding theory*, Graduate Texts in Mathematics **86**, Springer, Berlin, 1999.

[Morgan 1988] F. Morgan, "Area-minimizing surfaces, faces of Grassmannians, and calibrations", *Amer. Math. Monthly* **95**:9 (1988), 813–822.

[Morgan 1992] F. Morgan, "Calibrations and the size of Grassmann faces", *Aequationes Math.* **43**:1 (1992), 1–13.

[Müller 1966] C. Müller, *Spherical harmonics*, Lecture Notes in Mathematics **17**, Springer, Berlin, 1966.

[Pisier 1989] G. Pisier, *The volume of convex bodies and Banach space geometry*, Cambridge Tracts in Mathematics **94**, Cambridge University Press, Cambridge, 1989.

[Schneider 1993] R. Schneider, *Convex bodies: the Brunn–Minkowski theory*, Encyclopedia of Mathematics and its Applications **44**, Cambridge University Press, Cambridge, 1993.

[Schrijver 2003] A. Schrijver, *Combinatorial optimization: polyhedra and efficiency*, Algorithms and Combinatorics **24**, Springer, Berlin, 2003.

[Vyalyĭ 1995] M. N. Vyalyĭ, "On estimates for the values of a functional in polyhedra of the subgraph of least weight problem", pp. 27–43 in Комбинаторные модели и методы, Ross. Akad. Nauk Vychisl. Tsentr, Moscow, 1995. In Russian.

ALEXANDER BARVINOK
DEPARTMENT OF MATHEMATICS
UNIVERSITY OF MICHIGAN
ANN ARBOR, MI 48109-1109
UNITED STATES
   barvinok@umich.edu

GRIGORIY BLEKHERMAN
DEPARTMENT OF MATHEMATICS
UNIVERSITY OF MICHIGAN
ANN ARBOR, MI 48109-1109
UNITED STATES
   gblekher@umich.edu

Combinatorial and Computational Geometry
MSRI Publications
Volume **52**, 2005

The Hadwiger Transversal Theorem for Pseudolines

SAUGATA BASU, JACOB E. GOODMAN, ANDREAS HOLMSEN,
AND RICHARD POLLACK

ABSTRACT. We generalize the Hadwiger theorem on line transversals to
collections of compact convex sets in the plane to the case where the sets
are connected and the transversals form an arrangement of pseudolines.
The proof uses the embeddability of pseudoline arrangements in topological
affine planes.

Santaló [1940] showed, by an example, that Vincensini's proof [1935] of an
extension of Helly's theorem was incorrect. Vincensini claimed to have proved
that for any finite collection S of at least three compact convex sets in the plane,
any three of which are met by a line, there must exist a line meeting all the sets.
This would have constituted an extension of the planar Helly theorem [Helly
1923] to the effect that the same assertion holds if "line" is replaced by "point."
The Santaló example was later extended by Hadwiger and Debrunner [1964] to
show that even if the convex sets are disjoint the conclusion still may not hold.

In 1957, however, Hadwiger showed that the conclusion of the theorem *is*
valid if the hypothesis is strengthened by imposing a consistency condition on
the order in which the triples of sets are met by transversals:

THEOREM [Hadwiger 1957]. *If $B_1, \ldots, B_n$ is a family of disjoint compact convex
sets in the plane with the property that for any $1 \leq i < j < k \leq n$ there is a
line meeting each of B_i, B_j, B_k in that order, then there is a line meeting all the
sets B_i.*

In [Goodman and Pollack 1988] two of us gave a generalization of Hadwiger's
theorem to the case of hyperplane transversals, and this in turn was extended in
[Pollack and Wenger 1990; Wenger 1990], culminating in the following result:

Basu was supported in part by NSF grant CCR-0049070 and NSF Career Award 0133597.
Goodman was supported in part by NSA grant MDA904-03-I-0087 and PSC-CUNY grant
65440-0034. Holmsen was supported in part by the Mathematical Sciences Research Institute,
Berkeley. Pollack was supported in part by NSF grant CCR-9732101.

THEOREM [Anderson and Wenger 1996]. *Let A be a finite collection of connected sets in $\mathbb{R}^d$. A has a hyperplane transversal if and only if for some k with $0 \leq k < d$ there exists a rank $k+1$ acyclic oriented matroid structure on A such that every $k+2$ members of A are met by an oriented k-flat consistently with that oriented matroid structure.*

Our purpose in this paper is to extend the original Hadwiger theorem in a different direction — replacing "lines" by "pseudolines." A *pseudoline* in the affine plane is simply the homeomorphic image of a line. If that were all, the theorem would be true trivially: for any finite collection of sets there is a pseudoline meeting them in any prescribed order! (Of course this needs a suitable interpretation in the case where the sets are not mutually disjoint; see below.) But to reflect more accurately the properties of sets of lines in the plane, one insists that all the pseudolines under consideration form an *arrangement*, which means that they are finite in number, that any two meet exactly once, where they cross, and (for technical reasons) that they do not all pass through the same point.[1] (For examples of pseudoline arrangements that are not isomorphic, in a natural sense, to arrangements of straight lines, see, e.g., [Goodman 2004].) Furthermore, given a pseudoline arrangement A we say that a pseudoline l *extends* A if $A \cup \{l\}$ is also an arrangement of pseudolines. Thus the theorem we are going to prove is the following:

THEOREM 1. *Suppose $B_1, \ldots, B_n$ is a family of connected compact sets in the plane such that for each $1 \leq i < j < k \leq n$ there is a pseudoline l_{ijk} meeting each of B_i, B_j, B_k at points p_i, p_j, p_k, not necessarily distinct, contained in B_i, B_j, B_k, respectively, with p_j lying between p_i and p_k on l_{ijk}. Suppose further that the pseudolines l_{ijk} constitute an* arrangement A. *Then there exists a pseudoline l that extends the arrangement A and meets each set B_i.*

As in Wenger's generalization [1990], we do not assume the sets to be disjoint or even convex, merely connected. And in fact we will prove Theorem 1 by generalizing Wenger's proof, and by using the following result on topological planes:

THEOREM [Goodman et al. 1994]. *Any arrangement of pseudolines in the projective plane can be extended to a topological projective plane.*

Here a *topological projective plane* means $\mathbb{P}^2$, together with a distinguished collection $\mathcal{L}$ of pseudolines, one for each pair of points, varying continuously with the points, any two meeting (and crossing) exactly once. If we call a topological projective plane with one of its distinguished pseudolines removed a *topological affine plane* (TAP), the theorem above can trivially be modified to read: *Any*

---

[1]This is actually the definition of a "pseudoline arrangement" in the projective plane, while in the affine plane one allows pseudolines also to be "parallel"; in a finite arrangement, however, pseudolines can always be perturbed slightly to meet "at finite distance," and we will assume this whenever convenient.

arrangement of pseudolines in the affine plane can be extended to a TAP. We will use it in this form.

For background on pseudoline arrangements and on geometric transversal theory, the reader may consult the following surveys: [Eckhoff 1993; Goodman 2004; Goodman et al. 1993; Grünbaum 1972; Wenger 1999; Wenger 2004].

We now introduce some notions that will be used in the proof of the theorem.

Since $\mathbb{P}^2$ can be modeled by a closed circular disk Δ with antipodal points on the boundary $\partial\Delta$ identified, we will model our TAP by using int Δ, the interior of Δ, and call two pseudolines *parallel* if they meet on $\partial\Delta$. (From now on, whenever we speak of "pseudolines" in the TAP, we will mean members of the distinguished family of pseudolines constituting its "lines.") An *arrangement* of pseudolines is thus a finite set of Jordan arcs, each joining a pair of antipodal points of $\partial\Delta$, any two meeting (and crossing) exactly once, or possibly at their endpoints (the parallel case).

We will also speak of *directed* pseudolines, which corresponds to specifying one of the antipodal points where the pseudoline meets $\partial\Delta$. Thus it will make sense to say: let p be a point on $\partial\Delta$ and let l_p be a pseudoline in the direction p. Further, when we direct a pseudoline, we specify a positive and a negative open *half-space* bounded by that line, determined with respect to a fixed orientation of Δ. We denote these half-spaces by $H_+(l_p)$ and $H_-(l_p)$; see Figure 1.

Now let A and B be two connected compact sets in our TAP and let $p \in \partial\Delta$. If there is a pseudoline in the direction of p that contains points $a \in A$ and $b \in B$, with either $a = b$ or a preceding b on the pseudoline, we say that p is an (AB)-*transversal direction*. If there is a pseudoline l_p that strictly separates A and B such that $A \subset H_+(l_p)$ and $B \subset H_-(l_p)$, we say that p is a (AB)-*separating direction*.

Notice that a given direction can be both an (AB)-transversal direction and a (BA)-transversal direction; even the same pseudoline, in fact, can meet A before B and B before A in this sense.

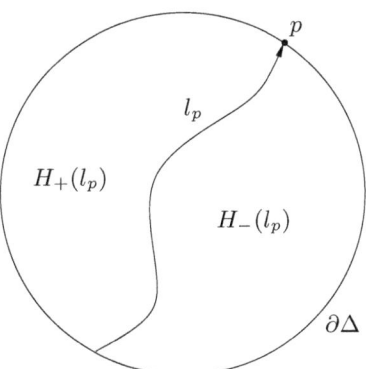

Figure 1.

Notice also that given a pair A, B, each direction p is either a transversal direction or a separating direction for A, B, but not both; this follows by a simple continuity argument, sweeping a pseudoline in direction p across the TAP.

Finally, notice that if there is an (AB)-separating direction p, no direction q can be both an (AB)-transversal direction and a (BA)-transversal direction. This follows from the fact that if two pseudolines have the same direction q, they must cross a given pseudoline l in direction p the same way: both from $H_+(l)$ to $H_-(l)$, or both from $H_-(l)$ to $H_+(l)$.

It then follows from the definition of a TAP and the compactness of our sets that the set T_{AB} of (AB)-transversal directions is a closed arc of $\partial\Delta$: If A and B have a point in common then clearly $T_{AB} = \partial\Delta$. If not, consider any two distinct directions $p_1, p_2 \in T_{AB}$. For $i = 1, 2$ choose points $a_i \in A$, $b_i \in B$ along a pseudoline l_i in direction p_i, with a_i preceding b_i, as well as a parametrized arc $a(t) \subset A$ from a_1 to a_2 and a parametrized arc $b(t) \subset B$ from b_1 to b_2. By continuity, the set of directions $\overrightarrow{a(t)b(t)}$ must contain one of the two arcs on $\partial\Delta$ joining p_1 and p_2. It follows that the set T_{AB} is itself an arc (possibly all of $\partial\Delta$), and this must be closed by the compactness of the sets A and B.

We have thus proved the following:

LEMMA 2. *Let A and B be connected compact sets in the plane. Then*

$$\partial\Delta = T_{AB} \cup S_{AB} \cup T_{BA} \cup S_{BA},$$

where $T_{AB} = -T_{BA}$ is the closed arc corresponding to the (AB)-transversal directions, and $S_{AB} = -S_{BA}$ is the open arc corresponding to the (AB)-separating directions. (Note that S_{AB} can be empty.)

To complete the proof of Theorem 1, we extend the arrangement $\mathcal{A}$ to a topological affine plane. We want to show first that there is a direction $p \in \partial\Delta$ that is a transversal direction for every pair B_i, B_j. For each pair B_i, B_j, let S_{ij} be the open arc of $(B_i B_j)$-separating directions. Now define the following antipodal sets:

$$\mathcal{S}_+ = \bigcup_{i<j} S_{ij} \quad , \quad \mathcal{S}_- = \bigcup_{i<j} S_{ji} \quad .$$

If there is no point $p \in \partial\Delta$ that is a transversal direction for every pair B_i, B_j then we must have $\partial\Delta = \mathcal{S}_+ \cup \mathcal{S}_-$. But since $\mathcal{S}_+$ and $\mathcal{S}_-$ are open sets that cover $\partial\Delta$ there must be a point $p \in \mathcal{S}_+ \cap \mathcal{S}_-$. But then we would have pseudolines l_1 and l_2, both directed toward p, and sets B_i, B_j, B_k, B_l with $i < j$ and $k < l$, such that $B_i \subset h_+(l_1)$, $B_j \subset h_-(l_1)$, $B_k \subset h_-(l_2)$, and $B_l \subset h_+(l_2)$. It is then easy to check that there would always be some triple that violates the transversal assumption; see Figure 2 for a typical case.

This means that there is a direction $q \in \partial\Delta$ that is a transversal direction for every pair B_i, B_j. It follows that q is not a separating direction for any pair B_i, B_j, so that a pseudoline in direction q sweeping through the TAP must pass simultaneously through all the sets B_i at some point. This completes the proof.

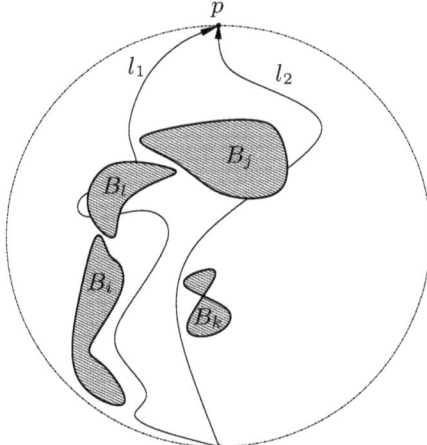

Figure 2. If $i < k$, there is no l_{ikl}; if $k < i$, there is no l_{kij}.

REMARKS. **1.** It is not hard to see that Theorem 1 is equivalent to the following.

THEOREM 3. *Suppose $\mathcal{L}$ is an arrangement of pseudolines in the affine plane. For each triple $i < j < k$ in $[1, n]$, select three (not necessarily distinct) points belonging to the same pseudoline of $\mathcal{L}$, and label them i, j, k, with the point labeled j between the other two (or possibly equal to one or both). Then there is a pseudoline l extending the arrangement $\mathcal{L}$ such that for each $i \in [1, n]$ there are points labeled i in both (closed) half-spaces bounded by l.*

2. As in the original Hadwiger theorem, one cannot strengthen the conclusion of Theorem 1 to include the assertion that the common transversal meets the sets in the order $1, 2, \ldots, n$ (see [Wenger 1990] for an example). But it is easily seen that, as in [Wenger 1990], that stronger assertion follows if we are willing to assume that every *six* of the sets are met in a consistent order; the argument is the same, *mutatis mutandis*.

THEOREM 4. *Suppose $B_1, \ldots, B_n$ is a family of at least six connected compact sets in the plane such that for each $1 \le f < g < h < i < j < k \le n$ there is a pseudoline l_{fghijk} meeting each of $B_f, B_g, B_h, B_i, B_j, B_k$ at points $p_f, p_g, p_h, p_i, p_j, p_k$, not necessarily distinct, contained in $B_f, B_g, B_h, B_i, B_j, B_k$, respectively, and occurring in that order on l_{fghijk}. Suppose further that the pseudolines l_{fghijk} constitute an arrangement $\mathcal{A}$. Then there exists a pseudoline l that extends the arrangement $\mathcal{A}$ and meets all of the sets $B_1, \ldots, B_n$ in that order.*

The example in [Wenger 1990] showing that the number 6 in the corresponding result for straight lines and convex sets is tight does not seem correct. Here is an example, however, showing that the result would fail for a collection $B_1, \ldots, B_6$ of convex sets if we assumed only that every five were met in a consistent order; here every five sets have a transversal meeting them in numerical order, but all six do not:

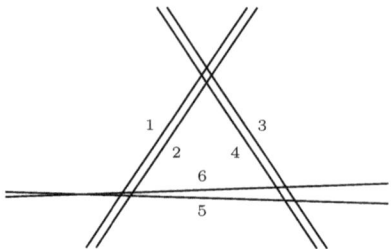

3. In the process of proving Theorem 1, we have actually proven the following (stronger) theorem about TAPs:

THEOREM 5. *If $B_1, \ldots, B_n$ is a family of connected compact sets in a topological affine plane $\mathcal{P}$ with the property that for any $1 \leq i < j < k \leq n$ there is a pseudoline of $\mathcal{P}$ meeting each of B_i, B_j, B_k in that order, then there is a pseudolineline of $\mathcal{P}$ meeting all the sets B_i.*

This raises the question: What other transversal theorems extend to TAPs?

4. Finally, what about higher dimensions? The notion of 'topological plane' extends only trivially to dimension ≥ 3, since, as is well-known, Desargues's theorem holds automatically in higher dimensions and any d-dimensional "topological projective space" is consequently isomorphic to the usual projective space $\mathbb{P}^d$. Nevertheless, one may ask: Does Theorem 1 extend in some way, in dimension > 2, to a result about (finite) arrangements of pseudohyperplane transversals?

References

[Anderson and Wenger 1996] L. Anderson and R. Wenger, "Oriented matroids and hyperplane transversals", *Adv. Math.* **119**:1 (1996), 117–125.

[Eckhoff 1993] J. Eckhoff, "Helly, Radon, and Carathéodory type theorems", pp. 389–448 in *Handbook of convex geometry*, vol. A, edited by P. M. Gruber and J. M. Wills, North-Holland, Amsterdam, 1993.

[Goodman 2004] J. E. Goodman, "Pseudoline arrangements", pp. 97–128 in *Handbook of discrete and computational geometry*, 2nd ed., edited by J. E. Goodman and J. O'Rourke, CRC, Boca Raton, FL, 2004.

[Goodman and Pollack 1988] J. E. Goodman and R. Pollack, "Hadwiger's transversal theorem in higher dimensions", *J. Amer. Math. Soc.* **1**:2 (1988), 301–309.

[Goodman et al. 1993] J. E. Goodman, R. Pollack, and R. Wenger, "Geometric transversal theory", pp. 163–198 in *New trends in discrete and computational geometry*, edited by J. Pach, Algorithms Combin. **10**, Springer, Berlin, 1993.

[Goodman et al. 1994] J. E. Goodman, R. Pollack, R. Wenger, and T. Zamfirescu, "Arrangements and topological planes", *Amer. Math. Monthly* **101**:9 (1994), 866–878.

[Grünbaum 1972] B. Grünbaum, *Arrangements and spreads*, American Mathematical Society, Providence, 1972.

[Hadwiger 1957] H. Hadwiger, "Ueber Eibereiche mit gemeinsamer Treffgeraden", *Portugal. Math.* **16** (1957), 23–29.

[Hadwiger and Debrunner 1964] H. Hadwiger and H. Debrunner, *Combinatorial geometry in the plane*, Holt, Rinehart and Winston, New York, 1964.

[Helly 1923] E. Helly, "Über Mengen konvexen Körper mit gemeinschaftlichen Punkten", *Jber. Deutsch. Math. Verein* **32** (1923), 175–176.

[Pollack and Wenger 1990] R. Pollack and R. Wenger, "Necessary and sufficient conditions for hyperplane transversals", *Combinatorica* **10**:3 (1990), 307–311.

[Santaló 1940] L. A. Santaló, "A theorem on sets of parallelepipeds with parallel edges", *Publ. Inst. Mat. Univ. Nac. Litoral* **2** (1940), 49–60.

[Vincensini 1935] P. Vincensini, "Figures convexes et variétés linéaires de l'espace euclidien à n dimensions", *Bull. Sci. Math.* **59** (1935), 163–174.

[Wenger 1990] R. Wenger, "A generalization of Hadwiger's transversal theorem to intersecting sets", *Discrete Comput. Geom.* **5**:4 (1990), 383–388.

[Wenger 1999] R. Wenger, "Progress in geometric transversal theory", pp. 375–393 in *Advances in discrete and computational geometry* (South Hadley, MA, 1996), edited by B. Chazelle et al., Contemp. Math. **223**, Amer. Math. Soc., Providence, RI, 1999.

[Wenger 2004] R. Wenger, "Helly-type theorems and geometric transversals", pp. 73–96 in *Handbook of discrete and computational geometry*, 2nd ed., edited by J. E. Goodman and J. O'Rourke, CRC, Boca Raton, FL, 2004.

SAUGATA BASU
SCHOOL OF MATHEMATICS
GEORGIA INSTITUTE OF TECHNOLOGY
ATLANTA, GA 30332
UNITED STATES
    saugata@math.gatech.edu

JACOB E. GOODMAN
DEPARTMENT OF MATHEMATICS
CITY COLLEGE, CUNY
NEW YORK, NY 10031
UNITED STATES
    jegcc@cunyvm.cuny.edu

ANDREAS HOLMSEN
DEPARTMENT OF MATHEMATICS
UNIVERSITY OF BERGEN
JOHANNES BRUNSGT. 12
5008 BERGEN
NORWAY
    andreash@mi.uib.no

RICHARD POLLACK
COURANT INSTITUTE, NYU
251 MERCER ST.
NEW YORK, NY 10012
USA
    pollack@courant.nyu.edu

Combinatorial and Computational Geometry
MSRI Publications
Volume **52**, 2005

Betti Number Bounds, Applications and Algorithms

SAUGATA BASU, RICHARD POLLACK, AND MARIE-FRANÇOISE ROY

ABSTRACT. Topological complexity of semialgebraic sets in $\mathbb{R}^k$ has been studied by many researchers over the past fifty years. An important measure of the topological complexity are the Betti numbers. Quantitative bounds on the Betti numbers of a semialgebraic set in terms of various parameters (such as the number and the degrees of the polynomials defining it, the dimension of the set etc.) have proved useful in several applications in theoretical computer science and discrete geometry. The main goal of this survey paper is to provide an up to date account of the known bounds on the Betti numbers of semialgebraic sets in terms of various parameters, sketch briefly some of the applications, and also survey what is known about the complexity of algorithms for computing them.

1. Introduction

Let R be a real closed field and S a semialgebraic subset of R^k, defined by a Boolean formula, whose atoms are of the form $P = 0$, $P > 0$, $P < 0$, where $P \in \mathcal{P}$ for some finite family of polynomials $\mathcal{P} \subset \mathrm{R}[X_1, \ldots, X_k]$. It is well known [Bochnak et al. 1987] that such sets are finitely triangulable. Moreover, if the cardinality of $\mathcal{P}$ and the degrees of the polynomials in $\mathcal{P}$ are bounded, then the number of topological types possible for S is finite [Bochnak et al. 1987]. (Here, two sets have the same topological type if they are semialgebraically homeomorphic). A natural problem then is to bound the topological complexity of S in terms of the various parameters of the formula defining S.

One measure of topological complexity are the various Betti numbers of S. The i-th Betti number of S (which we will denote by $b_i(S)$) is the rank of $H_i(S, \mathbb{Z})$. In case, R happens to be $\mathbb{R}$ then $H_i(S, \mathbb{Z})$ denotes the i-th singular homology group of S with integer coefficients. For semialgebraic sets defined

Mathematics Subject Classification: 14P10, 14P25.

Basu was supported in part by an NSF Career Award 0133597 and a Sloan Foundation Fellowship. Pollack was supported in part by NSF grant CCR-0098246.

over general real closed fields the definition of homology groups requires more care and several possibilities exists. For instance, if S is closed and bounded, then using the fact that S is finitely triangulable, $H_i(S, \mathbb{Z})$ can be taken to be the i-th simplicial homology group of S, and this definition agrees with the previous definition in case $R = \mathbb{R}$. For a general locally closed semialgebraic set, one can take for $H_i(S, \mathbb{Z})$ the i-th Borel–Moore homology groups, which are defined in terms of the simplicial homology groups of the one-point compactification of S, and which are known to be invariants under semialgebraic homeomorphisms [Bochnak et al. 1987]. Note that, even though some of the early results on bounding the Betti numbers of semialgebraic sets were stated only over $\mathbb{R}$, the bounds can be shown to hold over any real closed field by judicious applications of the Tarski–Seidenberg transfer principle. We refer the reader to [Basu et al. 2003] (Chapter 7) for more details.

2. Early Bounds

For a polynomial $P \in R[X_1, \ldots, X_k]$, we denote by $Z(P, \mathrm{R}^k)$ the set of zeros of P in R^k. The first results on bounding the Betti numbers of algebraic sets are due to Oleinik and Petrovsky [1949; 1951; 1949a; 1949b]. They considered the problem of bounding the Betti numbers of a nonsingular real algebraic hypersurface in $\mathbb{R}^k$ defined by a single polynomial equation of degree d. More precisely, they prove that the sum of the even Betti numbers, as well as the sum of the odd Betti numbers, of a nonsingular real algebraic hypersurface in R^k defined by a polynomial of degree d are each bounded by $\frac{1}{2}d^k +$ lower order terms. Independently, Thom [1965] proved a similar bound of $\frac{1}{2}d(2d - 1)^{k-1}$ on the sum of all the Betti numbers of $Z(P, \mathrm{R}^k)$, where P is only assumed to be nonnegative over R^k without the assumption that $Z(P, \mathrm{R}^k)$ is a nonsingular hypersurface. Milnor [1964] also proved the same bound in the case $Z(P, \mathrm{R}^k)$ is an arbitrary real algebraic subset. Moreover, he proved a bound of $(sd)(2sd - 1)^{k-1}$ on the sum of the Betti numbers of a basic semialgebraic set defined by the conjunction of s weak inequalities $P_1 \geq 0, \ldots, P_s \geq 0$, with $P_i \in \mathbb{R}[X_1, \ldots, X_k], \deg(P_i) \leq d$. Note that there is a cost for generality: the bounds of Thom and Milnor are slightly weaker (in the leading constant) than those proved by Oleinik and Petrovsky. Note also that these bounds on the sum of the Betti numbers of an algebraic set are tight, since the solutions to the system of equations,

$$(X_1 - 1)(X_1 - 2) \cdots (X_1 - d) = \cdots = (X_k - 1)(X_k - 2) \cdots (X_k - d) = 0,$$

or equivalently of the single equation

$$\left((X_1 - 1)(X_1 - 2) \cdots (X_1 - d)\right)^2 + \cdots + \left((X_k - 1)(X_k - 2) \cdots (X_k - d)\right)^2 = 0,$$

consist of d^k isolated points and the only nonzero Betti number of this set is $b_0 = d^k$.

The method used to obtain these bounds is based on a basic fact from Morse theory – that the sum of the Betti numbers of a compact, nonsingular, hyper-surface in $\mathbb{R}^k$ is at most the number of critical points of a well chosen projection. In case of a nonsingular real algebraic variety, the critical points of a projection map satisfy a simple system of algebraic equations obtaining by setting the poly-nomial defining the hypersurface, as well as $k-1$ different partial derivatives to zero. The number of solutions to such a system can be bounded from above by Bezout's theorem. The case of an arbitrary real algebraic variety (not neces-sarily compact and nonsingular) is reduced to the compact, nonsingular case by carefully using perturbation arguments.

Even though the bounds mentioned above are bounds on the sum of all the Betti numbers, in different combinatorial applications it suffices to have bounds only on the zero-th Betti number (that is the number of connected components). For instance, given a finite set of polynomials $\mathcal{P} \subset \mathrm{R}[X_1, \ldots, X_k]$, a natural question is how many of the 3^s sign conditions in $\{0, 1, -1\}^{\mathcal{P}}$ are actually realized at points in R^k. We define

$$
\begin{cases}
\operatorname{sign} x = 0 & \text{if and only if } x = 0, \\
\operatorname{sign} x = 1 & \text{if and only if } x > 0, \\
\operatorname{sign} x = -1 & \text{if and only if } x < 0.
\end{cases}
$$

Let $\mathcal{P} \subset \mathrm{R}[X_1, \ldots, X_k]$. A sign condition on $\mathcal{P}$ is an element of $\{0, 1, -1\}^{\mathcal{P}}$. A strict sign condition on $\mathcal{P}$ is an element of $\{1, -1\}^{\mathcal{P}}$. We say that $\mathcal{P}$ realizes the sign condition σ at $x \in \mathrm{R}^k$ if

$$
\bigwedge_{P \in \mathcal{P}} \operatorname{sign} P(x) = \sigma(P).
$$

The realization of the sign condition σ is

$$
\mathcal{R}(\sigma) = \Big\{ x \in \mathrm{R}^k \;\Big|\; \bigwedge_{P \in \mathcal{P}} \operatorname{sign} P(x) = \sigma(P) \Big\}.
$$

The sign condition σ is realizable if $\mathcal{R}(\sigma)$ is nonempty.

Warren [1968] proved a bound of $(4esd/k)^k$ on the number of strict sign conditions realized by a set of s polynomials in R^k whose degrees are bounded by d. Alon [1995] extended this result to all sign conditions by proving a bound of $(8esd/k)^k$. The fact that these bounds are polynomial in s (for fixed values of k) is important in many applications. Note that this bound is tight since it is an easy exercise to prove that the number of sign conditions realized by a family of linear polynomials in general position is

$$
\sum_{i=0}^{k} \sum_{j=0}^{k-i} \binom{s}{i} \binom{s-i}{j}; \tag{2-1}
$$

see for example [Basu et al. 2003].

3. Early Applications

One of the first applications of the bounds of Oleinik–Petrovsky, Thom and Milnor, was in proving lower bounds in theoretical computer science. The model for computation was taken to be algebraic decision trees. Given an input $x \in \mathbb{R}^k$, an algebraic decision tree decides membership of x in a certain fixed semialgebraic set $S \subset \mathbb{R}^k$. Starting from the root of the tree, at each internal node, v, of the tree, it evaluates a polynomial $f_v \in \mathbb{R}[X_1, \ldots, X_k]$ (where $\deg(f_v) \leq d$, for some fixed constant d), at the point $(x_1, \ldots, x_k)$ and branches according to the sign of the result. The leaf nodes of the tree are labelled as accepting or rejecting. On an input $x \in \mathbb{R}^k$, the algebraic decision tree accepts x if and only if the computation terminates at an accepting leaf node. Moreover, an algebraic decision tree tests membership in S, if it accepts x if and only if $x \in S$. The main idea behind using the Oleinik–Petrovsky, Thom and Milnor bounds in proving lower bounds for the problem of testing membership in a certain semialgebraic set $S \subset \mathbb{R}^k$ is that if the set S is topologically complicated, then an algebraic decision tree testing membership in it has to have large depth.

Ben-Or [1983] proved that the depth of an algebraic computation tree testing membership in S must be $\Omega(\log b_0(S))$. Several extensions of this result were proved by Yao [1995; 1997]. He proved that instead of $b_0(S)$ one could use in fact the Euler characteristic of S (which is the alternating sum of the Betti numbers), as well as the sum of the Betti numbers of S. This made the theorem useful for proving lower bounds for a wider class of problems by including sets with a single connected component but complicated topology [Montaña et al. 1991]. Another early application of the Oleinik–Petrovsky, Thom and Milnor bounds was in proving upper bounds on the number of order types of simple configurations of points in $\mathbb{R}^k$. Given an ordered set, S, of s points in $\mathbb{R}^k$, the order type of S is determined by the $\binom{s}{k+1}$ orientations of the $\binom{s}{k+1}$ oriented simplices spanned by $(k+1)$-tuples of points. A point configuration is simple if no $k+1$ of them are affinely dependent. Using Milnor's bound on the Betti numbers of basic semialgebraic sets Goodman and Pollack [1986b] proved an upper bound of s^{k^2} on the number of realizable simple order types of s points in $\mathbb{R}^k$ [Goodman and Pollack 1986a] rather than the trivial bound of 2^s. as well as on the number of combinatorial types of simple polytopes with s vertices in $\mathbb{R}^k$ [Goodman and Pollack 1986a]. In fact, Milnor's bound actually yields a bound on the number of isotopy classes of simple configurations of s points in $\mathbb{R}^k$. The isotopy class of a point configuration in $\mathbb{R}^k$ consists of all point configurations in $\mathbb{R}^k$ having the same order type which are reachable by continuous order type preserving deformations of the original point configuration. Alon [1995] extended these bounds to all configurations – not necessarily simple ones.

All of these applications are based on the simple observation that different strict sign conditions must belong to different connected components. Any situation where *geometric types* can be characterized by a sign condition gives an application of this type. Two other application in this spirit are bounds on the number of *weaving patterns* of lines [Pach et al. 1993] and the size of a grid which will support all order types of s points in the plane [Goodman et al. 1989; 1990].

4. Modern Bounds

Pollack and Roy [1993] proved a bound of $\binom{s}{k}O(d)^k$ on the number of connected components of the realizations of all realizable sign conditions of a family of s polynomials of degrees bounded by d. The proof was based on Oleinik–Petrovsky, Thom and Milnor's results for algebraic sets, as well as with deformation techniques and general position arguments.

From this bound one can deduce a tight bound on the number of isotopy classes of *all* point configurations in $\mathbb{R}^k$ (not just the simple ones). Note that Warren's bound mentioned before is a bound on the number of realizable strict sign conditions (extended by Alon to all sign conditions) but not on the number of connected components of their realizations. Thus, Warren's (or Alon's) bounds cannot be used to bound the number of isotopy classes (of simple or nonsimple configurations).

In some applications, notably in geometric transversal theory as well in bounding the complexity of the configuration space in robotics, it is useful to study the realizations of sign conditions of a family of s polynomials in $\mathrm{R}[X_1, \dots, X_k]$ restricted to a real variety $Z(Q, \mathrm{R}^k)$ where the real dimension of the variety $Z(Q, \mathrm{R}^k)$ can be much smaller than k. In [Basu et al. 1996] it was shown that the number of connected components of the realizations of all realizable sign condition of a family, $\mathcal{P} \subset \mathrm{R}[X_1, \dots, X_k]$ of s polynomials, restricted to a real variety of dimension k', where the degrees of the polynomials in $\mathcal{P} \cup \{Q\}$ are all bounded by d, is bounded by $\binom{s}{k'}O(d)^k$.

There are also results bounding the sum of the Betti numbers of semialgebraic sets defined by a conjunction of weak inequalities. Milnor [1964] proved a bound of $(sd)(2sd - 1)^{k-1}$ on the sum of the Betti numbers of a basic semialgebraic set defined by the conjunction of s weak inequalities $P_1 \geq 0$, ..., $P_s \geq 0$, with $P_i \in \mathbb{R}[X_1, \dots, X_k]$ such that $\deg(P_i) \leq d$. In another direction, Barvinok [1997] proved a bound of $k^{O(s)}$ on the sum of the Betti numbers of a basic, closed semialgebraic set defined by polynomials of degree at most 2. Unlike all previous bounds, this bound is polynomial in k for fixed values of s.

Extending such bounds to arbitrary semialgebraic sets is not trivial, because Betti numbers are not additive and the union of two topologically trivial semialgebraic sets can clearly have arbitrarily large higher Betti numbers. Basu [1999] proved a bound on the sum of the Betti numbers of a $\mathcal{P}$-closed semialgebraic set on a variety. A $\mathcal{P}$-closed semialgebraic set is one defined by a Boolean formula without negations whose atoms are of the form $P \geq 0$ or $P \leq 0$ with $P \in \mathcal{P}$. The bound is $s^{k'}O(d)^k$. Very recently Gabrielov and Vorobjov [$\geq$ 2005], succeeded in removing even the $\mathcal{P}$-closed assumption at the cost of a slightly worse bound.

They showed that the sum of the Betti numbers of an arbitrary semialgebraic set defined by a Boolean formula whose atoms are of the form $P = 0$, $P > 0$ or $P < 0$ with $P \in \mathcal{P}$, is bounded by $O(s^2d)^k$.

There have been recent refinements of the bounds on the Betti numbers of semialgebraic sets in another direction. All the bounds mentioned above are either bounds on the number of connected components or on the sums of all (or even or odd) Betti numbers. Basu [2003] proved different bounds (for each i) on the i-th Betti number of a basic, closed semialgebraic set on a variety. If S is a basic closed semialgebraic set defined by s polynomials in $\mathrm{R}[X_1, \ldots, X_k]$ of degree d, restricted to a real variety of dimension k' and defined by a polynomial of degree bounded by d, then $b_i(S)$ is bounded by $\binom{s}{k'-i}O(d)^k$. In the same paper, a bound of $s^\ell k^{O(\ell)}$ on the $(k - \ell)$-th Betti number of a basic, closed semialgebraic set defined by polynomials of degree at most 2 is proved. For fixed ℓ this bound is polynomial in both s and k. More recently, in [Basu et al. 2005] the authors bound (for each i) the sum of the i-th Betti number over all realizations of realizable sign conditions of a family of polynomials restricted to a variety of dimension k' by

$$\sum_{1 \leq j \leq k'-i} \binom{s}{j} 4^j d(2d - 1)^{k-1}.$$

This generalizes and makes more precise the bound in [Basu et al. 1996] which is the special case with $i = 0$. The technique of the proof uses a generalization of the Mayer–Vietoris exact sequence.

All the bounds on the Betti numbers of semialgebraic sets described above, depend on the degrees of the polynomials used in describing the semialgebraic set. However, it is well known that in the case of real polynomials of one variable, the number of real zeros can be bounded in terms of the number of monomials appearing in the polynomial (independent of the degree). This is an easy consequence of Descartes' law of signs [Basu et al. 2003]. Hence, it is natural to hope for a similar result in higher dimensions. Khovansky [1991] proved a bound of $2^{m^2}(mk)^k$ on the number of isolated real solutions of a system of k polynomial equations in k variables in which the number of monomials appearing with nonzero coefficients is bounded by m. Using this, one can obtain similar bounds on the sum of the Betti numbers of an algebraic set defined by a polynomial with at most m monomials in its support. The semialgebraic case requires some additional technique and it was shown in [Basu 1999] that the sum of the Betti numbers of a $\mathcal{P}$-closed semialgebraic set on a variety, is bounded by $s^{k'}2^{O(km^2)}$, where m is a bound on the number of monomials.

5. Modern Applications

Using [Pollack and Roy 1993] one immediately obtains reasonably tight bounds on the number of isotopy classes of not necessarily simple geometric objects such as the number of isotopy classes (with respect to order type) of configurations of n points in R^k or the number of isotopy classes (with respect to combinatorial type) of k−polytopes with n vertices.

Using [Basu et al. 1996], Goodman, Pollack, and Wenger [Goodman et al. 1996] were able to extend the known bounds on the number of *geometric permutations* (1-order types) induced by line transversals ($\ell = 1$) to the number of ℓ-order types induced by ℓ-flat transversals to n convex sets in R^3. As is the case for line transversals in $\mathbb{R}^3$, the lower bounds are about the square root of the upper bounds (in the plane, the corresponding result is tight [Edelsbrunner and Sharir 1990]). A much fuller discussion of Geometric Transversal Theory can be found in [Goodman et al. 1993].

6. Algorithms

A natural algorithmic problem is to design efficient algorithms for computing the Betti numbers of a given semialgebraic set. Clearly the problem of deciding whether a given semialgebraic set is empty is NP-hard, and counting its number of connected component is #P-hard. However, in view of the bounds described above we could hope for an algorithm having complexity polynomial in the number of polynomials and their degrees and singly exponential in the number of variables. This seems to be a very difficult problem in general and only partial results exist in this direction.

The cylindrical algebraic decomposition [Collins 1975] makes it possible to compute triangulations, and thus the number of connected components [Schwartz and Sharir 1983] as well as the higher Betti numbers in time polynomial in the number of polynomials and their degrees and doubly exponential in the number of variables (see [Basu et al. 2003]).

Various singly exponential time algorithms have been obtained for finding a point in every connected component of an algebraic set [Canny 1988b; Renegar 1992], of a semialgebraic set [Grigor'ev and Vorobjov 1988; Canny 1988b; Heintz et al. 1989; Renegar 1992], in every connected component of the sign conditions defined by a family of polynomials on a variety [Basu et al. 1997].

Computing the exact number of connected components in singly exponential time is a more difficult problem. The notion of a roadmap introduced by Canny [1988a] is the key to the solution. The basic algorithm has since been generalized and refined in several papers [Canny 1988a; 1993; Grigor'ev and Vorobjov 1992; Heintz et al. 1994; Gournay and Risler 1993; Basu et al. 2000] (see [Basu et al. 2003] for more details). Single exponential algorithms for computing the Euler–Poincaré characteristic (which is the alternating sum of the Betti numbers) of algebraic (as well as $\mathcal{P}$-closed semialgebraic) sets are described in [Basu 1999]. However, the problem of computing all the Betti numbers in single exponential time remains open.

References

[Alon 1995] N. Alon, "Tools from higher algebra", pp. 1749–1783 in *Handbook of combinatorics*, vol. 2, edited by R. L. Graham et al., Elsevier, Amsterdam, and MIT Press, Cambridge (MA), 1995.

[Barvinok 1997] A. I. Barvinok, "On the Betti numbers of semialgebraic sets defined by few quadratic inequalities", *Math. Z.* **225**:2 (1997), 231–244.

[Basu 1999] S. Basu, "On bounding the Betti numbers and computing the Euler characteristic of semi-algebraic sets", *Discrete Comput. Geom.* **22**:1 (1999), 1–18.

[Basu 2003] S. Basu, "Different bounds on the different Betti numbers of semi-algebraic sets", *Discrete Comput. Geom.* **30**:1 (2003), 65–85.

[Basu et al. 1996] S. Basu, R. Pollack, and M.-F. Roy, "On the number of cells defined by a family of polynomials on a variety", *Mathematika* **43**:1 (1996), 120–126.

[Basu et al. 1997] S. Basu, R. Pollack, and M.-F. Roy, "On computing a set of points meeting every cell defined by a family of polynomials on a variety", *J. Complexity* **13**:1 (1997), 28–37.

[Basu et al. 2000] S. Basu, R. Pollack, and M.-F. Roy, "Computing roadmaps of semi-algebraic sets on a variety", *J. Amer. Math. Soc.* **13**:1 (2000), 55–82.

[Basu et al. 2003] S. Basu, R. Pollack, and M.-F. Roy, *Algorithms in real algebraic geometry*, Algorithms and Computation in Mathematics **10**, Springer, Berlin, 2003.

[Basu et al. 2005] S. Basu, R. Pollack, and M.-F. Roy, "On the Betti numbers of sign conditions", *Proc. Amer. Math. Soc.* **133**:4 (2005), 965–974.

[Ben-Or 1983] M. Ben-Or, "Lower bounds for algebraic computation trees", pp. 80–86 in *Proceedings of the Fifteenth Annual ACM Symposium on Theory of Computing*, ACM Press, New York, 1983.

[Bochnak et al. 1987] J. Bochnak, M. Coste, and M.-F. Roy, *Géométrie algébrique réelle*, Ergebnisse der Mathematik (3) **12**, Springer, Berlin, 1987. Translated as *Real algebraic geometry*, Springer, Berlin, 1998.

[Canny 1988a] J. Canny, *The complexity of robot motion planning*, vol. 1987, ACM Doctoral Dissertation Awards, MIT Press, Cambridge, MA, 1988.

[Canny 1988b] J. Canny, "Some algebraic and geometric computations in PSPACE", pp. 460–467 in *Proceedings of the Twentieth Annual ACM Symposium on Theory of Computing* (Chicago, 1988), ACM Press, New York, 1988.

[Canny 1993] J. Canny, "Computing roadmaps of general semi-algebraic sets", *Comput. J.* **36**:5 (1993), 504–514.

[Collins 1975] G. E. Collins, "Quantifier elimination for real closed fields by cylindrical algebraic decomposition", pp. 134–183 in *Automata theory and formal languages* (Kaiserslautern, 1975), Lecture Notes in Comput. Sci. **33**, Springer, Berlin, 1975.

[Edelsbrunner and Sharir 1990] H. Edelsbrunner and M. Sharir, "The maximum number of ways to stab n convex nonintersecting sets in the plane is $2n-2$", *Discrete Comput. Geom.* **5**:1 (1990), 35–42.

[Gabrielov and Vorobjov $\geq$ 2005] A. Gabrielov and N. Vorobjov, "Betti numbers for quantifier-free formulae", *Discrete Comput. Geom.* To appear.

[Goodman and Pollack 1986a] J. E. Goodman and R. Pollack, "There are asymptotically far fewer polytopes than we thought", *Bull. Amer. Math. Soc. (N.S.)* **14**:1 (1986), 127–129.

[Goodman and Pollack 1986b] J. E. Goodman and R. Pollack, "Upper bounds for configurations and polytopes in $\mathbf{R}^d$", *Discrete Comput. Geom.* **1**:3 (1986), 219–227.

[Goodman et al. 1989] J. E. Goodman, R. Pollack, and B. Sturmfels, "Coordinate representation of order types requires exponential storage", pp. 405–410 in *Proceedings of the Twenty-First Annual ACM Symposium on Theory of Computing* (Seattle, 1989), ACM Press, New York, 1989.

[Goodman et al. 1990] J. E. Goodman, R. Pollack, and B. Sturmfels, "The intrinsic spread of a configuration in $\mathbf{R}^d$", *J. Amer. Math. Soc.* **3**:3 (1990), 639–651.

[Goodman et al. 1993] J. E. Goodman, R. Pollack, and R. Wenger, "Geometric transversal theory", pp. 163–198 in *New trends in discrete and computational geometry*, edited by J. Pach, Algorithms Combin. **10**, Springer, Berlin, 1993.

[Goodman et al. 1996] J. E. Goodman, R. Pollack, and R. Wenger, "Bounding the number of geometric permutations induced by k-transversals", *J. Combin. Theory Ser. A* **75**:2 (1996), 187–197.

[Gournay and Risler 1993] L. Gournay and J.-J. Risler, "Construction of roadmaps in semi-algebraic sets", *Appl. Algebra Engrg. Comm. Comput.* **4**:4 (1993), 239–252.

[Grigor'ev and Vorobjov 1988] D. Y. Grigor'ev and N. N. Vorobjov, Jr., "Solving systems of polynomial inequalities in subexponential time", *J. Symbolic Comput.* **5**:1-2 (1988), 37–64.

[Grigor'ev and Vorobjov 1992] D. Y. Grigor'ev and N. N. Vorobjov, Jr., "Counting connected components of a semialgebraic set in subexponential time", *Comput. Complexity* **2**:2 (1992), 133–186.

[Heintz et al. 1989] J. Heintz, M.-F. Roy, and P. Solernó, "On the complexity of semi-algebraic sets", pp. 293–298 in *Information processing* 89: *proceedings of the IFIP 11th World Computer Congress* (San Francisco, 1989), edited by G. X. Ritter, North-Holland, Amsterdam and New York, 1989.

[Heintz et al. 1994] J. Heintz, M.-F. Roy, and P. Solernó, "Single exponential path finding in semi-algebraic sets, II: The general case", pp. 449–465 in *Algebraic geometry and its applications* (West Lafayette, IN, 1990), edited by C. L. Bajaj, Springer, New York, 1994.

[Khovanskiĭ 1991] A. G. Khovanskiĭ, *Fewnomials*, Translations of Mathematical Monographs **88**, American Mathematical Society, Providence, RI, 1991.

[Milnor 1964] J. Milnor, "On the Betti numbers of real varieties", *Proc. Amer. Math. Soc.* **15** (1964), 275–280.

[Montaña et al. 1991] J. L. Montaña, L. M. Pardo, and T. Recio, "The nonscalar model of complexity in computational geometry", pp. 347–361 in *Effective methods in algebraic geometry* (Castiglioncello, 1990), edited by T. Mora and C. Traverso, Progr. Math. **94**, Birkhäuser, Boston, 1991.

[Oleĭnik 1949] O. A. Oleĭnik, "Some estimates for the Betti numbers of real algebraic surfaces", *Doklady Akad. Nauk SSSR (N.S.)* **67** (1949), 425–426.

[Oleĭnik 1951] O. A. Oleĭnik, "Estimates of the Betti numbers of real algebraic hypersurfaces", *Mat. Sbornik (N.S.)* **28(70)** (1951), 635–640.

[Pach et al. 1993] J. Pach, R. Pollack, and E. Welzl, "Weaving patterns of lines and line segments in space", *Algorithmica* **9**:6 (1993), 561–571.

[Petrovskiĭ and Oleĭnik 1949a] I. G. Petrovskiĭ and O. A. Oleĭnik, "On the topology of real algebraic surfaces", *Doklady Akad. Nauk SSSR (N.S.)* **67** (1949), 31–32.

[Petrovskiĭ and Oleĭnik 1949b] I. G. Petrovskiĭ and O. A. Oleĭnik, "On the topology of real algebraic surfaces", *Izvestiya Akad. Nauk SSSR. Ser. Mat.* **13** (1949), 389–402.

[Pollack and Roy 1993] R. Pollack and M.-F. Roy, "On the number of cells defined by a set of polynomials", *C. R. Acad. Sci. Paris Sér. I Math.* **316**:6 (1993), 573–577.

[Renegar 1992] J. Renegar, "On the computational complexity and geometry of the first-order theory of the reals, I", *J. Symbolic Comput.* **13**:3 (1992), 255–299.

[Schwartz and Sharir 1983] J. T. Schwartz and M. Sharir, "On the "piano movers" problem, II: General techniques for computing topological properties of real algebraic manifolds", *Adv. in Appl. Math.* **4**:3 (1983), 298–351.

[Thom 1965] R. Thom, "Sur l'homologie des variétés algébriques réelles", pp. 255–265 in *Differential and combinatorial topology: A symposium in honor of Marston Morse* (Princeton, 1964), Princeton Univ. Press, Princeton, N.J., 1965.

[Warren 1968] H. E. Warren, "Lower bounds for approximation by nonlinear manifolds", *Trans. Amer. Math. Soc.* **133** (1968), 167–178.

[Yao 1995] A. C.-C. Yao, "Algebraic decision trees and Euler characteristics", *Theoret. Comput. Sci.* **141**:1-2 (1995), 133–150.

[Yao 1997] A. C.-C. Yao, "Decision tree complexity and Betti numbers", *J. Comput. System Sci.* **55**:1 (1997), 36–43.

SAUGATA BASU
SCHOOL OF MATHEMATICS
GEORGIA INSTITUTE OF TECHNOLOGY
ATLANTA, GA 30332
UNITED STATES
saugata@math.gatech.edu

RICHARD POLLACK
COURANT INSTITUTE, NYU
251 MERCER ST.
NEW YORK, NY 10012
USA
pollack@courant.nyu.edu

MARIE-FRANÇOISE ROY
IRMAR (URA CNRS 305)
UNIVERSITÉ DE RENNES
CAMPUS DE BEAULIEU
35042 RENNES CEDEX
FRANCE
marie-francoise.roy@math.univ-rennes1.fr

Combinatorial and Computational Geometry
MSRI Publications
Volume **52**, 2005

Shelling and the h-Vector of the (Extra)ordinary Polytope

MARGARET M. BAYER

ABSTRACT. Ordinary polytopes were introduced by Bisztriczky as a (non-simplicial) generalization of cyclic polytopes. We show that the colex order of facets of the ordinary polytope is a shelling order. This shelling shares many nice properties with the shellings of simplicial polytopes. We also give a shallow triangulation of the ordinary polytope, and show how the shelling and the triangulation are used to compute the toric h-vector of the ordinary polytope. As one consequence, we get that the contribution from each shelling component to the h-vector is nonnegative. Another consequence is a combinatorial proof that the entries of the h-vector of any ordinary polytope are simple sums of binomial coefficients.

1. Introduction

This paper has a couple of main motivations. The first comes from the study of toric h-vectors of convex polytopes. The h-vector played a crucial role in the characterization of face vectors of simplicial polytopes [Billera and Lee 1981; McMullen and Shephard 1971; Stanley 1980]. In the simplicial case, the h-vector is linearly equivalent to the face vector, and has a combinatorial interpretation in a shelling of the polytope. The h-vector of a simplicial polytope is also the sequence of Betti numbers of an associated toric variety. In this context it generalizes to nonsimplicial polytopes. However, for nonsimplicial polytopes, we do not have a good combinatorial understanding of the entries of the h-vector. (Chan [1991] gives a combinatorial interpretation for the h-vector of cubical polytopes.)

This research was supported by the sabbatical leave program of the University of Kansas, and was conducted while the author was at the Mathematical Sciences Research Institute, supported in part by NSF grant DMS-9810361, and at Technische Universität Berlin, supported in part by Deutsche Forschungs-Gemeinschaft, through the DFG Research Center "Mathematics for Key Technologies" (FZT86) and the Research Group "Algorithms, Structure, Randomness" (FOR 13/1-1).

The definition of the (toric) h-vector for general polytopes (and even more generally, for Eulerian posets) first appeared in [Stanley 1987]. Already there Stanley raised the issue of computing the h-vector from a shelling of the polytope. Associated with any shelling, F_1, F_2, ..., F_n, of a polytope P is a partition of the faces of P into the sets $\mathcal{G}_j$ of faces of F_j not in $\bigcup_{i<j} F_i$. The h-vector can be decomposed into contributions from each set $\mathcal{G}_j$. When P is simplicial, the set $\mathcal{G}_j$ is a single interval $[G_j, F_j]$ in the face lattice of P, and the contribution to the h-vector is a single 1 in position $|G_j|$. For nonsimplicial polytopes, the set $\mathcal{G}_j$ is not so simple. It is not clear whether the contribution to the h-vector from $\mathcal{G}_j$ must be nonnegative, and, if it is, whether it counts something natural. (Tom Braden [2003] has announced a positive answer to this question, based on [Barthel et al. 2002; Karu 2002].) Another issue is the relation of the h-vector of a polytope P to the h-vector of a triangulation of P. This is addressed in [Bayer 1993; Stanley 1992].

A problem in studying nonsimplicial polytopes is the difficulty of generating examples with a broad range of combinatorial types. Bisztriczky [1997] discovered the fascinating "ordinary" polytopes, a class of generally nonsimplicial polytopes, which includes as its simplicial members the cyclic polytopes. These polytopes have been studied further in [Dinh 1999; Bayer et al. 2002; Bayer 2004]. The last of these articles showed that ordinary polytopes have surprisingly nice h-vectors, namely, the h-vector is the sum of the h-vector of a cyclic polytope and the shifted h-vector of a lower-dimensional cyclic polytope. These h-vectors were calculated from the flag vectors, and the calculation did not give a combinatorial explanation for the nice form that came out. So we were motivated to find a combinatorial interpretation for these h-vectors, most likely through shellings or triangulations of the polytopes.

This paper is organized as follows. In the second part of this introduction we give the main definitions. The brief Section 2 warms up with the natural triangulation of the multiplex. Section 3 is devoted to showing that the colex order of facets is a shelling of the ordinary polytope. The proof, while laborious, is constructive, explicitly describing the minimal new faces of the polytope as each facet is shelled on. We then turn in Section 4 to h-vectors of multiplicial polytopes in general, and of the ordinary polytope in particular. Here a "fake simplicial h-vector" arises in the shelling of the ordinary polytope. In Section 5, the triangulation of the multiplex is used to triangulate the boundary of the ordinary polytope. This triangulation is shown to have a shelling compatible with the shelling of Section 3. The shelling and triangulation together explain combinatorially the h-vector of the ordinary polytope.

About the title. Bisztriczky chose the name "ordinary polytope" to invoke the idea of ordinary curves. The name is, of course, a bit misleading, since it applies to a truly extraordinary class of polytopes. We feel that these polytopes are extraordinary because of their special structure, but we hope that they will also turn out to be extraordinary for their usefulness in understanding general convex polytopes.

Definitions. For common polytope terminology, refer to [Ziegler 1995].

The *toric h-vector* was defined by Stanley for Eulerian posets, including the face lattices of convex polytopes.

DEFINITION 1 [Stanley 1987]. Let P be a $(d-1)$-dimensional polytopal sphere. The *h*-vector and *g*-vector of P are encoded as polynomials:

$$h(P, x) = \sum_{i=0}^{d} h_i x^{d-i} \quad \text{and} \quad g(P, x) = \sum_{i=0}^{\lfloor d/2 \rfloor} g_i x^i,$$

with the relations $g_0 = h_0$ and $g_i = h_i - h_{i-1}$ for $1 \le i \le d/2$. Then the *h*-polynomial and *g*-polynomial are defined by the recursion

(i) $g(\varnothing, x) = h(\varnothing, x) = 1$, and

(ii) $h(P, x) = \displaystyle\sum_{\substack{G \text{ face of } P \\ G \neq P}} g(G, x)(x-1)^{d-1-\dim G}$.

It is easy to see that the *h*-vector depends linearly on the flag vector. In the case of simplicial polytopes, the formulas reduce to the well-known transformation between *f*-vector and *h*-vector.

DEFINITION 2 [Ziegler 1995]. Let $\mathcal{C}$ be a pure d-dimensional polytopal complex. If $d = 0$, a *shelling* of $\mathcal{C}$ is any ordering of the points of $\mathcal{C}$. If $d > 0$, a *shelling* of $\mathcal{C}$ is a linear ordering $F_1, F_2, \ldots, F_s$ of the facets of $\mathcal{C}$ such that for $2 \le j \le s$, the intersection $F_j \cap \left(\bigcup_{i<j} F_i \right)$ is nonempty and is the union of ridges (that is, $(d-1)$-dimensional faces) of $\mathcal{C}$ that form the initial segment of a shelling of F_j.

DEFINITION 3 [Bayer 1993]. A triangulation Δ of a polytopal complex $\mathcal{C}$ is *shallow* if and only if every face σ of Δ is contained in a face of $\mathcal{C}$ of dimension at most $2 \dim \sigma$.

THEOREM 1.1 [Bayer 1993]. *If Δ is a simplicial sphere forming a shallow triangulation of the boundary of the convex d-polytope P, then $h(\Delta, x) = h(P, x)$.*

Note: Theorem 4 in [Bayer 1993] gives $h(P, x) = h(\Delta, x)$ for a shallow subdivision Δ of the solid polytope P. The proof goes through for shallow subdivisions of the boundary, because it is based on the uniqueness of low-degree acceptable functions [Stanley 1987], which holds for lower Eulerian posets.

DEFINITION 4 [Bisztriczky 1996]. A d-dimensional *multiplex* is a polytope with an ordered list of vertices, x_0, x_1, ..., x_n, with facets F_0, F_1, ..., F_n given by

$$F_i = \text{conv}\{x_{i-d+1}, x_{i-d+2}, \ldots, x_{i-1}, x_{i+1}, x_{i+2}, \ldots, x_{i+d-1}\},$$

with the conventions that $x_i = x_0$ if $i < 0$, and $x_i = x_n$ if $i > n$.

Given an ordered set $V = \{x_0, x_1, \ldots, x_n\}$, a subset $Y \subseteq V$ is called a *Gale subset* if between any two elements of $V \setminus Y$ there is an even number of elements of Y. A polytope P with ordered vertex set V is a *Gale polytope* if the set of vertices of each facet is a Gale subset.

DEFINITION 5 [Bisztriczky 1997]. An *ordinary polytope* is a Gale polytope such that each facet is a multiplex with the induced order on the vertices.

Cyclic polytopes can be characterized as the simplicial Gale polytopes. Thus the only simplicial ordinary polytopes are cyclics. In fact, these are the only ordinary polytopes in even dimensions. However, the odd-dimensional, nonsimplicial ordinary polytopes are quite interesting.

We use the following notational conventions. Vertices are generally denoted by integers i rather than by x_i. Where it does not cause confusion, a face of a polytope or a triangulation is identified with its vertex set, and $\max F$ denotes the vertex of maximum index of the face F. Interval notation is used to denote sets of consecutive integers, so $[a, b] = \{a, a+1, \ldots, b-1, b\}$. If X is a set of integers and c is an integer, write $X + c = \{x + c : x \in X\}$.

2. Triangulating the Multiplex

Multiplexes have minimal triangulations that are particularly easy to describe.

THEOREM 2.1. *Let $M^{d,n}$ be a multiplex with ordered vertices 0, 1, ..., n. For $0 \leq i \leq n-d$, let T_i be the convex hull of $[i, i+d]$. Then $M^{d,n}$ has a shallow triangulation as the union of the $n-d+1$ d-simplices T_i.*

PROOF. The proof is by induction on n. For $n = d$, the multiplex $M^{d,d}$ is the simplex T_0 itself. Assume $M^{d,n}$ has a triangulation into simplices T_i, for $0 \leq i \leq n-d$. Consider the multiplex $M^{d,n+1}$ with ordered vertices 0, 1, ..., $n+1$. Then $M^{d,n+1} = \text{conv}(M^{d,n} \cup \{n+1\})$, where $n+1$ is a point beyond facet F_n of $M^{d,n}$, beneath the facets F_i for $0 \leq i \leq n-d+1$, and in the affine hulls of the facets F_i for $n-d+2 \leq i \leq n-1$. (See [Bisztriczky 1996].) Thus, $M^{d,n+1}$ is the union of $M^{d,n}$ and $\text{conv}(F_n \cup \{n+1\}) = T_{n+1-d}$, and $M^{d,n} \cap T_{n+1-d} = F_n$. By the induction assumption, the simplices T_i, with $0 \leq i \leq n+1-d$, form a triangulation of $M^{d,n+1}$.

The dual graph of the triangulation is simply a path. (The dual graph is the graph having a vertex for each d-simplex, and an edge between two vertices if the corresponding d-simplices share a $(d-1)$-face.) The ordering $T_0, T_1, T_2, \ldots,$

T_{n-d} is a shelling of the simplicial complex that triangulates $M^{d,n}$. So the h-vector of the triangulation is $(1, n-d, 0, 0, \ldots)$. This is the same as the g-vector of the boundary of the multiplex, which is the h-vector of the solid multiplex. So by [Bayer 1993], the triangulation is shallow. □

Note, however, that $M^{d,n}$ is not *weakly neighborly* for $n \geq d+2$ (as observed in [Bayer et al. 2002]). This means that it has nonshallow triangulations. This is easy to see because the vertices 0 and n are not contained in a common proper face of $M^{d,n}$.

Consider the induced triangulation of the boundary of $M^{d,n}$. For notational purposes we consider T_0 and T_n separately. All facets of T_0 except $[1, d]$ are boundary facets of $M^{d,n}$. Write $T_{0\backslash 0} = [0, d-1] = F_0$, and $T_{0\backslash j} = [0, d] \backslash \{j\}$ for $1 \leq j \leq d-1$. Write $T_{n-d\backslash n} = [n-d+1, n] = F_n$, and $T_{n\backslash j} = [n-d, n] \backslash \{j\}$ for $n-d+1 \leq j \leq n-1$. For $1 \leq i \leq n-d-1$, the facets of T_i are $T_{i\backslash j} = [i, i+d] \backslash \{j\}$. Two of these facets ($j = i$ and $j = i+d$) intersect the interior of $M^{d,n}$. For $1 \leq j \leq n-1$, the facet F_j is triangulated by $T_{i\backslash j}$ for $j-d+1 \leq i \leq j-1$ (and $0 \leq i \leq n-d$). The facet order $F_0, F_1, \ldots, F_n$, is a shelling of the multiplex $M^{d,n}$. The $(d-1)$-simplices $T_{i\backslash j}$ in the order $T_{0\backslash 0}, T_{0\backslash 1}, T_{0\backslash 2}, T_{1\backslash 2}, \ldots, T_{n-d-1\backslash n-2}$, $T_{n-d\backslash n-2}, T_{n-d\backslash n-1}, T_{n-d+1\backslash n}$ (increasing order of j and, for each j, increasing order of i), form a shelling of the triangulated boundary of $M^{d,n}$.

3. Shelling the Ordinary Polytope

Shelling is used to calculate the h-vector, and hence the f-vector of simplicial complexes (in particular, the boundaries of simplicial polytopes). This is possible because (1) the h-vector has a simple expression in terms of the f-vector and vice versa; (2) in a shelling of a simplicial complex, among the faces added to the subcomplex as a new facet is shelled on, there is a unique minimal face; (3) the interval from this minimal new face to the facet is a Boolean algebra; and (4) the numbers of new faces given by (3) match the coefficients in the f-vector/h-vector formula. These conditions all fail for shellings of arbitrary polytopes. However, some hold for certain shellings of ordinary polytopes.

As mentioned earlier, noncyclic ordinary polytopes exist only in odd dimensions. Furthermore, three-dimensional ordinary polytopes are quite different combinatorially from those in higher dimensions. We thus restrict our attention to ordinary polytopes of odd dimension at least five. It turns out that these are classified by the vertex figure of the first vertex.

THEOREM 3.1 [Bisztriczky 1997; Dinh 1999]. *For each choice of integers $n \geq k \geq d = 2m+1 \geq 5$, there is a unique combinatorial type of ordinary polytope $P = P^{d,k,n}$ such that the dimension of P is d, P has $n+1$ vertices, and the first vertex of P is on exactly k edges. The vertex figure of the first vertex of $P^{d,k,n}$ is the cyclic $(d-1)$-polytope with k vertices.*

We use the following description of the facets of $P^{d,k,n}$ by Dinh. For any subset $X \subseteq \mathbb{Z}$, let $\mathrm{ret}_n(X)$ (the *retraction* of X) be the set obtained from X by replacing every negative element by 0 and replacing every element greater than n by n.

THEOREM 3.2 [Dinh 1999]. *Let $\mathfrak{X}_n$ be the collection of sets*

$$X = [i, i+2r-1] \cup Y \cup [i+k, i+k+2r-1], \tag{3-1}$$

where $i \in \mathbb{Z}$, $1 \le r \le m$, Y is a paired $(d-2r-1)$-element subset of $[i+2r+1, i+k-2]$, and $|\mathrm{ret}_n(X)| \ge d$. The set of facets of $P^{d,k,n}$ is

$$\mathcal{F}(P^{d,k,n}) = \{\mathrm{ret}_n(X) : X \in \mathfrak{X}_n\}.$$

It is easy to check that when $n = k$, $|\mathrm{ret}_n(X)| = d$ for all $X \in \mathfrak{X}_n$, and that $\mathrm{ret}_n(\mathfrak{X}_n)$ is the set of d-element Gale subsets of $[0, k]$, that is, the facets of the cyclic polytope $P^{d,k,k}$.

Note that $\mathfrak{X}_{n-1} \subseteq \mathfrak{X}_n$. We wish to describe $\mathcal{F}(P^{d,k,n})$ in terms of $\mathcal{F}(P^{d,k,n-1})$; for this we need the following shift operations. If $F = \mathrm{ret}_{n-1}(X) \in \mathcal{F}(P^{d,k,n-1})$, let the right-shift of F be $\mathrm{rsh}(F) = \mathrm{ret}_n(X+1)$. Note that $\mathrm{rsh}(F)$ may or may not contain 0. In either case, $\mathrm{rsh}(F) \cap [1, n] = F+1$, so $|\mathrm{rsh}(F)| \ge |F| \ge d$, If $F = \mathrm{ret}_n(X) \in \mathcal{F}(P^{d,k,n})$, let the left-shift of F be $\mathrm{lsh}(F) = \mathrm{ret}_{n-1}(X-1)$. Note that $\mathrm{lsh}(F) \setminus \{0\} = (F-1) \cap [1, n]$; $\mathrm{lsh}(F)$ contains 0 if $0 \in F$ or $1 \in F$.

LEMMA 3.3. *If $n \ge k+1$ and $F \in \mathcal{F}(P^{d,k,n})$ with $\max F \ge k$, then $\mathrm{lsh}(F) \in \mathcal{F}(P^{d,k,n-1})$.*

PROOF. Let $F = \mathrm{ret}_n(X)$, with $X = [i, i+2r-1] \cup Y \cup [i+k, i+k+2r-1]$. Then $X-1$ also has the form of equation (3-1) (for $i-1$). The set $\mathrm{lsh}(F)$ is the vertex set of a facet of $P^{d,k,n-1}$ as long as $|\mathrm{lsh}(F)| \ge d$. We check this in three cases.

Case 1. If $k \le i+k+2r-1 \le n$, then $i+2r-1 \ge 0$, so $Y \subseteq [i+2r+1, i+k-2] \subseteq [2, i+k-2]$. Then

$$\mathrm{lsh}(F) \supseteq \max\{i+2r-2, 0\} \cup (Y-1) \cup [i+k-1, i+k+2r-2],$$

so $|\mathrm{lsh}(F)| \ge 1+(d-2r-1)+2r = d$.

Case 2. If $i+k \ge n$, then $i \ge n-k \ge 1$. Also, $|F| \ge d$ implies $\max Y \le n-1$. So

$$\mathrm{lsh}(F) = [i-1, i+2r-2] \cup (Y-1) \cup \{n-1\},$$

so $|\mathrm{lsh}(F)| = 2r+(d-2r-1)+1 = d$.

Case 3. If $i+k < n < i+k+2r-1$, then $i+2r-1 \ge n-k \ge 1$, and

$$F = [\max\{0, i\}, i+2r-1] \cup Y \cup [i+k, n],$$

so

$$|F| = (i+2r - \max\{0, i\}) + (d-2r-1) + (n-i-k+1)$$
$$= d+n-k - \max\{i, 0\} \ge d+1.$$

Then $|\mathrm{lsh}(F)| \ge |F|-1 \ge d$.

Thus, $\mathrm{lsh}(F)$ is a facet of $P^{d,k,n-1}$.   □

Identify each facet of the ordinary polytope $P^{d,k,n}$ with its ordered list of vertices. Then order the facets of $P^{d,k,n}$ in colex order. This means, if $F = i_1 i_2 \ldots i_p$ and $G = j_1 j_2 \ldots j_q$, then $F \prec_c G$ if and only if for some $t \geq 0$, $i_{p-t} < j_{q-t}$ while for $0 \leq s < t$, $i_{p-s} = j_{q-s}$.

LEMMA 3.4. *If $n \geq k+1$ and F_1 and F_2 are facets of $P^{d,k,n}$ with $\max F_i \geq k$, then $F_1 \prec_c F_2$ implies $\mathrm{lsh}(F_1) \prec_c \mathrm{lsh}(F_2)$.*

PROOF. Suppose $F_1 \prec_c F_2$, and let q be the maximum vertex in F_2 not in F_1. Then $\mathrm{lsh}(F_1) \prec_c \mathrm{lsh}(F_2)$ as long as $q \geq 2$, for in that case $q-1 \in \mathrm{lsh}(F_2) \backslash \mathrm{lsh}(F_1)$, while $[q, n-1] \cap \mathrm{lsh}(F_1) = [q, n-1] \cap \mathrm{lsh}(F_2)$. (If $q = 1$, then q shifts to 0 in $\mathrm{lsh}(F_2)$, but 0 may be in $\mathrm{lsh}(F_1)$ as a shift of a smaller element.) So we prove $q \geq 2$. Write

$$F_2 = \mathrm{ret}_n([i, i+2r-1] \cup Y \cup [i+k, i+k+2r-1])$$

and

$$F_1 = \mathrm{ret}_n([i', i'+2r'-1] \cup Y' \cup [i'+k, i'+k+2r'-1]).$$

Since $\max F_2 \geq k$, $i+2r-1 \geq 0$, so $Y \cup [i+k, i+k+2r-1] \subseteq [2, n]$, Thus, if $q \in Y \cup [i+k, i+k+2r-1]$, then $q \geq 2$. Otherwise

$$Y \cup [i+k, i+k+2r-1]) = Y' \cup [i'+k, i'+k+2r'-1]),$$

but $Y \neq Y'$. This can only happen when $Y \cup [i+k, i+k+2r-1])$ is an interval; in this case $i+k+2r-1 \geq n+1$. Then $q = i+2r-1 = (i+k+2r-1)-k \geq n+1-k \geq 2$.   □

PROPOSITION 3.5. *Let $n \geq k+1$. The facets of $P^{d,k,n}$ are*

$$\{F : F \in \mathcal{F}(P^{d,k,n-1}) \text{ and } \max F \leq n-2\}$$
$$\cup \{rsh(F) : F \in \mathcal{F}(P^{d,k,n-1}) \text{ and } \max F \geq n-2\}.$$

PROOF. If $\max X \leq n-2$, then $\mathrm{ret}_n(X) = \mathrm{ret}_{n-1}(X)$; in this case, letting $F = \mathrm{ret}_n(X)$, $F \in \mathcal{F}(P^{d,k,n-1})$ if and only if $F \in \mathcal{F}(P^{d,k,n})$. If $F \in \mathcal{F}(P^{d,k,n-1})$ with $\max F \geq n-2$, then $\mathrm{rsh}(F) \in \mathcal{F}(P^{d,k,n})$ with $\max \mathrm{rsh}(F) \geq n-1$. Now suppose that $G = \mathrm{ret}_n(X) \in \mathcal{F}(P^{d,k,n})$ with $\max G \geq n-1$. Let $F = \mathrm{lsh}(G) = \mathrm{ret}_{n-1}(X-1) \in \mathcal{F}(P^{d,k,n-1})$; then $\max F \geq n-2$. By definition, $\mathrm{rsh}(F) = \mathrm{ret}_n((X-1)+1) = \mathrm{ret}_n(X) = G$.   □

THEOREM 3.6. *Let F_1, F_2, $\ldots$, F_v be the facets of $P^{d,k,n}$ in colex order.*

(i) *F_1, F_2, $\ldots$, F_v is a shelling of $P^{d,k,n}$.*
(ii) *For each j there is a unique minimal face G_j of F_j not contained in $\bigcup_{i=1}^{j-1} F_i$.*
(iii) *For each j, $2 \leq j \leq v-1$, G_j contains the vertex of F_j of maximum index, and is contained in the $d-1$ highest vertices of F_j.*
(iv) *For each j, the interval $[G_j, F_j]$ is a Boolean lattice.*

Note that this theorem is not saying that the faces of $P^{d,k,n}$ in the interval $[G_j, F_j]$ are all simplices.

PROOF. We construct explicitly the faces G_j in terms of F_j. The reader may wish to refer to the example that follows the proof.

Cyclic polytopes. We start with the cyclic polytopes. (For the cyclics, the theorem is generally known, or at least a shorter proof based on [Billera and Lee 1981] is possible, but we will need the description of the faces G_j later.)

Let $F_1, F_2, \ldots, F_v$ be the facets, in colex order, of $P^{d,k,k}$, the cyclic d-polytope with vertex set $[0, k]$. Each facet F_j can be written as

$$F_j = I_j^0 \cup I_j^1 \cup I_j^2 \cup \cdots \cup I_j^p \cup I_j^k,$$

where I_j^0 is the interval of F_j containing 0, if $0 \in F_j$, and $I_j^0 = \varnothing$ otherwise; I_j^k is the interval of F_j containing k, if $k \in F_j$, and $I_j^k = \varnothing$ otherwise; and the I_j^ℓ are the other (even) intervals of F_j with the elements of I_j^ℓ preceding the elements of $I_j^{\ell+1}$. (For example, in $P^{7,9,9}$, $F_6 = \{0, 1, 2, 4, 5, 7, 8\}$, $I_6^0 = \{0, 1, 2\}$, $I_6^1 = \{4, 5\}$, $I_6^2 = \{7, 8\}$, and $I_6^9 = \varnothing$.) For the interval $[a, b]$, write $E([a, b])$ for the integers in the even positions in the interval, that is, $E([a, b]) = [a, b] \cap \{a + 2i + 1 : i \in \mathbb{N}\}$. Let $G_j = \bigcup_{\ell=1}^p E(I_j^\ell) \cup I_j^k$. Since $I_j^0 = F_j$ if and only if $j = 1$, $G_1 = \varnothing$, and for all $j > 1$, G_j contains the maximum vertex of F_j. Since F_j is a simplex, $[G_j, F_j]$ is a Boolean lattice.

To show that $F_1, F_2, \ldots, F_v$ is a shelling of $P^{d,k,k}$ we show that G_j is not in a facet before F_j and that every ridge of $P^{d,k,k}$ in F_j that does not contain G_j is contained in a previous facet. For $j > 0$ the face G_j consists of the right end-set I_j^k (if nonempty) and the set $\bigcup_{j=1}^p E(I_j^\ell)$ of singletons. Note that G_j satisfies condition (c) of the theorem (which here just says that the lowest vertex of F_j is not in G_j), unless $j = v$, in which case $G_v = F_v$. Any facet F of $P^{d,k,k}$ containing G_j must satisfy Gale's evenness condition and therefore must contain an integer adjacent to each element of $\bigcup_{j=1}^p E(I_j^\ell)$. If any element of the form $\max I_j^\ell + 1$ is in F, then F occurs after F_j in colex order. This implies that any F_i previous to F_j and containing G_j also contains $\bigcup_{\ell=1}^p I_j^\ell \cup I_j^k$. But F_j is the first facet in colex order that contains $\bigcup_{\ell=1}^p I_j^\ell \cup I_j^k$. So G_j is not in a facet before F_j.

Now let $g \in G_j$; we wish to show that $F_j \setminus \{g\}$ is in a previous facet. If $g \in E(I_j^\ell)$ for $\ell > 0$, let $F = F_j \setminus \{g\} \cup \{\min I_j^\ell - 1\}$. Then F satisfies Gale's evenness condition and is a facet before F_j. Otherwise $g \in I_j^k \setminus E(I_j^k)$; in this case let $F = F_j \setminus \{g\} \cup \{\max I_j^0 + 1\}$ (where we let $\max I_j^0 + 1 = 0$ if $I_j^0 = \varnothing$). Again F satisfies Gale's evenness condition and is a facet before F_j.

Thus the colex order of facets is a shelling order for the cyclic polytope $P^{d,k,k}$, and we have an explicit description for the minimal new face G_j as F_j is shelled on.

General ordinary. Now we prove the theorem for general $P^{d,k,n}$ by induction on $n \geq k$, for fixed k. Among the facets of $P^{d,k,n}$, first in colex order are those with maximum vertex at most $n-2$. These are also the first facets in colex order of $P^{d,k,n-1}$. Thus the induction hypothesis gives us that this initial segment is a partial shelling of $P^{d,k,n}$, and that assertions 2–4 hold for these facets.

Later facets. It remains to consider the facets of $P^{d,k,n}$ ending in $n-1$ or n. These facets come from shifting facets of $P^{d,k,n-1}$ ending in $n-2$ or $n-1$. Our strategy here will be to prove statement (b) of the theorem for these facets. The intersection of F_j with $\bigcup_{i=1}^{j-1} F_i$ is then the antistar of G_j in F_j, and so it is the union of $(d-2)$-faces that form an initial segment of a shelling of F_j. This will prove that the colex order $F_1, F_2, \ldots, F_v$ is a shelling of $P^{d,k,n}$.

Note that there is nothing to show for the last facet of $P^{d,k,n}$ in colex order. It is $F_v = [n-d+1, n]$, and is the only facet (other than the first) whose vertex set forms a single interval. Assume from now on that j is fixed, with $j \leq v-1$. Later we will describe recursively the minimal new face G_j as F_j is shelled on. It will always be the case that $\max F_j \in G_j$. We will prove that G_j is truly a new face (is not contained in a previous facet), and that every ridge not containing all of G_j is contained in a previous facet.

Ridges not containing the last vertex. It is convenient to start by showing that every ridge of $P^{d,k,n}$ contained in F_j and not containing $\max F_j$ is contained in an earlier facet. This case does not use the recursion needed for the other parts of the proof. Write

$$X = [i, i+2r-1] \cup Y \cup [i+k, i+k+2r-1]$$

and $F_j = \mathrm{ret}_n(X) = \{z_1, z_2, \ldots, z_p\}$ with $0 \leq z_1 < z_2 < \cdots < z_p \leq n$. The facet F_j is a $(d-1)$-multiplex, so its facets are of the form

$$F_j(\hat{z}_t) = \{z_\ell : 1 \leq \ell \leq p, \, 0 < |\ell - t| \leq d - 2\}$$

for $2 \leq t \leq p-1$, $F_j(\hat{z}_1) = \{z_1, z_2, \ldots, z_{d-1}\}$, and $F_j(\hat{z}_p) = \{z_{p-d+2}, \ldots, z_{p-1}, z_p\}$. If $F_j(\hat{z}_t)$ does not contain $\max F_j = z_p$, then $t \leq p-d+1$ and this implies $i \leq z_t \leq i+2r-1$. Consider such a z_t.

The first ridge. For $t = 1$, there are three cases to consider.

Case 1. Suppose $z_1 \geq 1$. Then $F_j(\hat{z}_1) = [i, i+2r-1] \cup Y$. Let I be the right-most interval of $F_j(\hat{z}_1)$. Let $Z = (I - k) \cup F_j(\hat{z}_1)$, and $F = \mathrm{ret}_n(Z)$. Since $i \geq 1$ and $\max F_j(\hat{z}_1) \leq i+k-2$, the interval $I - k$ contributes at least one new element to F, so $|F| \geq d$.

Case 2. Suppose $z_1 = 0$ and the right-most interval of $F_j(\hat{z}_1)$ is odd. In this case the left-most interval of F_j must also be odd, so $i < 0$, and $F_j(\hat{z}_1)$ contains $i+k$ but not $i+k-1$. Let $F = F_j(\hat{z}_1) \cup \{i+k-1\}$.

Case 3. Suppose $z_1 = 0$ and the right-most interval of $F_j(\hat{z}_1)$ is even (and then so is the left-most interval). Then $F_j(\hat{z}_1) = [0, i+2r-1] \cup Y \cup [i+k, k-1]$ (where the last interval is empty if $i = 0$). Let

$$F = F_j(\hat{z}_1) \cup \{i+2r\} = \{0\} \cup [1, i+2r] \cup Y \cup [i+k, k-1].$$

(When $i = 0$ and $r = (d-1)/2$, this gives $F = [0, d-1]$.) In all cases F is a facet of $P^{d,k,n}$ containing $F_j(\hat{z}_1)$. It does not contain $\max F_j$, so $F \prec_c F_j$.

Deleting a later vertex. Now assume $2 \le t \le p-d+1$; then $z_t \ge \max\{i+1, 1\}$. Here

$$F_j(\hat{z}_t) = [\max\{i, 0\}, z_t - 1] \cup [z_{t+1}, i+2r-1] \cup Y \cup [i+k, z_t - 1 + k],$$

and $|F_j(\hat{z}_t)| = z_t - \max\{i, 0\} + d - 2 \ge d - 1$. Also note that $z_t - 1 + k$ is the $(d-2)$nd element of $\{z_1, z_2, \ldots, z_p\}$ after z_t, so $z_t - 1 + k = z_{t+d-2} < z_p = \max F_j$.

Case 1. If $z_t - i$ is even, let $F = F_j(\hat{z}_t) \cup \{i+2r\}$. Then $F = \mathrm{ret}_n(Z)$, where

$$Z = [i, z_t - 1] \cup [z_t + 1, i+2r] \cup Y \cup [i+k, z_t - 1 + k],$$

and $|F| \ge d$.

Case 2. If $z_t - i$ is odd and $\max([i, i+2r-1] \cup Y) < i+k-2$, let $F = \mathrm{ret}_n(Z)$, where

$$Z = [i-1, z_t - 1] \cup [z_t + 1, i+2r-1] \cup Y \cup [i+k-1, z_t - 1 + k].$$

Then $F \supseteq F_j(\hat{z}_t) \cup \{i+k-1\}$, so $|F| \ge d$.

Case 3. Finally, suppose $z_t - i$ is odd and $\max Y = i+k-2$. Let $[q, i+k-2]$ be the right-most interval of Y, and let $F = \mathrm{ret}_n(Z)$, where

$$Z = [q-k, z_t - 1] \cup [z_t + 1, i+2r-1] \cup (Y \setminus [q, i+k-2]) \cup [q, z_t - 1 + k].$$

Then $F \supseteq F_j(\hat{z}_t) \cup \{i+k-1\}$, so $|F| \ge d$.

In all cases, F is a facet of $P^{d,k,n}$ containing $F_j(\hat{z}_t)$ and $\max F_j \notin F$, so F occurs before F_j in colex order.

Determining the minimal new face. We now describe the faces G_j recursively. (We are still assuming that $\max F_j \ge n-1$.) Let G be the face of $\mathrm{lsh}(F_j)$ that is the minimal new face when $\mathrm{lsh}(F_j)$ is shelled on, in the colex shelling of the polytope $P^{d,k,n-1}$. Let $G_j = G+1$; this is a subset of the last $d-1$ vertices of F_j and contains $\max F_j$. By [Bayer et al. 2002, Theorem 2.6] and [Bisztriczky 1996], G_j is a face of F_j. For any facet F_i of $P^{d,k,n}$, $G_j \subseteq F_i$ if and only if $G \subseteq \mathrm{lsh}(F_i)$. So by the induction hypothesis, G_j is not contained in a facet occurring before F_j in colex order.

Ridges in previous facets. It remains to show that any ridge of $P^{d,k,n}$ contained in F_j but not containing all of G_j is contained in a facet prior to F_j. Note that we have already dealt with those ridges not containing $\max F_j$. Now let $g \in G$, $g_j = g+1 \in G_j$, and assume $g_j \neq \max F_j$. The only ridge of $P^{d,k,n}$ contained in F_j, containing $\max F_j$, and not containing g_j is $F_j(\hat{g}_j)$.

Let H be the unique ridge of $P^{d,k,n-1}$ in $\mathrm{lsh}(F_j)$ containing $\max(\mathrm{lsh}(F_j))$, but not containing g. By the induction hypothesis, H is contained in a facet F of $P^{d,k,n-1}$ occurring before $\mathrm{lsh}(F_j)$ in colex order. Suppose $F_j(\hat{g}_j)$ is contained in a facet F_ℓ of $P^{d,k,n}$ occurring after F_j in colex order. Then H is contained in $\mathrm{lsh}(F_\ell)$. Thus the ridge H of $P^{d,k,n-1}$ is contained in three different facets: F (occurring before $\mathrm{lsh}(F_j)$ in colex order), $\mathrm{lsh}(F_j)$, and $\mathrm{lsh}(F_\ell)$ (occurring after $\mathrm{lsh}(F_j)$ in colex order). This contradiction shows that the ridge $F_j(\hat{g}_j)$ can only be contained in a facet of $P^{d,k,n}$ occurring before F_j in colex order.

Boolean intervals. Finally to verify assertion 4 of the theorem, observe that every facet F_j is a $(d-1)$-dimensional multiplex. The face G_j of F_j contains the maximum vertex u of F_j. The vertex figure of the maximum vertex in any multiplex is a simplex [Bisztriczky 1996]. The interval $[G_j, F_j]$ is an interval in $[u, F_j]$, which is the face lattice of a simplex, so $[G_j, F_j]$ is a Boolean lattice. $\square$

A nonrecursive description of the faces G_j, generalizing that for the cyclic case in the proof, is as follows. Write the facet F_j as a disjoint union, $F_j = A_j^0 \cup I_j^1 \cup I_j^2 \cup \cdots \cup I_j^p \cup I_j^n$, where I_j^n is the interval of F_j containing n if $n \in F_j$, and $I_j^n = \varnothing$ otherwise; the I_j^ℓ ($1 \le \ell \le p$) are even intervals of F_j written in increasing order; and A_j^0 is

- the interval containing 0, if $\max F_j \le k-1$;
- the union of the interval containing $\max F_j - k$ and the interval containing $\max F_j - k + 2$ (if the latter exists), if $k \le \max F_j \le n-1$;
- the interval containing $n-k$, if $\max F_j = n$ and $n-k \in F_j$;
- $\varnothing$, if $\max F_j = n$ and $n-k \notin F_j$.

Then $G_j = \bigcup_{\ell=1}^p E(I_j^\ell) \cup I_j^n$. The vertices of G_j are among the last d vertices of F_j and so are affinely independent [Bisztriczky 1996]; thus G_j is a simplex.

Example. Table 1 gives the faces F_j and G_j for the colex shelling of the ordinary polytope $P^{5,6,8}$.

Let us look at what happens when facet F_{13} is shelled on. The ridges of $P^{5,6,8}$ contained in F_{13} are 0123, 0236, 01367, 012678, 12378, 2368, and 3678. The first ridge, 0123, is contained in $F_1 = 01234$. The ridge 0236 is $F_{13}(\hat{z}_2) = F_{13}(\hat{1})$, and $\max([i, i+2r-1] \cup Y) = 3 < 4 = i+k-2$, so we find that 0236 is contained in $F_4 = 02356$. The ridge 01367 is $F_{13}(\hat{z}_3) = F_{13}(\hat{2})$, so we find that 01367 is contained in $F_6 = 013467$. This facet $F_{13} = 0123678$ is shifted from the facet 012567 of $P^{5,6,7}$, which in turn is shifted from the facet 01456 of the cyclic

j	F_j	G_j	j	F_j	G_j
1	01234	∅	9	23 56 8	68
2	012 45	5	10	3456 8	468
3	0 2345	35	11	1234 78	78
4	0 23 56	6	12	12 45 78	578
5	0 3456	46	13	0123 678	678
6	01 34 67	7	14	34 678	4678
7	01 4567	57	15	012 5678	5678
8	2345 8	8	16	45678	45678

Table 1. Shelling of $P^{5,6,8}$

polytope $P^{5,6,6}$. When 01456 occurs in the shelling of the cyclic polytope, its minimal new face is its right interval, 456. In $P^{5,6,8}$, then, the minimal new face when F_{13} is shelled on is 678. The other ridges of F_{13} not containing 678 are 12378 and 2368. The interval $[G_{13}, F_{13}]$ contains the triangle 678, the 3-simplex 3678, the 3-multiplex 012678, and F_{13} itself (which is a pyramid over 012678).

Note that for the multiplex, $M^{d,n} = P^{d,d,n}$, this theorem gives a shelling different from the one mentioned in Section 2. In the standard notation for the facets of the multiplex (see Definition 4), the colex shelling order is F_0, F_1, ..., F_{n-d}, F_{n-1}, F_{n-2}, ..., F_{n-d+1}, F_n. The statements of this section hold also for even-dimensional multiplexes.

4. The h-Vector from the Shelling

The h-vector of a simplicial polytope can be obtained easily from any shelling of the polytope. For P a simplicial polytope, and $\cup [G_j, F_j]$ the partition of a face lattice of P arising from a shelling, $h(P, x) = \sum_j x^{d-|G_j|}$. For general polytopes, the (toric) h-vector can also be decomposed according to the shelling partition. For a shelling, F_1, F_2, ..., F_n, of a polytope P, write $\mathcal{G}_j$ for the set of faces of F_j not in $\bigcup_{i<j} F_i$. Then $h(P, x) = \sum_{j=1}^n h(\mathcal{G}_j, x)$, where $h(\mathcal{G}_j, x) = \sum_{G \in \mathcal{G}_j} g(G, x)(x-1)^{d-1-\dim G}$. However, in general we do not know that the coefficients of $h(\mathcal{G}_j, x)$ count anything natural, nor even that they are nonnegative. Stanley raised this issue in [Stanley 1987, Section 6]. It has apparently been settled in [Braden 2003].

We turn now to h-vectors of ordinary polytopes. In [Bayer 2004] we used the flag vector of the ordinary polytope to compute its toric h-vector.

THEOREM 4.1 [Bayer 2004]. *For* $n \geq k \geq d = 2m+1 \geq 5$ *and* $1 \leq i \leq m$,

$$h_i(P^{d,k,n}) = \binom{k-d+i}{i} + (n-k)\binom{k-d+i-1}{i-1}.$$

We did not understand why the h-vector turned out to have such a nice form. Here we show how the h-vector can be computed from the colex shelling. Properties 2 and 4 of Theorem 3.6 are critical.

In [Bayer 2004] we showed that the flag vector of a multiplicial polytope depends only on the f-vector. However, for our purposes here it is more useful to write the h-vector in terms of the f-vector and the flag vector entries of the form f_{0i}. We introduce a modified f-vector. Let $\bar{f}_{-1} = f_{-1} = 1$, $\bar{f}_0 = f_0$, and $\bar{f}_{d-1} = f_{d-1} + (f_{0,d-1} - df_{d-1})$; and for $1 \leq j \leq d-2$, let

$$\bar{f}_j = f_j + (f_{0,j+1} - (j+2)f_{j+1}) + (f_{0,j} - (j+1)f_j).$$

(Thus, $\bar{f}_1 = f_1 + (f_{02} - 3f_2) + (f_{01} - 2f_1) = f_1 + (f_{02} - 3f_2)$.)

THEOREM 4.2. *If P is a multiplicial d-polytope, then*

$$h(P, x) = \sum_{i=0}^{d} h_i(P)x^{d-i} = \sum_{i=0}^{d} \bar{f}_{i-1}(P)(x-1)^{d-i}.$$

PROOF. As observed in the proof of Theorem 2.1, the g-polynomial of an e-dimensional multiplex M with $n+1$ vertices is $g(M, x) = 1 + (n-e)x$. So for a multiplicial d-polytope P,

$$h(P, x) = \sum_{\substack{G \text{ face of } P \\ G \neq P}} g(G, x)(x-1)^{d-1-\dim G}$$

$$= \sum_{\substack{G \text{ face of } P \\ G \neq P}} (1 + (f_0(G) - 1 - \dim G)x)(x-1)^{d-1-\dim G}$$

$$= \sum_{i=0}^{d} f_{i-1}(x-1)^{d-i} + \sum_{i=1}^{d-1} (f_{0i} - (i+1)f_i)x(x-1)^{d-1-i}$$

$$= \sum_{i=0}^{d} f_{i-1}(x-1)^{d-i} + \sum_{i=1}^{d-1} (f_{0i} - (i+1)f_i)[(x-1)^{d-i} + (x-1)^{d-1-i}]$$

$$= (x-1)^d + f_0(x-1)^{d-1}$$

$$\quad + \sum_{i=2}^{d-1} (f_{i-1} + (f_{0i} - (i+1)f_i) + (f_{0,i-1} - if_{i-1}))(x-1)^{d-i}$$

$$\quad + (f_{d-1} + (f_{0,d-1} - df_{d-1}))$$

$$= \sum_{i=0}^{d} \bar{f}_{i-1}(P)(x-1)^{d-i}. \qquad \square$$

Simplicial polytopes are a special case of multiplicial polytopes. Clearly, when P is simplicial, $\bar{f}(P) = f(P)$, and we recover the definition of the simplicial h-vector in terms of the f-vector. The multiplicial h-vector formula can be thought of as breaking into two parts: one involving the f-vector, and matching the simplicial h-vector formula; the other involving the "excess vertex counts," $f_{0,j} - (j+1)f_j$.

In the simplicial case the sum of the entries in the h-vector is the number of facets. For multiplicial polytopes $\sum_{i=0}^{d} h_i(P) = \bar{f}_{d-1}(P) = f_{d-1} + (f_{0,d-1} - df_{d-1})$.

In general, applying the simplicial h-formula to a nonsimplicial f-vector produces a vector with no (known) combinatorial interpretation. This vector is neither symmetric nor nonnegative in general. We will see that in the case of ordinary polytopes something special happens. Write $h'(P,x)$ for the h-polynomial that P would have if it were simplicial.

DEFINITION 6. The h'-*polynomial* of a multiplicial d-polytope P is given by

$$h'(P,x) = \sum_{i=0}^{d} h_i'(P)x^{d-i} = \sum_{i=0}^{d} f_{i-1}(P)(x-1)^{d-i}.$$

(The h'-vector is then the vector of coefficients of the h'-polynomial.)

THEOREM 4.3. *Let* $P^{d,k,n}$ *be an ordinary polytope. Let* $\bigcup_{j=1}^{v}[G_j, F_j]$ *be the partition of the face lattice of* $P^{d,k,n}$ *associated with the colex shelling of* $P^{d,k,n}$. *Then for all* i, $0 \leq i \leq d$, $h'(P^{d,k,n}, x) = \sum_{j=1}^{v} x^{d-|G_j|}$.

Furthermore, if $C^{d,k}$ *is the cyclic* d-*polytope with* $k+1$ *vertices, then for all* i, $0 \leq i \leq d$, $h_i'(P^{d,k,n}) \geq h_i(C^{d,k})$, *with equality for* $i > d/2$.

PROOF. Direct evaluation gives $h_0'(P) = h_d'(P) = 1$. Let $F_1, F_2, \ldots, F_v$ be the colex shelling of $P^{d,k,n}$. By Theorem 3.6, part 2, the set of faces of $P^{d,k,n}$ has a partition as $\bigcup_{j=1}^{v}[G_j, F_j]$. By Theorem 3.6, part 4, the interval $[G_j, F_j]$ has exactly

$$\binom{d-1-\dim G_j}{\ell - \dim G_j}$$

faces of dimension ℓ for $\dim G_j \leq \ell \leq d-1$. Let $k_i = |\{j : \dim G_j = i-1\}|$. Then $f_\ell = \sum_{i=0}^{\ell+1} \binom{d-i}{\ell-i+1} k_i$. These are the (invertible) equations that give f_ℓ in terms of h_i', so for all i, $h_i' = k_i = |\{j : \dim G_j = i-1\}|$.

The second part we prove by induction on $n \geq k$. We will also need the following statement, which we prove in the course of the induction as well. If F_j is a facet of $P^{d,k,n}$ with $\max F_j = n-2$, then $|G_j| \leq (d-1)/2$. The base case of the induction is the cyclic polytope, $C^{d,k} = P^{d,k,k}$. We need to show that if F_j is a facet of $C^{d,k}$ with $\max F_j = k-2$, then $|G_j| \leq (d-1)/2$. This follows from the description of G_j in the proof of Theorem 3.6, because in this case, in $F_j = I_j^0 \cup I_j^1 \cup I_j^2 \cup \cdots \cup I_j^p \cup I_j^k$, the set I_j^k is empty and $|G_j| = \frac{1}{2}|\bigcup_{\ell=1}^{p} I_j^\ell| \leq \frac{1}{2}(d-1)$ (since d is odd).

Recall from the proof of Theorem 3.6 that for each facet F_j of $P^{d,k,n}$, G_j is the same size as the minimum new face G of the corresponding facet of $P^{d,k,n-1}$; that facet is the same (as vertex set) as F_j, if $\max F_j \leq n-2$, and is $\mathrm{lsh}(F_j)$, if $\max F_j \geq n-1$. From Proposition 3.5 we see that each facet of $P^{d,k,n-1}$ with maximum vertex $n-2$ gives rise to two facets of $P^{d,k,n}$, while all others give rise to exactly one facet each. Thus, for all i,

$$h_i'(P^{d,k,n}) = h_i'(P^{d,k,n-1})$$
$$+ \left| \{ j : F_j \text{ is a facet of } P^{d,k,n} \text{ with } \max F_j = n-1 \text{ and } |G_j| = i \} \right|.$$

Thus, for all i, $h_i'(P^{d,k,n}) \geq h_i'(P^{d,k,n-1})$, so by induction, $h_i'(P^{d,k,n}) \geq h_i'(C^{d,k})$. Furthermore, if $\max F_j = n-1$, then $\max(\mathrm{lsh}(F_j)) = (n-1)-1$, so by the induction hypothesis, $|G_j| \leq (d-1)/2$. So for $i > d/2$, $h_i'(P^{d,k,n}) = h_i'(P^{d,k,n-1}) = h_i(C^{d,k})$. $\qquad\square$

Note that for the multiplex $M^{d,n}$ (d odd or even), $h'(M^{d,n}) = (1, n-d+1, 1, 1, \ldots, 1, 1)$, while $h(M^{d,n}) = (1, n-d+1, n-d+1, \ldots, n-d+1, 1)$.

Now for multiplicial polytopes, we consider the remaining part of the h-vector, coming from the parameters $f_{0,j} - (j+1)f_j$. This is

$$h(P, x) - h'(P, x)$$
$$= (f_{0,d-1} - d f_{d-1}) + \sum_{i=2}^{d-1} ((f_{0,i} - (i+1)f_i) + (f_{0,i-1} - i f_{i-1})) (x-1)^{d-i}.$$

So

$$h(P, x+1) - h'(P, x+1)$$
$$= (f_{0,d-1} - d f_{d-1}) + \sum_{i=2}^{d-1} ((f_{0,i} - (i+1)f_i) + (f_{0,i-1} - i f_{i-1})) x^{d-i}$$
$$= \sum_{i=2}^{d-1} (f_{0,i} - (i+1)f_i)(x+1)x^{d-1-i}.$$

So

$$\sum_{i=2}^{d-1} (h_i(P) - h_i'(P))(x+1)^{d-1-i} = \sum_{i=2}^{d-1} (f_{0,i} - (i+1)f_i)x^{d-1-i}.$$

For the ordinary polytope, this equation can be applied locally to give the contribution to $h(P^{d,k,n}, x) - h'(P^{d,k,n}, x)$ from each interval $[G_j, F_j]$ of the shelling partition. For each j, and each $i \geq \dim G_j$, let $b_{j,i} = \sum (f_0(H) - (i+1))$, where the sum is over all i-faces H in $[G_j, F_j]$. Let $b_j(x) = \sum_{i=\dim G_j}^{d-1} b_{j,i} x^{d-1-i}$. Write $b_j(x)$ in the basis of powers of $(x+1)$: $b_j(x) = \sum a_{j,i}(x+1)^{d-1-i}$. Then $a_{j,i} = h_i(\mathcal{G}_j) - h_i'(\mathcal{G}_j)$, the contribution to $h_i(P^{d,k,n}) - h_i'(P^{d,k,n})$ from faces in the interval $[G_j, F_j]$. Note that for fixed j, $\sum_i a_{j,i} = b_j(0) = f_0(F_j) - d$. We will return to the coefficients $a_{j,i}$ after triangulating the ordinary polytope.

Example. The h-vector of $P^{5,6,8}$ is $h(P^{5,6,8}) = (1, 4, 7, 7, 4, 1)$. The sum of the h_i is 24, which counts the 16 facets plus one for each of the four 6-vertex facets, plus two for each of the two 7-vertex facets. Referring to Table 1, we see that $h'(P^{5,6,8}) = (1, 4, 5, 3, 2, 1)$; from this we compute $f(P^{5,6,8}) = (9, 31, 52, 44, 16)$. The nonzero $a_{j,i}$ here are $a_{6,2} = a_{7,3} = a_{11,2} = a_{12,3} = 1$ and $a_{13,3} = a_{15,4} = 2$.

In this case each interval $[G_j, F_j]$ contributes to $h_i(P^{d,k,n}) - h_i'(P^{d,k,n})$ for at most one i, but this is not true in general.

5. Triangulating the Ordinary Polytope

Triangulations of polytopes or of their boundaries can be used to calculate the h-vector of the polytope if the triangulation is shallow [Bayer 1993]. The solid ordinary polytope need not have a shallow triangulation, but its boundary does have a shallow triangulation. The triangulation is obtained simply by triangulating each multiplex as in Section 2. This triangulation is obtained by "pushing" the vertices in the order 0, 1, ..., n. (See [Lee 1991] for pushing (placing) triangulations.)

THEOREM 5.1. *The boundary of the ordinary polytope $P^{d,k,n}$ has a shallow triangulation. The facets of one such triangulation are the Gale subsets of $[i, i+k]$ (where $0 \le i \le n-k$) of size d containing either 0 or n or the set $\{i, i+k\}$.*

PROOF. First we show that each such set is a consecutive subset of some facet of $P^{d,k,n}$. Suppose Z is a Gale subset of $[i, i+k]$ of size d containing $\{i, i+k\}$. Write $Z = [i, i+a-1] \cup Y \cup [i+k-b+1, i+k]$, where $a \ge 1$, $b \ge 1$, and

$$Y \cap \{i+a, i+k-b\} = \varnothing.$$

Since Z is a Gale subset, $|Y|$ is even; let $r = (d-1-|Y|)/2$. Since $|Z| = d$, $a+b = 2r+1$, so a and b are each at most $2r$. Define $X = [i+a-2r, i+a-1] \cup Y \cup [i+k-b+1, i+k-b+2r]$. Note that $i+k-b+1 = (i+a-2r)+k$. Then $\operatorname{ret}_n(X)$ is the vertex set of a facet of $P^{d,k,n}$, and Z is a consecutive subset of $\operatorname{ret}_n(X)$.

Now suppose that Z is a Gale subset of $[0, k]$ of size d containing 0, but not k. Write $Z = \{0\} \cup Y \cup [j-2r+1, j]$, where $j < k$, $r \ge 1$, and $j-2r \notin Y$. Then $|Y| = d-2r-1$, and $Z = \operatorname{ret}_n(X)$, where $X = [j-2r+1-k, j-k] \cup Y \cup [j-2r+1, j]$. So Z itself is the vertex set of a facet of $P^{d,k,n}$. The case of sets containing n but not $n-k$ works the same way.

Next we show that all consecutive d-subsets of facets F of $P^{d,k,n}$ are of one of these types. Let $F = \operatorname{ret}_n(X)$, where

$$X = [i, i+2r-1] \cup Y \cup [i+k, i+k+2r-1],$$

with Y a paired subset of size $d-2r-1$ of $[i+2r+1, i+k-2]$. Suppose first that $i+2r-1 \ge 0$ and $i+k \le n$. Let Z be a consecutive d-subset of F. Since $|Y| = d-2r-1$, $|[i, i+2r-1] \cap F| \le 2r$, and $|[i+k, i+k+2r-1] \cap F| \le 2r$, it follows that $i+2r-1$ and $i+k$ must both be in Z. Thus we can write $Z = [i+2r-a, i+2r-1] \cup Y \cup [i+k, i+k+b-1]$, with $a+b = 2r+1$, $i+2r-a \ge 0$, and $i+k+b-1 \le n$. Let $\ell = i+2r-a$. Then $i+k+b-1 = \ell+k$, so $0 \le \ell \le n-k$, and Z is a Gale subset of $[\ell, \ell+k]$ containing $\{\ell, \ell+k\}$.

If $i + 2r - 1 < 0$, then $i + k + 2r - 1 < k \le n$, and

$$F = \{0\} \cup Y \cup [i+k, i+k+2r-1].$$

Then $|F| = d$ and F itself is a Gale subset of $[0, k]$ of size d containing 0. Similarly for the case $i + k > n$.

The sets described are exactly the $(d-1)$-simplices obtained by triangulating each facet of $P^{d,k,n}$ according to Theorem 2.1. The fact that this triangulation is shallow follows from the corresponding fact for this triangulation of a multiplex.
$\square$

Let $\mathcal{T} = \mathcal{T}(P^{d,k,n})$ be this triangulation of $\partial P^{d,k,n}$. Since $\mathcal{T}$ is shallow, we have $h(P^{d,k,n}, x) = h(\mathcal{T}, x)$. We calculate $h(\mathcal{T}, x)$ by shelling $\mathcal{T}$.

THEOREM 5.2. *Let* $F_1, F_2, \ldots, F_v$ *be the colex order of the facets of* $P^{d,k,n}$. *For each* j, *if* $F_j = \{z_1, z_2, \ldots, z_{p_j}\}$ $(z_1 < z_2 < \cdots < z_{p_j})$, *and* $1 \le \ell \le p_j - d + 1$, *let* $T_{j,\ell} = \{z_\ell, z_{\ell+1}, \ldots, z_{\ell+d-1}\}$. *Then* $T_{1,1}, T_{1,2}, \ldots, T_{1,p_1-d+1}, T_{2,1},$ $\ldots, T_{2,p_2-d+1}, \ldots, T_{v,1}, \ldots, T_{v,p_v-d+1}$ *is a shelling of* $\mathcal{T}(P^{d,k,n})$.

Let $U_{j,\ell}$ *be the minimal new face when* $T_{j,\ell}$ *is shelled on. As vertex sets,* $U_{j,p_j-d+1} = G_j$.

PROOF. Throughout the proof, write $F_j = \{z_1, z_2, \ldots, z_{p_j}\}$ $(z_1 < z_2 < \cdots < z_{p_j})$. We first show that G_j is the unique minimal face of T_{j,p_j-d+1} not contained in $\bigcup_{i=1}^{j-1} \bigcup_{\ell=1}^{p_i-d+1} T_{i,\ell}) \cup (\bigcup_{\ell=1}^{p_j-d} T_{j,\ell})$. The set G_j is not contained in a facet of $P^{d,k,n}$ earlier than F_j. So G_j does not occur in a facet of $\mathcal{T}$ of the form $T_{i,\ell}$ for $i < j$. Also, $\max F_j \in G_j$, so G_j does not occur in a facet of $\mathcal{T}$ of the form $T_{j,\ell}$ for $\ell \le p_j - d$. Thus G_j does not occur in a facet of $\mathcal{T}$ before T_{j,p_j-d+1}.

We show that for $z_q \in G_j$, $T_{j,p_j-d+1} \setminus \{z_q\}$ is contained in a facet of $\mathcal{T}$ occurring before T_{j,p_j-d+1}. There is nothing to check for $j = v$, because $p_v - d + 1 = 1$ and so $T_{v,1} = F_v$ is the last simplex in the purported shelling order. So we may assume that $j < v$ and thus G_j is contained in the last $d-1$ vertices of F_j.

Case 1. If $p_j > d$ and $q = p_j$ (giving the maximal element of F_j), then $T_{j,p_j-d+1} \setminus \{z_{p_j}\} \subset T_{j,p_j-d}$.

Case 2. Suppose $p_j - d + 2 \le q \le p_j - 1$. Then

$$T_{j,p_j-d+1} \setminus \{z_q\} \subseteq \{z_{q-d+2}, \ldots, z_{q-1}, z_{q+1} \ldots, z_{p_j}\} = H.$$

This is a ridge of $P^{d,k,n}$ in F_j not containing G_j, and hence H is contained in a previous facet F_ℓ of $P^{d,k,n}$. Since H is a ridge in both F_j and F_ℓ, H is obtained from each facet by deleting a single element from a consecutive string of vertices in the facet. So $|H| \le |F_\ell \cap [z_{q-d+2}, z_{p_j}]| \le |H| + 1$, and so $d - 1 \le |F_\ell \cap [z_{p_j-d+1}, z_{p_j}]| \le d$. So $T_{j,p_j-d+1} \setminus \{z_q\}$ is contained in a consecutive set of d elements of F_ℓ, and hence in a $(d-1)$-simplex of $\mathcal{T}(P^{d,k,n})$ belonging to F_ℓ. This simplex occurs before T_{j,p_j-d+1} in the specified shelling order.

Case 3. Otherwise $p_j = d$ (so $p_j - d + 1 = 1$) and $q = d$. Then $T_{j,1} = F_j$ and $H = T_{j,1} \setminus \{z_d\}$ is a ridge of $P^{d,k,n}$ in F_j not containing $\max F_j$, so H is contained in a previous facet F_ℓ of $P^{d,k,n}$. As in Case 2, $d-1 \le |F_\ell \cap [z_1, z_{d-1}]| \le d$. So $T_{j,1} \setminus \{z_d\}$ is contained in a consecutive set of d elements of F_ℓ, and hence in a $(d-1)$-simplex of $\mathfrak{T}(P^{d,k,n})$ belonging to F_ℓ. This simplex occurs before T_{j,p_j-d+1} in the specified shelling order.

So in the potential shelling of $\mathfrak{T}$, G_j is the unique minimal new face as T_{j,p_j-d+1} is shelled on. Write $U_{j,p_j-d+1} = G_j$. At this point we need a clearer view of the simplex $T_{j,\ell}$. Recall that F_j is of the form $\mathrm{ret}_n(X)$, where $X = [i, i+2r-1] \cup Y \cup [i+k, i+k+2r-1]$, with Y a subset of size $d-2r-1$. If $i+2r-1 < 0$ or $i+k > n$, then $p_j = |F_j| = d$, and $T_{j,1} = T_{j,p_j-d+1} = F_j$; we have already completed this case. So assume $i+2r-1 \ge 0$ and $i+k \le n$. A consecutive string of length d in $\mathrm{ret}_n(X)$ must then be of the form $[i+s, i+2r-1] \cup Y \cup [i+k, i+k+s]$ for some s, $0 \le s \le 2r-1$. (All such strings—with appropriate Y—having $i+s \ge 0$ and $i+k+s \le n$ occur as $T_{j,\ell}$.) In particular, for $\ell < p_j - d+1$, $T_{j,\ell} = T_{j,\ell+1} \setminus \{\max T_{j,\ell+1}\} \cup \{\min T_{j,\ell+1} - 1\}$ and $\max T_{j,\ell} = \min T_{j,\ell} + k$.

Now define $U_{j,\ell}$ for $\ell \le p_j - d$ recursively by $U_{j,\ell} = U_{j,\ell+1} \setminus \{z\} \cup \{z-k, z-1\}$, where $z = \max T_{j,\ell+1}$. By the observations above, $U_{j,\ell} \subseteq T_{j,\ell}$. We prove by downward induction that $U_{j,\ell}$ is not contained in a facet F_i of $P^{d,k,n}$ before F_j, that $U_{j,\ell}$ is not contained in a facet of $\mathfrak{T}$ occurring before $T_{j,\ell}$, and that any ridge of $\mathfrak{T}$ in $T_{j,\ell}$ not containing all of $U_{j,\ell}$ is in an earlier facet of $\mathfrak{T}$. The base case of the induction is $\ell = p_j - d+1$, and this case has been handled above.

Note that $\{z-k, z-1\}$ is a diagonal of the 2-face $\{z-k-1, z-k, z-1, z\}$ of $P^{d,k,n}$ [Dinh 1999]. So if F_i is a facet of $P^{d,k,n}$ containing $U_{j,\ell}$, then F_i contains $\{z-k-1, z-k, z-1, z\}$. Thus F_i contains $U_{j,\ell+1}$, so, by the induction assumption, $i \ge j$. Therefore, for $i < j$, and any r, $T_{i,r}$ does not contain $U_{j,\ell}$. For $r < \ell$, $T_{j,r}$ does not contain $z-1 = \max T_{j,\ell}$, so $T_{j,r}$ does not contain $U_{j,\ell}$.

Now we will show that for any $g \in U_{j,\ell}$, $T_{j,\ell} \setminus \{g\}$ is in a previous facet of $\mathfrak{T}$.

Case 1. If $g = z-1 = \max T_{j,\ell}$ and $\ell \ge 2$, then $T_{j,\ell} \setminus \{g\} \subset T_{j,\ell-1}$.

Case 2. If $g = z-1 = \max T_{j,\ell}$ and $\ell = 1$, then $T_{j,\ell} \setminus \{g\}$ is the leftmost ridge of $P^{d,k,n}$ in F_j and, in particular, does not contain $\max F_j$. So $H = T_{j,\ell} \setminus \{g\}$ is contained in a previous facet F_e of $P^{d,k,n}$. As in the $\ell = p_j - d+1$ case, $F_e \cap [\min T_{j,\ell}, \max T_{j,\ell}]$ is contained in a consecutive set of d elements of F_e, and hence in a $(d-1)$-simplex of $\mathfrak{T}(P^{d,k,n})$ belonging to F_e. So $T_{j,\ell} \setminus \{g\}$ is contained in a previous facet of $\mathfrak{T}$.

Case 3. Suppose $g < z-1$ and $g \in U_{j,\ell} \cap U_{j,\ell+1}$. Since $\{z-1, z\} \subset T_{j,\ell+1}$, $T_{j,\ell+1}$ contains at most $d-3$ elements less than g. The ridge H of $P^{d,k,n}$ in F_j containing $T_{j,\ell+1} \setminus \{g\}$ consists of the $d-2$ elements of F_j below g and the (up to) $d-2$ elements of F_j above g. In particular, H contains $\min T_{j,\ell+1} - 1 = \min T_{j,\ell}$. So $T_{j,\ell} \setminus \{g\} \subset H$. Since $\dim T_{j,\ell} \setminus \{g\} = d-2$, H is the (unique) smallest face of $P^{d,k,n}$ containing $T_{j,\ell+1} \setminus \{g\}$. By the induction hypothesis $T_{j,\ell+1} \setminus \{g\}$ is contained in a previous facet $T_{i,r}$ of $\mathfrak{T}$; here $i < j$ because $\max T_{j,\ell+1} \in T_{j,\ell+1} \setminus \{g\}$. The $(d-2)$-simplex $T_{j,\ell+1} \setminus \{g\}$ is then contained in a ridge of $P^{d,k,n}$ contained in F_i, but this ridge must be H, by the uniqueness of H. So

$T_{j,\ell}\setminus\{g\} \subset H = F_i \cap F_j$. As in earlier cases, $F_i \cap [\min T_{j,\ell}, \max T_{j,\ell}]$ is contained in a consecutive set of d elements of F_i, and hence in a $(d-1)$-simplex of $\mathfrak{T}(P^{d,k,n})$ belonging to F_i. So $T_{j,\ell}\setminus\{g\}$ is contained in a previous facet of $\mathfrak{T}$.

Case 4. Finally, let $g = z - k$, which is $\min T_{j,\ell} + 1$. Then $T_{j,\ell}$ contains $d-2$ elements above g. Let H be the ridge of $P^{d,k,n}$ in F_j containing $T_{j,\ell}\setminus\{g\}$. Then $\max H = \max T_{j,\ell} < \max F_j$, so H does not contain G_j. So H is in a previous facet F_i of $P^{d,k,n}$. As in earlier cases, $F_i \cap [\min T_{j,\ell}, \max T_{j,\ell}]$ is contained in a consecutive set of d elements of F_i, and hence in a $(d-1)$-simplex of $\mathfrak{T}(P^{d,k,n})$ belonging to F_i. So $T_{j,\ell}\setminus\{g\}$ is contained in a previous facet of $\mathfrak{T}$.

Thus $T_{1,1}, T_{1,2}, \ldots, T_{1,p_1-d+1}, T_{2,1}, \ldots, T_{2,p_2-d+1}, \ldots, T_{v,1}, \ldots, T_{v,p_v-d+1}$ is a shelling of $\mathfrak{T}(P^{d,k,n})$. $\square$

COROLLARY 5.3. *Let $n \geq k \geq d = 2m+1 \geq 5$. Let $\cup [G_j, F_j]$ be the partition of the face lattice of $P^{d,k,n}$ from the colex shelling, and let $\cup [U_{j,\ell}, T_{j,\ell}]$ be the partition of the face lattice of $\mathfrak{T}(P^{d,k,n})$ from the shelling of Theorem 5.2. Then*

(i) *For each i, $h_i(P^{d,k,n}) \geq h'_i(P^{d,k,n})$.*
(ii) *The contribution to $h_i(P^{d,k,n}) - h'_i(P^{d,k,n})$ from the interval $[G_j, F_j]$ is*

$$a_{j,i} = \left|\{\ell : |U_{j,\ell}| = i, 1 \leq \ell \leq p_\ell - d\}\right| \geq 0.$$

PROOF. The h-vector of $\mathfrak{T}$ counts the sets $U_{j,\ell}$ of each size. Among these are all the sets G_j counted by the h'-vector of $P^{d,k,n}$. Thus

$$h_i(\mathfrak{T}(P^{d,k,n})) = \left|\{(j,\ell) : |U_{j,\ell}| = i\}\right|$$
$$\geq \left|\{(j,\ell) : |U_{j,\ell}| = i \text{ and } \ell = p_j - d+1\}\right| = h'_i(P^{d,k,n}).$$

Recall that we write $\mathcal{G}_j$ for the set of faces of F_j not in $\bigcup_{i<j} F_i$; here $\mathcal{G}_j$ is the set of faces in $[G_j, F_j]$. Write also $\mathfrak{T}G_j$ for the set of faces of $\mathfrak{T}$ that are contained in F_j but not in $\bigcup_{i<j} F_i$. By [Bayer 1993, Corollary 7], since $\mathfrak{T}$ is a shallow triangulation of $\partial P^{d,k,n}$, $g(G, x) = \sum (x-1)^{d-1-\dim \sigma}$, where the sum is over all faces σ of $\mathfrak{T}$ that are contained in G but not in any proper subface of G. Thus

$$h(\mathcal{G}_j, x) = \sum_{G \in [G_j, F_j]} g(G, x)(x-1)^{d-1-\dim G}$$
$$= \sum_{\sigma \in \mathfrak{T}G_j} (x-1)^{d-1-\dim \sigma} = \sum_{\ell=1}^{p_\ell-d+1} x^{d-|U_{j,\ell}|}$$

Since $h'(\mathcal{G}_j, x) = x^{d-|G_j|} = x^{d-|U_{j,p_j-d+1}|}$,

$$\sum_i a_{j,i} x^i = h(\mathcal{G}_j, x) - h'(\mathcal{G}_j, x) = \sum_{\ell=1}^{p_\ell-d} x^{d-|U_{j,\ell}|},$$

or

$$a_{j,i} = \left|\{\ell : |U_{j,\ell}| = i, 1 \leq \ell \leq p_\ell - d\}\right| \geq 0. \qquad \square$$

(j,ℓ)	$T_{j,\ell}$	$U_{j,\ell}$	(j,ℓ)	$T_{j,\ell}$	$U_{j,\ell}$
$1,1$	01234	∅	$11,1$	1234 7	27
$2,1$	012 45	5	$11,2$	234 78	78
$3,1$	0 2345	35	$12,1$	12 45 7	257
$4,1$	0 23 56	6	$12,2$	2 45 78	578
$5,1$	0 3456	46	$13,1$	0123 6	126
$6,1$	01 34 6	16	$13,2$	123 67	267
$6,2$	1 34 67	7	$13,3$	23 678	678
$7,1$	01 456	156	$14,1$	34 678	4678
$7,2$	1 4567	57	$15,1$	012 56	1256
$8,1$	2345 8	8	$15,2$	12 567	2567
$9,1$	23 56 8	68	$15,3$	2 5678	5678
$10,1$	3456 8	468	$16,1$	45678	45678

Table 2. Shelling of triangulation of $P^{5,6,8}$

Example. Table 2 gives the shelling of the triangulation of $P^{5,6,8}$. (Refer back to Table 1 for the shelling of $P^{5,6,8}$ itself.) Among the rows $(6,1)$, $(7,1)$, $(11,1)$, $(12,1)$, $(13,1)$, $(13,2)$, $(15,1)$, $(15,2)$ (rows (j,ℓ) that are not the last row for that j), count the $U_{j,\ell}$ of cardinality i to get $h_i(P^{5,6,8}) - h_i'(P^{5,6,8})$. Note that $U_{13,3} = G_{13}$ (from Table 1), and that $U_{13,2} = U_{13,3} \setminus \{8\} \cup \{2,7\}$. The ridges in $T_{13,2}$ are 1236, 1237, 1267, 1367, and 2367. The first ridge, 1236, falls under Case 1 of the proof of Theorem 5.2; it is contained in the previous facet, $T_{13,1}$. The next ridge, 1237, falls under Case 3; it is contained in the ridge 12378 of $P^{5,6,8}$ in $F_{13} = 0123678$, and 12378 also contains the ridge 2378 in $T_{13,3}$. The induction assumption says that 2378 is contained in an earlier facet, in this case $T_{11,2}$, and 12378 is contained in F_{11}. Finally, the ridge 1237 is contained in the simplex $T_{11,1}$, part of the triangulation of F_{11}. The last ridge of $T_{13,2}$ not containing 267 is 1367. It falls under Case 4. The set 1367 is contained in the ridge 01367 of $P^{5,6,8}$, contained in F_{13}. This ridge is also contained in the earlier facet F_6. The ridge 1367 of the triangulation is contained in the simplex $T_{6,2}$.

THEOREM 5.4. *Let* $n \geq d+k-1$. *For* $1 \leq i \leq d-1$, $h_i(P^{d,k,n}) - h_i(P^{d,k,n-1})$ *is the number of facets* $T_{j,\ell}$ *of* $\mathfrak{T}(P^{d,k,n})$ *such that* $\max F_j = n-1$ *and* $|U_{j,\ell}| = i$. *For* $1 \leq i \leq m$, *this is* $\binom{k-d+i-1}{i-1}$.

PROOF. Refer to Proposition 3.5 for a description of the facets of $P^{d,k,n}$ in terms of those of $P^{d,k,n-1}$. For $n \geq d+k-1$, for every facet $P^{d,k,n}$ ending in n, the translation $F-1$ is a facet of $P^{d,k,n-1}$. (For smaller n, a facet of $P^{d,k,n}$ may end in 0, in which case $\mathrm{lsh}(F)$ is a proper subset of $F-1$.) The same holds for the simplices $T_{j,\ell}$ triangulating these facets, and for the sets $U_{j,\ell}$. The facets of $P^{d,k,n}$ ending in $n-2$ are facets of $P^{d,k,n-1}$, and the same holds for the corresponding $T_{j,\ell}$ and $U_{j,\ell}$. The contributions to $h(P^{d,k,n})$ from facets ending in any element but $n-1$ thus total $h(P^{d,k,n-1})$. So for $1 \le i \le d-1$, $h_i(P^{d,k,n}) - h_i(P^{d,k,n-1})$ is the number of facets $T_{j,\ell}$ of $\mathcal{T}(P^{d,k,n})$ such that $\max F_j = n-1$ and $|U_{j,\ell}| = i$.

Now consider the set $\mathcal{S}$ of facets $T_{j,\ell}$ of $\mathcal{T}(P^{d,k,n})$ with $\max F_j = n-1$. For each $T \in \mathcal{S}$, T is a set of d elements occurring consecutively in some F_j with maximum element $n-1$. So T can be written as

$$T = [b, n-k-1] \cup [n-k+1, c] \cup Y \cup [e, b+k], \tag{5-1}$$

where

(i) $n-k-d+1 \le b \le n-k-1$;
(ii) $n-k \le c \le b+d-1$ and $c-n+k$ is even (here $c = n-k$ means $[n-k+1, c] = \varnothing$);
(iii) Y is a paired subset of $[c+2, e-1]$;
(iv) $e = b+k-1$ if $n-k-b$ is odd, and $e = b+k$ if $n-k-b$ is even; and
(v) $|T| = d$.

In these terms, the minimum new face U when T is shelled on is $U = [b+1, n-k-1] \cup E(Y) \cup \{b+k\}$.

We give a bijection between the facets T in $\mathcal{S}$ with $|U| = i$ (where $1 \le i \le m$) and the $(k-d)$-element subsets of $[1, k-d+i-1]$. Let T be as in Equation 5-1. Then $i = |U| = n-k-b+|Y|/2$. For each $x \ge c+1$, let $y(x)$ be the number of pairs in Y with both elements less than x. Let $a_1 = n-k-b = i-|Y|/2$. Write $[c+1, e-1] \setminus Y = \{x_1, x_2, \ldots, x_{k-d}\}$, with the x_ℓs increasing. (This set has $k-d$ elements because $d = (c-b) + |Y| + (b+k-e+1)$, so $|[c+1, e-1] \setminus Y| = e-c-1-|Y| = k-d$.) Set

$$A(T) = \{a_1 + y(x_\ell) + \ell - 1 : 1 \le \ell \le k-d\}.$$

To see that this is a subset of $[1, k-d+i-1]$, note that the elements of $A(T)$ form an increasing sequence with minimum element a_1 and maximum element $a_1 + y(x_{k-d}) + (k-d-1) \le a_1 + |Y|/2 + (k-d-1) = k-d+i-1$.

For the inverse of this map, write a $(k-d)$-element subset of $[1, k-d+i-1]$ as $A = \{a_1, a_2, \ldots, a_{k-d}\}$, with the a_ℓs increasing. Then $1 \le a_1 \le i$. Let

$$x_1 = n-k+d-2i+a_1-\chi(a_1 \text{ odd}).$$

Set

$$T(A) = [n-k-a_1, n-k-1] \cup [n-k+1, x_1-1] \cup Y \cup [n-a_1-\chi(a_1 \text{ odd}), n-a_1],$$

where

$$Y = ([x_1, n-a_1-1-\chi(a_1 \text{ odd})] \setminus \{x_1+2(a_\ell-a_1)-(\ell-1) : 1 \le \ell \le k-d\}).$$

We check that this gives a set of the required form.

(1) Since $1 \le a_1 \le i \le d-1$ $n-k-d+1 \le n-k-a_1 \le n-k-1$.
(2) $x_1-1-n+k = d-2i-1+(a_1-\chi(a_1 \text{ odd}))$, which is nonnegative and even; $x_1-1 = (n-k-a_1+d-1) - (2i-2a_1+\chi(a_1 \text{ odd})) \le n-k-a_1+d-1$.

(3) Y is clearly a subset of $[x_1+1, n-a_1-\chi(a_1 \text{ odd})-1]$. To see that Y is paired note that the difference between two consecutive elements in the removed set is
$$(x_1+2(a_{\ell+1}-a_1)-\ell)-(x_1+2(a_\ell-a_1)-(\ell-1)) = 2(a_{\ell+1}-a_\ell)-1.$$

(4) This condition holds by definition.

(5) To check the cardinality of $T(A)$, observe that
$$x_1+2(a_{k-d}-a_1)-(k-d-1) \le x_1+2(k-d+i-1)-2a_1-(k-d-1)$$
$$= x_1+k-d+2i-2a_1-1 = n-a_1-\chi(a_1 \text{ odd})-1.$$

So
$$\{x_1+2(a_\ell-a_1)-(\ell-1) : 1 \le \ell \le k-d\} \subseteq [x_1+1, n-a_1-1-\chi(a_1 \text{ odd})],$$

and
$$|Y| = (n-a_1-\chi(a_1 \text{ odd})-x_1)-(k-d) = 2i-2a_1.$$

So $|T(A)| = x_1-(n-k-a_1)+|Y|+\chi(a_1 \text{ odd}) = d$.

Also, in this case $U = [n-k-a_1+1, n-k-1] \cup E(Y) \cup \{n-a_1\}$, so $|U| = i$.

It is straightforward to check that these maps are inverses. The main point is that, if $a_\ell = a_1+y(x_\ell)+\ell-1$, then
$$x_1+2(a_\ell-a_1)-(\ell-1) = x_1+2(y(x_\ell)+\ell-1)-(\ell-1)$$
$$= x_1+2y(x_\ell)+\ell-1 = x_\ell. \qquad \square$$

Example. Consider the ordinary polytope $P^{7,9,15}$. There are six facets with maximum vertex 14; they are (with sets G_j underlined) $\{4,5,7,8,9,10,13,\underline{14}\}$, $\{4,5,7,8,10,\underline{11},13,\underline{14}\}$, $\{4,5,8,\underline{9},10,\underline{11},13,\underline{14}\}$, $\{2,3,4,5,7,8,11,\underline{12},13,\underline{14}\}$, $\{2,3,4,5,8,\underline{9},11,\underline{12},13,\underline{14}\}$, and $\{0,1,2,3,4,5,9,\underline{10},11,\underline{12},13,\underline{14}\}$. Among the 6-simplices occurring in the triangulation of these facets, six have $|U_{j,\ell}| = 3$. Table 3 gives the bijection from this set of simplices to the 2-element subsets of $[1,4]$.

$T_{j,\ell}$	b	c	e	Y	a_1	x_1, x_2	$y(x_i)$	$A(T_{j,\ell})$
$4,\underline{5},7,8,10,\underline{11},13$	4	8	13	$10,11$	2	$9,12$	$0,1$	$\{2,4\}$
$5,8,\underline{9},10,\underline{11},13,14$	5	6	13	$8,9,10,11$	1	$7,12$	$0,2$	$\{1,4\}$
$3,\underline{4},\underline{5},7,8,11,\underline{12}$	3	8	11	$\varnothing$	3	$9,10$	$0,0$	$\{3,4\}$
$4,\underline{5},7,8,11,\underline{12},13$	4	8	13	$11,12$	2	$9,10$	$0,0$	$\{2,3\}$
$5,8,\underline{9},11,\underline{12},13,14$	5	6	13	$8,9,11,12$	1	$7,10$	$0,1$	$\{1,3\}$
$5,9,\underline{10},11,\underline{12},13,14$	5	6	13	$9,10,11,12$	1	$7,8$	$0,0$	$\{1,2\}$

Table 3. Bijection with 2-element subsets of $\{1,2,3,4\}$

Again, the results of this section hold for even-dimensional multiplexes as well.

6. Afterword

The story of the combinatorics of simplicial polytopes is a beautiful one. There one finds an intricate interplay among the face lattice of the polytope, shellings, the Stanley–Reisner ring and the toric variety, tied together with the h-vector. The cyclic polytopes play a special role, serving as the extreme examples, and providing the environment in which to build representative polytopes for each h-vector (the Billera–Lee construction [Billera and Lee 1981]). In the general case of arbitrary convex polytopes, the various puzzle pieces have not interlocked as well. In this paper we made progress on putting the puzzle together for the special class of ordinary polytopes. Since the ordinary polytopes generalize the cyclic polytopes, a natural next step would be to mimic the Billera–Lee construction, or Kalai's extension of it [1988], on the ordinary polytopes, as a way of generating multiplicial flag vectors. It would also be interesting to see if there is a ring associated with these polytopes, particularly one having a quotient with Hilbert function equal to the h'-polynomial. Another open problem is to determine the best even-dimensional analogues of the ordinary polytopes. They may come from taking vertex figures of odd-dimensional ordinary polytopes, or from generalizing Dinh's combinatorial description of the facets of ordinary polytopes. Looking beyond ordinary and multiplicial polytopes, we should ask what other classes of polytopes have shellings with special properties that relate to the h-vector?

Acknowledgments

My thanks go to the folks at University of Washington, the Discrete and Computational Geometry program at MSRI and the Diskrete Geometrie group at TU-Berlin, who listened to me when it was all speculation. Particular thanks go to Carl Lee for helpful discussions.

References

[Barthel et al. 2002] G. Barthel, J.-P. Brasselet, K.-H. Fieseler, and L. Kaup, "Combinatorial intersection cohomology for fans", *Tohoku Math. J.* (2) **54**:1 (2002), 1–41.

[Bayer 1993] M. M. Bayer, "Equidecomposable and weakly neighborly polytopes", *Israel J. Math.* **81**:3 (1993), 301–320.

[Bayer 2004] M. M. Bayer, "Flag vectors of multiplicial polytopes", *Electron. J. Combin.* **11** (2004), Research Paper 65.

[Bayer et al. 2002] M. M. Bayer, A. M. Bruening, and J. D. Stewart, "A combinatorial study of multiplexes and ordinary polytopes", *Discrete Comput. Geom.* **27**:1 (2002), 49–63.

[Billera and Lee 1981] L. J. Billera and C. W. Lee, "A proof of the sufficiency of McMullen's conditions for f-vectors of simplicial convex polytopes", *J. Combin. Theory Ser. A* **31**:3 (1981), 237–255.

[Bisztriczky 1996] T. Bisztriczky, "On a class of generalized simplices", *Mathematika*
43:2 (1996), 274–285 (1997).

[Bisztriczky 1997] T. Bisztriczky, "Ordinary $(2m+1)$-polytopes", *Israel J. Math.* **102**
(1997), 101–123.

[Braden 2003] T. Braden, "g- and h-polynomials of non-rational polytopes: recent
progress", Abstract at meeting on topological and geometric combinatorics, Math-
ematisches Forschungsinstitut Oberwolfach, April 6–12 2003.

[Chan 1991] C. Chan, "Plane trees and H-vectors of shellable cubical complexes",
SIAM J. Discrete Math. **4**:4 (1991), 568–574.

[Dinh 1999] T. N. Dinh, *Ordinary polytopes*, Ph.D. thesis, The University of Calgary,
1999.

[Kalai 1988] G. Kalai, "Many triangulated spheres", *Discrete Comput. Geom.* **3**:1
(1988), 1–14.

[Karu 2002] K. Karu, "Hard Lefschetz Theorem for nonrational polytopes", version 4,
2002. Available at arXiv:math.AG/0112087.

[Lee 1991] C. W. Lee, "Regular triangulations of convex polytopes", pp. 443–456 in
Applied geometry and discrete mathematics: the Victor Klee festschrift, edited by P.
Gritzmann and B. Sturmfels, DIMACS Ser. Discrete Math. Theoret. Comput. Sci.
4, Amer. Math. Soc., Providence, RI, 1991.

[McMullen and Shephard 1971] P. McMullen and G. C. Shephard, *Convex polytopes
and the upper bound conjecture*, London Math. Soc. Lect. Note Series **3**, Cambridge
University Press, London, 1971.

[Stanley 1980] R. P. Stanley, "The number of faces of a simplicial convex polytope",
Adv. in Math. **35**:3 (1980), 236–238.

[Stanley 1987] R. Stanley, "Generalized H-vectors, intersection cohomology of toric va-
rieties, and related results", pp. 187–213 in *Commutative algebra and combinatorics*
(Kyoto, 1985), edited by M. Nagata and H. Matsumura, Adv. Stud. Pure Math. **11**,
North-Holland, Amsterdam and Tokyo, Kinokuniya, 1987.

[Stanley 1992] R. P. Stanley, "Subdivisions and local h-vectors", *J. Amer. Math. Soc.*
5:4 (1992), 805–851.

[Ziegler 1995] G. M. Ziegler, *Lectures on polytopes*, Graduate Texts in Mathematics
152, Springer-Verlag, New York, 1995.

MARGARET M. BAYER
DEPARTMENT OF MATHEMATICS
UNIVERSITY OF KANSAS
LAWRENCE, KS 66045-7523
UNITED STATES
bayer@math.ku.edu

Combinatorial and Computational Geometry
MSRI Publications
Volume **52**, 2005

On the Number of Mutually Touching Cylinders

ANDRÁS BEZDEK

ABSTRACT. In a three-dimensional arrangement of 25 congruent nonoverlapping infinite circular cylinders there are always two that do not touch each other.

1. Introduction

The following problem was posed by Littlewood [1968]:

What is the maximum number of congruent infinite circular cylinders that can be arranged in $\mathbb{R}^3$ *so that any two of them are touching? Is it 7?*

This problem is still open. The analogous problem concerning circular cylinders of finite length became known as a mathematical puzzle due to a the popular book [Gardner 1959]: Find an arrangement of 7 cigarettes so that any two touch each other. The question whether 7 is the largest such number is open. For constructions and for a more detailed account on both of these problems see the research problem collection [Moser and Pach ≥ 2005].

A very large bound for the maximal number of cylinders in Littlewood's original problem was found by the author in 1981 (an outline proof was presented at the Discrete Geometry meeting in Oberwolfach in that year). The bound was expressed in terms of various Ramsey constants, and so large that it merely showed the existence of a finite bound. In this paper we use a different approach to show that at most 24 cylinders can be arranged so that any two of them are touching:

THEOREM 1. *In an arrangement of* 25 *congruent nonoverlaping infinite circular cylinders there are always two that do not touch each other.*

Mathematics Subject Classification: 52C15, 52A40.

Keywords: packing, cylinders.

Partially supported by the Hungarian National Science Foundation, grant numbers T043520 and T038397.

In Section 2, we introduce the necessary terminology to talk about relative positions of the cylinders. In Section 3 we prove Theorem 1. We will describe a four-cylinder arrangement in which the cylinders cannot be mutually touching and show that in a family of 25 mutually touching cylinders there are always four cylinders of this type.

One of the needed lemmas can be stated and proved independently from the cylinder problem. To ease the description of the proof of Theorem 1 we place this lemma separately, in Section 4.

2. Terminology

The term *cylinder* will always refer to a circular cylinder infinite at both ends. More precisely, the *cylinder of radius r and axis l* is the set of those points in $\mathbb{R}^3$ that are at a distance of at most r from a given line l. If $r = 1$, we speak of *unit* cylinders. Two cylinders are *nonoverlapping* if they do not have common interior points. Two cylinders are *touching* if they do not overlap, but have at least one common boundary point.

Consider a family of mutually touching cylinders. For reference choose one of the cylinders, say c, and assign a positive direction to its axis l. We say that a cylinder *lies in front of* another cylinder with respect to the directed axis l if the first cylinder can be shifted parallel to l in the positive direction to infinity without crossing the other cylinder. This relation is not transitive, so it does not give rise to an ordering among the cylinders.

There is another natural way of describing a relative position among mutually touching cylinders. We say that a cylinder is (*clockwise*) *to the right* of another if a clockwise rotation by α (with $0 < \alpha \leq \pi$) around l takes the plane separating the second cylinder from c to the plane separating the first cylinder from c. To avoid ambiguity, we say that counterclockwise rotation around the axis l is the one which matches the right-hand rule with the thumb pointing in the positive direction of the axis l. The relation of "being to the right" clearly defines an order among cylinders that are touching c, in such a way that their contact points, if looked at from the direction of the axis of c, belong to a circular arc less than π. We will refer to this order as *the clockwise order* with respect to l.

3. Proof of Theorem 1

Assume we have an arrangement of 25 mutually touching cylinders so that one of the cylinders is c with directed axis l. Most likely the first thing one notices while studying cylinder arrangements is that no two of the cylinders are parallel. Otherwise the number of cylinders is at most four.

Most of our conclusions will come from studying the *front view*, which is what we see by looking at the cylinder packing from the positive direction of l. We intentionally use the term "front view" instead of "projection", since we would like to keep track of the relation of "being in front". Let the unit disc d be the image of cylinder c. The images of the other cylinders are strips of width 2, all touching disc d at different points. A simple integral averaging argument shows that among these 24 contact points in the front view one can choose 5 along an arc on the boundary of d with central angle at most $\pi/3$.

Label the corresponding cylinders c_1, c_2, c_3, c_4, c_5 in clockwise order, so that cylinder c_5 is rightmost.

LEMMA 1. *In any oriented complete graph with vertices labelled* $1, 2, 3, 4, 5$ *one can choose three vertices* $i < j < k$ *so that either* $i \to j \to k$ *or* $i \leftarrow j \leftarrow k$ *holds.*

PROOF. If the conclusion is not true, we may assume that $2 \to 3 \leftarrow 4$ or $2 \leftarrow 3 \to 4$ holds. Consider the first case: If $2 \leftarrow 4$, then either $1 \leftarrow 2 \leftarrow 4$ or $1 \to 2 \to 3$ holds, a contradiction. If $2 \to 4$ then either $3 \leftarrow 4 \leftarrow 5$ or $2 \to 4 \to 5$ holds, a contradiction. The second case is handled in the same way.   □

Consider the abstract complete graph whose vertices are the cylinders c_1, c_2, c_3, c_4, c_5. Orient the edges according to the "being in front" relation. According to Lemma 1 three of the cylinders, say c_1, c_2, c_3, are such that (i) c_1 is in front of c_2 which is in front of c_3, or (ii) c_1 is behind c_2 which is behind c_3.

We will show that cylinders c, c_1, c_2 and c_3 cannot be mutually touching. In this respect case (ii) can be reduced to case (i) by reflecting the cylinders along a plane passing through the axis of the cylinder c. Indeed such plane reflection preserves the relation of "being in front", but reverses the clockwise order. The impossibility of case (i) is stated as a separate lemma below. Its proof completes the proof of Theorem 1.

LEMMA 2 (A FORBIDDEN ARRANGEMENT OF FOUR CYLINDERS). *If a packing of four cylinders* c, c_1, c_2, c_3 *satisfies the conditions listed below, two of them must be disjoint.*

Contact condition: *Cylinders* c_1, c_2, c_3 *are touching* c *so that their contact points if looked at from the direction of the axis of* c *belong to a circular arc of length at most* $\pi/3$.

Clockwise order condition: *Cylinders* c_1, c_2, c_3 *are labelled according to their clockwise order with respect to the directed axis* l *of* c *so that* c_3 *is the rightmost one.*

"Being in front" condition: *Cylinder* c_1 *is in front of cylinder* c_2 *which is in front of cylinder* c_3 *with respect to the directed axis* l *of* c.

PROOF. Assume to the contrary that cylinders c, c_1, c_2, c_3 are mutually touching and satisfy all three conditions. Let strips s_1, s_2 and s_3 be the images of cylinders c_1, c_2 and c_3 in front view. Assume that strip s_3 is horizontal. Let the unit disc d with center O be the image of cylinder c. According to the contact condition and the clockwise order condition, the elevation angle of s_2 is positive and smaller than $\pi/3$. See Figure 1, left.

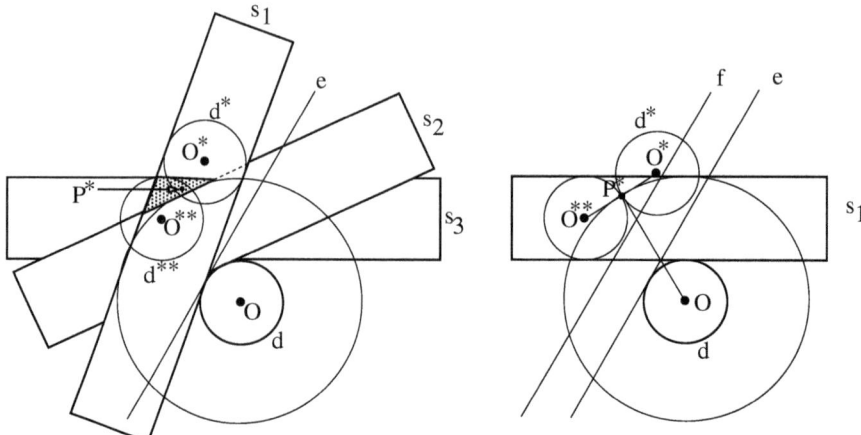

Figure 1.

Denote by P the contact point of cylinders c_1 and c_3, and by P^* the image of P in front view. P^* certainly belongs to both s_1 and s_3, but not to strip s_2, since c_2 is in front of c_3. Since strip s_1 is obtained from s_2 by a counterclockwise rotation around O, P^* lies to the left of strip s_2.

Let the unit discs d^* and d^{**} with centers O^* and O^{**} be the images in front view of the unit spheres inscribed in c_1 and c_3 respectively and containing P. Strip s_1 contains d^*, and is tangent to d. There are two such strips, but since P^* does not belong to s_2, the strip that is clockwise to the right of the other must be also to the right of s_2, thus it cannot be the same as s_1. Thus the position of d^* determines s_1.

Discs d^* and d^{**} are symmetrical with respect to point P^*. First fix P^* and move d^* horizontally to the right so that it has P^* on its boundary. Simultaneously move d^{**} so that P^* remains the symmetry center of d^* and d^{**}. Then move P^*, along with d^* and d^{**} horizontally to the right until P^* gets onto the circle centered at O of radius 3 (see Figure 1, right).

Notice that in the new position, (i) distance O^*O^{**} is 2 and the distance P^*O is 3, (ii) P^* is the midpoint of O^*O^{**} and (iii) O^{**} is on the left of the vertical line through O. Let e be the support line of d whose slope is $\sqrt{3}$. Lemma 3 of Section 4 states that in this new position, d^* lies to the left of line e, without touching e (except when $O^{**}O = 4$). This means that d^*, before it was moved, was to the left of line e, without touching e. Thus strip s_1 is obtained from s_3 by a counterclockwise rotation by an angle greater than $\pi/3$, contradicting Contact condition of Lemma 3. $\square$

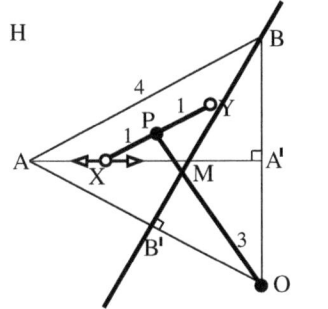 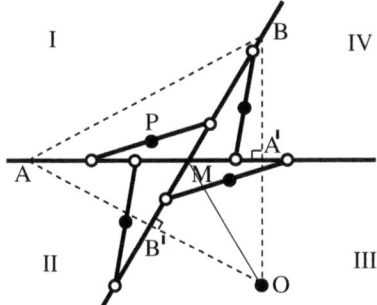

Figure 2.

4. T-linkages

By a *T-linkage* we will mean a mobile structure consisting of a bar of length 3 connected at its endpoint to the midpoint of a bar of length 2, so they can rotate about the contact point.

LEMMA 3. *Let AOB be an equilateral triangle of side length 4. Assume that a T-linkage is attached to O by the free endpoint of its longer bar (see Figure 2, left). As one endpoint of the shorter bar moves along the interior of median AA', the other endpoint of the shorter bar and A stay in the same open halfplane bounded by the line of median BB'.*

PROOF. Denote by H the open halfplane bounded by line BB' and containing A. Denote by M the intersection of AA' and BB'. A simple computation shows that when one endpoint of the shorter bar of the T-linkage coincides with M then the other one belongs to H. Thus, if Lemma 3 were not true then by a continuity argument the T-linkage would have a position with endpoints of the shorter bar on lines AA' and BB' respectively. We will prove that such a position does not exists. In fact we show more:

CLAIM. *If X is a point on line AA' different from both A and A' and if Y is a point on line BB' such that $XY = 2$, the distance from O to the midpoint of XY is smaller than 3.*

We distinguish four cases depending on which of the angles determined by lines of AA' and BB' contains the segment XY. Figure 2, right, shows how the angles are labelled I, II, III, IV. It suffices to check the cases when XY belongs to angles I or II. Indeed the cases of angles II and IV are the same by symmetry. Furthermore, if segment XY belongs to the angle III then reflecting XY around M we get a segment whose midpoint is farther from O than the midpoint of XY.

Case 1: *XY lies in angle I.* Let k be the circumcircle of the triangle XMY (see Figure 3, left). Since MO is the angle bisector of $\angle B'MA'$ the line of MO and

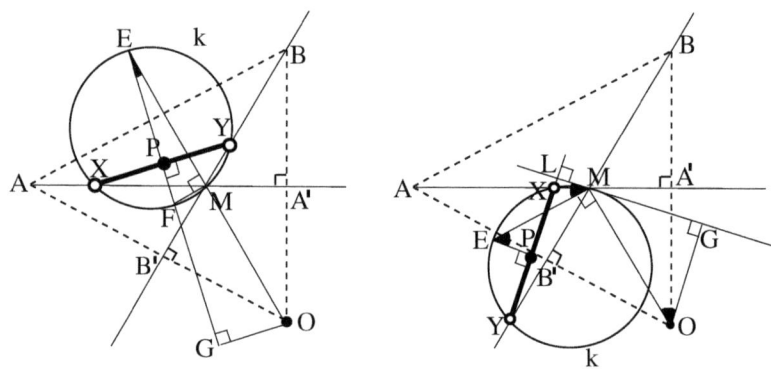

Figure 3.

the perpendicular bisector of XY intersect each other on k, say at E. Denote by F the diagonally opposite point of E on k.

Let G be the perpendicular projection of O onto line EF. Denote by P the midpoint of XY. We will express PO^2 in terms of the angle $\alpha = \angle PEM$ (with $-\pi/6 \leq \alpha \leq \pi/6$) and show that PO^2 is smaller than 9. Since $\angle XMY = 2\pi/3$ we have $EF = 4/\sqrt{3}$. Since $EP = \sqrt{3}$ and $MO = 4/\sqrt{3}$ we get

$$OE = EF \cos\alpha + MO = \frac{4}{\sqrt{3}}(\cos\alpha + 1).$$

Computing the parallel and perpendicular components of PO with respect to line EF we get

$$
\begin{aligned}
PO^2 &= OE^2 \sin^2\alpha + (OE^2 \cos\alpha - EP)^2 = OE^2 - 2\,OE \cos\alpha \sqrt{3} + 3 \\
&= \tfrac{16}{3}(\cos\alpha + 1)^2 - 8(\cos\alpha + 1)\cos\alpha + 3 = \tfrac{1}{3}(-8\cos^2\alpha + 8\cos\alpha + 25) \\
&= -\tfrac{1}{24}\left(\cos\alpha - \tfrac{1}{2}\right)^2 + 9 < 9,
\end{aligned}
$$

as claimed.

Case 2: *XY lies in angle II.* Let k be the circumcircle of triangle XMY (see Figure 3, right). The line perpendicular to MO and the perpendicular bisector of XY intersect each other on k, say at E. Let L be the perpendicular projection of M onto line XY. Let G be the perpendicular projection of O onto line LM.

Denote by P the midpoint of XY. We will express PO^2 in terms of the directed angle $\alpha = \angle PEM = \angle EML = \angle GOM$ (with $-\pi/3 \leq \alpha \leq \pi/3$) and show that PO^2 is smaller than 9. It is easy to see that $MO = 4/\sqrt{3}$, $EM = 4/\sqrt{3}\cos\alpha$ and $EP = 1/\sqrt{3}$. Computing the parallel and perpendicular components of PO with respect to line XY we get

$$PO^2 = (EM \cos \alpha - EP + MO \sin \alpha)^2 + (-EM \sin \alpha + MO \cos \alpha)^2$$
$$= \tfrac{1}{3}\big((4\cos^2 \alpha - 1 + 4\sin \alpha)^2 + (-4\cos \alpha \sin \alpha + 4\cos \alpha)^2\big)$$
$$= \tfrac{1}{3}(17 + 8\cos^2 \alpha - 8\sin \alpha) = \tfrac{1}{3}(25 - 8\sin^2 \alpha - 8\sin \alpha)$$
$$= -\tfrac{2}{3}(1 + 2\sin \alpha)^2 + 9 \leq 9.$$

Equality holds only if $\alpha = -\pi/6$, that is, when X coincides with A. Thus the Claim holds.   □

References

[Gardner 1959] M. Gardner, *Hexaflexagons and other mathematical diversions: The first Scientific American book of puzzles and games*, Simon & Schuster, New York, 1959.

[Littlewood 1968] J. E. Littlewood, *Some problems in real and complex analysis*, Heath Math. Monographs, Raytheon Education, Lexington (MA), 1968.

[Moser and Pach ≥ 2005] W. Moser and J. Pach, *Research problems in discrete geometry*, Academic Press. To appear.

ANDRÁS BEZDEK
DEPARTMENT OF MATHEMATICS
AUBURN UNIVERSITY
AUBURN, AL 36849-5310
UNITED STATES

RÉNYI INSTITUTE OF MATHEMATICS
HUNGARIAN ACADEMY OF SCIENCES
BUDAPEST H-1053
HUNGARY
   bezdean@auburn.edu

Combinatorial and Computational Geometry
MSRI Publications
Volume **52**, 2005

Edge-Antipodal 3-Polytopes

KÁROLY BEZDEK, TIBOR BISZTRICZKY, AND KÁROLY BÖRÖCZKY

ABSTRACT. A convex 3-polytope in E^3 is called edge-antipodal if any two vertices, that determine an edge of the polytope, lie on distinct parallel supporting planes of the polytope. We prove that the number of vertices of an edge-antipodal 3-polytope is at most eight, and that the maximum is attained only for affine cubes.

1. Introduction

Let X be a set of points in Euclidean d-space E^d. Then conv X and aff X denote, respectively, the convex hull and the affine hull of X.

Two points x and y are called *antipodal points* of X if there are distinct parallel supporting hyperplanes of conv X, one of which contains x and the other contains y. We say that X is an *antipodal set* if any two points of X are antipodal points of X. In the case that X is a convex d-polytope P, a related notion was recently introduced in [Talata 1999]. P is an *edge-antipodal d-polytope* if any two vertices of P, that lie on an edge of P, are antipodal points of P.

According to a well-known result of Danzer and Grünbaum [1962], conjectured independently by Erdős [1957] and Klee [1960], the cardinality of any antipodal set in E^d is at most 2^d. Talata [1999] conjectured that there exists a smallest positive integer m such that the cardinality of the vertex set of any edge-antipodal 3-polytope is at most m. In an elegant paper, Csikós [2003] showed that $m \leq 12$. In this paper, we prove that $m = 8$.

THEOREM. *The number of vertices of an edge-antipodal 3-polytope P is at most eight, with equality only if P is an affine cube.*

Mathematics Subject Classification: 52A40, 52B10, 52C10, 52C17.

Keywords: convex, polytope, edge-antipodal.

Bezdek and Böröczky were partially supported by the Hungarian National Science and Research Foundation OTKA T043556. Bezdek and Bisztriczky were supported by a Natural Sciences and Engineering Research Council of Canada Discovery Grant.

We remark that with some additional case analysis, it can be deduced from the proof of the Theorem that the vertex set of P is in fact antipodal. This is not the case for edge-antipodal d-polytopes P_d when $d \geq 4$ (see [Talata 1999] for $d = 4$), and thus, it seems highly challenging to determine the higher dimensional analogue of the Theorem. We note that Pór [2005] has shown that for each $d \geq 4$, there exists an integer $m(d)$, formula unknown, such that P_d has at most $m(d)$ vertices.

2. Proof of the Theorem

For sets $X_1, X_2, \ldots, X_n$ in E^3, let $[X_1, X_2, \ldots, X_n]$ be the convex hull of $X_1 \cup X_2 \cup \cdots \cup X_n$, and $\langle X_1, X_2, \ldots, X_n \rangle$ the affine hull of $X_1 \cup X_2 \cup \cdots \cup X_n$. For a point x, set $[x] = [\{x\}]$ and $\langle x \rangle = \langle \{x\} \rangle$.

For a point x and a line L in E^3, let $\ell(x, L)$ denote the line through x that is parallel to L. Likewise, if H is a plane in E^3, let $h(x, H)$ denote the plane through x that is parallel to L.

Let $P \subset E^3$ denote a (convex) 3-polytope with the set $\mathcal{V}(P)$ of vertices, the set $\mathcal{E}(P)$ of edges and the set $\mathcal{F}(P)$ of facets. We recall that by Euler's Theorem,

$$|\mathcal{V}(P)| - |\mathcal{E}(P)| + [\mathcal{F}(P)] = 2.$$

Let $v \in \mathcal{V}(P)$. Then v has *degree* k (deg $v = k$) if v is incident with exactly k edges of P. It is a consequence of Euler's Theorem (cf. [Fejes Tóth 1953]) that the average degree of a vertex of P is less than six, and thus,

REMARK 1. *Any 3-polytope contains a vertex of degree* k *with* $k \leq 5$.

Next, let

$$S = \{v_1, v_2, \ldots, v_n, v_{n+1} = v_1\} \subset \mathcal{V}(P),$$

where $n \geq 3$. We say that $[S]$ is a *contour section* of P if $\dim\langle S \rangle = 2$, $[S]$ is not a facet of P and $[v_i, v_{i+1}] \in \mathcal{E}(P)$ for $i = 1, \ldots, n$.

Finally, let v and w be antipodal vertices of P. When there is no danger of confusion, we denote by H_v^w and H_w^v, the distinct parallel supporting planes of P such that $v \in H_v^w$ and $w \in H_w^v$.

Henceforth, we assume that P is edge-antipodal. Thus, if $[v, w] \in \mathcal{E}(P)$ then v and w are antipodal.

We begin our arguments with some simple observations concerning a parallelogram $Q = [w, x, y, z]$ with sides $[w, x]$ and $[x, y]$:

REMARK 2. *If* $\{[w, x], [x, y]\} \subset \mathcal{E}(P)$ *then* $\langle w, z \rangle$ *and* $\langle y, z \rangle$ *are supporting lines of* P, *and* $\langle Q \rangle \cap P \subset Q$.

REMARK 3. *If* $[x, w, y, v] \subseteq Q \cap P$ *and* $[w, v] \in \mathcal{E}(P)$ *then* $v \in [y, z]$.

From these two remarks, we deduce:

REMARK 4. *Any facet or any contour section of P is a triangle or a parallelogram.*

We examine now P when it is nonsimplicial or simplicial, and determine when a subpolytope of P is necessarily edge-antipodal.

LEMMA 1. *Let $F = [w, x, y, z] \in \mathcal{F}(P)$ be a parallelogram with sides $[w, x]$ and $[x, y]$, and let H be a plane such that $H \cap F = [x, y]$ and $v \in (H \cap \mathcal{V}(P)) \setminus \{x, y\}$.*

1.1 *If $H \cap P$ is a contour section of P then $H \cap P$ is a parallelogram.*
1.2 *If $H \cap P$ is a facet of P then $h(v, \langle F \rangle)$ is a supporting plane of P.*

PROOF. We suppose that $H \cap P = [x, y, v]$ is a contour section, and seek a contradiction.

Let $L = \langle y, z \rangle$ and $R = [F, v, p]$ where p is the point on $\ell(v, L)$ such that $Q = [v, y, z, p]$ and $Q' = [v, x, w, p]$ are parallelograms. Next, $H \cap P \notin \mathcal{F}(P)$ implies that there is a $u \in \mathcal{V}(P)$ such that H separates u and R, and $[u, y] \in \mathcal{E}(P)$. We have now a contradiction by Remark 2. On the one hand; $\langle Q \rangle \cap P \subseteq Q$ and $\langle Q' \rangle \cap P \subseteq Q'$, and so $\ell(u, L)$ meets the relative interior of $H \cap P$. On the other hand; $\ell(u, L)$ is a supporting line of P.

Let $H \cap P \in \mathcal{F}(P)$. By Remark 4, $H \cap P$ is a parallelogram or a triangle.

If $H \cap P = [v, x, y, u]$ is a parallelogram with sides, say, $[v, x]$ and $[x, y]$ then

$$H_x^v \cap [v, x, y, u] = [x, y] \quad \text{and} \quad H_v^x \cap [v, x, y, u] = [v, u]$$

by Remark 2, and from this it follows that $h(v, \langle F \rangle)$ supports P. If $H \cap P = [v, x, y]$ then the assertion is immediate in the case that $H_x^v = \langle F \rangle$, and it is easy to check that $H_x^v \neq \langle F \rangle \neq H_y^v$ yields $h(v, \langle F \rangle) \cap P \subseteq \ell(v, L)$. □

LEMMA 2. *Let P be simplicial and $v \in \mathcal{V}(P)$. Then $\deg v \neq 5$.*

PROOF. We suppose that $[v, v_i, v_{i+1}] \in \mathcal{F}(P)$ for $i = 1, \dots, 5$ with $v_6 = v_1$, and seek a contradiction.

Let $\tilde{P} = [v, v_1, \dots, v_5]$. If $v_1, v_2, \dots, v_5$ are coplanar then $[v_1, \dots, v_5] \in \mathcal{F}(\tilde{P})$, $\mathcal{E}(\tilde{P}) \subset \mathcal{E}(P)$ and P is edge-antipodal; a contradiction by Remark 4.

Let, say, $[v_1, v_2, v_3, v_4] \in \mathcal{F}(\tilde{P})$. Then $H = \langle v_1, v_2, v_5 \rangle$ strictly separates v and $[v_3, v_4]$, and with $H \cap \langle v, v_j \rangle = \{u_j\}$ for $j \in \{3, 4\}$, $H \cap P$ is a pentagon with cyclically labelled vertices v_1, v_2, u_3, u_4, v_5. By Remark 2, $\ell(v_5, \langle v_1, v_2 \rangle)$ is a supporting line of $H \cap P$. Since v_1, v_2, v_3 and v_4 are coplanar, we obtain also from Remark 2 that $L' = \ell(v_3, \langle v_1, v_2 \rangle)$ is a supporting line of P. Then

$$\{[v, v_2, v_3], [v, v_3, v_4]\} \subset \mathcal{F}(P)$$

yields that $H' = \langle v, L' \rangle$ is a supporting plane of P, and $H \cap H'$ is a supporting line of $H \cap P$. Since $u_3 \in H \cap H'$ and the lines $H \cap H'$ and $\ell(v_5, \langle v_1, v_2 \rangle)$ are parallel, we obtain that $\{u_3, u_4, v_5\} \subset H'$ and v, v_3, v_4 and v_5 are coplanar; a contradiction.

Since $\tilde{P}$ is simplicial, there is an edge among the $[v_i, v_{i+1}]$'s such that neither $[v_{i-1}, v_i, v_{i+1}]$ nor $[v_i, v_{i+1}, v_{i+2}]$ is a face of $\tilde{P}$. Let, say,

$$[v_2, v_3, v_5] \in \mathcal{F}(\tilde{P}).$$

Then each of $\langle v_1, v_2, v_3 \rangle$ and $\langle v_2, v_3, v_4 \rangle$ strictly separates v and v_5, and we may assume that $H = \langle v_1, v_2, v_3 \rangle$ separates v and v_4. Hence, with $H \cap \langle v, v_j \rangle = \{u_j\}$ for $j \in \{4, 5\}$, the intersection $H \cap \tilde{P}$ is a pentagon with cyclically labelled vertices v_1, v_2, v_3, u_4, u_5. We apply now Remark 2 with $\langle v_1, v_2, v_3 \rangle$ and $\langle v_2, v_3, v_4 \rangle$, and obtain that $\ell(v_1, \langle v_2, v_3 \rangle)$ and $\ell(v_4, \langle v_2, v_3 \rangle)$ are supporting lines of $\tilde{P}$. This yields directly that $\ell(v_1, \langle v_2, v_3 \rangle)$ and $\ell(u_4, \langle v_2, v_3 \rangle)$ are parallel supporting lines of the pentagon $H \cap \tilde{P}$. Then v_1, u_4 and u_5 are collinear, and v, v_1, v_4 and v_5 are coplanar; a contradiction. □

LEMMA 3. *Let* $\{w, v_1, v_2, v_3, v_4, v_5 = v_1\} \subset \mathcal{V}(P)$ *such that* $[w, v_i, v_{i+1}] \in \mathcal{F}(P)$ *for* $i = 1, 2, 3, 4$. *Then* $P_w = [\mathcal{V}(P) \setminus \{w\}]$ *is edge-antipodal.*

PROOF. Since the assertion is immediate in the case that $\mathcal{E}(P_w) \subset \mathcal{E}(P)$, we may assume that the v_i's are not coplanar and that, say,

$$\mathcal{E}(P_w) \setminus \mathcal{E}(P) = \{[v_1, v_3]\}.$$

Let $H = \langle w, v_1, v_3 \rangle$, $U = \langle v_2, v_4 \rangle$, $Q = [w, v_1, v_3, p]$ be the parallelogram with sides $[w, v_1]$ and $[w, v_3]$, and H_w and H_1 be distinct parallel supporting planes of P such that $w \in H_w$ and $v_1 \in H_1$. We assume that $v_3 \notin H_w$ and observe that with $(v_2, v_4) = [v_2, v_4] \setminus \{v_2, v_4\}$:

(i) $H \cap U \in H \cap P \subseteq Q$ by Remark 2;
(ii) $H_w \cap Q = \{w\}$ and H_1 strictly separates v_3 and p;
(iii) $\langle w, v_1, u \rangle$ and $\langle w, v_3, u \rangle$ are supporting planes of P for each $u \in U \setminus (v_2, v_4)$;
(iv) $H \cap H_w$ and $H \cap H_1$ are supporting lines of the projection of P upon H along the direction of any line contained in H_w or H_1.

Let $H_w \cap U$ be the point $\bar{u}$, $\bar{U} = \langle w, \bar{u} \rangle$ and $\bar{P}$ be the projection of P upon H along $\bar{U}$.

Since $\bar{u} \in U \setminus (v_2, v_4)$, it follows from (iii) that $\langle w, v_1 \rangle$ and $\langle w, v_3 \rangle$ are supporting lines of $\bar{P}$. Since $\bar{U} \subset H_w$, it follows from (iv) that $H \cap H_1$ supports $\bar{P}$. But then $\langle v_1, p \rangle$ supports $\bar{P}$ by (ii), and consequently, $\langle w, v_3, \bar{u} \rangle$ and $\langle \ell(v_1, \bar{U}), p \rangle$ are parallel supporting planes of P, and hence of P_w.

In the case that $H_w \cap U = \varnothing$, letting figuratively $\bar{u} \in U$ tend to infinity yields that $\langle \ell(w, U), v_3 \rangle$ and $\langle \ell(v_1, U), p \rangle$ are parallel supporting planes of P, and hence of P_w. □

COROLLARY. *Let* P *be simplicial and* $w \in \mathcal{V}(P)$ *be such that* $\deg w \leq 4$. *Then* $P_w = [\mathcal{V}(P) \setminus \{w\}]$ *is edge-antipodal.*

We are now ready to proceed with the proof of the Theorem.

If P is not simplicial then by Remark 4, there is a parallelogram $F \in \mathcal{F}(P)$. By 1.2, there is a plane H, parallel to $\langle F \rangle$ and supporting P, that contains any vertex of $P \setminus F$ that is in an $F' \in \mathcal{F}(P)$ such that $F' \cap F \in \mathcal{E}(P)$. From this and Remark 2, it readily follows that H contains any vertex v of $P \setminus F$ such that $[v, x] \in \mathcal{E}(P)$ for some vertex x of F. Hence, $\mathcal{V}(P) \subset H \cup \langle F \rangle$ and $|\mathcal{V}(P)| \leq 8$ by Remark 4. We note that in this case, the degree of any vertex of P is at most four.

Let P be simplicial. If the degree of any vertex of P is at most four, we have

$$3 |\mathcal{F}(P)| = 2 |\mathcal{E}(P)| \leq 4 |\mathcal{V}(P)| ,$$

and it follows from Euler's Theorem that $|\mathcal{V}(P)| \leq 6$.

We suppose that there is a $w \in \mathcal{V}(P)$ such that $\deg w > 4$. Then $\deg w \geq 6$ by Lemma 2. From Remark 1, there is a $v_0 \in \mathcal{V}(P)$ such that $\deg v_0 \leq 4$. By the Corollary, $P_0 = [\mathcal{V}(P) \setminus \{v_0\}]$ is edge-antipodal. We note that $w \in \mathcal{V}(P_0)$ and $\deg w \geq 5$. Thus, P_0 is simplicial by the preceding, and $\deg w \geq 6$ by Lemma 2.

Since each iteration of the above yields a simplicial edge-antipodal subpolytope of P with w as a vertex, we have a contradiction.

Finally, we remark that if P is *strictly edge-antipodal* (meaning that whenever $[v, w] \in \mathcal{E}(P)$, there exist H_v^w and H_w^v such that $H_v^w \cap P = \{v\}$ and $H_w^v \cap P = \{w\}$), then $|\mathcal{V}(P)| \leq 5$. This follows from the Theorem (P is necessarily simplicial, $\mathcal{V}(P)$ is antipodal and $|\mathcal{V}(P)| \leq 6$) and the result of Grünbaum [1963] that there is no strictly antipodal set of six points in E^3.

References

[Csikós 2003] B. Csikós, "Edge-antipodal convex polytopes—a proof of Talata's conjecture", pp. 201–205 in *Discrete geometry*, edited by A. Bezdek, Pure Appl. Math. **253**, Dekker, New York, 2003.

[Danzer and Grünbaum 1962] L. Danzer and B. Grünbaum, "Über zwei Probleme bezüglich konvexer Körper von P. Erdős und von V. L. Klee", *Math. Z.* **79** (1962), 95–99.

[Erdős 1957] P. Erdős, "Some unsolved problems", *Michigan Math. J.* **4** (1957), 291–300.

[Fejes Tóth 1953] L. Fejes Tóth, *Lagerungen in der Ebene, auf der Kugel und im Raum*, Grundlehren der Mathematischen Wissenschaften **65**, Springer, Berlin, 1953.

[Grünbaum 1963] B. Grünbaum, "Strictly antipodal sets", *Israel J. Math.* **1** (1963), 5–10.

[Klee 1960] V. L. Klee, "Unsolved problems in intuitive geometry", Hectographical lecture notes, University of Washington, Seattle, 1960.

[Pór 2005] A. Pór, "On e-antipodal polytopes", 2005. Submitted to *Periodica Math. Hung.*

[Talata 1999] I. Talata, "On extensive subsets of convex bodies", *Period. Math. Hungar.* **38**:3 (1999), 231–246.

KÁROLY BEZDEK
DEPARTMENT OF MATHEMATICS AND STATISTICS
UNIVERSITY OF CALGARY
CALGARY, AB T2N 1N4
CANADA
  bezdek@math.ucalgary.ca

TIBOR BISZTRICZKY
DEPARTMENT OF MATHEMATICS AND STATISTICS
UNIVERSITY OF CALGARY
CALGARY, AB T2N 1N4
CANADA
  tbisztri@math.ucalgary.ca

KÁROLY BÖRÖCZKY
DEPARTMENT OF GEOMETRY
EŐTVŐS UNIVERSITY
BUDAPEST H-1117
HUNGARY
  boroczky@cs.elte.hu

Combinatorial and Computational Geometry
MSRI Publications
Volume **52**, 2005

A Conformal Energy for Simplicial Surfaces

ALEXANDER I. BOBENKO

ABSTRACT. A new functional for simplicial surfaces is suggested. It is invariant with respect to Möbius transformations and is a discrete analogue
of the Willmore functional. Minima of this functional are investigated. As
an application a bending energy for discrete thin-shells is derived.

1. Introduction

In the variational description of surfaces, several functionals are of primary
importance:

- The area $\mathcal{A} = \int dA$, where dA is the area element, is preserved by isometries.
- The total Gaussian curvature $\mathcal{G} = \int K \, dA$, where K is the Gaussian curvature,
  is a topological invariant.
- The total mean curvature $\mathcal{M} = \int H \, dA$, where H is the mean curvature,
  depends on the external geometry of the surface.
- The Willmore energy $\mathcal{W} = \int H^2 \, dA$ is invariant with respect to Möbius transformations.

Geometric discretizations of the first three functionals for simplicial surfaces are
well known. For the area functional the discretization is obvious. For the local
Gaussian curvature the discrete analog at a vertex v is defined as the angle defect

$$G(v) = 2\pi - \sum_i \alpha_i,$$

where the α_i are the angles of all triangles (see Figure 2) at vertex v. The total
Gaussian curvature is the sum over all vertices $G = \sum_v G(v)$. The local mean

Keywords: Conformal energy, Willmore functional, simplicial surfaces, discrete differential
geometry.

Partly supported by the DFG Research Center "Mathematics for key technologies" (FZT 86)
in Berlin.

135

curvature at an edge e is defined as

$$M(e) = l\theta,$$

where l is the length of the edge and θ is the angle between the normals to the adjacent faces at e (see Figure 6). The total mean curvature is the sum over all edges $M = \sum_e M(e)$. These discrete functionals possess the geometric symmetries of the smooth functionals mentioned above.

Until recently a geometric discretization of the Willmore functional was missing. In this paper we introduce a Möbius invariant energy for simplicial surfaces and show that it should be treated as a discrete Willmore energy.

2. Conformal Energy

Let S be a simplicial surface in 3-dimensional Euclidean space with set of vertices V, edges E and (triangular) faces F. We define a conformal energy for simplicial surfaces using the circumcircles of their faces. Each (internal) edge $e \in E$ is incident to two triangles. A consistent orientation of the triangles naturally induces an orientation of the corresponding circumcircles. Let $\beta(e)$ be the external intersection angle of the circumcircles of the triangles sharing e, which is the angle between the tangent vectors of the oriented circumcircles.

DEFINITION 1. The local conformal (discrete Willmore) energy at a vertex v is the sum

$$W(v) = \sum_{e \ni v} \beta(e) - 2\pi$$

over all edges incident on v. The conformal (discrete Willmore) energy of a simplicial surface S without boundary is the sum

$$W(S) = \frac{1}{2} \sum_{v \in V} W(v) = \sum_{e \in E} \beta(e) - \pi|V|,$$

over all vertices; here $|V|$ is the number of vertices of S.

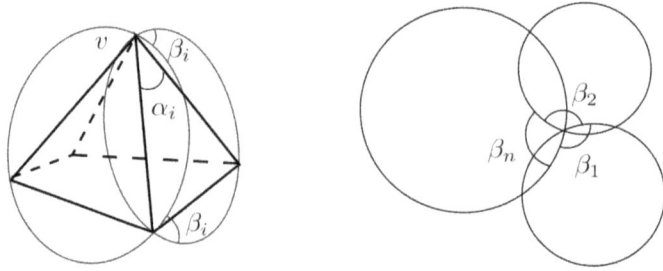

Figure 1. Definition of the conformal (discrete Willmore) energy.

Figure 1 presents two neighboring circles with their external intersection angle β_i as well as a view "from the top" at a vertex v showing all n circumcircles passing through v with the corresponding intersection angles $\beta_1, \ldots, \beta_n$. For simplicity we will consider only simplicial surfaces without boundary.

The energy $W(S)$ is obviously invariant with respect to Möbius transformations. This invariance is an important property of the classical Willmore energy defined for smooth surfaces (see below).

Also, $W(S)$ is well defined even for nonoriented simplicial surfaces, because changing the orientation of both circles preserves the angle $\beta(e)$.

The star $S(v)$ of the vertex v is the subcomplex of S comprised by the triangles incident with v. The vertices of $S(v)$ are v and all its neighbors. We call $S(v)$ *convex* if for any its face $f \in F(S(v))$ the star $S(v)$ lies to one side of the plane of F, and *strictly convex* if the intersection of $S(v)$ with the plane of f is f itself.

PROPOSITION 2. *The conformal energy is nonnegative*:

$$W(v) \geq 0.$$

It vanishes if and only if the star $S(v)$ is convex and all its vertices lie on a common sphere.

The proof is based on an elementary lemma:

LEMMA 3. *Let $\mathcal{P}$ be a (not necessarily planar) n-gon with external angles β_i. Choose a point P and connect it to all vertices of $\mathcal{P}$. Let α_i be the angles of the triangles at the tip P of the pyramid thus obtained (see Figure 2). Then*

$$\sum_{i=1}^{n} \beta_i \geq \sum_{i=1}^{n} \alpha_i,$$

and equality holds if and only if $\mathcal{P}$ is planar and convex and the vertex P lies inside $\mathcal{P}$.

The pyramid obtained is convex in this case; note that we distinguish between convex and strictly convex polygons (and pyramids). Some of the external angles β_i of a convex polygon may vanish. The corresponding side-triangles of the pyramid lie in one plane.

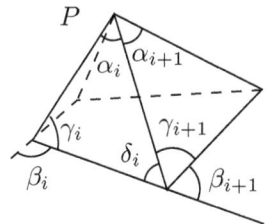

Figure 2. Toward the proof of Lemma 3.

PROOF. Denote by γ_i and δ_i the angles of the side-triangles at the vertices of $\mathcal{P}$ (see Figure 2). The claim of Lemma 3 follows from adding over all $i = 1, \ldots, n$ the two obvious relations

$$\beta_{i+1} \geq \pi - (\gamma_{i+1} + \delta_i), \qquad \pi - (\gamma_i + \delta_i) = \alpha_i.$$

All inequalities become equalities only in the case when $\mathcal{P}$ is planar, convex and contains P. $\qquad\square$

As a corollary we obtain a polygonal version of Fenchel's theorem [1929].

COROLLARY 4.

$$\sum_{i=1}^{n} \beta_i \geq 2\pi.$$

PROOF. For a given $\mathcal{P}$ choose the point P varying on a straight line encircled by $\mathcal{P}$. There always exist points P such that the star at P is not strictly convex, and thus $\sum \alpha_i \geq 2\pi$. $\qquad\square$

PROOF OF PROPOSITION 2. The claim of Proposition 2 is invariant with respect to Möbius transformations. Applying a Möbius transformation M that maps the vertex v to infinity, we make all circles passing through v into straight lines and arrive at the geometry shown in Figure 2, with $P = M(\infty)$. Now the result follows immediately from Corollary 4. $\qquad\square$

THEOREM 5. *Let S be a simplicial surface without boundary. Then*

$$W(S) \geq 0,$$

and equality holds if and only if S is a convex polyhedron inscribed in a sphere.

PROOF. Only the second statement needs to be proved. By, Proposition 2, the equality $W(S) = 0$ implies that all vertices and edges of S are convex (but not necessarily strictly convex). Deleting the edges that separate triangles lying in one plane one obtains a polyhedral surface S_P with circular faces and all strictly convex vertices and edges. Proposition 2 implies that for every vertex v there exists a sphere S_v with all vertices of the star $S(v)$ lying on it. For any edge (v_1, v_2) of S_P two neighboring spheres S_{v_1} and S_{v_2} share two different circles of their common faces. This implies $S_{v_1} = S_{v_2}$ and finally the coincidence of all the spheres S_v. $\qquad\square$

The discrete conformal energy W defined above is a discrete analogue of the Willmore energy [1993] for smooth surfaces, which is given by

$$W(S) = \frac{1}{4} \int_S (k_1 - k_2)^2 \, dA = \int_S H^2 \, dA - \int_S K \, dA.$$

Here dA is the area element, k_1, k_2 the principal curvatures, $H = \frac{1}{2}(k_1 + k_2)$ the mean curvature, $K = k_1 k_2$ the Gaussian curvature of the surface. Here we prefer a definition for $\mathcal{W}$ with a Möbius-invariant integrand. It differs from the one in the introduction by a topological invariant.

We mention two important properties of the Willmore energy:

- $\mathcal{W}(S) \geq 0$, and $\mathcal{W}(S) = 0$ if and only if S is the round sphere.
- $\mathcal{W}(S)$ (together with the integrand $(k_1 - k_2)^2 \, dA$) is Möbius-invariant [Blaschke 1929; Willmore 1993].

Whereas the first statement follows almost immediately from the definition, the second is a nontrivial property. We have shown that the same properties hold for the discrete energy W; in the discrete case Möbius invariance is built into the definition, and the nonnegativity of the energy is nontrivial.

In the same way one can define conformal (Willmore) energy for simplicial surfaces in Euclidean spaces of higher dimensions and space forms.

The discrete conformal energy is well defined for polyhedral surfaces with circular faces (not necessarily simplicial).

3. Computation of the Energy

Consider two triangles with a common edge. Let $a, b, c, d \in \mathbb{R}^3$ be their other edges, oriented as in Figure 3. Identifying vectors in $\mathbb{R}^3$ with imaginary quaternions $\operatorname{Im} \mathbb{H}$ one obtaines for the quaternionic product

$$ab = -\langle a, b \rangle + a \times b, \tag{3-1}$$

where $\langle a, b \rangle$ and $a \times b$ are the scalar and vector products in $\mathbb{R}^3$.

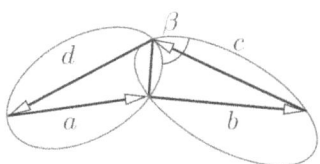

Figure 3. Formula for the angle between circumcircles.

PROPOSITION 6. *The external angle $\beta \in [0, \pi]$ between the circumcircles of the triangles in Figure 3 is given by one of the equivalent formulas:*

$$\cos(\beta) = -\frac{\operatorname{Re} q}{|q|} = -\frac{\operatorname{Re} abcd}{|abcd|} = \frac{\langle a, c \rangle \langle b, d \rangle - \langle a, b \rangle \langle c, d \rangle - \langle b, c \rangle \langle d, a \rangle}{|a|\,|b|\,|c|\,|d|},$$

where $q = ab^{-1}cd^{-1}$ is the cross-ratio of the quadrilateral.

PROOF. Since $\operatorname{Re} q$, $|q|$ and β are Möbius-invariant it is enough to prove the first formula for the planar case $a, b, c, d \in \mathbb{C}$, mapping all four vertices to a plane by a Möbius transformation. In this case q becomes the classical complex cross-ratio. Considering the arguments $a, b, c, d \in \mathbb{C}$ one easily arrives at $\beta = \pi - \arg q$.

The second representation follows from the identity $b^{-1} = -b/|b|$ for imaginary quaternions. Finally, applying (3–1) we obtain

$$\text{Re } abcd = \langle a, b \rangle \langle c, d \rangle - \langle a \times b, c \times d \rangle = \langle a, b \rangle \langle c, d \rangle + \langle b, c \rangle \langle d, a \rangle - \langle a, c \rangle \langle b, d \rangle. \quad \square$$

4. Minimizing Discrete Conformal Energy

Similarly to the smooth Willmore functional $\mathcal{W}$, minimizing the discrete conformal energy W makes the surface as round as possible.

Let $\boldsymbol{S}$ denote the combinatorial data of S. The simplicial surface S is called a geometric realization of the abstract simplicial surface $\boldsymbol{S}$.

DEFINITION 7. Critical points of $W(S)$ are called *simplicial Willmore surfaces*. The conformal (Willmore) energy of an abstract simplicial surface is the infimum over all geometric realizations

$$W(\boldsymbol{S}) = \inf_{S \in \boldsymbol{S}} W(S).$$

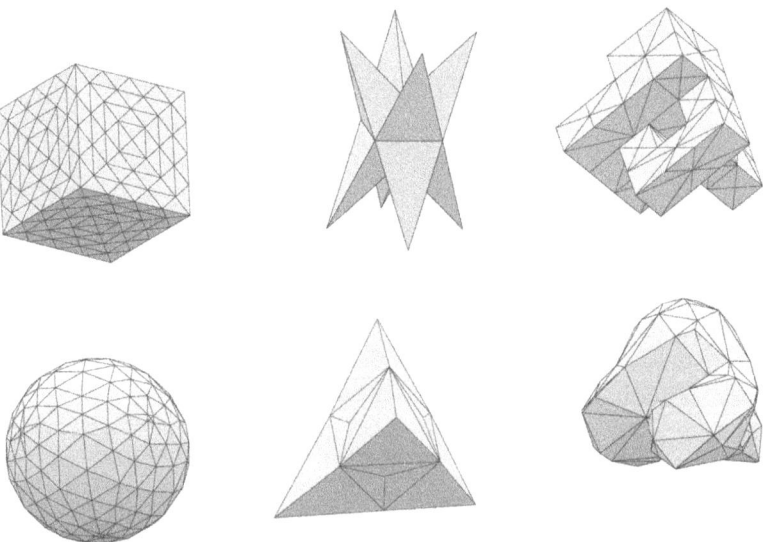

Figure 4. Discrete Willmore spheres of inscribable ($W = 0$) and noninscribable ($W > 0$) type, and discrete Boy surface.

Kevin Bauer implemented the proposed conformal functional with the Brakke's evolver [1992] and ran some numerical minimization experiments, whose results are exemplified in Figure 4. Corresponding entries in each row show initial configurations and the corresponding Willmore surfaces that minimize the conformal energy.

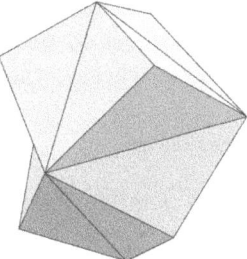

Figure 5. A discrete Willmore sphere of noninscribable type with 11 vertices and $W = 2\pi$.

Define the *discrete Willmore flow* as the gradient flow of the energy W. Under this flow the energy of the first simplicial sphere decreases to zero and the surface evolves into a convex polyhedron with all vertices lying on a sphere. The abstract simplicial surface of the central example is different and we obtain a simplicial Willmore sphere with positive conformal energy.

The rightmost example in the figure is a simplicial projective plane. The initial configuration is made from squares divided into triangles; see [Petit 1995]. We see that the minimum is close to the smooth Boy surface known (by [Karcher and Pinkall 1997]) to minimize the Willmore energy for projective planes.

The minimization of the conformal energy for simplicial spheres is related to a classical result of Steinitz [1928], who showed that there exist abstract simplicial 3-polytopes without geometric realizations all of whose vertices belong to a sphere. We call these combinatorial types *noninscribable*.

The noninscribable examples of Steinitz are constructed as follows [Grünbaum 2003]. Let $\boldsymbol{S}$ be an abstract simplicial sphere with vertices colored black and white. Denote the sets of white and black vertices by V_w and V_b respectively, so $V = V_w \cup V_b$. Assume that $|V_w| > |V_b|$ and that there are no edges connecting two white vertices. It is easy to see that $\boldsymbol{S}$ with these properties cannot be inscribed in a sphere. Indeed, assume that we have constructed such an inscribed convex polyhedron. Then the equality of the intersection angles at both ends of an edge (see left Figure 1) implies that

$$2\pi |V_b| \geq \sum_{e \in E} \beta(e) \geq 2\pi |V_w|.$$

This contradiction of the assumed inequality implies the claim.

To construct abstract polyhedra with $|V_w| > |V_b|$, take a polyhedron $\boldsymbol{P}$ whose number of vertices does not exceed the number of faces, $|\hat{F}| > |\hat{V}|$. Color all the vertices black, add white vertices at the faces and connect them to all black vertices of a face. This yields a polyhedron with black (original) edges and $|V_w| = |\hat{F}| > |V_b| = |\hat{V}|$. The example with minimal possible number of vertices $|V| = 11$ is shown in Figure 5. The starting polyhedron $\boldsymbol{P}$ here consists of two tetrahedra identified along a common face: $\hat{F} = 6$, $\hat{V} = 5$.

Hodgson, Rivin and Smith [Hodgson et al. 1992] have found a characterization of inscribable combinatorial types, based on a transfer to the Klein model of hyperbolic 3-space. It is not clear whether there exist noninscribable examples of non-Steinitz type.

Numerical experiments lead us to:

CONJECTURE 8. *The conformal energy of simplicial Willmore spheres is quantized*:

$$W = 2\pi N, \quad for\ N \in \mathbb{N}.$$

This statement belongs to differential geometry of discrete surfaces. It would be interesting to find a (combinatorial) meaning of the integer N. Compare also with the famous classification of smooth Willmore spheres by Bryant [1984], who showed that the energy of Willmore spheres is quantized by $\mathcal{W} = 4\pi N$, $N \in \mathbb{N}$.

The discrete Willmore energy is defined for ambient spaces ($\mathbb{R}^n$ or S^n) of any dimension. This leads to combinatorial Willmore energies

$$W_n(\boldsymbol{S}) = \inf_{S \in \boldsymbol{S}} W(S), \qquad S \subset S^n,$$

where the infimum is taken over all realizations in the n-dimensional sphere. Obviously these numbers build a nonincreasing sequence $W_n(\boldsymbol{S}) \geq W_{n+1}(\boldsymbol{S})$ that becomes constant for sufficiently large n.

Complete understanding of noninscribable simplicial spheres is an interesting mathematical problem. However the phenomenon of existence of such spheres might be seen as a problem in using of the conformal functional for applications in computer graphics, such as the fairing of surfaces. Fortunately the problem disappears after just one refinement step: all simplicial spheres become inscribable. Let $\boldsymbol{S}$ be an abstract simplicial sphere. Define its refinement $\boldsymbol{S}_R$ as follows: split every edge of $\boldsymbol{S}$ into two by putting additional vertices and connect these new vertices sharing a face of $\boldsymbol{S}$ by additional edges.

PROPOSITION 9. *The refined simplicial sphere $\boldsymbol{S}_R$ is inscribable, and thus $W(\boldsymbol{S}_R) = 0$.*

PROOF. Koebe's theorem (see [Ziegler 1995; Bobenko and Springborn 2004], for example) states that every abstract simplicial sphere $\boldsymbol{S}$ can be realized as a convex polyhedron S all of whose edges touch a common sphere S^2. Starting with this realization S it is easy to construct a geometric realization S_R of the refinement $\boldsymbol{S}_R$ inscribed in S^2. Indeed, choose the touching points of the edges of S with S^2 as additional vertices of S_R and project the original vertices of S (which lie outside of the sphere S^2) to S^2. One obtains a convex simplicial polyhedron S_R inscribed in S^2. $\square$

Another interesting variational problem involving the conformal energy is the optimization of triangulations of a given simplicial surface. Here one fixes the vertices and chooses an equivalent triangulation (abstract simplicial surface S) minimizing the conformal functional. The minimum

$$W(V) = \min_{S} W(S)$$

yields an "optimal" triangulation for a given vertex data. In the case of S^2 this optimal triangulation is well known.

PROPOSITION 10. *Let S be a simplicial surface with all vertices V on a two-dimensional sphere S^2. Then $W(S) = 0$ if and only if it is the Delaunay triangulation on the sphere, i.e., S is the boundary of the convex hull of V.*

In differential geometric applications such as the numerical minimization of the Willmore energy of smooth surfaces (see [Hsu et al. 1992]) it is not natural to preserve the triangulation by minimizing the energy, and one should also change the combinatorial type decreasing the energy.

The discrete conformal energy W is not just a discrete analogue of the Willmore energy. One can show that it approximates the smooth Willmore energy, although the smooth limit is very sensitive to the refinement method and must be chosen in a special way. A computation (to be published elsewhere) shows that if one chooses the vertices of a curvature line net of a smooth surface S for the vertices of S and triangularizes it, $W(S)$ converges to $\mathcal{W}(S)$ by natural refinement. On the other hand, the infinitesimal equilateral triangular lattice gives in the limit and energy half again higher. Possibly the minimization of the discrete Willmore energy with vertices on the smooth surface could be used for the computation of the curvature line net. We will be investigating this interesting and complicated phenomenon.

5. Bending of Simplicial Surfaces

An accurate model for the bending of discrete surfaces is important for modeling in virtual reality.

Let S_0 be a thin shell and S its deformation. The bending energy of smooth thin shells is given by the integral [Grinspun et al. 2003]

$$E = \int (H - H_0)^2 \, dA,$$

where H_0 and H are the mean curvatures of the original and deformed surface respectively. For $H_0 = 0$ it reduces to the Willmore energy.

To derive the bending energy for simplicial surfaces let us consider the limit of fine triangulation, i.e. of small angles between the normals of neighboring triangles. Consider an isometric deformation of two adjacent triangles. Let θ be the complement of the dihedral angle of the edge e, or, equivalently, the angle between the normals of these triangles (see Figure 6) and $\beta(\theta)$ the external intersection angle between the circumcircles of the triangles (see Figure 1) as a function of θ.

PROPOSITION 11. *Assume that the circumcenters of the circumcircles of two adjacent triangles do not coincide. In the limit of small angles $\theta \to 0$, the angle β between the circles behaves as*

$$\beta(\theta) = \beta(0) + \frac{l}{L}\theta^2 + o(\theta^3),$$

where l is the length of the edge and $L \neq 0$ is the distance between the centers of the circles.

This proposition and our definition of conformal energy for simplicial surfaces motivate to suggest

$$E = \sum_{e \in E} \frac{l}{L}\theta^2$$

for the bending energy of discrete thin-shells.

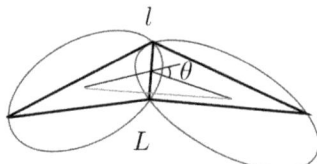

Figure 6. Toward the definition of the bending energy for simplicial surfaces.

In [Bridson et al. 2003; Grinspun et al. 2003] similar representations for the bending energy of simplicial surfaces were found empirically. They were demonstrated to give convincing simulations and good comparison with real processes. In [Grinspun et al. 2003] the distance between the barycenters is used for L in the energy expression but possible numerical advantages in using circumcenters are indicated.

Using the Willmore energy and Willmore flow is a hot topic in computer graphics. Applications include fairing of surfaces and surface restoration. We hope that our conformal energy will be useful for these applications and plan to work on them.

Acknowledgements

I thank Ulrich Pinkall for the discussion in which the idea of the discrete Willmore functional was born. I am also grateful to Günter Ziegler, Peter Schröder, Boris Springborn, Yuri Suris and Ekkerhard Tjaden for useful discussions and to Kevin Bauer for making numerical experiments with the conformal energy.

References

[Blaschke 1929] W. Blaschke, *Vorlesungen über Differentialgeometrie*, vol. III, Grundlehren der math. Wissenschaften **29**, Springer, Berlin, 1929.

[Bobenko and Springborn 2004] A. I. Bobenko and B. A. Springborn, "Variational principles for circle patterns and Koebe's theorem", *Trans. Amer. Math. Soc.* **356**:2 (2004), 659–689.

[Brakke 1992] K. A. Brakke, "The surface evolver", *Experiment. Math.* **1**:2 (1992), 141–165.

[Bridson et al. 2003] R. Bridson, S. Marino, and R. Fedkiw, "Simulation of clothing with folds and wrinkles", in *Eurographics/SIGGRAPH Symposium on Computer Animation* (San Diego, 2003), edited by D. Breen and M. Lin, 2003.

[Bryant 1984] R. L. Bryant, "A duality theorem for Willmore surfaces", *J. Differential Geom.* **20**:1 (1984), 23–53.

[Fenchel 1929] W. Fenchel, "Über Krümmung und Windung geschlossener Raumkurven", *Math. Ann.* **101** (1929), 238–252.

[Grinspun et al. 2003] E. Grinspun, A. N. Hirani, M. Desbrun, and P. Schröder, "Discrete shells", pp. 62–67 in *Eurographics/SIGGRAPH Symposium on Computer Animation* (San Diego, 2003), edited by D. Breen and M. Lin, 2003.

[Grünbaum 2003] B. Grünbaum, *Convex polytopes*, Graduate Texts in Mathematics **221**, Springer, New York, 2003.

[Hodgson et al. 1992] C. D. Hodgson, I. Rivin, and W. D. Smith, "A characterization of convex hyperbolic polyhedra and of convex polyhedra inscribed in the sphere", *Bull. Amer. Math. Soc. (N.S.)* **27**:2 (1992), 246–251.

[Hsu et al. 1992] L. Hsu, R. Kusner, and J. Sullivan, "Minimizing the squared mean curvature integral for surfaces in space forms", *Experiment. Math.* **1**:3 (1992), 191–207.

[Karcher and Pinkall 1997] H. Karcher and U. Pinkall, "Die Boysche Fläche in Oberwolfach", *Mitt. Dtsch. Math.-Ver.* no. 1 (1997), 45–47.

[Petit 1995] J.-P. Petit, *Das Topologikon*, Vieweg, Braunschweig, 1995.

[Steinitz 1928] E. Steinitz, "Über isoperimetrische Probleme bei konvexen Polyedern", *J. reine angew. math.* **159** (1928), 133–143.

[Willmore 1993] T. J. Willmore, *Riemannian geometry*, Oxford Univ. Press, New York, 1993.

[Ziegler 1995] G. M. Ziegler, *Lectures on polytopes*, Graduate Texts in Mathematics **152**, Springer, New York, 1995.

ALEXANDER I. BOBENKO
INSTITUT FÜR MATHEMATIK
TECHNISCHE UNIVERSITÄT BERLIN
STRASSE DES 17. JUNI 136
10623 BERLIN
GERMANY
bobenko@math.tu-berlin.de

Combinatorial and Computational Geometry
MSRI Publications
Volume **52**, 2005

On the Size of Higher-Dimensional Triangulations

PETER BRASS

ABSTRACT. I show that there are sets of n points in three dimensions, in general position, such that any triangulation of these points has only $O(n^{5/3})$ simplices. This is the first nontrivial upper bound on the MinMax triangulation problem posed by Edelsbrunner, Preparata and West in 1990: What is the minimum over all general-position point sets of the maximum size of any triangulation of that set? Similar bounds in higher dimensions are also given.

1. Introduction

In the plane, all triangulations of a set of points use the same number of triangles. This is a simple consequence of each triangle having an interior angle sum of π, and each interior point of the convex hull contributing an angle sum of 2π, which must be used up by the triangles.

Neither the constant size of triangulations nor the constant angle sum of simplices holds in higher dimensions. A classic example is the cube, which can be decomposed in two ways: into five simplices (cutting off alternate vertices) or into six simplices (which are even congruent; it is a well-known simple geometric puzzle to assemble six congruent simplices, copies of $\operatorname{conv}\big((000), (100), (010), (011)\big)$, into a cube).

For higher-dimensional cubes, the same problem was studied in a number of papers [Böhm 1989; Broadie and Cottle 1984; Haiman 1991; Hughes 1993; Hughes 1994; Lee 1985; Marshall 1998; Orden and Santos 2003; Sallee 1984; Smith 2000]. This suggest that one should be interested in the possible values of the numbers of simplices for arbitrary point sets.

It is well known that a triangulation of n points in d-dimensional space has size $\Omega(n)$ and $O(n^{\lceil d/2 \rceil})$. The lower bound is obvious (each point must go somewhere); and, at least in three-dimensional space, as upper bound one can use that

147

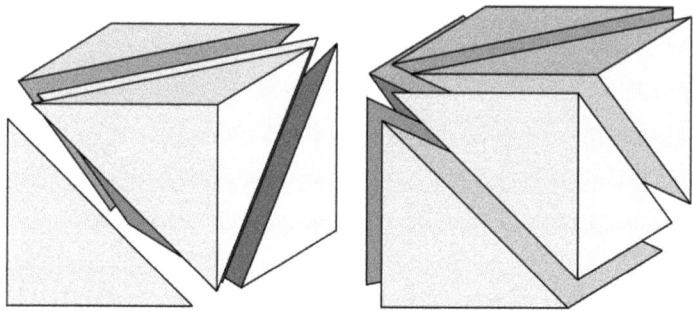

Figure 1. A cube can be triangulated with five or six simplices.

from each point the outer facets of the incident simplices can be viewed as faces
of a starshaped polytope with at most $n-1$ vertices, which is combinatorially
isomorphic to a convex polytope.

In more detail this problem was solved by Rothschild and Straus [1985], who
showed that the minimum number of simplices in any triangulation of any full-
dimensional set of n points in d-dimensional space is $n-d$. This is reached by
gluing simplices together along faces, such that each additional simplex generates
a new vertex, and all vertices are in convex position. Another method, without
the general position, would be to place $n-d+1$ points on a line, and $d-1$ points
off that line. They also showed that the maximum number of simplices in any
triangulation of any full-dimensional set of n points in d-dimensional space is
cyc_poly$(n+1, d, d+1)-(d+1) = \Theta(n^{\lceil d/2 \rceil})$, where cyc_poly$(n+1, d, d+1)$ is the
number of d-faces of the $d+1$-dimensional cyclic polytope on $n+1$ vertices. This
is a consequence of the upper bound theorem for simplicial d-spheres [Stanley
1983].

These were the maximum and minimum triangulation size, taken over all
sets of n points in d-dimensional space. As a next step, it would be interesting
to give bounds on the maximum and minimum triangulation size of a fixed
set [Rothschild and Straus 1985, Problem 6.2]. For that we have to make some
general-position assumption, no $d+1$ points collinear, otherwise there are always
point sets for which there is only a unique triangulation. The questions are:

MAXMIN PROBLEM. *What is the smallest number $f_d^{\mathrm{MaxMin}}(n)$, such that each
set of n points in d-dimensional space, no $d+1$ collinear, has a triangulation
with at most $f_d^{\mathrm{MaxMin}}(n)$ simplices?*

MINMAX PROBLEM. *What is the largest number $f_d^{\mathrm{MinMax}}(n)$, such that each set
of n points in d-dimensional space, no $d+1$ collinear, has a triangulation with
at least $f_d^{\mathrm{MinMax}}(n)$ simplices?*

This problem was considered in three-dimensional space by Edelsbrunner, Preparata and West [Edelsbrunner et al. 1990], who showed that $f_3^{\mathrm{MaxMin}}(n) \leq 3n-11$,
so every set of n point in general position in three-dimensional space has a

small triangulation. They also gave some bounds, if additionally the number of points of the convex hull is given. Together with the lower bound of Sleator, Tarjan and Thurston [Sleator et al. 1988], who constructed a convex polyhedron which requires $2n - 10$ simplices in any triangulation, this determines the exact minimum for n point sets in convex position, and leaves only a linear-sized gap in general.

For higher dimensions, the vertices of a cyclic polytope give a lower bound for $f_d^{\mathrm{MaxMin}}(n)$, since in any triangulation of the cyclic polytope, each facet must be facet of some simplex, and each simplex has only $d+1$ facets. Together with the above-mentioned general upper bound of [Rothschild and Straus 1985] on any triangulation this shows

$$\Omega\big(\mathrm{cyc\_poly}(n, d-1, d)\big) \le f_d^{\mathrm{MaxMin}}(n) \le O\big(\mathrm{cyc\_poly}(n+1, d, d+1)\big),$$

so

$$\Omega\big(n^{\lfloor d/2 \rfloor}\big) \le f_d^{\mathrm{MaxMin}}(n) \le O\big(n^{\lceil d/2 \rceil}\big).$$

For the MinMax-Problem, the situation is much worse, only constant-factor improvements for the trivial lower and upper bounds are known [Edelsbrunner et al. 1990; Urrutia 2003], so $\Omega(n) \le f_3^{\mathrm{MinMax}}(n) \le O(n^2)$; and although some other problems raised in [Edelsbrunner et al. 1990] were solved [Bern 1993], no progress on the growth rate of $f_3^{\mathrm{MinMax}}(n)$ was made since then. It is the aim of this paper to prove the first nontrivial upper bound.

THEOREM 1. $f_3^{\mathrm{MinMax}}(n) = O(n^{5/3})$.

This follows from

LEMMA 2. *Any triangulation of a point set in three-dimensional space that arises by a small perturbation from the $n^{1/3} \times n^{1/3} \times n^{1/3}$ lattice cube contains at most $O(n^{5/3})$ simplices.*

This upper bound is probably not sharp even in that class of perturbed lattice cubes. It is easy to construct a perturbed lattice cube that allows a triangulation of size $\Omega(n^{4/3})$, and that is probably the true maximum in that class.

The same argument works also in higher dimensions, unfortunately the improvement over the general upper bound of $O(n^{\lceil d/2 \rceil})$ on the number of simplices in any d-dimensional triangulation is very small, especially if compared with the only known (trivial) lower bound $f_d^{\mathrm{MinMax}}(n) = \Omega(n)$.

THEOREM 3. $f_d^{\mathrm{MinMax}}(n) = O\big(n^{(1/d)+(d-1)\lceil d/2 \rceil/d}\big)$ *for fixed dimension d.*

The improvement in the exponent is thus

$$\frac{1}{d}\left(\left\lceil \frac{d}{2} \right\rceil - 1\right) \approx \frac{1}{2}.$$

2. The Proof

Let X_n be a set of n points, which is obtained from the lattice cube

$$X_n^* = \left\{(x_1, x_2, x_3) \mid x_i \in \{1, \ldots, n^{1/3}\}\right\}$$

by a small perturbation. Any point $p \in X$ has a unique preimage $p^* \in X^*$ before the perturbation was applied, and any simplex $\{p_1, p_2, p_3, p_4\} \subset X$ has a preimage $\{p_1^*, p_2^*, p_3^*, p_4^*\} \subset X^*$, which is a possibly degenerate simplex (points coplanar or even collinear). Let $\mathcal{T}$ be the triangulation of X, then we partition $\mathcal{T} = \mathcal{T}_3 \cup \mathcal{T}_{\leq 2}$ by classifying the simplices $T \in \mathcal{T}$ according to the affine dimension of their preimage T^*; a simplex $T \in \mathcal{T}_3$ has a nondegenerate simplex T^* as preimage, a simplex $T \in \mathcal{T}_{\leq 2}$ has a coplanar, or even collinear, fourtuple T^* (degenerate simplex) as preimage.

We have less than $6n$ simplices in $\mathcal{T}_3$, since any nondegenerate simplex in X^* is a nondegenerate simplex with integer coordinates, so it has volume at least $\frac{1}{6}$; and the volume of $\text{conv}(X^*)$ is less than n.

The preimages T^* of simplices $T \in \mathcal{T}_3$ together partition the cube $\text{conv}(X^*)$ into nondegenerate simplices, and the vertices of these simplices are points of X^* so we can refine this partition to a triangulation $\mathcal{S}^*$ of X^*. Each face of a simplex T^*, $T \in \mathcal{T}_3$ of the partition is a union of faces of simplices from the triangulation $\mathcal{S}^*$. The triangulation $\mathcal{S}^*$ still contains at most $6n$ simplices.

The main problem is to bound $|\mathcal{T}_{\leq 2}|$, the number of almost-degenerate simplices in $\mathcal{T}$. Consider a simplex $T \in \mathcal{T}_{\leq 2}$, its preimage T^* is some coplanar fourtuple of points in X^*. Now T^* cannot intersect the interior of the preimage S^* of any of the full-dimensional simplices $S \in \mathcal{T}_3$. So each $T \in \mathcal{T}_{\leq 2}$ has a preimage T^* that is contained in the union of the faces of the S^*, $S \in \mathcal{T}_3$, so also in the union of faces of the S^*, $S^* \in \mathcal{S}^*$. Therefore each $T \in \mathcal{T}_{\leq 2}$ has a preimage T^* that is contained in a lattice plane of X^* spanned by a face of some S^* of the triangulation $\mathcal{S}^*$. Let $\{E_i\}_{i \in I}$ be the set of planes spanned by faces of simplices of the triangulation $S^* \in \mathcal{S}^*$, and let a_i be the number of simplices $S^* \in \mathcal{S}^*$ which have a face contained in the plane E_i. Since each of the $S^* \in \mathcal{S}^*$ contributes four faces, we have

$$\sum_{i \in I} a_i < 24n.$$

Since T^* is contained in the union of faces of simplices $S^* \in \mathcal{S}^*$, this holds also for the vertices of T^*; so they are either vertices of faces of the triangulation $\mathcal{S}^*$, or contained in the sides or relative interior of faces, which is not possible in a triangulation $\mathcal{S}^*$ of X^*. So each vertex of T^* is a vertex of some simplex S^*, and therefore the numbers b_i of points from $X^* \cap E_i$ that are vertices of T^* contained in E_i satisfies $\sum_{i \in I} b_i < 72n$. But also each b_i is at most $|X^* \cap E_i|$, so we have $b_i \leq n^{2/3}$ for each i. But these b_i points contained in E_i can generate only less than $O(b_i^2)$ simplices, since any set of b_i points can span at most $O(b_i^2)$

nonoverlapping simplices. So the total number of simplices in $\mathcal{T}_{\leq 2}$ is less than $\sum_{i \in I} C b_i^2$ for some C. Thus

$$|\mathcal{T}_{\leq 2}| \leq \max \left\{ \sum_{i \in I} C b_i^2 \,\bigg|\, \sum_{i \in I} b_i < 72n, \, 0 < b_i \leq n^{2/3} \right\} = O(n^{5/3}).$$

The d-dimensional version is proved in exactly the same way: the point set $X_{n,d}$ is any perturbation of the $n^{1/d} \times \cdots \times n^{1/d}$-lattice cube. Any triangulation $\mathcal{T}_{n,d}$ of such a set will contain at most $O(n)$ simplices with a full-dimensional preimage in the unperturbed lattice $X_{n,d}^*$, since any nondegenerate simplex with integer vertices has a volume at least $\frac{1}{d!}$. All the remaining simplices of the triangulation are near-degenerate, they have preimages which are contained in the union of faces of the full-dimensional simplices. The full-dimensional preimages of simplices partition the cube into nondegenerate simplices with vertices from $X_{n,d}^*$, and we can refine this to a triangulation $\mathcal{S}_{n,d}^*$ of $X_{n,d}^*$ with $O(n)$ simplices. The faces of this triangulation span a set of affine lattice subspaces. Each near-degenerate simplex has a preimage in one of these subspaces, and each vertex of that near-degenerate simplex has a preimage that is in $\mathcal{S}_{n,d}^*$ vertex of a simplex with a face that spans that affine subspace. The total number of pairs of vertices and incident faces in $\mathcal{S}_{n,d}^*$ is $O(n)$ and each of these pairs belongs to an affine lattice subspace, and can belong to the preimages of near-degenerate simplices only in that subspace. We sum now over all such subspaces, and count each point only for those subspaces where it is vertex with an incident face that spans the subspace. A subspace s that contains b_s points can contain only $O(b_s^{\lceil d/2 \rceil})$ preimages of near-degenerate simplices, since that is the maximum number of simplices that these b_s points can span. And each subspace contains at most $n^{(d-1)/d}$ points, since that is the maximum intersection of a proper affine subspace with the lattice cube. We now consider this just as an abstract optimization problem for the variables b_s, and get an upper bound of

$$\max \left\{ \sum_s O(b_s^{\lceil d/2 \rceil}) \,\bigg|\, \sum_s b_s = O(n), \, 0 < b_s \leq n^{(d-1)/d} \right\}.$$

This maximum is again reached if each nonvanishing b_s is as large as possible, so $b_s = n^{(d-1)/d}$ for $O(n^{1/d})$ variables b_s, which is the claimed bound.

3. Related Problems

The most important problem would be to get a nontrivial lower bound for $f_d^{\text{MinMax}}(n)$. It is still possible that there are point sets which allow only linear-sized triangulations. Perhaps it might help to compute some exact values and extremal configurations for small n; the first nontrivial values seem to be

$$f_3^{\text{MinMax}}(5) = 3 \quad \text{and} \quad f_3^{\text{MinMax}}(6) = 5,$$

both realized by points in convex position.

A good lower bound on $f_d^{\mathrm{MinMax}}(n)$ would also be interesting since it would imply an upper bound for the d-dimensional Heilbronn triangle problem. Let $g_d^{\mathrm{MinVol}}(n)$ be the maximum over all choices of n points from the unit cube of the minimum volume of a simplex spanned by this set, then

$$g_d^{\mathrm{MinVol}}(n) \leq \frac{1}{f_d^{\mathrm{MinMax}}(n)}.$$

For $d \geq 3$, the best upper bound we have on $g_d^{\mathrm{MinVol}}(n)$ is only slightly better than the trivial bound [Brass 2005]; for lower bounds see [Barequet 2001; Lefmann 2000].

It should be possible to determine the exact function for $f_3^{\mathrm{MaxMin}}(n)$, or at least the right multiplicative constant.

The problem of triangulating the d-cube with minimal number of simplices was already mentioned in the beginning. It does not quite fall in the model here, since the vertices of the cube are not in general position. The *maximum* number of simplices in any triangulation of the d-cube are $d!$, by the volume argument used above, and this number can be reached easily. The *minimum* number of simplices is known to be between

$$\frac{1}{2\sqrt{d+1}} \left(\frac{6}{d+1} \right)^{d/2} d! \qquad \text{and} \qquad (0.816)^d d!$$

(see [Smith 2000] and [Orden and Santos 2003], respectively); so the gap between upper an lower bound is still enormous, of order $2^{\Theta(d \log d)}$.

References

[Barequet 2001] G. Barequet, "A lower bound for Heilbronn's triangle problem in d dimensions", *SIAM J. Discrete Math.* **14**:2 (2001), 230–236.

[Bern 1993] M. Bern, "Compatible tetrahedralizations", pp. 281–288 in *Proceedings of the Ninth Annual Symposium on Computational Geometry* (San Diego, 1993), ACM Press, New York, 1993.

[Böhm 1989] J. Böhm, "Some remarks on triangulating a d-cube", *Beiträge Algebra Geom.* no. 29 (1989), 195–218.

[Brass 2005] P. Brass, "An upper bound for the d-dimensional Heilbronn triangle problem", *SIAM J. Discrete Math* (2005). To appear.

[Broadie and Cottle 1984] M. N. Broadie and R. W. Cottle, "A note on triangulating the 5-cube", *Discrete Math.* **52**:1 (1984), 39–49.

[Edelsbrunner et al. 1990] H. Edelsbrunner, F. P. Preparata, and D. B. West, "Tetrahedrizing point sets in three dimensions", *J. Symbolic Comput.* **10**:3-4 (1990), 335–347.

[Haiman 1991] M. Haiman, "A simple and relatively efficient triangulation of the n-cube", *Discrete Comput. Geom.* **6**:4 (1991), 287–289.

[Hughes 1993] R. B. Hughes, "Minimum-cardinality triangulations of the d-cube for $d = 5$ and $d = 6$", *Discrete Math.* **118**:1-3 (1993), 75–118.

[Hughes 1994] R. B. Hughes, "Lower bounds on cube simplexity", *Discrete Math.* **133**:1-3 (1994), 123–138.

[Lee 1985] C. W. Lee, "Triangulating the d-cube", pp. 205–211 in *Discrete geometry and convexity* (New York, 1982), edited by J. E. Goodman et al., Ann. New York Acad. Sci. **440**, New York Acad. Sci., New York, 1985.

[Lefmann 2000] H. Lefmann, "On Heilbronn's problem in higher dimension", pp. 60–64 in *Proceedings of the Eleventh Annual ACM-SIAM Symposium on Discrete Algorithms* (San Francisco, 2000), ACM, New York, 2000.

[Marshall 1998] T. H. Marshall, "Volume formulae for regular hyperbolic cubes", *Conform. Geom. Dyn.* **2** (1998), 25–28.

[Orden and Santos 2003] D. Orden and F. Santos, "Asymptotically efficient triangulations of the d-cube", *Discrete Comput. Geom.* **30**:4 (2003), 509–528.

[Rothschild and Straus 1985] B. L. Rothschild and E. G. Straus, "On triangulations of the convex hull of n points", *Combinatorica* **5**:2 (1985), 167–179.

[Sallee 1984] J. F. Sallee, "The middle-cut triangulations of the n-cube", *SIAM J. Algebraic Discrete Methods* **5**:3 (1984), 407–419.

[Sleator et al. 1988] D. D. Sleator, R. E. Tarjan, and W. P. Thurston, "Rotation distance, triangulations, and hyperbolic geometry", *J. Amer. Math. Soc.* **1**:3 (1988), 647–681.

[Smith 2000] W. D. Smith, "A lower bound for the simplexity of the n-cube via hyperbolic volumes", *European J. Combin.* **21**:1 (2000), 131–137.

[Stanley 1983] R. P. Stanley, *Combinatorics and commutative algebra*, Progress in Mathematics **41**, Birkhäuser, Boston, 1983.

[Urrutia 2003] J. Urrutia, "Coloraciones, tetraedralizaciones, y tetraedros vacios en coloraciones de conjuntos de puntos en $\mathbb{R}^3$", pp. 95–100 in *Proc. X Encuentros de Geometria Computacional* (Sevilla, 2003), edited by C. Grima et al., Universidad de Sevilla, 2003.

PETER BRASS
CITY COLLEGE OF NEW YORK, CUNY
DEPARTMENT OF COMPUTER SCIENCE
138TH STREET AT CONVENT AVE.
NEW YORK, NY 10031
peter@cs.ccny.cuny.edu

Combinatorial and Computational Geometry
MSRI Publications
Volume **52**, 2005

The Carpenter's Ruler Folding Problem

GRUIA CĂLINESCU AND ADRIAN DUMITRESCU

ABSTRACT. A carpenter's ruler is a ruler divided into pieces of different
lengths which are hinged where the pieces meet, which makes it possible to fold the ruler. The carpenter's ruler folding problem, originally
posed by Hopcroft, Joseph and Whitesides, is to determine the smallest
case (or interval on the line) into which the ruler fits when folded. The
problem is known to be NP-complete. The best previous approximation
ratio achieved, dating from 1985, is 2. We improve this result and provide a fully polynomial-time approximation scheme for this problem. In
contrast, in the plane, there exists a simple linear-time algorithm which
computes an exact (optimal) folding of the ruler in some convex case of
minimum diameter. This brings up the interesting problem of finding the
minimum area of a convex *universal case* (of unit diameter) for all rulers
whose maximum link length is one.

1. Introduction

The carpenter's ruler folding problem is: Given a sequence of rigid rods (links)
of various integral lengths connected end-to-end by hinges, to fold it so that its
overall folded length is minimum. It was first posed in [Hopcroft et al. 1985],
where the authors proved that the problem is NP-complete using a reduction
from the NP-complete problem PARTITION (see [Garey and Johnson 1979;
Cormen et al. 1990]). A simple linear-time factor 2 approximation algorithm,
as well as a pseudo-polynomial $O(L^2 n)$ time dynamic programming algorithm,
where L is the maximum link length, where presented in [Hopcroft et al. 1985]
(see also [Kozen 1992]). A physical ruler is idealized in the problem, so that the
ruler is allowed to fold onto itself and lie along a line segment whose length is
the size of the case, and thus no thickness results from the segments which lie
on top of each other.

The decision problem can be stated as follows. Given a ruler whose links have
lengths $l_1, l_2, \ldots, l_n$, can it be folded so that its overall folded length is at most
k? Note that different orderings of the links can result in different minimum case

---

Keywords: approximation scheme, carpenter's ruler, folding problems, universal case.

lengths. For example if the ruler has links of lengths 6, 6 and 3 in this order, the ruler can be folded into a case of length 6, but if the links occur in the order 6, 3 and 6, the optimal case-length is 9.

Our first result (Section 2) improves the 19-year old factor 2 approximation:

THEOREM 1. *There exists a fully polynomial-time approximation scheme for the carpenter's ruler folding problem.*

A fully polynomial-time approximation scheme (FPTAS) for a minimization problem is a family of algorithms A_ε, for all $\varepsilon > 0$, such that A_ε has running time polynomial in the size of the instance and $1/\varepsilon$, and the output of A_ε is at most $(1 + \varepsilon)$ times the optimum [Garey and Johnson 1979].

In Section 3, we study a natural, related question: the condition that the folding must lie on a line is relaxed, by considering foldings in the plane with the objective of minimizing the diameter of a convex case containing the folded ruler. Here foldings allow for a free reconfiguration of the joint angles, with the proviso that each link of the ruler maintains its length (the shape of the case is unconstrained). In contrast with the problem on the line, this variant admits an easy exact (optimal) solution which can be computed in linear time, using exact arithmetic.

This brings up the interesting problem of finding the minimum area of a convex case (of unit diameter) for all rulers whose maximum link length is one. A closed curve of unit diameter in the plane is said to be a *universal case* for all rulers whose maximum link length is one if each such ruler admits a planar folding inside the curve. Our results are summarized in:

THEOREM 2. *There exists an $O(n)$ algorithm for the carpenter's ruler folding problem in the plane with lengths $l_1, l_2, \ldots, l_n$, which computes a folding in a convex case of minimum diameter $L = \max(l_1, \ldots, l_n)$. The minimum area A of a convex universal case (of unit diameter) for all rulers whose maximum link length is one satisfies*

$$\frac{3}{8} \le A \le \frac{\pi}{3} - \frac{\sqrt{3}}{4}.$$

The lower bound is $\frac{3}{8} = 0.375$ and the upper bound is ≈ 0.614. We believe the latter is closer to the truth.

Other folding problems with links allowed to cross have been studied, for example in [Hopcroft et al. 1984; Kantabutra 1992; Kantabutra 1997; Kantabutra and Kosaraju 1986; van Kreveld et al. 1996], while linkage folding problems for noncrossing links have been investigated for example in [Connelly et al. 2003; Streinu 2000]. For other *universal cover* problems, such as the *worm problem*, see [Croft et al. 1991; Klee and Wagon 1991] and the references therein.

2. Proof of Theorem 1

We present two approximation schemes: one based on trimming the solution space and one based on rounding and scaling. We start with notation and observations which apply to both algorithms.

A folding F of the ruler can be specified by the position on the line of the first (free) endpoint of the ruler (i.e., the free endpoint of the first link) and a binary string of length n in which the i-th bit is -1 or 1 depending on whether the i-th segment is folded to the left or right of its fixed endpoint (view this as a sequential process). We call this binary string the *folding vector*.

For a given folding F, let the interval $I_F = [a_F, b_F]$ be the smallest closed interval which contains it (i.e., it contains all the segments of the ruler). We refer to it as the *folding interval*. See also Figure 1.

Figure 1. A carpenter's ruler with segments of length 1, 3, 2 and 4 folded so that it fits into a case of length 5 (left). Its folding vector is $(-1, 1, 1 - 1)$. Another folding into a case of same length (right). Its folding vector is $(1, -1, 1 - 1)$.

Denote by OPT the minimum folded length for a ruler whose lengths are $l_1, l_2, \ldots, l_n$. A trivial lower bound — on which the 2-approximation algorithm is based — is OPT $\geq L$, where $L = \max(l_1, l_2, \ldots, l_n)$ is the maximum rod length. We further exploit this observation and the 2-approximation algorithm given in [Hopcroft et al. 1985].

OBSERVATION 1. *An optimal solution can be computed by fixing the first segment at $[0, l_1]$ (with the free endpoint of the first link at 0), and then computing all foldings that extend it, whose intervals have length at most $2L$ (thus are included in the interval $[-2L + l_1, 2L]$).*

PROOF. Consider an optimal solution. Clearly the first segment can be fixed at any given position of its free endpoint and at any of the two possible orientations. Since there exist approximate solutions whose folding intervals have length at most $2L$, foldings with larger intervals do not need to be considered (are not optimal). □

One can also see that the observation can be somewhat strengthened, since in fact, any of the links can be fixed at a given position and orientation.

OBSERVATION 2. *An optimal solution can be computed by first fixing one segment of length L (if more exist, select one arbitrarily) at $[0, L]$ and then computing all foldings that extend it, whose intervals have length at most $2L$ (thus are included in the interval $[-L, 2L]$).*

Consider a folding F, whose vector is $(\varepsilon_1, \ldots, \varepsilon_n)$, for a given ruler $l_1, l_2, \ldots, l_n$. For $i = 1, \ldots, n$, let the *partial folding* F_i of the ruler $l_1, l_2, \ldots, l_i$ be that whose folding vector is $(\varepsilon_1, \ldots, \varepsilon_i)$.

For a folding F whose interval is $[a, b]$, clearly the endpoint x of the last segment also lies in the same interval, i.e., $x \in [a, b]$. We say that F has *parameters* a, b and x, or that F is given by a, b and x.

A FPTAS based on trimming the solution space. We now describe the first algorithm which we note, has some similarity features with the fully polynomial-time approximation scheme for the subset-sum problem [Ibarra and Kim 1975] (see also [Cormen et al. 1990] for a more accessible presentation). Let ε be the approximation parameter, where $0 < \varepsilon < 1$. For simplicity assume that $m = 8n/\varepsilon$ is an integer. Set $\delta = L\varepsilon/(2n)$. Consider the partition of the interval $[-2L, 2L]$ into m *elementary intervals* of length δ, given by $[-2L + j\delta, -2L + (j+1)\delta)$, for $j = 0, \ldots, m-1$, except that the last interval in this sequence, for $j = m-1$, is closed at both ends. For simplicity of exposition, we consider the interval $[-2L, 2L]$ instead of the interval $[-2L+l_1, 2L]$ mentioned in Observation 1 (and then the expression of m above is an overestimate). An *interval triplet* denoted (I_a, I_b, I_x), is any of the m^3 ordered triples of elementary intervals.

The algorithm iteratively computes a set of partial foldings $\mathcal{F}_i$ of the ruler $l_1, l_2, \ldots, l_i$, for $i = 1, \ldots, n$, so that at most one partial folding per interval triplet is maintained at the end of the i-th iteration. A partial folding whose folding interval is $[a, b]$, and the endpoint of the last segment at x is associated with the interval triplet (I_a, I_b, I_x), where $a \in I_a$, $b \in I_b$ and $x \in I_x$. If at step i more partial foldings per interval triplet are computed, all but one of them are discarded; the one selected for the next step is chosen arbitrarily from those computed.

$\mathcal{F}_1$ consists of one (partial) folding, given by $a_1' = 0$, $b_1' = l_1$, $x_1' = l_1$. Let $i \geq 2$. In the i-th iteration, the algorithm computes from the set $\mathcal{F}_{i-1}$ of partial foldings of the first $i-1$ links, all the partial foldings of the first i links that extend foldings in $\mathcal{F}_{i-1}$, and whose intervals are included in the interval $[-2L, 2L]$ (there are at most $2|\mathcal{F}_{i-1}|$ of these). It then "trims" this set to obtain $\mathcal{F}_i$, so that if an interval triplet has more partial foldings associated with it, exactly one is maintained for the next iteration. Clearly, $|\mathcal{F}_i| \leq m^3$ at the end of the i-th iteration, for any $i = 1, \ldots, n$. Note that this bound holds during the execution of each iteration as well. After the last iteration n, the algorithm outputs a folding of the ruler (one in $\mathcal{F}_n$) whose interval has minimum length.

Let now F be an optimal folding as specified in Observation 1, whose vector is $(\varepsilon_1, \ldots, \varepsilon_n)$. We have $\varepsilon_1 = 1$. For $i = 1, \ldots, n$, let the partial folding F_i have the (folding) interval $[a_i, b_i]$ and the endpoint of the last segment at $x_i \in [a_i, b_i]$. We have $a_1 = 0$, $b_1 = l_1$ and $x_1 = l_1$, and also $x_i = \sum_{j=1}^{i} \varepsilon_j l_j$, for $i = 1, \ldots, n$.

LEMMA 1. *For $i = 1, \ldots, n$, the algorithm computes a partial folding $F_i' \in \mathcal{F}_i$ of the ruler $l_1, l_2, \ldots, l_i$, whose interval is $[a_i', b_i']$ and the endpoint of the last segment is at x_i', so that*

$$\text{(A)} \qquad |a_i - a_i'| \leq i\delta,$$
$$\text{(B)} \qquad |b_i - b_i'| \leq i\delta,$$
$$\text{(X)} \qquad |x_i - x_i'| \leq i\delta.$$

PROOF. We proceed by induction. The basis $i = 1$ is clear. Let $i \geq 2$, and assume that a partial folding F_{i-1}' of the ruler $l_1, l_2, \ldots, l_{i-1}$, is computed by the algorithm after $i - 1$ iterations, as specified. We thus have

$$|a_{i-1} - a_{i-1}'| \leq (i-1)\delta,$$
$$|b_{i-1} - b_{i-1}'| \leq (i-1)\delta,$$
$$|x_{i-1} - x_{i-1}'| \leq (i-1)\delta.$$

The partial folding F_i (corresponding to F) has parameters

$$a_i = \min(a_{i-1}, x_{i-1} + \varepsilon_i l_i),$$
$$b_i = \max(b_{i-1}, x_{i-1} + \varepsilon_i l_i),$$
$$x_i = x_{i-1} + \varepsilon_i l_i.$$

Consider the partial folding F_i'' obtained from F_{i-1}' (i.e., which extends F_{i-1}') so that its i-th bit in the folding vector is ε_i (the same as in F_i). Note that the algorithm computes F_i'' in the first part of iteration i (before trimming). Its parameters are

$$a_i'' = \min(a_{i-1}', x_{i-1}' + \varepsilon_i l_i),$$
$$b_i'' = \max(b_{i-1}', x_{i-1}' + \varepsilon_i l_i),$$
$$x_i'' = x_{i-1}' + \varepsilon_i l_i.$$

Let the interval triplet which contains F_i'' be (I_a, I_b, I_x). The algorithm discards all but one partial folding in this interval triplet, say F_i', with parameters a_i', b_i', x_i'. This implies that

$$|a_i' - a_i''| \leq \delta,$$
$$|b_i' - b_i''| \leq \delta,$$
$$|x_i' - x_i''| \leq \delta.$$

The lemma follows once we show that

$$\text{(A')} \qquad |a_i - a_i''| \leq (i-1)\delta,$$
$$\text{(B')} \qquad |b_i - b_i''| \leq (i-1)\delta,$$
$$\text{(X')} \qquad |x_i - x_i''| \leq (i-1)\delta,$$

since then, the partial folding F_i' which is computed by the algorithm, satisfies the imposed conditions after step i, e.g. for (A),

$$|a_i - a_i'| \leq |a_i - a_i''| + |a_i'' - a_i'| \leq (i-1)\delta + \delta = i\delta.$$

(B) and (X) follow in a similar way.

We will show that (A') holds by examining four cases, depending on how the minimums for a_i and for a_i'' are achieved. The proof of (B') is very similar (with max taking the place of min) and will be omitted.

To prove (A'), recall that

$$|a_i'' - a_i| = |\min(a_{i-1}', x_{i-1}' + \varepsilon_i l_i) - \min(a_{i-1}, x_{i-1} + \varepsilon_i l_i)|.$$

Put $\Delta = |a_i'' - a_i|$. We distinguish four cases.

Case 1: $\min(a_{i-1}', x_{i-1}' + \varepsilon_i l_i) = a_{i-1}'$ and $\min(a_{i-1}, x_{i-1} + \varepsilon_i l_i) = a_{i-1}$. Then using the induction hypothesis,

$$\Delta = |a_{i-1}' - a_{i-1}| \leq (i-1)\delta.$$

Case 2: $\min(a_{i-1}', x_{i-1}' + \varepsilon_i l_i) = x_{i-1}' + \varepsilon_i l_i$ and $\min(a_{i-1}, x_{i-1} + \varepsilon_i l_i) = x_{i-1} + \varepsilon_i l_i$. Similarly, the induction hypothesis yields

$$\Delta = |x_{i-1}' - x_{i-1}| \leq (i-1)\delta.$$

Case 3: $\min(a_{i-1}', x_{i-1}' + \varepsilon_i l_i) = a_{i-1}'$ and $\min(a_{i-1}, x_{i-1} + \varepsilon_i l_i) = x_{i-1} + \varepsilon_i l_i$. Note that in this case $\varepsilon_i = -1$. We have two subcases.

Case 3.1: $x_{i-1} - l_i \leq a_{i-1}'$. Recall that $a_{i-1}' \leq x_{i-1}' - l_i$. We have

$$x_{i-1} - l_i \leq a_{i-1}' \leq x_{i-1}' - l_i.$$

Then

$$\Delta = |a_{i-1}' - (x_{i-1} - l_i)| \leq |x_{i-1}' - l_i - (x_{i-1} - l_i)| \leq (i-1)\delta,$$

where the last in the chain of inequalities above is implied by the induction hypothesis.

Case 3.2: $a_{i-1}' \leq x_{i-1} - l_i$. Recall that $x_{i-1} - l_i \leq a_{i-1}$. We have

$$a_{i-1}' \leq x_{i-1} - l_i \leq a_{i-1}.$$

Then

$$\Delta = |a_{i-1}' - (x_{i-1} - l_i)| \leq |a_{i-1}' - a_{i-1}| \leq (i-1)\delta,$$

again by the induction hypothesis.

Case 4: $\min(a_{i-1}', x_{i-1}' + \varepsilon_i l_i) = x_{i-1}' + \varepsilon_i l_i$ and $\min(a_{i-1}, x_{i-1} + \varepsilon_i l_i) = a_{i-1}$. Note that in this case $\varepsilon_i = -1$. Thus $x_{i-1}' - l_i \leq a_{i-1}'$ and $a_{i-1} \leq x_{i-1} - l_i$. We have two subcases.

Case 4.1: $x'_{i-1} - l_i \le a_{i-1}$. Then

$$\Delta = |a_{i-1} - (x'_{i-1} - l_i)| \le |x_{i-1} - l_i - (x'_{i-1} - l_i)| \le (i-1)\delta.$$

Case 4.2: $a_{i-1} \le x'_{i-1} - l_i$. Then

$$\Delta = |(x'_{i-1} - l_i) - a_{i-1}| \le |a'_{i-1} - a_{i-1}| \le (i-1)\delta.$$

This concludes the proof of (A').

We also clearly have

$$|x_i - x''_i| = |(x_{i-1} + \varepsilon_i l_i) - (x'_{i-1} + \varepsilon_i l_i)| = |x_{i-1} - x'_{i-1}| \le (i-1)\delta,$$

which proves (X') and concludes the proof of the lemma. □

Lemma 1 for $i = n$ implies that the algorithm computes a folding F' of the ruler whose interval is $[a', b']$, so that if F is an optimal folding whose interval is $[a, b]$,

$$|a - a'| \le n\delta = L\varepsilon/2,$$
$$|b - b'| \le n\delta = L\varepsilon/2.$$

Since the algorithm selects in the end a folding whose interval length is minimum, it outputs one whose interval length is not more than

$$|b' - a'| \le |b - a| + \varepsilon L \le (1+\varepsilon)\text{OPT}.$$

The last in the chain of inequalities above follows from the lower bound $b - a = \text{OPT} \ge L$.

It takes $O(\log L)$ time to compute the three parameters for each partial folding, and $O(\log L)$ space to store this information. Since there are n iterations, and each takes $O(m^3 \log L)$ time, the total running time is $O(nm^3 \log L)) = O(n^4 (1/\varepsilon)^3 \log L)$. As each (partial) folding can be stored in $O(n \log L)$ space, the total space is also $O(n^4 (1/\varepsilon)^3 \log L)$.

REMARK 1. Using Observation 2, one can modify the algorithm so that $m = 6n/\varepsilon$ (versus $m = 8n/\varepsilon$), which leads to maintaining a somewhat smaller number of interval triplets.

A FPTAS based on rounding and scaling. We apply the rounding and scaling technique, inspired by the method used to obtain an approximation scheme for Knapsack (from [Ibarra and Kim 1975]; see also [Garey and Johnson 1979, pages 135–137]). The algorithm is:

(i) Set

$$\bar{l}_i = \left\lfloor \frac{l_i}{L} 4n \frac{1}{\varepsilon} \right\rfloor.$$

Call the new instance of the carpenter's ruler folding problem with lengths $\bar{l}_i$ the *reduced* instance.

(ii) Use the pseudo-polynomial algorithm in [Hopcroft et al. 1985] to solve exactly the reduced instance. Output the same folding vector.

Note that the maximum length of the reduced instance is $\bar{L} = \lfloor 4n\frac{1}{\varepsilon} \rfloor$ and therefore the running time of the algorithm is

$$O(n \log L + \bar{L}^2 n) = O(n \log L + n^3 (1/\varepsilon)^2).$$

We refine the notation as follows: given folding F whose vector is $(\varepsilon_1^F, \ldots, \varepsilon_n^F)$, set $x_0^F = 0$ and for $i = 1, \ldots, n$, set $x_i^F = \sum_{j=1}^{i} \varepsilon_j^F l_j$. As before $a_F = \min_{i=0}^{n} x_i^F$ and $b_F = \max_{i=0}^{n} x_i^F$, and note that the length of F is $b_F - a_F$. Define $\bar{x}_i^F$, $\bar{a}_F$ and $\bar{b}_F$ in the same way using the length function $\bar{l}$ instead of l. Let

$$q_i := \frac{l_i}{L} 4n \frac{1}{\varepsilon} - \left\lfloor \frac{l_i}{L} 4n \frac{1}{\varepsilon} \right\rfloor.$$

Note that $0 \leq q_i < 1$ and $l_i = (\bar{l}_i + q_i) L\varepsilon/(4n)$.

Let A be any folding for the original instance and B be an optimum folding for the reduced instance. We have:

$$x_i^B = \sum_{j=1}^{i} \varepsilon_i^B l_i = \sum_{j=1}^{i} \varepsilon_i^B (\bar{l}_i + q_i) \frac{L\varepsilon}{4n} = \frac{L\varepsilon}{4n} \left(\bar{x}_i^B + \sum_{j=1}^{i} \varepsilon_i^B q_i \right).$$

Using $0 \leq q_i < 1$, we obtain

$$x_i^B \leq \frac{L\varepsilon}{4n} (\bar{x}_i^B + n) \leq \frac{L\varepsilon}{4n} \bar{x}_i^B + \frac{L\varepsilon}{4},$$

and therefore

$$b_B \leq \frac{L\varepsilon}{4n} \bar{b}_B + \frac{L\varepsilon}{4}.$$

Similarly we have:

$$x_i^B \geq \frac{L\varepsilon}{4n} \bar{x}_i^B - \frac{L\varepsilon}{4},$$

and consequently

$$a_B \geq \frac{L\varepsilon}{4n} \bar{a}_B - \frac{L\varepsilon}{4}.$$

Using the fact that B has optimum length for $\bar{l}$, and the inequality $b_A - a_A \geq L$, we get:

$$b_B - a_B \leq \frac{L\varepsilon}{4n} (\bar{b}_B - \bar{a}_B) + \frac{L\varepsilon}{2} \leq \frac{L\varepsilon}{4n} (\bar{b}_A - \bar{a}_A) + \frac{\varepsilon}{2} (b_A - a_A). \qquad (2\text{--}1)$$

Further:

$$\bar{x}_i^A = \sum_{j=1}^{i} \varepsilon_i^A \bar{l}_i = \sum_{j=1}^{i} \varepsilon_i^A \left(\frac{l_i}{L} 4n \frac{1}{\varepsilon} - q_i \right) = \frac{4n}{L\varepsilon} x_i^A - \sum_{j=1}^{i} \varepsilon_i^A q_i \leq \frac{4n}{L\varepsilon} x_i^A + n,$$

and therefore

$$\bar{b}_A \leq \frac{4n}{L\varepsilon} b_A + n. \qquad (2\text{--}2)$$

Similarly we have:

$$\bar{x}_i^A = \frac{4n}{L\varepsilon} x_i^A - \sum_{j=1}^{i} \varepsilon_i^A q_i \geq \frac{4n}{L\varepsilon} x_i^A - n,$$

and consequently

$$\bar{a}_A \geq \frac{4n}{L\varepsilon}a_A - n. \tag{2–3}$$

Plugging Equations (2–2) and (2–3) into (2–1) and using again the inequality $b_A - a_A \geq L$, we obtain

$$b_B - a_B \leq \frac{L\varepsilon}{4n}\left(\frac{4n}{L\varepsilon}b_A + n - \left(\frac{4n}{L\varepsilon}a_A - n\right)\right) + \frac{\varepsilon}{2}(b_A - a_A) \leq (b_A - a_A)(1 + \varepsilon).$$

If we now let A be an optimal folding for the original instance, we find that $b_B - a_B \leq (1 + \varepsilon)\mathrm{OPT}$; this completes the second proof of Theorem 1.

3. Folding in the Plane: Proof of Theorem 2

For the purposes of this section, a folding of the ruler is a polygonal chain of n segments (links), numbered from 1 to n, lying in the plane. Let q_0 be the free endpoint of the first link, and q_1 be its other endpoint. Call $v_1 = \overline{q_0 q_1}$ the vector of link 1. Inductively define ($q_2, \ldots, q_n$ and) $v_2, \ldots, v_n$, the vectors of links $2, \ldots, n$. The *joint angle* between links i and $i + 1$ is the angle $\in [0, \pi]$ between v_i and v_{i+1}. The angle is counterclockwise if it describes a left turn, and clockwise if it describes a right turn. Angles of 0 and π are considered both left and right turns.

It is obvious that the diameter of any convex case in which the ruler is folded is at least L, where L is the maximum link length. The following simple linear-time algorithm computes a folding of the ruler, so that all joint angles in $(0, \pi]$ are clockwise (or counterclockwise). The algorithm is certainly implicit in [Hopcroft et al. 1985], where an extensive analysis of reconfiguration problems for rulers confined in discs is made.

Fix arbitrarily a disk D of diameter L, whose boundary is the circle C. Fix the first free endpoint of the ruler (i.e., the free endpoint of the first link) at some point p_0 of C. For $i = 1, \ldots, n$, iteratively fix the next point of the ruler (i.e., the next endpoint of its i-th link) at one of the at most two intersection points of C with the circle with center at p_{i-1} and radius l_i. One can also select the appropriate intersection point at each step, so that all joint angles in $(0, \pi]$ are clockwise (or counterclockwise). An illustration appears in Figure 2. Consider now the closed convex curve R, of unit diameter, obtained from a *Reuleaux triangle*, by replacing one of the circular arcs with a straight segment, as in Figure 3. (A Reuleaux triangle can be obtained from an equilateral triangle ABC by joining each pair of its vertices by a circular arc whose center is at the third vertex; see [Yaglom and Boltyanskiĭ 1961].) The above algorithm can be modified to compute a folding of a ruler with maximum link length 1 inside R: Fix the first free endpoint of the ruler at some point p_0 of the circular arc AB. Iteratively fix the next point of the ruler at some intersection point (it exists!) with the open curve BAC. The area of R is $\frac{1}{3}\pi - \frac{1}{4}\sqrt{3} \approx 0.614$, as claimed.

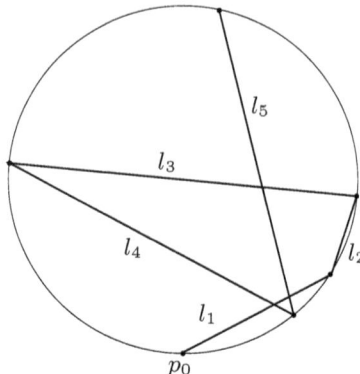

Figure 2. A carpenter's ruler with five links folded so that it fits in a circular case of diameter L, where $L = l_3$ is the maximum link-length. All joint angles in this folding are counterclockwise (i.e., left turns).

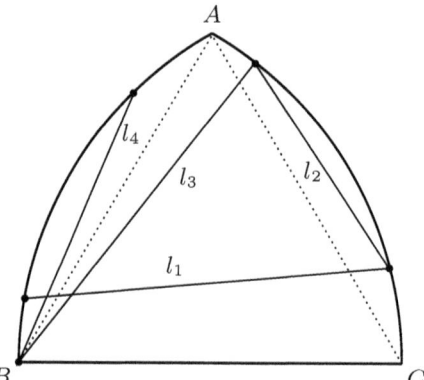

Figure 3. The closed curve R obtained from a Reuleaux triangle, and a ruler with four links folded inside; the length of l_3 is 1.

It remains to prove the lower bound in Theorem 2. Consider a 3-link ruler $ABCD$ with lengths $AB = 1$, $BC = x < 1$ and $CD = 1$, where the choice of the length $x = \frac{1}{2}(\sqrt{7} - 1) \approx 0.8229$ of the middle link is explained below. We will show that the area of any convex case for it is at least $\frac{3}{8}$. In any folding in which the unit length links do not intersect, the diameter of the case exceeds one. Assume therefore that they intersect (see Figure 4). The area of $BCAD$ (i.e., the convex hull of the four endpoints of the links) is

$$\frac{ab\sin\alpha}{2} + \frac{(1-a)(1-b)\sin\alpha}{2} + \frac{a(1-b)\sin\alpha}{2} + \frac{(1-a)b\sin\alpha}{2} = \frac{\sin\alpha}{2},$$

where $\alpha = \widehat{BOD}$.

For a given x, the area is minimized when either $A = D$ so that the folding forms an isosceles triangle (small x), or when AD is parallel to BC and $AD = 1$

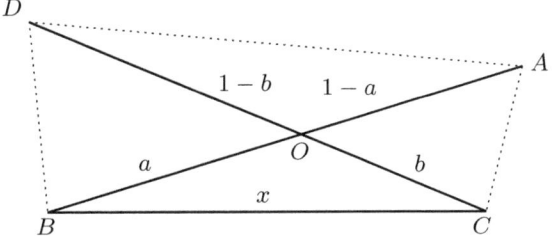

Figure 4. A ruler with link-lengths 1, x and 1.

(large x). The area of the isosceles triangle is

$$\sqrt{\left(1+\frac{x}{2}\right)\left(\frac{x}{2}\right)\left(\frac{x}{2}\right)\left(1-\frac{x}{2}\right)}.$$

The area of the trapezoid $BCAD$ is

$$\frac{1+x}{2}\sqrt{1-\left(\frac{1+x}{2}\right)^2}.$$

Now choose x to balance the two areas. A routine calculation gives

$$x = \frac{\sqrt{7}-1}{2},$$

and the corresponding area is $3/8$. This completes the proof of Theorem 2.

We conclude with these questions: Is the curve R a convex universal case of minimum area? If not, what is the minimum area of such a universal case? Does convexity of the case make any difference?

Acknowledgement

We thank the anonymous referee (of an earlier version) who suggested the modification of the Reuleaux triangle in Figure 3.

References

[Connelly et al. 2003] R. Connelly, E. D. Demaine, and G. Rote, "Straightening polygonal arcs and convexifying polygonal cycles", *Discrete Comput. Geom.* **30**:2 (2003), 205–239.

[Cormen et al. 1990] T. H. Cormen, C. E. Leiserson, and R. L. Rivest, *Introduction to algorithms*, MIT Press, Cambridge, MA, 1990.

[Croft et al. 1991] H. T. Croft, K. J. Falconer, and R. K. Guy, *Unsolved problems in geometry*, Problem Books in Mathematics, Springer, New York, 1991.

[Garey and Johnson 1979] M. R. Garey and D. S. a. Johnson, *Computers and intractability*, W. H. Freeman and Co., San Francisco, 1979.

[Hopcroft et al. 1984] J. Hopcroft, D. Joseph, and S. Whitesides, "Movement problems for 2-dimensional linkages", *SIAM J. Comput.* **13**:3 (1984), 610–629.

[Hopcroft et al. 1985] J. Hopcroft, D. Joseph, and S. Whitesides, "On the movement of robot arms in 2-dimensional bounded regions", *SIAM J. Comput.* **14**:2 (1985), 315–333.

[Ibarra and Kim 1975] O. H. Ibarra and C. E. Kim, "Fast approximation algorithms for the knapsack and sum of subset problems", *J. Assoc. Comput. Mach.* **22**:4 (1975), 463–468.

[Kantabutra 1992] V. Kantabutra, "Motions of a short-linked robot arm in a square", *Discrete Comput. Geom.* **7**:1 (1992), 69–76.

[Kantabutra 1997] V. Kantabutra, "Reaching a point with an unanchored robot arm in a square", *Internat. J. Comput. Geom. Appl.* **7**:6 (1997), 539–549.

[Kantabutra and Kosaraju 1986] V. Kantabutra and S. R. Kosaraju, "New algorithms for multilink robot arms", *J. Comput. System Sci.* **32**:1 (1986), 136–153.

[Klee and Wagon 1991] V. Klee and S. Wagon, *Old and new unsolved problems in plane geometry and number theory*, vol. 11, The Dolciani Mathematical Expositions, Math. Assoc. America, Washington (DC), 1991.

[Kozen 1992] D. C. Kozen, *The design and analysis of algorithms*, Texts and Monographs in Computer Science, Springer, New York, 1992.

[van Kreveld et al. 1996] M. van Kreveld, J. Snoeyink, and S. Whitesides, "Folding rulers inside triangles", *Discrete Comput. Geom.* **15**:3 (1996), 265–285.

[Streinu 2000] I. Streinu, "A combinatorial approach to planar non-colliding robot arm motion planning", pp. 443–453 in *41st Annual Symposium on Foundations of Computer Science* (Redondo Beach, CA, 2000), IEEE Comput. Soc. Press, Los Alamitos, CA, 2000.

[Yaglom and Boltyanskiĭ 1961] I. M. Yaglom and V. G. Boltyanskiĭ, *Convex figures*, vol. 4, Library of the mathematical circle, Holt, Rinehart and Winston, New York, 1961.

GRUIA CĂLINESCU
DEPARTMENT OF COMPUTER SCIENCE
ILLINOIS INSTITUTE OF TECHNOLOGY
CHICAGO, IL 60616
UNITED STATES
calinesc@iit.edu

ADRIAN DUMITRESCU
COMPUTER SCIENCE
UNIVERSITY OF WISCONSIN–MILWAUKEE
3200 N. CRAMER STREET
MILWAUKEE, WI 53211
UNITED STATES
ad@cs.uwm.edu

Combinatorial and Computational Geometry
MSRI Publications
Volume **52**, 2005

A Survey of Folding and Unfolding in Computational Geometry

ERIK D. DEMAINE AND JOSEPH O'ROURKE

ABSTRACT. We survey results in a recent branch of computational geometry: folding and unfolding of linkages, paper, and polyhedra.

CONTENTS

1. Introduction   168
2. Linkages   168
  2.1. Definitions and fundamental questions   168
  2.2. Fundamental questions in 2D   171
  2.3. Fundamental questions in 3D   175
  2.4. Fundamental questions in 4D and higher dimensions   181
  2.5. Protein folding   181
3. Paper   183
  3.1. Categorization   184
  3.2. Origami design   185
  3.3. Origami foldability   189
  3.4. Flattening polyhedra   191
4. Polyhedra   193
  4.1. Unfolding polyhedra   193
  4.2. Folding polygons into convex polyhedra   196
  4.3. Folding nets into nonconvex polyhedra   199
  4.4. Continuously folding polyhedra   200
5. Conclusion and Higher Dimensions   201
Acknowledgements   202
References   202

Demaine was supported by NSF CAREER award CCF-0347776. O'Rourke was supported by
NSF Distinguished Teaching Scholars award DUE-0123154.

1. Introduction

Folding and unfolding problems have been implicit since Albrecht Dürer [1525], but have not been studied extensively in the mathematical literature until recently. Over the past few years, there has been a surge of interest in these problems in discrete and computational geometry. This paper gives a brief survey of most of the work in this area. Related, shorter surveys are [Connelly and Demaine 2004; Demaine 2001; Demaine and Demaine 2002; O'Rourke 2000]. We are currently preparing a monograph on the topic [Demaine and O'Rourke ≥ 2005].

In general, we are interested in how objects (such as linkages, pieces of paper, and polyhedra) can be moved or reconfigured (folded) subject to certain constraints depending on the type of object and the problem of interest. Typically the process of *unfolding* approaches a more basic shape, whereas *folding* complicates the shape. We define the *configuration space* as the set of all configurations or states of the object permitted by the folding constraints, with paths in the space corresponding to motions (foldings) of the object.

This survey is divided into three sections corresponding to the type of object being folded: linkages, paper, or polyhedra. Unavoidably, areas with which we are more familiar or for which there is a more extensive literature are covered in more detail. For example, more problems have been explored in linkage and paper folding than in polyhedron folding, and our corresponding sections reflect this imbalance. On the other hand, this survey cannot do justice to the wealth of research on protein folding, so only a partial survey appears in Section 2.5.

2. Linkages

2.1. Definitions and fundamental questions. A *linkage* or *framework* consists of a collection of rigid line segments (*bars* or *links*) joined at their endpoints (*vertices* or *joints*) to form a particular graph. A linkage can be *folded* by moving the vertices in $\mathbb{R}^d$ in any way that preserves the length of every bar. Unless otherwise specified, we assume the vertices to be universal joints, permitting the full angular range of motions. Restricted angular motions will be discussed in Section 2.5.2.

Linkages have been studied extensively in the case that bars are permitted to cross; see, for example, [Hopcroft et al. 1984; Jordan and Steiner 1999; Kapovich and Millson 1995; Kempe 1876; Lenhart and Whitesides 1995; Sallee 1973; Whitesides 1992]. Such linkages can be very complex, even in the plane. Kempe [1876] suggested an incomplete argument to show that a planar linkage can be built so that a vertex traces an arbitrary polynomial curve — there is a linkage that can "sign your name." It was not until recently that Kempe's claim was established rigorously by Kapovich and Millson [2002]. Hopcroft, Joseph, and Whitesides [Hopcroft et al. 1984] showed that deciding whether a planar linkage can reach a particular configuration is PSPACE-complete. Jordan and Steiner [1999] proved that there is a linkage whose configuration space is homeomorphic to an arbitrary compact real algebraic variety with Euclidean topology, and thus planar linkages are equivalent to the theory of the reals (solving systems of polynomial inequalities over reals). On the other hand, for a linkage whose graph is just a cycle, all configurations can be reached in Euclidean space of any dimension greater than 2 by a sequence of simple motions [Lenhart and Whitesides 1995; Sallee 1973], and in the plane there is a simple restriction characterizing which polygons can be inverted in orientation [Lenhart and Whitesides 1995].

Recently there has been much work on the case that the linkage must remain *simple*, that is, never have two bars cross.[1] The remainder of this survey assumes this noncrossing constraint. Such linkage folding has applications in hydraulic tube bending [O'Rourke 2000] and motion planning of robot arms. There are also connections to protein folding in molecular biology, which we touch upon in Section 2.5. See also [Connelly et al. 2003; O'Rourke 2000; Toussaint 1999a] for other surveys on linkage folding without crossings.

Perhaps the most fundamental question one can ask about folding linkages is whether it is possible to fold between any two configurations. That is, is there a folding between any two simple configurations of the same linkage (with matching graphs, combinatorial embeddings, and bar lengths) while preserving the bar lengths and not crossing any bars during the folding? Because folding motions can be reversed and concatenated, this fundamental question is equivalent to whether every simple configuration can be folded into some *canonical configuration*, a configuration whose definition depends on the type of linkage under consideration.

We concentrate here on allowing all continuous motions that maintain simplicity, but we should mention that different applications often further constrain the permissible motions in various ways. For example, hydraulic tube bending allows only one joint to bend at any one time, and moreover the joint angle can never reverse direction. Such constraints often drastically alter what is possible. See, for example, [Arkin et al. 2003].

In the context of linkages whose edges cannot cross, three general types of linkages are commonly studied, characterized by the structure of their associated graphs (see Figure 1): a *polygonal arc* or *open polygonal chain* (a single path); a *polygonal cycle*, *polygon*, or *closed polygonal chain* (a single cycle); and a *polygonal tree* (a single tree).[2] The canonical configuration of an arc is the *straight configuration*, all vertex angles equal to 180°. A canonical configuration

---

[1] Typically, bars are allowed to touch, provided they do not properly cross. However, insisting that bars only touch at common endpoints does not change the results.

[2] More general graphs have been studied largely in the context of allowing bars to cross, exploring either aspects of the configurations space (e.g., the Kempe work mentioned earlier), or the conditions which render the graph rigid. Graph rigidity is a rich topic, not detailed here, which also plays a role in the noncrossing-bar scenario in Section 2.2.1.

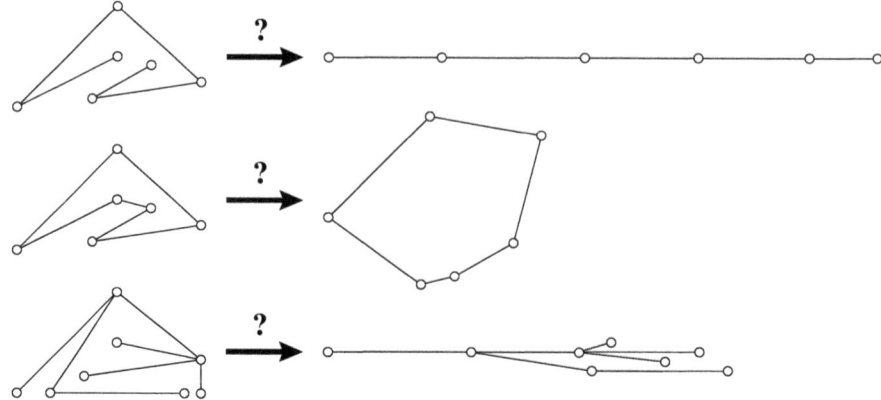

Figure 1. The three common types of linkages and their associated canonical configurations. From top to bottom, a polygonal arc $\overset{?}{\to}$ the straight configuration, a polygonal cycle $\overset{?}{\to}$ a convex configuration, and a polygonal tree $\overset{?}{\to}$ a (nearly) flat configuration.

of a cycle is a *convex configuration*, planar and having all interior vertex angles less than or equal to 180°. It is relatively easy to show that convex configurations are indeed "canonical" in the sense that any one can be folded into any other, a result that first appeared in [Aichholzer et al. 2001]. Finally, a canonical configuration of a tree is a *flat configuration*: all vertices lie on a horizontal line, and all bars point "rightward" from a common root. Again it is easy to fold any flat configuration into any other [Biedl et al. 2002b].

The fundamental questions thus become whether every arc can be straightened, every cycle can be convexified, and every tree can be flattened. The answers to these questions depend on the dimension of the space in which the linkage starts, and the dimension of the space in which the linkage may be folded. Over the past few years, this collection of questions has been completely resolved:

Can all arcs be straightened?

2D: Yes [Connelly et al. 2003]
3D: No [Cantarella and Johnston 1998; Biedl et al. 2001]
4D+: Yes [Cocan and O'Rourke 2001]

Can all cycles be convexified?

2D: Yes [Connelly et al. 2003]
3D: No [Cantarella and Johnston 1998; Biedl et al. 2001]
4D+: Yes [Cocan and O'Rourke 2001]

Can all trees be flattened?

2D: No [Biedl et al. 2002b]
3D: No (from arcs)
4D+: Yes [Cocan and O'Rourke 2001]

The answers for arcs and cycles are analogous to the existence of knots tied from one-dimensional string: nontrivial knots exist only in 3D. In contrast, the situation for trees presents an interesting difference in 2D: while trees in the plane are topologically unknotted, they can be geometrically locked. This observation is some evidence for the belief that the fundamental problems are most difficult in 2D.

The next three subsections describe the historical progress of these results and other results closely related to the fundamental questions. Along the way, Sections 2.3.1–2.3.4 describe several special forms of linkage folding arising out of a problem posed by Erdős in 1935; and Section 2.3.8 considers the generalization of multiple chains. Finally, Section 2.5 discusses the connections between linkage folding and protein folding, and describes the most closely related results and open problems.

2.2. Fundamental questions in 2D. Section 2.2.1 describes the development of the theorems for straightening arcs and convexifying cycles in 2D. Section 2.2.2 discusses the contrary result that not all trees can be flattened.

2.2.1. The carpenter's rule problem: polygonal chains in 2D. The questions of whether every polygonal arc can be straightened and every polygonal cycle can be convexified in the plane have arisen in many contexts over the last quarter of a century.[3] In the discrete and computational geometry community, the arc-straightening problem has become known as the *carpenter's rule problem* because a carpenter's rule folds like a polygonal arc.

Most people's initial intuition is that the answers to these problems are YES, but describing a precise general motion proved difficult. It was not until 2000 that the problems were solved by Connelly, Demaine, and Rote [Connelly et al. 2003], with an affirmative answer. Figure 2 shows an example of the motion resulting from this theorem.

More generally, the result in [Connelly et al. 2003] shows that a collection of nonintersecting polygonal arcs and cycles in the plane may be simultaneously folded so that the outermost arcs are straightened and the outermost cycles are convexified. The "outermost" proviso is necessary because arcs and cycles cannot always be straightened and convexified when they are contained in other cycles. The key idea for the solution, introduced by Günter Rote, is to look for *expansive* motions in which no vertex-to-vertex distance decreases. Bars cannot cross before getting closer, so expansiveness allows us to ignore the difficult nonlocal constraint that bars must not cross. Expansiveness brings the problem into the areas of rigidity theory and tensegrity theory, which study frameworks of rigid bars, unshrinkable *struts*, and unexpandable *cables*. Tools from these

---

[3]Posed independently by Stephen Schanuel and George Bergman in the early 1970's, Ulf Grenander in 1987, William Lenhart and Sue Whitesides in 1991, and Joseph Mitchell in 1992; see [Connelly et al. 2003].

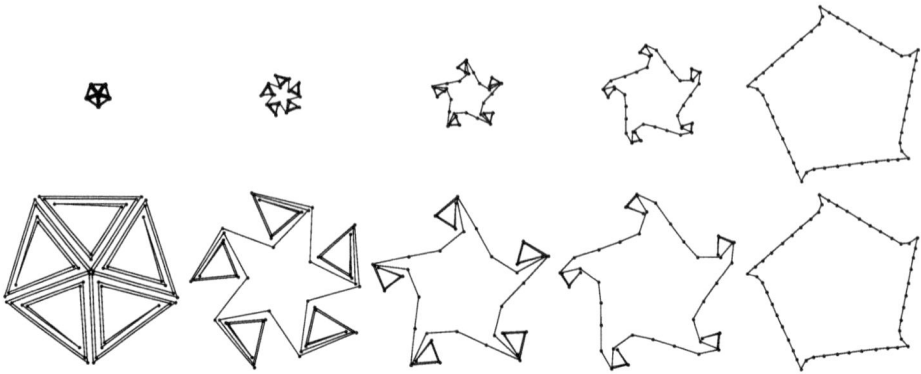

Figure 2. Two views of convexifying a "doubled tree" linkage. The top snapshots are all scaled the same, and the bottom snapshots are rescaled to improve visibility.

areas helped show that, *infinitesimally*, arcs and cycles can be unfolded expansively. These infinitesimal motions are combined by flowing along a vector field defined implicitly by an optimization problem. As a result, the motion is piecewise-differentiable (C^1). In addition, any symmetries present in the initial configuration of the linkage are preserved throughout the motion. Similar techniques show that the area of each cycle increases by this motion and furthermore by any expansive motion [Connelly et al. 2003].

Since the original theorem, two additional algorithms have been developed for unfolding polygonal chains. Figure 3 provides a visual comparison of all three algorithms.

Ileana Streinu [2000] demonstrated another expansive motion for straightening arcs and convexifying polygons that is piecewise-algebraic, composed of a polynomial-length sequence of *mechanisms*, each with a single degree of freedom. In this sense the motion is easier to implement "mechanically." It is also possible to compute the algebraic curves involved, though the running time is exponential in n. This method also elucidates an interesting combinatorial structure to 2D linkage unfolding through "pseudotriangulations," which have subsequently received much attention in computational geometry (see [O'Rourke 2002; Rote 2003], for example).

Cantarella, Demaine, Iben, and O'Brien [Cantarella et al. 2004] gave an energy-based algorithm for straightening arcs and convexifying polygons. This algorithm follows the downhill gradient of an appropriate energy function, corresponding roughly to the intuition of filling the polygon with air. The resulting motion is not expansive, essentially averaging out the strut constraints. On the other hand, the existence of the downhill gradient relies on the existence of expansive motions from [Connelly et al. 2003], by showing that the latter decrease energy. The motion avoids self-intersection not through expansiveness but by designing the energy function to approach $+\infty$ near an intersecting configura-

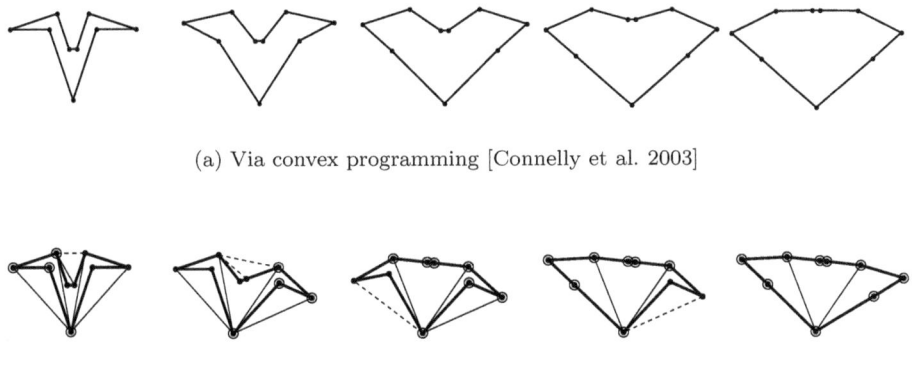

(a) Via convex programming [Connelly et al. 2003]

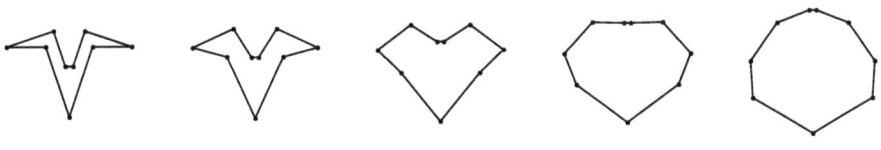

(b) Via pseudotriangulations [Streinu 2000]. Pinned vertices are circled.

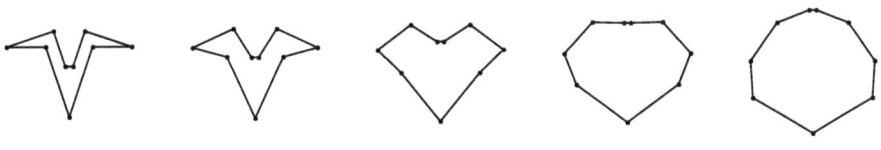

(c) Via energy minimization [Cantarella et al. 2004].

Figure 3. Convexifying a common polygon via all three convexification methods.

tion; any downhill flow avoids such spikes. The result is a C^∞ motion, easily computed as a piecewise-linear motion in angle space. The number of steps in the piecewise-linear motion is polynomial in two quantities: in the number of vertices n, and in the ratio between the maximum edge length and the initial minimum distance between a vertex and an edge.

2.2.2. Trees in 2D. It was shown in [Biedl et al. 2002b] that not all trees can be flattened in the plane. The example there consists of at least 5 *petals* connected at a central high-degree vertex. The version shown in Figure 4 uses 8 petals. Each petal is an arc of three bars, the last of which is "wedged" into the center vertex.

Intuitively, the argument that the tree is locked is as follows. No petal can be straightened unless enough angular room has been made. But no petal can be reduced to occupy less angular space by more than a small positive number unless the petal has already been straightened. This circular dependence implies that no petal can be straightened, so the tree is locked. The details of this argument, in particular obtaining suitable tolerances for closeness, are somewhat intricate [Biedl et al. 2002b]. The key is that each petal occupies a wedge of space whose angle is less than $90°$, which is why at least 5 petals are required.

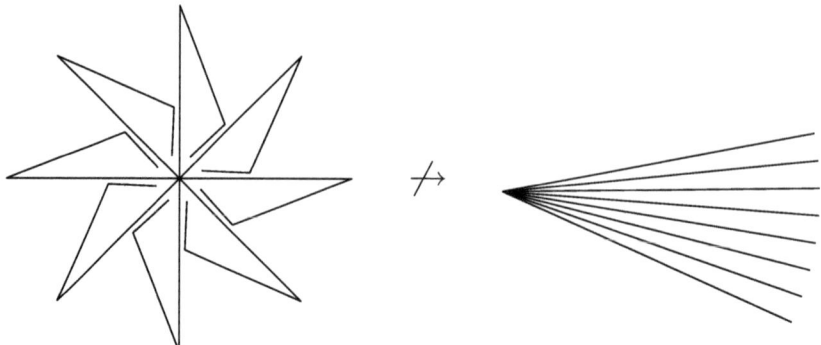

Figure 4. The locked tree on the left, from [Biedl et al. 2002b], cannot be reconfigured into the nearly flat configuration on the right. (Figure 1 of [Biedl et al. 2002b].)

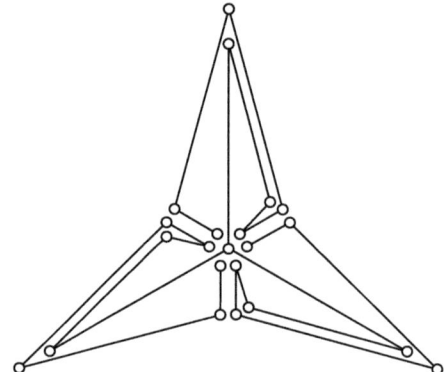

Figure 5. The locked tree from [Connelly et al. 2002]. Based on Figure 1(c) of [Connelly et al. 2002].

This tree remains locked if we replace the central degree-5 (or higher) vertex with multiple degree-3 vertices connected by very short bars [Biedl et al. 2002b, full version]. Connelly, Demaine, and Rote [Connelly et al. 2002] showed that the tree in Figure 5, with a single degree-3 vertex and the remaining vertices having degrees 1 and 2, is locked, proving tightness of the arc-and-cycle result in [Connelly et al. 2003]. In [Connelly et al. 2002] an extension to rigidity/tensegrity theory is given that permits establishing via linear programming that many classes of planar linkages (e.g., trees) are locked. In particular, this method is used to give short proofs that the tree in Figure 4 and the tree with one degree-3 vertex are *strongly locked*, in the sense that sufficiently small perturbations of the vertex positions and bar lengths result in a tree that cannot be moved more than ε in the configuration space for any $\varepsilon > 0$.

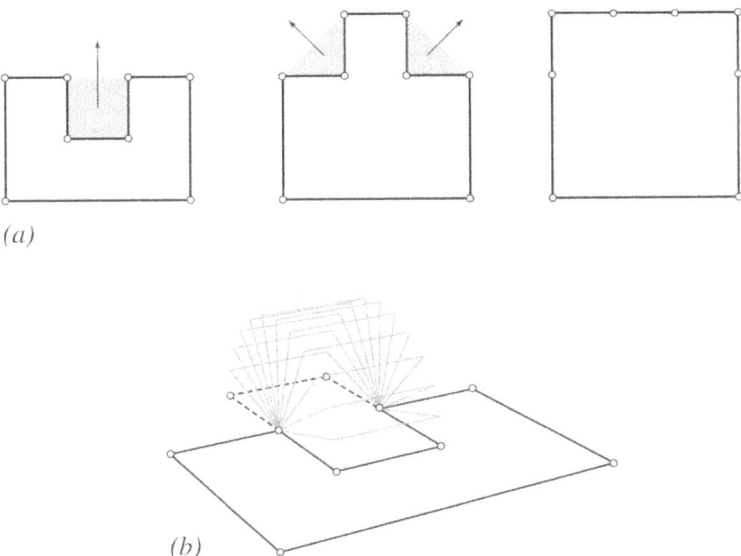

(a)

(b)

Figure 6. (a) Flipping a polygon until it is convex. Pockets are shaded. (b) The first flip shown in three dimensions.

2.3. Fundamental questions in 3D.
Linkage folding in 3D was initiated earlier, by Paul Erdős [1935]. His problem and its solution are described in Section 2.3.1. Sections 2.3.2–2.3.4 consider various extensions of this problem. All of this work deals with linkages that start in the plane, but fold through 3D. The more general situation, an arbitrary linkage starting in 3D, is addressed in Section 2.3.6. As this problem proves unsolvable in general, additional special cases are addressed in Section 2.3.7. Finally, Section 2.3.8 considers the generalized problem of multiple interlocking chains.

2.3.1. Flips for planar polygons in 3D.
The roots of linkage folding go back to [Erdős 1935], a problem posed in the *American Mathematics Monthly*. Define a *pocket* of a polygon to be a region bounded by a subchain of the polygon edges, and define the *lid* of the pocket to be the edge of the convex hull connecting the endpoints of that subchain. Every nonconvex polygon has at least one pocket. Erdős defined a *flip* as a rotation of a pocket's chain of edges into 3D about the pocket lid by 180°, landing the subchain back in the plane of the polygon, such that the polygon remains simple (i.e., non-self-intersecting); see Figure 6. He asked whether every polygon may be convexified by a finite number of simultaneous pocket flips.

The answer was provided in a later issue of the *Monthly* [Nagy 1939]. First, Nagy observed that flipping several pockets at once could lead to self-crossing; see Figure 7b. However, restricting to one flip at a time, Nagy proved that a finite number of flips suffice to convexify any polygon; see Figure 6 for a three-step example. This beautiful result has been rediscovered and reproved several

Figure 7. Flipping multiple pockets simultaneously can lead to crossings [Nagy 1939].

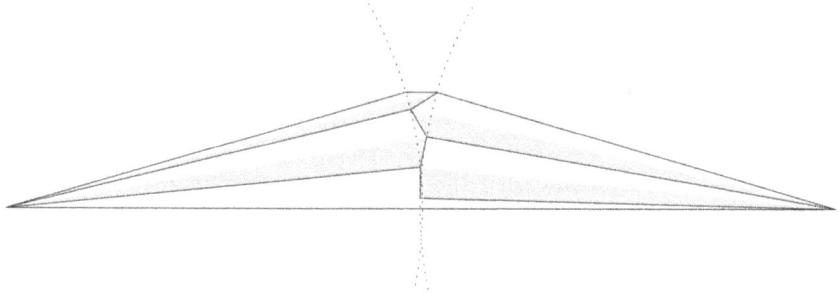

Figure 8. Quadrangles can require arbitrarily many flips to convexify [Grünbaum 1995; Toussaint 1999b; Biedl et al. 2001].

times, as uncovered by Grünbaum and Toussaint and detailed in their histories of the problem [Grünbaum 1995; Toussaint 1999b]; only recently has a subtle oversight in Nagy's proof been corrected.

Unfortunately, the number of required flips can be arbitrarily large in terms of the number of vertices, even for a quadrangle. This fact was originally proved by Joss and Shannon (1973); see [Grünbaum 1995; Toussaint 1999b; Biedl et al. 2001]. Figure 8 shows the construction. By making the vertical edge of the quadrangle very short and even closer to the horizontal edge, the angles after the first flip approach the mirror image of the original quadrangle, and hence the number of required flips approaches infinity.

Mark Overmars[4] posed the still-open problem of bounding the number of flips in terms of natural measures of geometric closeness such as the diameter (maximum distance between two vertices), sharpest angle, or the minimum feature size (minimum distance between two nonincident edges).

Another open problem is to determine the complexity of finding the shortest or longest sequence of flips to convexify a given polygon. Weak NP-hardness has been established for the related problem of finding the longest sequence of *flipturns* [Aichholzer et al. 2002].

---

[4]Personal communication, February 1998.

2.3.2. Flips in nonsimple polygons. Flips can be generalized to apply to nonsimple polygons: consider two vertices adjacent along the convex hull of the polygon, splitting the polygon into two chains, and rotate one (either) chain by 180° with respect to the other chain about the axis through the two vertices. Simplicity may not be preserved throughout the motion, just as it may not hold in the initial or final configuration. The obvious question is whether every nonsimple polygon can be convexified by a finite sequence of such flips. Grünbaum and Zaks [1998] proved that if at each step we choose the flip that maximizes the resulting sum of distances between all pairs of vertices, then this metric increases at each flip, and the polygon becomes convex after finitely many flips. Without sophisticated data structures, computing these flips requires $\Omega(n^2)$ time per flip. Toussaint [1999b] proved that a different sequence of flips convexifies a nonsimple polygon, and this sequence can be computed in $O(n)$ time per flip. More recently, it has been established[5] that every sequence of flips eventually convexifies a nonsimple polygon. We expect that each flip can be executed in polylogarithmic amortized time using dynamic convex-hull data structures as in [Aichholzer et al. 2002].[6]

2.3.3. Deflations. A *deflation* [Fevens et al. 2001; Wegner 1993; Toussaint 1999b] is the reverse of a flip, in the sense that a deflation of a polygon should result in a simple polygon that can be flipped into the original polygon. More precisely, a deflation is a rotation by 180° about a line meeting the polygon at two vertices and nowhere else, thus separating the chain into two subchains, such that the rotation does not cause any intersections. Hence, after the deflation, this line becomes a line of support (a line extending a convex-hull edge). Wegner [1993] proposed the notion of deflations, and their striking similarity to flips led him to conjecture that every polygon can be deflated only a finite number of times. Surprisingly, this is not true: Fevens, Hernandez, Mesa, Soss, and Toussaint [Fevens et al. 2001] characterized a class of quadrangles whose unique deflation leads to another quadrangle in the class, thus repeating ad infinitum.

2.3.4. Other variations. Erdős flips have inspired several directions of research on related notions, including pivots, pops, and flipturns. See [Toussaint 1999b] for a survey of this area, with more recent work on flipturns in [Ahn et al. 2000; Aichholzer et al. 2002; Biedl 2005].

2.3.5. Efficient algorithms for planar linkages in 3D. Motivated by the inefficiency of the flip algorithm, Biedl et al. [2001] developed an algorithm to convexify planar polygons by motions in 3D using a linear number of simple moves. The essence of this algorithm is to lift the polygon, bar by bar, at all times maintaining a convex chain (or *arch*) lying in a plane orthogonal to the plane containing the polygon; see Figure 9. The details of the algorithm are significantly more involved than the overarching idea.

---

[5]Personal communication with Therese Biedl, May 2001.

[6]Personal communication with Jeff Erickson.

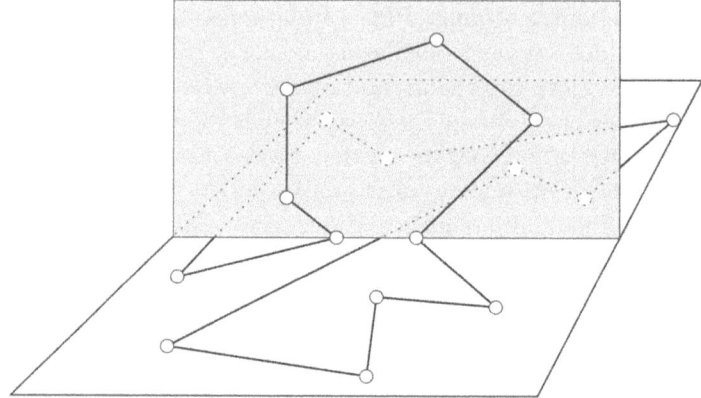

Figure 9. A planar polygon partially lifted into a convex arch lying in a vertical plane (shaded). (Based on Figure 6 of [Biedl et al. 2001].)

A second linear-time algorithm, which is in some ways conceptually simpler, was developed by Aronov, Goodman, and Pollack [Aronov et al. 2002]. Their algorithm at all times maintains the arch as a convex quadrilateral. At each step, the algorithm lifts two edges, forming a "twisted trapezoid," incorporates the trapezoid into the arch, makes the arch planar, and reduces it back to a quadrilateral. Avoiding intersections during the lifting phase requires a delicate argument.

In contrast to convexifying a cycle, it is relatively easy to straighten a polyg-onal arc lying in a plane, or on the surface of a convex polyhedron, by motions in 3D [Biedl et al. 2001]. For an arc in a plane, the basic idea is to pull the arc up into a vertical line. For a convex surface, the same idea is followed, but with the orientation of the line changing to remain normal to the surface. The algorithm lifts each bar in turn, from one end of the arc to the other, at all times maintaining a prefix of the arc in a line normal to the current facet of the polyhedron. Each lifting motion causes two joint angles to rotate, so that the lifted prefix remains normal to the facet at all times, while the remainder of the chain remains in its original position. Whenever the algorithm reaches a vertex that bridges between two adjacent facets, it rotates the prefix to bring it normal to the next facet. This algorithm also generalizes to flattening planar trees and trees on the surface of a convex polyhedron, via motions in 3D.

2.3.6. Almost knots. What if the linkage starts in an arbitrary position in 3D instead of in a plane? In general, a polygonal arc or an unknotted polygonal cycle in 3D cannot always be straightened or convexified [Cantarella and Johnston 1998; Toussaint 2001; Biedl et al. 2001] (page 170). Figure 10 shows an example of a locked arc in 3D. Provided that each of the two end bars is longer than the sum s of the middle three bar lengths, the ends of the chain cannot get close enough to the middle bars to untangle the chain (sometimes called the "knitting

Figure 10. A locked polygonal arc in 3D with 5 bars [Cantarella and Johnston 1998; Biedl et al. 2001].

needles" example). More precisely, because the ends of the chain remain outside a sphere with radius s and centered at one of the middle vertices, we can connect the ends of the chain with an unknotted flexible cord outside the sphere, and any straightening motion unties the resulting knot, which is impossible without crossings [Biedl et al. 2001].

Alt, Knauer, Rote, and Whitesides [Alt et al. 2004] proved that it is PSPACE-hard to decide whether a 3D polygonal arc (or a 2D polygonal tree) can be reconfigured between two specified configurations. On the other hand, it remains open to determine the complexity of deciding whether a polygonal arc can be straightened. The next two sections describe special cases of 3D chains, more general than planar chains, that can be straightened and convexified.

2.3.7. Simple projection. The "almost knottedness" of the example in Section 2.3.6 suggests that polygonal chains having simple orthogonal projections can always be straightened or convexified. This fact is established by two papers [Biedl et al. 2001; Calvo et al. 2001]. In addition, there is a polynomial-time algorithm to decide whether a polygonal chain has a simple projection, and if so find a suitable plane for projection [Bose et al. 1999].

For a polygonal arc with a simple orthogonal projection, the straightening method is relatively straightforward [Biedl et al. 2001]. The basic idea is to process the arc from one end to the other, accumulating bars into a compact "accordion" (x-monotone chain) lying in a plane orthogonal to the projection plane, in which each bar is nearly vertical. Once this accumulation is complete, the planar accordion is unfolded joint-by-joint into a straight arc. We observe that a similar algorithm can be used to fold a polygonal tree with a simple orthogonal projection into a generalized accordion, which can then be folded into a flat configuration.

For a polygonal cycle with a simple orthogonal projection, the convexification method is based on two steps [Calvo et al. 2001]. First, the projection of the polygon is convexified via the results described in Section 2.2.1, by folding the 3D polygon to track the shadow, keeping constant the ascent of each bar. Second, Calvo, Krizanc, Morin, Soss, and Toussaint [Calvo et al. 2001] develop an algorithm for convexifying a polygon with convex projection. The basic idea is to reconfigure the convex projection into a triangle, and stretch each accordion formed by an edge in the projection. In linear time they show how to compute a motion for the second step that consists of $O(n)$ simple moves, each changing at most seven vertex angles.

2.3.8. Interlocked chains in 3D. Although we have settled on page 170 the question of when *one* chain can lock (only in 3D), the conditions that permit pairs of chains to "interlock" are largely unknown. This line of investigation was prompted by a question posed by Anna Lubiw [Demaine and O'Rourke 2001]: into how many pieces must an n-bar 3D chain be cut (at vertices) so that the pieces can be separated and straightened? It is now known that the chain need be fractured into no more than $\lceil n/2 \rceil - 1$ pieces [Demaine et al. 2002b] but this upper bound is likely not tight: the only lower bound known is $\lfloor (n-1)/4 \rfloor$.

A collection of disjoint, noncrossing chains can be *separated* if, for any distance d, there is a non-self-crossing motion that results in every pair of points on different chains being separated by at least d. If a collection cannot be separated, its chains are *interlocked*. Which collections of relatively short chains can interlock was investigated in several papers [Demaine et al. 2003c; Demaine et al. 2002b]. Three typical results (all for chains with universal joints) are as follows:

(i) No pair of 3-bar open chains can interlock, even with an arbitrary number of additional 2-bar open chains.
(ii) A 3-bar open chain can interlock with a 4-bar closed chain. (See Figure 11.)
(iii) A 3-bar open chain can interlock with a 4-bar open chain.

The proof of the first result (for just a pair of 3-bar chains) identifies a plane parallel to and separating the middle bars of each chain, and then nonuniformly scales the coordinate system to straighten the other links while avoiding intersections. The second result uses a topological argument based on "links" (multicomponent knots), in a manner similar to the use of knots in the proof that the chain in Figure 10 is locked. The proof of the third listed result is quite intricate, relying on ad hoc geometric arguments [Demaine et al. 2002b]. There are many open problems here, one of the most intriguing being this: what is the smallest k that permits a k-bar open chain to interlock with a 2-bar open chain? (See [Glass et al. 2004].)

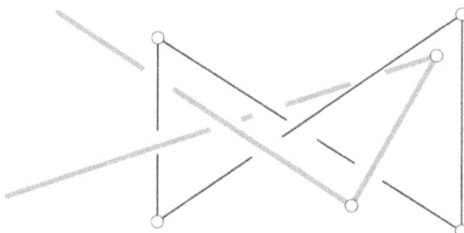

Figure 11. A 3-bar open chain (grey) interlocked with a 4-bar closed chain (black).

2.4. Fundamental questions in 4D and higher dimensions. In all dimensions higher than 3, it is known that all knots are trivial; analogously, all polygonal arcs can be straightened, all polygonal cycles can be convexified, and all polygonal trees can be flattened [Cocan and O'Rourke 2001] (page 170). Intuitively, this result holds because the number of degrees of freedom of any vertex is at least two higher than the dimensionality of the obstacles imposed by any bar. This property allows Cocan and O'Rourke [2001] to establish, for example, that the last bar of a polygonal arc can be unfolded by itself to any target position that is simple.

Cocan and O'Rourke [2001] show how to straighten an arc using $O(n)$ simple moves that can be computed in $O(n^2)$ time and $O(n)$ space. On the other hand, their method for convexifying a polygon requires $O(n^6)$ simple moves and $O(n^6 \log n)$ time to compute.

2.5. Protein folding. Protein folding [Chan and Dill 1993; Hayes 1998; Merz and Le Grand 1994] is an important problem in molecular biology because it is generally believed that the folded structure of a protein (the fundamental building block of life) determines its function and behavior.

2.5.1. Connection to linkages. A protein can be modeled by a linkage in which the vertices represent amino acids and the bars represent bonds connecting them. The bars representing bonds are typically close in length, within a factor less than two. Depending on the level of detail, the protein can be modeled as a tree (more precise) or as a chain (less precise).

An amazing property of proteins is that they fold quickly and consistently to a minimum-energy configuration. Understanding this motion has immediate connections to linkage folding in 3D. A central unsolved theoretical question [Biedl et al. 2001] arising in this context is whether every equilateral polygonal arc in 3D can be straightened. Cantarella and Johnston [Cantarella and Johnston 1998] proved that this is true for arcs of at most 5 bars. More generally, can every equilateral polygonal tree in 3D be flattened?

2.5.2. Fixed-angle linkages. A more accurate mathematical model of foldings of proteins is not by linkages whose vertices are universal joints, but rather by *fixed-angle* linkages in which each vertex forms a fixed angle between its incident bars. This angular constraint roughly halves the number of degrees of freedom in the linkage; the basic motion is rotating a portion of the linkage around a bar of the linkage. Foldings of such linkages have been explored extensively by Soss and Toussaint [Soss and Toussaint 2000; Soss 2001]. For example, they prove in [Soss and Toussaint 2000] that it is NP-complete to decide whether a fixed-angle polygonal arc can be flattened (reconfigured to lie a plane), and in [Soss 2001] that it is NP-complete to decide whether a fixed-angle polygonal arc can be folded into its mirror image.

More positive results analyze the polynomial complexity of determining the maximum extent of a rotation around a bar: Soss and Toussaint [Soss and Toussaint 2000; Soss 2001] prove an $O(n^2)$ upper bound, and Soss, Erickson, and Overmars [Soss 2001; Soss et al. 2003] give a 3SUM-hardness reduction, suggesting an $\Omega(n^2)$ lower bound.

Another line of investigation on fixed-angle chains was opened in [Aloupis et al. 2002a; Aloupis et al. 2002b]. Define a linkage X to be *flat-state connected* if, for each pair of its flat realizations x_1 and x_2, there is a reconfiguration from x_1 to x_2 that avoids self-intersection throughout. In general this motion alters the linkage to nonflat configurations in $\mathbb{R}^3$ intermediate between the two flat states. The main question is to determine whether every fixed-angle open chain is flat-state connected. It has been established that the answer is YES for chains all of whose fixed angles between consecutive bars are nonacute [Aloupis et al. 2002a], and although other special cases have been settled [Aloupis et al. 2002b], the main question remains open.

2.5.3. Producible chains. A connection between fixed-angle nonacute chains and a model of protein production was recently established in [Demaine et al. 2003b]. Here the ribosome — the "machine" that creates protein chains in biological cells — is modeled as a cone, with the fixed-angled chain produced bar-by-bar inside and emerging through the cone's apex. A configuration of a chain is said to be α-*producible* if there exists a continuous motion of the chain as it is created by the above model from within a cone of half-angle $\alpha \le \pi/2$. The main result of [Demaine et al. 2003b] is a theorem that identifies producible with flattenable chains, in this sense: a configuration of a chain whose fixed angles are $\ge \pi - \alpha$, for $\alpha \le \pi/2$, is α-producible if and only if it is flattenable. For example, for $\alpha = 45°$, this theorem says that a fixed-135°-angle chain (which is nonacute) is producible within a 90° cone if and only if that configuration is flattenable.

The proof uses a coiled cannonical configuration of the chain, which can be obtained by time-reversal of the production steps, winding the chain inside the cone. This canonical form establishes that all α-producible chains can be reconfigured to one another. Then it is shown how to produce any flat configuration by rolling the cone around on the plane into which the flat chain is produced. Because locked chains are not flattenable, the equivalence of producible and flattenable configurations shows that cone production cannot lead to locked configurations. This result in turn leads to the conclusion that the producible chains are rare, in a technical sense, suggesting that the entire configuration space for folding proteins might not need to be searched.

2.5.4. The H-P model. So far in this section we have not considered the forces involved in protein folding in nature. There are several models of these forces.

One of the most popular models of protein folding is the hydrophobic-hydrophilic (H-P) model [Chan and Dill 1993; Dill 1990; Hayes 1998], which defines both a geometry and a quality metric of foldings. This model represents a protein as a chain of amino acids, distinguished into two categories, hydrophobic (H) and hydrophilic (P). A folding of such a protein chain in this model is an embedding along edges of the square lattice in 2D or the cubic lattice in 3D without self-intersection. The optimum or minimum-energy folding maximizes the number of hydrophobic (H) nodes that are adjacent in the lattice. Intuitively, this metric causes hydrophobic amino acids to avoid the surrounding water.

This combinatorial model is attractive in its simplicity, and already seems to capture several essential features of protein folding such as the tendency for the hydrophobic components to fold to the center of a globular protein [Chan and Dill 1993]. While a 3D H-P model most naturally matches the physical world, in fact it is more realistic as a 2D model for computationally feasible problem sizes. The reason for this is that the perimeter-to-area ratio of a short 2D chain is a close approximation to the surface-to-volume ratio of a long 3D chain [Chan and Dill 1993; Hayes 1998].

Much work has been done on the H-P model [Berger and Leighton 1998; Chan and Dill 1991; Chan and Dill 1990; Crescenzi et al. 1998; Hart and Istrail 1996; Lau and Dill 1989; Lau and Dill 1990; Lipman and Wilber 1991; Unger and Moult 1993a; Unger and Moult 1993b; Unger and Moult 1993c]. Recently, on the computational side, Berger and Leighton [Berger and Leighton 1998] proved NP-completeness of finding the optimal folding in 3D, and Crescenzi et al. [Crescenzi et al. 1998] proved NP-completeness in 2D. Hart and Istrail [Hart and Istrail 1996] have developed a 3/8-approximation in 3D and a 1/4-approximation in 2D for maximizing the number of hydrophobic-hydrophobic adjacencies.

Aichholzer, Bremner, Demaine, Meijer, Sacristan, and Soss [Aichholzer et al. 2003] have begun exploring an important yet potentially more tractable aspect of protein folding: can we design a protein that folds stably into a desired shape? In the H-P model, a protein folds *stably* if it has a unique minimum-energy configuration. So far, Aichholzer et al. [Aichholzer et al. 2003] have proved the existence of stably folding proteins of all lengths divisible by 4, and for closed chains of all possible (even) lengths. It remains open to characterize the possible shapes (connected subsets of the square grid) attained by stable protein foldings.

3. Paper

Paper folding (origami) has led to several interesting mathematical and computational questions over the past fifteen years or so. A piece of paper, normally a (solid) polygon such as a square or rectangle, can be folded by a continuous motion that preserves the distances on the surface and does not cause the paper to properly self-intersect. Informally, paper cannot tear, stretch, or cross itself, but may otherwise bend freely. (There is a contrast here to folding other materials, such as sheet metal, that must remain piecewise planar throughout the folding process.) Formally, a folding is a continuum of isometric embeddings of the piece of paper in $\mathbb{R}^3$. However, the use of the term "embedding" is weak:

paper is permitted to touch itself provided it does not properly cross itself. In particular, a *flat folding* folds the piece of paper back into the plane, and so the paper must necessarily touch itself. We frequently ignore the continuous motion of a folding and instead concentrate on the final folded state of the paper; in the case of a flat folding, the flat folded state is called a *flat origami*. This concentration on the final folded state was recently justified by a proof that there always exists a continuous motion from a planar polygonal piece of paper to any "legal" folded state [Demaine et al. 2004].

Some of the pioneering work in origami mathematics (see Section 3.3.1) studies the *crease pattern* that results from unfolding a flat origami, that is, the graph of edges on the paper that fold to edges of a flat origami. Stated in reverse, what crease patterns have flat foldings? Various necessary conditions are known [Hull 1994; Justin 1994; Kawasaki 1989], but there is little hope for a polynomial characterization: Bern and Hayes [Bern and Hayes 1996] have shown that this decision problem is NP-hard.

A more recent trend, as in [Bern and Hayes 1996], is to explore *computational origami*, the algorithmic aspects of paper folding. This field essentially began with Robert Lang's work on algorithmic origami design [Lang 1996], starting around 1993. Since then, the field of computational origami has grown significantly, in particular in the past two years by applying computational geometry techniques. This section surveys this work. See also [Demaine and Demaine 2002].

3.1. Categorization.

Most results in computational origami fall under one or more of three categories: universality results, efficient decision algorithms, and computational intractability results. This categorization applies more generally to folding and unfolding, but is particularly useful for results in computational origami.

A *universality result* shows that, subject to a certain model of folding, everything is possible. For example, any tree-shaped origami base (Section 3.2.2), any polygonal silhouette (Section 3.2.1), and any polyhedral surface (Section 3.2.1) can be folded out of a sufficiently large piece of paper. Universality results often come with efficient algorithms for finding the foldings; pure existence results are rare.

When universality results are impossible (some objects cannot be folded), the next-best result is an *efficient decision algorithm* to determine whether a given object is foldable. Here "efficient" normally means "polynomial time." For example, there is a polynomial-time algorithm to decide whether a "map" (grid of creases marked mountain and valley) can be folded by a sequence of "simple folds" (Section 3.3.4).

Not all paper-folding problems have efficient algorithms, and this can be proved by a *computational intractability result*. For example, it is NP-hard to tell whether a given crease pattern folds into some flat origami (Section 3.3.2),

even when folds are restricted to simple folds (Section 3.3.4). These results imply
that there are no polynomial-time algorithms for these problems, unless some
of the hardest computational problems known can also be solved in polynomial
time, which is generally deemed unlikely.

We further distinguish computational origami results as addressing either
origami design or *origami foldability*. In origami design, some aspects of the
target configuration are specified, and the goal is to design a suitable detailed
folded state that can be folded out of paper. In origami foldability, the target configuration is unspecified and arbitrary; rather, the initial configuration
is specified, in particular the crease pattern, possibly marked with mountains
and valleys, and the goal is to fold something (anything) using precisely those
creases. While at first it may seem that understanding origami foldability is a
necessary component for origami design, the results indicate that in fact origami
design is easier to solve than origami foldability, which is usually intractable.

Our survey of computational origami is divided accordingly into Section 3.2
(origami design) and Section 3.3 (origami foldability).

3.2. Origami design. We define *origami design* loosely as, given a piece of
paper, fold it into an object with certain desired properties, e.g., a particular
shape. The natural theoretical version of this problem is to ask for an origami
with a specific silhouette or three-dimensional shape; this problem can be solved
in general (Section 3.2.1), although the algorithms developed so far do not lead
to practical foldings. A specific form of this problem has been solved for practical purposes by Lang's tree method (Section 3.2.2), which has brought modern
origami design to a new level of complexity. Related to this work is the problem
of folding a piece of paper to align a prescribed graph (Section 3.2.3), which can
be used for a magic trick involving folding and one complete straight cut.

3.2.1. Silhouettes and polyhedra. A direct approach to origami design is to
specify the exact final shape that the paper should take. More precisely, suppose
we specify a particular flat silhouette, or a three-dimensional polyhedral surface,
and desire a folding of a sufficiently large square of paper into precisely this
object, allowing coverage by multiple layers of paper. For what polyhedral shapes
is this possible? This problem is implicit throughout origami design, and was
first formally posed in [Bern and Hayes 1996]. The surprising answer is "always,"
as established by Demaine, Demaine, and Mitchell in 1999 [Demaine et al. 1999c;
2000d].

The basic idea of the approach is to fold the piece of paper into a thin strip,
and then wrap this strip around the desired shape. This wrapping can be done
particularly efficiently using methods in computational geometry. Specifically,
three algorithms are described in [Demaine et al. 2000d] for this process. One
algorithm optimizes paper usage: the amount of paper required can be made
arbitrarily close to the surface area of the shape, but only at the expense of increasing the aspect ratio of the rectangular paper. Another algorithm maximizes

Figure 12. A flat folding of a square of paper, black on one side and white on the other side, designed by John Montroll [Montroll 1991, pp. 94–103]. (Figure 1(b) of [Demaine et al. 2000d].)

the width of the strip subject to some constraints. A third algorithm places the visible *seams* of the paper in any desired pattern forming a decomposition of the sides into convex polygons. In particular, the number and total length of seams can be optimized in polynomial time in most cases [Demaine et al. 2000d].

All of these algorithms allow an additional twist: the paper may be colored differently on both sides, and the shape may be two-colored according to which side should be showing. In principle, this allows the design of two-color models similar to the models in Montroll's *Origami Inside-Out* [Montroll 1993]. An example is shown in Figure 12.

Because of the use of thin strips, none of these methods lead to practical foldings, except for small examples or when the initial piece of paper is a thin strip. Nonetheless, the universality results of [Demaine et al. 2000d] open the door to many new problems. For example, how small a square can be folded into a desired object, e.g., a $k \times k$ chessboard? This optimization problem remains open even in this special case, as do many other problems about finding efficient, practical foldings of silhouettes, two-color patterns, and polyhedra.

3.2.2. Tree method. The *tree method of origami design* is a general approach for "true" origami design (in contrast to the other topics that we discuss, which involve less usual forms of origami). In short, the tree method enables design of efficient and practical origami within a particular class of three-dimensional shapes, most useful for origami design. Some components of this method, such as special cases of the constituent molecules and the idea of disk packing, as well as other methods for origami design, have been explored in the Japanese technical origami community, in particular by Jun Maekawa, Fumiaki Kawahata, and Toshiyuki Meguro. This work has led to several successful designs, but a full survey is beyond the scope of this paper; see [Lang 2003; Lang 1998]. It suffices to say that the explosion in origami design over the last 30 years, during which the majority of origami models have been designed, may largely be due to an understanding of these general techniques.

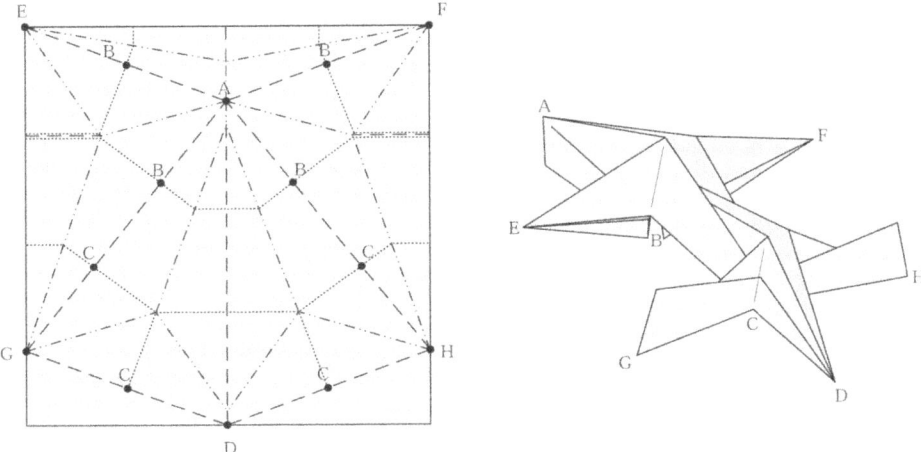

Figure 13. Lang's TreeMaker applied to an 8-vertex tree to produce a lizard base. (Figure 2.1.11 of [Lang 1998].)

Here we concentrate on Robert Lang's work [1994a; 1994b; 1996; 1998; 2003], which is the most extensive. Over the past decade, starting around 1993, Lang has developed the tree method to the point where an algorithm and computer program have been explicitly defined and implemented. Anyone with a Macintosh computer can experiment with the tree method using Lang's program TreeMaker [Lang 1998].

The tree method allows one to design an origami *base* in the shape of a specified tree with desired edge lengths, which can then be folded and shaped into an origami model. See Figure 13 for an example. More precisely, the tree method designs a *uniaxial base* [Lang 1996], which must have the following properties: the base lies above and on the xy-plane, all facets of the base are perpendicular to the xy-plane, the projection of the base to the xy plane is precisely where the base comes in contact with the xy-plane, and this projection is a one-dimensional tree.

It is known that every metric tree (unrooted tree with prescribed edge lengths) is the projection of a uniaxial base that can be folded from, e.g., a square. The tree method gives an algorithm to find the folding that is optimal in the sense that it folds the uniaxial base with the specified projection using the smallest possible square piece of paper (or more generally, using the smallest possible scaling of a given convex polygon). These foldings have led to many impressive origami designs; see [Lang 2003] in particular.

There are two catches to this result. First, it is currently unknown whether the prescribed folding self-intersects, though it is conjectured that self-intersection does not arise, and this conjecture has been verified on extensive examples. Second, the optimization problem is difficult, a fairly general form of nonlinear constrained optimization. So while optimization is possible in principle in finite

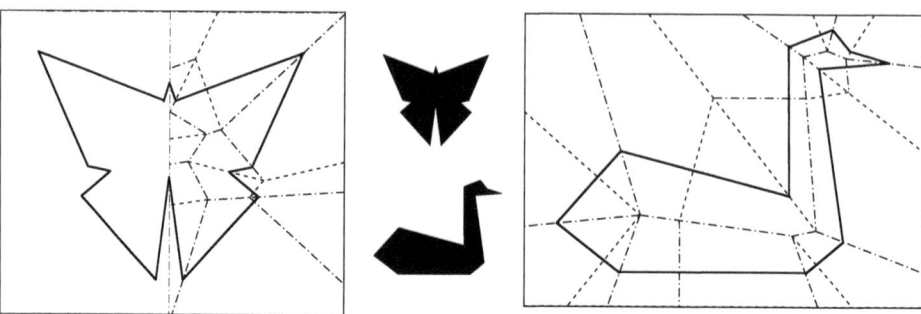

Figure 14. Crease patterns for folding a rectangle of paper flat so that one complete straight cut makes a butterfly (left) or a swan (right), based on [Demaine et al. 2000c; Demaine et al. 1999b].

time, in practice heuristics must be applied; fortunately, such heuristics frequently yield good, practical solutions. Indeed, additional practical constraints can be imposed, such as symmetry in the crease pattern, or the constraint that angles of creases are integer multiples of some value (e.g., 22.5°) subject to some flexibility in the metric tree.

3.2.3. One complete straight cut. Take a piece of paper, fold it flat, make one complete straight cut, and unfold the pieces. What shapes can result? This *fold-and-cut* problem was first formally stated by Martin Gardner [1960], but goes back much further, to a Japanese puzzle book [Sen 1721] and perhaps to Betsy Ross in 1777 [Harper's 1873]; see also [Houdini 1922, pp. 176–177]. A more detailed history can be found in [Demaine et al. 2000c].

More formally, given a planar graph drawn with straight edges on a piece of paper, can the paper be folded flat so as to map the entire graph to a common line, and map nothing else to that line? The surprising answer is that this is always possible, for any collection of line segments in the plane, forming nonconvex polygons, adjoining polygons, nested polygons, etc. There are two solutions to the problem. The first (partial) solution [Demaine et al. 2000c; Demaine et al. 1999b] is based on a structure called the straight skeleton, which captures the symmetries of the graph, thereby exploiting a more global structure of the problem. This solution applies to a large class of instances, which we do not describe in detail here. See Figure 14 for two examples. The second (complete) solution [Bern et al. 2002] is based on disk packing to make the problem more local, and achieves efficient bounds on the number of creases.

While this problem may not seem directly connected to pure paper folding because of the one cut, the equivalent problem of folding a piece of paper to line up a given collection of edges is in fact closely connected to origami design. Specifically, one subproblem that arises in TreeMaker (Section 3.2.2) is that the piece of paper is decomposed into convex polygons, and the paper must be folded flat so as to line up all the edges of the convex polygons, and place the interior

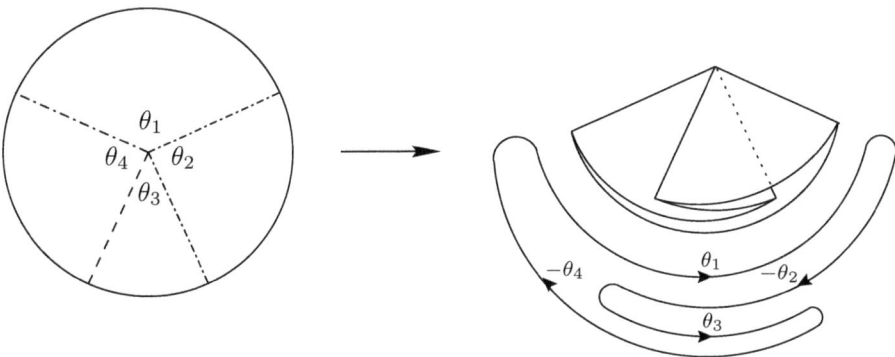

Figure 15. A locally flat-foldable vertex: $\theta_1 + \theta_3 + \cdots = \theta_2 + \theta_4 + \cdots = 180°$.

of these polygons above this line. The fold-and-cut problem is a generalization of this situation to arbitrary graphs: nonconvex polygons, nested polygons, etc. In TreeMaker, there are important additional constraints in how the edges can be lined up, called path constraints, which are necessary to enforce the desired geometric tree. These constraints lead to additional components in the solution called *gussets*.

3.3. Origami foldability. We distinguish origami design from *origami foldability* in which the starting point is a given crease pattern and the goal is to fold an origami that uses precisely these creases. (Arguably, this is a special case of our generic definition of origami design, but we find it a useful distinction.) The most common case studied is when the resulting origami should be flat, i.e., lie in a plane.

3.3.1. Local foldability. For crease patterns with a single vertex, it is relatively easy to characterize flat foldability. Without specified crease directions, a single-vertex crease pattern is flat-foldable precisely if the alternate angles around the vertex sum to 180°; see Figure 15. This is known as Kawasaki's theorem [Bern and Hayes 1996; Hull 1994; Justin 1994; Kawasaki 1989]. When the angle condition is satisfied, a characterization of valid mountain-valley assignments and flat foldings can be found in linear time [Bern and Hayes 1996; Justin 1994], using Maekawa's theorem [Bern and Hayes 1996; Hull 1994; Justin 1994] and another theorem of Kawasaki [Bern and Hayes 1996; Hull 1994; Kawasaki 1989] about constraints on mountains and valleys. In particular, Hull has shown that the number of distinct mountain-valley assignments of a vertex can be computed in linear time [Hull 2003].

A crease pattern is called *locally foldable* if there is a mountain-valley assignment so that each vertex locally folds flat, i.e., a small disk around each vertex folds flat. Testing local foldability is nontrivial because each vertex has flexibility in its assignment, and these assignments must be chosen consistently: no crease should be assigned both mountain and valley by the two incident vertices.

Bern and Hayes [Bern and Hayes 1996] proved that consistency can be resolved efficiently when it is possible: local foldability can be tested in linear time.

3.3.2. Existence of folded states. Given a crease pattern, does it have a flat folded state? Bern and Hayes [Bern and Hayes 1996] have proved that this decision problem is NP-hard, and thus computationally intractable. Because local foldability is easy to test, the only difficult part is global foldability, or more precisely, computing a valid *overlap order* of the crease faces that fold to a common portion of the plane. Indeed, Bern and Hayes [Bern and Hayes 1996] prove that, given a crease pattern and a mountain-valley assignment that definitely folds flat, finding the overlap order of a flat folded state is NP-hard.

3.3.3. Equivalence to continuous folding process. In the previous section we have alluded to the difference between two models of folding: the final folded state (specified by a crease pattern, mountain-valley or angle assignment, and overlap order) and a continuous motion to bring the paper to that folded state. Basically all results, in particular those described so far, have focused on the former model: proving that a folded state exists with the desired properties. Intuitively, by appropriately flexing the paper, any folded state can be reached by a continuous motion, so the two models should be equivalent. Only recently has this been proved, initially for rectangular pieces of paper [Demaine and Mitchell 2001], and recently for general polygonal pieces of paper [Demaine et al. 2004] but overall the number of creases is uncountably infinite. An interesting open problem is whether a finite crease pattern suffices.

The only other paper of which we are aware that explicitly constructs continuous folding processes is [Demaine and Demaine 1997]. This paper proves that every convex polygon can be folded into a uniaxial base via Lang's universal molecule [Lang 1998] without gussets. Furthermore, unlike [Demaine and Mitchell 2001], no additional creases are introduced during the motion, and each crease face remains flat. This result can be used to animate the folding process.

3.3.4. Map folding: sequence of simple folds. In contrast to the complex origami folds arising from reaching folded states [Demaine and Demaine 1997; Demaine and Mitchell 2001], we can consider the less complex model of simple folds. A *simple fold* (or *book fold*) is a fold by $\pm 180°$ along a single line. Examples are shown in Figure 16. This model is closely related to "pureland origami", introduced by Smith [1976; 1980; 1988; 1993].

We can ask the same foldability questions for a sequence of simple folds. Given a crease pattern, can it be folded flat via a sequence of simple folds? What if a particular mountain-valley assignment is imposed?

An interesting special case of these problems is *map folding* (see Figure 16): given a rectangle of paper with horizontal and vertical creases, each marked mountain or valley, can it be folded flat via a sequence of simple folds? Traditionally, map folding has been studied from a combinatorial point of view; see,

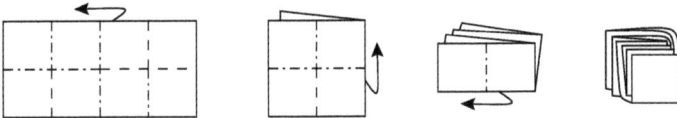

Figure 16. Folding a 2×4 map via a sequence of 3 simple folds.

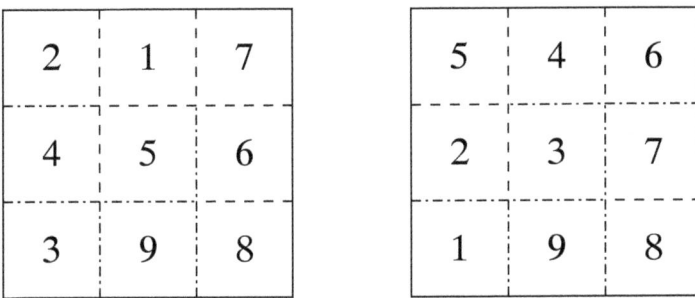

Figure 17. Two maps that cannot be folded by simple folds, but can be folded flat. (These are challenging puzzles.) The numbering indicates the overlap order of faces. (Figure 12 of [Arkin et al. 2004].)

e.g., [Lunnon 1968; Lunnon 1971]. Arkin, Bender, Demaine, Demaine, Mitchell, Sethia, and Skiena [Arkin et al. 2004] have shown that deciding foldability of a map by simple folds can be solved in polynomial time. If the simple folds are required to fold all layers at once, the running time is at most $O(n \log n)$, and otherwise the running time is linear.

Surprisingly, slight generalizations of map folding are (weakly) NP-complete [Arkin et al. 2004]. Deciding whether a rectangle with horizontal, vertical, and diagonal ($\pm 45°$) creases can be folded via a sequence of simple folds is NP-complete. Alternatively, if the piece of paper is more general, a polygon with horizontal and vertical sides, and the creases are only horizontal and vertical, the same problem is NP-complete.

These hardness results are *weak* in the sense that they leave open the existence of a *pseudopolynomial-time* algorithm, whose running time is polynomial in the total length of creases. Another intriguing open problem, posed by Jack Edmonds, is the complexity of deciding whether a map has some flat folded state, as opposed to a folding by a sequence of simple folds. Examples of maps in which these two notions of foldability differ are shown in Figure 17.

3.4. Flattening polyhedra. When one flattens a cardboard box for recycling, generally the surface is cut open. Suppose instead of allowing cuts to a polyhedral surface in order to flatten it, we treat it as a piece of paper and fold as in origami. We run into the same dichotomy as in Section 3.3.2: do we want a continuous motion of the polyhedron, or does a description of the final folded state suffice? If we start with a convex polyhedron, and each face of the crease pattern must remain rigid during the folding, then Connelly's extension [Connelly 1980] of

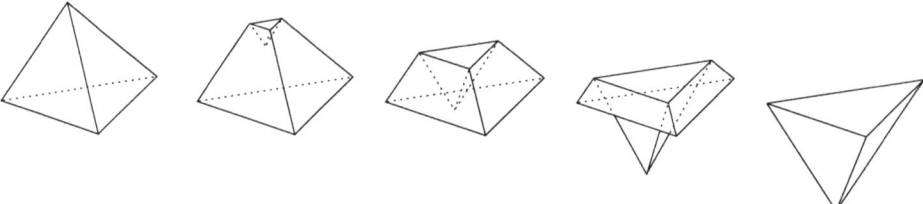

Figure 18. Inverting a tetrahedral cone by a continuous isometric motion. Based on Figure 2.5 of [Connelly 1993].

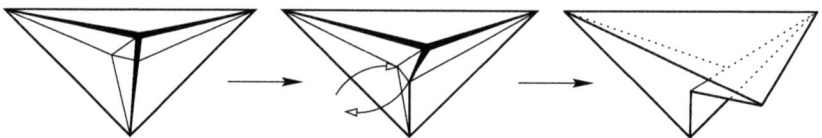

Figure 19. Flattening a tetrahedron, from left to right. Note that the faces are not flat in the middle picture.

Cauchy's rigidity theorem [1813] (see also [Cromwell 1997, pp. 219–247]) says that the polyhedron cannot fold at all. Even if we start with a nonconvex polyhedron and keep each face of the crease pattern rigid, the Bellows Theorem [Connelly et al. 1997] says that the volume of the polyhedron cannot change, so foldings are limited. However, if we allow the paper to curve (e.g., introduce new creases) during the motion, as in origami, then folding becomes surprisingly flexible. For example, a cone can be inverted [Connelly 1993]; see Figure 18.

A natural question is whether every polyhedron can be *flattened*: folded into a flat origami. Intuitively, this can be achieved by applying force to the polyhedral model, but in practice this can easily lead to tearing. There is an interesting connection of this problem to a higher-dimensional version of the fold-and-cut problem from Section 3.2.3. Given any polyhedral complex, can $\mathbb{R}^3$ be folded (through $\mathbb{R}^4$) "flat" into $\mathbb{R}^3$ so that the surface of the polyhedral complex maps to a common plane, and nothing else maps to that plane? While the applicability of four dimensions is difficult to imagine, the problem's restriction to the surface of the complex is quite practical, e.g., in packaging: *flatten* the polyhedral complex into a flat folded state, without cutting or stretching the paper.

The flattening problem remains open if we desire a continuous folding process into the flat state. If we instead focus on the existence of a flat folded state of a polyhedron, then much more is known. Demaine, Demaine, and Lubiw[7] have shown how to flatten several classes of polyhedra, including convex polyhedra and orthogonal polyhedra. See Figure 19 for an example. Recently, Demaine,

Demaine, Hayes, and Lubiw[8] have shown that all polyhedra have flat folded states. They conjecture further that every polyhedral complex can be flattened.

A natural question is whether the methods of Demaine and Mitchell [Demaine and Mitchell 2001] and [Demaine et al. 2004] described in Section 3.3.3 can be generalized to show that these folded states induce continuous folding motions as in Figure 18.

4. Polyhedra

A standard method for building a model of a polyhedron is to cut out a flat *net* or *unfolding*, fold it up, and glue the edges together so as to make precisely the desired surface. Given the polyhedron of interest, a natural problem is to find a suitable unfolding. On the other hand, given a polygonal piece of paper, we might ask whether it can be folded and its edges can be glued together so as to form a convex polyhedron. These two questions are addressed in Sections 4.1 and 4.2, respectively. Section 4.3 extends different forms of the latter question to nonconvex polyhedra. Section 4.4 connects these problems to linkage and paper folding.

4.1. Unfolding polyhedra. A classic open problem is whether (the surface of) every convex polyhedron can be cut along some of its edges and unfolded into one flat piece without overlap [Shephard 1975; O'Rourke 2000]. Such *edge-unfoldings* go back to Dürer [1525], and have important practical applications in manufacturing, such as sheet-metal bending [O'Rourke 2000; Wang 1997]. It seems folklore that the answer to this question should be YES, but the evidence for a positive answer is actually slim. Only very simple classes of polyhedra are known to be edge-unfoldable; for example, pyramids, prisms, "prismoids,"[9] and other more specialized classes [Demaine and O'Rourke ≥ 2005]. In contrast, experiments by Schevon [Schevon 1989; O'Rourke 2000] suggest that a random edge-unfolding of a random polytope overlaps with probability 1. Of course, such a result would not preclude, for every polytope, the existence of at least one nonoverlapping edge-unfolding, or even that a large but subconstant fraction of the polytope's edge-unfoldings do not overlap. However, the unlikeliness of finding an unfolding by chance makes the search more difficult.

An easier version of this edge-unfolding problem is the *fewest-nets* problem: prove an upper bound on the number of pieces required by a multipiece non-overlapping edge unfolding of a convex polyhedron. The obvious upper bound is the number F of faces in the polyhedron; the original problem asks whether an upper bound of 1 is possible. The first bound of cF for $c < 1$ was obtained by

---

[7]Manuscript, March 2001.

[8]Manuscript in preparation.

[9]The convex hull of two equiangular convex polygons, oriented so that corresponding edges are parallel.

Michael Spriggs,[10] who established $c = 2/3$. The smallest value of c obtained so far[11] is $1/2$. Proving an upper bound that is sublinear in F would be a significant advancement.

We can also examine to what extent edge unfoldings can be generalized to nonconvex polyhedra. In particular, define a polyhedron to be *topologically convex* if its 1-skeleton (graph) is the 1-skeleton of a convex polyhedron. Does every topologically convex polyhedron have an edge-unfolding? In particular, every polyhedron composed of convex faces and homeomorphic to a sphere is topologically convex; can they all be edge-unfolded? This problem was posed by Schevon [Schevon 1987].

Bern, Demaine, Eppstein, Kuo, Mantler, and Snoeyink [Bern et al. 2003] have shown that the answer to both of these questions is NO: there is a polyhedron composed of triangles and homeomorphic to a sphere that has no (one-piece, nonoverlapping) edge-unfolding. The polyhedron is shown in Figure 20. It consists of four "hats" glued to the faces of a regular tetrahedron, such that only the peaks of the hats have positive curvature, that is, have less than 360° of incident material. This property limits the unfoldings significantly, because (1) any set of cuts must avoid cycles in order to create a one-piece unfolding, and (2) a leaf in a forest of cuts can only lie at a positive-curvature vertex of the polyhedron: a leaf at a negative-curvature vertex (more than 360° of incident material) would cause local overlap.

The complexity of deciding whether a given topologically convex polyhedron can be edge-unfolded remains open.

Another intriguing open problem in this area is whether every polyhedron homeomorphic to a sphere has *some* one-piece unfolding, not necessarily using cuts along edges. It is known that every convex polyhedron has an unfolding in this model, allowing cuts across the faces of the polytope. Specifically, the *star unfolding* [Agarwal et al. 1997; Aronov and O'Rourke 1992] cuts the shortest paths from a common source point to each vertex of the polytope, and the *source unfolding* [Mitchell et al. 1987] cuts the points with more than one shortest path to a common source. Both of these unfoldings avoid overlap, the star unfolding being the more difficult case to establish [Aronov and O'Rourke 1992]. The source unfolding (but not the star unfolding) also generalizes to unfold convex polyhedra in higher dimensions [Miller and Pak 2003].

But many nonconvex polyhedra also have such unfoldings. For example, Figure 20 illustrates one for the polyhedron described above. Biedl, Demaine, Demaine, Lubiw, Overmars, O'Rourke, Robbins, and Whitesides [Biedl et al. 1998] have shown how to unfold many orthogonal polyhedra, even with holes and knotted topology, although it remains open whether all orthogonal polyhedra

---

[10]Personal communication, August 2003.

[11]Personal communication from Vida Dujmović, Pat Morin, and David Wood, February 2004.

Figure 20. (Left) Simplicial polyhedron with no edge-unfolding. (Right) An unfolding when cuts are allowed across faces.

can be unfolded. The only known scenario that prevents unfolding altogether [Bern et al. 2003] is a polyhedron with a single vertex of negative curvature (see Figure 21), but this requires the polyhedron to have boundary (edges incident to only one face).

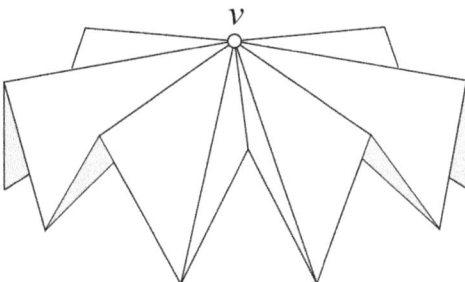

Figure 21. A polyhedron with boundary that has no one-piece unfolding even when cuts are allowed across faces. Vertex v has negative curvature, that is, more than 360° of incident material. (Based on Figure 9 of [Bern et al. 2003].)

A recent approach to unfolding both convex and nonconvex polyhedra in any dimension is the notion of "vertex-unfolding" [Demaine et al. 2003a]; see Figure 22. Specifically, a *vertex-unfolding* may cut only along edges of the polyhedron (like an edge-unfolding) but permits the facets to remain connected only at vertices (instead of along edges as in edge-unfolding). Thus, a vertex-unfolding is connected, but its interior may be disconnected, "pinching" at a vertex. This notion also generalizes to polyhedra in any dimension. Demaine, Eppstein, Erickson, Hart, and O'Rourke [Demaine et al. 2003a] proved that every simplicial

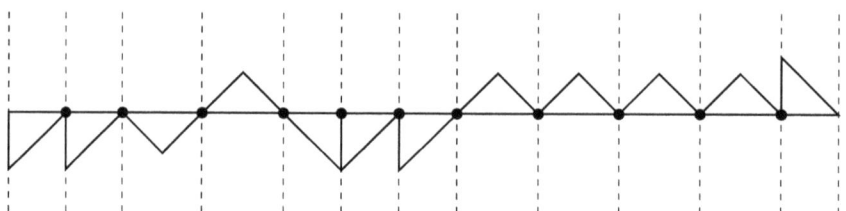

Figure 22. Vertex-unfolding of a triangulated cube with hinge points aligned. (Based on Figure 2 of [Demaine et al. 2003a].)

manifold in any dimension has a nonoverlapping vertex-unfolding. In particular, this result covers triangulated polyhedra in 3D, possibly with boundary, but it remains open to what extent vertex-unfoldings exist for polyhedra with nontriangular faces. For example, does every convex polyhedron in 3D have a vertex-unfolding?

4.2. Folding polygons into convex polyhedra. In addition to unfolding polyhedra into simple planar polygons, we can consider the reverse problem of folding polygons into polyhedra. More precisely, when can a polygon have its boundary glued together, with each portion gluing to portions of matching length, and the resulting topological object be folded into a *convex* polyhedron? (There is almost too much flexibility with nonconvex polyhedra for this problem, but see Section 4.3 for related problems of interest in this context.) A particular kind of gluing is an *edge-to-edge* gluing, in which each entire edge of the polygon is glued to precisely one other edge of the polygon. The existence of such a gluing requires a perfect pairing of edges with matching lengths.

4.2.1. Edge-to-edge gluings. Introducing this area, Lubiw and O'Rourke [Lubiw and O'Rourke 1996] showed how to test in polynomial time whether a polygon has an edge-to-edge gluing that can be folded into a convex polyhedron, and how to list all such edge-to-edge gluings in exponential time. A key tool in their work is a theorem of A. D. Aleksandrov [Alexandrov 1950]. The theorem states that a topological gluing can be realized geometrically by a convex polyhedron precisely if the gluing is topologically a sphere, and at most 360° of material is glued to any one point — that is, every point should have nonnegative curvature.

Based on this tool, Lubiw and O'Rourke use dynamic programming to develop their algorithms. There are $\Omega(n^2)$ subproblems corresponding to gluing subchains of the polygon, assuming that the two ends of the subchain have already been glued together. These subproblems are additionally parameterized by how much angle of material remains at the point to which the two ends of the chain glue in order to maintain positive curvature. It is this parameterization that forces enumeration of all gluings to take exponential time. But for the decision problem of the existence of any gluing, the remaining angle at the ends only needs to be bounded, and only polynomially many subproblems need to be considered, resulting in an $O(n^3)$ algorithm.

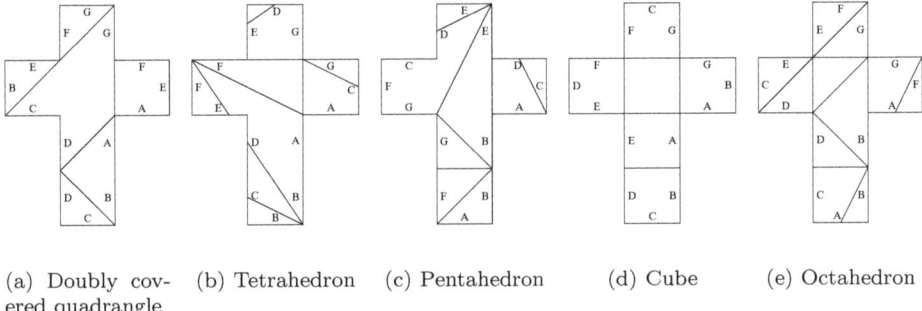

(a) Doubly covered quadrangle   (b) Tetrahedron   (c) Pentahedron   (d) Cube   (e) Octahedron

Figure 23. The five edge-to-edge gluings of the Latin cross [Lubiw and O'Rourke 1996].

A particularly surprising discovery from this work [Lubiw and O'Rourke 1996] is that the well-known "Latin cross" unfolding of the cube can be folded into exactly five convex polyhedra by edge-to-edge gluing: a doubly covered (flat) quadrangle, an (irregular) tetrahedron, a pentahedron, the cube, and an (irregular) octahedron. See Figure 23 for crease patterns and gluing instructions. These foldings are the subject of a video [Demaine et al. 1999a].

4.2.2. Non-edge-to-edge gluings. More recently, Demaine, Demaine, Lubiw, and O'Rourke [Demaine et al. 2000b; Demaine et al. 2002a] have extended this work in various directions, in particular to non-edge-to-edge gluings.

In contrast to edge-to-edge gluings, any convex polygon can be glued into a continuum of distinct convex polyhedra, making it more difficult for an algorithm to enumerate all gluings of a given polygon. Fortunately, there are only finitely many *combinatorially distinct* gluings of any polygon. For convex polygons, there are only polynomially many combinatorially distinct gluings, and they can be enumerated for a given convex polygon in polynomial time. This result generalizes to any polygon in which there is a constant bound on the sharpest angle. For general nonconvex polygons, there can be exponentially many ($2^{\Theta(n)}$) combinatorially distinct gluings, but only that many. Again this corresponds to an algorithm running in $2^{O(n)}$ time. Because of the exponential worst-case lower bound on the number of combinatorially distinct gluings, we are justified both here and in the enumeration algorithm of [Lubiw and O'Rourke 1996] to spend exponential time. It remains open whether there is an output-sensitive algorithm, whose running time is polynomial in the number of resulting gluings, or in the number of gluings desired by the user. For non-edge-to-edge gluings, it even remains open whether there is a polynomial-time algorithm to decide whether a gluing exists.

The algorithms for enumerating all non-edge-to-edge gluings have been implemented independently by Anna Lubiw (July 2000) and by Koichi Hirata [Hirata 2000] (June 2000). These programs have been applied to the example of the

Latin cross. There are surprisingly many more, but still finitely many, non-edge-to-edge gluings: a total of 85 distinct gluings (43 modulo symmetry). A manual reconstruction of the polyhedra resulting from these gluings reveals 23 distinct shapes: the cube, seven different tetrahedra, three different pentahedra, four different hexahedra, six different octahedra, and two flat quadrangles [Demaine et al. 2000a; Demaine and O'Rourke ≥ 2005].

Alexander, Dyson, and O'Rourke [Alexander et al. 2002] performed a case study of all the gluings of the square, reconstructing all the incongruent polyhedra that result. This situation is complicated by the existence of entire continua of gluings and polyhedra. Nonetheless, the entire configuration space of the polyhedra can be characterized, as shown in Figure 24. Although in this case it is connected, there are convex polygons of n vertices whose space of all gluings into polyhedra has $\Omega(n^2)$ connected components [Demaine and O'Rourke ≥ 2005]. Although it is almost certain that all of these gluings lead to distinct polyhedra, it seems difficult to establish this property without a method for reconstructing the three-dimensional structure, the topic of the next section.

4.2.3. Constructing polyhedra. Another intriguing open problem in this area [Demaine et al. 2002a] remains relatively unexplored: Aleksandrov's theorem implies that any valid gluing (homeomorphic to a sphere and having nonnegative curvature everywhere) can be folded into a unique convex polyhedron, but how efficiently can this polyhedron be constructed? The key difficulty here is to determine the dihedral angles of the polyhedron, that is, by how much each crease is folded. Finding a (superset of) the creases is straightforward:[12] every edge of the polyhedron is a shortest path between two positive-curvature vertices, so compute all-pairs shortest paths in the polyhedral metric defined by the gluing [Chen and Han 1996; Kaneva and O'Rourke 2000; Kapoor 1999].

Sabitov [Sabitov 1996] recently presented a finite algorithm for this reconstruction problem, reducing the problem to finding roots of a collection of polynomials of exponentially high degree. The algorithm is based on another his results [Sabitov 1998; Sabitov 1996] that expresses the volume of a triangulated polyhedron as the root of a polynomial in the edge lengths, independent of how the polyhedron is geometrically embedded in 3-space. (This result was also used to settle the famous Bellows Conjecture [Connelly et al. 1997].) Sabitov's algorithm was recently extended and its bounds improved by Fedorchuk and Pak [Fedorchuk and Pak 2004] to express the internal vertex-to-vertex diagonal lengths as roots of a polynomial of degree 4^m for a polyhedron of m edges. The polyhedron can easily be reconstructed from these diagonal lengths.

---

[12]Personal communication with Boris Aronov, June 1998. The essence of the argument is also present in [Alexandrov 1941].

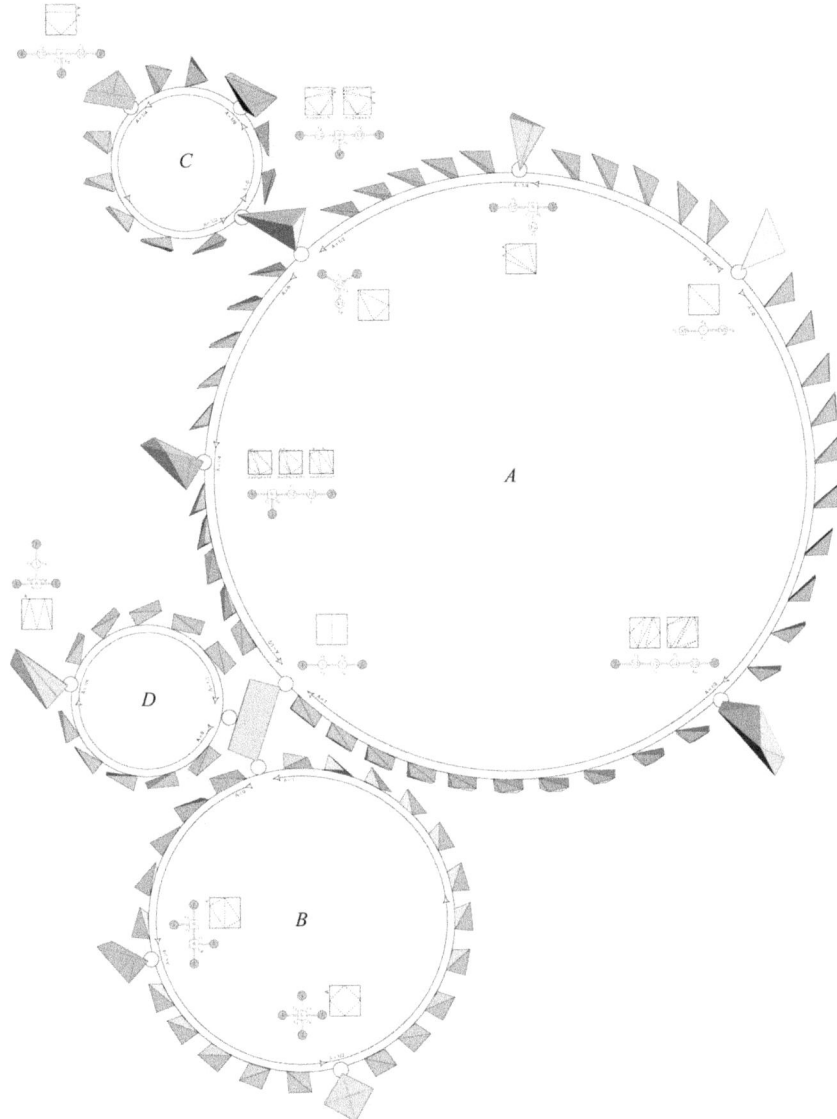

Figure 24. The continua of polyhedra foldable from a square. (Figure 2 of [Alexander et al. 2002].)

4.3. Folding nets into nonconvex polyhedra.

Define a *net* to be a connected edge-to-edge gluing of polygons to form a tree structure, the edges shared by polygons denoting creases. An open problem mentioned in Section 4.2.3 is deciding whether a given net can be folded into a convex polyhedron using only the given creases. More generally, we can ask whether a given net folds into a nonconvex polyhedron. Now Aleksandrov's theorem and Cauchy's rigidity theorem do not apply, so for a given gluing we are no longer easily guaranteed existence or uniqueness.

Given the dihedral angles associated with creases in the net, it is easy to decide foldability in polynomial time [Biedl et al. 1999b; Sun 1999]: we only need to check that edges match up and no two faces cross. Without the dihedral angles, when does a given net fold into any polyhedron? Biedl, Lubiw, and Sun [Biedl et al. 1999b; Sun 1999] proved a closely related problem to be weakly NP-complete: does a given orthogonal net (each face is an orthogonal polygon) fold into an orthogonal polyhedron? The difference with this problem is that it constrains each dihedral angle to be ±90°. It remained open whether this constraint actually restricted what polyhedra could be folded, even for this particular reduction. More generally, is there a nonorthogonal polyhedron (i.e., one that has at least one dihedral angle not a multiple of 90°) having orthogonal faces and that is homeomorphic to a sphere? The answer to this question (posed in [Biedl et al. 1999b]) turns out to be NO, as proved by Donoso and O'Rourke [Donoso and O'Rourke 2002]. The answer is YES, however, if the polyhedron is allowed to have genus 6 or larger; on the other hand, the answer remains NO for genus up to 2 [Biedl et al. 2002a]. It remains open whether such nonorthogonal polyhedra with orthogonal faces exist with genus 3, 4, or 5.

4.4. Continuously folding polyhedra. The results described so far for polyhedron folding and unfolding are essentially about folded or unfolded states, and not about the continuous process of reaching such states. In the context of paper folding, we saw in Section 3.3.3 that these two notions are largely equivalent. In the context of linkages, we saw that the two notions can differ, particularly in 3D. Relatively little has been studied in the context of polyhedron folding.

One special case that has been explored is orthogonal polyhedra. Specifically, Biedl, Lubiw, and Sun [Biedl et al. 1999b; Sun 1999] have proved that there is an edge-unfolding of an orthogonal polyhedron (which is an orthogonal net) that cannot be folded into the orthogonal polyhedron by a continuous motion that keeps the faces rigid and avoids self-intersection. The basis for their example is the locked polygonal arc in 3D (Figure 10), converted into an orthogonal locked polygonal arc in 3D, and then "thickened" into an orthogonal tube. A single chain of faces in the unfolding is what prevents the continuous foldability.

One would expect, analogous to the results described in Section 3.3.3 [Demaine and Mitchell 2001], that collections of polygons hinged together into a tree can be folded into all possible configurations if we allow additional creases during the motion. However, this extension (equivalent to a polygonal piece of paper) remains open. A particularly interesting version of this question, posed in [Biedl et al. 1999b], is whether a finite number of additional creases suffice.

An interesting collection of open questions arise when we consider polyhedron foldings with creases only at polyhedron edges. For example, do all convex

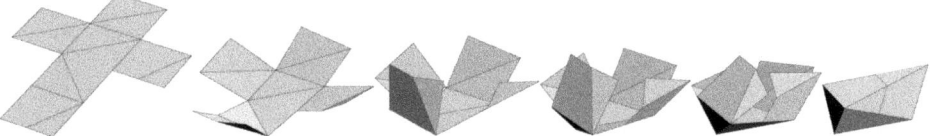

Figure 25. Folding the Latin cross into an octahedron, according to the crease pattern in Figure 23(e), by affinely interpolating all dihedral angles. (Figure 2 of [Demaine et al. 1999a].)

polyhedra have *continuous* edge-unfoldings? (This question may be easier to answer negatively than the classic edge-unfolding problem.) Figure 25 shows a simple example of such a folding, taken from a longer video [Demaine et al. 1999a], based on the simple rule of affinely interpolating each dihedral angle from start to finish. Connelly, as reported in [Miller and Pak 2003], asked whether the source unfolding can be *continuously bloomed*, i.e., unfolded so that all dihedral angles increase monotonically. Although an affirmative answer to this question has just been obtained,[13] it remains open whether every general unfolding can be executed continuously.

5. Conclusion and Higher Dimensions

Our goal has been to survey the results in the newly developing area of folding and unfolding, which offers many beautiful mathematical and computational problems. Much progress has been made recently in this area, but many important problems remain open. For example, most aspects of unfolding polyhedra remain unsolved, and we highlight two key problems in this context: can all convex polyhedra be edge-unfolded, and can all polyhedra be generally unfolded? Another exciting new direction is the developing connection between linkage folding and protein folding.

Finally, higher dimensions are just beginning to be explored. We mentioned in Section 2.4 that 1D (one-dimensional) linkages in higher dimensions have been explored. But 2D "linkages" in 4D — and higher-dimensional analogs — have received less attention. One model is 2D polygons hinged together at their edges to form a chain. Such a hinged chain has fewer degrees of freedom than a 1D linkage in 3D; for example, a hinged chain can be forced to fold like a planar linkage by extruding the linkage orthogonal to the plane. See Figure 26. Biedl, Lubiw, and Sun [Biedl et al. 1999b; Sun 1999] showed that even hinged chains of rectangles do not have connected configuration spaces, by considering an orthogonal version of Figure 10. It would be interesting to explore these chains of rectangles in 4D.

Turning to the origami context, one natural open problem is a generalization of the fold-and-cut problem: given a polyhedral complex drawn on a *d*-

---

[13]Personal communication with Stefan Langerman et al., February 2004.

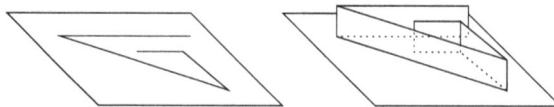

Figure 26. Extruding a linkage into an equivalent collection of polygons (rectangles) hinged together at their edges.

dimensional piece of paper, is it always possible to fold the paper flat (into d-space) while mapping the $(d-1)$-dimensional facets of the complex to a common $(d-1)$-dimensional hyperplane? What if our goal is to map all k-dimensional faces to a common k-dimensional flat, for all $k = 0, 1, \ldots, d$?

Salvador Dali's famous painting ("Christ") of Christ on an unfolded 4D hypercube suggests the possibilities for unfolding higher-dimensional polyhedra. All of the unsolved problems related to unfolding 3D to 2D are equally unsolved in their higher-dimensional analogs. We mentioned in Section 4.1 a rare exception: the vertex-unfolding algorithm generalizes to unfold simplicial manifolds without overlap in arbitrary dimensions. Miller and Pak [2003] have established that the source unfolding generalizes to higher dimensions to yield nonoverlapping unfoldings, but that the most natural generalization of the star unfolding does not even suffice to unfold, let alone without overlap. Nevertheless, with one general unfolding available, the natural analog of the edge-unfolding question remains: Does every convex d-polytope have a *ridge unfolding*, a cutting of $(d-2)$-dimensional faces that unfolds the polytope into $\mathbb{R}^{d-1}$ without overlap?

Acknowledgements

We appreciate the helpful remarks of Joseph Mitchell and the suggestions of the referees.

References

[Agarwal et al. 1997] P. K. Agarwal, B. Aronov, J. O'Rourke, and C. A. Schevon, "Star unfolding of a polytope with applications", *SIAM J. Comput.* **26**:6 (1997), 1689–1713.

[Ahn et al. 2000] H.-K. Ahn, P. Bose, J. Czyzowicz, N. Hanusse, E. Kranakis, and P. Morin, "Flipping your lid", *Geombinatorics* **10**:2 (2000), 57–63.

[Aichholzer et al. 2001] O. Aichholzer, E. D. Demaine, J. Erickson, F. Hurtado, M. Overmars, M. A. Soss, and G. T. Toussaint, "Reconfiguring convex polygons", *Comput. Geom.* **20**:1–2 (2001), 85–95.

[Aichholzer et al. 2002] O. Aichholzer, C. Cortés, E. D. Demaine, V. Dujmović, J. Erickson, H. Meijer, M. Overmars, B. Palop, S. Ramaswami, and G. T. Toussaint, "Flipturning polygons", *Discrete Comput. Geom.* **28**:2 (2002), 231–253.

[Aichholzer et al. 2003] O. Aichholzer, D. Bremner, E. D. Demaine, H. Meijer, V. Sacristán, and M. Soss, "Long proteins with unique optimal foldings in the H-P model", *Comput. Geom.* **25**:1–2 (2003), 139–159.

[Alexander et al. 2002] R. Alexander, H. Dyson, and J. O'Rourke, "The convex polyhedra foldable from a square", pp. 31–32 in *Proceedings of the Japan Conference on Discrete and Computational Geometry* (Tokyo, 2002), 2002.

[Alexandrov 1941] A. D. Alexandrov, "Existence of a given polyhedron and of a convex surface with a given metric", *Doklady Acad. Sci. URSS (N.S.)* **30** (1941), 103–106. In Russian; translation as pp. 169–173 in *Selected Works*, Part I, edited by Yu. G. Reshetnyak and S. S. Kutateladze, Gordon and Breach, Amsterdam, 1996.

[Alexandrov 1950] A. D. Aleksandrov, *Vypuklye mnogogranniki*, Gosudarstv. Izdat. Tehn.-Teor. Lit., Moscow-Leningrad, 1950.

[Aloupis et al. 2002a] G. Aloupis, E. D. Demaine, V. Dujmović, J. Erickson, S. Langerman, H. Meijer, I. Streinu, J. O'Rourke, M. Overmars, M. Soss, and G. T. Toussaint, "Flat-state connectivity of linkages under dihedral motions", pp. 369–380 in *Proceedings of the 13th Annual International Symposium on Algorithms and Computation* (Vancouver, 2002), edited by P. Bose and P. Morin, Lecture Notes in Computer Science **2518**, 2002.

[Aloupis et al. 2002b] G. Aloupis, E. D. Demaine, H. Meijer, J. O'Rourke, I. Streinu, and G. Toussaint, "Flat-state connectedness of fixed-angle chains: Special acute chains", pp. 27–30 in *Proceedings of the 14th Canadian Conference on Computational Geometry* (Lethbridge, AB), 2002.

[Alt et al. 2004] H. Alt, C. Knauer, G. Rote, and S. Whitesides, "On the complexity of the linkage reconfiguration problem", pp. 1–13 in *Towards a theory of geometric graphs*, Contemp. Math. **342**, Amer. Math. Soc., Providence, RI, 2004.

[Arkin et al. 2003] E. M. Arkin, S. P. Fekete, and J. S. B. Mitchell, "An algorithmic study of manufacturing paperclips and other folded structures", *Comput. Geom.* **25**:1-2 (2003), 117–138.

[Arkin et al. 2004] E. M. Arkin, M. A. Bender, E. D. Demaine, M. L. Demaine, J. S. B. Mitchell, S. Sethia, and S. S. Skiena, "When can you fold a map?", *Comput. Geom.* **29**:1 (2004), 23–46.

[Aronov and O'Rourke 1992] B. Aronov and J. O'Rourke, "Nonoverlap of the star unfolding", *Discrete Comput. Geom.* **8**:3 (1992), 219–250.

[Aronov et al. 2002] B. Aronov, J. E. Goodman, and R. Pollack, "Convexification of planar polygons in $\mathbb{R}^3$", 2002.

[Berger and Leighton 1998] B. Berger and T. Leighton, "Protein folding in the hydrophobic-hydrophilic (HP) model is NP-complete", *Journal of Computational Biology* **5**:1 (1998), 27–40.

[Bern and Hayes 1996] M. Bern and B. Hayes, "The complexity of flat origami", pp. 175–183 in *Proceedings of the 7th Annual ACM–SIAM Symposium on Discrete Algorithms* (Atlanta, 1996), ACM, New York, 1996.

[Bern et al. 2002] M. Bern, E. Demaine, D. Eppstein, and B. Hayes, "A disk-packing algorithm for an origami magic trick", pp. 17–28 in *Origami³: Proceedings of the 3rd International Meeting of Origami Science, Math, and Education* (Asilomar, CA, 2001), edited by T. Hull, A K Peters, Natick, MA, 2002.

[Bern et al. 2003] M. Bern, E. D. Demaine, D. Eppstein, E. Kuo, A. Mantler, and J. Snoeyink, "Ununfoldable polyhedra with convex faces", *Comput. Geom.* **24**:2 (2003), 51–62.

[Biedl 2005] T. Biedl, "Polygons needing many flipturns", *Discrete Comput. Geom* (2005). To appear.

[Biedl et al. 1998] T. Biedl, E. Demaine, M. Demaine, A. Lubiw, M. Overmars, J. O'Rourke, S. Robbins, and S. Whitesides, "Unfolding some classes of orthogonal polyhedra", pp. 70–71 in *Proceedings of the 10th Canadian Conference on Computational Geometry* (Montreal), 1998. Available at http://cgm.cs.mcgill.ca/cccg98/proceedings/.

[Biedl et al. 1999a] T. Biedl, E. Demaine, M. Demaine, S. Lazard, A. Lubiw, J. O'Rourke, M. Overmars, S. Robbins, I. Streinu, G. Toussaint, and S. Whitesides, "Locked and unlocked polygonal chains in 3D", Technical Report 060, Smith College, Northampton (MA), 1999. Available at arXiv:cs.CG/9910009.

[Biedl et al. 1999b] T. Biedl, A. Lubiw, and J. Sun, "When can a net fold to a polyhedron?", pp. 70–71 in *Proceedings of the 11th Canadian Conference on Computational Geometry* (Vancouver, 1999), 1999. Available at http://www.cs.ubc.ca/conferences/CCCG/elec_proc/fp16.ps.gz.

[Biedl et al. 2000] T. Biedl, E. Demaine, M. Demaine, S. Lazard, A. Lubiw, J. O'Rourke, S. Robbins, I. Streinu, G. Toussaint, and S. Whitesides, "On reconfiguring tree linkages: Trees can lock", Technical Report SOCS-00.7, McGill University, Montreal, 2000. Available at arXiv:cs.CG/9910024.

[Biedl et al. 2001] T. Biedl, E. Demaine, M. Demaine, S. Lazard, A. Lubiw, J. O'Rourke, M. Overmars, S. Robbins, I. Streinu, G. Toussaint, and S. Whitesides, "Locked and unlocked polygonal chains in three dimensions", *Discrete Comput. Geom.* **26**:3 (2001), 269–281. Full version: [Biedl et al. 1999a].

[Biedl et al. 2002a] T. Biedl, T. M. Chan, E. D. Demaine, M. L. Demaine, P. Nijjar, R. Uehara, and M. Wang, "Tighter bounds on the genus of nonorthogonal polyhedra built from rectangles", pp. 105–108 in *Proceedings of the 14th Canadian Conference on Computational Geometry* (Lethbridge, AB), 2002.

[Biedl et al. 2002b] T. Biedl, E. Demaine, M. Demaine, S. Lazard, A. Lubiw, J. O'Rourke, S. Robbins, I. Streinu, G. Toussaint, and S. Whitesides, "A note on reconfiguring tree linkages: trees can lock", *Discrete Appl. Math.* **117**:1-3 (2002), 293–297. Full version: [Biedl et al. 2000].

[Bose et al. 1999] P. Bose, F. Gomez, P. Ramos, and G. Toussaint, "Drawing nice projections of objects in space", *Journal of Visual Communication and Image Representation* **10**:2 (1999), 155–172.

[Calvo et al. 2001] J. A. Calvo, D. Krizanc, P. Morin, M. Soss, and G. Toussaint, "Convexifying polygons with simple projections", *Inform. Process. Lett.* **80**:2 (2001), 81–86.

[Cantarella and Johnston 1998] J. Cantarella and H. Johnston, "Nontrivial embeddings of polygonal intervals and unknots in 3-space", *J. Knot Theory Ramifications* **7**:8 (1998), 1027–1039.

[Cantarella et al. 2004] J. H. Cantarella, E. D. Demaine, H. N. Iben, and J. F. O'Brien, "An energy-driven approach to linkage unfolding", pp. 134–143 in *Proceedings of the*

20th Annual ACM Symposium on Computational Geometry (Brooklyn, NY, 2004), ACM, New York, 2004.

[Cauchy 1813] A. L. Cauchy, "Deuxième mémoire sur les polygones et polyèdres", *Journal de l'École Polytechnique* **9** (1813), 87–98.

[Chan and Dill 1990] H. S. Chan and K. A. Dill, "Origins of structure in globular proteins", *Proceedings of the National Academy of Sciences USA* **87** (1990), 6388–6392.

[Chan and Dill 1991] H. S. Chan and K. A. Dill, "Polymer principles in protein structure and stability", *Annual Reviews of Biophysics and Biophysical Chemistry* **20** (1991), 447–490.

[Chan and Dill 1993] H. S. Chan and K. A. Dill, "The protein folding problem", *Physics Today* **46**:2 (February 1993), 24–32.

[Chen and Han 1996] J. Chen and Y. Han, "Shortest paths on a polyhedron, I: Computing shortest paths", *Internat. J. Comput. Geom. Appl.* **6**:2 (1996), 127–144.

[Cocan and O'Rourke 2001] R. Cocan and J. O'Rourke, "Polygonal chains cannot lock in 4D", *Comput. Geom.* **20**:3 (2001), 105–129.

[Connelly 1980] R. Connelly, "The rigidity of certain cabled frameworks and the second-order rigidity of arbitrarily triangulated convex surfaces", *Adv. in Math.* **37**:3 (1980), 272–299.

[Connelly 1993] R. Connelly, "Rigidity", pp. 223–271 in *Handbook of convex geometry*, vol. A, North-Holland, Amsterdam, 1993.

[Connelly and Demaine 2004] R. Connelly and E. D. Demaine, "Geometry and topology of polygonal linkages", pp. 197–218 in *Handbook of discrete and computational geometry*, 2nd ed., edited by J. E. Goodman and J. O'Rourke, CRC, Boca Raton, FL, 2004.

[Connelly et al. 1997] R. Connelly, I. Sabitov, and A. Walz, "The bellows conjecture", *Beiträge Algebra Geom.* **38**:1 (1997), 1–10.

[Connelly et al. 2002] R. Connelly, E. D. Demaine, and G. Rote, "Infinitesimally locked self-touching linkages with applications to locked trees", pp. 287–311 in *Physical knots: knotting, linking, and folding geometric objects in* $\mathbb{R}^3$ (Las Vegas, 2001), edited by J. Calvo et al., Contemp. Math. **304**, Amer. Math. Soc., Providence, RI, 2002.

[Connelly et al. 2003] R. Connelly, E. D. Demaine, and G. Rote, "Straightening polygonal arcs and convexifying polygonal cycles", *Discrete Comput. Geom.* **30**:2 (2003), 205–239.

[Crescenzi et al. 1998] P. Crescenzi, D. Goldman, C. Papadimitriou, A. Piccolboni, and M. Yannakakis, "On the complexity of protein folding", *Journal of Computational Biology* **5**:3 (1998), 423–465.

[Cromwell 1997] P. R. Cromwell, *Polyhedra*, Cambridge University Press, Cambridge, 1997.

[Demaine 2001] E. D. Demaine, "Folding and unfolding linkages, paper and polyhedra", pp. 113–124 in *Discrete and computational geometry* (JCDCG, Tokyo, 2000: Revised papers), edited by M. U. Jin Akiyama, Mikio Kano, Lecture Notes in Comput. Sci. **2098**, Springer, Berlin, 2001.

[Demaine and Demaine 1997] E. D. Demaine and M. L. Demaine, "Computing extreme origami bases", Technical Report CS-97-22, University of Waterloo, Waterloo, ON, 1997.

[Demaine and Demaine 2002] E. D. Demaine and M. L. Demaine, "Recent results in computational origami", pp. 3–16 in *Origami³: Proceedings of the 3rd International Meeting of Origami Science, Math, and Education* (Asilomar, CA, 2001), edited by T. Hull, A K Peters, Natick, MA, 2002.

[Demaine and Mitchell 2001] E. D. Demaine and J. S. B. Mitchell, "Reaching folded states of a rectangular piece of paper", pp. 73–75 in *Proceedings of the 13th Canadian Conference on Computational Geometry* (Waterloo, ON), 2001. Available at http://compgeo.math.uwaterloo.ca/~cccg01/proceedings/.

[Demaine and O'Rourke 2001] E. D. Demaine and J. O'Rourke, "Open problems from CCCG 2000", in *Proceedings of the 13th Canadian Conference on Computational Geometry* (Waterloo, ON), 2001. Available at http://compgeo.math.uwaterloo.ca/~cccg01/proceedings.

[Demaine and O'Rourke ≥ 2005] E. D. Demaine and J. O'Rourke, *Geometric folding algorithms: linkages, origami, and polyhedra*, Cambridge University Press, New York. To appear.

[Demaine et al. 1999a] E. Demaine, M. Demaine, A. Lubiw, J. O'Rourke, and I. Pashchenko, "Metamorphosis of the cube", pp. 409–410 in *8th Annual Video Review of Computational Geometry, Proceedings of the 15th Annual ACM Symposium on Computational Geometry* (Miami Beach, 1999), 1999. Available at http://theory.csail.mit.edu/~edemaine/metamorphosis/.

[Demaine et al. 1999b] E. D. Demaine, M. L. Demaine, and A. Lubiw, "Folding and one straight cut suffice", pp. 891–892 in *Proceedings of the 10th Annual ACM-SIAM Symposium on Discrete Algorithms* (Baltimore, MD), 1999.

[Demaine et al. 1999c] E. D. Demaine, M. L. Demaine, and J. S. B. Mitchell, "Folding flat silhouettes and wrapping polyhedral packages: new results in computational origami", pp. 105–114 in *Proceedings of the 15th Annual Symposium on Computational Geometry* (Miami Beach, 1999), ACM, New York, 1999.

[Demaine et al. 2000a] E. Demaine, M. Demaine, A. Lubiw, and J. O'Rourke, "The 85 foldings of the Latin cross", 2000. Available at http://theory.csail.mit.edu/~edemaine/aleksandrov/cross/.

[Demaine et al. 2000b] E. Demaine, M. Demaine, A. Lubiw, and J. O'Rourke, "Examples, counterexamples, and enumeration results for foldings and unfoldings between polygons and polytopes", Technical Report 069, Smith College, Northampton, MA, 2000. Available at arXiv:cs.CG/0007019.

[Demaine et al. 2000c] E. D. Demaine, M. L. Demaine, and A. Lubiw, "Folding and cutting paper", pp. 104–117 in *Discrete and computational geometry* (JCDCG, Tokyo, 1998: Revised papers), edited by M. K. J. Akiyama and M. Urabe, Lecture Notes in Comput. Sci. **1763**, Springer, Berlin, 2000.

[Demaine et al. 2000d] E. D. Demaine, M. L. Demaine, and J. S. B. Mitchell, "Folding flat silhouettes and wrapping polyhedral packages: new results in computational origami", *Comput. Geom.* **16**:1 (2000), 3–21.

[Demaine et al. 2002a] E. D. Demaine, M. L. Demaine, A. Lubiw, and J. O'Rourke, "Enumerating foldings and unfoldings between polygons and polytopes", *Graphs Combin.* **18**:1 (2002), 93–104. Available at arXiv:cs.CG/0107024.

[Demaine et al. 2002b] E. D. Demaine, S. Langerman, J. O'Rourke, and J. Snoeyink, "Interlocked open linkages with few joints", pp. 189–198 in *Proceedings of the 18th Annual ACM Symposium on Computational Geometry* (Barcelona), 2002.

[Demaine et al. 2003a] E. D. Demaine, D. Eppstein, J. Erickson, G. W. Hart, and J. O'Rourke, "Vertex-unfoldings of simplicial manifolds", pp. 215–228 in *Discrete geometry: in honor of W. Kuperberg's 60th birthday*, edited by A. Bezdek, Pure Appl. Math. **253**, Dekker, New York, 2003. Available at http://www.arXiv.org/abs/cs.CG/0110054.

[Demaine et al. 2003b] E. D. Demaine, S. Langerman, and J. O'Rourke, "Geometric restrictions on producible polygonal protein chains", pp. 395–404 in *Algorithms and computation* (14th Annual Symposium, Kyoto, 2003), edited by H. O. Toshihide Ibaraki, Naoki Katoh, Lecture Notes in Comput. Sci. **2906**, Springer, Berlin, 2003. To appear in *Algorithmica*.

[Demaine et al. 2003c] E. D. Demaine, S. Langerman, J. O'Rourke, and J. Snoeyink, "Interlocked open and closed linkages with few joints", *Comput. Geom.* **26**:1 (2003), 37–45.

[Demaine et al. 2004] E. D. Demaine, S. L. Devadoss, J. S. B. Mitchell, and J. O'Rourke, "Continuous foldability of polygonal paper", pp. 64–67 in *Proceedings of the 16th Canadian Conference on Computational Geometry* (Montreal), 2004.

[Dill 1990] K. A. Dill, "Dominant forces in protein folding", *Biochemistry* **29**:31 (1990), 7133–7155.

[Donoso and O'Rourke 2002] M. Donoso and J. O'Rourke, "Nonorthogonal polyhedra built from rectangles", 2002. Available at arXiv:cs.CG/0110059.

[Dürer 1525] A. Dürer, *Unterweysung der Messung mit dem Zirkel und Richtscheyt, in Linien Ebnen und gantzen Corporen*, 1525. Reprinted 2002, Verlag Alfons Uhl, Nördlingen; translated as *The Painter's Manual*, Abaris Books, New York, 1977.

[Erdős 1935] P. Erdős, "Problem 3763", *American Mathematical Monthly* **42** (1935).

[Fedorchuk and Pak 2004] M. Fedorchuk and I. Pak, "Rigidity and polynomial invariants of convex polytopes", preprint, 2004. Available at http://www-math.mit.edu/~pak/pp16.pdf.

[Fevens et al. 2001] T. Fevens, A. Hernandez, A. Mesa, P. Morin, M. Soss, and G. Toussaint, "Simple polygons with an infinite sequence of deflations", *Beiträge zur Algebra und Geometrie* **42**:2 (2001), 307–311.

[Gardner 1960] M. Gardner, "Mathematical games", *Scientific American* (June 1960). Reprinted as "Paper cutting", chapter 5, pp. 58–69 in *New mathematical diversions* (revised edition), Math. Assoc. of America, Washington, DC, 1995.

[Glass et al. 2004] J. Glass, S. Langerman, J. O'Rourke, J. Snoeyink, and J. K. Zhong, "A 2-chain can interlock with a *k*-chain", 2004. Available at www.arXiv.org/cs.CG/0410052.

[Grünbaum 1995] B. Grünbaum, "How to convexify a polygon", *Geombinatorics* **5** (1995), 24–30.

[Grünbaum and Zaks 1998] B. Grünbaum and J. Zaks, "Convexification of polygons by flips and by flipturns", Technical Report 6/4/98, Department of Mathematics, University of Washington, Seattle, 1998.

[Harper's 1873] Harper's, "National standards and emblems", *Harper's New Monthly Magazine* **47** (July 1873), 171–181. Available at http://memory.loc.gov/ammem/ndlpcoop/moahtml/title/lists/harp_V47I278.html.

[Hart and Istrail 1996] W. E. Hart and S. Istrail, "Fast protein folding in the hydrophobic-hydrophilic model within three-eighths of optimal", *Journal of Computational Biology* **3**:1 (1996), 53–96.

[Hayes 1998] B. Hayes, "Prototeins", *American Scientist* **86** (1998), 216–221.

[Hirata 2000] K. Hirata, "Polytope2", 2000. Available at http://weyl.ed.ehime-u.ac.jp/cgi-bin/WebObjects/Polytope2. Online software.

[Hopcroft et al. 1984] J. Hopcroft, D. Joseph, and S. Whitesides, "Movement problems for 2-dimensional linkages", *SIAM J. Comput.* **13**:3 (1984), 610–629.

[Houdini 1922] H. Houdini, *Paper magic*, Dutton, 1922.

[Hull 1994] T. Hull, "On the mathematics of flat origamis", *Congr. Numer.* **100** (1994), 215–224.

[Hull 2003] T. Hull, "Counting mountain-valley assignments for flat folds", *Ars Combin.* **67** (2003), 175–187.

[Jordan and Steiner 1999] D. Jordan and M. Steiner, "Configuration spaces of mechanical linkages", *Discrete Comput. Geom.* **22**:2 (1999), 297–315.

[Justin 1994] J. Justin, "Towards a mathematical theory of origami", pp. 15–29 in *Proceedings of the 2nd International Meeting of Origami Science and Scientific Origami* (Otsu, Japan, 1994), edited by K. Miura, 1994.

[Kaneva and O'Rourke 2000] B. Kaneva and J. O'Rourke, "An implementation of Chen & Han's shortest paths algorithm", pp. 139–146 in *Proceedings of the 12th Canadian Conference on Computational Geometry* (Fredericton, NB, 2000), 2000. Available at http://www.cs.unb.ca/conf/cccg/eProceedings/8.ps.gz.

[Kapoor 1999] S. Kapoor, "Efficient computation of geodesic shortest paths", pp. 770–779 in *Proceedings of the 32nd Annual ACM Symposium on Theory of Computing* (Atlanta, GA, 1999), ACM, New York, 1999.

[Kapovich and Millson 1995] M. Kapovich and J. Millson, "On the moduli space of polygons in the Euclidean plane", *J. Differential Geom.* **42**:1 (1995), 133–164.

[Kapovich and Millson 2002] M. Kapovich and J. J. Millson, "Universality theorems for configuration spaces of planar linkages", *Topology* **41**:6 (2002), 1051–1107.

[Kawasaki 1989] T. Kawasaki, "On the relation between mountain-creases and valley-creases of a flat origami", pp. 229–237 in *Proceedings of the 1st International Meeting of Origami Science and Technology* (Ferrara, Italy, 1989), edited by H. Huzita, 1989. An unabridged Japanese version appeared in *Sasebo College of Technology Report* **27** (1990), 153–157.

[Kempe 1876] A. B. Kempe, "On a general method of describing plane curves of the nth degree by linkwork", *Proceedings of the London Mathematical Society* **7** (1876), 213–216.

[Lang 1994a] R. J. Lang, "Mathematical algorithms for origami design", *Symmetry Cult. Sci.* **5**:2 (1994), 115–152.

[Lang 1994b] R. J. Lang, "The tree method of origami design", pp. 73–82 in *Proceedings of the 2nd International Meeting of Origami Science and Scientific Origami* (Otsu, Japan, 1994), edited by K. Miura, 1994.

[Lang 1996] R. J. Lang, "A computational algorithm for origami design", pp. 98–105 in *Proceedings of the 12th Annual ACM Symposium on Computational Geometry* (Philadelphia, 1996), ACM, New York, 1996.

[Lang 1998] R. J. Lang, "TreeMaker 4.0: A Program for Origami Design", 1998. Available at http://origami.kvi.nl/programs/treemaker/trmkr40.pdf.

[Lang 2003] R. J. Lang, *Origami design secrets: Mathematical methods for an ancient art*, A K Peters, Natick, MA, 2003.

[Lau and Dill 1989] K. F. Lau and K. A. Dill, "A lattice statistical mechanics model of the conformation and sequence spaces of proteins", *Macromolecules* **22** (1989), 3986–3997.

[Lau and Dill 1990] K. F. Lau and K. A. Dill, "Theory for protein mutability and biogenesis", *Proceedings of the National Academy of Sciences USA* **87** (1990), 638–642.

[Lenhart and Whitesides 1995] W. J. Lenhart and S. H. Whitesides, "Reconfiguring closed polygonal chains in Euclidean *d*-space", *Discrete Comput. Geom.* **13**:1 (1995), 123–140.

[Lipman and Wilber 1991] D. J. Lipman and W. J. Wilber, "Modelling neutral and selective evolution of protein folding", *Proceedings of The Royal Society of London, Series B* **245**:1312 (1991), 7–11.

[Lubiw and O'Rourke 1996] A. Lubiw and J. O'Rourke, "When can a polygon fold to a polytope?", Technical Report 048, Smith College, Northampton, MA, 1996.

[Lunnon 1968] W. F. Lunnon, "A map-folding problem", *Math. Comp.* **22** (1968), 193–199.

[Lunnon 1971] W. F. Lunnon, "Multi-dimensional map-folding", *Comput. J.* **14** (1971), 75–80.

[Merz and Le Grand 1994] K. M. Merz and S. M. Le Grand (editors), *The protein folding problem and tertiary structure prediction*, edited by K. M. Merz and S. M. Le Grand, Birkhäuser, Boston, 1994.

[Miller and Pak 2003] E. Miller and I. Pak, "Metric combinatorics of convex polyhedra: cut loci and nonoverlapping unfoldings", preprint, 2003. Available at http://www.math.umn.edu/~ezra/Foldout/fold.pdf.

[Mitchell et al. 1987] J. S. B. Mitchell, D. M. Mount, and C. H. Papadimitriou, "The discrete geodesic problem", *SIAM J. Comput.* **16**:4 (1987), 647–668.

[Montroll 1991] J. Montroll, *African animals in origami*, Dover, New York, 1991.

[Montroll 1993] J. Montroll, *Origami inside-out*, Dover, New York, 1993.

[Nagy 1939] B. Nagy, "Solution to problem 3763", *American Mathematical Monthly* **46** (March 1939), 176–177.

[O'Rourke 2000] J. O'Rourke, "Folding and unfolding in computational geometry", pp. 258–266 in *Discrete and computational geometry* (JCDCG, Tokyo, 1998: Revised papers), edited by M. U. Jin Akiyama, Mikio Kano, Lecture Notes in Comput. Sci. **1763**, Springer, Berlin, 2000.

[O'Rourke 2002] J. O'Rourke, "Computational geometry column 43", *Internat. J. Comput. Geom. Appl.* **12**:3 (2002), 263–265. Also in *SIGACT News* **33**:1 (2002), 58-60.

[Rote 2003] G. Rote, "Pseudotriangulations: A survey and recent results", 2003. Available at page.mi.fu-berlin.de/~rote/Papers/slides/Pseudotriangulations-JGA2003. Course notes, *Journées de Géométrie Algorithmique* 2003.

[Sabitov 1996] I. K. Sabitov, "The volume of a polyhedron as a function of its metric", *Fundam. Prikl. Mat.* **2**:4 (1996), 1235–1246. In Russian.

[Sabitov 1998] I. K. Sabitov, "The volume as a metric invariant of polyhedra", *Discrete Comput. Geom.* **20**:4 (1998), 405–425.

[Sallee 1973] G. T. Sallee, "Stretching chords of space curves", *Geometriae Dedicata* **2** (1973), 311–315.

[Schevon 1987] C. Schevon, "Unfolding polyhedra", Usenet article on sci.math, 1987. Available at http://www.ics.uci.edu/~eppstein/gina/unfold.html.

[Schevon 1989] C. Schevon, Ph.D. thesis, Johns Hopkins University, Baltimore, MD, 1989.

[Sen 1721] K. C. Sen, *Wakoku chiyekurabe (Mathematical contests)*, 1721. Excerpts available from http://theory.csail.mit.edu/~edemaine/foldcut/sen_book.html.

[Shephard 1975] G. C. Shephard, "Convex polytopes with convex nets", *Math. Proc. Cambridge Philos. Soc.* **78**:3 (1975), 389–403.

[Smith 1976] J. S. Smith, "Origami profiles", *British Origami* **58** (June 1976).

[Smith 1980] J. S. Smith, *Pureland origami*, BOS **14**, British Origami Society, Stockport, 1980.

[Smith 1988] J. S. Smith, *Pureland origami 2*, BOS **29**, British Origami Society, Stockport, 1988.

[Smith 1993] J. S. Smith, *Pureland origami 3*, BOS **43**, British Origami Society, Stockport, 1993.

[Soss 2001] M. Soss, *Geometric and computational aspects of molecular reconfiguration*, Ph.D. thesis, School of Computer Science, McGill University, Montreal, 2001.

[Soss and Toussaint 2000] M. Soss and G. T. Toussaint, "Geometric and computational aspects of polymer reconfiguration", *J. Math. Chem.* **27**:4 (2000), 303–318.

[Soss et al. 2003] M. A. Soss, J. Erickson, and M. H. Overmars, "Preprocessing chains for fast dihedral rotations is hard or even impossible", *Comput. Geom.* **26**:3 (2003), 235–246.

[Streinu 2000] I. Streinu, "A combinatorial approach to planar non-colliding robot arm motion planning", pp. 443–453 in *41st Annual Symposium on Foundations of Computer Science* (Redondo Beach, CA, 2000), IEEE Comput. Soc. Press, Los Alamitos, CA, 2000.

[Sun 1999] J. Sun, *Folding orthogonal polyhedra*, Master's thesis, Department of Computer Science, University of Waterloo, Waterloo, ON, 1999.

[Toussaint 1999a] G. Toussaint, "Computational polygonal entanglement theory", pp. 269–278 in *Proceedings of the VIII Encuentros de Geometría Computacional* (Castellon, Spain, 1999), 1999.

[Toussaint 1999b] G. Toussaint, "The Erdős–Nagy theorem and its ramifications", pp. 1–13 in *Proceedings of the 11th Canadian Conference on Computational Geometry* (Vancouver, 1999), 1999. Available at http://www.cs.ubc.ca/conferences/CCCG/elec_proc/fp19.ps.gz.

[Toussaint 2001] G. Toussaint, "A new class of stuck unknots in Pol6", *Beiträge Algebra Geom.* **42**:2 (2001), 301–306.

[Unger and Moult 1993a] R. Unger and J. Moult, "Finding the lowest free energy conformation of a protein is a NP-hard problem: Proof and implications", *Bulletin of Mathematical Biology* **55**:6 (1993), 1183–1198.

[Unger and Moult 1993b] R. Unger and J. Moult, "A genetic algorithm for 3D protein folding simulations", pp. 581–588 in *Proceedings of the 5th International Conference on Genetic Algorithms* (San Mateo, CA, 1993), 1993.

[Unger and Moult 1993c] R. Unger and J. Moult, "Genetic algorithms for protein folding simulations", *Journal of Molecular Biology* **231**:1 (1993), 75–81.

[Wang 1997] C.-H. Wang, *Manufacturability-driven decomposition of sheet metal*, Ph.D. thesis, Robotics Institute, Carnegie Mellon University, Pittsburgh, PA, 1997. Issued as technical report CMU-RI-TR-97-35.

[Wegner 1993] B. Wegner, "Partial inflation of closed polygons in the plane", *Beiträge Algebra Geom.* **34**:1 (1993), 77–85.

[Whitesides 1992] S. Whitesides, "Algorithmic issues in the geometry of planar linkage movement", *Australian Computer Journal* **24**:2 (1992), 42–50.

ERIK D. DEMAINE
MIT COMPUTER SCIENCE AND ARTIFICIAL INTELLIGENCE LABORATORY
32 VASSAR STREET
CAMBRIDGE, MA 02139
UNITED STATES
edemaine@mit.edu

JOSEPH O'ROURKE
DEPARTMENT OF COMPUTER SCIENCE
SMITH COLLEGE
NORTHAMPTON, MA 01063
UNITED STATES
orourke@cs.smith.edu

Combinatorial and Computational Geometry
MSRI Publications
Volume **52**, 2005

On the Rank of a Tropical Matrix

MIKE DEVELIN, FRANCISCO SANTOS, AND BERND STURMFELS

ABSTRACT. This is a foundational paper in tropical linear algebra, which is
linear algebra over the min-plus semiring. We introduce and compare three
natural definitions of the rank of a matrix, called the Barvinok rank, the
Kapranov rank and the tropical rank. We demonstrate how these notions
arise naturally in polyhedral and algebraic geometry, and we show that
they differ in general. Realizability of matroids plays a crucial role here.
Connections to optimization are also discussed.

1. Introduction

The rank of a matrix M is one of the most important notions in linear algebra.
This number can be defined in many different ways. In particular, the following
three definitions are equivalent:

- *The rank of M is the smallest integer r for which M can be written as the
  sum of r rank one matrices. A matrix has rank 1 if it is the product of a
  column vector and a row vector.*
- *The rank of M is the smallest dimension of any linear space containing the
  columns of M.*
- *The rank of M is the largest integer r such that M has a nonsingular $r \times r$
  minor.*

Our objective is to examine these familiar definitions over an algebraic struc-
ture which has no additive inverses. We work over the *tropical semiring* $(\mathbb{R}, \oplus, \odot)$,
whose arithmetic operations are

$$a \oplus b := \min(a, b) \quad \text{and} \quad a \odot b := a + b.$$

This work was conducted during the Discrete and Computational Geometry semester at
M.S.R.I., Berkeley. Mike Develin held the AIM Postdoctoral Fellowship 2003-2008 and Bernd
Sturmfels held the MSRI-Hewlett Packard Professorship 2003/2004. Francisco Santos was
partially supported by the Spanish Ministerio de Ciencia y Tecnología (grant BFM2001-1153).
Bernd Sturmfels was partially supported by National Science Foundation grant DMS-0200729.

The set $\mathbb{R}^d$ of real d-vectors and the set $\mathbb{R}^{d \times n}$ of real $d \times n$-matrices are semi-modules over the semiring $(\mathbb{R}, \oplus, \odot)$. The operations of matrix addition and matrix multiplication are well defined. All our definitions of rank make sense over the tropical semiring $(\mathbb{R}, \oplus, \odot)$:

DEFINITION 1.1. The *Barvinok rank* of a matrix $M \in \mathbb{R}^{d \times n}$ is the smallest integer r for which M can be written as the tropical sum of r matrices, each of which is the tropical product of a $d \times 1$-matrix and a $1 \times n$-matrix.

DEFINITION 1.2. The *Kapranov rank* of a matrix $M \in \mathbb{R}^{d \times n}$ is the smallest dimension of any tropical linear space (to be defined in Definition 3.2) containing the columns of M.

DEFINITION 1.3. A square matrix $M = (m_{ij}) \in \mathbb{R}^{r \times r}$ is *tropically singular* if the minimum in

$$\det M := \bigoplus_{\sigma \in \mathcal{S}_r} m_{1\sigma_1} \odot m_{2\sigma_2} \odot \cdots \odot m_{r\sigma_r}$$

$$= \min \left\{ m_{1\sigma_1} + m_{2\sigma_2} + \cdots + m_{r\sigma_r} : \sigma \in \mathcal{S}_r \right\}$$

is attained at least twice. Here $\mathcal{S}_r$ denotes the symmetric group on $\{1, 2, \ldots, r\}$. The *tropical rank* of a matrix $M \in \mathbb{R}^{d \times n}$ is the largest integer r such that M has a nonsingular $r \times r$ minor.

These three definitions are easily seen to agree for $r = 1$, but in general they are not equivalent:

THEOREM 1.4. *If M is a matrix with entries in the tropical semiring $(\mathbb{R}, \oplus, \odot)$,*

$$\text{tropical rank}(M) \leq \text{Kapranov rank}(M) \leq \text{Barvinok rank}(M). \qquad (1\text{--}1)$$

Both of these inequalities can be strict.

The proof of Theorem 1.4 consists of Propositions 3.6, 4.1, 7.2 and Theorem 7.3 in this paper. As we go along, several alternative characterizations of the Barvinok, Kapranov and tropical ranks will be offered. One of them arises from the fact that every $d \times n$-matrix M defines a tropically linear map $\mathbb{R}^n \to \mathbb{R}^d$. The image of M is a polyhedral complex in $\mathbb{R}^d$. Following [Develin and Sturmfels 2004], we identify this polyhedral complex with its image in the tropical projective space $\mathbb{TP}^{d-1} = \mathbb{R}^d / \mathbb{R}(1, 1, \ldots, 1)$. This image is the *tropical convex hull* of (the columns of) M as in [Develin and Sturmfels 2004]. Equivalently, this *tropical polytope* is the set of all tropical linear combinations of the columns of M. We show in Section 4 that the tropical rank of M equals the dimension of this tropical polytope plus one, thus justifying the definition of the vanishing of the determinant given in Definition 1.3.

The discrepancy between Definition 1.3 and Definition 1.2 comes from the crucial distinction between tropical polytopes and tropical linear spaces, as explained in [Richter-Gebert et al. 2005, § 1]. The latter are described in [Speyer

and Sturmfels 2004] where it is shown that they are parametrized by the tropical
Grassmannian. That the two inequalities in Theorem 1.4 can be strict corresponds to two facts about tropical geometry which are unfamiliar from classical
geometry. Strictness of the first inequality corresponds to the fact that a point
configuration in tropical space can have a d-dimensional convex hull but not lie
in any d-dimensional affine subspace. Strictness of the second inequality corresponds to the fact that a point configuration in a d-dimensional subspace need
not lie in the convex hull of $d + 1$ points.

We start out in Section 2 by studying the Barvinok rank (Definition 1.1). This
notion of rank arises in the context of combinatorial optimization [Barvinok et al.
1998; Butkovič 2003; Çela et al. 1998]. In Section 3 we study the Kapranov rank
(Definition 1.2). This notion is the most natural one from the point of view of
algebraic geometry, where tropical arithmetic arises as the "tropicalization" of
arithmetic in a power series ring. It has good algebraic and geometric properties
but is difficult to characterize combinatorially; for instance, it depends on the
base field of the power series ring, which here we take to be the complex numbers
$\mathbb{C}$, unless otherwise stated.

In Section 4 we study the tropical rank (Definition 1.3). This is the best notion
of rank from a geometric and combinatorial perspective. For instance, it can be
expressed in terms of regular subdivisions of products of simplices [Develin and
Sturmfels 2004]. In Section 5, we use this characterization to show that the
tropical and Kapranov ranks agree when either of them is equal to $\min(d, n)$.

Section 6 is devoted to another case where the Kapranov and tropical ranks
agree, namely when either of them equals two. The set of $d \times n$-matrices enjoying
this property is the space of trees with d leaves and n marked points. This space
is studied in the companion paper [Develin 2004].

The second inequality of Theorem 1.4 is strict for many matrices (see Proposition 2.2 for examples), but it requires more effort to find matrices for which the
first inequality is strict. Such matrices are constructed in Section 7 by relating
Kapranov rank to realizability of matroids.

Our definition of "tropically nonsingular" is equivalent to what is called
"strongly regular" in the literature on the min-plus algebra [Butkovič and Hevery
1985; Cuninghame-Green 1979]. The resulting notion of tropical rank, as well
as the notion of Barvinok rank, have previously appeared in this literature. In
fact, linear algebra in the tropical semiring has been called "the linear algebra of
combinatorics" [Butkovič 2003]. In the final section of the paper we revisit some
of that literature, which is concerned mainly with algorithmic issues, and relate
it to our results. We also point out several (mostly algorithmic) open questions.

Summing up, the three definitions of rank studied in this paper generally
disagree, and they have different flavors (combinatorial, algebraic, geometric).
But they all share some of the familiar properties of matrix rank over a field.
The following properties are easily checked for each of the three definitions of
rank: the rank of a matrix and its transpose are the same; the rank of a minor

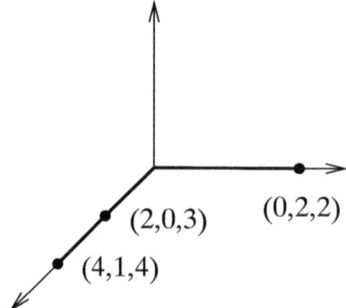

Figure 1. A tropical line in $\mathbb{TP}^2$, and a one-dimensional tropical polytope.

cannot exceed that of the whole matrix; the rank is invariant under (tropical) multiplication of rows or columns by constants, and under insertion of a row or column obtained as a combination of others; the rank of $M \oplus N$ is at most the sum of the ranks of M and N; the rank of $(M \mid N)$ is at least the ranks of M and of N and at most the sum of their ranks; and the rank of $M \odot N$ is at most the minimum of the ranks of M and N.

2. The Barvinok Rank

The Traveling Salesman Problem can be solved in polynomial time if the distance matrix is the tropical sum of r matrices of tropical rank one (with $\oplus$ as "max" instead of "min"). This result was proved by Barvinok, Johnson and Woeginger [Barvinok et al. 1998], building on earlier work of Barvinok. This motivates our definition of *Barvinok rank* as the smallest r for which $M \in \mathbb{R}^{d \times n}$ is expressible in this fashion. Since matrices of tropical rank one are of the form $X \odot Y^T$, for two column vectors $X \in \mathbb{R}^d$ and $Y \in \mathbb{R}^n$, this is equivalent to saying that M has a representation

$$M \quad = \quad X_1 \odot Y_1^T \ \oplus \ X_2 \odot Y_2^T \ \oplus \ \cdots \ \oplus \ X_r \odot Y_r^T. \tag{2-1}$$

For example, the following equation shows a 3×3-matrix which has Barvinok rank two:

$$\begin{pmatrix} 0 & 4 & 2 \\ 2 & 1 & 0 \\ 2 & 4 & 3 \end{pmatrix} \quad = \quad \begin{pmatrix} 0 \\ 2 \\ 2 \end{pmatrix} \odot (0,4,2) \quad \oplus \quad \begin{pmatrix} 3 \\ 0 \\ 3 \end{pmatrix} \odot (2,1,0).$$

This matrix also has tropical rank 2 and Kapranov rank 2 because the matrix is tropically singular. The column vectors lie on the tropical line in $\mathbb{TP}^2 = \mathbb{R}^3/\mathbb{R}(1,1,1)$ defined by $2 \odot x_1 \oplus 3 \odot x_2 \oplus 0 \odot x_3$, depicted in Figure 1. Their convex hull, darkened, is a subset of the line and thus one-dimensional.

We next present two reformulations of the definition of Barvinok rank: in terms of tropical convex hulls as introduced in [Develin and Sturmfels 2004], and via a "tropical morphism" between matrix spaces.

PROPOSITION 2.1. *Let M be a real $d \times n$-matrix. The following properties are equivalent:*

(a) *M has Barvinok rank at most r.*

(b) *The columns of M lie in the tropical convex hull of r points in $\mathbb{TP}^{d-1}$.*

(c) *There are matrices $X \in \mathbb{R}^{d \times r}$ and $Y \in \mathbb{R}^{r \times n}$ such that $M = X \odot Y$. Equivalently, M lies in the image of the following tropical morphism, which is defined by matrix multiplication:*

$$\phi_r : \mathbb{R}^{d \times r} \times \mathbb{R}^{r \times n} \to \mathbb{R}^{d \times n}, \quad (X, Y) \mapsto X \odot Y. \qquad (2\text{-}2)$$

PROOF. Let $M_1, \ldots, M_n \in \mathbb{R}^d$ be the column vectors of M. Let $X_1, \ldots, X_r \in \mathbb{R}^d$ and $Y_1, \ldots, Y_r \in \mathbb{R}^n$ be the columns of two unspecified matrices $X \in \mathbb{R}^{d \times r}$ and $Y \in \mathbb{R}^{n \times r}$. Let Y_{ij} denote the jth coordinate of Y_i. The following three algebraic identities are easily seen to be equivalent:

(a) $M = X_1 \odot Y_1^T \oplus X_2 \odot Y_2^T \oplus \cdots \oplus X_r \odot Y_r^T$,

(b) $M_j = Y_{1j} \odot X_1 \oplus Y_{2j} \odot X_2 \oplus \cdots \oplus Y_{rj} \odot X_r$ for all $j = 1, \ldots, n$, and

(c) $M = X \odot Y^T$.

Statement (b) says that each column vector of M lies in the tropical convex hull of $X_1, \ldots, X_r$. The entries of the matrix Y are the multipliers in that tropical linear combination. This shows that the three conditions (a), (b) and (c) in the statement of the proposition are equivalent. $\qquad \square$

Part (b) of Proposition 2.1 suggests that the Barvinok rank of a tropical matrix is more an analogue of the nonnegative rank of a matrix than of the usual rank. Recall (from [Cohen and Rothblum 1993], for instance) that the *nonnegative rank* of a real nonnegative matrix $M \in \mathbb{R}^{d \times n}$ is the smallest r for which M can be written as a product of nonnegative matrices of format $d \times r$ and $r \times n$. Equivalently, it is the smallest r for which the columns (or rows) of M lie in the positive hull of r nonnegative vectors. Compare this with the formulation of Barvinok rank given in Proposition 2.1 (b); this closer connection comes from the fact that tropical linear combinations yield an object more analogous to a "positive span" or "convex hull" [Develin and Sturmfels 2004; Richter-Gebert et al. 2005] than a linear span. For more information on nonnegative rank see [Cohen and Rothblum 1993], and for the connection to rank over other semigroup rings see [Gregory and Pullman 1983].

By Proposition 2.1, the set of all Barvinok matrices of rank $\leq r$ is the image of the tropical morphism ϕ_r. In particular, this set is a polyhedral fan in $\mathbb{R}^{d \times n}$. This fan has interesting combinatorial structure, even for $r = 2$. These fans are discussed in more detail in [Ardila 2004] and [Develin 2004].

We next present an example of a matrix which shows that the Barvinok rank can be much larger than the other two notions of rank. The matrix to be considered is the *classical identity matrix*

$$C_n \quad = \quad \begin{pmatrix} 1 & 0 & 0 & \cdots & 0 \\ 0 & 1 & 0 & \cdots & 0 \\ 0 & 0 & 1 & \cdots & 0 \\ \vdots & \vdots & \vdots & \ddots & \vdots \\ 0 & 0 & 0 & \cdots & 1 \end{pmatrix}. \tag{2-3}$$

This looks like the unit matrix (in classical arithmetic) but it is far from being a unit matrix in tropical arithmetic, where 0 is the neutral element for $\odot$ and ∞ is the neutral element for $\oplus$. After obtaining the following result, we learned that the same calculation had already been done in [Çela et al. 1998].

PROPOSITION 2.2. *The Barvinok rank of the matrix C_n is the smallest integer r such that*

$$n \le \binom{r}{\lfloor \frac{r}{2} \rfloor}.$$

PROOF. Let r be an integer and assume that $n \le \binom{r}{\lfloor r/2 \rfloor}$. We first show that Barvinok rank$(C_n) \le r$. Let $S_1, \ldots, S_n$ be distinct subsets of $\{1, \ldots, r\}$ each having cardinality $\lfloor r/2 \rfloor$. For each $k \in 1, \ldots, r$, we define an $n \times n$-matrix $X_k = (x_{ij}^k)$ with entries in $\{0, 1, 2\}$ as follows:

$$x_{ij}^k = \begin{cases} 0 & \text{if } k \in S_i \backslash S_j, \\ 2 & \text{if } k \in S_j \backslash S_i, \\ 1 & \text{otherwise.} \end{cases}$$

The matrix X_k has tropical rank one. To see this, let $V_k \in \{0, 1\}^n$ denote the vector with ith coordinate equal to one or zero depending on whether k is an element of S_i or not. Then

$$X_k \quad = \quad V_k^T \odot (1 \odot (-V_k)).$$

To prove Barvinok rank$(C_n) \le r$, it now suffices to establish the identity

$$C_n \quad = \quad X_1 \oplus X_2 \oplus \cdots \oplus X_r.$$

Indeed, all diagonal entries of the matrices on the right hand side are 1, and the off-diagonal entries (for $i \ne j$) of the right hand side are $\min(x_{ij}^1, x_{ij}^2, \ldots, x_{ij}^r) = 0$, because $S_i \backslash S_j$ is nonempty.

To prove the converse direction, we consider an arbitrary representation

$$C_n \quad = \quad Y_1 \oplus Y_2 \oplus \cdots \oplus Y_r$$

where the matrices $Y_k = (y_{ij}^k)$ have tropical rank one. For each k we set $T_k := \{(i, j) : y_{ij}^k = 0\}$. Since the matrices Y_k are nonnegative and have tropical rank one, it follows that each T_k is a product $I_k \times J_k$, where I_k and J_k are subsets of

$\{1, \ldots, n\}$. Moreover, we have $I_k \cap J_k = \varnothing$ because the diagonal entries of Y_k are not zero. For each $i = 1, \ldots, n$ we set

$$S_i := \{k : i \in I_k\} \subseteq \{1, \ldots, r\}.$$

We claim that no two of the sets $S_1, \ldots, S_n$ are contained in one another. Sperner's Theorem [Aigner and Ziegler 1998] then proves that $n \leq \binom{r}{\lfloor r/2 \rfloor}$. To prove the claim, observe that if $S_i \subset S_j$ then the entry $y_{i,j}^k$ cannot be zero for any k. Indeed, if $k \in S_i \subseteq S_j$ then $j \in I_k$ implies $j \notin J_k$. And if $k \notin S_i$ then $i \notin I_k$.   □

For example, C_6 has Barvinok rank 4, as the following decomposition shows:

$$C_6 = \begin{pmatrix} 1 & 1 & 1 & 2 & 2 & 2 \\ 1 & 1 & 1 & 2 & 2 & 2 \\ 1 & 1 & 1 & 2 & 2 & 2 \\ 0 & 0 & 0 & 1 & 1 & 1 \\ 0 & 0 & 0 & 1 & 1 & 1 \\ 0 & 0 & 0 & 1 & 1 & 1 \end{pmatrix} \oplus \begin{pmatrix} 1 & 1 & 0 & 0 & 0 & 1 \\ 1 & 1 & 0 & 0 & 0 & 1 \\ 2 & 2 & 1 & 1 & 1 & 2 \\ 2 & 2 & 1 & 1 & 1 & 2 \\ 2 & 2 & 1 & 1 & 1 & 2 \\ 1 & 1 & 0 & 0 & 0 & 1 \end{pmatrix} \oplus \begin{pmatrix} 1 & 0 & 1 & 0 & 1 & 0 \\ 2 & 1 & 2 & 1 & 2 & 1 \\ 1 & 0 & 1 & 0 & 1 & 0 \\ 2 & 1 & 2 & 1 & 2 & 1 \\ 1 & 0 & 1 & 0 & 1 & 0 \\ 2 & 1 & 2 & 1 & 2 & 1 \end{pmatrix} \oplus \begin{pmatrix} 1 & 2 & 2 & 2 & 1 & 1 \\ 0 & 1 & 1 & 1 & 0 & 0 \\ 0 & 1 & 1 & 1 & 0 & 0 \\ 0 & 1 & 1 & 1 & 0 & 0 \\ 1 & 2 & 2 & 2 & 1 & 1 \\ 1 & 2 & 2 & 2 & 1 & 1 \end{pmatrix}.$$

Similarly, C_{36} has Barvinok rank 8, even though all its 35×35 minors have Barvinok rank 7 (and its 8×8 minors have Barvinok rank at most 5). Asymptotically,

$$\text{Barvinok rank}(C_n) \sim \log_2 n.$$

We will see in Examples 3.5 and 4.4 that the Kapranov rank and tropical rank of C_n are both two.

3. The Kapranov Rank

The tropical semiring has a strong connection to power series rings and their algebraic geometry. We review the basic setup from [Speyer and Sturmfels 2004; Sturmfels 2002]. Let $K = \mathbb{C}\{\{t\}\}$ be the field of Puiseux series with complex coefficients. The elements in K are formal power series $f = c_1 t^{a_1} + c_2 t^{a_2} + \cdots$, where $a_1 < a_2 < \cdots$ are rational numbers that have a common denominator. Let $\deg : K^* \to \mathbb{Q}$ be the natural valuation sending a nonzero Puiseux series f to its degree a_1. For any two elements $f, g \in K$, we have $\deg(fg) = \deg f + \deg g = \deg f \odot \deg g$. In general we also have $\deg(f + g) = \min(\deg f, \deg g) = \deg f \oplus \deg g$, unless there is a cancellation of leading terms. Thus the tropical arithmetic is naturally induced from ordinary arithmetic in power series fields.

The field $K = \mathbb{C}\{\{t\}\}$ is algebraically closed of characteristic zero. If I is any ideal in $K[x_1, \ldots, x_d]$ then we write $V(I)$ for its variety in the d-dimensional algebraic torus $(K^*)^d$. Thus the elements of $V(I)$ are vectors $x(t) = (x_1(t), \ldots, x_d(t))$ where each $x_i(t)$ is a Puiseux series and $f(x(t)) = 0$ for each polynomial $f \in I$. Let us now enlarge the field K and allow all formal power series $f = c_1 t^{a_1} + c_2 t^{a_2} + \cdots$ where the a_i can be real numbers, not just rationals. We denote this larger field by $\tilde{K}$ and we write $\tilde{V}(I)$ for the variety in $(\tilde{K}^*)^d$ defined by I.

The degree map can be applied coordinatewise, giving rise to a map which takes vectors of nonzero power series into $\mathbb{R}^d$:

$$\deg : (\tilde{K}^*)^d \to \mathbb{R}^d, \quad (f_1(t), \ldots, f_d(t)) \mapsto (\deg f_1, \ldots, \deg f_d).$$

We define the *tropical variety* of I, denoted $\mathcal{T}(I) \subset \mathbb{R}^d$, to be the image of $\tilde{V}(I)$ under the map deg. In [Speyer and Sturmfels 2004; Sturmfels 2002], the following alternative description of the tropical variety is given:

THEOREM 3.1. *The tropical variety $\mathcal{T}(I)$ is the set of vectors $w \in \mathbb{R}^n$ such that the initial ideal $\mathrm{in}_w(I) = \langle \mathrm{in}_w(f) : f \in I \rangle$ contains no monomial. The dimension of $\mathcal{T}(I)$ is the (topological) dimension of $V(I)$.*

The first statement in Theorem 3.1 is due to Misha Kapranov (in the special case when I is a principal ideal) and the third author (for arbitrary ideals I, in [Sturmfels 2002]). A complete proof can be found in [Speyer and Sturmfels 2004]. The second statement in Theorem 3.1 is due to Bieri and Groves [1984]. An elementary proof of this result, and the fact that $\mathcal{T}(I)$ is a polyhedral fan, appears in [Sturmfels 2002, § 9].

We defined Kapranov rank to be the smallest dimension of any tropical linear space containing the columns of M; now, we can make this precise by defining tropical linear spaces.

DEFINITION 3.2. A *tropical linear space* in $\mathbb{R}^d$ is any subset $\mathcal{T}(I)$ where I is an ideal generated by affine-linear forms $a_1 x_1 + \cdots + a_d x_d + b$ in $\tilde{K}[x] = \tilde{K}[x_1, \ldots, x_d]$. Its *dimension* is its topological dimension, which is equal to d minus the number of minimal generators of I.

Note that here the scalars $a_1, \ldots, a_n, b$ are power series in t with complex coefficients, the choice of the complex numbers being crucial. If I is the principal ideal generated by one affine-linear form $a_1 x_1 + \cdots + a_n x_n + b$, then $\mathcal{T}(I)$ is a *tropical hypersurface*. Tropical linear spaces were studied in [Speyer and Sturmfels 2004], where it was shown that they are parametrized by the *tropical Grassmannian*. Every tropical linear space L is a finite intersection of tropical hyperplanes, but not conversely, and the number of tropical hyperplanes needed is generally larger than the codimension of L.

Recall from Definition 1.2 that the Kapranov rank of a matrix $M \subset \mathbb{R}^{d \times n}$ is the smallest dimension of any tropical linear space containing the columns of M. It is not completely apparent in this definition that the Kapranov rank of a matrix and its transpose are the same, but this follows from our next result. Let J_r denote the ideal generated by all the $(r+1) \times (r+1)$-subdeterminants of a $d \times n$-matrix of indeterminates (x_{ij}). This is a prime ideal of dimension $rd + rn - r^2$, and the generating determinants form a Gröbner basis. The variety $V(J_r)$ consists of all $d \times n$-matrices with entries in K^* whose (classical) rank is at most r.

THEOREM 3.3. *For a real matrix* $M = (m_{ij}) \in \mathbb{R}^{d \times n}$ *the following statements are equivalent:*

(a) *The Kapranov rank of M is at most r.*
(b) *The matrix M lies in the tropical determinantal variety $\mathcal{T}(J_r)$.*
(c) *There exists a $d \times n$-matrix $F = (f_{ij}(t))$ with nonzero entries in the field $\tilde{K}$ such that the rank of F is less than or equal to r and $\deg(f_{ij}) = m_{ij}$ for all i and j.*

The power series matrix F in part (c) is called a *lift* of M. We abbreviate this as $\deg F = M$.

PROOF. The equivalence of (b) and (c) is simply our definition of tropical variety applied to the ideal J_r since, over the field $\tilde{K}$, lying in the variety of the determinantal ideal J_r is equivalent to having rank at most r. To see that (c) implies (a), consider the linear subspace of $\tilde{K}^d$ spanned by the columns of F. This is an r-dimensional linear space over a field, so it is defined by an ideal I generated by $d - r$ linearly independent linear forms in $\tilde{K}[x_1, \ldots, x_d]$. The tropical linear space $\mathcal{T}(I)$ contains all the column vectors of $M = \deg F$.

Conversely, suppose that (a) holds, and let L be a tropical linear space of dimension r containing the columns of M. Pick a linear ideal I in $\tilde{K}[x_1, \ldots, x_d]$ such that $L = \mathcal{T}(I)$. By applying the definition of tropical variety to the ideal I, we see that each column vector of M has a preimage in $\tilde{V}(I) \subset (\tilde{K}^*)^d$ under the degree map. Let F be the $d \times n$-matrix over $\tilde{K}$ whose columns are these preimages. Then the column space of F is contained in the variety defined by I, so we have $\mathrm{rank}(F) \leq r$, and $\deg F = M$ as desired. $\square$

COROLLARY 3.4. *The Kapranov rank of a matrix $M \in \mathbb{R}^{d \times n}$ is the smallest rank of any lift of M.*

The ideal J_1 is generated by the 2×2-minors $x_{ij}x_{kl} - x_{il}x_{kj}$ of the $d \times n$-matrix (x_{ij}). Therefore, a matrix of Kapranov rank one must certainly satisfy the linear equations $m_{ij} + m_{kl} = m_{il} + m_{kj}$. This happens if and only if there exist real vectors $X = (x_1, \ldots, x_d)$ and $Y = (y_1, \ldots, y_n)$ with

$$m_{ij} = x_i + y_j \text{ for all } i, j \quad \Longleftrightarrow \quad m_{ij} = x_i \odot y_j \text{ for all } i, j \quad \Longleftrightarrow \quad M = X^T \odot Y.$$

Conversely, if such X and Y exist, we can lift M to a matrix of rank one by substituting $t^{m_{ij}}$ for m_{ij}. Therefore, a matrix M has Kapranov rank one if and only if it has Barvinok rank one. In general, the Kapranov rank can be much smaller than the Barvinok rank, as the following example shows.

EXAMPLE 3.5. Let $n \geq 3$ and consider the classical identity matrix C_n. It does not have Kapranov rank one, so it has Kapranov rank at least two. Let

$a_3, a_4, \ldots, a_n$ be distinct nonzero complex numbers. Consider the matrix

$$
F_n = \begin{pmatrix}
t & 1 & t+a_3 & t+a_4 & \cdots & t+a_n \\
1 & t & 1+a_3 t & 1+a_4 t & \cdots & 1+a_n t \\
t-a_3 & 1 & t & t-a_3+a_4 & \cdots & t-a_3+a_n \\
t-a_4 & 1 & t-a_4+a_3 & t & \cdots & t-a_4+a_n \\
\vdots & \vdots & \vdots & \vdots & \ddots & \vdots \\
t-a_n & 1 & t-a_n+a_3 & t-a_n+a_4 & \cdots & t
\end{pmatrix}.
$$

The matrix F_n has rank 2 because the i-th column (for $i \geq 3$) equals the first column plus a_i times the second column. Since $\deg F_n = C_n$, we conclude that C_n has Kapranov rank two.

The two-dimensional tropical plane containing the columns of C_n is the two-dimensional fan L in $\mathbb{R}^n$ which consists of the n cones $\{x_i \geq x_1 = \cdots = x_{i-1} = x_{i+1} = \cdots = x_n\}$; this is the tropical variety defined by the ideal in $K[x_1, \ldots, x_n]$ generated by $n-2$ linear forms with generic coefficients in $\mathbb{C}$. Its image in $\mathbb{TP}^{n-1}$ is the line all of whose tropical Plücker coordinates are zero [Speyer and Sturmfels 2004].

The following proposition establishes half of Theorem 1.4.

PROPOSITION 3.6. *Every matrix $M \in \mathbb{R}^{d \times n}$ satisfies Kapranov rank$(M) \leq$ Barvinok rank(M), and this inequality can be strict.*

PROOF. Suppose that M has Barvinok rank r. Write $M = M_1 \oplus \cdots \oplus M_r$ where each M_i has Barvinok rank one. Then M_i has Kapranov rank one, so there exists a rank one matrix F_i over $\tilde{K}$ such that $\deg F_i = M_i$. Moreover, by multiplying the matrices F_i by random complex numbers, we can choose F_i such that there is no cancellation of leading terms in t when we form the matrix $F = F_1 + \cdots + F_r$. This means $\deg F = M$. Clearly, the matrix F has rank at most r. Theorem 3.3 implies that M has Kapranov rank at most r. Example 3.5 shows that the inequality can be strict.   □

A general algorithm for computing the Kapranov rank of a matrix M involves computing a Gröbner basis of the determinantal ideal J_r. Suppose we wish to decide whether a given real $d \times n$-matrix $M = (m_{ij})$ has Kapranov rank $> r$. To decide this question, we fix any term order $\prec_M$ on the polynomial ring $\mathbb{C}[x_{ij}]$ which refines the partial ordering on monomials given assigning weight m_{ij} to the variable x_{ij}, and we compute the reduced Gröbner basis $\mathcal{G}$ of J_r in the term order $\prec_M$. For each polynomial g in $\mathcal{G}$, we consider its leading form $\mathrm{in}_M(g)$ with respect to the partial ordering coming from M. As noted in [Sturmfels 1996, §1], we have $\mathrm{in}_{\prec_M}(\mathrm{in}_M(g)) = \mathrm{in}_{\prec_M}(g)$ for all $g \in \mathcal{G}$.

The ideal generated by the set of leading forms $\{\mathrm{in}_M(g) : g \in \mathcal{G}\}$ is the initial ideal $\mathrm{in}_M(J_r)$. Let x^{all} denote the product of all dn unknowns x_{ij}. The second step in our algorithm is to compute the saturation of the initial ideal with respect to the coordinate hyperplanes:

$$\left(\text{in}_M(J_r) : \langle x^{\text{all}} \rangle^\infty \right) = \{ f \in \mathbb{C}[x_{ij}] : f \cdot (x^{\text{all}})^s \in J_r \text{ for some } s \in \mathbb{N} \}. \quad (3\text{--}1)$$

Computing such an ideal quotient, given the generators $\text{in}_M(g)$, is a standard operation in computational commutative algebra. It is a built-in command in software systems such as CoCoA [CoCoA 2000–], Macaulay 2 [Grayson and Stillman 1993–] or Singular [Greuel et al. 2001]. Here is a direct consequence of Theorems 3.1 and 3.3:

COROLLARY 3.7. *The matrix M has Kapranov rank $> r$ if and only if* (3–1) *is the unit ideal $\langle 1 \rangle$.*

In view of this, the (combinatorial) Theorem 5.5, Theorem 6.5 and Corollary 7.4 have the following commutative algebra implications. Recall from [Richter-Gebert et al. 2005] that a finite generating set S of an ideal I is a *tropical basis* if, for every weight vector $w \in \mathbb{R}^n$ for which the initial ideal $\text{in}_w(I)$ contains a monomial, there is an $f \in S$ such that $\text{in}_w(f)$ is a monomial. Every ideal I in $K[x_1, \ldots, x_n]$ has a tropical basis but tropical bases are often much larger than minimal generating sets.

COROLLARY 3.8. *The 3×3-minors of a matrix of indeterminates form a tropical basis. The same holds for the maximal minors of a matrix, but it does not hold for the 4×4-minors of a 7×7-matrix.*

We have defined Kapranov rank in terms of power series arithmetic over the complex field $\mathbb{C}$, which is a canonical choice for doing algebraic geometry. However, the same definition works over any field k. One can consider the Puiseux series field $K = k\{\{t\}\}$ with either rational or real exponents. Note that the former is not algebraically closed if k is algebraically closed of characteristic p, but this need not concern us. We denote the latter by $\tilde{K}$ as before. All we need is the degree map $(\tilde{K}^*)^d \to \mathbb{R}^d$. We make the following analogous definitions.

DEFINITION 3.9. Let $K = k\{\{t\}\}$. A *tropical linear space over k* is the image under "deg" of any linear subspace of the $\tilde{K}$-vector space $\tilde{K}^d$. Its *dimension* is equal to the dimension of that linear subspace. The *Kapranov rank over k* of a matrix $M \in \mathbb{R}^{n \times d}$ is the smallest dimension of a tropical linear space containing the columns of M.

Unless otherwise stated, we will concern ourselves only with Kapranov rank over the complex numbers. In the general setting, Theorem 3.3 is true over all fields, but Proposition 3.6 is true only over infinite fields because in its proof we needed to take random coefficients. Indeed, Example 6.6 in Section 6 shows a matrix whose Kapranov rank over the 2-element field $\mathbb{F}_2$ is greater than the Barvinok rank. Even over algebraically closed fields, the Kapranov rank of a matrix may depend on the characteristic of the field. We will discuss this further and give examples in Section 7.

4. The Tropical Rank

We begin by proving the first inequality in Theorem 1.4. To complete the proof of Theorem 1.4, it remains to be seen that the inequality can be strict. This will be done in Section 7.

PROPOSITION 4.1. *Every matrix* $M \in \mathbb{R}^{d \times n}$ *satisfies*

$$tropical\ rank\,(M) \leq Kapranov\ rank\,(M).$$

PROOF. If the matrix M has a tropically nonsingular $r \times r$ minor, then any lift of M to the power series field $\tilde{K}$ must have the corresponding $r \times r$-minor nonsingular over $\tilde{K}$, since the leading exponent of its determinant occurs only once in the sum. Consequently, no lift of M to $\tilde{K}$ can have rank less than r. By Theorem 3.3, this means that the Kapranov rank of M must be at least r.   □

The set of all tropical linear combinations of a set of n vectors in $\mathbb{R}^d$ is a polyhedral complex. It has a 1-dimensional lineality space, spanned by the vector $(1, \ldots, 1)$, but upon quotienting out by this 1-dimensional space, we get a bounded subset in *tropical projective space* $\mathbb{TP}^{d-1} = \mathbb{R}^d / \mathbb{R}(1, \ldots, 1)$. This set is the *tropical convex hull* of the n given points in $\mathbb{TP}^{d-1}$, and it was investigated in depth in [Develin and Sturmfels 2004]. We review some relevant definitions and facts.

We fix a subset $V = \{v_1, \ldots, v_n\} \subseteq \mathbb{R}^d$. Given a point $x \in \mathbb{R}^d$, its *type* is the d-tuple of sets $S = (S_1, \ldots, S_d)$, where each $S_j \subset \{1, \ldots, n\}$ and $i \in S_j$ if $x_j - v_{ij} \geq x_k - v_{ik}$ for all $k \in \{1, \ldots, n\}$. Let X_S be the region consisting of points with type S; then according to [Develin and Sturmfels 2004, Theorem 15], the tropical convex hull of V equals the union of the bounded regions X_S, which are precisely those regions for which each S_j is nonempty. (If x is a point in the tropical convex hull with type S, then expressing x as a linear combination of the v_i's, we have $i \in S_j$ if the contribution of v_i is responsible for the j-th coordinate of x.) Indeed, (the topological closures of) these regions provide a polytopal decomposition of the tropical convex hull of V. Note that by definition, any type has the property that each $i \in \{1, \ldots, n\}$ is in some S_j.

The dimension of a particular cell X_S of the tropical polytope can be easily computed from the combinatorics of the d-tuple S: let G_S be the graph which has vertex set $1, \ldots, d$, with i and j connected by an edge if $S_i \cap S_j$ is nonempty. The dimension of X_S is one less than the number of connected components of the graph G_S.

Recall from Definition 1.3 that the tropical rank of a matrix is the size of the largest nonsingular square minor, and that an $r \times r$ matrix M is nonsingular if $\bigodot_{i=1}^r M_{\sigma(i),i} = \sum_{i=1}^r M_{\sigma(i),i}$ achieves its minimum only once as σ ranges over the symmetric group $\mathcal{S}_r$. Here is another characterization.

THEOREM 4.2. *Let $M \subset \mathbb{R}^{d \times n}$ be a matrix. Then the tropical rank of M is equal to one plus the dimension of the tropical convex hull of the columns of M, viewed as a tropical polytope in $\mathbb{TP}^{d-1}$.*

PROOF. Let $V = \{v_1, \dots, v_n\}$ be the set of columns of M, and let $P = \text{tconv}(V)$ be their tropical convex hull in $\mathbb{TP}^{d-1}$. Suppose that r is the tropical rank of M, that is, there exists a tropically nonsingular $r \times r$-submatrix of M, but all larger square submatrices are tropically singular.

We first show that $\dim P \geq r - 1$. We fix a nonsingular $r \times r$-submatrix M' of M. Deleting the rows outside M' means projecting P into $\mathbb{TP}^{r-1}$, and deleting the columns outside M' means passing to a tropical subpolytope P' of the image. Both operations can only decrease the dimension, so it suffices to show $\dim P' \geq r - 1$. Hence, we can assume that M is itself a tropically nonsingular $r \times r$-matrix. Also, without loss of generality, we can assume that the minimum over $\sigma \in \mathcal{S}_r$ of

$$f(\sigma) \;=\; \sum_{i=1}^{r} v_{i,\sigma(i)} \tag{4-1}$$

is uniquely achieved when σ is the identity element $e \in \mathcal{S}_r$. We now claim that the cell $X_{(\{1\},\dots,\{r\})}$ exists; to do this, we need to demonstrate that there exists a point with type $(\{1\}, \dots, \{r\})$.

The inequalities which must be valid on this cell are $x_k - x_j \leq v_{jk} - v_{jj}$ for $j \neq k$. We claim that these inequalities define a full-dimensional region. Suppose not; then, by Farkas' Lemma, there exists a nonnegative linear combination of the inequalities $x_k - x_j \leq v_{jk} - v_{jj}$ which equals $0 \leq c$ for some nonpositive real c. This linear combination would imply that some other $\sigma \in \mathcal{S}_r$ has $f(\sigma) \leq f(e)$, a contradiction. So this cell is full-dimensional; it follows immediately that picking a point in its interior yields a point with type $(\{1\}, \dots, \{r\})$, since because these inequalities are all strict, no other type-inducing inequalities can hold.

For the converse, suppose that $\dim P \geq r$. Pick a region X_S of dimension r, and assume by translating the points (which adds a constant to each row of X_S, not changing the rank of the matrix) that $(0, \dots, 0)$ is in X_S, so that the only inequalities valid on 0 are those given by S. The graph G_S has $r + 1$ connected components, so we can pick $r + 1$ elements of $\{1, \dots, n\}$ of which no two appear in a common S_j. Assume without loss of generality that this set is $\{1, \dots, r+1\}$, and again without loss of generality rearrange the labeling of the coordinates so that $i \in S_j$ if and only if $i = j$, for $1 \leq i, j \leq r + 1$.

We now claim that the square submatrix consisting of the first $r + 1$ rows and columns of M is tropically nonsingular. Indeed, we have (using the definition of $f(\sigma)$ given in (4-1)):

$$f(\sigma) - f(e) \;=\; \sum_{i=1}^{r+1} v_{i,\sigma(i)} - \sum_{i=1}^{r+1} v_{ii} \;=\; \sum_{i=1}^{r+1} \big(v_{i,\sigma(i)} - v_{ii}\big),$$

but whenever $\sigma(i) \neq i$, $v_{i,\sigma(i)} - v_{ii} > 0$ since $i \in S_i$ and $i \notin S_{\sigma(i)}$ for the point 0. Therefore, if σ is not the identity, we have $f(\sigma) - f(e) > 0$, and e is the unique permutation in $\mathcal{S}_{r+1}$ minimizing the expression (4–1). So M has tropical rank at least $r + 1$. This is a contradiction, and we conclude that $\dim P = r - 1$. $\square$

We next present a combinatorial formula for the tropical rank of a zero-one matrix, or any matrix which has only two distinct entries. We define the *support* of a vector in tropical space $\mathbb{R}^d$ as the set of its zero coordinates. We define the *support poset* of a matrix M to be the set of all unions of supports of column vectors of M. This set is partially ordered by inclusion.

PROPOSITION 4.3. *The tropical rank of a zero-one matrix with no column of all ones equals the maximum length of a chain in its support poset.*

The assumption that there is no column of all ones is needed for the statement to hold because a column of zeroes and a column of ones represent the same point in tropical projective space $\mathbb{TP}^{d-1}$.

PROOF. There is no loss of generality in assuming that every union of supports of columns of M is actually the support of a column. Indeed, the tropical sum of a set of columns gives a column whose support is the union of supports, and appending this column to M does not change the tropical rank since the tropical convex hull of the columns remains the same. Therefore, if there is a chain with r elements in the support poset we may assume that there is a set of r columns with supports contained in one another. Since there is no column of ones, from this we can easily extract an $r \times r$ minor with zeroes on and below the diagonal and 1's above the diagonal, which is tropically nonsingular.

    Reciprocally, suppose there is a tropically nonsingular $r \times r$ minor N. We claim that the support poset of N has a chain of length r, from which it follows that the support poset of M also has a chain of length r. Assume without loss of generality that the unique minimum permutation sum is obtained in the diagonal. This minimum sum cannot be more than one, because if n_{ii} and n_{jj} are both 1 then changing them for n_{ij} and n_{ji} does not increase the sum. If the minimum is zero, orienting an edge from i to j if entry ij of N is zero yields an acyclic digraph, which admits an ordering. Rearranging the rows and columns according to this ordering yields a matrix with 1's above the diagonal and 0's on and below the diagonal. The tropical sum of the last i columns (which corresponds to union of the corresponding supports) then produces a vector with 0's exactly in the last i positions. Hence, there is a proper chain of supports of length r.

    If the minimum permutation sum in N is 1, then let n_{ii} be the unique diagonal entry equal to 1. The i-th row in N must consist of all 1's: if n_{ij} is zero, then changing n_{ij} and n_{ji} for n_{ii} and n_{jj} does not increase the sum. Changing this row of ones to a row of zeroes does not affect the support poset of N (it just adds an element to every support), and yields a nonsingular zero-one matrix with minimum sum zero to which we can apply the argument in the previous paragraph. □

EXAMPLE 4.4. The tropical rank of the classical identity matrix C_n equals two (for all n), since all of its 3×3 minors are tropically singular, while the principal 2×2 minors are not. The supports of its columns are all the sets of cardinality $n - 1$ and the support poset consists of them and the whole set $\{1, \ldots, n\}$. The maximal chains in the poset have indeed length two.

As with the matrices of Barvinok rank r, the $d \times n$ matrices of tropical rank at most r form a polyhedral fan given as the intersection of the tropical hypersurfaces $\mathcal{T}(f)$ where f runs over the **set of** $(r + 1) \times (r + 1)$-subdeterminants of a $d \times n$-matrix of unknowns (x_{ij}). Note that this is very similar to the Kapranov rank; by Theorem 3.3, the set of $d \times n$ matrices of tropical rank is the intersection of the tropical hypersurfaces $\mathcal{T}(f)$ where f runs over the **ideal generated by the** $(r + 1) \times (r + 1)$-subdeterminants of a $d \times n$-matrix of unknowns (x_{ij}).

However, these are not equal; matrices can have Kapranov rank strictly bigger than their tropical rank, as will be seen in Section 7. In this sense, the subdeterminants of a given size $r \geq 4$ do not form a tropical basis for the ideal they generate.

5. Mixed Subdivisions and Corank One

A useful tool in tropical convexity is the computation of tropical convex hulls by means of mixed subdivisions of the Minkowski sum of several copies of a simplex. We recall the definition of mixed subdivisions, adapted to the case of interest to us. See [Santos 2003] for more details.

DEFINITION 5.1. Let Δ^{d-1} be the standard $(d - 1)$-simplex in $\mathbb{R}^d$, with vertex set $A = \{e_1, \ldots, e_d\}$. Let $n\Delta^{d-1}$ denote its dilation by a factor of n, which we regard as the convex hull of the Minkowski sum $A + A + \cdots + A$ (n times). Let $M = (v_{ij}) \subset \mathbb{R}^{d \times n}$ be a matrix. Consider the lifted simplices

$$P_i := \text{conv}\{(e_1, v_{1i}), \ldots, (e_d, v_{di})\} \subset \mathbb{R}^{d+1} \quad \text{for } i = 1, 2, \ldots, n.$$

The *regular mixed subdivision* of $n\Delta^{d-1}$ induced by M is the set of projections of the lower faces of the Minkowski sum $P_1 + \cdots + P_n$. Here, a face is called lower if its outer normal cone contains a vector with last coordinate negative.

It was shown in [Develin and Sturmfels 2004, § 4] that there is a bijection between the cells X_S in the convex hull of the columns of M and the interior cells in the regular subdivision of a product of simplices induced by M. Via the Cayley trick [Santos 2003], the latter biject to interior cells in the regular mixed subdivision defined above. Here we provide a short direct proof of the composition of these two bijections:

LEMMA 5.2. *Let $M \subset \mathbb{R}^{d \times n}$ and let $S = (S_1, \ldots, S_d)$, where each S_j is a subset of $\{1, \ldots, n\}$. Then, the following properties are equivalent:*

(1) *There exists a point in $\mathbb{R}^d$ of type S relative to the n points given by the columns of M.*

(2) *There is a nonnegative matrix M' such that M' is obtained from M by adding constants to rows or columns of M, and such that $M'_{ji} = 0$ precisely when $i \in S_j$.*

(3) *The regular mixed subdivision of $n\Delta^{d-1}$ induced by M has as a cell the Minkowski sum $\tau_1 + \cdots + \tau_n$ where $\tau_i = \mathrm{conv}(\{e_j : i \in S_j\})$.*

Moreover, if this happens, the cells referred to in parts (1) and (3) have complementary dimensions.

PROOF. Adding a constant to a row of M amounts to translating the set of n points in $\mathbb{TP}^{n-1}$, while adding a constant to a column leaves the point set unchanged. Consider a cell X_S in the tropical convex hull, let x be any point in the relative interior of X_S and let M' be the (unique) matrix obtained by translating the point set by a vector $-x$ and normalizing every column by adding a scalar so that its minimum coordinate equals 0. Conversely, for a matrix M' as in (2), consider the point x whose coordinates are the amounts added to the columns of M to obtain M'. The point x is in the tropical convex hull of the columns of M. Let S be its type. Then the modified matrix M' has zeroes precisely in entries (j, i) with $i \in S_j$, proving the equivalence of (1) and (2).

For the equivalence of (2) and (3), observe that adding a constant to a row or column of M does not change the mixed subdivision of $\sum P_i$. For a nonnegative matrix M' with at least a zero in every column, the positions of the zero entries define the face of $\sum P_i$ in the negative vertical direction. Conversely, for every cell of the regular mixed subdivision, we can apply a linear transformation changing only the last coordinate to give that cell height zero and all other vertices positive height (this is what it means to be in the lower envelope.) The resulting height function is precisely the matrix M' in (2), which proves the equivalence of (2) and (3). The assertion on dimensions is easy to prove. $\square$

This lemma implies that the tropical convex hull is dual to the regular mixed subdivision.

COROLLARY 5.3. *Given a matrix M, the poset of types in the tropical convex hull of its columns and the poset of interior cells of the corresponding regular mixed subdivision are antiisomorphic.*

PROOF. From the proof of Lemma 5.2, it is clear that the poset of types (under $S < T$ if $S_j \subset T_j$ for each j) and the poset of cells in the regular mixed subdivision are antiisomorphic. Meanwhile, a type S is in the tropical convex hull of its columns if and only if each S_j is nonempty; this is the same condition categorizing when the corresponding cell is contained in the boundary of the mixed subdivision (which occurs whenever there exists a vertex appearing in no summand.) □

COROLLARY 5.4. *Let $M \subset \mathbb{R}^{d \times n}$. The tropical rank of M equals d minus the minimal dimension of an interior cell in the regular mixed subdivision of $n\Delta^{d-1}$ induced by M.*

We can use these tools to prove that the tropical and Kapranov ranks of a matrix coincide if the latter is maximal.

THEOREM 5.5. *If a $d \times n$ matrix M has Kapranov rank equal to d, then it has tropical rank equal to d as well.*

PROOF. By Corollary 5.4, M has tropical rank d if and only if the corresponding regular mixed subdivision has an interior vertex. The theorem then follows from the next two lemmas. □

LEMMA 5.6. *A $d \times n$-matrix M has Kapranov rank less than d if and only if the corresponding regular mixed subdivision has a cell that intersects all facets of $n\Delta^{d-1}$.*

PROOF. If M has Kapranov rank less than d, then its column vectors lie in a tropical hyperplane. Since all tropical hyperplanes are translates of one another, there is no loss of generality in assuming that it is the hyperplane defined by $x_1 \oplus \cdots \oplus x_d$. That is, after normalization, all columns of M are nonnegative and have at least two zeroes. Then, by Lemma 5.2, the zero entries of M define a cell B in the regular mixed subdivision none of whose Minkowski summands are single vertices. In particular, for every facet F of Δ^{d-1} and for every $i \in \{1, \ldots, n\}$, the i-th summand of B is at least an edge and hence it intersects F. Hence, B intersects all facets of $n\Delta^{d-1}$. For the converse suppose the regular mixed subdivision has a cell B which intersects all facets of $n\Delta^{d-1}$. We may assume that M gives height zero to the points in that cell and positive height to all the others. The intersection of B with the j-th facet is given by the zero entries in M after deletion of the j-th row. In particular, B intersects the j-th facet if and only if every column has a zero entry outside of the j-th row, and so B intersects all facets if and only if all columns of M have at least two zeroes, implying that these all lie in the hyperplane defined by $x_1 \oplus \cdots \oplus x_d$. □

The cell in the preceding statement need not be unique. For example, if a tetrahedron is sliced by planes parallel to two opposite edges, then each maximal cell meets all the facets of the tetrahedron.

LEMMA 5.7. *In every polyhedral subdivision of a simplex which has no interior vertices, but arbitrarily many vertices on the boundary, there is a cell that intersects all of the facets.*

PROOF. Observe that there is no loss of generality in assuming that the polyhedral subdivision S is a triangulation. For a triangulation, we use Sperner's

Lemma [Aigner and Ziegler 1998]: "if the vertices of a triangulation of Δ are labeled so that (1) the vertices of Δ receive different labels and (2) the vertices in any face F of Δ receive labels among those of the vertices of F, then there is a fully labeled simplex".

Our task is to give our triangulation a Sperner labeling with the property that every vertex labeled i lies in the i-th facet of the simplex. The way to obtain this is: the vertex opposite to facet i is labeled $i + 1$. More generally, the label i of a vertex v is taken so that v is contained in facet i but not on facet $i - 1$. All labels are modulo d. $\square$

6. Matrices of Rank Two

By Theorem 4.2, if a matrix has tropical rank two, then the tropical convex hull of its columns is one-dimensional. Since it is contractible [Develin and Sturmfels 2004], this tropical polytope is a tree. Another way of showing this is via the corresponding regular mixed subdivision. Tropical rank 2 means that all the interior cells have codimension zero or one. Hence, the subdivision is constructed by slicing the simplex via a certain number of hyperplanes (which do not meet inside the simplex) and its dual graph is a tree. The special case when the matrix has Barvinok rank two is characterized by the following proposition.

PROPOSITION 6.1. *The following are equivalent for a matrix M*:

(1) *It has Barvinok rank 2.*
(2) *All its 3×3 minors have Barvinok rank 2.*
(3) *The tropical convex hull of its columns is a path.*

PROOF. $(1) \Longrightarrow (2)$ is trivial (the Barvinok rank of a minor cannot exceed that of the whole matrix) and $(3) \Longrightarrow (1)$ is easy: if a tropical polytope is a path, then it is the tropical convex hull of its two endpoints. Proposition 2.1 then implies that the Barvinok rank is two.

For $(2) \Longrightarrow (3)$ first observe that the case where M is 3×3 again follows from Proposition 2.1. We next prove the case where M is $d \times 3$ by contradiction: since the tropical convex hulls of rows and of columns of a matrix are isomorphic as cell complexes [Develin and Sturmfels 2004, Theorem 23], assume that the tropical convex hull of the *rows* of M is not a path. Then, there are three rows whose tropical convex hull is not a path, and their 3×3 minor has Barvinok rank 3. Finally, if M is of arbitrary size $d \times n$ and the tropical convex hull of its columns is not a path, consider three columns whose tropical convex hull is not a path and apply the previous case to them. $\square$

Our goal in this section is to show that if M has tropical rank 2 then it has Kapranov rank 2. Following Theorem 3.3 (c), this is done by constructing an explicit lift to a rank 2 matrix over $\tilde{K}$.

LEMMA 6.2. *Let M be a matrix of tropical rank two. Let x be a point in the tropical convex hull of the columns of M. Let M' be the matrix obtained by adding $-x$ to every column and then normalizing columns to have zero as their minimal entry. After possibly reordering the rows and columns, M' has the following block structure:*

$$M' := \begin{pmatrix} 0 & 0 & 0 & \cdots & 0 \\ 0 & A_1 & 0 & \cdots & 0 \\ 0 & 0 & A_2 & \cdots & 0 \\ \vdots & \vdots & \vdots & \ddots & \vdots \\ 0 & 0 & 0 & \cdots & A_k \end{pmatrix},$$

where the matrices A_i have all entries positive and every 2×2 minor has the property that the minimum of its four entries is achieved twice. Each $\mathbf{0}$ represents a matrix of zeroes of the appropriate size, and the first row and column blocks of M' may have size zero. Moreover, the tropical convex hull of the columns of M' is the union of the tropical convex hulls of the column vectors of the blocks augmented by the zero vector $\mathbf{0}$, and two of these k trees meet only at the point $\mathbf{0}$.

PROOF. First, adjoin the column x to our matrix if it does not already exist; since x is in the convex hull of M, this will not change the tropical convex hull of the columns of M. We can then simply remove it at the end, when it is transformed into a column of all zeroes. Thus, we can assume that one of the columns of the matrix M' consists of all zeroes.

The asserted block decomposition means that any two given columns of M' have either equal or disjoint cosupports, where the cosupport of a column is the set of positions where it does not have a zero. To prove that this holds, just observe that if it didn't then M' would have the following minor, where $+$ denotes a strictly positive entry. (Recall that each column has a zero in it.)

$$\begin{pmatrix} 0 & + & + \\ 0 & 0 & + \\ 0 & ? & 0 \end{pmatrix}$$

But this 3×3-matrix is tropically nonsingular. The assertion of the 2×2 minors follows from the fact that the nonnegative matrix

$$\begin{pmatrix} 0 & a & b \\ 0 & c & d \\ 0 & 0 & 0 \end{pmatrix}$$

is tropically singular if and only if the minimum of a, b, c and d is achieved twice.

Finally, the assertion about the convex hulls is trivial, since any linear combination of column vectors from a given block will have all zero entries except in the coordinates corresponding to that block. Any path joining two such points from different blocks will pass through the origin. $\square$

We next introduce a technical lemma for making a power series lifting sufficiently generic.

LEMMA 6.3. *Let A be a nonnegative matrix with no zero column and suppose that the smallest entry in A occurs most frequently in the first column. Let $\tilde{A}$ be the matrix*

$$\begin{pmatrix} 0 & \mathbf{0} \\ \mathbf{0} & A \end{pmatrix}$$

obtained by adjoining a row and a column of zeroes. If $\tilde{A}$ has Kapranov rank two, then $\tilde{A}$ has a rank-2 lift $F \in \tilde{K}^{d \times n}$ in which every 2×2 minor is nonsingular and the i-th column can be written as a linear combination $\lambda_i u_1 + \mu_i u_2$ of the first two columns u_1 and u_2, with $\deg \lambda_i \geq \deg \mu_i = 0$.

PROOF. Starting with an arbitrary rank-2 lift $\tilde{F}$ of $\tilde{A}$, let F be obtained by adding to every column a $\tilde{K}$-linear combination of the first column of $\tilde{F}$ with coefficients of sufficiently high degree (so as to not change the degrees of the entries) but otherwise generic. This preserves the degree of every entry and thus the rank of the lift, but makes every 2×2 minor of F nonsingular; by "generic," all we require is that the ratio between the coefficients of two columns is not equal to the ratio between those two columns if they are scalar multiples of each other. No column of $\tilde{F}$ is a scalar multiple of its first column since no column of $\tilde{A}$ aside from the first is constant, so no column of F is a scalar multiple of the first column either.

Since the lift has rank two and the first two columns are linearly independent, the i-th column of F is now a $\tilde{K}$-linear combination $\lambda_i u_1 + \mu_i u_2$ of the first two columns. If the degrees of λ_i and μ_i are different, then their minimum must be zero in order to get a degree zero element in the first entry of column i. But then $\deg \mu_i > \deg \lambda_i = 0$ is impossible, because it would make the i-th column of A all zero. Hence $\deg \lambda_i > \deg \mu_i = 0$.

If the degrees are equal, then they are nonpositive in order to get degree zero for the first entry in $\lambda_i u_1 + \mu_i u_2$. But they cannot be equal and negative, or otherwise entries of positive degree in u_2 would produce entries of negative degree in u_i. Hence, $\deg \lambda_i = \deg \mu_i = 0$ in this case. $\square$

COROLLARY 6.4. *Let A and B be nonnegative matrices. Assume that the two matrices*

$$\tilde{A} := \begin{pmatrix} A & \mathbf{0} \\ \mathbf{0} & 0 \end{pmatrix} \quad and \quad \tilde{B} := \begin{pmatrix} 0 & \mathbf{0} \\ \mathbf{0} & B \end{pmatrix}$$

have Kapranov rank equal to 2. Then, the matrix

$$M := \begin{pmatrix} A & \mathbf{0} & \mathbf{0} \\ \mathbf{0} & 0 & \mathbf{0} \\ \mathbf{0} & \mathbf{0} & B \end{pmatrix}$$

has Kapranov rank equal to 2 as well.

PROOF. We may assume that neither A nor B has a zero column. Hence Lemma 6.3 applies to both of them. We number the rows of M from $-k$ to k' and its columns from $-l$ to l', where $k \times l$ and $k' \times l'$ are the dimensions of

A and B respectively. In this way, A (respectively B) is the minor of negative (respectively, positive) indices. The row and column indexed zero consist of all zeroes. To further exhibit the symmetry between A and B the columns and rows in $\tilde{A}$ will be referred to "in reverse". That is to say, the first and second columns of it are the ones indexed 0 and -1 in M.

We now construct a lifting $F = (a_{i,j}) \in \mathbb{C}\{\{t\}\}^{d \times n}$ of M. We assume that we are given rank-2 lifts of $\tilde{A}$ and $\tilde{B}$ which satisfy the conditions of the previous lemma. Furthermore, we assume that the lift of the entry $(0,0)$ is the same in both, which can be achieved by scaling the first row in one of them.

We use exactly those lifts of $\tilde{A}$ and $\tilde{B}$ for the upper-left and bottom-right corner minors of M. Our task is to complete that with an entry $a_{i,j}$ for every i, j with $ij < 0$, such that $\deg(a_{i,j}) = 0$ and the whole matrix still has rank 2. We claim that it suffices to choose the entry $a_{-1,1}$ of degree zero and sufficiently generic. That this choice fixes the rest of the matrix is easy to see: The entry $a_{1,-1}$ is fixed by the fact that the 3×3 minor

$$\begin{pmatrix} a_{-1,-1} & a_{-1,0} & a_{-1,1} \\ a_{0,-1} & a_{0,0} & a_{0,1} \\ a_{1,-1} & a_{1,0} & a_{1,1} \end{pmatrix}$$

needs to have rank 2. All other entries $a_{i,-1}$ and $a_{i,1}$ are fixed by the fact that the entries $a_{i,-1}$, $a_{i,0}$ and $a_{i,1}$ (two of which come from either $\tilde{A}$ or $\tilde{B}$) must satisfy the same dependence as the three columns of the minor above. For each $j = -l, \ldots, -2$ (respectively $j = 2, \ldots, l'$), let λ_j and μ_j be the coefficients in the expression of the j-th column of $\tilde{A}$ (respectively, of $\tilde{B}$) as $\lambda_j u_0 + \mu_j u_{-1}$ (respectively, $\lambda_j u_0 + \mu_j u_1$). Then, $a_{i,j} = \lambda_j a_{i,0} + \mu_j a_{i,-1}$ (respectively, $a_{i,j} = \lambda_j a_{i,0} + \mu_j a_{i,1}$).

What remains to be shown is that if $a_{-1,1}$ is of degree zero and sufficiently generic, all the new entries are of degree zero too. For this, observe that if $j \in \{-l', \ldots, 2\}$ then $a_{i,j}$ is of degree zero as long as the coefficient of degree zero in $a_{i,-1}$ are different from the degree zero coefficients in the quotient $-\lambda_j a_{i,0}/\mu_j$ (here we are using the assumption that $\deg \lambda_j \geq \deg \mu_j \geq 0$). The same is true for $j \in \{2, \ldots, l\}$, with $a_{i,1}$ instead of $a_{i,-1}$. In terms of the choice of $a_{-1,1}$, this translates to the following determinant having nonzero coefficient in degree zero:

$$\begin{pmatrix} a_{i,-1} & a_{i,0} & -\lambda_j a_{i,0}/\mu_j \\ a_{-1,-1} & a_{-1,0} & a_{-1,1} \\ a_{0,-1} & a_{0,0} & a_{0,1} \end{pmatrix} \quad \text{or} \quad \begin{pmatrix} a_{0,-1} & a_{0,0} & a_{0,1} \\ a_{1,-1} & a_{1,0} & a_{1,1} \\ -\lambda_j a_{i,0}/\mu_j & a_{i,0} & a_{i,1} \end{pmatrix},$$

respectively for $j \in \{-l', \ldots, 2\}$ or $j \in \{2, \ldots, l\}$. That $a_{-1,1}$ and $a_{1,-1}$ sufficiently generic imply nonsingularity of these matrices follows from the fact that the following 2×2 minors come from the given lifts of $\tilde{A}$ and $\tilde{B}$, hence they are nonsingular:

$$\begin{pmatrix} a_{i,-1} & a_{i,0} \\ a_{0,-1} & a_{0,0} \end{pmatrix}, \quad \begin{pmatrix} a_{0,0} & a_{0,1} \\ a_{i,0} & a_{i,1} \end{pmatrix}. \qquad \square$$

THEOREM 6.5. *Let M be a matrix of tropical rank 2. Then its Kapranov rank equals 2 as well.*

PROOF. The Kapranov rank of M is always at least the tropical rank, so we need only show that the Kapranov rank is less than or equal to 2. If the tropical convex hull P of the columns of M is a path, then M has Barvinok rank 2 (by Proposition 6.1) and thus Kapranov rank 2. Otherwise, let x be a node of degree at least three in the tree P. We apply the method of Lemma 6.2. Since x has degree at least three, it follows that there are at least three blocks A_i. In particular, M has at least three columns. We induct on the number of columns of M. If M has exactly three columns, then each block A_i is a single column, and every row of M has at most one positive entry. It is easy to construct an explicit lift of rank 2: in each row, lift the positive entry α as $-t^\alpha$ and the zero entries as -1 and $1 + t^\alpha$. If there are rows of zeroes, lift them as $(-1, -1, 2)$, for example.

Next, suppose that M has $m \geq 4$ columns. The two blocks with the smallest number of combined columns have at least 2 and at most $m - 2$ rows all together. Possibly after adding a row and column of zeroes, this provides a decomposition of our matrix as

$$M = \begin{pmatrix} 0 & \mathbf{0} & \mathbf{0} \\ \mathbf{0} & A & \mathbf{0} \\ \mathbf{0} & \mathbf{0} & B \end{pmatrix},$$

where both A and B have at least two columns (A is the union of these two blocks, B the union of all other blocks.) It then follows that the minors

$$\begin{pmatrix} 0 & \mathbf{0} \\ \mathbf{0} & A \end{pmatrix} \quad \text{and} \quad \begin{pmatrix} 0 & \mathbf{0} \\ \mathbf{0} & B \end{pmatrix}$$

both have fewer columns than the original matrix. By the inductive hypothesis they have Kapranov rank 2. Applying Corollary 6.4 completes the inductive step of the theorem. □

In the proof of Lemma 6.3 we again required the ability to pick generic field elements. Thus, Theorem 6.5 holds over any infinite coefficient field, but it may fail over finite fields. This is illustrated by the following example. Proposition 4.1 and Theorem 5.5 fail here too, as does the fact that Kapranov rank is invariant under insertion of a tropical combination of existing columns.

EXAMPLE 6.6. The matrix

$$M = \begin{pmatrix} 1 & 0 & 0 \\ 0 & 1 & 0 \\ 0 & 0 & 0 \end{pmatrix} = \begin{pmatrix} 1 & 0 & 0 \\ 2 & 1 & 1 \\ 1 & 0 & 0 \end{pmatrix} \oplus \begin{pmatrix} 1 & 2 & 1 \\ 0 & 1 & 0 \\ 0 & 1 & 0 \end{pmatrix}.$$

has Barvinok and tropical ranks equal to 2, but Kapranov rank 3 over the two-element field $\mathbb{F}_2$.

7. Matrices Constructed from Matroids

One of the important properties of rank in usual linear algebra is that it produces a matroid. Unfortunately, the definitions of tropical rank, Kapranov rank, and Barvinok rank all fail to do this. Consider the configuration of four points in the tropical plane $\mathbb{TP}^2$ given by the columns of

$$M = \begin{pmatrix} 0 & 0 & 0 & 0 \\ 0 & 0 & 1 & 2 \\ 1 & 0 & 0 & -1 \end{pmatrix}.$$

By any of our three definitions of rank, the maximal independent sets of columns are $\{1,2\}$, $\{1,3,4\}$, and $\{2,3,4\}$. These do not all have the same size, and so they cannot be the bases of a matroid. The central obstruction here is that the sets $\{1,2,3\}$ and $\{1,2,4\}$ are (tropically) collinear, but the set $\{1,2,3,4\}$ is not. Despite this failure, there is a strong connection between tropical linear algebra and matroids.

The results in Sections 5 and 6 imply that any matrix whose tropical and Kapranov ranks disagree must be at least of size 5×5. The smallest example we know is 7×7. It is based on the *Fano matroid*. To explain the example, and to show how to construct many others, we prove a theorem about tropical representations of matroids. The reader is referred to [Oxley 1992] for matroid basics.

DEFINITION 7.1. Let $\mathcal{M}$ be a matroid. The *cocircuit matrix of* $\mathcal{M}$, denoted $\mathcal{C}(\mathcal{M})$, has rows indexed by the elements of the ground set of $\mathcal{M}$ and columns indexed by the cocircuits of $\mathcal{M}$. It has a 0 in entry (i,j) if the i-th element is in the j-th cocircuit and a 1 otherwise.

In other words, $\mathcal{C}(\mathcal{M})$ is the zero-one matrix whose columns have the cocircuits of $\mathcal{M}$ as supports. (As before, the support of a column is its set of zeroes.) As an example, the matrix C_n of Section 2 is the cocircuit matrix of the uniform matroid of rank 2 with n elements. Similarly, the cocircuit matrix of the uniform matroid $U_{n,r}$ has size $n \times \binom{n}{r-1}$ and its columns are all the zero-one vectors with exactly $r - 1$ ones. The following results show that its tropical and Kapranov ranks equal r. The tropical polytopes defined by these matrices are the *tropical hypersimplices* studied in [Joswig 2005].

PROPOSITION 7.2. *The tropical rank of the cocircuit matrix* $\mathcal{C}(\mathcal{M})$ *is the rank of the matroid* $\mathcal{M}$.

PROOF. This is a special case of Proposition 4.3 because the rank of $\mathcal{M}$ is the maximum length of a chain of nonzero covectors, and the supports of covectors are precisely the unions of supports of cocircuits. Note that $\mathcal{C}(\mathcal{M})$ cannot have a column of ones because every cocircuit is nonempty. □

THEOREM 7.3. *If the Kapranov rank of $\mathcal{C}(\mathcal{M})$ over the ground field k is equal to the rank of $\mathcal{M}$, then $\mathcal{M}$ is representable over k. If k is an infinite field, then the converse also holds.*

PROOF. Let $\mathcal{M}$ be a matroid of rank r on $\{1, \ldots, d\}$ which has n cocircuits and suppose that $F \in \tilde{K}^{d \times n}$ is a rank r lift of the cocircuit matrix $\mathcal{C}(\mathcal{M})$. For each row f_i of F, let $v_i \in k^d$ be the vector of constant terms in $f_i \in \tilde{K}^d$. We claim that $V = \{v_1, \ldots, v_d\}$ is a representation of $\mathcal{M}$. First note that V has rank at most r since every $\tilde{K}$-linear relation among the vectors f_i translates into a k-linear relation among the v_i. Our claim says that $\{i_1, \ldots, i_r\}$ is a basis of $\mathcal{M}$ if and only if $\{v_{i_1}, \ldots, v_{i_r}\}$ is a basis of V. Suppose $\{i_1, \ldots, i_r\}$ is a basis of $\mathcal{M}$. Then, as in the proof of Proposition 4.3, we can find a square submatrix of $\mathcal{C}(\mathcal{M})$ using rows $i_1, \ldots, i_r$ with 0's on and below the diagonal and 1's above it. This means that the lifted submatrix of constant terms is lower-triangular with nonzero entries along the diagonal. It implies that $v_{i_1}, \ldots, v_{i_r}$ are linearly independent, and, since $\operatorname{rank}(V) \leq r$, they must be a basis. We also conclude $\operatorname{rank}(V) = r$. If $\{i_1, \ldots, i_r\}$ is not a basis in $\mathcal{M}$, there exists a cocircuit containing none of them; this means that some column of $\mathcal{C}(\mathcal{M})$ has all 1's in rows $i_1, \ldots, i_r$. Therefore, $f_{i_1}, \ldots, f_{i_r}$ all have zero constant term in that coordinate, which means that $v_{i_1}, \ldots, v_{i_r}$ are all 0 in that coordinate. Since the cocircuit is not empty, not all vectors v_j have an entry of 0 in that coordinate, and so $\{v_{i_1}, \ldots, v_{i_r}\}$ cannot be a basis. This shows that V represents $\mathcal{M}$ over k, which completes the proof of the first statement in Theorem 7.3.

For the second statement, let us assume that $\mathcal{M}$ has no loops. This is no loss of generality because a loop corresponds to a row of 1's in $\mathcal{C}(\mathcal{M})$, which does not increase the Kapranov rank because every column has at least a zero. Assume $\mathcal{M}$ is representable over k and fix a $d \times n$-matrix $A \in k^{d \times n}$ such that the rows of A represent $\mathcal{M}$ and the sets of nonzero coordinates along the columns of A are the cocircuits of $\mathcal{M}$. Suppose $\{1, \ldots, r\}$ is a basis of $\mathcal{M}$ and let A' be the submatrix of A consisting of the first r rows. Write

$$ A = \begin{pmatrix} \mathbf{I}_r \\ C \end{pmatrix} \cdot A' $$

where $\mathbf{I}_r$ is the identity matrix and $C \in k^{(d-r) \times r}$. Observe that A, hence C, cannot have a row of zeroes (because $\mathcal{M}$ has no loops). Since k is an infinite field, there exists a matrix $B' \in k^{r \times n}$ such that all entries of the $d \times r$-matrix $\begin{pmatrix} \mathbf{I}_r \\ C \end{pmatrix} \cdot B'$ are nonzero. We now define

$$ F = \begin{pmatrix} \mathbf{I}_r \\ C \end{pmatrix} \cdot (A' + tB') \in \tilde{K}^{d \times n}. $$

This matrix has rank r and $\deg F = \mathcal{C}(\mathcal{M})$. This completes the proof of Theorem 7.3. $\qquad \square$

If k is representable over a finite field, its Kapranov rank (with respect to that field) may still exceed its tropical rank. It is easy to find examples; for instance, the matroid represented by $\{(0,1),(1,0),(1,1),(0,0)\}$ over $\mathbb{F}_2$ will work.

COROLLARY 7.4. *Let $\mathcal{M}$ be a matroid which is not representable over a given field k. Then the Kapranov rank with respect to k of the tropical matrix $\mathcal{C}(\mathcal{M})$ exceeds its tropical rank.*

This corollary furnishes many examples of matrices whose Kapranov rank exceeds their tropical rank. Consider, for example, the Fano and non-Fano matroids, depicted in Figure 2. They both have rank three and seven elements. The

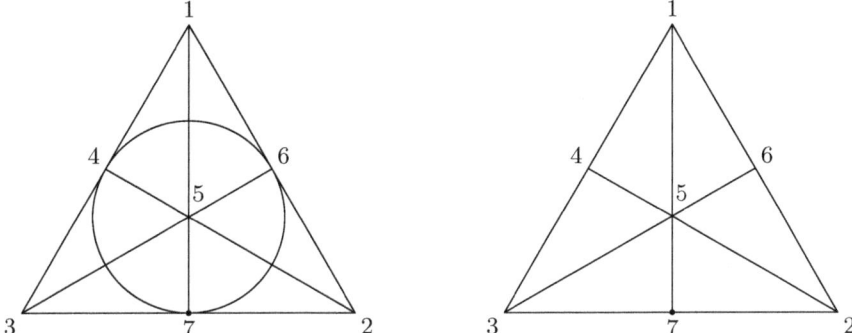

Figure 2. The Fano (left) and non-Fano (right) matroids.

first is only representable over fields of characteristic two, the second only over fields of characteristic different from two. In particular, Corollary 7.4 applied to these two matroids implies that over every field there are matrices with tropical rank equal to three and Kapranov rank larger than that. Also, it shows that the Kapranov rank of a matrix may be different over different fields k and k', even if k and k' are assumed to be algebraically closed. This is a more significant discrepancy than that of Example 6.6, which used a finite field.

More explicitly, the cocircuit matrix of the Fano matroid is

$$\mathcal{C}(\mathcal{M}) = \begin{pmatrix} 1 & 1 & 0 & 1 & 0 & 0 & 0 \\ 0 & 1 & 1 & 0 & 1 & 0 & 0 \\ 0 & 0 & 1 & 1 & 0 & 1 & 0 \\ 0 & 0 & 0 & 1 & 1 & 0 & 1 \\ 1 & 0 & 0 & 0 & 1 & 1 & 0 \\ 0 & 1 & 0 & 0 & 0 & 1 & 1 \\ 1 & 0 & 1 & 0 & 0 & 0 & 1 \end{pmatrix}.$$

This matrix is the smallest known example of a matrix whose Kapranov rank over $\mathbb{C}$ (four) is strictly larger than its tropical rank (three). Put differently, the seven columns of this matrix (in $\mathbb{TP}^6$) have as their tropical convex hull a two-dimensional cell complex which does not lie in any two-dimensional linear subspace of $\mathbb{TP}^6$, a feature decidedly absent from ordinary geometry.

Applied to nonrepresentable matroids, such as the *Vamos matroid* (rank 4, 8 elements, 41 cocircuits) or the *non-Pappus matroid* (rank 3, 9 elements, 20 cocircuits) [Oxley 1992], Corollary 7.4 yields matrices with different Kapranov and tropical ranks over *every* field. One can also get examples in which the difference of the two ranks is arbitrarily large. Indeed, given matrices A and B, we can construct the matrix

$$M := \begin{pmatrix} A & \infty \\ \infty' & B \end{pmatrix},$$

where ∞ and ∞' denote matrices of the appropriate dimensions and whose entries are sufficiently large. Appropriate choices of these large values (pick the extra columns to be points in the tropical convex hull of the columns of A and B and add large constants to each column) will ensure that the tropical and Kapranov ranks of M are the sums of those of A and of B. The difference between the Kapranov and tropical ranks of M is equal to the sum of this difference for A and for B.

The construction in Theorem 7.3 is closely related to the *Bergman complex* of the matroid $\mathcal{M}$. Ardila and Klivans [≥ 2005] showed that this complex is triangulated by the order complex of the lattice of flats of $\mathcal{M}$. Since flats correspond to unions of cocircuits, the following result is easily derived:

PROPOSITION 7.5. *The Bergman complex of the matroid $\mathcal{M}$ is equal to the tropical convex hull of the rows of the modified cocircuit matrix $\mathcal{C}'(\mathcal{M})$, where the 1's in $\mathcal{C}(\mathcal{M})$ are replaced by ∞'s.*

For the Fano matroid, the Bergman complex is the cone over the incidence graph of points and lines in the matroid. It consists of 15 vertices, 35 edges and 21 triangles.

8. Related Work and Open Questions

As mentioned in the introduction, our definition of nonsingular square matrix corresponds to the notion of "strongly regular" in the literature on the max-plus (or min-plus) algebra. The definition of "regular matrix" in [Butkovič 1995; Butkovič and Hevery 1985; Cuninghame-Green 1979] is the following one, for which we prefer to use a different name:

DEFINITION 8.1. A square matrix M is *positively tropically regular* if, in the formula for its tropical determinant, the minimum over all even permutations equals the minimum over odd permutations. The *positive tropical rank* of a matrix is the maximum size of a positively tropically regular minor.

The reason for this terminology is that M is positively tropically regular if it lies outside the positive tropical variety defined by the determinant. For basics on positive tropical varieties and a detailed study of the positive tropical Grassmannian see [Speyer and Williams 2003]. The positive tropicalization of

determinantal varieties leads also to a notion of *positive Kapranov rank* that
satisfies the inequalities

$$\text{pos. tropical rank}\,(M) \leq \text{pos. Kapranov rank}\,(M) \leq \text{Barvinok rank}\,(M).$$

Of course, the tropical and Kapranov ranks are less than or equal to their positive
counterparts.

Our notion of tropical rank, however, appears in [Butkovič and Hevery 1985;
Cuninghame-Green 1979] under a different name. Proposition 8.3 below was
previously proved in [Butkovič and Hevery 1985]:

DEFINITION 8.2. The columns of a matrix $M \in \mathbb{R}^{d \times n}$ are *strongly linearly
independent* if there is a column vector $b \in \mathbb{R}^d$ such that the tropical linear
system $M \odot x = b$ has a unique solution $x \in \mathbb{R}^n$. A square matrix is *strongly
regular* if its columns are strongly linearly independent.

PROPOSITION 8.3. *Strongly regular and tropically nonsingular are equivalent,
for a square matrix.*

PROOF. Suppose an $r \times r$ matrix M is tropically nonsingular; then there is some
$(r-1)$-dimensional cell X_S in the tropical convex hull of its columns in $\mathbb{TP}^{r-1}$.
After relabeling we have $S_i = \{i\}$ for $i = 1, 2, \ldots, r$. Then taking a point in the
relative interior of X_S yields a vector $b \in \mathbb{R}^r$ for which $M \odot x = b$ has a unique
solution, each x_i being necessarily equal to $b_i - m_{ii}$.

Conversely, suppose the columns of an $r \times r$ matrix M are strongly linearly
independent. Pick $b \in \mathbb{R}^r$ such that $M \odot x = b$ has a unique solution. Then, for
each x_j, there exists a b_i for which the expression $\sum M_{ik}x_k$ is uniquely minimized
for $k = j$ (otherwise we could increase x_j and get the same value for $M \odot x$).
This is equivalent to b having type S, where $S_j = \{i\}$.   □

COROLLARY 8.4. *The tropical rank of a matrix equals the largest size of a
strongly linearly independent subset of its columns.*

We now discuss some algorithmic issues. Apart from Corollary 8.4, the main
result in [Butkovič and Hevery 1985] is an $O(n^3)$ algorithm to check strong (i.e.,
tropical) regularity of an $n \times n$ matrix. The key step is to find a permutation that achieves the minimum in the determinantal tropical sum, which is the
assignment problem in combinatorial optimization [Papadimitriou and Steiglitz
1982]. Similarly, it is shown in [Butkovič 1995] that the problem of testing positive tropical regularity of square matrices is equivalent to the problem of testing
existence of even cycles in directed graphs.

For the Barvinok rank, we quote some results from [Çela et al. 1998]:

PROPOSITION 8.5. *The computation of the Barvinok rank of a matrix $M \in
\{0,1\}^{d \times n}$ is an NP-complete problem. Deciding whether a matrix has Barvinok
rank 2 can be done in time $O(dn)$.*

NP-completeness is proved by a reduction to the problem of covering a bipartite graph by complete bipartite subgraphs. For the case of rank 2, an algorithm is derived from the fact that matrices of Barvinok rank 2 are permuted *Monge matrices*. Çela et al. also prove that a matrix has Barvinok rank 2 if and only if all its 3×3 minors do (our Proposition 6.1) and that the Barvinok rank is bounded below by the maximum size of a strongly regular minor (i.e., by the tropical rank).

We finish by listing some open questions, most of them with an algorithmic flavor:

(1) Singularity of a single minor can be tested in polynomial time. But a naive algorithm to compute the tropical rank would need to check an exponential number of them. Can the tropical rank of a matrix be computed in polynomial time? In other words, is there a tropical analogue of Gauss elimination?

(2) Fix an integer k. The number of square minors of size at most $k+1$ of a $d \times n$ matrix M is polynomial in dn. Hence, there is a polynomial time algorithm for deciding whether M has tropical rank smaller or equal to k. Is the same true for the Barvinok rank? It is even open whether Barvinok rank equal to 3 can be tested in polynomial time.

(3) For a fixed k, a positive answer to either of the following two questions would imply a positive answer to the previous one:

(i) Is there a number $N(k)$ such that if all minors of M of size at most $N(k)$ have Barvinok rank at most k then M itself has Barvinok rank at most k? Proposition 2.2 shows that

$$ N(k) \geq \binom{k+1}{\lfloor \frac{k+1}{2} \rfloor}. $$

(ii) Is there a polynomial time algorithm for the Barvinok rank of matrices with tropical rank bounded by k? (This is open even for $k = 2$).

(iii) Can we obtain a bound on the Kapranov rank given the tropical rank? That is, given a positive integer r, can we find a bound $N(r)$ so that all matrices of tropical rank r have Kapranov rank at most $N(r)$? The example of the classical identity matrix shows that the same cannot be done for Barvinok rank.

(4) Can the Barvinok rank of a matrix M be defined in terms of the regular mixed subdivision of $n\Delta^{d-1}$ produced by M? Ideally, we would like a "nice and simple" characterization such as the one given for the tropical rank in Corollary 5.4. But the question we pose is whether matrices producing the same mixed subdivision have necessarily the same Barvinok rank.

(5) All the questions above are open for the Kapranov rank, too.

(6) Is there a 5×5-matrix having tropical rank 3 but Kapranov rank 4?

Acknowledgment

We thank Günter Rote for helpful discussions and for pointing us to references [Butkovič 1995; Butkovič and Hevery 1985].

References

[Aigner and Ziegler 1998] M. Aigner and G. M. Ziegler, *Proofs from The Book*, Springer-Verlag, Berlin, 1998. Second edition, 2001; third edition, 2004.

[Ardila 2004] F. Ardila, "A tropical morphism related to the hyperplane arrangement of the complete bipartite graph", 2004. http://www.arxiv.org/math.CO/0404287. To appear in *Discrete Comput. Geom.*

[Ardila and Klivans ≥ 2005] F. Ardila and C. Klivans, "The Bergman complex of a matroid and phylogenetic trees", to appear in *J. Comb. Theory B.* http://www.arxiv.org/math.CO/0311370.

[Barvinok et al. 1998] A. Barvinok, D. S. Johnson, G. J. Woeginger, and R. Woodroofe, "The maximum traveling salesman problem under polyhedral norms", pp. 195–201 in *Integer programming and combinatorial optimization* (Houston, 1998), edited by R. E. Bixby et al., Lecture Notes in Comput. Sci. **1412**, Springer, Berlin, 1998.

[Bieri and Groves 1984] R. Bieri and J. R. J. Groves, "The geometry of the set of characters induced by valuations", *J. Reine Angew. Math.* **347** (1984), 168–195.

[Butkovič 1995] P. Butkovič, "Regularity of matrices in min-algebra and its time-complexity", *Discrete Appl. Math.* **57**:2-3 (1995), 121–132.

[Butkovič 2003] P. Butkovič, "Max-algebra: the linear algebra of combinatorics?", *Linear Algebra Appl.* **367** (2003), 313–335.

[Butkovič and Hevery 1985] P. Butkovič and F. Hevery, "A condition for the strong regularity of matrices in the minimax algebra", *Discrete Appl. Math.* **11**:3 (1985), 209–222.

[Çela et al. 1998] E. Çela, R. Rudolf, and G. Woeginger, "On the Barvinok rank of matrices", 1998. Some results were subsequently sharpened in collaboration with G. Rote, 1998.

[CoCoA 2000–] CoCoA, "CoCoA: a system for doing computations in commutative algebra", 2000–. http://cocoa.dima.unige.it.

[Cohen and Rothblum 1993] J. E. Cohen and U. G. Rothblum, "Nonnegative ranks, decompositions, and factorizations of nonnegative matrices", *Linear Algebra Appl.* **190** (1993), 149–168.

[Cuninghame-Green 1979] R. Cuninghame-Green, *Minimax algebra*, vol. 166, Lecture Notes in Economics and Mathematical Systems, Springer-Verlag, Berlin, 1979.

[Develin 2004] M. Develin, "The space of n points on a tropical line in d-space", preprint, 2004. http://www.arxiv.org/math.CO/0401224. To appear in *Collectanea Math.*

[Develin and Sturmfels 2004] M. Develin and B. Sturmfels, "Tropical convexity", *Doc. Math.* **9** (2004), 1–27.

[Grayson and Stillman 1993–] D. R. Grayson and M. E. Stillman, "Macaulay 2, a software system for research in algebraic geometry", 1993–. http://www.math.uiuc.edu/Macaulay2/.

[Gregory and Pullman 1983] D. A. Gregory and N. J. Pullman, "Semiring rank: Boolean rank and nonnegative rank factorizations", *J. Combin. Inform. System Sci.* **8**:3 (1983), 223–233.

[Greuel et al. 2001] G. M. Greuel, G. Pfister, and H. Schönemann, "Singular 2.0, a system for polynomial computations", 2001. http://www.singular.uni-kl.de.

[Joswig 2005] M. Joswig, "Tropical half-spaces", 2005.

[Oxley 1992] J. G. Oxley, *Matroid theory*, Oxford University Press, New York, 1992.

[Papadimitriou and Steiglitz 1982] C. H. Papadimitriou and K. Steiglitz, *Combinatorial optimization: algorithms and complexity*, Prentice-Hall., Englewood Cliffs, NJ, 1982.

[Richter-Gebert et al. 2005] J. Richter-Gebert, B. Sturmfels, and T. Theobald, "First steps in tropical geometry", in *Idempotent mathematics and mathematical physics* (Vienna, 2003), edited by G. L. Litvinov and V. P. Maslov, Amer. Math. Society, Providence, 2005. http://www.arxiv.org/math.AG/0306366.

[Santos 2003] F. Santos, "The Cayley Trick and triangulations of products of simplices", Technical report, 2003. http://www.arxiv.org/math.CO/0312069. To appear in *Proceedings of the Joint Summer Research Conference on Integer Points in Polyhedra: geometry, number theory, algebra, and optimization*, edited by A. Barvinok et al., Contemporary Mathematics, Amer. Math. Soc.

[Speyer and Sturmfels 2004] D. Speyer and B. Sturmfels, "The tropical Grassmannian", *Adv. Geom.* **4**:3 (2004), 389–411.

[Speyer and Williams 2003] D. Speyer and L. Williams, "The tropically totally positive Grassmannian", 2003. http://www.arxiv.org/math.CO/0312297. To appear in *J. Alg. Combinatorics*.

[Sturmfels 1996] B. Sturmfels, *Gröbner bases and convex polytopes*, University Lecture Series **8**, Amer. Math. Soc., Providence, 1996.

[Sturmfels 2002] B. Sturmfels, *Solving systems of polynomial equations*, CBMS Regional Conference Series in Mathematics **97**, Amer. Math. Soc., Providence, and CBMS, Washington, DC, 2002.

MIKE DEVELIN
AMERICAN INSTITUTE OF MATHEMATICS
360 PORTAGE AVE.
PALO ALTO, CA 94306
UNITED STATES
develin@post.harvard.edu

FRANCISCO SANTOS
DEPTO. DE MATEMÁTICAS, ESTADÍSTICA Y COMPUTACIÓN
UNIVERSIDAD DE CANTABRIA
E-39005 SANTANDER
SPAIN
francisco.santos@unican.es

BERND STURMFELS
DEPARTMENT OF MATHEMATICS
UNIVERSITY OF CALIFORNIA
BERKELEY, CA 94720
UNITED STATES
bernd@math.berkeley.edu

Combinatorial and Computational Geometry
MSRI Publications
Volume **52**, 2005

The Geometry of Biomolecular Solvation

HERBERT EDELSBRUNNER AND PATRICE KOEHL

ABSTRACT. Years of research in biology have established that all cellular functions are deeply connected to the shape and dynamics of their molecular actors. As a response, structural molecular biology has emerged as a new line of experimental research focused on revealing the structure of biomolecules. The analysis of these structures has led to the development of computational biology, whose aim is to predict from molecular simulation properties inaccessible to experimental probes.

Here we focus on the representation of biomolecules used in these simulations, and in particular on the hard sphere models. We review how the geometry of the union of such spheres is used to model their interactions with their environment, and how it has been included in simulations of molecular dynamics.

In parallel, we review our own developments in mathematics and computer science on understanding the geometry of unions of balls, and their applications in molecular simulation.

1. Introduction

The molecular basis of life rests on the activity of biological macro-molecules, mostly nucleic acids and proteins. A perhaps surprising finding that crystallized over the last handful of decades is that geometric reasoning plays a major role in our attempt to understand these activities. In this paper, we address this connection between biology and geometry, focusing on hard sphere models of biomolecules.

The biomolecular revolution. Most living organisms are complex assemblies of cells, the building blocks for life. Each cell can be seen as a small chemical factory, involving thousands of different players with a large range of size and function. Among them, biological macro-molecules hold a special place. These usually large molecules serve as storage for the genetic information (the

---

Keywords: Molecular simulations, implicit solvent models, space-filling diagrams, spheres, balls, surface area, volume, derivatives.

Research of the two authors is partially supported by NSF under grant CCR-00-86013.

nucleic acids such as DNA and RNA), and as key actors of cellular functions (the proteins). Biochemistry, the field that studies these biomolecules, is currently experiencing a major revolution. In hope of deciphering the rules that define cellular functions, large scale experimental projects are performed as collaborative efforts involving many laboratories in many countries. The main aims of these projects are to provide maps of the genetic information of different organisms (the *genome projects*), to derive as much structural information as possible on the products of the corresponding genes (the *structural genomics projects*), and to relate these genes to the function of their products, usually deduced from their structure (the *functional genomics projects*). The success of these projects is completely changing the landscape of research in biology. As of February 2004, more than 170 whole genomes have been sequenced, corresponding to a database of over a million gene sequences. The need to store this data efficiently and to analyze its contents has led to the emergence of a collaborative effort between computer science and biology, referred to as *bio-informatics*. In parallel, the repository of biomolecular structures [Bernstein et al. 1977; Berman et al. 2000] contains more than 24,000 structures of proteins and nucleic acids. The similar need to organize and analyze the structural information contained in this database is leading to the emergence of another partnership between computer science and biology, namely *biogeometry*.

Significance of shape. Molecular structure or shape and chemical reactivity are highly correlated as the latter depends on the positions of the nuclei and electrons within the molecule. Indeed, chemists have long used three-dimensional plastic and metal models to understand the many subtle effects of structure on reactivity and have invested in experimentally determining the structure of important molecules. The same applies to biochemistry, where structural genomics projects are based on the premise that the structure of biomolecules implies their function. This premise rests on a number of specific and quantifiable correlations:

- enzymes fold into unique structures and the three-dimensional arrangement of their side-chains determines their catalytic activity;
- there is theoretical evidence that the mechanisms underlying protein complex formation depend on the shapes of the biomolecules involved [Levy et al. 2004];
- the folding rate of many small proteins correlates with a gross topological parameter that quantifies the difference between distance in space and along the main-chain [Plaxco et al. 1998; Alm and Baker 1999; Muñoz and Eaton 1999; Alm et al. 2002].

There is also evidence that the geometry of a protein plays a major role in defining its tolerance to mutation [Koehl and Levitt 2002]. We note in passing that structural biologists often refer to the 'topology' of a biomolecule when they mean the 'geometry' or 'shape' of the same. A common concrete model representing this shape is a union of balls, in which each ball corresponds to an atom. Properties of the biomolecule are then expressed in terms of properties of the union. For example, the potential active sites are detected as cavities [Liang et al. 1998c; Edelsbrunner et al. 1998; Liang et al. 1998b] and the interaction with the environment is quantified through the surface area and/or volume of the union of balls [Eisenberg and McLachlan 1986; Ooi et al. 1987; Liang et al. 1998a]. In what follows, we discuss in detail the geometric properties of union of balls, and relate them to the physical properties of the biomolecules they represent.

Outline. Section 2 describes biomolecules, and surveys their different levels of representation, focusing on the hard sphere models used in nearly all molecular simulation. Section 3 describes the relationship between the geometry of a biomolecule and its energetics. Section 4 surveys analytical and approximate methods used in biomolecular simulations for computing the area and volume of a molecule, and their derivatives with respect to the atomic coordinates. Section 5 develops the mathematical background needed to give compact formulas for geometric measurements. Section 6 discusses implementations of these formulas and presents experimental results. Section 7 concludes the paper with a discussion of future research directions.

2. Biomolecules

Following the Greek philosopher Democritus, who proclaimed that all matter is an assemblage of atoms, we can build a hierarchy that relates life to atoms. All living organisms can be described as arrangements of cells, the smallest units capable of carrying functions important for life. Cells can be divided into organelles, which are themselves assemblies of biomolecules. These biomolecules are usually polymers of smaller subunits, whose atomic structures are known from standard chemistry. There are many remarkable aspects to this hierarchy, one of them being that it is ubiquitous to all life forms, from unicellular organisms to complex multicellular species like us. Unraveling the secrets behind this hierarchy has become one of the major challenges of the twentieth and now twenty-first centuries. While physics and chemistry have provided significant insight into the structure of the atoms and their arrangements in small chemical structures, the focus now is set on understanding the structure and function of biomolecules, mainly nucleic acids and proteins. Our presentation of these molecules follow the general dogma in biology that states that the genetic information contained in DNA is first transcribed to RNA molecules which are then translated into proteins.

DNA. Deoxyribonucleic acid is a long polymer built from four different building blocks, the nucleotides. The sequence in which the nucleotides are arranged contains the entire information required to describe cells and their functions.

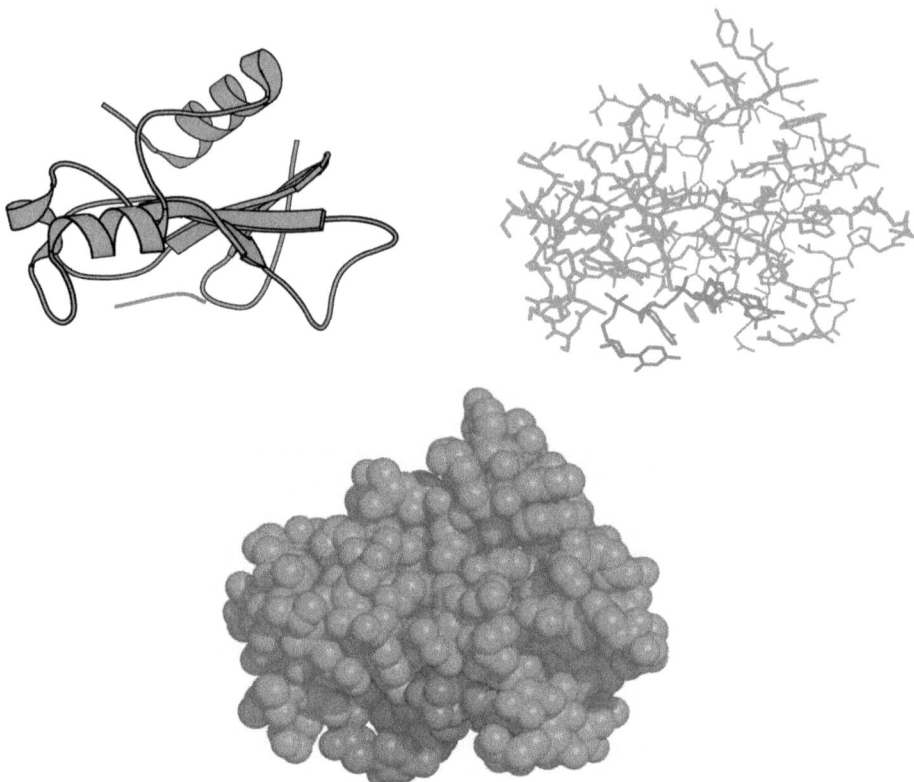

Figure 1. Visualizing protein-ligand interaction. Barnase is a small protein of 110 residues which has an endonuclease activity—it is able to cleave DNA fragments. Here we show the complex it forms with the small DNA fragment d(CGAC) [Buckle and Fersht 1994], using three different types of visualization. The coordinates are taken from the PDB file 1BRN. The protein is shown in green, and the DNA fragment in red.

Top left: Cartoon. This representation provides a high level view of the local organization of the protein in secondary structures, shown as idealized helices and strands. The DNA is shown as a short rod. This view highlights the position of the binding site where the DNA sits.

Top right: Skeletal model. This representation uses lines to represent bonds; atoms are located at their endpoints where the lines meet. It emphasizes the chemical nature of both molecules: for example, the four aromatic rings of the nucleotides of the DNA molecule are clearly visible.

Bottom: Space-filling diagram. Atoms are represented as balls centered at the atoms, with radii equal to the van der Waals radii of the atoms. This representation shows the tight binding between the protein and the ligand, that was not obvious from the other diagrams. Each of the representations is complementary to the others, and usually the biochemist uses all three when studying a protein, alone or, as illustrated here, in interaction with a ligand. The top panels were drawn using MOLSCRIPT [Kraulis 1991] and the bottom one with Pymol (http://www.pymol.org).

Despite this essential role in cellular functions, DNA molecules adopt surprisingly simple structures. Each nucleotide contains two parts, a backbone consisting of a deoxyribose and a phosphate, and an aromatic base, of which there are four types: adenine (A), thymine (T), guanine (G) and cytosine (C). The nucleotides are capable of being linked together to form a long chain, called a *strand*. Cells contain strands of DNA in pairs that are exact mirrors of each other. When correctly aligned, A pairs with T, G pairs with C, and the two strands form a double helix [Watson and Crick 1953]. The geometry of this helix is surprisingly uniform, with only small, albeit important, structural differences between regions of different sequences. The order in which the nucleotides appear in one DNA strand defines its sequence. Some stretches of the sequence contain information that can be translated first into an RNA molecule and then into a protein. These stretches are called *genes*; the ensemble of all genes of an organism constitutes its *genome* or *genetic information*. The remainder is *junk DNA*, which is assumed to correspond to fragments of genes that have been lost over the course of evolution. The DNA strands can stretch for millions of nucleotides. The size of the strands, as well as the fraction of junk DNA vary greatly between organisms and do not necessarily reflect differences in the complexity of the organisms. For example, the wheat genome contains approximately $1.6 \cdot 10^{10}$ bases, which is close to five times the size of the human genome. For a complete list of the genomes, see http://wit.integratedgenomics.com/GOLD/ [Bernal et al. 2001]. The whole DNA molecules of more than 170 organisms have been sequenced in the existing genome projects, and many others are underway. There are more than a million genes that have been extracted from the DNA sequences and are collected in databases; see http://www.ebi.ac.uk/embl.

RNA. Ribonucleic acid molecules are very similar to DNA, being formed as sequences of four types of nucleotides, namely A, G, C, and uracil (U), which is a derivative of thymine. The sugar in the nucleotides of RNA is a ribose, which includes an extra oxygen compared to deoxyribose. The presence of this bulky extra oxygen prevents the formation of long and stable double helices. The single-stranded RNA can adopt a large variety of conformations, which remain difficult to predict based on its sequence. Interestingly, RNA is considered an essential molecule in the early steps of the origin of life. It is generally accepted now that before the appearance of living cells, the assemblies of self-replicating molecules were RNAs. In this early world, a single type of molecule performed both the function of active agents and the repository of its own description [Gilbert 1986; Gesteland and Atkins 1993; Cech 1993]. The activity of the RNA was related to its three-dimensional shape, while the coding corresponded to its linear sequence. This single molecule world had limitations since any modification of the RNA meant to improve its catalytic function could lead to a loss of its coding capabilities. Cellular life has evolved from this primary world by separating the two functions. RNA molecules now mainly serve as templates that are used to synthesize the active molecules, namely the proteins. The information needed to synthesize the RNA is read from the genes coded by the DNA. It is assumed that DNA molecules evolved as a more stable, and consequently more reliable form of RNAs for storage purpose.

Proteins. While all biomolecules play an important part in life, there is something special about proteins, which are the products of the information contained in the genes. They are the active elements of life whose chemical activities regulate all cellular activities. According to Jacques Monod, it is in the protein that lies the secret of life: "C'est à ce niveau d'organisation chimique que gît, s'il y en a un, le secret de la vie" [Monod 1973]. As a consequence, studies of their sequence and structure occupy a central role in biology.

Proteins are heteropolymer chains of amino acids, often referred to as *residues*. This term comes from chemistry and describes the material found at the bottom of a reaction tube once a protein has been cut into pieces in order to determine its composition. There are twenty types of amino acids, which share a common *backbone* and are distinguished by their chemically diverse *side-chains*, which range in size from a single hydrogen atom to large aromatic rings and can be charged or include only nonpolar saturated hydrocarbons; see Table 1. The order

Type	Amino acids
nonpolar	glycine, alanine, valine, leucine, isoleucine, proline, methionine, tryptophan, phenylalanine
polar (neutral)	serine, threonine, asparagine, glutamine, cysteine, tyrosine
polar (acidic)	aspartic acid, glutamic acid
polar (basic)	lysine, arginine, histidine

Table 1. Classification of the 20 amino acids according to the chemical properties of their side-chains [Timberlake 1992]. Nonpolar amino acids do not have concentration of electric charges and are usually not soluble in water. Polar amino acids carry local concentration of charges, and are either globally neutral, negatively charged (acidic), or positively charged (basic). Acidic and basic amino acids are classically referred to as electron acceptors and electron donors, respectively, which can associate to form salt bridges in proteins.

in which amino acids appear defines the *primary sequence* of the protein. In its native environment, the polypeptide chain adopts a unique three-dimensional shape, referred to as the *tertiary* or *native structure* of the protein. In this structure, nonpolar amino acids have a tendency to re-group and form the core, while polar amino acids remain accessible to the solvent. The backbones are connected in sequence forming the protein *main-chain*, which frequently adopts canonical local shapes or *secondary structures*, such as α-helices and β-strands.

The former is a right handed helix with 3.6 amino acids per turn, while the latter is an approximately planar layout of the backbone. In the tertiary structure, β-strands are usually paired in parallel or anti-parallel arrangements, to form β-sheets. On average, the protein main-chain consists of about 25% in α-helix formation, 25% in β-strands, with the rest adopting less regular structural arrangements [Brooks et al. 1988]. From the seminal work of Anfinsen [1973], we know that the sequence fully determines the three-dimensional structure of the protein, which itself defines its function. While the key to the decoding of the information contained in genes was found more than fifty years ago (the genetic code), we have not yet found the rules that relate a protein sequence to its structure [Koehl and Levitt 1999; Baker and Sali 2001]. Our knowledge of protein structure therefore comes from years of experimental studies, either using X-ray crystallography or NMR spectroscopy. The first protein structures to be solved were those of hemoglobin and myoglobin [Kendrew et al. 1960; Perutz et al. 1960]. Currently, there are more than 16,000 protein structures in the database of biomolecular structures [Bernstein et al. 1977; Berman et al. 2000]; see http://www.rcsb.org.

Visualization. The need for visualizing biomolecules is based on the early understanding that their shape determines their function. Early crystallographers who studied proteins and nucleic acids could not rely—as it is common nowadays—on computers and computer graphics programs for representation and analysis. They had developed a large array of finely crafted physical models that allowed them to have a feeling for these molecules. These models, usually made out of painted wood, plastic, rubber and/or metal were designed to highlight different properties of the molecule under study. In the *space-filling models*, such as CPK [Corey and Pauling 1953; Koltun 1965], atoms are represented as spheres, whose radii are the atoms' van der Waals radii. They provide a volumetric representation of the biomolecules, and are useful to detect cavities and pockets that are potential active sites. In the *skeletal models*, chemical bonds are represented by rods, whose junctions define the position of the atoms. These models were used for example in [Kendrew et al. 1960], which studied myoglobin. They are useful to the chemists by highlighting the chemical reactivity of the biomolecules and, consequently, their potential activity. With the introduction of computer graphics to structural biology, the principles of these models have been translated into software such that molecules could be visualized on the computer screen. Figure 1 shows examples of computer visualizations of a protein-DNA interaction, including space-filling and skeletal representations.

3. Biomolecular Modeling

While the structural studies provide the necessary data on biomolecules, the key to their success lies in unraveling the connection between structure and function. A survey of the many modeling initiatives motivated by this question is beyond the scope of this paper; detailed descriptions of biomolecular simulation techniques and their applications can be found in [Leach 2001; Becker et al. 2001]. We shall focus here on those in which geometry plays an essential role, mainly in the definition and computation of the energy of the biomolecule.

The apparition of computers, and the rapid increase of their power has given hope that theoretical methods can play a significant role in biochemistry. Computer simulations are expected to predict molecular properties that are inaccessible to experimental probes, as well as how these properties are affected by a change in the composition of a molecular system. For example, thermodynamics and kinetics play an important role in most functions of proteins. Proteins have to fold into a stable conformation in order to be active. Improper folding leads to inactive proteins that can accumulate and lead to disease (such as the prion proteins). Many proteins also adopt slightly different conformations in different environments. The cooperative rearrangement of hemoglobin upon binding of oxygen, for example, is essential for oxygen transport and release [Perutz 1990]. Predicting the equilibrium conformation of a protein in solution remains a "holy grail" in structural biology. In addition, while a few experimental probes exist to monitor protein dynamics events, such as hydrogen exchange experiments in NMR and small angle scattering of x-rays or neutrons, they remain elusive mainly because of the huge hierarchy of time-scale they involve. Biomolecular simulations have been designed to solve some of these problems. In particular, their aims are to describe the thermodynamic equilibrium properties of the system under study, through sampling of its free energy surface, as well as its dynamical properties.

Energy function. The state of a biomolecule is usually described in terms of its energy landscape. The native state corresponds to a large basin in this landscape, and it is mostly the structure of this basin that is of interest. Theoretically, the laws of quantum mechanics completely determine the wave function of any given molecule, and, in principle, we can compute the energy eigenvalues by solving Schrödinger's equation. In practice, however, only the simplest systems such as the hydrogen atom have an exact, explicit solution to this equation and modelers of large molecular systems must rely on approximations. Simulations of biomolecules are based on a space-filling representation of the molecule, in which the atoms are modeled by hard spheres that interact through empirical forces. A typical, semi-empirical energy function used in classical molecular simulation has the form

$$U = \sum_b k_b \, (r_b - r_b^0)^2 + \sum_b k_a \, (\theta_a - \theta_a^0)^2 + \sum_t k_t \big(1 + \cos n(\phi_t - \phi_t^0)\big)$$

$$+ \sum_{i<j} \left(\frac{A_{ij}}{r_{ij}^{12}} - \frac{B_{ij}}{r_{ij}^6} + \frac{q_i q_j}{r_{ij}} \right)$$

[Levitt et al. 1995; Liwo et al. 1997a; 1997b; MacKerell et al. 1998; Kaminski et al. 2001; Price and Brooks 2002]. The terms in the first three sums represent bonded interactions: covalent bonds, valence angles, and torsions around bonds. The two terms in the last sum represent nonbonded interactions: a Lennard-Jones potential for van der Waals forces and the Coulomb potential for electrostatics. This sum usually excludes pairs of atoms separated by one, or two covalent bonds. The force constants, k, the minima, r^0, θ^0 and ϕ^0, the Lennard Jones parameters, A and B, and the atomic charges q define the force field. They are derived from data on small organic molecules, from both experiments and *ab initio* quantum calculations.

Note that U given above corresponds to the *internal energy* of the molecule, while we really need its *free energy* to describe its thermodynamic state. In thermodynamics, the term *free energy* denotes the total amount of energy in a system which can be converted to work. For a molecule, "work" is the transfer of energy related to organized motion. The free energy F is the difference between the internal energy of the molecule, and its *entropy*, where the entropy is a measure of disorder:

$$F = U - TS,$$

where T is the temperature of the system. Ideally, F is minimum when U is minimum and S is maximum. These two conditions however cannot be satisfied simultaneously by a molecule: U is minimum when there are many favorable contacts, leading to a single compact conformation for the molecule, while S is maximum when there is no privileged conformation for the molecule. In general, the termodynamic equilibrium is reached through a compromise between these two terms. To get an estimate of the free energy of a molecule, we need to compute its internal energy, and sample the conformational space it can access. This sampling is performed through simulations, which are discussed below.

Simulation algorithms. There are three main types of algorithms used in this field, which we now describe.

Molecular dynamics simulations proceed by solving the classical equations of motions for the positions, velocities and accelerations of all atoms and molecules of the system under study. A state of the system is either described in cartesian or internal coordinates and the solution is computed numerically. In early work, macromolecules were simulated *in vacuo*, and only heavy (no hydrogen) atoms were included [McCammon et al. 1977]. This has changed as modern computers are now sufficiently powerful to simulate biomolecules in atomic detail using all-atom representations [Levitt and Sharon 1988]. The strengths of molecular dynamics are that it efficiently samples the states accessible to a system around its energy minimum, and that it provides kinetic data on the transitions between these states [Cheatham and Kollman 2000; Karplus and McCammon 2002]. The weakness of molecular dynamics is an inability to access long time-scales (on the order of one microsecond or more even for small biomolecules [Duan and Kollman 1998].)

Monte Carlo techniques applied to biomolecular studies use stochastic moves, corresponding to rotation, translation, insertion or deletion of whole molecules, to sample the conformational space available to the molecule under study, and to calculate ensemble averages of physical or geometric quantities of interest, such as energy, or the fluctuation of some specific inter-atomic distances. In the limit of long Monte Carlo simulations, these ensemble averages correspond to thermodynamics equilibrium properties. A strength of Monte Carlo simulations is that they can be adapted to explore unfavorable regions of the energy landscape. This has been used to sample conformations of small simplified models of proteins, yielding a full characterization of the thermodynamics of their folding process [Hao and Scheraga 1994a; Hao and Scheraga 1994b].

A *molecular mechanics* study is not really a simulation as such, rather a mechanical investigation of the properties of one or more molecules. A good example would be finding the minimum of the potential energy U of a molecule. Note that U does not include entropic effects. Thus, the conformation of a molecule obtained through minimization of U does not necessarily correspond to the thermodynamic equilibrium state, which corresponds to the minimum of the free energy.

Protein solvation. Soluble biomolecules adopt their stable conformation in water, and are unfolded in the gas phase. It is therefore essential to account for water in any modeling experiment. Molecular dynamics simulation that include a large number of solvent molecules are the state of the art in this field, but they are inefficient as most of the computing time is spent on updating the position of the water molecule. It should further be noted that it is not always possible to account for the interaction with the solvent explicitly. For example, energy minimization of a system including both a protein and water molecules does not account for the entropy of water, which would behave like ice with respect to the protein. An alternative approach takes the effect of the solvent implicitly into account. In such an implicit solvent model, the effects of water is included in an effective solvation potential, $W = W_{\text{elec}} + W_{\text{np}}$, in which the first term accounts for the molecule-solvent electrostatics polarization, and the second for the molecule-solvent van der Waals interactions and for the formation of a cavity in the solvent. There is a large body of work that focuses on computing W_{elec}. A survey of the corresponding models is beyond the scope of this paper and we refer the reader to the excellent review [Simonson 2003] for more information.

Here we focus on computing W_{np}, the nonpolar effect of water on the biomolecule, sometimes referred to as the *hydrophobic effect*. Biomolecules contain both hydrophilic and hydrophobic parts. In their folded states, the hydrophilic parts are usually at the surface, where they can interact with water, and the

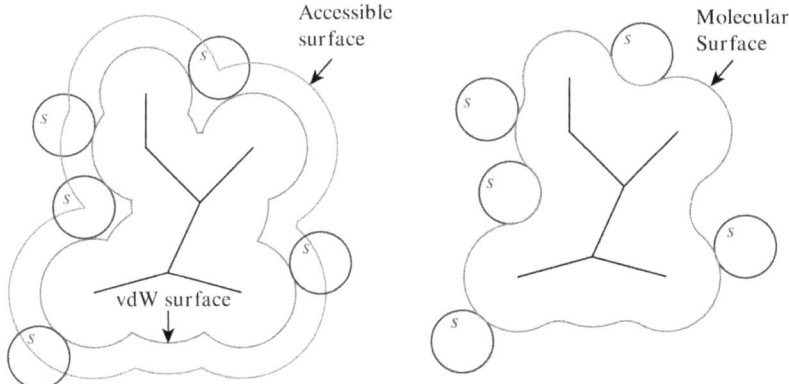

Figure 2. Different notions of protein surface. The *van der Waals surface* of a molecule (shown in red) is the surface of the union of balls representing all atoms, with radii set to the van der Waals radii. The *accessible surface* of the same molecule (shown in green) is the surface generated by the center of a solvent sphere (marked *S*) rolling on the van der Waals surface. The radius of the solvent sphere is usually set to 1.4 Å, the approximate radius of a water molecule. The accessible surface is also the obtained after expanding the radius of the atomic spheres by the radius of the solvent sphere. The *molecular surface* (shown in magenta) is the envelope generated by the rolling sphere. It differs from the van der Waals surface by covering portions of the volume inaccessible to the rolling sphere.

hydrophobic parts are buried in the interior, where they form an "oil drop with a polar coat" [Kauzmann 1959].

Quantifying the hydrophobic effect. In order to quantify the hydrophobic effect, Lee and Richards introduced the concept of the solvent-accessible surface [Lee and Richards 1971], illustrated in Figure 2. They computed the accessible area of each atom in both the folded and extended state of a protein, and found that the decrease in accessible area between the two states is greater for hydrophobic than for hydrophilic atoms. These ideas were further refined by Eisenberg and McLachlan [1986], who introduced the concept of a solvation free energy, computed as a weighted sum of the accessible areas A_i of all atoms i of the biomolecule:

$$W_{\mathrm{np}} = \sum_i \alpha_i A_i,$$

where α_i is the atomic solvation parameter. It is not clear, however, which surface area should be used to compute the solvation energy [Wood and Thompson 1990; Tunon et al. 1992; Simonson and Brünger 1994]. There is also some evidence that for small solute, the hydrophobic term W_{np} is not proportional to the surface area [Simonson and Brünger 1994], but rather to the solvent excluded volume of the molecule [Lum et al. 1999]. A volume-dependent solvation term was originally introduced by Gibson and Scheraga [1967] as the hydration shell model. Note that the ambiguity in the choice of the definition of the surface of a protein extends to the choice of its volume definition. Within this debate on the exact form of the solvation energy, there is however a consensus that it depends on the geometry of the biomolecule under study. Inclusion of W_{np} in a molecular simulation therefore requires the calculation of accurate surface areas and volumes. If the simulations rely on minimization, or integrate the equations of motion, the derivatives of the solvation energy are also needed. The calculation of second derivatives is also of interest in studying the normal modes of a biomolecule in a continuum solvent.

4. Computing Volumes and Areas

In this section, we review existing approaches to computing the surface area and/or volume of a biomolecule represented as a union of balls. The original approach of Lee and Richards [1971] computed the accessible surface area by first cutting the molecule with a set of parallel planes. The intersection of a plane with an atomic ball, if it exists, is a circle which can be partitioned into accessible arcs on the boundary and occluded arcs in the interior of the union. The accessible surface area of atom i is the sum of the contributions of all its accessible arcs, computed approximately as the product of the arc length and the spacing between the plane. This method was originally implemented in the program ACCESS [Lee and Richards 1971] and later in NACCESS (http://wolf.bms.umist.ac.uk/naccess/). Shrake and Rupley [1973] refined Lee and Richards' method and proposed a Monte Carlo numerical integration of the accessible surface area. Their method placed 92 points on each atomic sphere, and determined which points were accessible to solvent (not inside any other sphere). Efficient implementations of this method include applications of lookup tables [Legrand and Merz 1993], of vectorized algorithm [Wang and Levinthal 1991] and of parallel algorithms [Futamura et al. 2004]. Similar numerical methods have been developed for computing the volume of a union of balls [Rowlinson 1963; Pavani and Ranghino 1982; Gavezzotti 1983].

The surface area and/or volume computed by numerical integration over a set of points, even if closely spaced, is not accurate and cannot be readily differentiated. To improve upon the numerical methods, analytical approximations to the accessible surface area have been developed, which either treat multiple overlapping balls probabilistically [Wodak and Janin 1980; Hasel et al. 1988; Cavallo et al. 2003] or ignore them altogether [Street and Mayo 1998; Weiser et al. 1999a]. Better analytical methods describe the molecule as a union of pieces of balls, each defined by their center, radius, and arcs forming their boundary, and subsequently apply analytical geometry to compute the surface area and volume [Richmond 1984; Connolly 1985; Dodd and Theodorou 1991; Petitjean 1994; Irisa 1996]. Pavani and Ranghino [1982] proposed a method for computing the volume of a molecule by inclusion-exclusion. In their implementation, only intersections of up to three balls were considered. Petitjean however noticed that practical situations for proteins frequently involve simultaneous overlaps of up to six balls [Petitjean 1994]. Subsequently, Pavani and Ranghino's idea was generalized to any number of simultaneous overlaps by Gibson and Scheraga [Gibson and Scheraga 1987] and by Petitjean [Petitjean 1994], applying a theorem that states that higher-order overlaps can always be reduced to lower-order overlaps [Kratky 1978]. Doing the reduction correctly remains however computationally difficult and expensive. The Alpha Shape Theory solves this problem using Delaunay triangulations and their filtrations, as described by Edelsbrunner [Edelsbrunner 1995]. It will be presented in greater detail in the next section.

The distinction between approximate and exact computation also applies to existing methods for computing the derivatives of the volume and surface area of a molecule with respect to its atomic coordinates [Kundrot et al. 1991; Gogonea and Osawa 1994; Gogonea and Osawa 1995; Cossi et al. 1996]. In the case of the derivatives of the surface area, computationally efficient methods were implemented in the MSEED software by Perrot et al. [1992] and in the SASAD software by Sridharan et al. [1994]. All these methods introduce approximations to deal with singularities caused by numerical errors or by discontinuities in the derivatives [Gogonea and Osawa 1995]. There is also an inherent difficulty in using a potential based on surface area or volume in biomolecular simulations. Although the area and volume are continuous in the position of the atoms, their derivatives are not. This problem of discontinuities was studied in more details for surface area calculation [Perrot et al. 1992; Wawak et al. 1994].

The complexity of the computation of the area and volume of a union of balls, the problems of singularities encountered when computing their derivatives, and the inherent existence of discontinuities have led to the development of alternative geometric representations of molecules. Here we mention the Gaussian description of molecular shape, that allows for easy analytical computation of surface area, volume and derivatives [Grant and Pickup 1995; Weiser et al. 1999b], and the molecular skin, which will be described in the next section.

5. Alpha Shape Theory

In this section, we discuss in some detail the inclusion-exclusion approach to computing area, volume, and their derivatives. It is based on the concept of alpha complexes [Edelsbrunner et al. 1983; Edelsbrunner and Mücke 1994], which are sub-complexes of the Delaunay triangulation [Delaunay 1934] of a set of spheres.

Voronoi decomposition and dual complex. Consider a finite set of spheres S_i with centers $z_i \in \mathbb{R}^3$ and radii $r_i \in \mathbb{R}$ and let B_i be the ball bounded by S_i. To allow for varying radii, we measure square distance of a point x from S_i using $\pi_i(x) = \|x - z_i\|^2 - r_i^2$. The *Voronoi region* of S_i consists of all points

x at least as close to S_i as to any other sphere: $V_i = \{x \in \mathbb{R}^3 \mid \pi_i(x) \leq \pi_j(x)\}$. As illustrated in Figure 3, the Voronoi region of S_i is a convex polyhedron obtained as the common intersection of finitely many closed half-spaces, one per sphere $S_j \neq S_i$. If S_i and S_j intersect in a circle then the plane bounding the corresponding half-space passes through that circle. It follows that the Voronoi regions decompose the union of balls B_i into convex regions of the form $B_i \cap V_i$. The boundary of each such region consists of spherical patches on S_i and planar patches on the boundary of V_i. The spherical patches separate the inside from the outside and the planar patches decompose the inside of the union. The

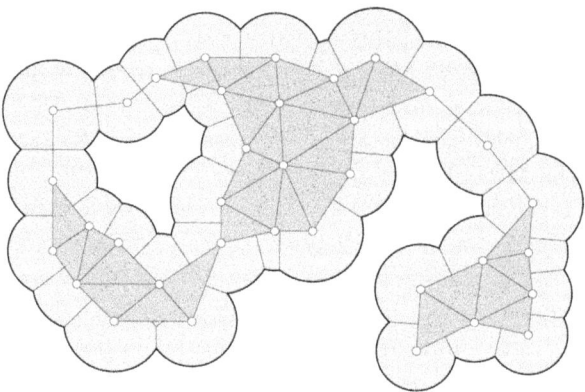

Figure 3. Voronoi decomposition and dual complex. Given a finite set of disks, the Voronoi diagram decomposes the plane into regions in which one circle minimizes the square distance measured as $\|x - z_i\|^2 - r_i^2$. In the drawing, we restrict the Voronoi diagram to within the portion of the plane covered by the disks and get a decomposition of the union into convex regions. The dual Delaunay triangulation is obtained by drawing edges between circle centers of neighboring Voronoi regions. To draw the dual complex of the disks we limit ourselves to edges and triangles between centers whose corresponding restricted Voronoi regions have a nonempty common intersection.

Delaunay triangulation is the dual of the Voronoi diagram, obtained by drawing an edge between the centers of S_i and S_j if the two corresponding Voronoi regions share a common face. Furthermore, we draw a triangle connecting z_i, z_j and z_k if V_i, V_j and V_k intersect in a common line segment, and we draw a tetrahedron connecting z_i, z_j, z_k and z_ℓ if V_i, V_j, V_k and V_ℓ meet at a common point. Assuming general position of the spheres, there are no other cases to be considered. We refer to this as the *generic case* but hasten to mention that because of limited precision it is rare in practice. Nevertheless, we can simulate a perturbation in our algorithm [Edelsbrunner and Mücke 1990], which is an effective method to consistently unfold potentially complicated degenerate cases to nondegenerate ones.

Suppose we limit the construction of the dual triangulation to within the union of balls, as illustrated in Figure 3. In other words, we draw a dual edge between z_i and z_j only if $B_i \cap V_i$ and $B_j \cap V_j$ share a common face, and similarly for triangles and tetrahedra. The result is a sub-complex of the Delaunay triangulation which we refer to as the *dual complex* $K = K_0$ of the set of spheres. For various reasons, including the definition of pockets in biomolecules [Edelsbrunner et al. 1998], it is useful to alter the spheres by increasing or decreasing their radii. We do this in a way that leaves the Voronoi diagram invariant. Modeling growth with a positive real number and shrinkage with a positive real multiple of the imaginary unit, both denoted as α, we obtain a real number α^2 that may be positive or negative. For each i let $S_i(\alpha)$ be the sphere with center z_i and radius $\sqrt{r_i^2 + \alpha^2}$. Interpreting spheres with imaginary radii as empty, the *alpha complex* K_α of the spheres S_i is the dual complex of the spheres $S_i(\alpha)$. If we increase α^2 continuously from $-\infty$ to $+\infty$ we get a continuous nested sequence of unions of balls and a discrete nested sequence of alpha complexes.

Area and volume formulas. A simplex τ in the dual complex can be interpreted abstractly as a collection of balls, one ball if it is a vertex, two if it is an edge, etc. In this interpretation, the dual complex is a system of sets of balls, and because every face of a simplex in K also belongs to K, this system is closed under containment. It now makes sense to write $\mathrm{vol} \bigcap \tau$ for the volume of the intersection of the balls in τ. This is the kind of term we would see in an inclusion-exclusion formula for the volume of the union of balls, $\bigcup_i B_i$. As proved in [Edelsbrunner 1995], the inclusion-exclusion formula that corresponds to the dual complex gives indeed the correct volume.

VOLUME THEOREM:

$$\mathrm{vol} \bigcup_i B_i = \sum_{\tau \in K} (-1)^{\dim \tau} \, \mathrm{vol} \bigcap \tau.$$

Here $\dim \tau = \mathrm{card} \, \tau - 1$ is the dimension of the simplex. This result overcomes past difficulties by implicitly reducing higher-order to lower-order overlaps. An added advantage of this formula is that the balls in each term form a unique geometric configuration so that the analytic calculation of the volume can be done without case analysis. Specifically, the balls in a simplex $\tau \in K$ are *independent* in the sense that for every face $\upsilon \subseteq \tau$ there exists a point that lies inside all balls that belong to υ and outside all balls that belong to τ but not to υ.

A similar formula can be derived for the area of the boundary of the union of balls. One way to arrive at this formula is to consider a sphere S_i and to observe that its contribution is the area of the entire sphere, $4\pi r_i^2$, minus the portion covered by caps of the form $S_i \cap B_j$, for $j \neq i$. The configuration of caps on S_i is but a spherical version of a configuration of disks, and computing its area is the same problem as computing the volume of a set of balls, only one dimension lower. To express that area as an alternating sum we need its dual complex, but this is nothing other than the link of S_i in K, consisting of all simplices υ that do not contain B_i but are faces of simplices that contain B_i: $B_i \not\subseteq \upsilon$ and $\upsilon \cup \{B_i\} \in K$. Specifically, the area contribution of S_i is the area of the sphere minus the sum of $(-1)^{\dim \upsilon} \mathrm{area}\, (S_i \cap \bigcap \upsilon)$. We collect all these contributions and combine terms to get the final result.

AREA THEOREM:

$$\mathrm{area} \bigcup_i B_i = \sum_{\tau \in K} (-1)^{\dim \tau}\, \mathrm{area} \bigcap \tau.$$

We see that the principle of inclusion-exclusion is quite versatile, which is important for applications in which we might want to measure aspects of the union of balls that are similar to but different from its volume and surface area. Examples are

- the total length of arcs in the boundary;
- voids of empty space surrounded by the union;
- weighted versions of the above.

Of the three extensions, the least obvious is how to measure voids. The other two are needed to express the derivative of the weighted volume and area, which are discussed next.

Area and volume derivatives. We are interested in the derivatives of the area and the volume of a union of n balls with respect to their positions in space. Since we keep the radii fixed, we may specify the configuration by the vector $\boldsymbol{z} \in \mathbb{R}^{3n}$ of center coordinates. The area thus becomes a function $f : \mathbb{R}^{3n} \to \mathbb{R}$, and similar for the volume. The *derivative* of f at $\boldsymbol{z}$ is the best linear approximation at that configuration, $\mathrm{D}f_{\boldsymbol{z}} : \mathbb{R}^{3n} \to \mathbb{R}$. This linear function is completely specified by the gradient $\boldsymbol{a} = \nabla f(\boldsymbol{z})$, namely

$$\mathrm{D}f_{\boldsymbol{z}}(\boldsymbol{t}) = \langle \boldsymbol{a}, \boldsymbol{t} \rangle,$$

in which $\boldsymbol{t} \in \mathbb{R}^{3n}$ is the motion vector. In [Edelsbrunner and Koehl 2003; Bryant et al. 2004] we gave formulas for the derivatives by specifying the gradient in terms of simple parameters readily computable from the input spheres. To state the result for the area, let $\zeta_{ij} = \|z_i - z_j\|$ be the distance between the two centers and write $u_{ij} = (z_i - z_j)/\zeta_{ij}$ for the unit vector in the direction of the connecting line. For each $k \neq i, j$ let

$$w_{ijk} = u_{ik} - \langle u_{ik}, u_{ij} \rangle \cdot u_{ij}$$

be the component of u_{ik} normal to u_{ij}, and let $u_{ijk} = w_{ijk}/\|w_{ijk}\|$ be the unit vector in that normal direction. Finally, let r_{ijk} be half the distance between the two points at which the spheres S_i, S_j, and S_k meet. For completeness, we state the result for the case in which the area contribution is weighted by the constant α_i, the corresponding atomic solvation parameter.

WEIGHTED AREA DERIVATIVE THEOREM: *The gradient $\boldsymbol{a} \in \mathbb{R}^{3n}$ of the weighted area derivative at a configuration of balls $\boldsymbol{z} \in \mathbb{R}^{3n}$ is*

$$
\begin{bmatrix} \boldsymbol{a}_{3i+1} \\ \boldsymbol{a}_{3i+2} \\ \boldsymbol{a}_{3i+3} \end{bmatrix} = \sum_j \left(s_{ij} \cdot a_{ij} + \sum_k b_{ijk} \cdot a_{ijk} \right),
$$

$$
a_{ij} = \pi \left((\alpha_i r_i + \alpha_j r_j) - (\alpha_i r_i - \alpha_j r_j) \frac{r_i^2 - r_j^2}{\zeta_{ij}^2} \right) \cdot u_{ij},
$$

$$
a_{ijk} = 2 r_{ijk} \frac{\alpha_i r_i - \alpha_j r_j}{\zeta_{ij}} \cdot u_{ijk},
$$

for $0 \leq i < n$. The sums are over all boundary edges $z_i z_j$ and their triangles $z_i z_j z_k$ in K.

The geometrically interesting terms in the formula are s_{ij}, the fraction of the circle $S_i \cap S_j$ that belongs to the boundary of the union, and b_{ijk}, the fraction of the line segment connecting the point pair $S_i \cap S_j \cap S_k$ that belongs to the Voronoi segment $V_i \cap V_j \cap V_k$. A remarkable aspect of the formula is the existence of terms that depend on three rather than just two spheres. These terms vanish in the unweighted case if all radii are the same. We can reuse some of the notation to state the result for the volume. We again state the result for the case in which the volume of $B_i \cap V_i$ is weighted by the constant α_i.

WEIGHTED VOLUME DERIVATIVE THEOREM: *The gradient $\boldsymbol{v} \in \mathbb{R}^{3n}$ of the weighted volume derivative of a configuration of balls $\boldsymbol{z} \in \mathbb{R}^{3n}$ is*

$$
\begin{bmatrix} \boldsymbol{v}_{3i+1} \\ \boldsymbol{v}_{3i+2} \\ \boldsymbol{v}_{3i+3} \end{bmatrix} = \sum_j b_{ij} r_{ij}^2 \pi (y_{ij} \cdot u_{ij} + x_{ij} \cdot v_{ij}),
$$

$$
y_{ij} = \frac{\alpha_i + \alpha_j}{2} + \frac{(\alpha_j - \alpha_i)(r_i^2 - r_j^2)}{2\zeta_{ij}^2},
$$

$$
x_{ij} = \frac{2(\alpha_i - \alpha_j)}{3\zeta_{ij}},
$$

for $0 \leq i < n$. The sum is over all edges $z_i z_j$ in K.

Here r_{ij} is the radius of the disk spanned by the circle $S_i \cap S_j$ and b_{ij} is the fraction of this disk that belongs to the corresponding Voronoi polygon, $V_i \cap V_j$. The most interesting term in this formula is the average vector v_{ij} from the center of the disk to the boundary of its intersection with the Voronoi polygon. In computing the average, we weight each point on this boundary by the area of the infinitesimal triangle it defines with the center. This vector is used to express gain and loss of weighted volume as the disk rotates and trades off contributions of the two balls it separates. In the unweighted case, we gain as much as we lose which explains why x_{ij} vanishes and thus cancels any effect v_{ij} would have.

Continuity of the derivative. If considered over all configurations, the derivative of f is a function $Df : \mathbb{R}^{3n} \times \mathbb{R}^{3n} \to \mathbb{R}$. As described earlier, for each state $z \in \mathbb{R}^{3n}$, this is a linear function $\mathbb{R}^{3n} \to \mathbb{R}$ completely specified by the gradient at z. It is convenient to introduce another function $\nabla f : \mathbb{R}^{3n} \to \mathbb{R}^{3n}$ such that $Df(z, t) = \langle \nabla f(z), t \rangle$. For the purpose of simulating molecular motion, it is important that ∇f be continuous, at least mostly, and if there are discontinuities, that we are able to recognize and predict them. Unfortunately, the derivatives of the weighted area and the weighted volume are both not everywhere continuous. The good news is that the formulas in the two Derivative Theorems permit a complete analysis.

Interestingly, a configuration at which ∇f is not continuous is necessarily a configuration at which the dual complex is ambiguous, and this is true for the area and the volume. For example, the area derivative has a discontinuity at configurations that contain two spheres touching in a point that belongs to the boundary of the union. The set of configurations z that contain such spheres is a $(3n-1)$-dimensional subset of $\mathbb{R}^{3n}$. In contrast, the volume derivative has discontinuities only at configurations that contain two equal spheres or three spheres that meet in a common circle, both in the weighted and the unweighted case. The set of such configurations is a $(3n-3)$-dimensional subset of $\mathbb{R}^{3n}$. A molecular dynamics simulation has to do extra work to compensate for the missing information whenever it runs into a discontinuity of the derivative [Carver 1978; Gear and Østerby 1984]. This occurs less often for the volume than for the area, firstly because the dimension of such configurations is less and secondly because the specific structure of these configurations makes them physically unlikely.

Voids and pockets. A *void* V is a maximal connected subset of space that is disjoint from and completely surrounded by the union of balls. Its surface area is easily computed by identifying the sphere patches on the boundary of the union that also bound the void. It helps to know that there is a deformation retraction from $\bigcup_i B_i$ to the dual complex [Edelsbrunner 1995]. Similarly, there is a corresponding void in K represented by a connected set of simplices in the Delaunay triangulation, that do not belong to K. This set U is open and its boundary (the simplices added by closure) forms what one may call the dual complex of the boundary of V. We use normalized angles to select the relevant portions of the intersections of balls. To define this concept, let v be a face of a simplex τ and consider a sufficiently small sphere in the affine hull of τ whose center is in the interior of v. The *normalized angle* $\varphi_{v,\tau}$ is the fraction of the sphere contained in τ. For example, if τ is a tetrahedron then we get the solid angle at a vertex, the dihedral angle at an edge, and $\frac{1}{2}$ at a triangle.

VOID AREA THEOREM:

$$\text{area } V = \sum_{v \subseteq \tau} (-1)^{\dim v} \varphi_{v,\tau} \text{ area} \bigcap v.$$

The sum is over all faces $v \in K$ of simplices $\tau \in U$.

The correctness of the formula is not immediate and relies on an identity for simplices proved in [Edelsbrunner 1995]. Similarly, we can use U to compute the volume of V.

VOID VOLUME THEOREM:

$$\mathrm{vol}\, V = \mathrm{vol}\, U - \sum_{v \subseteq \tau} (-1)^{\dim v} \varphi_{v,\tau} \,\mathrm{vol} \bigcap v.$$

The sum is over all faces $v \in K$ of simplices $\tau \in U$.

Here, vol U is simply the sum of volumes of the tetrahedra of U. There are similar angle-weighted formulas for the entire union of balls. It would be interesting to generalize the Void Area and Volume Theorems to pockets as defined in [Edelsbrunner et al. 1998]. In contrast to a void, a *pocket* is not completely surrounded but connected to the outside through narrow channels. Again we have a corresponding set of simplices in the Delaunay triangulation that do not belong to the dual complex, but this set is partially closed at the places the pocket connects to the outside. The inclusion-exclusion formulas still apply, but there are cases in which the cancellation of terms near the connecting channel is not complete and leads to slightly incorrect measurements.

Alternative geometric representations. The sensitivity of simulation software to discontinuities in the derivative suggests that we approximate the surface area by another function. For example, we may use a *shell representation* and approximate area by the volume in that shell. This can be done with uniform thickness everywhere, or with variable thickness that depends on the radii, such as $\bigcup_i B_i(\varepsilon) - \bigcup_i B_i(-\varepsilon)$, where the small positive ε affects the radii as formulated in the definition of the alpha complex. The latter lends itself to fast computation because both the outer and the inner union have their dual complex in the same Delaunay triangulation and measuring both takes barely more time than measuring one. Another alternative to the union of balls is the *molecular surface* explained in Figure 2. Here we roll a sphere with fixed radius r about a union of balls. The rolling motion is captured by the boundary of another union in which all balls grow by r in radius. For each patch, arc, and vertex in this boundary the molecular surface contains a (smaller) sphere patch, a torus patch, and a (reversed) sphere patch. We can therefore collect all patches of the molecular surface using the dual complex of the grown balls and get the surface area by accumulation. At rare occasions, the patches form self-intersections which leads to slightly incorrect measurements. Computing these self-intersections can be rather involved analytically [Bajaj et al. 1997]. A similar alternative to unions of balls is the *molecular skin* as defined in [Edelsbrunner 1999]. Instead of torus patches, it uses hyperboloids of one and two sheets to blend between the spheres; see Figure 4. The surface is decomposed into simple patches by a mix of the

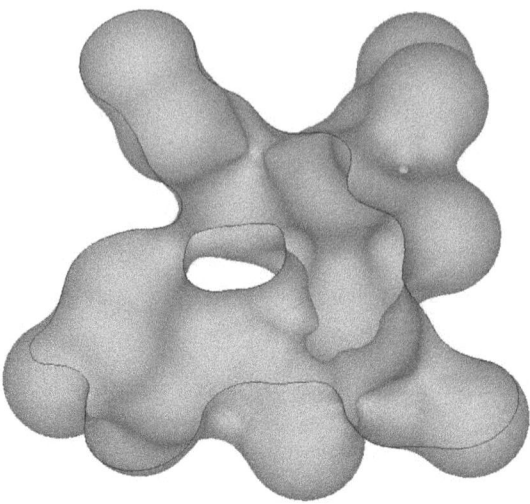

Figure 4. Molecular skin in cut-away view. Half the surface of a small molecule of about forty atoms.

Voronoi diagram and the Delaunay triangulation. These patches are free of self-intersections and the area can be computed by accumulation, as before but without running the risk of making mistakes. At this time, there is no complete analysis of the volume and area derivatives available, neither for the molecular surface nor the molecular skin.

6. Algorithm and Implementation

We have written a new version of the Alpha Shape software [Edelsbrunner and Mücke 1994], specific to molecular simulation applications, implementing the weighted surface area, the weighted volume, and the derivatives of both. The software is distributed as and Open Source program under the name AlphaVol at http://biogeometry.duke.edu/software/proshape.

Overview. The software takes as input a set of spheres S_i in $\mathbb{R}^3$, each specified by the coordinates of its center z_i and its radius r_i. Such a set representing a protein can for example be extracted from the corresponding PDB file using one of several standard sets of van der Waals radii. The computation is performed through three successive tasks:

1. Construct the Delaunay triangulation.
2. Extract the dual complex.
3. Measure the union using inclusion-exclusion.

The main difference to the old Alpha Shapes software is the speed resulting from an improvement of all steps by about two orders of magnitude. We achieve this through careful redesign of low-level computations (determinants in Task 1 and term management in Task 3) and the limitation in scope (dual complex instead of filtration of alpha complexes in Task 2). We review all three steps, focusing on nonobvious implementation details that have an impact on the correctness and running time of the software.

Delaunay triangulation. Our implementation of the Delaunay triangulation is based on the randomized incremental algorithm described in [Edelsbrunner and Shah 1996]. Following the paper's recommendation, we use a minimalist approach to storing the triangulation in a linear array of tetrahedra. For each tetrahedron, we store the indices of its four vertices, the indices of the four neighboring tetrahedra, a label, and the position of the opposite vertex in the vertex list of each neighboring tetrahedron. For each vertex we use four double-precision real numbers for the coordinates and the radius of the corresponding sphere. The triangles and edges are implicit in this representation.

The triangulation is constructed incrementally, by adding one sphere at a time. Before starting the construction, we re-index such that $S_1, S_2, \ldots, S_n$ is a random permutation of the input spheres. To reduce the number of cases, we choose four additional spheres with their centers at infinity so that all input spheres are contained in the tetrahedron they define. Let D_i be the Delaunay triangulation of the four spheres at infinity together with $S_1, S_2, \ldots, S_i$. The algorithm proceeds by iterating three steps:

For $i = 1$ to n,
1.1. find the tetrahedron $\tau \in D_{i-1}$ that contains z_i;
1.2. add z_i to decompose τ into four tetrahedra;
1.3. flip locally non-Delaunay triangles in the link of z_i.

Step 1.1 is implemented using the jump-and-walk technique proposed by Mücke et al. [1999]. Here we choose a small random sample of the vertices in the current triangulation and walk from the vertex closest to z_i to τ. In this walk, we repeatedly test whether z_i is inside a tetrahedron v and whether v remains in the current Delaunay triangulation. These tests are decided by computing the signs of the determinants of four 4-by-4 matrices, which place z_i relative to the faces of v, and the sign of one 5-by-5 matrix. By noticing that any two of the 4-by-4 matrices share three rows (corresponding to z_i and the vertices of a shared edge) we find that 28 multiplications suffice to compute all five determinants. In Step 1.2, the sphere S_i is sometimes discarded without decomposing τ, namely when its Voronoi region is empty. This usually does not happen for molecular data. A flip in Step 1.3 replaces two tetrahedra by three or three by two. We are also prepared to remove a sphere by replacing four tetrahedra by one, but this again is usually not necessary for molecular data. The fact that any arbitrary ordering of the flips will successfully repair the Delaunay triangulation is nontrivial but has been established in [Edelsbrunner and Shah 1996]. The numerical tests needed to decide which flips to make compute again signs of determinants of 4-by-4 and 5-by-5 matrices. As before, we save time by recognizing common rows and reusing partial results in the form of shared minors. An important ingredient in this context is the treatment of singularities. Inexact versions of the numerical tests are vulnerable to roundoff errors and can lead to wrong output. Following work in computational geometry [Fortune and VanWyk 1996], we implemented both tests using a so-called floating-point filter that first evaluates the tests approximately, using floating-points arithmetic, and if the results cannot be trusted, switches to exact arithmetic. As a side-benefit, we can now correctly recognize degenerate cases and use a simulated perturbation to consistently reduce them to general cases [Edelsbrunner and Mücke 1990].

Dual complex. Given the Delaunay triangulation D of the input spheres, we construct the dual complex $K \subseteq D$ by labeling the Delaunay simplices. Specifically, for each simplex $\tau \in D$ there is a threshold α_τ such that $\tau \in K_\alpha$ iff $\alpha_\tau^2 \leq \alpha^2$. Hence τ belongs to the dual complex iff $\alpha_\tau^2 \leq 0$. We call τ a *critical* simplex if α_τ separates the case in which the balls $B_i(\alpha)$ defining τ have an empty common intersection from the case in which they have a nonempty common intersection. These simplices are characterized by the fact that all other balls are further than orthogonal from the smallest sphere orthogonal to all balls B_i defining τ. (Two balls B_i and B_j of centers z_i and z_j and radii r_i and r_j, respectively, are orthogonal iff $\|z_i - z_j\|^2 = r_i^2 + r_j^2$.) All other simplices are *regular* and need a critical simplex they are face of to be included in the dual complex. To label the Delaunay simplices, we therefore need to be able to recognize critical simplices and to decide the signs of their square thresholds. Both tests can be expressed in terms of the signs of the determinants of small matrices whose entries are center coordinates and square radii of the input spheres. Detailed expressions for these tests can be found in [Edelsbrunner 1992; Edelsbrunner and Mücke 1994].

We evaluate these tests with the same care for singularities and numerical uncertainties as used in the construction of the Delaunay triangulation. Specifically, we apply filters and repeat the computation in exact arithmetic unless we can be sure that the initial floating-point computation gives the correct sign.

Weighted surface area and volume. We compute the weighted volume of a union of balls using the Volume Theorem in Section 4. The weights are worked into the formula by decomposing each term, $\text{vol} \bigcap \tau$, into $\dim \tau + 1$ terms using the bisector planes also used in the Voronoi diagram. This decomposition is natural since it is the easiest way to compute the volume of $\bigcap \tau$ in the first place, even in the unweighted case.

We could do the same for the weighted area, effectively reducing the formula in the Area Theorem further to an alternating sum in which every term is the area of the intersection of a sphere with up to three half-spaces. Simple analytic formulas for the area of such an intersection can be found in [Edelsbrunner and Fu 1994]. We choose an alternative path deriving a similar formula (yielding the same result) from the angle-weighted formula given in the Void Area Theorem. An adaptation of this formula to an entire union of balls gives

$$\text{area} \bigcup_i B_i = \sum_v (-1)^{\dim v} \varphi_v \text{ area} \bigcap v,$$

where the sum is over all simplices v in the boundary of K and φ_v is the normalized angle around v not covered by simplices that contain v as a face. As before, we further decompose each term into the intersection of a sphere and a small number of half-spaces. The above sum is usually shorter than that in the straight Area Theorem, which has a term for every simplex in the dual complex. Another difference is that each term is the intersection of at most three balls as opposed to at most four in the Area Theorem. The two differences compensate for the extra effort of computing normalized angles and more, leading to code that for proteins is about twice as fast as that based on the straight Area Theorem.

Derivatives. We now explain how we compute the geometric ingredients in the two Derivative Theorems stated in Section 5. For the area derivative, these are the fractions s_{ij} and b_{ijk}. Both can be computed using inclusion-exclusion over links inside the dual complex. Recall that s_{ij} is the fraction of the circle $S_i \cap S_j$ that belongs to the boundary of the union of balls. Equivalently, it is the fraction of the circle not covered by arcs of the form $S_i \cap S_j \cap B_k$. We may interpret these arcs as one-dimensional balls and measure their union using inclusion-exclusion, not unlike the formula in the Volume Theorem. We find the same symmetry in dimension in the corresponding combinatorial complexes. Specifically, the (one-dimensional) dual complex of the arcs is isomorphic to the link of the edge $z_i z_j$ in the dual complex of the balls. The link of this edge in the Delaunay triangulation is a cycle and in K is a sub-complex consisting of vertices z_k and edges $z_k z_\ell$. Writing s_{ij}^k and $s_{ij}^{k\ell}$ for the fractions of the circle inside B_k and inside $B_k \cap B_\ell$, we have

$$s_{ij} = 1 - \sum_k s_{ij}^k + \sum_{k,\ell} s_{ij}^{k\ell},$$

where the sums range over the link of the edge $z_i z_j$ in K. The computation of b_{ijk} is similar but simpler because the dimension of the link of a triangle is only zero, consisting of at most two vertices. Consider the line segment connecting the two points at which S_i, S_j and S_k meet and note that all points x on this line segment have the same distance to the three spheres: $\pi_i(x) = \pi_j(x) = \pi_k(x)$. Writing b_{ijk}^ℓ for the fraction of points x whose square distance from S_ℓ is less than from the three defining spheres we get $b_{ijk} = 1 - \sum_\ell b_{ijk}^\ell$, where the sum is over the vertices z_ℓ in the link of the triangle. For further details refer to [Bryant et al. 2004]. The same quantity but one dimension higher appears in the volume derivative. Specifically, b_{ij} is the fraction of the disk B_{ij} spanned by the circle $S_i \cap S_j$ that belongs to the corresponding Voronoi polygon. Let B_{ij}^k

be the subset of points x in this disk whose square distance to S_k is less than to the two defining spheres: $\pi_k(x) < \pi_i(x) = \pi_j(x)$. Similarly, let $B_{ij}^{k\ell} = B_{ij}^k \cap B_{ij}^\ell$ and write b_{ij}^k and $b_{ij}^{k\ell}$ for the respective fractions of the disk they define. Then $b_{ij} = 1 - \sum_k b_{ij}^k + \sum_{k,\ell} b_{ij}^{k\ell}$. Finally consider the average vector v_{ij} from the center of the disk to the boundary of its intersection with the Voronoi polygon. Its computation follows the same pattern of inclusion-exclusion over the link of the edge, $v_{ij} = 0 - \sum_k v_{ij}^k + \sum_{k,\ell} v_{ij}^{k\ell}$, where v_{ij}^k is the average vectors to the arc minus the average vector to the line segment in the boundary of B_{ij}^k, and similarly $v_{ij}^{k\ell}$ is the difference between the two average vectors of $B_{ij}^{k\ell}$. For further details refer to [Edelsbrunner and Koehl 2003].

Performance. We discuss the actual performance of AlphaVol. We have computed the weighted surface areas and volumes, as well as their derivatives with respect to atomic coordinates, of 2,868 proteins varying in size from 17 to 500 residues. These proteins contain between 124 and 4,063 atoms. Computing times for AlphaVol on an Intel 1600 MHz Pentium IV computer are shown in Figure 5.

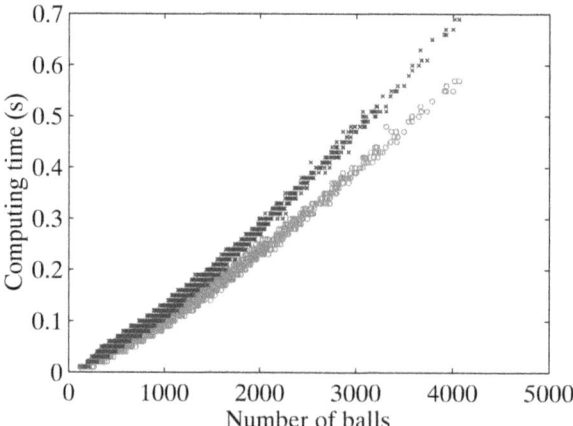

Figure 5. Performance of AlphaVol. The running time (in seconds) required by AlphaVol to compute the weighted volume and weighted surface area of a protein, with (x) and without (o) derivative is plotted against the number of atoms of the protein. The running times are measured on an Intel 1600 MHz Pentium IV computer, running Linux. AlphaVol is written in Fortran, and was compiled using ifc, the Intel Fortran compiler for Linux.

As described above, AlphaVol first computes the Delaunay triangulation of the n input spheres. Although in the worst case this takes quadratic time for constructing a quadratic number of simplices, for protein data the running time is typically $O(n \log n)$ for constructing $O(n)$ simplices. The time for constructing the dual complex and measuring the union of balls is linear in the number of simplices in the Delaunay triangulation and therefore typically in $O(n)$. The experimentally observed total running time of AlphaVol is compatible with a complexity of $O(n \log n)$, both with and without derivatives, for up to 4,000 balls. Approximately 45% of the total running time is spent on the Delaunay triangulation, 10% on the dual complex, and 45% on the weighted area and volume. Computing the derivatives of both adds another 20%.

Applications. AlphaVol exists as a stand-alone program that can be used to compute the solvation energy of a biomolecule. We have also inserted AlphaVol into the molecular dynamics software ENCAD [Levitt et al. 1995] and GROMACS [Lindahl et al. 2001], but it is too early to say anything about the corresponding results. Recall from Section 3 that AlphaVol accounts for the nonpolar effect of water on a biomolecule, W_{np}, which is only one element of the effective solvation potential W to be used in simulations with implicit solvent. While there is a large body of work on computing the other part, W_{elec} [Simonson 2003], there is not yet any concensus on the model to be used for simulation. We have recently started a project on this specific problem.

7. Discussion

The Alpha Shape Theory with the two Derivative Theorems provides a fast, accurate and robust method for computing the interaction of water with a bio-molecule in an implicit solvent model. To our knowledge, the corresponding software, AlphaVol, is the only program that deals explicitly with the problem of discontinuities of the derivatives, which are detected as singularities in the construction of the dual complex [Edelsbrunner and Koehl 2003; Bryant et al. 2004]. We conclude this paper with a short discussion of two immediate applications of this work.

Macro-molecular machinery. Recent advances in structural biology have produced an abundance of data on large macro-molecular complexes; see for example the myosin motors at http://www.proweb.org/myosin/index.html, the RNA polymerase transcription complexes [Cramer et al. 2001; Bushnell and Kornberg 2003], and the ribosome complexes [Wimberly et al. 2000; Yusupov et al. 2001; Ban et al. 2002]. Modeling the dynamics of such large systems is as important as modeling smaller proteins. It becomes impractical, however, to consider all atoms of the molecular machinery, and we need to introduce approximations that consider the system at coarser levels of detail. One possible approach is to represent the macro-molecular complex with a small number of spheres, supplemented with a model for their interactions that captures the physics of the underlying atomic model. These interactions will include an internal potential, and a potential to account for the solvent environment of the system. We expect the latter to resemble the solvation potential described in Section 2, in which the software AlphaVol will prove useful.

Normal modes. Collective motions in which substantial parts move as units relative to the rest play an important role in defining the function of a biomolecule. Examples include domain motions during catalytic activities (e.g. citrate synthase [Remington et al. 1982]), as well as the transition from one conformation to another for proteins that have more than one functionally distinct state. These processes involve the correlated motion of many atoms and are slower than local vibrations. They are difficult and costly to detect using classical molecular dynamics simulations, which motivates the use of normal modes dynamics as an alternative approach to detecting these collective motions [Go et al. 1983; Levitt et al. 1983; Brooks and Karplus 1983]. The normal modes are found by assuming that the potential energy can be approximated as a quadratic function of its variables and solving an eigenvalue problem to give a closed analytical description of the motion. The eigenvalues give the frequencies of the modes and the eigenvectors give the details of the corresponding motions. At a local minimum, the quadratic approximation is obtained by a Taylor expansion to the second order of the total potential energy. Computing normal modes therefore requires computing the second derivatives of the energy function. However, it is difficult to define a meaningful energy minimum for a system involving a large biomolecule in the midst of small water molecules since their geometric and physical properties are so different. We believe that this difficulty can be circumvented by using an implicit solvent model. Computing the Taylor expansion of the energy function including an implicit solvent model would then require the second derivatives of the weighted surface area and/or volume of the biomolecule. We have recently applied the mathematical tools described in this paper to derive formulas for both (manuscript in preparation).

References

[Alm and Baker 1999] E. Alm and D. Baker, "Prediction of protein-folding mechanisms from free energy landscapes derived from native structures", *Proc. Natl. Acad. Sci. (USA)* **96** (1999), 11305–11310.

[Alm et al. 2002] E. Alm, A. V. Morozov, T. Kortemme, and D. Baker, "Simple physical models connect theory and experiments in protein folding kinetics", *J. Mol. Biol.* **322** (2002), 463–476.

[Anfinsen 1973] C. B. Anfinsen, "Principles that govern protein folding", *Science* **181** (1973), 223–230.

[Bajaj et al. 1997] C. Bajaj, H. Y. Lee, R. Merkert, and V. Pascucci, "NURBS based B-rep models from macromolecules and their properties", pp. 217–228 in *Proc. 4th Sympos. Solid Modeling Appl.*, 1997.

[Baker and Sali 2001] D. Baker and A. Sali, "Protein structure prediction and structural genomics", *Science* **294** (2001), 93–96.

[Ban et al. 2002] N. Ban, P. Nissen, J. Hansen, P. B. Moore, and T. A. Steitz, "The complete atomic structure of the large ribosomal subunit at 2.4 angstrom resolution", *Science* **289** (2002), 905–920.

[Becker et al. 2001] O. M. Becker, A. D. McKerell, B. Roux, and M. Watanabe (editors), *Computational biochemistry and biophysics*, edited by O. M. Becker et al., Marcel Dekker Inc., New York, 2001.

[Berman et al. 2000] H. M. Berman, J. Westbrook, Z. Feng, G. Gilliland, T. N. Bhat, H. Weissig, et al., "The Protein Data Bank", *Nucl. Acids. Res.* **28** (2000), 235–242.

[Bernal et al. 2001] A. Bernal, U. Ear, and N. Kyrpides, "Genomes OnLine Database (GOLD): a monitor of genome projects world-wide", *Nucl. Acids. Res.* **29** (2001), 126–127.

[Bernstein et al. 1977] F. C. Bernstein, T. F. Koetzle, G. William, D. J. Meyer, M. D. Brice, J. R. Rodgers, et al., "The protein databank: a computer-based archival file for macromolecular structures", *J. Mol. Biol.* **112** (1977), 535–542.

[Brooks and Karplus 1983] B. R. Brooks and M. Karplus, "Harmonic dynamics of proteins: normal modes and fluctuations in bovine pancreatic trypsin inhibitor", *Proc. Natl. Acad. Sci. (USA)* **80** (1983), 3696–3700.

[Brooks et al. 1988] C. Brooks, M. Karplus, and M. Pettitt, "Proteins: a theoretical perspective of dynamics, structure and thermodynamics", *Adv. Chem. Phys.* **71** (1988), 1–259.

[Bryant et al. 2004] R. Bryant, H. Edelsbrunner, P. Koehl, and M. Levitt, "The area derivative of a space-filling diagram", *Discrete Comput. Geom.* **32** (2004), 293–308.

[Buckle and Fersht 1994] A. M. Buckle and A. R. Fersht, "Subsite binding in an RNase: structure of a barnase-tetranucleotide complex at 1.76 angstrom resolution", *Biochemistry* **33** (1994), 1644–1653.

[Bushnell and Kornberg 2003] D. A. Bushnell and R. D. Kornberg, "Complete, 12-subunit RNA polymerase II at 4.1-angstrom resolution: implications for the initiation of transcription", *Proc. Natl. Acad. Sci. (USA)* **100** (2003), 6969–6973.

[Carver 1978] M. B. Carver, "Efficient integration over discontinuities in ordinary differential equation simulators", *Math. Comput. Simul.* **20** (1978), 190–196.

[Cavallo et al. 2003] L. Cavallo, J. Kleinjung, and F. Fraternali, "POPS: a fast algorithm for solvent accessible surface areas at atomic and residue level", *Nucl. Acids. Res.* **31** (2003), 3364–3366.

[Cech 1993] T. R. Cech, "The efficiency and versality of catalytic RNA: implications for an RNA world", *Gene* **135** (1993), 33–36.

[Cheatham and Kollman 2000] T. E. Cheatham and P. A. Kollman, "Molecular dynamics simulation of nucleic acids", *Ann. Rev. Phys. Chem.* **51** (2000), 435–471.

[Connolly 1985] M. L. Connolly, "Computation of molecular volume", *J. Am. Chem. Soc.* **107** (1985), 1118–1124.

[Corey and Pauling 1953] R. B. Corey and L. Pauling, "Molecular models of amino acids, peptides and proteins", *Rev. Sci. Instr.* **24** (1953), 621–627.

[Cossi et al. 1996] M. Cossi, B. Mennucci, and R. Cammi, "Analytical first derivatives of molecular surfaces with respect to nuclear coordinates", *J. Comp. Chem.* **17** (1996), 57–73.

[Cramer et al. 2001] P. Cramer, D. A. Bushnell, and R. D. Kornberg, "Structural basis of transcription: RNA polymerase II at 2.8 angstrom resolution", *Science* **292** (2001), 1863–1876.

[Delaunay 1934] B. Delaunay, "Sur la sphère vide", *Izv. Akad. Nauk SSSR, Otdelenie Matematicheskii i Estestvennyka Nauk* **7** (1934), 793–800.

[Dodd and Theodorou 1991] L. R. Dodd and D. N. Theodorou, "Analytical treatment of the volume and surface area of molecules formed by an arbitrary collection of unequal spheres intersected by planes", *Mol. Phys.* **72** (1991), 1313–1345.

[Duan and Kollman 1998] Y. Duan and P. A. Kollman, "Pathways to a protein folding intermediate observed in a 1-microsecond simulation in aqueous solution", *Science* **282** (1998), 740 – 744.

[Edelsbrunner 1992] H. Edelsbrunner, "Weighted alpha shapes", Technical Report UIUC-CS-R-92-1760, Comput. Sci. Dept., Univ. Illinois, Urbana, Illinois, 1992.

[Edelsbrunner 1995] H. Edelsbrunner, "The union of balls and its dual shape", *Discrete Comput. Geom.* **13** (1995), 415–440.

[Edelsbrunner 1999] H. Edelsbrunner, "Deformable smooth surface design", *Discrete Comput. Geom.* **21** (1999), 87–115.

[Edelsbrunner and Fu 1994] H. Edelsbrunner and P. Fu, "Measuring space filling diagrams and voids", Technical Report UIUC-BI-MB-94-01, Beckman Inst., Univ. Illinois, Urbana, Illinois, 1994.

[Edelsbrunner and Koehl 2003] H. Edelsbrunner and P. Koehl, "The weighted-volume derivative of a space-filling diagram", *Proc. Natl. Acad. Sci. (USA)* **100** (2003), 2203–2208.

[Edelsbrunner and Mücke 1990] H. Edelsbrunner and E. P. Mücke, "Simulation of simplicity: a technique to cope with degenerate cases in geometric algorithms", *ACM Trans. Graphics* **9** (1990), 66–104.

[Edelsbrunner and Mücke 1994] H. Edelsbrunner and E. P. Mücke, "Three-dimensional alpha shapes", *ACM Trans. Graphics* **13** (1994), 43–72.

[Edelsbrunner and Shah 1996] H. Edelsbrunner and N. R. Shah, "Incremental topo-logical flipping works for regular triangulations", *Algorithmica* **15** (1996), 223–241.

[Edelsbrunner et al. 1983] H. Edelsbrunner, D. G. Kirkpatrick, and R. Seidel, "On the shape of a set of points in the plane", *IEEE Trans. Inform. Theory* **IT-29** (1983), 551–559.

[Edelsbrunner et al. 1998] H. Edelsbrunner, M. A. Facello, and J. Liang, "On the definition and construction of pockets in macromolecules", *Discrete Appl. Math.* **88** (1998), 83–102.

[Eisenberg and McLachlan 1986] D. Eisenberg and A. D. McLachlan, "Solvation energy in protein folding and binding", *Nature (London)* **319** (1986), 199–203.

[Fortune and VanWyk 1996] S. Fortune and C. J. VanWyk, "Static analysis yields efficient exact integer arithmetic for computational geometry", *ACM Trans. Graph.* **15** (1996), 223–248.

[Futamura et al. 2004] N. Futamura, S. Alura, D. Ranjan, and B. Hariharan, "Efficient parallel algorithms for solvent accessible surface area of proteins", *IEEE Trans. Parallel Dist. Syst.* **13** (2004), 544–555.

[Gavezzotti 1983] A. Gavezzotti, "The calculation of molecular volumes and the use of volume analysis in the investigation of structured media and of solid-state organic reactivity", *J. Am. Chem. Soc.* **105** (1983), 5220–5225.

[Gear and Østerby 1984] C. W. Gear and O. Østerby, "Solving ordinary differential equations with discontinuities", *ACM Trans. Math. Softw.* **10** (1984), 23–24.

[Gesteland and Atkins 1993] R. F. Gesteland and J. A. Atkins, *The RNA world: the nature of modern RNA suggests a prebiotic RNA world*, Cold Spring Harbor Laboratory Press, Plainview, NY, 1993.

[Gibson and Scheraga 1967] K. D. Gibson and H. A. Scheraga, "Minimization of polypeptide energy, I: Preliminary structures of bovine pancreatic ribonuclease S-peptide", *Proc. Natl. Acad. Sci. (USA)* **58** (1967), 420–427.

[Gibson and Scheraga 1987] K. D. Gibson and H. A. Scheraga, "Exact calculation of the volume and surface-area of fused hard-sphere molecules with unequal atomic radii", *Mol. Phys.* **62** (1987), 1247–1265.

[Gilbert 1986] W. Gilbert, "The RNA world", *Nature* **319** (1986), 618.

[Go et al. 1983] N. Go, T. Noguti, and T. Nishikawa, "Dynamics of a small globular protein in terms of low-frequency vibrational modes", *Proc. Natl. Acad. Sci. (USA)* **80** (1983), 3696–3700.

[Gogonea and Osawa 1994] V. Gogonea and E. Osawa, "Implementation of solvent effect in molecular mechanics, 3: The first and second-order analytical derivatives of excluded volume.", *J. Mol. Struct. (Theochem)* **311** (1994), 305–324.

[Gogonea and Osawa 1995] V. Gogonea and E. Osawa, "An improved algorithm for the analytical computation of solvent-excluded volume: the treatment of singularities in solvent-accessible surface-area and volume functions", *J. Comp. Chem.* **16** (1995), 817–842.

[Grant and Pickup 1995] J. A. Grant and B. T. Pickup, "A Gaussian description of molecular shape", *J. Phys. Chem.* **99** (1995), 3503–3510.

[Hao and Scheraga 1994a] M. H. Hao and H. A. Scheraga, "Monte Carlo Simulation of a first order transition for protein folding", *J. Phys. Chem.* **98** (1994), 4940–4948.

[Hao and Scheraga 1994b] M. H. Hao and H. A. Scheraga, "Statistical thermodynamics of protein folding – sequence dependence", *J. Phys. Chem.* **98** (1994), 9882–9893.

[Hasel et al. 1988] W. Hasel, T. F. Hendrikson, and W. C. Still, "A rapid approximation to the solvent accessible surface areas of atoms", *Tetrahed. Comp. Method.* **1** (1988), 103–106.

[Irisa 1996] M. Irisa, "An elegant algorithm of the analytical calculation for the volume of fused spheres with different radii", *Comp. Phys. Comm.* **98** (1996), 317–338.

[Kaminski et al. 2001] G. A. Kaminski, R. A. Friesner, J. Tirado-Rives, and W. L. Jorgensen, "Evaluation and reparametrization of the OPLS-AA force field for proteins via comparison with accurate quantum chemical calculations on peptides", *J. Phys. Chem. B.* **105** (2001), 6474–6487.

[Karplus and McCammon 2002] M. Karplus and J. A. McCammon, "Molecular dynamics simulations of biomolecules", *Nature Struct. Biol.* **9** (2002), 646–652.

[Kauzmann 1959] W. Kauzmann, "Some factors in the interpretation of protein denaturation", *Adv. Protein Chem.* **14** (1959), 1–63.

[Kendrew et al. 1960] J. Kendrew, R. Dickerson, B. Strandberg, R. Hart, D. Davies, and D. Philips, "Structure of myoglobin: a three dimensional Fourier synthesis at 2 angstrom resolution", *Nature* **185** (1960), 422–427.

[Koehl and Levitt 1999] P. Koehl and M. Levitt, "A brighter future for protein structure prediction", *Nature Struct. Biol.* **6** (1999), 108–111.

[Koehl and Levitt 2002] P. Koehl and M. Levitt, "Protein topology and stability defines the space of allowed sequences", *Proc. Natl. Acad. Sci. (USA)* **99** (2002), 1280–1285.

[Koltun 1965] W. L. Koltun, "Precision space-filling atomic models", *Biopolymers* **3** (1965), 665–679.

[Kratky 1978] K. W. Kratky, "Area of intersection of *n* equal circular disks", *J. Phys. A.: Math. Gen.* **11** (1978), 1017–1024.

[Kraulis 1991] P. J. Kraulis, "MOLSCRIPT: a program to produce both detailed and schematic plots of protein structures", *J. Appl. Crystallo.* **24** (1991), 946–950.

[Kundrot et al. 1991] C. E. Kundrot, J. W. Ponder, and F. M. Richards, "Algorithms for calculating excluded volume and its derivatives as a function of molecular-conformation and their use in energy minimization", *J. Comp. Chem.* **12** (1991), 402–409.

[Leach 2001] A. R. Leach, *Molecular modelling: principles and applications*, 2nd ed., Prentice Hall, 2001.

[Lee and Richards 1971] B. Lee and F. M. Richards, "Interpretation of protein structures: estimation of static accessibility", *J. Mol. Biol.* **55** (1971), 379–400.

[Legrand and Merz 1993] S. M. Legrand and K. M. Merz, "Rapid approximation to molecular-surface area via the use of boolean logic and look-up tables", *J. Comp. Chem.* **14** (1993), 349–352.

[Levitt and Sharon 1988] M. Levitt and R. Sharon, "Accurate simulation of protein dynamics in solution", *Proc. Natl. Acad. Sci. (USA)* **85** (1988), 7557–7561.

[Levitt et al. 1983] M. Levitt, C. Sander, and P. S. Stern, "Protein normal-mode dynamics: trypsin inhibitor, crambin, ribonuclease and lysozyme", *J. Mol. Biol.* **181** (1983), 423–447.

[Levitt et al. 1995] M. Levitt, M. Hirshberg, R. Sharon, and V. Daggett, "Potential-energy function and parameters for simulations of the molecular-dynamics of proteins and nucleic-acids in solution", *Comp. Phys. Comm.* **91** (1995), 215–231.

[Levy et al. 2004] Y. Levy, P. G. Wolynes, and J. N. Onuchic, "Protein topology determines binding mechanism", *Proc. Natl. Acad. Sci. (USA)* **101** (2004), 511–516.

[Liang et al. 1998a] J. Liang, H. Edelsbrunner, P. Fu, P. V. Sudhakar, and S. Subramaniam, "Analytical shape computation of macromolecules, I: Molecular area and volume through alpha shape", *Proteins: Struct. Func. Genet.* **33** (1998), 1–17.

[Liang et al. 1998b] J. Liang, H. Edelsbrunner, P. Fu, P. V. Sudhakar, and S. Subramaniam, "Analytical shape computation of macromolecules, II: Inaccessible cavities in proteins", *Proteins: Struct. Func. Genet.* **33** (1998), 18–29.

[Liang et al. 1998c] J. Liang, H. Edelsbrunner, and C. Woodward, "Anatomy of protein pockets and cavities: measurement of binding site geometry and implications for ligand design", *Prot. Sci.* **7** (1998), 1884–1897.

[Lindahl et al. 2001] E. Lindahl, B. Hess, and D. van der Spoel, "GROMACS 3.0: a package for molecular simulation and trajectory analysis", *J. Molec. Mod.* **7** (2001), 306–317.

[Liwo et al. 1997a] A. Liwo, S. Oldziej, M. R. Pincus, R. J. Wawak, S. Rackovsky, and H. A. Scheraga, "A united-residue force field for off-lattice protein-structure simulations, 1: Functional forms and parameters of long-range side-chain interaction potentials from protein crystal data", *J. Comp. Chem.* **18** (1997), 849–873.

[Liwo et al. 1997b] A. Liwo, M. R. Pincus, R. J. Wawak, S. Rackovsky, S. Oldziej, and H. A. Scheraga, "A united-residue force field for off-lattice protein-structure simulations, 2: Parameterization of short-range interactions and determination of weights of energy terms by Z-score optimization", *J. Comp. Chem.* **18** (1997), 874–887.

[Lum et al. 1999] K. Lum, D. Chandler, and J. D. Weeks, "Hydrophobicity at small and large length scales", *J. Phys. Chem. B.* **103** (1999), 4570–4577.

[MacKerell et al. 1998] A. D. MacKerell, D. Bashford, M. Bellott, R. L. Dunbrack, J. D. Evanseck, M. J. Field, et al., "All-atom empirical potential for molecular modeling and dynamics studies of proteins", *J. Phys. Chem. B.* **102** (1998), 3586–3616.

[McCammon et al. 1977] J. A. McCammon, B. R. Gelin, and M. Karplus, "Dynamics of folded proteins", *Nature* **267** (1977), 585–590.

[Monod 1973] J. Monod, *Le hasard et la nécessité*, Le Seuil, Paris, France, 1973.

[Mücke et al. 1999] E. P. Mücke, I. Saias, and B. Zhu, "Fast randomized point location without preprocessing in two- and three-dimensional Delaunay triangulations", *Comput. Geom.: Theory Appl.* **12** (1999), 63–83.

[Muñoz and Eaton 1999] V. Muñoz and W. A. Eaton, "A simple model for calculating the kinetics of protein folding from three-dimensional structures", *Proc. Natl. Acad. Sci. (USA)* **96** (1999), 11311–11316.

[Ooi et al. 1987] T. Ooi, M. Oobatake, G. Nemethy, and H. A. Scheraga, "Accessible surface-areas as a measure of the thermodynamic parameters of hydration of peptides", *Proc. Natl. Acad. Sci. (USA)* **84** (1987), 3086–3090.

[Pavani and Ranghino 1982] R. Pavani and G. Ranghino, "A method to compute the volume of a molecule", *Computers and Chemistry* **6** (1982), 133–135.

[Perrot et al. 1992] G. Perrot, B. Cheng, K. D. Gibson, J. Vila, K. A. Palmer, A. Nayeem, et al., "MSEED: a program for the rapid analytical determination of accessible surface-areas and their derivatives", *J. Comp. Chem.* **13** (1992), 1–11.

[Perutz 1990] M. F. Perutz, "Mechanisms regulating the reactions of human hemoglobin with oxygen and carbon monoxide", *Annu. Rev. Physiol.* **52** (1990), 1–25.

[Perutz et al. 1960] M. Perutz, M. Rossmann, A. Cullis, G. Muirhead, G. Will, and A. North, "Structure of hemoglobin: a three-dimensional Fourier synthesis at 5.5 angstrom resolution, obtained by X-ray analysis", *Nature* **185** (1960), 416–422.

[Petitjean 1994] M. Petitjean, "On the analytical calculation of van-der-Waals surfaces and volumes: some numerical aspects", *J. Comp. Chem.* **15** (1994), 507–523.

[Plaxco et al. 1998] K. W. Plaxco, K. T. Simons, and D. Baker, "Contact order, transition state placement and the refolding rates of single domain proteins", *J. Mol. Biol.* **277** (1998), 985–994.

[Price and Brooks 2002] D. J. Price and C. L. Brooks, "Modern protein force fields behave comparably in molecular dynamics simulations", *J. Comp. Chem.* **23** (2002), 1045–1057.

[Remington et al. 1982] S. Remington, G. Weigand, and R. Huber, "Crystallographic refinement and atomic models of two different forms of citrate synthase at 2.7 and 1.7 angstrom resolution", *J. Mol. Biol.* **158** (1982), 111–152.

[Richmond 1984] T. J. Richmond, "Solvent accessible surface-area and excluded volume in proteins: analytical equations for overlapping spheres and implications for the hydrophobic effect", *J. Mol. Biol.* **178** (1984), 63–89.

[Rowlinson 1963] J. S. Rowlinson, "The triplet distribution function in a fluid of hard spheres", *Mol. Phys.* **6** (1963), 517–524.

[Shrake and Rupley 1973] A. Shrake and J. A. Rupley, "Environment and exposure to solvent of protein atoms: lysozyme and insulin", *J. Mol. Biol.* **79** (1973), 351–371.

[Simonson 2003] T. Simonson, "Electrostatics and dynamics of proteins", *Rep. Prog. Phys* **66** (2003), 737–787.

[Simonson and Brünger 1994] T. Simonson and A. T. Brünger, "Solvation free-energies estimated from macroscopic continuum theory: an accuracy assessment", *J. Phys. Chem.* **98** (1994), 4683–4694.

[Sridharan et al. 1994] S. Sridharan, A. Nicholls, and K. A. Sharp, "A rapid method for calculating derivatives of solvent accessible surface areas of molecules", *J. Comp. Chem.* **16** (1994), 1038–1044.

[Street and Mayo 1998] A. G. Street and S. L. Mayo, "Pairwise calculation of protein solvent-accessible surface areas", *Folding & Design* **3** (1998), 253–258.

[Timberlake 1992] K. C. Timberlake, *Chemistry*, 5th ed., Harper Collins, New York, 1992.

[Tunon et al. 1992] I. Tunon, E. Silla, and J. L. Pascual-Ahuir, "Molecular-surface area and hydrophobic effect", *Protein Eng.* **5** (1992), 715–716.

[Wang and Levinthal 1991] H. Wang and C. Levinthal, "A vectorized algorithm for calculating the accessible surface area of macromolecules", *J. Comp. Chem.* **12** (1991), 868–871.

[Watson and Crick 1953] J. D. Watson and F. H. C. Crick, "A Structure for Deoxyribose Nucleic Acid", *Nature* **171** (1953), 737–738.

[Wawak et al. 1994] R. J. Wawak, K. D. Gibson, and H. A. Scheraga, "Gradient discontinuities in calculations involving molecular-surface area", *J. Math. Chem.* **15** (1994), 207–232.

[Weiser et al. 1999a] J. Weiser, P. S. Shenkin, and W. C. Still, "Approximate atomic surfaces from linear combinations of pairwise overlaps (LCPO)", *J. Comp. Chem.* **20** (1999), 217–230.

[Weiser et al. 1999b] J. Weiser, P. S. Shenkin, and W. C. Still, "Optimization of Gaussian surface calculations and extension to solvent accessible surface areas", *J. Comp. Chem.* **20** (1999), 688–703.

[Wimberly et al. 2000] B. T. Wimberly, D. E. Brodersen, W. M. Clemons Jr., R. J. Morgan-Warren, A. P. Carter, C. Vonrhein, et al., "Structure of the 30S ribosomal subunit", *Nature* **407** (2000), 327–339.

[Wodak and Janin 1980] S. J. Wodak and J. Janin, "Analytical approximation to the accessible surface-area of proteins", *Proc. Natl. Acad. Sci. (USA)* **77** (1980), 1736–1740.

[Wood and Thompson 1990] R. H. Wood and P. T. Thompson, "Differences between pair and bulk hydrophobic interactions", *Proc. Natl. Acad. Sci. (USA)* **87** (1990), 946–949.

[Yusupov et al. 2001] M. M. Yusupov, G. Z. Yusupova, A. Baucom, K. Lieberman, T. N. Earnest, J. H. D. Cate, and H. F. Noller, "Crystal structure of the ribosome at 5.5 angstrom resolution", *Science* **292** (2001), 883–896.

HERBERT EDELSBRUNNER
DEPARTMENT OF COMPUTER SCIENCE
DUKE UNIVERSITY
DURHAM, NC 27708
UNITED STATES
edels@cs.duke.edu

PATRICE KOEHL
DEPARTMENT OF COMPUTER SCIENCE AND GENOME CENTER
UNIVERSITY OF CALIFORNIA
DAVIS, CA 95616
UNITED STATES
koehl@cs.ucdavis.edu

Combinatorial and Computational Geometry
MSRI Publications
Volume **52**, 2005

Inequalities for Zonotopes

RICHARD EHRENBORG

Dedicated to Louis Billera on his sixtieth birthday

ABSTRACT. We present two classes of linear inequalities that the flag f-vectors of zonotopes satisfy. These inequalities strengthen inequalities for polytopes obtained by the lifting technique of Ehrenborg.

1. Introduction

The systematic study of flag f-vectors of polytopes was initiated by Bayer and Billera [1985]. Billera then suggested the study of flag f-vectors of zonotopes; see the dissertation of his student Liu [1995]. The essential computational results of the field appeared in two papers by Billera, Ehrenborg and Readdy [Billera et al. 1997; 1998]. Here we present two classes of linear inequalities for the flag f-vectors of zonotopes. These classes are motivated by our recent results for polytopes [Ehrenborg 2005].

The flag f-vector of a convex polytope contains all the enumerative incidence information between the faces of the polytope. For an n-dimensional polytope the flag f-vector consists of 2^n entries; in other words, the flag f-vector lies in the vector space $\mathbb{R}^{2^n}$. Bayer and Billera [1985] showed that the flag vectors of n-dimensional polytopes span a subspace of $\mathbb{R}^{2^n}$, called the generalized Dehn–Sommerville subspace and denoted by GDSS_n. Bayer and Klapper [1991] proved that GDSS_n is naturally isomorphic to the n-th homogeneous component of the noncommutative ring $\mathbb{R}\langle c, d \rangle$, where the grading is given by $\deg c = 1$ and $\deg d = 2$. Hence, the flag f-vector of a polytope P can be encoded by a noncommutative polynomial $\Psi(P)$ in the variables c and d, called the cd-index.

The next essential step is to consider linear inequalities that the flag f-vector of polytopes satisfy. The known linear inequalities are: the nonnegativity of the

Research partially supported by National Science Foundation grant 0200624.

toric g-vector [Kalai 1987; Karu 2001; Stanley 1987], inequalities obtained by the Kalai convolution [Kalai 1988], and that the $\mathbf{cd}$-index is minimized coefficientwise on the n-dimensional simplex Σ_n [Billera and Ehrenborg 2000]. Recently we introduced in [Ehrenborg 2005] a lifting technique that allows one to use lower dimensional inequalities to obtain higher-dimensional inequalities. Here is a special case of this lifting technique:

THEOREM 1.1. *Let u, q and v be three $\mathbf{cd}$-monomials such that the sum of the degrees of u, q and v is n and the degree of q is k. Let Δ_q denote the coefficient of the $\mathbf{cd}$-monomial q in the $\mathbf{cd}$-index of a k-dimensional simplex Σ_k. Then for all n-dimensional polytopes P we have*

$$\langle u \cdot (q - \Delta_q \cdot \mathbf{c}^k) \cdot v \mid \Psi(P) \rangle \geq 0,$$

where the bracket $\langle \cdot \mid \cdot \rangle$ is the standard inner product on $\mathbb{R}\langle \mathbf{c}, \mathbf{d} \rangle$.

The purpose of this paper is to improve Theorem 1.1 for zonotopes.

Recall that a zonotope is a polytope obtained as the Minkowski sum of line segments. The flag f-vectors of n-dimensional zonotopes lie in the subspace GDSS_n. Billera, Ehrenborg and Readdy [Billera et al. 1998] proved that they do not lie in any proper subspace of GDSS_n. They also showed that among all n-dimensional zonotopes (and more generally, the dual of the lattice of regions of oriented matroids), the n-dimensional cube minimizes the $\mathbf{cd}$-index coefficient-wise [Billera et al. 1997]. This is the zonotopal analogue of Stanley's Gorenstein* lattice conjecture [Stanley 1994b, Conjecture 2.7].

We continue this vein of research by introducing further classes of linear inequalities for flag f-vectors of zonotopes. We develop two sharper versions of the inequality appearing in Theorem 1.1. For an n-dimensional zonotope we show that the expression in Theorem 1.1 is at least the value obtained by the n-dimensional cube C_n; see Theorem 3.1. The second improvement is the case when $u = 1$. We can replace the factor Δ_q by a larger factor, the coefficient of q in the $\mathbf{cd}$-index of the k-dimensional cube C_k; see Theorem 3.6.

2. Preliminaries

For standard terminology for posets, see [Stanley 1986]. A partially ordered set (poset) P is ranked if there is a rank function $\rho : P \to \mathbb{Z}$ such that when x is covered by y then $\rho(y) = \rho(x) + 1$. The poset P is graded of rank n if it is ranked and has a minimal element $\hat{0}$ and a maximal element $\hat{1}$ such that $\rho(\hat{0}) = 0$ and $\rho(\hat{1}) = n$. Define the interval $[x, y]$ to be the subposet $\{z \in P : x \leq z \leq y\}$. Observe that the interval $[x, y]$ is also a graded poset of rank $\rho(y) - \rho(x)$.

Let P be a graded poset of rank $n + 1$. For $S = \{s_1 < s_2 < \cdots < s_k\}$ a subset of $\{1, \ldots, n\}$, define f_S to be the number of chains $\hat{0} = x_0 < x_1 < \cdots < x_{k+1} = \hat{1}$, where the rank of the element x_i is s_i for $1 \leq i \leq k$. These 2^n values constitute the *flag f-vector* of the poset P. Define the *flag h-vector* of P by the two equivalent relations $h_S = \sum_{T \subseteq S} (-1)^{|S-T|} f_T$ and $f_S = \sum_{T \subseteq S} h_T$. There has been a lot of recent work in understanding the flag f-vectors of graded posets and Eulerian posets. For example, see [Bayer 2001; Bayer and Hetyei 2001; Billera and Hetyei 2000].

For S a subset of $\{1, \ldots, n\}$ define the monomial $u_S = u_1 u_2 \cdots u_n$, where $u_i = \boldsymbol{a}$ if $i \notin S$ and $u_i = \boldsymbol{b}$ if $i \in S$. Define the $\boldsymbol{ab}$-*index* of a graded poset P of rank $n + 1$ to be the sum

$$\Psi(P) = \sum_S h_S \cdot u_S.$$

A poset P is Eulerian if every interval $[x, y]$, where $x \neq y$, has the same number of elements of odd rank as the number of elements of even rank. This condition states that every interval $[x, y]$ satisfies the Euler–Poincaré relation. The condition of being Eulerian is equivalent to the condition that the Möbius function $\mu(x, y)$ is $(-1)^{\rho(x,y)}$. The two main examples of Eulerian posets are the strong Bruhat order and face lattices of convex polytopes.

The following result was conjectured by Fine and proved by Bayer and Klapper [1991]. It states that the generalized Dehn–Sommerville subspace GDSS_n is naturally isomorphic to the space of $\boldsymbol{cd}$-polynomials of degree n.

THEOREM 2.1. *The $\boldsymbol{ab}$-index of an Eulerian poset P, $\Psi(P)$, can be written in terms of $\boldsymbol{c} = \boldsymbol{a} + \boldsymbol{b}$ and $\boldsymbol{d} = \boldsymbol{a} \cdot \boldsymbol{b} + \boldsymbol{b} \cdot \boldsymbol{a}$.*

When $\Psi(P)$ is expressed in terms of $\boldsymbol{c}$ and $\boldsymbol{d}$ it is called the $\boldsymbol{cd}$-*index* of the poset P. There exist several proofs of this result in the literature; see [Bayer and Klapper 1991; Billera and Liu 2000; Ehrenborg 2001; Ehrenborg and Readdy 2002; Stanley 1994a]. The $\boldsymbol{cd}$-index has been extraordinarily useful for flag vector computations; see [Bayer and Ehrenborg 2000; Billera et al. 1997; Ehrenborg and Readdy 1998]. Moreover, this basis is now emerging as a key tool for obtaining linear inequalities for the entries of the flag f-vector; see [Billera and Ehrenborg 2000; Ehrenborg 2005; Ehrenborg and Fox 2003; Stanley 1994a].

Define an inner product $\langle \cdot \mid \cdot \rangle$ on $\mathbb{R}\langle \boldsymbol{c}, \boldsymbol{d} \rangle$ by $\langle u \mid v \rangle = \delta_{u,v}$ for all $\boldsymbol{cd}$-monomials u and v, and extend this relation by linearity. Using this notation any linear inequality on the flag f-vector of an n-dimensional polytope can be expressed as $\langle H \mid \Psi(P) \rangle \geq 0$, where H is homogeneous $\boldsymbol{cd}$-polynomial of degree n.

In the remainder of this section we will focus upon the $\boldsymbol{cd}$-index of zonotopes. However, all the results carry over to oriented matroids. In order to keep the statements of the results explicit, we will use the geometric language of zonotopes and their hyperplane arrangements.

A zonotope Z is a polytope obtained by the Minkowski sum of line segments, that is, $Z = [\boldsymbol{0}, \boldsymbol{v}_1] + \cdots + [\boldsymbol{0}, \boldsymbol{v}_m]$. For each line segment $[\boldsymbol{0}, \boldsymbol{v}_i]$ let H_i be the hyperplane through the origin that is orthogonal to $\boldsymbol{v}_i$. The collection of these hyperplanes $\mathcal{H} = \{H_1, \ldots, H_m\}$ is the central hyperplane arrangement associated to the zonotope Z. The intersection lattice L of the arrangement $\mathcal{H}$

is the collection of all the intersections of the hyperplanes $H_1, \ldots, H_m$ ordered by reverse inclusion.

Let ω be the linear map from $\mathbb{R}\langle a, b \rangle$ to $\mathbb{R}\langle c, d \rangle$ defined on an ab-monomial by replacing each occurrence of ab with $2d$ and then replacing the remaining variables by c. Here is the fundamental theorem for computing the cd-index of a zonotope:

THEOREM 2.2 [Billera et al. 1997]. *Let Z be a zonotope (and more generally, let Z be the dual of the lattice of regions of an oriented matroid). Let L be the intersection lattice of the associated central hyperplane arrangement $\mathcal{H}$ and $\Psi(L)$ the ab-index of the lattice L. Then the cd-index of the zonotope and the sum of the cd-indices of all the vertex figures of the zonotope are given by*

$$\Psi(Z) = \omega(a \cdot \Psi(L)),$$

$$\sum_v \Psi(Z/v) = 2 \cdot \omega(\Psi(L)),$$

where v ranges over all vertices of the zonotope Z.

The first of these identities is [Billera et al. 1997, Theorem 3.1]. The second follows from the first by using the linear map h defined in Section 8 of the same reference.

It remains to compute the ab-index of the intersection lattice L. We do this using R-labelings. For more details, see [Billera et al. 1997, Section 7] and [Björner 1980; Stanley 1974; 1986]. Linearly order the hyperplanes in the arrangement $\mathcal{H}$ as $\mathcal{H} = \{H_1, \ldots, H_m\}$. Mark each edge $x \prec y$ in the Hasse diagram of the lattice L with the smallest (in the given linear order) hyperplane H such that intersecting x with H gives y. That is,

$$\lambda(x, y) = \min\{i \,:\, x \cap H_i = y\}.$$

For a maximal chain $c = \{\hat{0} = x_0 \prec x_1 \prec \cdots \prec x_n = \hat{1}\}$ in the intersection lattice L define its *descent set* $D(c)$ by

$$D(c) = \{i \,:\, \lambda(x_{i-1}, x_i) > \lambda(x_i, x_{i+1})\}.$$

THEOREM 2.3 [Billera et al. 1997, Section 7]. *The ab-index of intersection lattice L is given by*

$$\Psi(L) = \sum_c u_{D(c)},$$

where the sum ranges over all maximal chains c in the lattice L.

3. Inequalities for Zonotopes

In this section we will improve Theorem 1.1 for zonotopes. Let C_n denote the n-dimensional cube.

THEOREM 3.1. *Let Z be an n-dimensional zonotope (and more generally, let Z be the dual of the lattice of regions of an oriented matroid). Let q be a **cd**-monomial of degree k that contains at least one $\boldsymbol{d}$. Then the **cd**-index $\Psi(Z)$ satisfies the inequality*

$$\langle u \cdot (q - \Delta_q \cdot \boldsymbol{c}^k) \cdot v \mid \Psi(Z) - \Psi(C_n) \rangle \geq 0.$$

*for any two **cd**-monomials u and v such that $\deg u + \deg v = n - k$.*

DEFINITION 3.2. *Let q be a **cd**-monomial of degree k that contains at least one $\boldsymbol{d}$. For two **cd**-polynomials z and w define the order relation $z \preceq_q w$ if the inequality $\langle u \cdot (q - \Delta_q \cdot \boldsymbol{c}^k) \cdot v \mid w - z \rangle \geq 0$ holds for all **cd**-monomials u and v.*

In this notation Theorem 3.1 becomes $\Psi(Z) \succeq_q \Psi(C_n)$ and that of Theorem 1.1 becomes $\Psi(P) \succeq_q 0$. Note that this order relation differs slightly from the order relation used in [Ehrenborg 2005].

LEMMA 3.3. *Let z and w be nonnegative **cd**-polynomials such that $z \succeq_q 0$ and $w \succeq_q 0$. Then we have $z \cdot \boldsymbol{d} \cdot w \succeq_q 0$.*

PROOF. Without loss of generality, we may assume that z and w are homogeneous polynomials. We would like to prove that

$$\langle u \cdot (q - \Delta_q \cdot \boldsymbol{c}^k) \cdot v \mid z \cdot \boldsymbol{d} \cdot w \rangle \geq 0,$$

for all **cd**-monomials u and v such that $\deg u + \deg v = \deg(zdw) - k$, where k is the degree of q. We do this in three cases. The first case is $\deg(u\boldsymbol{c}^k) \leq \deg z$. Try to factor $v = v_1 \cdot v_2$ such that $\deg(u\boldsymbol{c}^k v_1) = \deg z$. If such factoring is not possible, both sides of the inequality are equal to zero. If factoring is possible then $\langle u(q - \Delta_q \boldsymbol{c}^k)v \mid zdw \rangle = \langle u(q - \Delta_q \boldsymbol{c}^k)v_1 \mid z \rangle \cdot \langle v_2 \mid dw \rangle \geq 0$. The second case is $\deg u \geq \deg(zd)$, which is symmetric to the first case.

The third is $\deg(u\boldsymbol{c}^k) > \deg z$ and $\deg u < \deg(zd)$. Since z and w have nonnegative coefficients we have $\langle uqv \mid zdw \rangle \geq 0$. Moreover, $\langle u\boldsymbol{c}^k v \mid zdw \rangle = 0$. This completes the third case.   □

PROPOSITION 3.4. *Let Z be an n-dimensional zonotope and let Z' be the zonotope obtained by taking the Minkowski sum of Z with a line segment in the affine span of Z. Then we have $\Psi(Z') \succeq_q \Psi(Z)$.*

PROOF. Let $\mathcal{H}$ and $\mathcal{H}'$ be the associated hyperplane arrangements and let H be the new hyperplane. Let $\mathcal{H}'$ inherit the linear order of $\mathcal{H}$ with the new hyperplane H inserted at the end of the linear order. Similarly, let L and L' be the corresponding intersection lattices. Observe that every maximal chain in L is also a maximal chain in L'. Also observe that there is no maximal chain in L' whose last label is H. Hence the difference in the $\boldsymbol{ab}$-indices between the two intersection lattices is

$$\Psi(L') - \Psi(L) = \sum_c u_{D(c)}$$

$$= \sum_{\hat{0} < x \prec y} \Psi([\hat{0}, x]) \cdot \boldsymbol{ab} \cdot \Psi([y, \hat{1}]) + \sum_{\hat{0} = x \prec y} \boldsymbol{b} \cdot \Psi([y, \hat{1}]),$$

where the sum on the first line is over all maximal chains c containing the label H and the sums on the second line are over edges $x \prec y$ in the Hasse diagram of L' having the label H. Applying the map $w \longmapsto \omega(\boldsymbol{a} \cdot w)$ we obtain

$$\Psi(Z') - \Psi(Z) = \sum_{\hat{0} < x \prec y} \omega(\boldsymbol{a} \cdot \Psi([\hat{0}, x])) \cdot 2\boldsymbol{d} \cdot \omega(\Psi([y, \hat{1}])) + \sum_{\hat{0} \prec y} 2\boldsymbol{d} \cdot \omega(\Psi([y, \hat{1}])). \quad (3.1)$$

The term $\omega(\boldsymbol{a} \cdot \Psi([\hat{0}, x]))$ is the $\boldsymbol{cd}$-index of a zonotope and hence is nonnegative in the order $\succeq_q$ by Theorem 1.1. Similarly, the term $\omega(\Psi([y, \hat{1}]))$ is one half of the sum of $\boldsymbol{cd}$-indices of the vertex figures of a zonotope and hence is also $\succeq_q$-nonnegative. The result now follows by Lemma 3.3 and the property that the order $\succeq_q$ is preserved under addition. □

PROOF OF THEOREM 3.1. Observe that any n-dimensional zonotope is obtained from the n-dimensional cube C_n by Minkowski adding line segments. Thus the result follows from Proposition 3.4. □

The second improvement of the zonotopal inequalities is when comparing the coefficients of $\boldsymbol{c}^k v$ and qv, that is, when u is equal to 1. Let $\square_q$ denote the coefficient of the monomial q in the $\boldsymbol{cd}$-index of the k-dimensional cube C_k, that is, $\square_q = \langle q \,|\, \Psi(C_k) \rangle$. For ease in notation, we introduce a second order relation.

DEFINITION 3.5. *Let q be a $\boldsymbol{cd}$-monomial of degree k that contains at least one $\boldsymbol{d}$ and let z and w be two $\boldsymbol{cd}$-polynomials. Define the order relation $z \preceq'_q w$ on the $\boldsymbol{cd}$-polynomials z and w by $\langle (q - \square_q \cdot \boldsymbol{c}^k) \cdot v \,|\, w - z \rangle \geq 0$ for all $\boldsymbol{cd}$-monomials v.*

THEOREM 3.6. *Let Z be an n-dimensional zonotope (and more generally, let Z be the dual of the lattice of regions of an oriented matroid). Let q be a $\boldsymbol{cd}$-monomial of degree k that contains at least one $\boldsymbol{d}$. Then the $\boldsymbol{cd}$-index $\Psi(Z)$ satisfies the inequality $\Psi(Z) \succeq'_q \Psi(C_n)$. That is, for all $\boldsymbol{cd}$-monomials v of degree $n - k$ we have*

$$\langle (q - \square_q \cdot \boldsymbol{c}^k) \cdot v \,|\, \Psi(Z) - \Psi(C_n) \rangle \geq 0.$$

The proof of Theorem 3.6 consists of the following lemma and two propositions.

LEMMA 3.7. *Let z and w be two nonnegative $\boldsymbol{cd}$-polynomials such that $z \succeq'_q 0$. Then we have $z \cdot \boldsymbol{d} \cdot w \succeq'_q 0$. Furthermore if $\deg q \leq \deg z$ we have that $z \cdot w \succeq'_q 0$.*

PROOF. We want to show that $\langle (q - \square_q \boldsymbol{c}^k) v \,|\, z \boldsymbol{d} w \rangle \geq 0$ for all $\boldsymbol{cd}$-monomials v, where $k = \deg q$. Consider first the case when $k \leq \deg z$. Try to write $v = v_1 \cdot v_2$

such that $k + \deg v_1 = \deg z$. If this is not possible both sides are equal to zero. If this is possible we have $\langle (q - \square_q \boldsymbol{c}^k) v \mid z \boldsymbol{d} w \rangle = \langle (q - \square_q \boldsymbol{c}^k) v_1 \mid z \rangle \cdot \langle v_2 \mid \boldsymbol{d} w \rangle \geq 0$. The second case is $k > \deg z$. Then right away we have $\langle \boldsymbol{c}^k v \mid z \boldsymbol{d} w \rangle = 0$. Also $\langle qv \mid z \boldsymbol{d} w \rangle \geq 0$, since both z and w have nonnegative coefficients. The second statement of the lemma is proved by similar reasoning, where there is only the case $\langle (q - \square_q \boldsymbol{c}^k) v \mid z w \rangle = \langle (q - \square_q \boldsymbol{c}^k) v_1 \mid z \rangle \cdot \langle v_2 \mid w \rangle \geq 0$. $\square$

PROPOSITION 3.8. *The $\boldsymbol{cd}$-index of the n-dimensional cube C_n satisfies*

$$\Psi(C_n) \succeq'_q 0.$$

PROOF. The proof is by induction on n. Observe that when $n < \deg q$ there is nothing to prove. When $n = \deg q$ the result is directly true. The induction step is based on the Purtill recursion for the $\boldsymbol{cd}$-index of the n-dimensional cube; see [Ehrenborg and Readdy 1996; Purtill 1993] or [Ehrenborg and Readdy 1998, Proposition 4.2]:

$$\Psi(C_{n+1}) = \Psi(C_n) \cdot \boldsymbol{c} + \sum_{i=0}^{n-1} 2^{n-i} \cdot \binom{n}{i} \cdot \Psi(C_i) \cdot \boldsymbol{d} \cdot \Psi(\Sigma_{n-i-1}).$$

By Lemma 3.7 we observe that all the terms in this expression are greater than 0 in the order $\succeq'_q$. $\square$

PROPOSITION 3.9. *Let Z be an n-dimensional zonotope and let Z' be the zonotope obtained by taking the Minkowski sum of Z with a line segment in the affine span of Z. Assume that all zonotopes W of dimension $n-1$ and less satisfy the relation $0 \preceq'_q \Psi(W)$. Then the order relation $\Psi(Z) \preceq'_q \Psi(Z')$ holds.*

PROOF. The proof follows the same outline as the proof of Proposition 3.4. By Lemma 3.7 each term in equation (3.1) is nonnegative in the order $\preceq'_q$. Since the property of being nonnegative is preserved under addition, the result follows. $\square$

PROOF OF THEOREM 3.6. We work by induction. The case $n = 0$ is straightforward. For the induction step assume that every zonotope W of dimension k less than n satisfies the inequality $\Psi(C_k) \preceq'_q \Psi(W)$. Especially, we know that the $\boldsymbol{cd}$-index of a lower dimensional zonotope is nonnegative in the order $\preceq'_q$. Thus by Proposition 3.9 we know that $\Psi(Z) \preceq'_q \Psi(Z')$ holds for n-dimensional zonotopes. Now the theorem follows from Propositions 3.8. $\square$

4. Concluding Remarks

In the view of the lifting technique in [Ehrenborg 2005], it is natural to consider the following conjecture.

CONJECTURE 4.1. *Let H be a $\boldsymbol{cd}$-polynomial homogeneous of degree k such that $\langle H \mid \Psi(P) \rangle \geq 0$ for all k-dimensional polytopes P. Then for all n-dimensional* *zonotopes (and more generally, the dual of the lattice of regions of an oriented matroid) the inequality*

$$\langle u \cdot H \cdot v \mid \Psi(Z) - \Psi(C_n) \rangle \geq 0$$

*holds for all **cd**-monomials u and v such that the sum of their degrees is $n - k$, u does not end with c and v does not begin with c.*

Conjecture 4.1 is the zonotopal analogue of Conjecture 6.1 in [Ehrenborg 2005]. Theorem 3.1 is the verification of Conjecture 4.1 in the case when $H = q - \Delta_q \cdot c^k$. Moreover, in the light of Theorem 3.6 we also suggest the next conjecture.

CONJECTURE 4.2. *Let H be a **cd**-polynomial homogeneous of degree k such that for all k-dimensional zonotopes Z (and more generally, the dual of the lattice of regions of an oriented matroid) the inequality $\langle H \mid \Psi(Z) - \Psi(C_k) \rangle \geq 0$ holds. Then for all n-dimensional zonotopes (oriented matroids) the inequality*

$$\langle H \cdot v \mid \Psi(Z) - \Psi(C_n) \rangle \geq 0$$

*holds for all **cd**-monomials v of degree $n - k$.*

There are other natural questions that arise. For instance, is there a way to interpolate between Theorems 3.1 and 3.6? Such an interpolation would let the factor vary between the constants Δ_q and $\square_q$, depending on the degree of the monomial u. Another inequality to consider is the following multiplicative version of Theorem 3.1:

CONJECTURE 4.3. *The **cd**-index of a zonotope Z (and more generally, the dual of the lattice of regions of an oriented matroid) satisfies the inequality*

$$\frac{\langle uqv \mid \Psi(Z) \rangle}{\langle uc^k v \mid \Psi(Z) \rangle} \geq \frac{\langle uqv \mid \Psi(C_n) \rangle}{\langle uc^k v \mid \Psi(C_n) \rangle}.$$

More linear inequalities for the flag f-vector of zonotopes can be obtained by the Kalai convolution [1988]. That is, if the two inequalities $\langle H_1 \mid \Psi(Z) \rangle \geq 0$ and $\langle H_2 \mid \Psi(P) \rangle \geq 0$ hold for all m-dimensional zonotopes, respectively all n-dimensional polytopes, then the inequality $\langle H_1 * H_2 \mid \Psi(Z) \rangle \geq 0$ holds for all $(m+n+1)$-dimensional zonotopes. For an explicit description of the convolution on **cd**-polynomials, see [Ehrenborg 2005, Proposition 2.2].

Finally, another class of linear inequalities for the flag f-vector of zonotopes have been obtained by Varchenko and Liu; see [Fukuda et al. 1991; Liu 1995; Varchenko 1988]. Recently, this class has been sharpened by Stenson [2003].

Acknowledgements

I would like to thank the MIT Mathematics Department, where this research was initiated while the author was a Visiting Scholar, for their kind support. The author also thanks Margaret Readdy for many helpful discussions.

References

[Bayer 2001] M. M. Bayer, "Signs in the *cd*-index of Eulerian partially ordered sets", *Proc. Amer. Math. Soc.* **129**:8 (2001), 2219–2225.

[Bayer and Billera 1985] M. M. Bayer and L. J. Billera, "Generalized Dehn–Sommerville relations for polytopes, spheres and Eulerian partially ordered sets", *Invent. Math.* **79**:1 (1985), 143–157.

[Bayer and Ehrenborg 2000] M. M. Bayer and R. Ehrenborg, "The toric *h*-vectors of partially ordered sets", *Trans. Amer. Math. Soc.* **352**:10 (2000), 4515–4531.

[Bayer and Hetyei 2001] M. M. Bayer and G. Hetyei, "Flag vectors of Eulerian partially ordered sets", *European J. Combin.* **22**:1 (2001), 5–26.

[Bayer and Klapper 1991] M. M. Bayer and A. Klapper, "A new index for polytopes", *Discrete Comput. Geom.* **6**:1 (1991), 33–47.

[Billera and Ehrenborg 2000] L. J. Billera and R. Ehrenborg, "Monotonicity of the cd-index for polytopes", *Math. Z.* **233**:3 (2000), 421–441.

[Billera and Hetyei 2000] L. J. Billera and G. Hetyei, "Linear inequalities for flags in graded partially ordered sets", *J. Combin. Theory Ser. A* **89**:1 (2000), 77–104.

[Billera and Liu 2000] L. J. Billera and N. Liu, "Noncommutative enumeration in graded posets", *J. Algebraic Combin.* **12**:1 (2000), 7–24.

[Billera et al. 1997] L. J. Billera, R. Ehrenborg, and M. Readdy, "The *c-2d*-index of oriented matroids", *J. Combin. Theory Ser. A* **80**:1 (1997), 79–105.

[Billera et al. 1998] L. J. Billera, R. Ehrenborg, and M. Readdy, "The *cd*-index of zonotopes and arrangements", pp. 23–40 in *Mathematical essays in honor of Gian-Carlo Rota* (Cambridge, MA, 1996), edited by B. E. Sagan and R. P. Stanley, Progr. Math. **161**, Birkhäuser, Boston, 1998.

[Björner 1980] A. Björner, "Shellable and Cohen–Macaulay partially ordered sets", *Trans. Amer. Math. Soc.* **260**:1 (1980), 159–183.

[Ehrenborg 2001] R. Ehrenborg, "*k*-Eulerian posets", *Order* **18**:3 (2001), 227–236.

[Ehrenborg 2005] R. Ehrenborg, "Lifting inequalities for polytopes", *Adv. Math.* **193** (2005), 205–222.

[Ehrenborg and Fox 2003] R. Ehrenborg and H. Fox, "Inequalities for **cd**-indices of joins and products of polytopes", *Combinatorica* **23**:3 (2003), 427–452.

[Ehrenborg and Readdy 1996] R. Ehrenborg and M. Readdy, "The **r**-cubical lattice and a generalization of the **cd**-index", *European J. Combin.* **17**:8 (1996), 709–725.

[Ehrenborg and Readdy 1998] R. Ehrenborg and M. Readdy, "Coproducts and the *cd*-index", *J. Algebraic Combin.* **8**:3 (1998), 273–299.

[Ehrenborg and Readdy 2002] R. Ehrenborg and M. Readdy, "Homology of Newtonian coalgebras", *European J. Combin.* **23**:8 (2002), 919–927.

[Fukuda et al. 1991] K. Fukuda, S. Saito, and A. Tamura, "Combinatorial face enumeration in arrangements and oriented matroids", *Discrete Appl. Math.* **31**:2 (1991), 141–149.

[Kalai 1987] G. Kalai, "Rigidity and the lower bound theorem, I", *Invent. Math.* **88**:1 (1987), 125–151.

[Kalai 1988] G. Kalai, "A new basis of polytopes", *J. Combin. Theory Ser. A* **49**:2 (1988), 191–209.

[Karu 2001] K. Karu, "Hard Lefschetz theorem for nonrational polytopes", Technical report, 2001. Available at math.AG/0112087.

[Liu 1995] N. Liu, *Algebraic and combinatorial methods for face enumeration in polytopes*, Ph.D. thesis, Cornell University, Ithaca, NY, 1995.

[Purtill 1993] M. Purtill, "André permutations, lexicographic shellability and the *cd*-index of a convex polytope", *Trans. Amer. Math. Soc.* **338**:1 (1993), 77–104.

[Stanley 1974] R. P. Stanley, "Finite lattices and Jordan-Hölder sets", *Algebra Universalis* **4** (1974), 361–371.

[Stanley 1986] R. P. Stanley, *Enumerative combinatorics*, vol. I, Wadsworth and Brooks/Cole, Monterey, CA, 1986.

[Stanley 1987] R. Stanley, "Generalized *H*-vectors, intersection cohomology of toric varieties, and related results", pp. 187–213 in *Commutative algebra and combinatorics* (Kyoto, 1985), edited by M. Nagata and H. Matsumura, Adv. Stud. Pure Math. **11**, North-Holland, Amsterdam and Kinokuniya, Tokyo, 1987.

[Stanley 1994a] R. P. Stanley, "Flag *f*-vectors and the *cd*-index", *Math. Z.* **216**:3 (1994), 483–499.

[Stanley 1994b] R. P. Stanley, "A survey of Eulerian posets", pp. 301–333 in *Polytopes: abstract, convex and computational* (Scarborough, ON, 1993), edited by T. Bisztriczky et al., NATO Adv. Sci. Inst. Ser. C Math. Phys. Sci. **440**, Kluwer Acad. Publ., Dordrecht, 1994.

[Stenson 2003] C. Stenson, "Families of tight inequalities for polytopes", Technical report, 2003.

[Varchenko 1988] A. N. Varchenko, "The numbers of faces of a configuration of hyperplanes", *Dokl. Akad. Nauk SSSR* **302**:3 (1988), 527–530. In Russian; translation in *Soviet Math. Dokl.* **38** (1989), 291–295.

RICHARD EHRENBORG
DEPARTMENT OF MATHEMATICS
UNIVERSITY OF KENTUCKY
LEXINGTON, KY 40506
UNITED STATES
jrge@ms.uky.edu

Combinatorial and Computational Geometry
MSRI Publications
Volume **52**, 2005

Quasiconvex Programming

DAVID EPPSTEIN

ABSTRACT. We define *quasiconvex programming*, a form of generalized linear programming in which one seeks the point minimizing the pointwise maximum of a collection of quasiconvex functions. We survey algorithms for solving quasiconvex programs either numerically or via generalizations of the dual simplex method from linear programming, and describe varied applications of this geometric optimization technique in meshing, scientific computation, information visualization, automated algorithm analysis, and robust statistics.

1. Introduction

Quasiconvex programming is a form of geometric optimization, introduced in [Amenta et al. 1999] in the context of mesh improvement techniques and since applied to other problems in meshing, scientific computation, information visualization, automated algorithm analysis, and robust statistics [Bern and Eppstein 2001; 2003; Chan 2004; Eppstein 2004]. If a problem can be formulated as a quasiconvex program of bounded dimension, it can be solved algorithmically in a linear number of constant-complexity primitive operations by generalized linear programming techniques, or numerically by generalized gradient descent techniques. In this paper we survey quasiconvex programming algorithms and applications.

1.1. Quasiconvex functions. Let Y be a totally ordered set, for instance the real numbers $\mathbb{R}$ or integers $\mathbb{Z}$ ordered numerically. For any function $f : X \mapsto Y$, and any value $\lambda \in Y$, we define the *lower level set*

$$f^{\leq \lambda} = \{x \in X \mid f(x) \leq \lambda\}.$$

A function $q : X \mapsto Y$, where X is a convex subset of $\mathbb{R}^d$, is called *quasiconvex* [Dharmadhikari and Joag-Dev 1988] when its lower level sets are all convex. A one-dimensional quasiconvex function is more commonly called *unimodal*, and

This research was supported in part by NSF grant CCR-9912338.

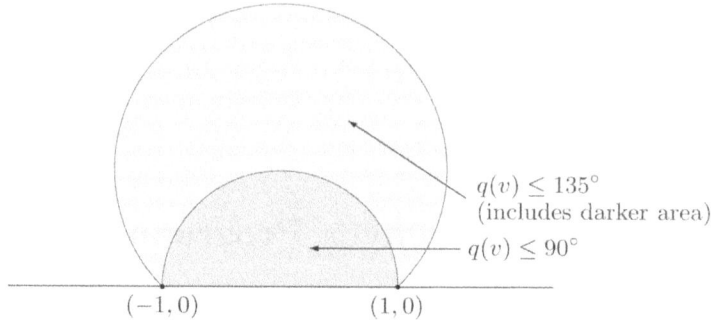

Figure 1. Level sets of the quasiconvex function $q(v) = 180° - \angle uvw$, for $u = (-1, 0)$ and $w = (1, 0)$, restricted to the half-plane $y \geq 0$.

another way to define a quasiconvex function is that it is unimodal along any line through its domain.

As an example, let $H = \{(x, y) \mid y > 0\}$ be the upper half-plane in $\mathbb{R}^2$, let $u = (-1, 0)$ and $w = (1, 0)$, and let q measure the angle complementary to the one subtended by segment uw from point v: thus $q(v) = 180° - \angle uvw$. Each level set $q^{\leq \lambda}$ consists of the intersection with H of a disk having u and w on its boundary (Figure 1). Since these sets are all convex, q is quasiconvex.

Quasiconvex functions are a generalization of the well-known set of *convex functions*, which are the functions $\mathbb{R}^d \mapsto \mathbb{R}$ satisfying the inequality

$$f\big(p\bar{x} + (1 - p)\bar{y}\big) \leq p\, f(\bar{x}) + (1 - p)f(\bar{y})$$

for all $\bar{x}, \bar{y} \in \mathbb{R}^d$ and all $0 \leq p \leq 1$: it is a simple consequence of this inequality that any convex function has convex lower level sets. However, there are many functions that are quasiconvex but not convex; for instance, the complementary angle function q defined above is not convex, as can be seen from the fact that its values are bounded above by $180°$. As another example, the function $\chi_K(\bar{x})$ that takes the value 0 within a convex set K and 1 outside K has as its lower level sets K and $\mathbb{R}^d$, so it is quasiconvex, but not convex.

If r is convex or quasiconvex and $f : Y \mapsto Z$ is monotonically nondecreasing, then $q(\bar{x}) = f(r(\bar{x}))$ is quasiconvex; for instance the function χ_K above can be factored in this way into the composition of a convex function $d_K(\bar{x})$ measuring the Euclidean distance from $\bar{x}$ to K with a monotonic function f mapping 0 to itself and all larger values to 1. In the other direction, given a quasiconvex function $q : X \mapsto Y$, one can often find a monotonic function $f : Y \mapsto \mathbb{R}$ that, when composed with q, turns it into a convex function. However this sort of convex composition is not always possible. For instance, in the case of the step function χ_K described above, any nonconstant composition of χ_K remains two-valued and hence cannot be convex.

1.2. Nested convex families. Quasiconvex functions are closely related to *nested convex families*. Following [Amenta et al. 1999], we define a nested convex family to be a map $\kappa : Y \mapsto K(\mathbb{R}^d)$, where Y is a totally ordered set and $K(\mathbb{R}^d)$ denotes the family of compact convex subsets of $\mathbb{R}^d$, and where κ is further required to satisfy the following two axiomatic requirements (the second of which is a slight generalization of the original definition that allows Y to be discrete):

(i) For every $\lambda_1, \lambda_2 \in Y$ with $\lambda_1 < \lambda_2$ we have $\kappa(\lambda_1) \subseteq \kappa(\lambda_2)$.
(ii) For all $\lambda \in Y$ such that $\lambda = \inf\{\lambda' \mid \lambda' > \lambda\}$ we have $\kappa(\lambda) = \bigcap_{\lambda' > \lambda} \kappa(\lambda')$.

If Y has the property that every subset of Y has an infimum (for instance, $Y = \mathbb{R} \cup \{\infty, -\infty\}$), then from any nested convex family $\kappa : Y \mapsto K(\mathbb{R}^d)$ we can define a function $q_\kappa : \mathbb{R}^d \mapsto Y$ by the formula

$$q_\kappa(\bar{x}) = \inf\{\lambda \mid \bar{x} \in \kappa(\lambda)\}.$$

LEMMA 1.1. *For any nested convex family $\kappa : Y \mapsto K(\mathbb{R}^d)$ and any $\lambda \in Y$,* $q_\kappa^{\leq \lambda} = \kappa(\lambda)$.

PROOF. The lower level sets of q_κ are

$$q_\kappa^{\leq \lambda} = \{\bar{x} \in R^d \mid q_\kappa(\bar{x}) \leq \lambda\} = \{\bar{x} \in R^d \mid \inf\{\lambda' \mid \bar{x} \in \kappa(\lambda')\} \leq \lambda\}.$$

For any $\bar{x} \in \kappa(\lambda)$ we have $\lambda \in \{\lambda' \mid \bar{x} \in \kappa(\lambda')\}$ so the infimum of this set can not be greater than λ and $\bar{x} \in q_\kappa^{\leq \lambda}$. For any $\bar{x} \notin \kappa(\lambda)$, $\inf\{\lambda' \mid \bar{x} \in \kappa(\lambda')\} \geq \lambda^+ > \lambda$ by the second property of nested convex families, so $\bar{x} \notin q_\kappa^{\leq \lambda}$. Therefore, $q_\kappa^{\leq \lambda} = \kappa(\lambda)$.   □

In particular, q_κ has convex lower level sets and so is quasiconvex.

Conversely, suppose that q is quasiconvex and has bounded lower level sets. Then we can define a nested convex family

$$\kappa_q(\lambda) = \begin{cases} \bigcap_{\lambda' > \lambda} \mathrm{cl}(q^{\leq \lambda'}) & \text{if } \lambda = \inf\{\lambda' \mid \lambda' > \lambda\}, \\ \mathrm{cl}(q^{\leq \lambda}) & \text{otherwise}, \end{cases}$$

where cl denotes the topological closure operation.

If q does not have bounded lower level sets, we can still form a nested convex family by restricting our attention to a compact convex subdomain $K \subset \mathbb{R}^d$:

$$\kappa_{q,K}(\lambda) = \begin{cases} \bigcap_{\lambda' > \lambda} \mathrm{cl}(K \cap q^{\leq \lambda'}) & \text{if } \lambda = \inf\{\lambda' \mid \lambda' > \lambda\}, \\ \mathrm{cl}(K \cap q^{\leq \lambda}) & \text{otherwise}. \end{cases}$$

This restriction to a compact subdomain is necessary to handle linear functions and other functions without bounded level sets within our mathematical framework.

The following two theorems allow us to use nested convex families and quasiconvex functions interchangeably for each other for most purposes: more specifically, a nested convex family conveys exactly the same information as a continuous quasiconvex function with bounded lower level sets. Thus, later, we will use whichever of the two notions is more convenient for the purposes at hand, using these theorems to replace an object of one type for an object of the other in any algorithms or lemmas needed for our results.

THEOREM 1.2. *For any nested convex family κ, we have $\kappa = \kappa_{q_\kappa}$.*

PROOF. If λ is not an infimum of larger values, then $q_\kappa(x) \le \lambda$ if and only if $x \in \kappa(\lambda)$. So $\kappa_{q_\kappa}(\lambda) = \mathrm{cl}(q_\kappa^{\le\lambda}) = \{x \mid q_\kappa(x) \le \lambda\} = \kappa(\lambda)$.

Otherwise, by Lemma 1.1,

$$\kappa_{q_\kappa}(\lambda) = \bigcap_{\lambda' > \lambda} \mathrm{cl}(\kappa(\lambda')).$$

The closure operation does not modify the set $\kappa(\lambda')$, because it is already closed, so we can replace $\mathrm{cl}(\kappa(\lambda'))$ above by $\kappa(\lambda'))$, giving

$$\kappa_{q_\kappa}(\lambda) = \bigcap_{\lambda' > \lambda} \kappa(\lambda').$$

The intersection on the right-hand side of the equation further simplifies to $\kappa(\lambda)$ by the second property of nested convex families. □

THEOREM 1.3. *If $q : X \mapsto \mathbb{R}$ is a continuous quasiconvex function with bounded lower level sets, then $q_{\kappa_q} = q$.*

PROOF. By Lemma 1.1, $q_{\kappa_q}^{\le\lambda} = \kappa_q(\lambda)$. Assume first that $\lambda = \inf\{\lambda' \mid \lambda' > \lambda\}$. Expanding the definition of κ_q, we get

$$q_{\kappa_q}^{\le\lambda} = \bigcap_{\lambda' > \lambda} \mathrm{cl}(q^{\le\lambda'}).$$

If q is continuous, its level sets are closed, so we can simplify this to

$$q_{\kappa_q}^{\le\lambda} = \bigcap_{\lambda' > \lambda} q^{\le\lambda'}.$$

Suppose the intersection on the right-hand side of the formula is nonempty, and let $\bar{x}$ be any point in it. We wish to show that $q(\bar{x}) \le \lambda$, so suppose for a contradiction that $q(\bar{x}) > \lambda$. But then there is a value λ' strictly between λ and $q(\bar{x})$ (else λ would not be the infimum of all greater values), and $\bar{x} \notin q^{\le\lambda'}$, contradicting the assumption that $\bar{x}$ is in the intersection. Therefore, $q(\bar{x})$ must be at most equal to λ.

As we have now shown that $q(\bar{x}) \le \lambda$ for any $\bar{x}$ in $q_{\kappa_q}^{\le\lambda}$, it follows that $q_{\kappa_q}^{\le\lambda}$ cannot contain any points outside $q^{\le\lambda}$. On the other hand, $q_{\kappa_q}^{\le\lambda}$ is formed by intersecting a collection of supersets of $q^{\le\lambda}$, so it contains all points inside $q^{\le\lambda}$. Therefore, the two sets are equal.

If $\lambda \ne \inf\{\lambda' > \lambda\}$, the same equality can be seen even more simply to be true, since we have no intersection operation to eliminate. Since q_{κ_q} and q have the same level sets, they are the same function. □

Thanks to these two theorems, we do not lose any information by using the function q_κ in place of the nested convex family κ, or by using the nested convex family $\kappa_{q_\kappa} = \kappa$ in place of a quasiconvex function that is of the form $q = q_\kappa$ or in place of a continous quasiconvex function with bounded lower level sets. In most situations quasiconvex functions and nested convex families can be treated as equivalent and interchangeable: if we are given a quasiconvex function q and need a nested convex family, we can use the family κ_q, and if we are given a nested convex family κ and need a quasiconvex function, we can use the function q_κ or $q_{\kappa,K}$. Our quasiconvex programs' formal definition will involve inputs that are nested convex families only, but in our applications of quasiconvex programming we will describe inputs that are quasiconvex functions, and which will be assumed to be converted to nested convex families as described above.

1.3. Quasiconvex programs. If a finite set of functions q_i are all quasiconvex and have the same domain and range, then the function $Q(\bar{x}) = \max_{i \in S} q_i(\bar{x})$ is also quasiconvex, and it becomes of interest to find a point where Q achieves its minimum value. For instance, in Section 2.2 below we discuss in more detail the smallest enclosing ball problem, which can be defined by a finite set of functions q_i, each of which measures the distance to an input site; the minimum of Q marks the center of the smallest enclosing ball of the sites. Informally, we use *quasiconvex programming* to describe this search for the point minimizing the pointwise maximum of a finite set of quasiconvex functions.

More formally, Amenta et al. [1999] originally defined a *quasiconvex program* to be formed by a set of nested convex families $S = \{\kappa_1, \kappa_2, \ldots \kappa_n\}$; the task to be solved is finding the value

$$\Lambda(S) = \inf \left\{ (\lambda, \bar{x}) \;\middle|\; \bar{x} \in \bigcap_{\kappa_i \in S} \kappa_i(\lambda) \right\}$$

where the infimum is taken in the lexicographic ordering, first by λ and then by the coordinates of $\bar{x}$. However, we can simplify the infimum operation in this definition by replacing it with a minimum; that is, it is always true that the set defined on the right-hand side of the definition has a least point $\Lambda(S)$. To prove this, suppose that $(\lambda, \bar{x})$ is the infimum, that is, there is a sequence of pairs $(\lambda_j, \bar{x}_j)$ in the right-hand side intersection that converges to $(\lambda, \bar{x})$, and $(\lambda, \bar{x})$ is the smallest pair with this property. Clearly, each $\lambda_j \geq \lambda$ (else $(\lambda_j, \bar{x}_j)$ would be a better solution) and it follows from the fact that the sets κ_i are closed and nested that we can take each $\bar{x}_j = \bar{x}$. But then, it follows from the second property of nested convex families that $\bar{x} \in \kappa_i(\lambda)$ for all $\kappa_i \in S$.

In terms of the quasiconvex functions defining a quasiconvex program, we would like to say that the value of the program consists of a pair $(\lambda, \bar{x})$ such that, for each input function q_i, $q_i(\bar{x}) \leq \lambda$, and that no other pair with the same property has a smaller value of λ. However, $\max_i q_i(\bar{x})$ may not equal λ if at least one of the input quasiconvex functions is discontinuous. For instance, consider a one-dimensional quasiconvex program with two functions $q_0(x) = |x|$, $q_1(x) = 1$ for $x \geq 0$, and $q_1(x) = 0$ for $x < 0$. This program has value $(0,0)$, but $\max\{q_0(0), q_1(0)\} = 1$. The most we can say in general is that there exists a sequence of points $\bar{x}_j$ converging to x with $\lim_{j\to\infty} \max_i q_i(\bar{x}_j) = \lambda$. This technicality is, however, not generally a problem in our applications.

In subsequent sections we explore various examples of quasiconvex programs, algorithms for quasiconvex programming, and applications of those algorithms.

(The expression *quasiconvex programming* has also been applied to the problem of minimizing a single quasiconvex function over a convex domain; see [Kiwiel 2001; Xu 2001], for example. The two formulations are easily converted to each other using the ideas described in Section 2.6. For the applications described in this survey, we prefer the formulation involving minimizing the pointwise maximum of multiple quasiconvex functions, as it places greater emphasis on combinatorial algorithms and less on numerical optimization.)

2. Examples

We begin our study of quasiconvex programming by going through some simple examples of geometric optimization problems, and showing how they may be formulated as low-dimensional quasiconvex programs.

2.1. Sighting point. When we introduced the definition of quasiconvex functions, we used as an example the complementary angle subtended by a line segment from a point: $q(v) = 180° - \angle uvw$. If we have a collection of line segments forming a star-shaped polygon, and form a quasiconvex program from the functions corresponding to each line segment, then the point v that minimizes the maximum function value must lie in the kernel of the polygon. If we define the *angular resolution* of the polygon from v to be the minimum angle formed by any two consecutive vertices as seen from v, then this choice of v makes the angular resolution be as large as possible.

This problem of maximizing the angular resolution was used by Matoušek et al. [1996] as an example of an LP-type problem that does not form a convex program. It can also be viewed as a special case of the mesh smoothing application described below in Section 4.1.

McKay [1989] had asked about a similar problem in which one wishes to choose a viewpoint maximizing the angular resolution of an unordered set of points that is not connected into a star-shaped polygon. However, it does not seem possible to form a quasiconvex program from this version of the problem: for star-shaped polygons, we know on which side of each line segment the optimal point must lie, so we can use quasiconvex functions with level sets that are intersections of disks and half-planes, but for point sets, without knowing where the viewpoint lies with respect to the line through any pair of points, we need to use the absolute value $|q(v)|$ of the angle formed at v by each pair of points. This modification

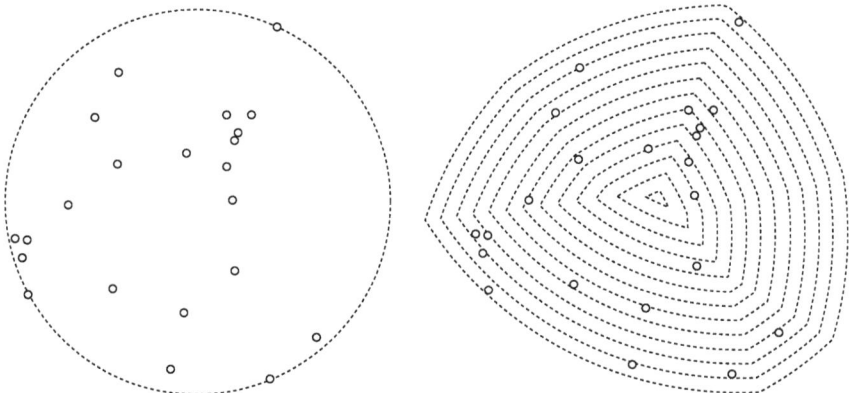

Figure 2. Smallest enclosing ball of a set of points (left), and the level sets of $\max_i q_i(x)$ for the distance functions q_i defining the quasiconvex program for the smallest enclosing ball (right).

leads to nonquasiconvex functions with level sets that are unions or intersections of two disks. It remains open whether an efficient algorithm for McKay's sighting point problem exists.

2.2. Smallest enclosing ball. Consider the problem of finding the minimum radius Euclidean sphere that encloses all of a set of points $S = \{\bar{p}_i\} \subset \mathbb{R}^d$ (Figure 2, left). As we show below, this smallest enclosing ball problem can easily be formulated as a quasiconvex program. The smallest enclosing ball problem has been well studied and linear time algorithms are known in any fixed dimension [Dyer 1984; Fischer et al. 2003; Gärtner 1999; Megiddo 1983; Welzl 1991], so the quasiconvex programming formulation does not lead to improved solutions for this problem, but it provides an illuminating example of how to find such a formulation more generally, and in later sections we will use the smallest enclosing ball example to illustrate our quasiconvex programming algorithms.

Define the function $q_i(\bar{x}) = d(\bar{x}, \bar{p}_i)$ where d is the Euclidean distance. Then the level set $q_i^{\leq \lambda}$ is simply a Euclidean ball of radius λ centered at $\bar{p}_i$, so q_i is quasiconvex (in fact, it is convex). The function $q_S(\bar{x}) = \max_i q_i(\bar{x})$ (the level sets of which are depicted in Figure 2, right) measures the maximum distance from $\bar{x}$ to any of the input points, so a Euclidean ball of radius $q_S(\bar{x})$ centered at $\bar{x}$ will enclose all the points and is the smallest ball centered at $\bar{x}$ that encloses all the points.

If we form a quasiconvex program from the functions q_i, the solution to the program consists of a pair $(\lambda, \bar{x})$ where $\lambda = q_S(\bar{x})$ and λ is as small as possible. That is, the ball with radius λ centered at $\bar{x}$ is the smallest enclosing ball of the input points.

Any smallest enclosing ball problem has a *basis* of at most $d + 1$ points that determine its value. More generally, it will turn out that any quasiconvex program's value is similarly determined by a small number of the input functions;

this phenomenon will prove central in our ability to apply generalized linear programming algorithms to solve quasiconvex programs.

If we generalize each q_i to be the Euclidean distance to a convex set K_i, the resulting quasiconvex program finds the smallest sphere that touches or encloses each K_i. In a slightly different generalization, if we let $q_i(\bar{x}) = d(\bar{x}, \bar{p}_i) + r_i$, a sphere centered at $\bar{x}$ with radius $q_i(\bar{x})$ or larger will contain the sphere centered at $\bar{p}_i$ with radius r_i. So, solving the quasiconvex program with this family of functions q_i will find the smallest enclosing ball of a family of balls [Megiddo 1989; Gärtner and Fischer 2003].

2.3. Hyperbolic smallest enclosing ball.

Although we have defined quasiconvex programming in terms of Euclidean space $\mathbb{R}^n$, the definition involves only concepts such as convexity that apply equally well to other geometries such as hyperbolic space $\mathbb{H}^n$. Hyperbolic geometry (e.g. see [Iversen 1992]) may be defined in various ways; for instance by letting $\mathbb{H}^n$ consist of the unit vectors of $\mathbb{R}^{n+1}$ according to the inner product $\langle \bar{x}, \bar{y} \rangle = \sum_{i<n}(x_i y_i) - x_n y_n$, and defining the distance $d(\bar{x}, \bar{y}) = \cosh^{-1} \langle \bar{x}, \bar{y} \rangle$. Angles, congruence, lines, hyperplanes, and other familiar Euclidean concepts can also be defined in a straightforward way for hyperbolic space. Hyperbolic geometry satisfies many of the same axioms as Euclidean geometry, but not the famous *parallel postulate*: in the hyperbolic plane $\mathbb{H}^2$, given a line ℓ and a point $p \notin \ell$, there will be infinitely many lines through p that do not meet ℓ. A hyperbolic convex set K is defined as in Euclidean space to be one in which, for any two points $\{p, q\} \subset K$, all points on the line segment connecting p to q also belong to K. Similarly, a quasiconvex function $\mathbb{H}^n \mapsto \mathbb{R}$ is one for which all lower level sets are convex, or equivalently one that is unimodal on any line in $\mathbb{H}^n$. As in the Euclidean case we may define a hyperbolic quasiconvex program to be the problem of searching for the point minimizing the pointwise maximum of a collection of hyperbolic quasiconvex functions.

There are several standard ways of representing the points and other geometric objects of Hyperbolic space within a Euclidean space, of which the two best known are the Poincaré and Klein models (Figure 3). In the Poincaré model, the points of $\mathbb{H}^n$ are represented as Euclidean points interior to an n-dimensional unit ball or half-space, and lines of $\mathbb{H}^n$ are represented as arcs of circles that meet the boundary of this unit ball or half-space perpendicularly. In this model, the hyperbolic angle between two objects in $\mathbb{H}^n$ is equal to the Euclidean angle between the models of those objects, and hyperbolic circles and spheres are modeled by Euclidean circles and spheres; however, hyperbolic distances do not equal distances within the Poincaré model, and objects that are straight or flat hyperbolically may have curved models. In the Klein model, again, points of $\mathbb{H}^n$ are represented as Euclidean points interior to an n-dimensional unit ball, but the hyperbolic line connecting two points is represented as the restriction to the ball of the Euclidean line connecting the models of those points. In this model,

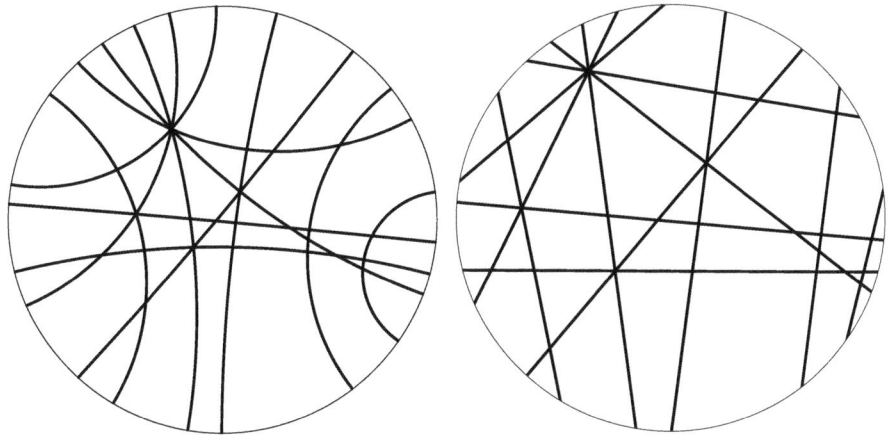

Figure 3. Poincaré (left) and Klein (right) models of the hyperbolic plane. Both models show the same hyperbolic arrangement of lines; analogous models exist for any higher dimensional hyperbolic space. Figure taken from [Bern and Eppstein 2001].

angles and distances may be distorted but straightness is preserved: a straight or flat hyperbolic object will have a straight or flat model. In particular, since the definition of convexity involves only straight line segments, a convex hyperbolic object will have a convex Klein model and vice versa. The Poincaré and Klein models for a hyperbolic space are not uniquely defined, as one may choose any hyperbolic point to be modeled by the center of the Euclidean unit ball, and that ball may rotate arbitrarily around its center.

If we let k be a function mapping $\mathbb{H}^n$ to a Klein model in $\mathbb{R}^n$, and if each $q_i(\bar{x})$ is a hyperbolic quasiconvex function, then $\hat{q}_i(\bar{x}) = q_i(k^{-1}(\bar{x}))$ is a Euclidean quasiconvex function. More, $\hat{q}_i$ has bounded lower levels sets since they are all subsets of the unit ball. Let $(\lambda, \bar{x})$ be the solution to the Euclidean quasiconvex program defined by the set of functions $\hat{q}_i$. Then, if $\bar{x}$ is interior to the unit ball defining the Klein model, $(\lambda, k^{-1}(\bar{x}))$ is the solution to the hyperbolic quasiconvex program defined by the original functions q_i. On the other hand, $\bar{x}$ may be on the boundary of the Klein model; if so, $\bar{x}$ may be viewed as an infinite point of the hyperbolic space, and is the limit of sequence of points within the space with monotonically decreasing values. The latter possibility, of an infinite solution to the quasiconvex program, can only occur if some of the hyperbolic quasiconvex functions have unbounded lower level sets. Therefore, as noted in [Bern and Eppstein 2001], hyperbolic quasiconvex programs may in general be solved as easily as their Euclidean counterparts.

As an example, consider the problem of finding the hyperbolic ball of minimum radius containing all of a collection of hyperbolic points $\bar{p}_i$. As in the Euclidean case, we can define $q_i(\bar{x})$ to be the (hyperbolic) distance from $\bar{x}$ to $\bar{p}_i$; this function has convex hyperbolic balls as its level sets, so it is quasiconvex. And, just as in the Euclidean case, the solution to the quasiconvex program defined by the functions q_i is the pair $(\lambda, \bar{x})$ where the hyperbolic ball of radius λ centered at $\bar{x}$ is the minimum enclosing ball of the points $\bar{p}_i$.

2.4. Optimal illumination. Suppose that we have a room (modeled as a possibly nonconvex three-dimensional polyhedron) and wish to place a point source of light in order to light up the whole room as brightly as possible: that is, we wish to maximize the minimum illumination received on any point of the room's surface. The quasiconvex programs we are studying solve min-max rather than max-min problems, but that is easily handled by negating the input functions.

So, let $q_i(\bar{x})$ be the negation of the intensity of light received at point i of the room's surface, as a function of $\bar{x}$, the position of the light source. It is not hard to see that, within any face of the polyhedron, the light intensity is least at some vertex of the face, since those are the points at maximal distance from the light source and with minimal angle to it. Therefore, we need only consider a finite number of possibilities for i: one for each pair (f, v) where f is a face of the polyhedron and v is a vertex of f. For each such pair, we can compute q_i via a simple formula of optics, $q_i(\bar{x}) = -\bar{u} \cdot (\bar{x} - v)/d(\bar{x}, v)^3$, where d is as usual the Euclidean distance, and u is a unit vector facing inwards at a perpendicular angle to f. In this formula, one factor $\bar{u} \cdot (\bar{x} - v)/d(\bar{x}, v)$ accounts for the angle of incidence of light from the source onto the part of face v near vertex f, while the other factor $1/d(\bar{x}, v)^2$ accounts for the inverse-square rule for falloff of light from a point source in three-dimensional space. Note that we can neglect occlusions from other faces in this formula, because, if some face is occluded, then at least one other face will be facing away from the light source and entirely unilluminated; this unilluminated face will dominate the occluded one in our min-max optimization.

In [Amenta et al. 1999], as part of a proof of quasiconvexity of a more complex function used for smoothing three-dimensional meshes by solid angles, we showed that the function q_i defined above is quasiconvex; more precisely, we showed that $(-q_i(\bar{x}))^{-1/2}$ is a convex function of $\bar{x}$ by using *Mathematica* to calculate the principal determinants of its Hessian, and by showing from the structure of the resulting formulae that these determinants are always nonnegative. Therefore, we can express the problem of finding an optimal illumination point as a quasiconvex program.

2.5. Longest intersecting prefix. This example is due to Chan [2004]. Suppose we are given an ordered sequence of convex sets K_i, $0 \leq i < n$, that are all subsets of the same compact convex set $X \subset \mathbb{R}^d$. We would like to find the maximum value ℓ such that $\bigcap_{i<\ell} K_i \neq \varnothing$. That is, we would like to find the longest prefix of the input sequence, such that the convex sets in this prefix have a nonempty intersection (Figure 4).

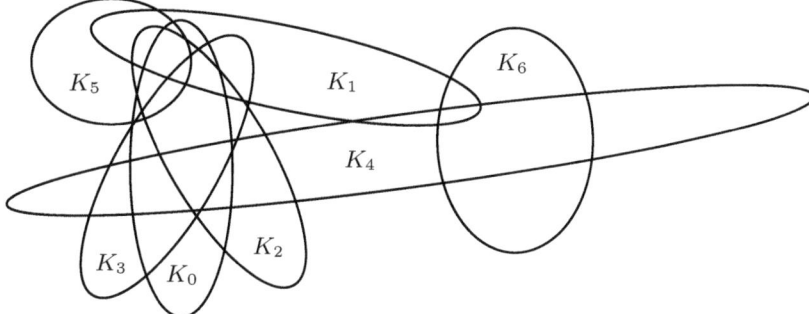

Figure 4. Instance of a longest intersecting prefix problem. The longest inter-secting prefix is (K_0, K_1, K_2, K_3).

To represent this as a quasiconvex program, define a nested convex family $\kappa_i : \mathbb{Z} \mapsto K(\mathbb{R}^d)$ for each set K_i in the sequence, as follows:

$$\kappa_i(\lambda) = \begin{cases} K_i, & \text{if } \lambda < -i \\ X, & \text{otherwise.} \end{cases}$$

The optimal value $(\lambda, \bar{x})$ for the quasiconvex program formed by this set of nested convex families has $\bar{x} \in \kappa_i(\lambda) = K_i$ for all $i < -\lambda$, so the prefix of sets with index up to (but not including) $-\lambda$ has a nonempty intersection containing $\bar{x}$. Since the quasiconvex program solution minimizes λ, $-\lambda$ is the maximum value with this property. That is, the first $-\lambda$ values of the sequence K_i form its longest intersecting prefix.

More generally, the same technique applies equally well when each of the convex sets K_i has an associated value k_i, and we must find the maximum value ℓ such that $\bigcap_{k_i < \ell} K_i \neq \varnothing$. The longest intersecting prefix problem can be seen as a special case of this problem in which the values k_i form a permutation of the integers from 0 to $n - 1$. We will see an instance of this generalized longest intersecting prefix problem, in which the values k_i are integers with some repeated values, when we describe Chan's solution to the Tukey median problem.

2.6. Linear, convex, quasiconvex. There are many ways of modeling linear programs, but one of the simplest is the following: a linear program is the search for a vector $\bar{x}$ that satisfies all of a set of closed linear inequalities $\bar{a}_i \cdot \bar{x} \geq b_i$ and that, among all such feasible vectors, minimizes a linear objective function $f(x) = \bar{c} \cdot \bar{x}$. The vectors $\bar{x}$, $\bar{a}_i$, and $\bar{c}$ all have the same dimension, which we call the dimension of the linear program. We typically use the symbol n to denote the number of inequalities in the linear program. It is often useful to generalize such programs somewhat, by keeping the linear constraints but allowing the objective function $f(x)$ to be convex instead of linear; such a generalization is known as

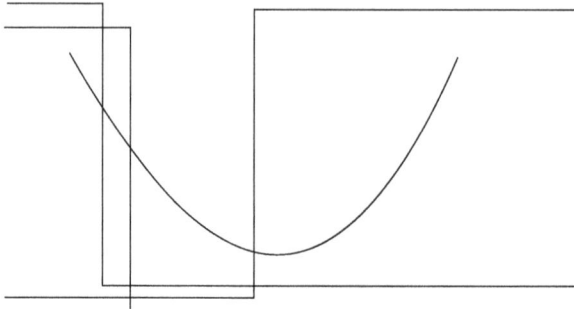

Figure 5. Conversion of convex program into quasiconvex program, by treating each half-space constraint as a quasiconvex step function.

a convex program, and many linear programming algorithms can be adapted to handle the convex case as well.

For instance, consider the following geometric problem, which arises in collision detection algorithms for maintaining simulations of virtual environments: we are given as input two k-dimensional convex bodies P and Q, specified as intersections of half-spaces $P = \bigcap P_i$ and $Q = \bigcap Q_i$; we wish to find the closest pair of points $\bar{p}, \bar{q}$ with $\bar{p} \in P$ and $\bar{q} \in Q$. If we view $\bar{p}, \bar{q}$ as forming a $2k$-dimensional vector $\bar{x}$, then each constraint $\bar{p} \in P_i$ or $\bar{q} \in Q_i$ is linear in $\bar{x}$, but the objective function $d(\bar{p}, \bar{q})$ is nonlinear: evaluating the distance using the Pythagorean formula results in a formula that is the square root of a sum of squares of differences of coordinates. We can square the formula to eliminate the square root, but what remains is a convex quadratic function. Thus, the closest distance problem can be expressed as a convex program; similar formulations are also possible when P and Q are expressed as convex hulls of their vertex sets [Matoušek et al. 1996].

These formulations seem somewhat different from our quasiconvex programming framework: in the linear and convex programming formulations above, we have a large set of constraints and a single objective function, while in quasiconvex programming we have many input functions that take a role more analogous to objectives than constraints. Nevertheless, as we now show, any linear or convex program can be modeled as a quasiconvex program. Intuitively, the idea is simply to treat each half-space constraint as a quasiconvex step function, and include them together with the convex objective functions in the set of quasiconvex functions defining a quasiconvex program (Figure 5).

THEOREM 2.1. *Suppose a convex program is specified by n linear inequalities $\bar{a}_i \cdot \bar{x} \geq b_i$ and a convex objective function $f(\bar{x})$, and suppose that the solution of this convex program is known to lie within a compact convex region K. Then we can find a set of $n+1$ nested convex families $\kappa_i(\lambda)$ such that the solution $(\lambda, \bar{x})$ of the quasiconvex program formed by these nested convex families is an optimal solution to the convex program, with $\lambda = f(\bar{x})$.*

PROOF. For each inequality $\bar{a}_i \cdot \bar{x} \geq b_i$ form a nested convex family

$$\kappa_i(\lambda) = K \cap \{\bar{x} \mid \bar{a}_i \cdot \bar{x} \geq b_i\};$$

that is, κ_i ignores its argument λ and produces a constant compact convex set of the points satisfying the ith inequality. Also form a nested convex family $\kappa_n = \kappa_{f,K}$ representing the objective function.

If $(\lambda, \bar{x})$ is the optimal solution to the quasiconvex program defined by the nested convex families κ_i, then $\bar{a}_i \cdot \bar{x} \geq b_i$ (else $\bar{x}$ would not be contained in $\kappa_i(\lambda)$) and $\lambda = f(\bar{x})$ (else either $\bar{x}$ would be outside $\kappa_n(\lambda)$ or the pair $(f(\bar{x}), \bar{x})$ would be a better solution). There could be no $\bar{y}$ satisfying all constraints $\bar{a}_i \cdot \bar{y} \geq b_i$ with $f(\bar{y}) < \lambda$, else $(f(\bar{y}), \bar{y})$ would be a better solution than $(\lambda, \bar{x})$ for the quasiconvex program. Therefore, $\bar{x}$ provides the optimal solution to the convex program as the result states.   □

The region K is needed for this result as a technicality, because our quasiconvex programming formulation requires the nested convex families to be compact. In practice, though, it is not generally difficult to find K; for instance, in the problem of finding closest distances between convex bodies, we could let K be a bounding box defined by extreme points of the convex bodies in each axis-aligned direction.

3. Algorithms

We now discuss techniques for solving quasiconvex programs, both numerically and combinatorially.

3.1. Generalized linear programming. Although linear programs can be solved in polynomial time, regardless of dimension [Karmarkar 1984; Khachiyan 1980], known results in this direction involve time bounds that depend not just on the number and dimension of the constraints, but also on the magnitude of the coordinates used to specify the constraints. In typical computational geometry applications the dimension is bounded but these magnitudes may not be, so there has been a long line of work on linear programming algorithms that take a linear amount of time in terms of the number of constraints, independent of the magnitude of coordinates, but possibly with an exponential dependence on the dimension of the problem [Adler and Shamir 1993; Chazelle and Matoušek 1993; Clarkson 1986; 1987; 1995; Dyer and Frieze 1989; Matoušek et al. 1996; Megiddo 1983; Megiddo 1984; 1991]. In most cases, these algorithms can be interpreted as *dual simplex methods*: as they progress, they maintain a *basis* of d constraints, and the point $\bar{x}$ optimizing the objective function subject to the constraints in the basis. At each step, the basis is replaced by another one with a worse value of $\bar{x}$; when no more basis replacement steps are possible, the correct solution has been found.

Very quickly, workers in this area realized that similar techniques could also be applied to certain nonlinear programs such as the minimum enclosing ball problem [Adler and Shamir 1993; Amenta 1994; Chazelle and Matoušek 1993; Clarkson 1995; Dyer 1984; 1992; Fischer et al. 2003; Gärtner 1995; 1999; Matoušek et al. 1996; Megiddo 1983; Post 1984; Welzl 1991]. One of the most popular and general formulations of this form of generalized linear program is the class of *LP-type problems* defined by Matoušek et al. [1996]; we follow the description of this formulation from [Amenta et al. 1999].

An LP-type problem consists of a finite set S of *constraints* and an *objective function* f mapping subsets of S to some totally ordered space and satisfying the following two properties:

(i) For any $A \subset B$, $f(A) \leq f(B)$.
(ii) For any A, p, and q,

$$f(A) = f(A \cup \{p\}) = f(A \cup \{q\}) \implies f(A) = f(A \cup \{p, q\}).$$

The problem is to compute $f(S)$ using only evaluations of f on small subsets of S.

For instance, in linear programming, S is a set of half-spaces and $f(S)$ is the point in the intersection of the half-spaces at which some linear function takes its minimum value. In the smallest enclosing ball problem, S consists of the points themselves, and $f(A)$ is the smallest enclosing ball of A, where the total ordering on balls is given by their radii. It is not hard to see that this system satisfies the properties above: removing points can only make the radius shrink or stay the same, and if a ball contains the additional points p and q separately it contains them both together.

A *basis* of an LP-type problem is a set B such that $f(A) < f(B)$ for any $A \subsetneq B$. Thus, due to the first property of an LP-type problem, the value of the overall problem is the same as the value of the *optimal basis*, the basis B that maximizes $f(B)$. The *dimension* of an LP-type problem is the maximum cardinality of any basis; although we have not included it above, a requirement that this dimension be bounded is often included in the definition of an LP-type problem. The dimension of an LP-type problem may differ from the dimension of some space $\mathbb{R}^d$ that may be associated in some way with the problem; for instance, for smallest enclosing balls in $\mathbb{R}^d$, the dimension of the LP-type problem turns out to be $d + 1$ instead of d.

As described in [Matoušek et al. 1996], efficient and simple randomized algorithms for bounded-dimension LP-type problems are known, with running time $O(dnT + t(d)E \log n)$ where n is the number of constraints, T measures the time to test whether $f(B) = f(B \cup \{x\})$ for some basis B and element $x \in S$, $t(d)$ is exponential or subexponential, and E is the time to perform a *basis-change operation* in which we must find the basis of a constant-sized subproblem and use it to replace the current basis. It is also possible with certain additional assumptions to solve these problems deterministically in time linear in n [Chazelle and Matoušek 1993].

As shown in [Amenta et al. 1999], quasiconvex programs can be expressed as LP-type problems, in such a way that the dimension of the LP-type problem is not much more than the dimension of the domain of the quasiconvex functions; therefore, quasiconvex programs can be solved in a linear number of function evaluations and a sublinear number of basis-change operations.

In order to specify the LP-type dimension of these problems, we need one additional definition: suppose we have a nested convex family κ_i. If $\kappa_i(\lambda)$ does not depend on λ, we say that κ_i is *constant*; such constant families arose, for instance, in our treatment of convex programs. Otherwise, suppose κ_i is associated with a quasiconvex function q_i. If there is no open set S such that q_i is constant over S, and if $\kappa(t')$ is contained in the interior of $\kappa(t)$ for any $t' < t$, we say that κ is *continuously shrinking*. We note that this property is different from the related and more well-known property of *strict quasiconvexity* (a quasiconvex function is strictly quasiconvex if, whenever it is constant on a line segment, it remains constant along the whole line containing the segment): L_1 distance from the origin (in $\mathbb{R}^d$, $d > 1$) is continuously shrinking but not strictly quasiconvex. On the other hand, the function

$$f(x, y) = \min\{r \mid x^2 + (y - r)^2 \leq r^2\}$$

(on the closed upper half-plane $y \geq 0$) is strictly quasiconvex but not continuously shrinking, since the origin is on the boundary of all its level sets.

We repeat the analysis of [Amenta et al. 1999], showing that quasiconvex programs are LP-type problems, below.

THEOREM 3.1. *Any quasiconvex program forms an LP-type problem of dimension at most $2d + 1$. If each κ_i in the quasiconvex program is either constant or continuously shrinking, the dimension is at most $d + 1$.*

PROOF. We form an LP-type problem in which the set S consists of the nested convex families defining the quasiconvex program, and the objective function $\Lambda(T)$ gives the value of the quasiconvex program defined by the nested convex families in T. Then, property 1 of LP-type problems is obvious: adding another nested convex family to the input can only further constrain the solution values and increase the min-max solution. To prove property 2, recall that $\Lambda(T)$ is defined as the minimum point of the intersection $\{(\lambda, \bar{x}) \mid \bar{x} \in \kappa_i(\lambda)\}$ (the intersection is nonempty by the remark in Section 1.3 about replacing infima by minima). If this point belongs to the intersection for sets A, $A \cup \{\kappa_i\}$, and $A \cup \{\kappa_k\}$, then clearly it belongs to the intersection for $A \cup \{\kappa_i, \kappa_j\}$. It remains only to show the stated bounds on the dimension.

First we prove the dimension bound for the general case, where we do not assume continuous shrinking of the families in S. Let $(\lambda, \bar{x}) = \Lambda(S)$. For any

$\lambda' < \lambda$,

$$\bigcap_{i \in S} \kappa_i(\lambda') = \varnothing,$$

so by Helly's theorem some $(d+1)$-tuple of sets $\kappa_i(\lambda')$ has empty intersection. If there is some $\lambda'' < \lambda$ for which this $(d+1)$-tuple's intersection becomes nonempty, replace λ' by λ'', find another $(d+1)$-tuple with empty intersection for the new λ', and repeat until this replacement process terminates. There are only finitely many possible $(d+1)$-tuples of nested convex families, and each replacement increases λ', so the replacement process must terminate and we eventually find a $(d+1)$-tuple B^- of nested convex families that has empty intersection for all $\lambda' < \lambda$.

With this choice of B^-, $\Lambda(B^-) = (\lambda, \bar{y})$ for some $\bar{y}$, so the presence of B^- forces the LP-type problem's solution to have the correct value of λ. We must now add further nested convex families to our basis to force the solution to also have the correct value of $\bar{x}$. Recall that

$$\bar{x} \in L = \bigcap_{i \in S} \kappa_i(\lambda),$$

and $\bar{x}$ is the minimal point in L. By Helly's theorem again, the location of this minimal point is determined by some d-tuple B^+ of the sets $\kappa_i(\lambda)$. Then $\Lambda(B^- \cup B^+) = \Lambda(S)$, so some basis of S is a subset of $B^- \cup B^+$ and has cardinality at most $2d + 1$.

Finally, we must prove the improved dimension bound for well-behaved nested convex families, so suppose each $\kappa_i \in S$ is constant or continuously shrinking. Our strategy will be to again find a tuple B^- that determines λ, and a tuple B^+ that determines $\bar{x}$, but we will use continuity to make the sizes of these two tuples add to at most $d + 1$.

The set L defined above has empty interior: otherwise, we could find an open region X within L, and a nested family $\kappa_i \in S$ such that $\kappa_i(\lambda') \cap X = \varnothing$ for any $\lambda' < \lambda$, violating the assumption that κ_i is constant or continuously shrinking. If the interior of some $\kappa_i(\lambda)$ contains a point of the affine hull of L, we say that κ_i is *slack*; otherwise we say that κ_i is *tight*. The boundary of a slack $\kappa_i(\lambda)$ intersects L in a subset of measure zero (relative to the affine hull of L), so we can find a point $\bar{y}$ in the relative interior of L and not on the boundary of any slack κ_i. Form the projection $\pi : \mathbb{R}^d \mapsto \mathbb{R}^{d - \dim L}$ onto the orthogonal complement of L.

For any ray r in $\mathbb{R}^{d - \dim L}$ starting at the point $\pi(L)$, we can lift that ray to a ray $\hat{r}$ in $\mathbb{R}^d$ starting at $\bar{y}$, and find a hyperplane containing L and separating the interior of some $\kappa_i(\lambda)$ from $\hat{r} \setminus \{\bar{y}\}$. This separated κ_i must be tight (because it has $\bar{y}$ on its boundary as the origin of the ray) so the separating hyperplane must contain the affine hull of L (otherwise some point in L within a small neighborhood of $\bar{x}$ would be interior to κ_i). Therefore the hyperplane is projected by π to a lower dimensional hyperplane separating $\pi(\kappa_i(\lambda))$ from $\pi(L)$. Since one can find such a separation for any ray, $\bigcap_{\text{tight } \kappa_i} \pi(\kappa_i(\lambda))$ can not contain any points of any such ray and must consist of the single point $\pi(L)$. At least one tight κ_j must be continuously shrinking (rather than constant), since otherwise $\bigcap_{\kappa_i \in S} \kappa_i(\lambda')$ would be nonempty for some $\lambda' < \lambda$. The intersection of the interior of $\pi(\kappa_j(\lambda))$ with the remaining projected tight constraints $\pi(\kappa_i(\lambda))$ is empty, so by Helly's theorem, we can find a $(d - \dim L + 1)$-tuple B^- of these convex sets having empty intersection, and the presence of B^- forces the LP-type problem's solution to have the correct value of λ. Similarly, we can reduce the size of the set B^+ determining $\bar{x}$ from d to $\dim L$, so the total size of a basis is at most $(d - \dim L + 1) + \dim L = d + 1$. □

This result provides theoretically efficient combinatorial algorithms for quasiconvex programs, and allows us to claim $O(n)$ time randomized algorithms for most quasiconvex programming problems in the standard computational model for computational geometry, in which primitives of constant description complexity may be assumed to be solved in constant time. For certain well-behaved sets of quasiconvex functions (essentially, the family of sets $S_{\bar{x}, \lambda} = \{\kappa \in S \mid \bar{x} \in \kappa(\lambda)\}$ should have bounded Vapnik–Chervonenkis dimension) the technique of Chazelle and Matousek [1993] applies and these problems can be solved deterministically in $O(n)$ time.

However, there are some difficulties with this approach in practice. In particular, although the basis-change operations have constant description complexity, it may not always be clear how to implement them efficiently. Therefore, in Section 3.3 we discuss alternative numerical techniques for solving quasiconvex programs directly, based only on simpler operations (function and gradient evaluation). It may be of interest to combine the two approaches, by using numerical techniques to solve the basis change operations needed for the LP-type approach; however, we do not have any theory describing how the LP-type algorithms might be affected by approximate numerical results in the basis-change steps.

3.2. Implicit quasiconvex programming. In some circumstances we may have a set of n inputs that leads to a quasiconvex program with many more than n quasiconvex functions; for instance, there may be one such function per pair of inputs. If we directly apply an LP-type algorithm, we will end up with a running time much larger than the $O(n)$ input size. Chan [2004] showed that, in such circumstances, the time for solving the quasiconvex program can often be sped up to match the time for a *decision algorithm* that merely tests whether a given pair $(\lambda, \bar{x})$ provides a feasible solution to the program.

As a simple example, consider a variation of the smallest enclosing ball problem. Suppose that we wish to place a center that minimizes the maximum sum of distances to any k-tuple of sites, rather than (as in the smallest enclosing ball problem) minimizing the maximum distance to a single site. This can be expressed again as a quasiconvex program: the sum of distances to any k-tuple of sites is quasiconvex, as it is a sum of convex functions. There are $O(n^k)$ such functions, so the problem can be solved in $O(n^k)$ time by the methods discussed already. However, the quality of any fixed center can easily be evaluated much more quickly, in $O(n)$ time, and Chan's technique provides an automatic method for turning this fast evaluation algorithm into a fast optimization algorithm for choosing the best center location.

Chan's result applies more generally to LP-type problems, but we state it here as it applies to implicit quasiconvex programming.

THEOREM 3.2. *Let $\mathfrak{Q}$ be a space of quasiconvex functions, $\mathcal{P}$ be a space of input values, and $f : 2^{\mathcal{P}} \mapsto 2^{\mathfrak{Q}}$ map sets of input values to sets of functions in $\mathfrak{Q}$. Further, suppose that $\mathcal{P}$, f, and $\mathcal{S}$ satisfy the following properties:*

- *There exists a constant-time subroutine for solving quasiconvex programs of the form $f(B)$ for any $B \subset \mathcal{P}$ with $|B| = O(1)$.*
- *There exists a decision algorithm that takes as input a set $P \subset \mathcal{P}$ and a pair $(\lambda, \bar{x})$, and returns yes if and only if $\bar{x} \in \kappa(\lambda)$ for all $\kappa \in f(P)$. The running time of the decision algorithm is bounded by $D(|P|)$, where there exists a constant $\varepsilon > 0$ such that $D(n)/n^{\varepsilon}$ is monotone increasing.*
- *There are constants α and r such that, for any input set $P \subset \mathcal{P}$, we can find in time at most $D(|P|)$ a collection of sets P_i, $0 \le i < r$, each of size at most $\alpha|P|$, for which $f(P) = \bigcup_i f(P_i)$.*

Then for any $P \subset \mathcal{P}$ we can solve the quasiconvex program $f(P)$, where $|P| = n$, in randomized expected time $O(D(n))$.

The proof involves solving a slightly more general problem in which we are given, not just a single input P, but a set of inputs $P_1, \ldots, P_d$, where d is the dimension of the LP-type problems coming from $\mathfrak{Q}$, and must solve the quasiconvex program $\cup f(P_i)$. Given any such problem, we partition each input P_i into r^i subproblems $P_{i,j}$ of size at most $\alpha^i n$ for an appropriately chosen i, by repeatedly subdividing large subproblems into smaller ones. We then view the subproblems $P_{i,j}$ as being constraints for an LP-type problem in which the objective function is the solution to the quasiconvex program $\bigcup_{P_{i,j} \in S} f(P_{i,j})$. This new LP-type problem turns out to have the same dimension as the quasiconvex programs with which we started, and the result follows by applying a standard LP-type algorithm to this problem and solving the divide-and-conquer recurrence that results.

The first and last conditions of the theorem are easily met when $f(P)$ produces one or a constant number of quasiconvex functions per k-tuple of inputs for some constant k (as in our example of optimizing the sum of k distances): then, constant sized input sets lead to constant sized quasiconvex programs, and if the input is partitioned into $k+1$ equal-sized subsets, the complements of these subsets provide the sets P_i needed for the last condition. For such problems, the main difficulty in applying this theorem is finding an appropriate decision algorithm. For our example of minimizing the maximum sum of k distances, the decision algorithm is also straightforward (select and add the k largest distances

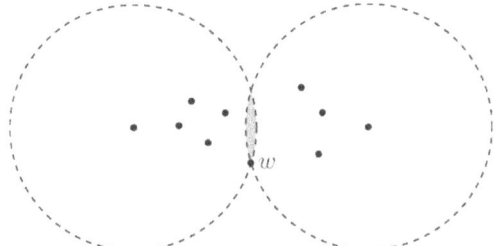

Figure 6. Example showing the difficulty of applying standard gradient descent methods to quasiconvex programming. The function to be minimized is the maximum distance to any point; only points within the narrow shaded intersection of circles have function values smaller than the value at point w. Figure taken from [Eppstein 2004].

from the given center to the sites) and so we can apply Chan's result to solve this problem in $O(n)$ time.

Chan's implicit quasiconvex programming algorithm is important in the robust statistics application described later. This algorithm has also been applied to problems of inverse parametric minimum spanning tree computation [Chan 2004; Eppstein 2003a] and facility location [Eppstein and Wortman 2005].

3.3. Smooth quasiconvex programming. If all functions $q_i(\bar{x})$ are quasiconvex, the function $q(\bar{x}) = \max_i q_i(\bar{x})$ is itself quasiconvex, so we can apply hill-climbing procedures to find its infimum. Such hill climbing procedures may be desirable in preference to the combinatorial algorithms for LP-type problems, as they avoid the difficulty of describing and implementing an appropriate exact basis change procedure. In addition, a hill climbing information that uses only numerical evaluation of function values (or possibly also function gradient evaluations) can be implemented in a generic way that does not depend on the specific form of the quasiconvex functions given to it as input.

However, many of the known nonlinear optimization techniques require the function being optimized to satisfy some smoothness conditions. In many of our applications the individual functions q_i are smooth, but their maximum q may not be smooth, so it is difficult to apply standard gradient descent techniques. The difficulty may be seen, for instance, in the smallest enclosing ball problem in the plane (Figure 6). A basis for this problem may consist of either two or three points. If a point set has only two points in its basis, and our hill climbing procedure for circumradius has reached a point w equidistant from these two points and near but not on their midpoint, then improvements to the function value $q(w)$ may be found only by moving w in a narrow range of directions towards the midpoint. Standard gradient descent algorithms may have a difficult time finding such an improvement direction.

To avoid these difficulties, we introduced in [Eppstein 2004] the following algorithm, which we call *smooth quasiconvex programming*, and which can be viewed as a generalization of Zoutendijk's [1960] method of feasible directions for convex programming. If a quasiconvex function q_i is differentiable, and w is a point where q_i is not minimal, then one can find a point with a smaller value by moving a sufficiently small distance from x along any direction having negative dot product with the gradient of q_i at w. Thus, we can improve $q(w)$ by moving in a direction that is negative with respect to all the gradients of the functions that determine the value of $q(w)$.

We formalize this notion and generalize it to nondifferentiable functions as follows. Assume for the purposes of this algorithm that, for each of the input quasiconvex functions q_i, and each $\bar{x}$ that is not the minimum point of q_i, we also can compute a vector-valued function $q_i^*(\bar{x})$, satisfying the following properties:

(i) If $q_i(\bar{y}) < q_i(\bar{x})$, then $(\bar{y} - \bar{x}) \cdot q_i^*(\bar{x}) > 0$, and
(ii) If $q_i^*(\bar{x}) \cdot \bar{y} > 0$, then for all sufficiently small $\varepsilon > 0$, $q_i(\bar{x} + \varepsilon\bar{y}) < q_i(\bar{x})$.

Less formally, any vector $\bar{y}$ is an improving direction for $q_i(\bar{x})$ if and only if it has positive inner product with $q_i^*(\bar{x})$.

If the level set $q_i^{\leq\lambda}$ is a *smooth* convex set (one that has at each of its boundary points a unique tangent plane), then the vector $q_i^*(\bar{x})$ should be an inward-pointing normal vector to the tangent plane to $q_i^{\leq q(\bar{x})}$ at $\bar{x}$. For example, in the smallest enclosing ball problem, the level sets are spheres, having tangent planes perpendicular to the radii, and q_i^* should point inwards along the radii of these spheres. If q_i is differentiable then q_i^* can be computed as the negation of the gradient of q_i, but the functions q_i^* also exist for discontinuous functions with smooth level sets.

Our smooth quasiconvex programming algorithm begins by selecting an initial value for $\bar{x}$, and a desired output tolerance. Once these values are selected, we repeat the following steps:

(i) Compute the set of vectors $q_i^*(\bar{x})$, for each i such that $q_i(\bar{x})$ is within the desired tolerance of $\max_i q_i(\bar{x})$.
(ii) Find an improving direction $\bar{y}$; that is, a vector such that $\bar{y} \cdot q_i^*(\bar{x}) > 0$ for each vector $q_i^*(\bar{x})$ in the computed set. If no such vector exists, $q(\bar{x})$ is within the tolerance of its optimal value and the algorithm terminates.
(iii) Search for a value ε for which $q(\bar{x} + \varepsilon\bar{y}) \leq q(\bar{w})$, and replace $\bar{x}$ by $\bar{x} + \varepsilon\bar{y}$.

The search for a vector $\bar{y}$ in step 2 can be expressed as a linear program. However, when the dimension of the quasiconvex functions' domain is at most two (as in the planar smallest enclosing ball problem) it can be solved more simply by sorting the vectors $q_i^*(\bar{x})$ radially around the origin and choosing $\bar{y}$ to be the average of two extreme vectors.

In step 3, it is important to choose ε carefully. It would be natural, for instance, to choose ε as large as possible while satisfying the inequality in that step; such a value could be found by a simple doubling search. However, such a choice could lead to situations where the position of $\bar{x}$ oscillates back and forth across the true optimal location. Instead, it may be appropriate to reduce the resulting ε by a factor of two before replacing $\bar{x}$.

We do not have any theory regarding the convergence rate of the smooth quasiconvex programming algorithm, but we implemented it and applied it successfully in the automated algorithm analysis application discussed below [Eppstein 2004]. Our implementation appeared to exhibit linear convergence: each iteration increased the number of bits of precision of the solution by a constant. Among numerical algorithms techniques, the sort of gradient descent we perform here is considered naive and inefficient compared to other techniques such as conjugate gradients or Newton iteration, and it would be of interest to see how well these more sophisticated methods could be applied to quasiconvex programming.

4. Applications

We have already described some simple instances of geometric optimization problems that can be formulated as quasiconvex programs. Here we describe some more complex applications of geometric optimization, in which quasiconvex programming plays a key role.

4.1. Mesh smoothing. An important step in many scientific computation problems, in which differential equations describing airflow, heat transport, stress, global illumination, or similar quantities are simulated, is *mesh generation* [Bern and Eppstein 1995; Bern and Plassmann 2000]. In this step, a complex two- or three-dimensional domain is partitioned into simpler regions, called elements, such as triangles or quadrilaterals in the plane or tetrahedra or cuboids in three dimensions. Once these elements are formed, one can then set up simple equations relating the values of the quantity of interest in each of the elements, and solve the equations to produce the results of the simulation. In this section we are particularly concerned with *unstructured mesh generation*, in which the pattern of connections from element to element does not form a regular grid; we will consider a problem in structured mesh generation in a later section.

In meshing problems, it is important to find a mesh that has small elements in regions of fine detail, but larger elements elsewhere, so that the total number of elements is minimized; this allows the system of equations derived from the mesh to be solved quickly. It is also important for the accuracy of the simulation that the mesh elements be *well shaped*; typically this means that no element should have very sharp angles or angles very close to 180°. To achieve a high quality mesh, it is important not only to find a good initial placement of mesh vertices (the main focus of most meshing papers) but then to modify the mesh by changing its topology and moving vertices until no further quality increase can be achieved. We here concentrate on the problem of moving mesh vertices

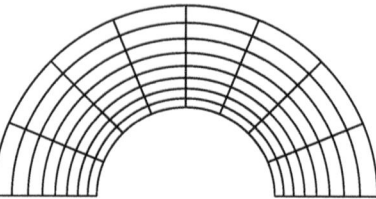

Figure 7. Mesh of an arched domain. Too much Laplacian smoothing can lead to invalid placements of the internal vertices beyond the boundaries of the arch.

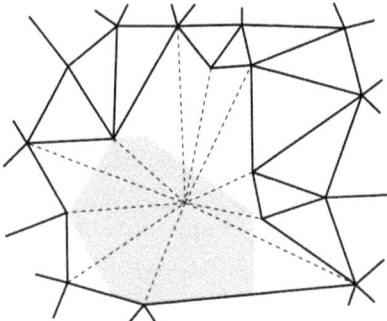

Figure 8. Optimization-based smoothing of a triangular mesh in $\mathbb{R}^2$. At each step we remove a vertex from the mesh, leaving a star-shaped polygon, then add a new vertex within the kernel (shaded) of the star-shaped region and retriangulate. Figure taken from [Amenta et al. 1999].

while retaining a fixed mesh topology, known as *mesh smoothing* [Amenta et al. 1999; Bank and Smith 1997; Canann et al. 1998; Djidjev 2000; Freitag 1997; Freitag et al. 1995; 1999; Freitag and Ollivier-Gooch 1997; Vollmer et al. 1999].

Two approaches to mesh smoothing have commonly been used, although they may sometimes be combined [Canann et al. 1998; Freitag 1997]: In *Laplacian smoothing*, all vertices are moved towards the centroid of their neighbors. Although this is easy and works well for many instances, it has some problems; for instance in a regular mesh on an arched domain (Figure 7), repeated Laplacian smoothing can cause the vertices at the top of the arch to sag downwards, eventually moving them to invalid positions beyond the boundaries of the domain.

Instead, *optimization-based smoothing* takes a more principled approach, in which we decide on a measure of element quality that best fits our application, and then seek the vertex placement that optimizes that quality measure. However, since simultaneous global optimization of all vertex positions seems a very difficult problem, we instead cycle through the vertices optimizing their positions a single vertex at a time. At each step (Figure 8), we select a vertex and remove it from the mesh, leaving a star-shaped region consisting of the elements incident to that vertex. Then, we place a new vertex within the kernel of the

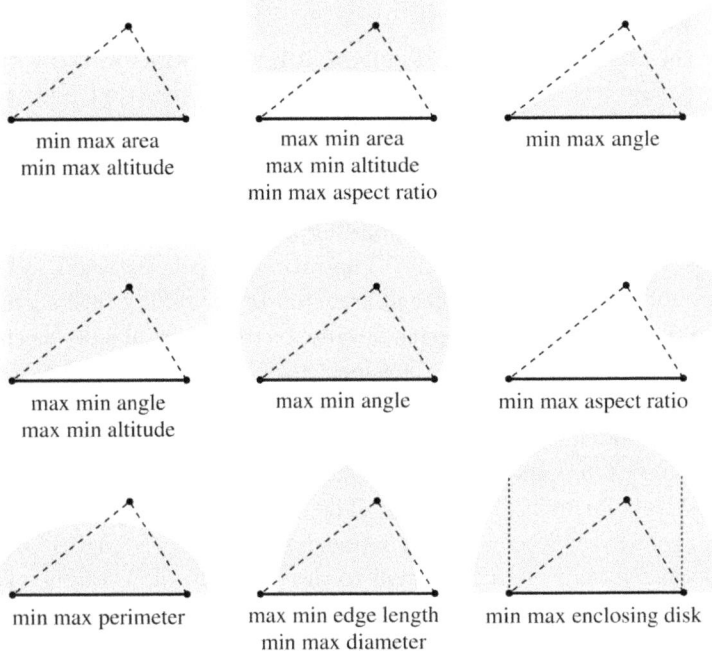

Figure 9. Level set shapes for various mesh element quality measures. Figure modified from one in [Amenta et al. 1999].

star-shaped region, and form a mesh again by connecting the new vertex to the boundary of the region. Each step improves the overall mesh quality, so this optimization process eventually converges to a locally optimal placement, but we have no guarantees about its quality with respect to the globally optimal placement.

However, in the individual vertex placement steps we need accept no such compromises with respect to global optimization. As we showed in [Amenta et al. 1999], for many natural measures $q_i(\bar{x})$ of the quality of an element incident to vertex $\bar{x}$ (with smaller numbers indicating better quality), the problem of finding a mesh minimizing the maximum value of q_i can be expressed as a quasiconvex program. Figure 9 illustrates the level set shapes resulting from various of these quasiconvex optimization-based mesh smoothing problems. For shape-based quality measures, such as maximizing the minimum angle, the optimal vertex placement will naturally land in the interior of the kernel of the region formed by the removal of the previous vertex placement. For some other quality measures, such as minimizing the maximum perimeter, it may be appropriate to also include constant quasiconvex functions, forcing the vertex to stay within the kernel, similar to the functions used in our transformation of convex programs to quasiconvex programs. It would also be possible to handle multiple quality measures simultaneously by including quasiconvex functions of more than one type in the optimization problem.

In most of the cases illustrated in Figure 9, it is straightforward to verify that the quality measure has level sets of the convex shape illustrated. One possible exception is the problem of minimizing the maximum aspect ratio (ratio of the longest side length to shortest altitude) of any element. To see that this forms a quasiconvex optimization problem, Amenta et al. [1999] consider separately the ratios of the three sides to their corresponding altitudes; the maximum of these three will give the overall aspect ratio. The ratio of a side external to the star to its corresponding altitude has a feasible region (after taking into account the kernel constraints) forming a half-space parallel to the external side, as shown in Figure 9 (top center). To determine the aspect ratio on one of the other two sides of a triangle Δ_i, normalize the triangle coordinates so that the replaced point has coordinates (x, y) and the other two have coordinates $(0, 0)$ and $(1, 0)$. The side length is then $\sqrt{x^2 + y^2}$, and the altitude is $y/\sqrt{x^2 + y^2}$, so the overall aspect ratio has the simple formula $(x^2 + y^2)/y$. The locus of points for which this is a constant b is given by $x^2 + y^2 = by$, or equivalently $x^2 + (y - (b/2))^2 = (b/2)^2$. Thus the feasible region is a circle tangent to the fixed side of Δ_i at one of its two endpoints (Figure 9, center right). Another nontrivial case is that of minimizing the smallest enclosing ball of the element, shown in the bottom right of the figure; in that case the level set boundary consists of curves of two types, according to whether, for placements in that part of the level set, the enclosing ball touches two or three of the element vertices, but the curves meet at a common tangent point to form a smooth convex level set.

Bank and Smith [1997] define yet another measure of the quality of a triangle, computed by dividing the triangle's area by the sum of the squares of its edge lengths. This gives a dimensionless quantity which Bank and Smith normalize to be one for the equilateral triangle (and less than one for any other triangle). As Bank and Smith show, the lower level sets for this mesh quality measure form circles centered on the perpendicular bisector of the two fixed points of the mesh element, so, as with the other measures, finding the placement optimizing Bank and Smith's measure can be expressed as a quasiconvex program.

We have primarily discussed triangular mesh smoothing here, but the same techniques apply with little modification to many natural element quality measures for quadrilateral and tetrahedral mesh smoothing. Smoothing of cubical meshes is more problematic, though, as moving a single vertex may cause the faces of one of the cuboid elements to become significantly warped. Several individual quasiconvex quality measures for quadrilateral and tetrahedral meshes, and the shapes of their level sets, are discussed in more detail in [Amenta et al. 1999]. The most interesting of these from the mathematical viewpoint is the problem of maximizing the minimum solid angle of any tetrahedral element, as

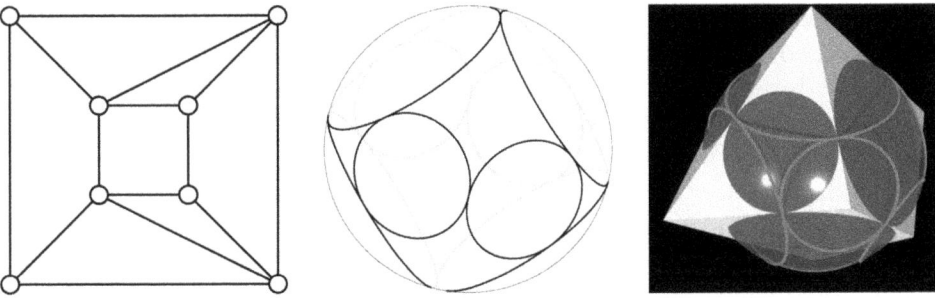

Figure 10. Planar graph (left), its representation as a set of tangent disks on a sphere (center), and the corresponding polyhedral representation (right). Left and center images taken from [Bern and Eppstein 2001].

measured at its vertices, which with some difficulty we were able to show leads to a quasiconvex objective function.

4.2. Graph drawing. The Koebe–Thurston–Andreev embedding theorem [Brightwell and Scheinerman 1993; Koebe 1936; Sachs 1994] states that any planar graph embedding can be transformed into a collection of disks with disjoint interiors on the surface of a sphere, one disk per vertex, such that two disks are tangent if and only if the corresponding two vertices are adjacent (Figure 10, left and center). The representation of the graph as such a collection of tangent disks is sometimes called a *coin graph*. For maximal planar graphs, this coin graph representation is unique up to *Möbius transformations* (the family of transformations of the sphere that transform circles to circles), and for nonmaximal graphs it can be made unique by adding a new vertex within each face of the embedding, adjacent to all vertices of the face, and finding a disk representation of the resulting augmented maximal planar graph.

Given a coin graph representation, the graph itself can be drawn on the sphere, say by placing a vertex at the center of each circle and connecting two vertices by edges along an arc of a great circle; similar drawings are also possible in the plane by using polar projection to map the circles in the sphere onto circles in the plane [Hliněný 1997]. Coin graphs can also be used to form a three-dimensional *polyhedral representation* of the graph, as follows: embed the sphere in space, and, for each disk, form a cone in space that is tangent to the sphere at the disk's boundary; then, form a polyhedron by taking the convex hull of the cone apexes. The resulting polyhedron's skeleton is isomorphic to the original graph, and its edges are tangent to the sphere (Figure 10, right).

In order to use these techniques for visualizing graphs, we would like to choose a coin graph representation that leads to several desirable properties identified as standard within the graph drawing literature [di Battista et al. 1999], including the display of as many as possible of the symmetries of the original graph, and the separation of vertices as far apart from each other as possible. In [Bern and

Eppstein 2001] we used quasiconvex programming to formalize the search for a drawing based on these objectives.

In order to understand this formalization, we need some more background knowledge about Möbius transformations and their relation to hyperbolic geometry. We can identify the unit sphere that the Möbius transformations transform as being the boundary of a Poincaré or Klein model of hyperbolic space $\mathbb{H}^3$. The points on the sphere can be viewed as "infinite" points that do not belong to $\mathbb{H}^3$ but are the limit points of certain sequences of points within $\mathbb{H}^3$. With this identification, circles on the sphere become the limit points of hyperplanes in $\mathbb{H}^3$. Any isometry of $\mathbb{H}^3$ takes hyperplanes to hyperplanes, and therefore can be extended to a transformation of the sphere that takes circles to circles, and the converse turns out to be true as well. We can determine an isometry of $\mathbb{H}^3$ by specifying which point of $\mathbb{H}^3$ is mapped to the center of the Poincaré or Klein model, and then by specifying a spatial rotation around that center point. The rotation component of this isometry does not change the shape of objects on the sphere, so whenever we seek the Möbius transformation that optimizes some quality measure of a transformed configuration of disks on the sphere, we can view the problem more simply as one of seeking the optimal center point of the corresponding isometry in $\mathbb{H}^3$.

To see how we apply this technique to our graph drawing problem, first consider a version of the problem in which we seek a disk representation maximizing the radius of the smallest disk. More generally, given any collection of circles on the sphere, we wish to transform the circles in order to maximize the minimum radius. Thus, let $q_i(\bar{x})$ measure the (negation of the) transformed radius of the ith circle, as a function of the transformed center point $\bar{x} \in \mathbb{H}^3$. If we let H_i denote the hyperplane in $\mathbb{H}^3$ that has the ith circle as its set of limit points, then the transformed radius is maximized when the circle is transformed into a great circle; that is, when $\bar{x} \in H_i$. If we choose a center point $\bar{x}$ away from H_i, the transformed radius will be smaller, and due to the uniform nature of hyperbolic space the radius can be written as a function only of the distance from $\bar{x}$ to H_i, not depending in any other way on the location of $\bar{x}$. That is, the level sets of q_i are the convex hyperbolic sets within some distance R of the hyperplane H_i. Therefore, q_i is a quasiconvex hyperbolic function. In fact, the quasiconvex program defined by the functions q_i can be viewed as a hyperbolic version of a generalized minimum enclosing ball problem, in which we seek the center $\bar{x}$ of the smallest ball that touches each of the convex sets H_i. The two-dimensional version of this problem, in which we seek the smallest disk touching each of a collection of hyperbolic lines, is illustrated in Figure 11. If we form a Klein or Poincaré model with the resulting optimal point $\bar{x}$ at the center of the model, the corresponding Möbius transformation of the model's boundary maximizes the minimum radius of our collection of circles.

Further, due to the uniqueness of quasiconvex program optima, the resulting disk representation must display all the symmetries possible for the original pla-

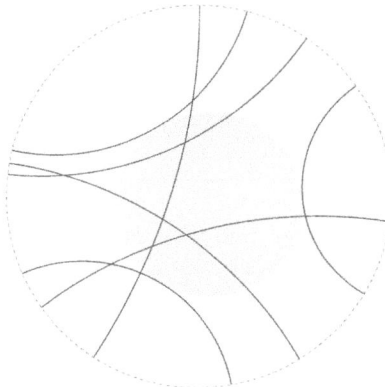

Figure 11. Two-dimensional analogue of max-min radius transform problem: find the smallest disk touching all of a collection of hyperbolic lines.

nar graph embedding; for, if not all symmetries were displayed, one could use an undisplayed symmetry to relabel the vertices of the disk representation, achieving a second disk representation with equal quality to the first. For instance, in Figure 10, the disk representation shown has three planes of mirror symmetry while the initial drawing has only one mirror symmetry axis.

Bern and Eppstein [2001] then consider an alternative version of the graph drawing problem, in which the objective is to maximize the minimum distance between certain pairs of vertices on the sphere surface. For instance, one could consider only pairs of vertices that are adjacent in the graph, or instead consider all pairs; in the latter case we can reduce the number of pairs that need be examined by the algorithm by using the Delaunay triangulation in place of the complete graph. The problem of maximizing the minimum spherical distance among a set of pairs of vertices can be formulated as a quasiconvex program by viewing each pair of vertices as the two limit points of a hyperbolic line in $\mathbb{H}^3$, finding the center $\bar{x}$ of the smallest ball in $\mathbb{H}^3$ that touches each of these hyperbolic lines, and using this choice of center point to transform the sphere.

Möbius transformations can also be performed on the augmented plane $\mathbb{R}^2 \cup \{\infty\}$ instead of on a sphere, and act on lines and circles within that plane; a line can be viewed as a limiting case of a circle that passes through the special point ∞. Multiplication of each coordinate of each point by the same constant k forms a special type of Möbius transformation, which (if $k > 1$) increases every distance, so it does not make sense to look for an unrestricted Möbius transformation of the plane that maximizes the minimum Euclidean distance among a collection of pairs of points. However, Bern and Eppstein were able to show, given a collection of points within the unit ball in the plane, that seeking the Möbius transformation that takes that disk to itself and maximizes the minimum distances between certain pairs of the points can again be expressed

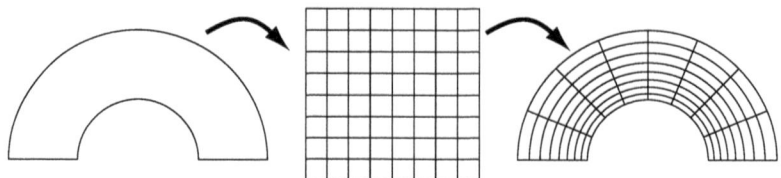

Figure 12. Conformal meshing: transform domain to a more simply shaped region with a known mesh, then invert the transformation to transform the mesh back to the original domain.

as a two-dimensional quasiconvex program. The proof of quasiconvexity is more complex and involves simultaneously treating the unit ball as a Poincaré model of $\mathbb{H}^2$ and the entire plane as the boundary of a Poincaré model of $\mathbb{H}^3$.

Along with these coin graph based drawing methods, Bern and Eppstein also considered a different graph drawing question, more directly involving hyperbolic geometry. The Poincaré and Klein models of projective geometry have been considered by several authors as a way of achieving a "fish-eye" view of a large graph, so that a local neighborhood in the graph is visible in detail near the center of the view while the whole graph is spread out on a much smaller scale at the periphery [Lamping et al. 1995; Munzner 1997; Munzner and Burchard 1995]. Bern and Eppstein [Bern and Eppstein 2001] found quasiconvex programming formulations of several versions of the problem of selecting an initial viewpoint for these hyperbolic drawings, in order for the whole graph to be visible in as large a scale as possible. For instance, a natural version of this problem would be to choose a viewpoint minimizing the maximum hyperbolic distance to any vertex, which is just the hyperbolic smallest enclosing ball problem again. One question in this area that they left open is whether one can use quasiconvex programming to find a Klein model of a given graph that maximizes the minimum Euclidean distance between adjacent vertices.

4.3. Conformal mesh generation. The ideas of mesh generation and optimal Möbius transformation coincide in the problem of conformal mesh generation [Bern and Eppstein 2001]. In this problem, we wish to generate a mesh for a simply-connected domain in $\mathbb{R}^2$ by using a *conformal transformation* (that is, a transformation that preserves angles of incidence between transformed curves) to map the shape into some easy-to-mesh domain such as a square, then invert the transformation to map the meshed square back into the original domain (Figure 12). There has been much work on algorithms for finding conformal maps [Driscoll and Vavasis 1998; Howell 1990; Smith 1991; Stenger and Schmidtlein 1997; Trefethen 1980] and conformal meshes have significant advantages: the orthogonality of the angles at mesh vertices means that one can avoid certain additional terms in the definition of the partial differential equation to be solved [Bern and Plassmann 2000; Thompson et al. 1985].

If we replace the square in Figure 12 by a disk, the Riemann mapping theorem tells us that a conformal transformation always exists and is, moreover, unique up to Möbius transformations that transform the disk to itself; any such transformation preserves conformality. Thus, we have several degrees of freedom for controlling the size of the mesh elements produced by the conformal method: we can use a larger or smaller grid on the disk or square, but we can also use a Möbius transformation in order to enlarge certain portions of the domain and shrink others before meshing it. We would like to use these degrees of freedom to construct a mesh that has small elements in regions of the domain where fine detail is desired, and large elements elsewhere, in order to limit the total number of elements of the resulting mesh.

Bern and Eppstein [2001] formalized the problem by assuming an input domain in which certain interior points p_i are marked with a desired element size s_i. If we find a conformal map f from the domain to a disk, the gradient of f maps the marked element sizes to desired sizes s_i' in the transformed disk: $s_i' = \|f'(p_i)\|$. We can then choose a structured mesh with element size $\min s_i'$ in the disk, and transform it back to a mesh of the original domain. The goal is to choose our conformal map in a way that maximizes $\min s_i'$, so that we can use a structured mesh with as few elements as possible. Another way of interpreting this is that s_i' can be seen as the radius of a small disk at $f(p_i)$. What we seek is the transformation that maximizes the minimum of these radii. This is not quite the same as the max-min radius graph drawing problem of the previous section, because the circles to be optimized belong to $\mathbb{R}^2$ instead of to a sphere, but as in the previous section we can view the unit disk as being a Poincaré model of $\mathbb{H}^2$ (using the fact that circles in $\mathbb{H}^2$ are mapped by the Poincaré model into circles in the unit disk), and seek a hyperbolic isometry that maps $\mathbb{H}^2$ into itself and optimizes the circle radii. The transformed radius of a circle is a function only of the distance from that circle to the center point of the transformed model, so the level sets of the functions representing the transformed radii are themselves circles and the functions are quasiconvex.

The quasiconvex conformal meshing technique of Bern and Eppstein does not account for two remaining degrees of freedom: first, it is possible to rotate the unit disk around its center point and, while that will not change the element size as measured by Bern and Eppstein's formalization, it will change the element orientations. This is more important if we also consider the second degree of freedom, which is that instead of using a uniform grid on a square, we could use a rectangle with arbitrary aspect ratio. Bern and Eppstein leave as an open question whether we can efficiently compute the optimal choice of conformal map to a high-aspect-ratio rectangle to maximize the minimum desired element size.

4.4. Brain flat mapping. In [Hurdal et al. 1999] methods are described for visualizing the complicated structure of the brain by stretching its surface onto a flat plane. This stretching is done via conformal maps: surfaces of major brain components such as the cerebellum are simply connected, so there exists a conformal map from these surfaces onto a Euclidean unit disk, sphere, or hyperbolic plane. The authors approximate this conformal map by using a fine triangular mesh to represent the brain surface, and forming the Koebe disk representation of this mesh. Each triangle from the brain surface can then be mapped to the triangle connecting the corresponding three disk centers. As in the conformal meshing example, there is freedom to modify the conformal map by means of a Möbius transformation, so Bern and Eppstein [2001] suggested that the optimal Möbius transformation technique described in the previous two sections could also be useful in this application.

Although conformal transformation preserves angles, it distorts other important geometric information such as area. Bern and Eppstein proposed to ameliorate this distortion by using an optimal Möbius transformation to find the conformal transformation minimizing the maximum ratio a/a' where a is the area of a triangle in the initial three-dimensional map, and a' is the area of its image in the flat map.

Unfortunately it has not yet been possible to prove that this optimization problem leads to quasiconvex optimization problems. Bern and Eppstein formalized the difficulty in the following open question: Let T be a triangle in the unit disk or on the surface of a sphere, and let C be the set of center points for Poincaré models (of $\mathbb{H}^2$ in the disk case or $\mathbb{H}^3$ in the sphere case) such that the Möbius transformations corresponding to center points in C transform T into a triangle of area at least A. Is C necessarily convex? Note that, at least in the spherical case, the area of the transformed triangle is the same as the hyperbolic solid angle of T as viewed from the center point, so this question seems strongly reminiscent of the difficult problem of proving quasiconvexity for tetrahedral mesh smoothing to maximize the minimum Euclidean solid angle, discussed in the initial subsection of this section. A positive answer would allow the quasiconvex programming technique to be applied to this brain flat mapping application.

4.5. Optimized color gamuts. Tiled projector systems [Humphreys and Hanrahan 1999; Li et al. 2000; Raskar et al. 1999] are a recent development in computer display technology, in which the outputs of multiple projectors are combined into large seamless displays for collaborative workspaces. There are many difficult research issues involved in achieving this seamlessness: how to move the data quickly enough to all the screens, how to maintain physical alignment of the projectors, how to handle the radial reduction in brightness (vignetting) common to many projector systems, and so on. Here we concentrate on one small piece of this puzzle: matching colors among the outputs of mul-

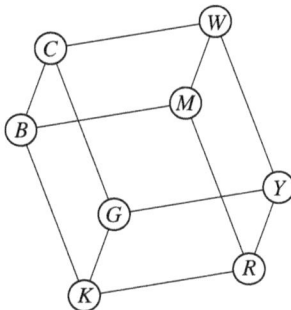

Figure 13. An additive color gamut, with vertices labeled by colors: K = black, R = red, G = green, B = blue, C = cyan, M = magenta, Y = yellow, W = white.

tiple projectors. Any imaging device has a *gamut*, the set of colors that it can produce. However, two projectors, even of the same model, will have somewhat different gamuts due to factors such as color filter batches and light bulb ages. We seek a common gamut of colors that can be produced by all the projectors, and a coordinate system for that gamut so that we can display color images in a seamless fashion across multiple projectors [Bern and Eppstein 2003; Majumder et al. 2000; Stone 2001].

Most projectors, and most computer graphics software, use an *additive color* system in which colors are produced by adding signals of three primary colors, typically red, green, and blue. If we view the gamuts as sets of points in a linear three-dimensional device-independent color space, additive color systems produce gamuts that are the Minkowski sums of three line segments, one per color signal, and therefore have the geometric form of parallelepipeds (Figure 13). The color spaces representing human vision are three-dimensional, so these parallelepipeds have twelve degrees of freedom: three for the *black point* of the projector (representing the color of light it projects when it is given a zero input signal) and three each for the three primary colors (that is, the color that the projector produces when given an input signal with full strength in one primary color channel and zero in the other two color channels). The black point and the three primary colors form four of the eight parallelepiped vertices; the other four are the secondary colors cyan, yellow, and magenta, and the white point produced when all three input color channels are saturated.

The computational task of finding a common color gamut, then, can be represented as a twelve-dimensional geometric optimization problem in which we seek the best parallelepiped to use as our gamut, according to some measure of gamut quality, while constraining our output parallelepiped to lie within the intersection of a collection of input parallelepipeds, one per projector of our system.

To represent this problem as a quasiconvex program, Bern and Eppstein [2003] suppose that we are given eight quasiconvex functions d_K, d_R, d_G, d_B, d_C, d_M,

d_Y, and d_W, where each $d_X : \mathbb{R}^3 \mapsto \mathbb{R}$ measures the distance of a color from the ideal location of corner X of the color cube (here each capital letter is the initial of one of the colors at the color cube corners, except for K which by convention stands for black). This formulation allows different distance functions to be used for each color; for instance, we might want to weight d_K and d_W more strongly than the other six color distances. We also form eight functions $f_X : \mathbb{R}^{12} \mapsto \mathbb{R}^3$ mapping our twelve-dimensional parametrization of color gamuts into the color values of each of the gamut corners. If we parametrize a gamut by the black point and three primary colors, then f_K, f_R, f_G, and f_B are simply coordinate projections, while the other four functions are simple linear combinations of the coordinates. For each of the eight colors X, define $q_X(\bar{x}) = d_X(f_X(\bar{x}))$. The level sets of q_X are simply Cartesian products of the three dimensional level sets of d_X with complementary nine-dimensional subspaces of $\mathbb{R}^{12}$, so they are convex and each q_X is quasiconvex.

It remains to formulate the requirement that our output gamut lie within the intersection of the input gamuts. If we are given n input gamuts, form a half-space $H_{i,j}$ (with $0 \le i < n$ and $0 \le j < 6$) for each of the six facets of each of these parallelepipeds, and for each color X form a nested convex family $\kappa_{i,j,X}(\lambda) = \{\bar{x} \in \mathbb{R}^{12} \mid f_X(\bar{x}) \in H_{i,j}\}$ that ignores its argument λ and returns a constant half-space. We can then represent the problem of finding a feasible gamut that minimizes the maximum distance from one of its corners to the corner's ideal location as the quasiconvex program formed by the eight quasiconvex functions q_X together with the $48n$ nested convex families $\kappa_{i,j,X}$.

4.6. Analysis of backtracking recurrences. In this section we discuss another application of quasiconvex programming, in the automated analysis of algorithms, from our paper [Eppstein 2004]. There has been much research on exponential-time exact algorithms for problems that are NP-complete (so that no polynomial time solution is expected); see [Beigel 1999; Byskov 2003; Dantsin and Hirsch 2000; Eppstein 2001a; 2001b; 2003b; Gramm et al. 2000; Paturi et al. 1998; Schöning 1999] for several recent papers in this area. Although other techniques are known, many of these algorithms use a form of backtracking search in which one repeatedly performs some case analysis to find an appropriate structure in the problem instance, and then uses that structure to split the problem into several smaller subproblems which are solved by recursive calls to the algorithm.

For example, as part of a graph coloring algorithm [Eppstein 2001b] we used the following subroutine for listing all maximal independent sets of a graph G that have at most k vertices in the maximum independent set (we refer to such a set as a k-*MIS*). The subroutine consists of several different cases, and applies the first of the cases which is found to be present in the input graph G:

- If G contains a vertex v of degree zero, recursively list each $(k-1)$-MIS in $G \setminus \{v\}$ and append v to each listed set.

$T(n, h) \leq \max$

$\begin{cases}
T(n+3, h-2)+T(n+3, h-1)+T(n+4, h-2)+T(n+5, h-2), \\
T(n, h+1)+T(n+1, h+2), \\
2\,T(n+2, h)+2\,T(n+3, h), \\
2\,T(n+2, h)+2\,T(n+3, h), \\
T(n+3, h-2)+T(n+3, h-1)+T(n+5, h-3)+T(n+5, h-2), \\
T(n+1, h)+T(n+3, h-1)+3\,T(n+3, h+3), \\
T(n+3, h-2)+2\,T(n+3, h-1)+T(n+7, h-2), \\
T(n+1, h)+2\,T(n+4, h-2), \\
3\,T(n+1, h+2)+2\,T(n+1, h+5), \\
2\,T(n+2, h)+T(n+3, h+1)+T(n+4, h)+T(n+4, h+1), \\
T(n+1, h-1)+T(n+4, h-1), \\
T(n+1, h+3)+2\,T(n+2, h)+T(n+3, h), \\
2\,T(n+2, h-1), \\
T(n, h+3)+T(n+1, h+2)+T(n+2, h), \\
T(n+1, h-1)+T(n+4, h-1), \\
2\,T(n+1, h+1)+T(n+2, h+1), \\
9\,T(n+2, h+3), \\
T(n+1, h)+T(n+1, h+1), \\
9\,T(n+9, h-5)+9\,T(n+9, h-4), \\
T(n+3, h-2)+T(n+3, h-1)+T(n+5, h-2)+2\,T(n+6, h-3), \\
T(n+1, h-1)+T(n+4, h)+T(n+4, h+1), \\
2\,T(n+2, h)+T(n+3, h)+T(n+4, h)+T(n+5, h), \\
T(n+1, h)+2\,T(n+2, h+1), \\
T(n+1, h-1), \\
2\,T(n+2, h+1)+T(n+3, h-2)+T(n+3, h), \\
T(n+1, h+1)+T(n+1, h+2)+T(n+2, h), \\
2\,T(n+2, h)+2\,T(n+3, h), \\
T(n+1, h+2)+T(n+2, h-1)+T(n+2, h+1), \\
T(n+1, h), \\
T(n+2, h+1)+T(n+3, h-2)+T(n+4, h-3), \\
T(n-1, h+2), \\
3\,T(n+4, h)+7\,T(n+4, h+1), \\
T(n+2, h-1)+2\,T(n+3, h-1), \\
T(n+2, h-1)+T(n+2, h)+T(n+2, h+1), \\
T(n+3, h-2)+T(n+3, h)+2\,T(n+4, h-2), \\
T(n+1, h)+T(n+3, h-1)+T(n+3, h+3)+T(n+5, h)+T(n+6, h-1), \\
2\,T(n+1, h+4)+3\,T(n+3, h+1)+3\,T(n+3, h+2), \\
3\,T(n+3, h+1)+T(n+3, h+2)+3\,T(n+3, h+3)+3\,T(n+4, h), \\
T(n+2, h-1)+T(n+3, h-1)+T(n+4, h-2), \\
T(n, h+1), \\
T(n+1, h+2)+T(n+3, h-2)+T(n+3, h-1), \\
2\,T(n+3, h-1)+T(n+3, h+2)+T(n+5, h-2)+T(n+5, h-1)+T(n+5, h)+2\,T(n+7, h-3), \\
T(n+2, h+2)+2\,T(n+3, h)+3\,T(n+3, h+1)+T(n+4, h), \\
T(n+3, h-2)+T(n+3, h-1)+T(n+5, h-3)+T(n+6, h-3)+T(n+7, h-4), \\
T(n+1, h-1), \\
T(n+1, h)+2\,T(n+3, h), \\
4\,T(n+3, h+1)+5\,T(n+3, h+2), \\
4\,T(n+2, h+3)+3\,T(n+4, h)+3\,T(n+4, h+1), \\
T(n+3, h-2)+2\,T(n+3, h-1)+T(n+6, h-3), \\
4\,T(n+2, h+3)+6\,T(n+3, h+2), \\
T(n, h+1)+T(n+4, h-3), \\
T(n+1, h-1)+2\,T(n+3, h+2), \\
2\,T(n+2, h+1)+3\,T(n+2, h+3)+2\,T(n+2, h+4), \\
2\,T(n+2, h)+2\,T(n+2, h+3), \\
2\,T(n+2, h)+T(n+2, h+3)+T(n+3, h+2)+T(n+4, h)+T(n+4, h+1), \\
2\,T(n, h+2), \\
T(n+2, h)+T(n+3, h-2)+T(n+3, h-1), \\
T(n+3, h-2)+2\,T(n+4, h-2)+T(n+5, h-3), \\
T(n+1, h)+T(n+5, h-4)+T(n+5, h-3), \\
T(n+1, h+2)+T(n+2, h-1)+T(n+3, h-1), \\
T(n+2, h-1)+T(n+2, h)+T(n+4, h-1), \\
10\,T(n+3, h+2), \\
6\,T(n+2, h+2), \\
T(n+2, h)+T(n+3, h), \\
2\,T(n+3, h-1)+T(n+3, h+2)+T(n+5, h-2)+T(n+5, h-1)+T(n+5, h)+T(n+6, h-2)+T(n+7, h-2), \\
6\,T(n+3, h+1), \\
3\,T(n, h+3), \\
T(n+2, h-1)+T(n+2, h)+T(n+4, h-2), \\
2\,T(n+5, h-3)+5\,T(n+5, h-2), \\
2\,T(n+2, h)+T(n+2, h+1)+T(n+4, h-1), \\
8\,T(n+1, h+4), \\
T(n+3, h-2)+T(n+3, h-1)+T(n+5, h-3)+T(n+5, h-2)+T(n+7, h-3), \\
T(n+1, h-1)+T(n+2, h+2), \\
5\,T(n+2, h+2)+2\,T(n+2, h+3)
\end{cases}$

Table 1. A recurrence arising from unpublished work with J. Byskov on graph coloring algorithms, taken from [Eppstein 2004].

- If G contains a vertex v of degree one, with neighbor u, recursively list each $(k-1)$-MIS in $G \setminus N(u)$ and append u to each listed set. Then, recursively list each $(k-1)$-MIS in $G \setminus \{u, v\}$ and append v to each listed set.
- If G contains a path v_1-v_2-v_3 of degree-two vertices, then, first, recursively list each $(k-1)$-MIS in $G \setminus N(v_1)$ and append v_1 to each listed set. Second, list each $(k-1)$-MIS in $G \setminus N(v_2)$ and append v_2 to each listed set. Finally, list each $(k-1)$-MIS in $G \setminus (\{v_1\} \cup N(v_3))$ and append v_3 to each listed set. Note that, in the last recursive call, v_1 may belong to $N(v_3)$ in which case the number of vertices is only reduced by three.
- If G contains a vertex v of degree three or more, recursively list each k-MIS in $G \setminus \{v\}$. Then, recursively list each $(k-1)$-MIS in $G \setminus N(v)$ and append v to each listed set.

Clearly, at least one case is present in any nonempty graph, and it is not hard to verify that any k-MIS will be generated by one of the recursive calls made from each case. Certain of the sets generated by this algorithm as described above may not be maximal, but if these nonmaximal outputs cause difficulties they can be removed by an additional postprocessing step. We can bound the worst-case number of output sets produced by this algorithm as the solution to the following recurrence in the variables n and k:

$$
T(n,k) = \max \begin{cases} T(n-1, k-1) \\ 2T(n-2, k-1) \\ 3T(n-3, k-1) \\ T(n-1, k) + T(n-4, k-1) \end{cases}
$$

As base cases, $T(0,0) = 1$, $T(n,-1) = 0$, and $T(n,k) = 0$ for $k > n$. Each term in the overall maximization of the recurrence comes from a case in the case analysis; the recurrence uses the maximum of these terms because, in a worst-case analysis, the algorithm has no control over which case will arise. Each summand in each term comes from a recursive subproblem called for that case. It turns out that, for the range of parameters of interest $n/4 \le k \le n/3$, the recurrence above is dominated by its last two terms, and has the solution $T(n,k) = (4/3)^n (3^4/4^3)^k$. We can also find graphs having this many k-MISs, so the analysis given by the recurrence is tight. Similar but somewhat more complicated multivariate recurrences have arisen in our algorithm for 3-coloring [Eppstein 2001a] with variables counting 3- and 4-value variables in a constraint satisfaction instance, and in our algorithm for the traveling salesman problem in cubic graphs [Eppstein 2003b] with variables counting vertices, unforced edges, forced edges, and 4-cycles of unforced edges. Another such recurrence, of greater complexity but with the same general form, is depicted in Table 1.

We would like to perform this type of analysis algorithmically: if we are given as input a recurrence such as the ones discussed above, can we efficiently determine its asymptotic solution, and determine which of the cases in the analysis are the critical ones for the performance of the backtracking algorithm that generated the recurrence? We showed in [Eppstein 2004] that these questions can be answered automatically by a quasiconvex programming algorithm, as follows.

Let $\bar{x}$ denote a vector of arguments to the input recurrence, and for each term in the input recurrence define a univariate linear recurrence, by replacing $\bar{x}$ with a weighted linear combination $\xi = \bar{w} \cdot \bar{x}$ throughout. For instance, in the k-bounded maximal independent set recurrences, the four terms in the recurrence lead to four linear recurrences

$$t_1(\xi) = t_1(\xi - \bar{w} \cdot (1,1)),$$
$$t_2(\xi) = 2t_2(\xi - \bar{w} \cdot (2,1)),$$
$$t_3(\xi) = 3t_3(\xi - \bar{w} \cdot (3,1)),$$
$$t_4(\xi) = t_4(\xi - \bar{w} \cdot (1,0)) + t_4(\xi - \bar{w} \cdot (4,1)).$$

We can solve each of these linear recurrences to find constants c_i such that $t_i(\xi) = O(c_i^{\xi})$; it follows that, for any weight vector $\bar{w}$, $T(\bar{x}) = O(\max c_i^{\bar{w} \cdot \bar{x}})$.

This technique only yields a valid bound when each linear recurrence is solvable; that is, when each term on the right-hand side of each linear recurrence has a strictly smaller argument than the term on the left hand side. In addition, different choices of $\bar{w}$ in this upper bound technique will give us different bounds.

To get the tightest possible upper bound from this technique, for $\bar{x} = n\bar{t}$ where $\bar{t}$ is a fixed *target vector*, constrain $\bar{w} \cdot \bar{t} = 1$ (this is a normalizing condition since multiplying $\bar{w}$ by a scalar does not affect the overall upper bound), and express c_i as a function $c_i = q_i(\bar{w})$ of the weight vector $\bar{w}$; set $c_i = +\infty$ whenever the corresponding linear inequality has a right-hand side term with argument greater than or equal to that on the left hand side. We show in [Eppstein 2004] that these functions q_i are quasiconvex, as their level sets can be expressed by the formula

$$q_i^{\leq \lambda} = \left\{ \bar{w} \ \middle| \ \sum_j \lambda^{-\bar{w} \cdot \delta_{i,j}} \leq 1 \right\},$$

where the right-hand side describes a level set of a sum of convex functions of $\bar{w}$. Therefore, we can find the vector $\bar{w}$ minimizing $\max_i q_i(w)$ as a quasiconvex program. The value λ of this quasiconvex program gives us an upper bound $T(n\bar{t}) = O(\lambda^n)$ on our input recurrence.

In the same paper, we also show a lower bound $T(n\bar{t}) = \Omega(\lambda^n n^{-c})$, so the upper bound is tight to within a factor that is polylogarithmic compared to the overall solution. The lower bound technique involves relating the recurrence solution to the probability that a random walk in a certain infinite directed graph reaches the origin, where the sets of outgoing edges from each vertex in the graph are also determined randomly with probabilities determined from the gradients surrounding the optimal solution of the quasiconvex program for the upper bound.

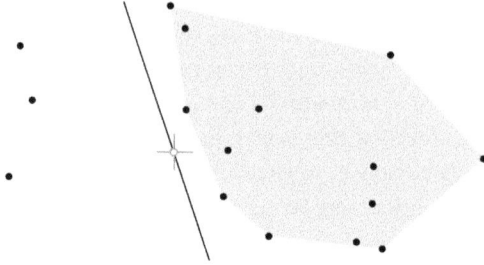

Figure 14. The Tukey depth of the point marked with the $+$ sign is three, since there is a half-plane containing it and only three sample points; equivalently, three points can be removed from the sample set to place the test point outside the convex hull of the remaining points (shaded).

4.7. Robust statistics. If one has a set of n observations $x_i \in \mathbb{R}$, and wishes to summarize them by a single number, the average or mean is a common choice. However, it is sensitive to *outliers*: replacing a single observation by a value far from the mean can change the mean to an arbitrarily chosen value. In contrast, if one uses the median in place of the mean, at least $n/2$ observations need to be corrupted before the median can be changed to an arbitrary value; if fewer than $n/2$ observations are corrupted, the median will remain within the interval spanned by the uncorrupted values. In this sense, the median is *robust* while the mean is not. More generally, we define a statistic to be robust if its *breakdown point* (the number of observations that must be corrupted to cause it to take an arbitrary value) is at least cn for some constant $c > 0$.

If one has observations $\bar{x}_i \in \mathbb{R}^d$, it is again natural to attempt to summarize them by a single point $\bar{x} \in \mathbb{R}^d$. In an attempt to generalize the success of the median in the one-dimensional problem, statisticians have devised many notions of the *depth* of a point, from which we can define a generalized median as being the point of greatest depth [Gill et al. 1992; Hodges 1955; Liu 1990; Liu et al. 1999; Mahalanobis 1936; Oja 1983; Tukey 1975; Zuo and Serfling 2000]. Of these definitions, the most important and most commonly used is the *Tukey depth* [Hodges 1955; Tukey 1975], also known as *half-space depth* or *location depth*. According to this definition, the depth of a point $\bar{x}$ (which need not be one of our sample points) is the minimum number of sample points contained in any half-space that contains $\bar{x}$ (Figure 14). The *Tukey median* is any point of maximum depth. It follows by applying Helly's theorem to the system of half-spaces containing more than $dn/(d+1)$ observations that, for observations in $\mathbb{R}^d$, the Tukey median must have depth at least $n/(d+1)$. This depth is also its breakdown point, so the Tukey median is robust, and it has other useful statistical properties as well, such as invariance under affine transformations and the ability to form a center-outward ordering of the observations based on their depths.

There has been much research on the computation of Tukey medians, and of other points with high Tukey depth [Chan 2004; Clarkson et al. 1996; Cole 1987; Cole et al. 1987; Jadhav and Mukhopadhyay 1994; Langerman and Steiger 2000; 2001; 2003; Matoušek 1992; Naor and Sharir 1990; Rousseeuw and Ruts 1998; Struyf and Rousseeuw 2000]. Improving on many previously published algorithms, Chan [Chan 2004] found the best bound known for Tukey median construction, $O(n \log n + n^{d-1})$ randomized expected time, using his implicit quasiconvex programming technique.

Let B be a bounding box of the sample points. Each d-tuple t of sample points that are in general position in $\mathbb{R}^d$ defines a hyperplane that bounds two closed half-spaces, H_t^+ and H_t^-. If we associate with each such half-space a number k_t^+ or k_t^- that counts the number of sample points in the corresponding half-space, then the pairs $(B \cap H_t^\pm, -k_t^\pm)$ can be used to form a generalized longest intersecting prefix problem, as defined in Section 2.5; borrowing the terminology of LP-type problems, call any such pair a *constraint*. The solution to the quasiconvex program defined by this set of constraints is a pair $(k, \bar{x})$, where k is minimal and every half-space with more than k samples contains $\bar{x}$. If a half-space H contains fewer than $n - k$ samples, therefore, it does not contain $\bar{x}$, so the depth of $\bar{x}$ is at least $n - k$. Any point of greater depth would lead to a better solution to the problem, so $\bar{x}$ must be a Tukey median of the samples, and we can express the problem of finding a Tukey median as a quasiconvex program. This program, however, has $O(n^d)$ constraints, larger than Chan's claimed time bound. To find Tukey medians more quickly, Chan applies his implicit quasiconvex programming technique: we need to be able to solve constant sized subproblems in constant time, solve decision problems efficiently, and partition large problems into smaller subproblems.

It is tempting to perform the partition step as described after Theorem 3.2, by dividing the set of samples arbitrarily into $d+1$ equal-sized subsets and using the complements of these subsets. However, this idea does not seem to work well for the Tukey median problem: the difficulty is that the numbers $k_t^\pm$ do not depend only on the subset, but on the whole original set of sample points.

Instead, Chan modifies the generalized longest intersecting prefix problem (in a way that doesn't change its optimal value) by including a constraint for every possible half-space, not just those half-spaces bounded by d-tuples of samples. There are infinitely many such constraints but that will not be problematic as long as we can satisfy the requirements of the implicit quasiconvex programming technique. To perform the partition step for this technique, we use a standard tool for divide and conquer in geometric algorithms, known as ε-cuttings. We form the projective dual of the sample points, which is an arrangement of hyperplanes in $\mathbb{R}^d$; each possible constraint boundary is dual to a point in $\mathbb{R}^d$ somewhere in this arrangement, and the number $k_t^\pm$ for the constraint equals the number of arrangement hyperplanes above or below this dual point. We then partition the arrangement into a constant number of simplices, such that each simplex is crossed by at most εn hyperplanes. For each simplex we form a subproblem, consisting of the sample points corresponding to hyperplanes that cross the simplex, together with a constant amount of extra information: the simplex itself and the numbers of hyperplanes that pass above and below it. Each such subproblem corresponds to a set of constraints dual to points in the simplex. When recursively dividing a subproblem already of this form into even smaller sub-subproblems, we intersect the sub-subproblem simplices with the subproblem simplex and partition the resulting polytopes into smaller simplices; this increases the number of sub-subproblems by a constant factor. In this way we fulfill the condition of Theorem 3.2 that we can divide a large problem into a constant number of subproblems, each described by an input of size a constant fraction of the original.

Subproblems of constant size may be solved by constructing and searching the arrangement dual to the samples within the simplex defining the subproblem. It remains to describe how to perform the decision algorithm needed for Theorem 3.2. Decision algorithms for testing the Tukey depth of a point were already known [Rousseeuw and Ruts 1996; Rousseeuw and Struyf 1998], but here we need to solve a slightly more general problem due to the extra information associated with each subproblem. Given k, $\bar{x}$, and a subproblem of our overall problem, we must determine whether there exists a *violated constraint*; that is, a half-space that is dual to a point in the simplex defined by the subproblem, and that contains more than k sample points but does not contain $\bar{x}$. Let H be the hyperplane dual to $\bar{x}$, and Δ be the simplex defining the subproblem. If there exists a violated constraint dual to a point $h \in \Delta$, we can assume without loss of generality that either $h \in H$ or h is on the boundary of Δ; for, if not, we could find another half-space containing as many or more samples by moving h along a vertical line segment until it reaches either H or the boundary. Within H and each boundary plane of the simplex, we can construct the $(d-1)$-dimensional arrangement formed by intersecting this plane with the planes dual to the sample points, in time $O(n \log n + n^{d-1})$. Within each face of these arrangements, all points are dual to half-spaces that contain the same number of samples, and as we move from face to face, the number of sample points contained in the half-spaces changes by ± 1, so we can compute these numbers in constant time per face as we construct these arrangements. By searching all faces of these arrangements we can find a violated constraint, if one exists.

To summarize, by applying the implicit quasiconvex programming technique of Theorem 3.2 to a generalized longest intersecting prefix problem, using ε-cuttings to partition problems into subproblems and $(d-1)$-dimensional arrangements to solve the decision algorithm as described above, Chan [2004] shows how to find the Tukey median of any point set in randomized expected time $O(n \log n + n^{d-1})$.

5. Conclusions

We have introduced quasiconvex programming as a formalization for geometric optimization intermediate in expressivity between linear and convex programming on the one hand, and LP-type problems on the other. Quasiconvex programs are capable of expressing a wide variety of geometric optimization problems and applications, but are still sufficiently concrete that they can be solved both by rapidly converging numeric local improvement techniques and (given the assumption of constant-time primitives for solving constant-sized subproblems) by strongly-polynomial combinatorial optimization algorithms. The power of this approach is demonstrated by the many and varied applications in which quasiconvex programming arises.

References

[Adler and Shamir 1993] I. Adler and R. Shamir, "A randomization scheme for speeding up algorithms for linear and convex quadratic programming problems with a high constraints-to-variables ratio", *Mathematical Programming* **61** (1993), 39–52.

[Amenta 1994] N. Amenta, "Helly-type theorems and generalized linear programming", *Discrete Comput. Geom.* **12** (1994), 241–261.

[Amenta et al. 1999] N. Amenta, M. W. Bern, and D. Eppstein, "Optimal point placement for mesh smoothing", *J. Algorithms* **30**:2 (1999), 302–322. Available at http://www.arxiv.org/cs.CG/9809081.

[Bank and Smith 1997] R. E. Bank and R. K. Smith, "Mesh smoothing using a posteriori error estimates", *SIAM J. Numerical Analysis* **34**:3 (1997), 979–997.

[di Battista et al. 1999] G. di Battista, P. Eades, R. Tamassia, and I. G. Tollis, *Graph drawing: algorithms for the visualization of graphs*, Prentice-Hall, 1999.

[Beigel 1999] R. Beigel, "Finding maximum independent sets in sparse and general graphs", pp. S856–S857 in *Proc. 10th ACM-SIAM Symp. Discrete Algorithms*, 1999. Available at http://www.eecs.uic.edu/~beigel/papers/mis-soda.PS.gz.

[Bern and Eppstein 1995] M. W. Bern and D. Eppstein, "Mesh generation and optimal triangulation", pp. 47–123 in *Computing in euclidean geometry*, second ed., edited by D.-Z. Du and F. K.-M. Hwang, Lecture Notes Series on Computing **4**, World Scientific, 1995. Available at http://www.ics.uci.edu/~eppstein/pubs/BerEpp-CEG-95.pdf.

[Bern and Eppstein 2001] M. W. Bern and D. Eppstein, "Optimal Möbius transformations for information visualization and meshing", pp. 14–25 in *Proc. 7th Worksh. Algorithms and Data Structures*, edited by F. Dehne et al., Lecture Notes in Computer Science **2125**, Springer, 2001. Available at http://www.arxiv.org/cs.CG/0101006.

[Bern and Eppstein 2003] M. W. Bern and D. Eppstein, "Optimized color gamuts for tiled displays", pp. 274–281 in *Proc. 19th ACM Symp. Computational Geometry*, 2003. Available at http://www.arxiv.org/cs.CG/0212007.

[Bern and Plassmann 2000] M. W. Bern and P. E. Plassmann, "Mesh generation", Chapter 6, pp. 291–332 in *Handbook of computational geometry*, edited by J.-R. Sack and J. Urrutia, Elsevier, 2000.

[Brightwell and Scheinerman 1993] G. R. Brightwell and E. R. Scheinerman, "Representations of planar graphs", *SIAM J. Discrete Math.* **6**:2 (1993), 214–229.

[Byskov 2003] J. M. Byskov, "Algorithms for k-colouring and finding maximal independent sets", pp. 456–457 in *Proc. 14th ACM-SIAM Symp. Discrete Algorithms*, 2003.

[Canann et al. 1998] S. A. Canann, J. R. Tristano, and M. L. Staten, "An approach to combined Laplacian and optimization-based smoothing for triangular, quadrilateral, and quad-dominant meshes", pp. 479–494 in *Proc. 7th Int. Meshing Roundtable*, Sandia Nat. Lab., 1998. Available at http://www.andrew.cmu.edu/user/sowen/ abstracts/Ca513.html.

[Chan 2004] T. M.-Y. Chan, "An optimal randomized algorithm for maximum Tukey depth", pp. 423–429 in *Proc. 15th ACM-SIAM Symp. Discrete Algorithms*, 2004.

[Chazelle and Matoušek 1993] B. Chazelle and J. Matoušek, "On linear-time deterministic algorithms for optimization problems in fixed dimensions", pp. 281–290 in *Proc. 4th ACM-SIAM Symp. Discrete Algorithms*, 1993.

[Clarkson 1986] K. L. Clarkson, "Linear programming in $O(n \times 3^{d^2})$ time", *Information Processing Letters* **22** (1986), 21–24.

[Clarkson 1987] K. L. Clarkson, "New applications of random sampling in computational geometry", *Discrete Comput. Geom.* **2** (1987), 195–222.

[Clarkson 1995] K. L. Clarkson, "Las Vegas algorithms for linear and integer programming when the dimension is small", *J. ACM* **42**:2 (1995), 488–499.

[Clarkson et al. 1996] K. L. Clarkson, D. Eppstein, G. L. Miller, C. Sturtivant, and S.-H. Teng, "Approximating center points with iterated Radon points", *Int. J. Computational Geometry & Applications* **6**:3 (1996), 357–377.

[Cole 1987] R. Cole, "Slowing down sorting networks to obtain faster sorting algorithms", *J. ACM* **34** (1987), 200–208.

[Cole et al. 1987] R. Cole, M. Sharir, and C. K. Yap, "On k-hulls and related problems", *SIAM J. Computing* **16** (1987), 61–77.

[Dantsin and Hirsch 2000] E. Dantsin and E. A. Hirsch, "Algorithms for k-SAT based on covering codes", Preprint 1/2000, Steklov Inst. of Mathematics, St. Petersburg, 2000. Available at ftp://ftp.pdmi.ras.ru/pub/publicat/preprint/2000/01-00.ps.gz.

[Dharmadhikari and Joag-Dev 1988] S. Dharmadhikari and K. Joag-Dev, *Unimodality, convexity and applications*, Academic Press, 1988.

[Djidjev 2000] H. N. Djidjev, "Force-directed methods for smoothing unstructured triangular and tetrahedral meshes", pp. 395–406 in *Proc. 9th Int. Meshing Roundtable*, Sandia Nat. Lab., 2000. Available at http://www.andrew.cmu.edu/user/sowen/ abstracts/Dj763.html.

[Driscoll and Vavasis 1998] T. A. Driscoll and S. A. Vavasis, "Numerical conformal mapping using cross-ratios and Delaunay triangulation", *SIAM J. Sci. Computation* **19**:6 (1998), 1783–1803. Available at ftp://ftp.cs.cornell.edu/pub/vavasis/ papers/crdt.ps.gz.

[Dyer 1984] M. E. Dyer, "On a multidimensional search procedure and its application to the Euclidean one-centre problem", *SIAM J. Computing* **13** (1984), 31–45.

[Dyer 1992] M. E. Dyer, "On a class of convex programs with applications to computational geometry", pp. 9–15 in *Proc. 8th ACM Symp. Computational Geometry*, 1992.

[Dyer and Frieze 1989] M. E. Dyer and A. M. Frieze, "A randomized algorithm for fixed-dimensional linear programming", *Mathematical Programming* **44** (1989), 203–212.

[Eppstein 2001a] D. Eppstein, "Improved algorithms for 3-coloring, 3-edge-coloring, and constraint satisfaction", pp. 329–337 in *Proc. 12th ACM-SIAM Symp. Discrete Algorithms*, 2001. Available at http://www.arxiv.org/cs.DS/0009006.

[Eppstein 2001b] D. Eppstein, "Small maximal independent sets and faster exact graph coloring", pp. 462–470 in *Proc. 7th Worksh. Algorithms and Data Structures*, edited by F. Dehne et al., Lecture Notes in Computer Science **2125**, Springer, 2001. Available at http://www.arxiv.org/cs.DS/0011009.

[Eppstein 2003a] D. Eppstein, "Setting parameters by example", *SIAM J. Computing* **32**:3 (2003), 643–653. Available at http://dx.doi.org/10.1137/S0097539700370084.

[Eppstein 2003b] D. Eppstein, "The traveling salesman problem for cubic graphs", pp. 307–318 in *Proc. 8th Worksh. Algorithms and Data Structures* (Ottawa, 2003), edited by F. Dehne et al., Lecture Notes in Computer Science **2748**, Springer, 2003. Available at http://www.arxiv.org/cs.DS/0302030.

[Eppstein 2004] D. Eppstein, "Quasiconvex analysis of backtracking algorithms", pp. 781–790 in *Proc. 15th ACM-SIAM Symp. Discrete Algorithms*, 2004. Available at http://www.arxiv.org/cs.DS/0304018.

[Eppstein and Wortman 2005] D. Eppstein and K. Wortman, "Minimum dilation stars", in *Proc. 21st ACM Symp. Computational Geometry*, 2005. To appear.

[Fischer et al. 2003] K. Fischer, B. Gärtner, and M. Kutz, "Fast smallest-enclosing-ball computation in high dimensions", pp. 630–641 in *Proc. 11th Eur. Symp. Algorithms*, Lecture Notes in Computer Science **2832**, Springer, 2003.

[Freitag 1997] L. A. Freitag, "On combining Laplacian and optimization-based mesh smoothing techniques", pp. 37–43 in *Proc. Symp. Trends in Unstructured Mesh Generation*, Amer. Soc. Mechanical Engineers, 1997. Available at ftp://info.mcs.anl.gov/pub/tech_reports/plassman/lori_combined.ps.Z.

[Freitag and Ollivier-Gooch 1997] L. A. Freitag and C. F. Ollivier-Gooch, "Tetrahedral mesh improvement using face swapping and smoothing", *Int. J. Numerical Methods in Engineering* **40**:21 (1997), 3979–4002. Available at ftp://info.mcs.anl.gov/pub/tech_reports/plassman/lori_improve.ps.Z.

[Freitag et al. 1995] L. A. Freitag, M. T. Jones, and P. E. Plassmann, "An efficient parallel algorithm for mesh smoothing", pp. 47–58 in *Proc. 4th Int. Meshing Roundtable*, Sandia Nat. Lab., 1995.

[Freitag et al. 1999] L. A. Freitag, M. T. Jones, and P. E. Plassmann, "A parallel algorithm for mesh smoothing", *SIAM J. Scientific Computing* **20**:6 (1999), 2023–2040.

[Gärtner 1995] B. Gärtner, "A subexponential algorithm for abstract optimization problems", *SIAM J. Computing* **24** (1995), 1018–1035.

[Gärtner 1999] B. Gärtner, "Fast and robust smallest enclosing balls", pp. 325–338 in *Proc. 7th Eur. Symp. Algorithms*, Lecture Notes in Computer Science **1643**, Springer, 1999.

[Gärtner and Fischer 2003] B. Gärtner and K. Fischer, "The smallest enclosing ball of balls: combinatorial structure and algorithms", pp. 292–301 in *Proc. 19th ACM Symp. Computational Geometry*, 2003.

[Gill et al. 1992] J. Gill, W. Steiger, and A. Wigderson, "Geometric medians", *Discrete Math.* **108** (1992), 37–51.

[Gramm et al. 2000] J. Gramm, E. A. Hirsch, R. Niedermeier, and P. Rossmanith, "Better worst-case upper bounds for MAX-2-SAT", in *Proc. 3rd Worksh. on the Satisfiability Problem*, 2000. Available at http://ssor.twi.tudelft.nl/~warners/SAT2000abstr/hirsch.html.

[Hliněný 1997] P. Hliněný, "Touching graphs of unit balls", pp. 350–358 in *Proc. 5th Int. Symp. Graph Drawing*, Lecture Notes in Computer Science **1353**, Springer, 1997.

[Hodges 1955] J. L. Hodges, "A bivariate sign test", *Ann. Mathematical Statistics* **26** (1955), 523–527.

[Howell 1990] L. H. Howell, *Computation of conformal maps by modified Schwarz–Christoffel transformations*, Ph.D. thesis, MIT, 1990. Available at http://gov/www.llnl.CASC/people/howell/lhhphd.ps.gz.

[Humphreys and Hanrahan 1999] G. Humphreys and P. Hanrahan, "A distributed graphics system for large tiled displays", in *Proc. IEEE Visualization '99*, 1999. Available at http://graphics.stanford.edu/papers/mural_design/mural_design.pdf.

[Hurdal et al. 1999] M. K. Hurdal, P. L. Bowers, K. Stephenson, D. W. L. Summers, K. Rehm, K. Shaper, and D. A. Rottenberg, "Quasi-conformally flat mapping the human cerebellum", Technical Report FSU-99-05, Florida State Univ., Dept. of Mathematics, 1999. Available at http://www.math.fsu.edu/~aluffi/archive/paper98.ps.gz.

[Iversen 1992] B. Iversen, *Hyperbolic geometry*, London Math. Soc. Student Texts **25**, Cambridge Univ. Press, 1992.

[Jadhav and Mukhopadhyay 1994] S. Jadhav and A. Mukhopadhyay, "Computing a centerpoint of a finite planar set of points in linear time", *Discrete Comput. Geom.* **12** (1994), 291–312.

[Karmarkar 1984] N. Karmarkar, "A new polynomial-time algorithm for linear programming", *Combinatorica* **4** (1984), 373–395.

[Khachiyan 1980] L. G. Khachiyan, "Polynomial algorithm in linear programming", *U.S.S.R. Comput. Math. and Math. Phys.* **20** (1980), 53–72.

[Kiwiel 2001] K. C. Kiwiel, "Convergence and efficiency of subgradient methods for quasiconvex minimization", *Mathematical Programming* **90**:1 (2001), 1–25.

[Koebe 1936] P. Koebe, "Kontaktprobleme der konformen Abbildung", *Ber. Verh. Sächs. Akad. Wiss. Leipzig Math.-Phys. Kl.* **88** (1936), 141–164.

[Lamping et al. 1995] J. Lamping, R. Rao, and P. Pirolli, "A focus+context technique based on hyperbolic geometry for viewing large hierarchies", pp. 401–408 in *Proc. ACM Conf. Human Factors in Computing Systems*, 1995. Available at http://www.parc.xerox.com/istl/projects/uir/pubs/pdf/UIR-R-1995-04-Lamping-CHI95-FocusContext.pdf.

[Langerman and Steiger 2000] S. Langerman and W. Steiger, "Computing a maximal depth point in the plane", pp. 46 in *Proc. 4th Japan Conf. Discrete Comput. Geom.*, edited by J. Akiyama et al., Lecture Notes in Computer Science **2098**, Springer, 2000.

[Langerman and Steiger 2001] S. Langerman and W. Steiger, "Computing a high depth point in the plane", pp. 227–233 in *Developments in Robust Statistics: Proc. ICORS 2001*, edited by R. Dutter et al., 2001.

[Langerman and Steiger 2003] S. Langerman and W. Steiger, "Optimization in arrangements", pp. 50–61 in *Proc. STACS 2003: 20th Ann. Symp. Theoretical Aspects of Computer Science*, edited by H. Alt and M. Habib, Lecture Notes in Computer Science **2607**, Springer, 2003.

[Li et al. 2000] K. Li, H. Chen, Y. Chen, D. W. Clark, P. Cook, S. Damianakis, G. Essl, A. Finkelstein, T. Funkhouser, A. Klein, Z. Liu, E. Praun, R. Samanta, B. Shedd, J. P. Singh, G. Tzanetakis, and J. Zheng, "Early experiences and challenges in building and using a scalable display wall system", *IEEE Computer Graphics & Appl.* **20**:4 (2000), 671–680. Available at http://www.cs.princeton.edu/omnimedia/papers/cga00.pdf.

[Liu 1990] R. Liu, "On a notion of data depth based on random simplices", *Ann. Statistics* **18** (1990), 405–414.

[Liu et al. 1999] R. Liu, J. M. Parelius, and K. Singh, "Multivariate analysis by data depth: descriptive statistics, graphics and inference", *Ann. Statistics* **27** (1999), 783–858.

[Mahalanobis 1936] P. C. Mahalanobis, "On the generalized distance in statistics", *Proc. Nat. Acad. Sci. India* **12** (1936), 49–55.

[Majumder et al. 2000] A. Majumder, Z. He, H. Towles, and G. Welch, "Achieving color uniformity across multi-projector displays", in *Proc. IEEE Visualization 2000*, 2000. Available at http://www.cs.unc.edu/~welch/media/pdf/vis00_color.pdf.

[Matoušek 1992] J. Matoušek, "Computing the center of a planar point set", pp. 221–230 in *Discrete and computational geometry: Papers from the DIMACS Special Year*, edited by J. E. Goodman et al., Amer. Math. Soc., 1992.

[Matoušek et al. 1996] J. Matoušek, M. Sharir, and E. Welzl, "A subexponential bound for linear programming", *Algorithmica* **16** (1996), 498–516.

[McKay 1989] J. McKay, "Sighting point", Message to sci.math bulletin board, April 1989. Available at http://www.ics.uci.edu/~eppstein/junkyard/maxmin-angle.html.

[Megiddo 1983] N. Megiddo, "Linear time algorithms for linear programming in $\mathbb{R}^3$ and related problems", *SIAM J. Computing* **12** (1983), 759–776.

[Megiddo 1984] N. Megiddo, "Linear programming in linear time when the dimension is fixed", *J. ACM* **31** (1984), 114–127.

[Megiddo 1989] N. Megiddo, "On the ball spanned by balls", *Discrete Comput. Geom.* **4** (1989), 605–610.

[Munzner 1997] T. Munzner, "Exploring large graphs in 3D hyperbolic space", *IEEE Comp. Graphics Appl.* **18**:4 (1997), 18–23. Available at http://graphics.stanford.edu/papers/h3cga/.

[Munzner and Burchard 1995] T. Munzner and P. Burchard, "Visualizing the structure of the world wide web in 3D hyperbolic space", pp. 33–38 in *Proc. VRML '95*, ACM, 1995. Available at http://www.geom.umn.edu/docs/research/webviz/webviz/.

[Naor and Sharir 1990] N. Naor and M. Sharir, "Computing a point in the center of a point set in three dimensions", pp. 10–13 in *Proc. 2nd Canad. Conf. Comput. Geom.*, 1990.

[Oja 1983] H. Oja, "Descriptive statistics for multivariate distributions", *Stat. Prob. Lett.* **1** (1983), 327–332.

[Paturi et al. 1998] R. Paturi, P. Pudlák, M. E. Saks, and F. Zane, "An improved exponential-time algorithm for k-SAT", pp. 628–637 in *Proc. 39th IEEE Symp. Foundations of Computer Science*, 1998. Available at http://www.math.cas.cz/ ~pudlak/ppsz.ps.

[Post 1984] M. J. Post, "Minimum spanning ellipsoids", pp. 108–116 in *Proc. 16th ACM Symp. Theory of Computing*, 1984.

[Raskar et al. 1999] R. Raskar, M. S. Brown, R. Yang, W.-C. Chen, G. Welch, H. Towles, B. Seales, and H. Fuchs, "Multi-projector displays using camera-based registration", in *Proc. IEEE Visualization '99*, 1999. Available at http://www.cs.unc.edu/ Research/ootf/stc/Seamless/.

[Rousseeuw and Ruts 1996] P. J. Rousseeuw and I. Ruts, "Algorithm AS 307: bivariate location depth", *J. Royal Statistical Soc., Ser. C: Applied Statistics* **45** (1996), 516–526.

[Rousseeuw and Ruts 1998] P. J. Rousseeuw and I. Ruts, "Constructing the bivariate Tukey median", *Statistica Sinica* **8**:3 (1998), 827–839.

[Rousseeuw and Struyf 1998] P. J. Rousseeuw and A. Struyf, "Computing location depth and regression depth in higher dimensions", *Statistics and Computing* **8** (1998), 193–203.

[Sachs 1994] H. Sachs, "Coin graphs, polyhedra, and conformal mapping", *Discrete Math.* **134**:1–3 (1994), 133–138.

[Schöning 1999] U. Schöning, "A probabilistic algorithm for k-SAT and constraint satisfaction problems", pp. 410–414 in *Proc. 40th IEEE Symp. Foundations of Computer Science*, 1999.

[Seidel 1991] R. Seidel, "Low dimensional linear programming and convex hulls made easy", *Discrete Comput. Geom.* **6** (1991), 423–434.

[Smith 1991] W. D. Smith, "Accurate circle configurations and numerical conformal mapping in polynomial time", preprint, 1991. Available at http://citeseer.nj.nec.com/ smith94accurate.html.

[Stenger and Schmidtlein 1997] F. Stenger and R. Schmidtlein, "Conformal maps via sinc methods", pp. 505–549 in *Proc. Conf. Computational Methods in Function Theory*, edited by N. Papamichael et al., World Scientific, 1997. Available at http:// www.cs.utah.edu/~stenger/PAPERS/stenger-sinc-comformal-maps.ps.

[Stone 2001] M. C. Stone, "Color and brightness appearance issues for tiled displays", *IEEE Computer Graphics & Appl.* **21**:5 (2001), 58–66. Available at http:// graphics.stanford.edu/papers/tileddisplays/.

[Struyf and Rousseeuw 2000] A. Struyf and P. J. Rousseeuw, "High-dimensional computation of deepest location", *Computational Statistics and Data Analysis* **34** (2000), 415–426.

[Thompson et al. 1985] J. F. Thompson, Z. U. A. Warsi, and C. W. Mastin, *Numerical grid generation: foundations and applications*, North-Holland, 1985.

[Trefethen 1980] L. N. Trefethen, "Numerical computation of the Schwarz-Christoffel transformation", *SIAM J. Sci. Stat. Comput.* **1**:1 (1980), 82–102.

[Tukey 1975] J. W. Tukey, "Mathematics and the picturing of data", pp. 523–531 in *Proc. Int. Congress of Mathematicians*, vol. 2, edited by R. D. James, Canadian Mathematical Congress, Vancouver, 1975.

[Vollmer et al. 1999] J. Vollmer, R. Mencl, and H. Müller, "Improved Laplacian smoothing of noisy surface meshes", in *Proc. 20th Conf. Eur. Assoc. for Computer Graphics (EuroGraphics '99)*, 1999.

[Welzl 1991] E. Welzl, "Smallest enclosing disks (balls and ellipsoids)", pp. 359–370 in *New results and new trends in computer science*, edited by H. Maurer, Lecture Notes in Computer Science **555**, Springer, 1991.

[Xu 2001] H. Xu, "Level function method for quasiconvex programming", *J. Optimization Theory and Applications* **108**:2 (2001), 407–437.

[Zoutendijk 1960] G. Zoutendijk, *Methods of feasible directions: A study in linear and non-linear programming*, Elsevier, 1960.

[Zuo and Serfling 2000] Y. Zuo and R. Serfling, "General notions of statistical depth function", *Ann. Statistics* **28**:2 (2000), 461–482.

DAVID EPPSTEIN
COMPUTER SCIENCE DEPARTMENT
DONALD BREN SCHOOL OF INFORMATION AND COMPUTER SCIENCES
UNIVERSITY OF CALIFORNIA, IRVINE
IRVINE, CA 92697-3424
UNITED STATES
eppstein@uci.edu

Combinatorial and Computational Geometry
MSRI Publications
Volume **52**, 2005

De Concini–Procesi Wonderful Arrangement Models: A Discrete Geometer's Point of View

EVA MARIA FEICHTNER

ABSTRACT. This article outlines the construction of De Concini–Procesi arrangement models and describes recent progress in understanding their significance from the algebraic, geometric, and combinatorial point of view. Throughout the exposition, strong emphasis is given to the combinatorial and discrete geometric data that lie at the core of the construction.

CONTENTS

1. An Invitation to Arrangement Models   333
2. Introducing the Main Character   334
3. The Combinatorial Core Data: A Step Beyond Geometry   342
4. Returning to Geometry   348
5. Adding Arrangement Models to the Geometer's Toolbox   353
Acknowledgments   358
References   359

1. An Invitation to Arrangement Models

The complements of coordinate hyperplanes in a real or complex vector space are easy to understand: The coordinate hyperplanes in $\mathbb{R}^n$ dissect the space into 2^n open orthants; removing the coordinate hyperplanes from $\mathbb{C}^n$ leaves the complex torus $(\mathbb{C}^*)^n$. Arbitrary subspace arrangements, i.e., finite families of linear subspaces, have complements with far more intricate combinatorics in the real case, and far more intricate topology in the complex case. Arrangement models improve this complicated situation locally — constructing an arrangement model means to alter the ambient space so as to preserve the complement and to replace the arrangement by a divisor with normal crossings, i.e., a collection of smooth hypersurfaces which locally intersect like coordinate hyperplanes. Almost a decade ago, De Concini and Procesi provided a canonical construction of arrangement models — *wonderful arrangement models* — that had significant impact in various fields of mathematics.

Why should a discrete geometer be interested in this model construction?

Because there is a wealth of *wonderful* combinatorial and discrete geometric structure lying at the heart of the matter. Our aim here is to bring these discrete pearls to light.

First, combinatorial data plays a descriptive role in various places: The combinatorics of the arrangement fully prescribes the model construction and a natural stratification of the resulting space. We will see details and examples in Section 2. In fact, the rather coarse combinatorial data reflect enough of the situation so as to, for instance, determine algebraic-topological invariants of the arrangement models (compare the topological interpretation of the algebra $D(\mathcal{L}, \mathcal{G})$ in Section 4.2).

Secondly, the combinatorial data put forward in the study of arrangement models invites purely combinatorial generalizations. We discuss such generalizations in Section 3 and show in Section 4 how this combinatorial generalization opens unexpected views when related back to geometry.

Finally, in Section 5 we propose arrangement models as a tool for resolving group actions on manifolds. Again, it is an open eye for discrete core data that enables the construction.

We have kept the exposition self-contained and illustrated it with many examples. We invite discrete geometers to discover an algebro-geometric context in which familiar discrete structures play a key role. We hope that yet many more bridges will be built between algebraic and discrete geometry — areas that, despite the differences in terminology, concepts, and methods, share what has inspired and driven mathematicians for centuries: a passion for geometry.

2. Introducing the Main Character

2.1. Basics on arrangements. We first need to fix some basic terminology, in particular as it concerns the combinatorial data of an arrangement. We suggest that the reader, who is not familiar with the setting, reads through the first part of this Section and compares the notions to the illustrations given for braid arrangements in Example 2.1.

An *arrangement* $\mathcal{A} = \{U_1, \ldots, U_n\}$ is a finite family of linear subspaces in a real or complex vector space V. The topological space most obviously associated to such an arrangement is its *complement* in the ambient space, $\mathcal{M}(\mathcal{A}) := V \setminus \bigcup \mathcal{A}$.

Having arrangements in real vector spaces in mind, the topology of $\mathcal{M}(\mathcal{A})$ does not look very interesting: the complement is a collection of open polyhedral cones, and apart from their number there is no significant associated topological data. In the complex case, however, already a single hyperplane in $\mathbb{C}^1$, the origin, has a nontrivial complement: it is homotopy equivalent to S^1, the 1-dimensional sphere. The complement of two (for instance, coordinate) hyperplanes in $\mathbb{C}^2$ is homotopy equivalent to the torus $S^1 \times S^1$.

The combinatorial data associated with an arrangement is usually recorded in the *intersection lattice* $\mathcal{L} = \mathcal{L}(\mathcal{A})$, which is the set of intersections of subspaces in $\mathcal{A}$, partially ordered by reversed inclusion. We adopt terminology from the theory of partially ordered sets and often denote the unique minimum in $\mathcal{L}(\mathcal{A})$ (corresponding to the empty intersection, i.e., the ambient space V) by $\hat{0}$ and the unique maximum of $\mathcal{L}(\mathcal{A})$ (the overall intersection of subspaces in $\mathcal{A}$) by $\hat{1}$. In many situations, the elements of the intersection lattice are labeled by the codimension of the corresponding intersection. For arrangements of hyperplanes, this information is recorded in the rank function of the lattice — the codimension of an intersection X is the number of elements in a maximal chain in the half-open interval $(\hat{0}, X]$ in $\mathcal{L}(\mathcal{A})$.

As with any poset, we can consider the *order complex* $\Delta(\bar{\mathcal{L}})$ of the proper part, $\bar{\mathcal{L}} := \mathcal{L} \setminus \{\hat{0}, \hat{1}\}$, of the intersection lattice, i.e., the abstract simplicial complex formed by the linearly ordered subsets in $\bar{\mathcal{L}}$,

$$\Delta(\bar{\mathcal{L}}) = \{ X_1 < \ldots < X_k \mid X_i \in \mathcal{L} \setminus \{\hat{0}, \hat{1}\}\}.$$

The topology of $\Delta(\bar{\mathcal{L}})$ plays a prominent role for describing the topology of arrangement complements. For instance, it is the crucial ingredient for the explicit description of cohomology groups of $\mathcal{M}(\mathcal{A})$ by Goresky and MacPherson [1988, Part III].

For hyperplane arrangements, the homotopy type of $\Delta(\bar{\mathcal{L}})$ is well known: the complex is homotopy equivalent to a wedge of spheres of dimension equal to the codimension of the total intersection of $\mathcal{A}$. The number of spheres can as well be read from the intersection lattice, it is the absolute value of its Möbius function. For subspace arrangements however, the barycentric subdivision of any finite simplicial complex can appear as the order complex of the intersection lattice.

Besides $\Delta(\bar{\mathcal{L}})$, we will often refer to the cone over $\Delta(\bar{\mathcal{L}})$ obtained by extending the linearly ordered sets in $\bar{\mathcal{L}}$ by the maximal element $\hat{1}$ in $\mathcal{L}$. We will denote this complex by $\Delta(\mathcal{L} \setminus \{\hat{0}\})$ or $\Delta(\mathcal{L}_{>\hat{0}})$.

In order to have a standard example at hand, we briefly discuss braid arrangements. This class of arrangements has figured prominently in many places and has helped develop lots of arrangement theory over the last decades.

EXAMPLE 2.1 (BRAID ARRANGEMENTS). The arrangement $\mathcal{A}_{n-1}$ given by the hyperplanes

$$H_{ij} : \quad x_i = x_j, \quad \text{for } 1 \leq i < j \leq n,$$

in real n-dimensional vector space is called the (*real*) *rank* $n-1$ *braid arrangement*. There is a complex version of this arrangement. It consists of hyperplanes H_{ij} in $\mathbb{C}^n$ given by the *same* linear equations. We denote the arrangement by $\mathcal{A}_{n-1}^{\mathbb{C}}$. Occasionally, we will use the analogous $\mathcal{A}_{n-1}^{\mathbb{R}}$ if we want to stress the real

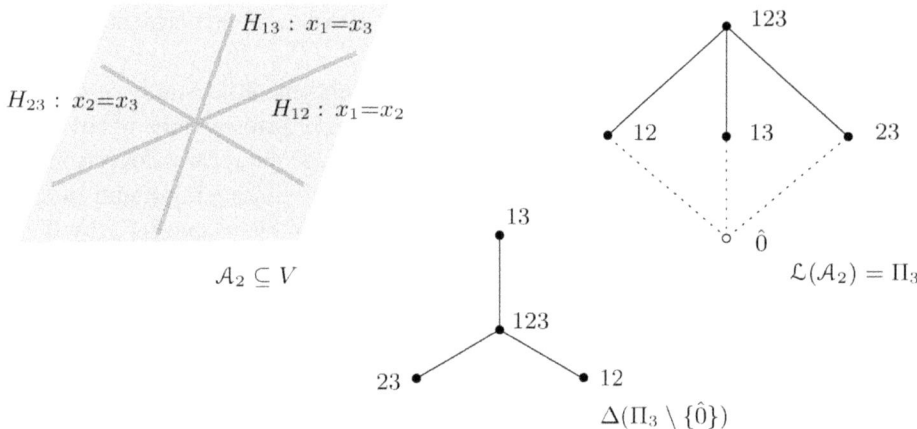

Figure 1. The rank 2 braid arrangement $\mathcal{A}_2$, its intersection lattice Π_3, and the order complex $\Delta(\Pi_3 \setminus \{\hat{0}\})$.

setting. In many situations a similar reasoning applies to the real and to the complex case. To simplify notation, we then use $\mathbb{K}$ to denote $\mathbb{R}$ or $\mathbb{C}$.

Observe that the diagonal $\Delta = \{x \in \mathbb{K}^n \mid x_1 = \ldots = x_n\}$ is the overall intersection of hyperplanes in $\mathcal{A}_{n-1}$. Without loosing any relevant information on the topology of the complement, we will often consider $\mathcal{A}_{n-1}$ as an arrangement in complex or real $(n-1)$-dimensional space $V = \mathbb{K}^n/\Delta \cong \{x \in \mathbb{K}^n \mid \sum x_i = 0\}$. This explains the indexing for braid arrangements, which may appear unusual at first sight.

The complement $\mathcal{M}(\mathcal{A}_{n-1}^{\mathbb{R}})$ is a collection of $n!$ polyhedral cones, corresponding to the $n!$ linear orders on n pairwise noncoinciding coordinate entries. The complement $\mathcal{M}(\mathcal{A}_{n-1}^{\mathbb{C}})$ is the classical configuration space of the complex plane

$$F(\mathbb{C}, n) = \{(x_1, \ldots, x_n) \in \mathbb{C}^n \mid x_i \neq x_j \text{ for } i \neq j\}.$$

This space is the classifying space of the pure braid group, which explains the occurrence of the term "braid" for this class of arrangements.

As the intersection lattice of the braid arrangement $\mathcal{A}_{n-1}$ we recognize the *partition lattice* Π_n, i.e., the set of set partitions of $\{1, \ldots, n\}$ ordered by reversed refinement. The correspondence to intersections in the braid arrangement can be easily described: The blocks of a partition correspond to sets of coordinates with identical entries, thus to the set of points in the corresponding intersection of hyperplanes.

The order complex $\Delta(\overline{\Pi}_n)$ is a pure, $(n-1)$-dimensional complex that is homotopy equivalent to a wedge of $(n-1)!$ spheres of dimension $n-1$.

In Figure 1 we depict the real rank 2 braid arrangement $\mathcal{A}_2$ in $V = \mathbb{R}^3/\Delta$, its intersection lattice Π_3, and the order complex $\Delta(\Pi_3 \setminus \{\hat{0}\})$. We denote partitions in Π_3 by their nontrivial blocks. The complex depicted is a cone over $\Delta(\overline{\Pi}_3)$, a union of three points, which indeed is the wedge of two 0-dimensional spheres.

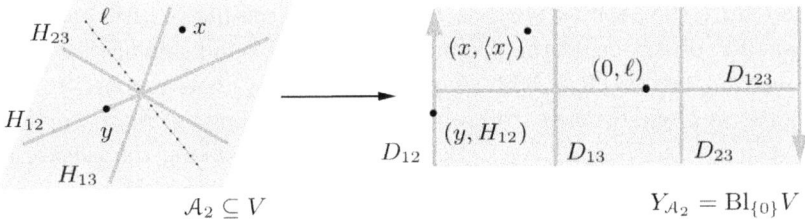

Figure 2. The maximal wonderful model for $\mathcal{A}_2$.

2.2. The model construction. We provide two alternative definitions for De Concini–Procesi arrangement models: the first one describes the models as closures of open embeddings of the arrangement complements. It comes in handy for technical purposes. By contrast, the second definition, which describes arrangement models as results of sequences of blowups, is much more intuitive and suitable for inductive constructions and proofs.

DEFINITION 2.2 (MODEL CONSTRUCTION I). Let $\mathcal{A}$ be an arrangement of real or complex linear subspaces in V. Consider the map

$$\Psi : \quad \mathcal{M}(\mathcal{A}) \to V \times \prod_{X \in \mathcal{L}_{>\hat{0}}} \mathbb{P}(V/X) \tag{2-1}$$

$$x \mapsto \left(x, (\langle x, X \rangle / X)_{X \in \mathcal{L}_{>\hat{0}}} \right);$$

it encodes the relative position of each point in the arrangement complement $\mathcal{M}(\mathcal{A})$ with respect to the intersection of subspaces in $\mathcal{A}$. The map Ψ is an open embedding; the closure of its image is called the *(maximal) De Concini–Procesi wonderful model for $\mathcal{A}$* and is denoted by $Y_{\mathcal{A}}$.

DEFINITION 2.3 (MODEL CONSTRUCTION II). Let $\mathcal{A}$ be an arrangement of real or complex linear subspaces in V. Let $X_1, \ldots, X_t$ be a linear extension of the opposite order $\mathcal{L}^{\mathrm{op}}_{>\hat{0}}$ on $\mathcal{L}_{>\hat{0}}$. The *(maximal) De Concini–Procesi wonderful model for $\mathcal{A}$* is the result $Y_{\mathcal{A}}$ of successively blowing up subspaces $X_1, \ldots, X_t$, respectively their proper transforms.

To avoid confusion with spherical blowups that have been appearing in model constructions as well [Gaiffi 2003], let us emphasize here that, also in the real setting, we think about blowups as substituting points by projective spaces. Before we list the main properties of arrangement models let us look at a first example.

EXAMPLE 2.4 (THE ARRANGEMENT MODEL $Y_{\mathcal{A}_2}$). Consider the rank 2 braid arrangement $\mathcal{A}_2$ in $V = \mathbb{R}^3/\Delta$. Following the description in Definition 2.3 we obtain $Y_{\mathcal{A}_2}$ by a single blowup of V at $\{0\}$. The result is an open Möbius band; the exceptional divisor $D_{123} \cong \mathbb{R}\mathbb{P}^1$ in $Y_{\mathcal{A}_2}$ intersects transversally with the proper transforms D_{ij} of the hyperplanes H_{ij}, $1 \le i < j \le 3$. We illustrate the blowup in Figure 2.

To recognize the Möbius band as the closure of the image of Ψ according to Definition 2.2, observe that the product on the right-hand side of (2–1) consists of two relevant factors, $V \times \mathbb{RP}^1$. A point x in $\mathcal{M}(\mathcal{A}_2)$ gets mapped to $(x, \langle x \rangle)$ and we observe a one-to-one correspondence between points in $\mathcal{M}(\mathcal{A}_2)$ and points in $Y_{\mathcal{A}_2} \setminus (D_{123} \cup D_{12} \cup D_{13} \cup D_{23})$. Points added when taking the closure are of the form (y, H_{ij}) for $y \in H_{ij} \setminus \{0\}$ and $(0, \ell)$ for ℓ some line in V.

Observe that the triple intersection of hyperplanes in V has been replaced by double intersections of hypersurfaces in $Y_{\mathcal{A}_2}$. Without changing the topology of the arrangement complement, the arrangement of hyperplanes has been replaced by a normal crossing divisor. Moreover, note that the irreducible divisor components D_{12}, D_{13}, D_{23}, and D_{123} intersect if and only if their indexing lattice elements form a chain in $\mathcal{L}(\mathcal{A}_2)$.

The observations we made for $Y_{\mathcal{A}_2}$ are special cases of the main properties of (maximal) De Concini–Procesi models that we list in the following:

THEOREM 2.5 [De Concini and Procesi 1995a, Theorems in § 3.1 and § 3.2]. (1) *The arrangement model Y_A as defined in 2.2 and 2.3 is a smooth variety with a natural projection map to the original ambient space, $\pi : Y_A \longrightarrow V$, which is one-to-one on the arrangement complement $\mathcal{M}(\mathcal{A})$.*

(2) *The complement of $\pi^{-1}(\mathcal{M}(\mathcal{A}))$ in Y_A is a divisor with normal crossings; its irreducible components are the proper transforms D_X of intersections X in $\mathcal{L}$,*

$$Y_A \setminus \pi^{-1}(\mathcal{M}(\mathcal{A})) = \bigcup_{X \in \mathcal{L}_{>\hat{0}}} D_X.$$

(3) *Irreducible components D_X for $X \in \mathcal{S} \subseteq \mathcal{L}_{>\hat{0}}$ intersect if and only if $\mathcal{S}$ is a linearly ordered subset in $\mathcal{L}_{>\hat{0}}$. If we think about Y_A as stratified by the irreducible components of the normal crossing divisor and their intersections, then the poset of strata coincides with the face poset of the order complex $\Delta(\mathcal{L}_{>\hat{0}})$.*

EXAMPLE 2.6 (THE ARRANGEMENT MODEL $Y_{\mathcal{A}_3}$). We now consider a somewhat larger and more complicated example, the rank 3 braid arrangement $\mathcal{A}_3$ in $V \cong \mathbb{R}^4/\Delta$. First note that the intersection lattice of $\mathcal{A}_3$ is the partition lattice Π_4, which we depict in Figure 3 for later reference. Again, we denote partitions by their nontrivial blocks.

Following again the description of arrangement models given in Definition 2.3, the first step is to blow up V at $\{0\}$. We obtain a line bundle over $\mathbb{RP}^2$; in Figure 4 we depict the exceptional divisor $D_{1234} \cong \mathbb{RP}^2$ stratified by the intersections of proper transforms of hyperplanes in $\mathcal{A}_3$.

This first step is now followed by the blowup of triple, respectively double intersections of proper transforms of hyperplanes in arbitrary order. In each such intersection the situation locally corresponds to the blowup of a 2-dimensional real vector space in a point as discussed in Example 2.4. Topologically, the arrangement model $Y_{\mathcal{A}_3}$ is a line bundle over a space obtained from a 7-fold punc-

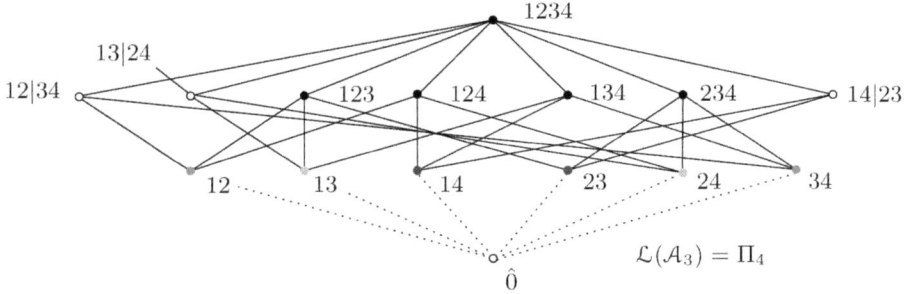

Figure 3. The intersection lattice of $\mathcal{A}_3$.

tured $\mathbb{RP}^2$ by gluing 7 Möbius bands along their boundaries into the boundary components.

We can easily check the statements of Theorem 2.5 for $Y_{\mathcal{A}_3}$. In particular, we see that intersections of irreducible divisors in $Y_{\mathcal{A}_3}$ are nonempty if and only if the corresponding index sets form a chain in $\mathcal{L}_{>\hat{0}}$. For instance, the 0-dimensional stratum of the divisor stratification that is encircled in Figure 4 corresponds to the chain $14 < 134 < 1234$ in $\Pi_4 \setminus \{\hat{0}\}$. For comparison, we depict the order complex of $\Pi_4 \setminus \{\hat{0}\}$ in Figure 5. Recall that the complex is a pure 2-dimensional cone with apex 1234 over $\Delta(\overline{\Pi}_4)$; we only draw its base.

If our only objective was to construct a model for $\mathcal{M}(\mathcal{A}_3)$ with a normal crossing divisor, it would be enough to blow up $\mathrm{Bl}_{\{0\}}V$ in the 4 triple intersections. The result would be a line bundle over a 4-fold punctured $\mathbb{RP}^2$ with 4 Möbius bands glued into boundary components.

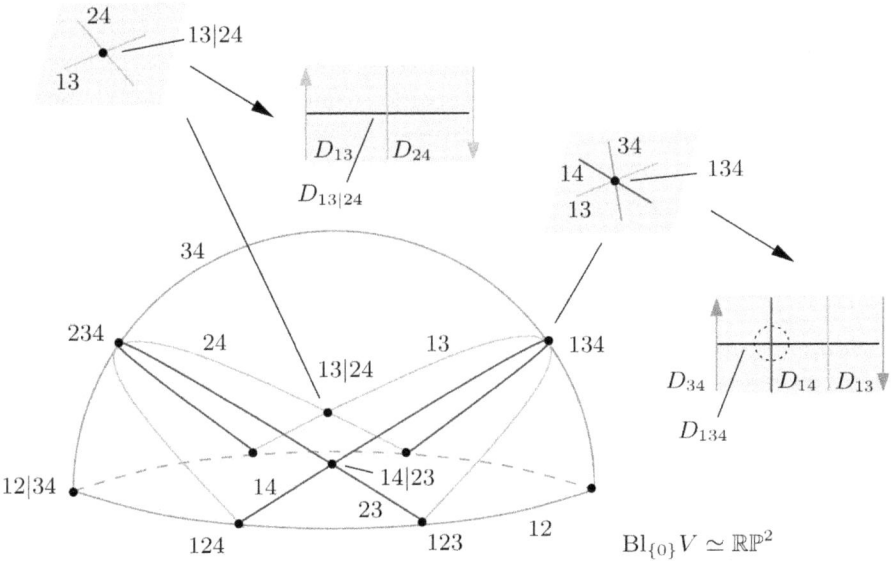

Figure 4. The construction of $Y_{\mathcal{A}_3}$.

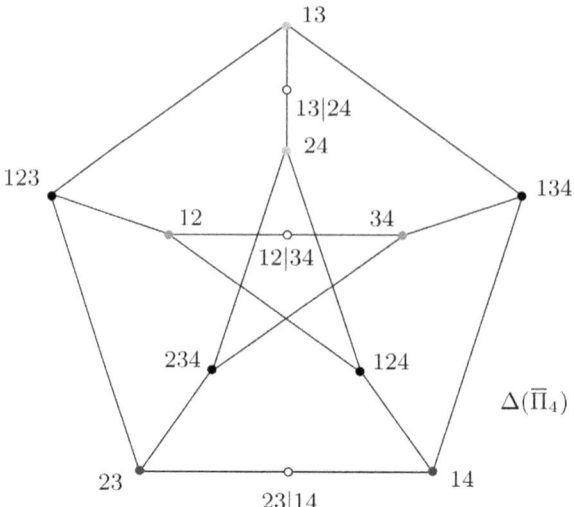

Figure 5. The order complex $\Delta(\overline{\Pi}_4)$.

This observation leads to a generalization of the model construction presented so far: it is enough to do successive blowups on a specific *subset* of intersections in $\mathcal{A}$ to obtain a model with similar properties as those summarized in Theorem 2.5. In fact, appropriate subsets of intersections lattices, so-called *building sets*, were specified in [De Concini and Procesi 1995a]; all give rise to wonderful arrangement models in the sense of Theorem 2.5. The only reservation being that the order complex $\Delta(\mathcal{L}_{>\hat{0}})$ is no longer indexing nonempty intersections of irreducible divisors: chains in $\mathcal{L}_{>\hat{0}}$ are replaced by so-called *nested sets* — subsets of building sets that again form an abstract simplicial complex.

We will not give the original definitions of De Concini and Procesi for building sets and nested sets in this survey. Instead, we will present a generalization of these notions for arbitrary meet-semilattices in Section 3.1. This combinatorial abstraction has proved useful in many cases beyond arrangement model constructions. Its relation to the original geometric context will be explained in Section 4.1.

2.3. Some remarks on history. Before we proceed, we briefly sketch the historic background of De Concini–Procesi arrangement models. Moreover, we outline an application to a famous problem in arrangement theory that, among other issues, served as a motivation for the model construction.

Compactifications of configuration spaces due to Fulton and MacPherson [1994] have prepared the scene for wonderful arrangement models. Their work is concerned with classical configurations spaces $F(X,n)$ of smooth algebraic varieties X, i.e., spaces of n-tuples of pairwise distinct points in X:

$$F(X, n) = \{ (x_1, \ldots, x_n) \in X^n \mid x_i \neq x_j \text{ for } i \neq j \}.$$

A compactification $X[n]$ of $F(X, n)$ is constructed in which the complement of the original configuration space is a normal crossing divisor; in fact, $X[n]$ has properties analogous to those listed for arrangement models in Theorem 2.5. The relation to the arrangement setting can be summarized by saying that, on the one hand, the underlying spaces in the configuration space setting are incomparably more complicated — smooth algebraic varieties X rather than real or complex linear space; the combinatorics, on the other hand, is far simpler — it is the combinatorics of our basic Examples 2.4 and 2.6, the partition lattice Π_n. The notion of building sets and nested sets, which constitutes the defining combinatorics of arrangement models, has its roots in the Fulton–MacPherson construction for configuration spaces, hence is inspired by the combinatorics of Π_n.

Looking along the time line in the other direction, De Concini–Procesi arrangement models have triggered a number of more general constructions with similar spirit: compactifications of conically stratified complex manifolds by MacPherson and Procesi [1998], and model constructions for mixed real subspace and halfspace arrangements and real stratified manifolds by Gaiffi [2003] that use spherical rather than classical blowups.

As a first impact, the De Concini–Procesi model construction has yielded substantial progress on a longstanding open question in arrangement theory [De Concini and Procesi 1995a, Section 5], the question being whether combinatorial data of a complex subspace arrangement determines the cohomology algebra of its complement. For arrangements of hyperplanes, there is a beautiful description of the integral cohomology algebra of the arrangement complement in terms of the intersection lattice — the *Orlik–Solomon algebra* [1980]. Also, a prominent application of Goresky and MacPherson's Stratified Morse Theory states that cohomology of complements of (complex and real) subspace arrangements, as graded groups over $\mathbb{Z}$, are determined by the intersection lattice and its codimension labelling. In fact, there is an explicit description of cohomology groups in terms of homology of intervals in the intersection lattice [Goresky and MacPherson 1988, Part III]. However, whether *multiplicative* structure is determined as well remained an open question 20 years after it had been answered for arrangements of hyperplanes (see [Feichtner and Ziegler 2000; Longueville 2000] for results on particular classes of arrangements).

The De Concini–Procesi construction allows to apply Morgan's theory on rational models for complements of normal crossing divisors [Morgan 1978] to arrangement complements and to conclude that their *rational* cohomology algebras indeed are determined by the combinatorics of the arrangement. A key step in the description of the Morgan model is the presentation of cohomology of divisor components and their intersections in purely combinatorial terms [De Concini and Procesi 1995a, 5.1, 5.2]. For details on this approach to arrangement cohomology, see [De Concini and Procesi 1995a, 5.3].

Unfortunately, the Morgan model is fairly complicated even for small arrangements, and the approach is bound to rational coefficients. The model has been considerably simplified in work of Yuzvinsky [2002; 1999]. In [Yuzvinsky 2002] explicit presentations of cohomology algebras for certain classes of arrangements were given. However, despite an explicit conjecture of an integral model for arrangement cohomology in [Yuzvinsky 2002, Conjecture 6.7], extending the result to integral coefficients remained out of reach. Only years later, the question has been fully settled to the positive in work of Deligne, Goresky and MacPherson [Deligne et al. 2000] with a sheaf-theoretic approach, and parallely by de Longueville and Schultz [2001] using rather elementary topological methods: Integral cohomology algebras of complex arrangement complements are indeed determined by combinatorial data.

3. The Combinatorial Core Data: A Step Beyond Geometry

We will now abandon geometry for a while and in this section fully concentrate on combinatorial and algebraic gadgets that are inspired by De Concini–Procesi arrangement models.

We first present a combinatorial analogue of De Concini–Procesi resolutions on purely order theoretic level following [Feichtner and Kozlov 2004, Sections 2 and 3]. Based on the notion of building sets and nested sets for arbitrary lattices proposed therein, we define a family of commutative graded algebras for any given lattice.

The next Section then will be devoted to relate these objects to geometry — to the original context of De Concini–Procesi arrangement models and, more interestingly so, to different seemingly unrelated contexts in geometry.

3.1. Combinatorial resolutions. We will state purely combinatorial definitions of *building sets* and *nested sets*. Recall that, in the context of model constructions, building sets list the strata that are to be blown up in the construction process, and nested sets describe beforehand the nonempty intersections of irreducible divisor components in the final resolution.

Let $\mathcal{L}$ be a finite meet-semilattice, i.e., a finite poset such that any pair of elements has a unique maximal lower bound. In particular, such a meet-semilattice has a unique minimal element that we denote with $\hat{0}$. We will talk about semilattices for short. As a basic reference on partially ordered sets we refer to [Stanley 1997, Chapter 3].

DEFINITION 3.1 (COMBINATORIAL BUILDING SETS). A subset $\mathcal{G} \subseteq \mathcal{L}_{>\hat{0}}$ in a finite meet-semilattice $\mathcal{L}$ is called a *building set* if for any $X \in \mathcal{L}_{>\hat{0}}$ and $\max \mathcal{G}_{\leq X} = \{G_1, \ldots, G_k\}$ there is an isomorphism of posets

$$\varphi_X : \prod_{j=1}^{k} [\hat{0}, G_j] \xrightarrow{\cong} [\hat{0}, X] \qquad (3\text{--}1)$$

with $\varphi_X(\hat{0}, \ldots, G_j, \ldots, \hat{0}) = G_j$ for $j = 1, \ldots, k$. We call $F_{\mathcal{G}}(X) := \max \mathcal{G}_{\leq X}$ the *set of factors* of X in $\mathcal{G}$.

There are two extreme examples of building sets for any semilattice: we can take the full semilattice $\mathcal{L}_{>\hat{0}}$ as a building set. On the other hand, the set of elements X in $\mathcal{L}_{>\hat{0}}$ which do not allow for a product decomposition of the lower interval $[\hat{0}, X]$ form the unique minimal building set (see Example 3.3 below).

Intuitively speaking, building sets are formed by elements in the semilattice that are the perspective factors of product decompositions.

Any choice of a building set $\mathcal{G}$ in $\mathcal{L}$ gives rise to a family of so-called *nested sets*. These are, roughly speaking, subsets of $\mathcal{G}$ whose antichains are sets of factors with respect to the chosen building set. Nested sets form an abstract simplicial complex on the vertex set $\mathcal{G}$. This simplicial complex plays the role of the order complex for arrangement models more general than the maximal models discussed in Section 2.2.

DEFINITION 3.2 (NESTED SETS). Let $\mathcal{L}$ be a finite meet-semilattice and $\mathcal{G}$ a building set in $\mathcal{L}$. A subset $\mathcal{S}$ in $\mathcal{G}$ is called *nested* (or $\mathcal{G}$-*nested* if specification is needed) if, for any set of incomparable elements $X_1, \ldots, X_t$ in $\mathcal{S}$ of cardinality at least two, the join $X_1 \vee \ldots \vee X_t$ exists and does not belong to $\mathcal{G}$. The $\mathcal{G}$-nested sets form an abstract simplicial complex $\mathcal{N}(\mathcal{L}, \mathcal{G})$, the *nested set complex* with respect to $\mathcal{L}$ and $\mathcal{G}$.

Observe that if we choose the full semilattice as a building set, then a subset is nested if and only if it is linearly ordered in $\mathcal{L}$. Hence, the nested set complex $\mathcal{N}(\mathcal{L}, \mathcal{L}_{>\hat{0}})$ coincides with the order complex $\Delta(\mathcal{L}_{>\hat{0}})$.

EXAMPLE 3.3 (BUILDING SETS AND NESTED SETS FOR THE PARTITION LATTICE). Choosing the maximal building set in the partition lattice Π_n, we obtain the order complex $\Delta((\Pi_n) \setminus \{\hat{0}\})$ as the associated complex of nested sets. Topologically, it is a cone over a wedge of $(n-1)!$ spheres of dimension $n-1$.

The minimal building set $\mathcal{G}_{\min}$ in Π_n is given by partitions with at most one block of size larger or equal 2, the so-called modular elements in Π_n. We can identify these partitions with subsets of $\{1, \ldots, n\}$ of size larger or equal 2. A collection of such subsets is nested, if and only if none of the pairs of subsets have a nontrivial intersection, i.e., for any pair of subsets they are either disjoint or one is contained in the other. Referring to a naive picture of such containment relation explains the choice of the term *nested*—it appeared first in the work of Fulton and MacPherson [1994] on compactifications of classical configuration spaces. As we noted earlier, the combinatorics they are concerned with is indeed the combinatorics of the partition lattice.

For the rank 3 partition lattice Π_3, maximal and minimal building sets coincide, $\mathcal{G} = \Pi_3 \setminus \{\hat{0}\}$. The nested set complex $\mathcal{N}(\Pi_3, \mathcal{G})$ is the order complex $\Delta(\Pi_3 \setminus \{\hat{0}\})$ depicted in Figure 1.

For the rank 4 partition lattice Π_4, we have seen the nested set complex for the maximal building set $\mathcal{N}(\Pi_4, \mathcal{G}_{\max})$ in Figure 5. The nested set complex associated with the minimal building set $\mathcal{G}_{\min}$ in Π_4 is depicted in Figure 6. Again, $\mathcal{N}(\Pi_4, \mathcal{G}_{\min})$ is a cone with apex 1234, and we only draw its base, $\mathcal{N}(\overline{\Pi}_4, \mathcal{G}_{\min})$.

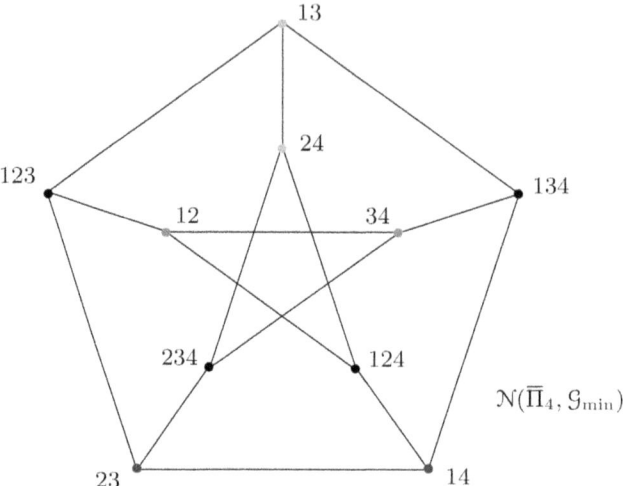

Figure 6. The nested set complex $\mathcal{N}(\overline{\Pi}_4, \mathcal{G}_{\min})$.

Adding one or two 2-block partitions to $\mathcal{G}_{\min}$ yields all the other building sets for Π_4. The corresponding nested set complexes are subdivisions of $\mathcal{N}(\Pi_4, \mathcal{G}_{\min})$.

When studying the (maximal) wonderful model Y_{A_3} in Example 2.6 we had observed that, if we only wanted to achieve a model with normal crossing divisors, it would have been enough to blow up the overall and the triple intersections. This selection of strata, respectively elements in $\mathcal{L}(\mathcal{A}_3) = \Pi_4$, exactly corresponds to the minimal building set $\mathcal{G}_{\min}$ in Π_4 — a geometric motivation for Definition 3.1.

We can also get a glimpse on the geometry that motivates the definition of nested sets: comparing simplices in $\mathcal{N}(\Pi_4, \mathcal{G}_{\min})$ with intersections of irreducible divisor components in the arrangement model resulting from blowups along subspaces in $\mathcal{G}_{\min}$, we see that there is a 1-1 correspondence. For instance, $\{12, 34\}$ is a nested set with respect to $\mathcal{G}_{\min}$, and divisor components D_{12} and D_{34} intersect in the model (compare Figure 4).

It is not a coincidence that, in the example above, one nested set complex is a subdivision of the other if one building set contains the other. In fact, the following holds:

THEOREM 3.4 [Feichtner and Müller 2005, Proposition 3.3, Theorem 4.2]. *For any finite meet-semilattice $\mathcal{L}$, and $\mathcal{G}$ a building set in $\mathcal{L}$, the nested set complex*

$\mathcal{N}(\mathcal{L}, \mathcal{G})$ *is homotopy equivalent to the order complex of* $\mathcal{L}_{>\hat{0}}$,

$$\mathcal{N}(\mathcal{L}, \mathcal{G}) \simeq \Delta(\mathcal{L}_{>\hat{0}}).$$

Moreover, if $\mathcal{L}$ *is atomic, i.e., any element is a join of a set of atoms, and* $\mathcal{G}$ *and* $\mathcal{H}$ *are building sets with* $\mathcal{G} \supseteq \mathcal{H}$, *then the nested set complex* $\mathcal{N}(\mathcal{L}, \mathcal{G})$ *is obtained from* $\mathcal{N}(\mathcal{L}, \mathcal{H})$ *by a sequence of stellar subdivisions. In particular, the complexes are homeomorphic.*

We now propose a construction on semilattices producing new semilattices: the *combinatorial blowup* of a semilattice in an element.

DEFINITION 3.5 (COMBINATORIAL BLOWUP). For a semilattice $\mathcal{L}$ and an element X in $\mathcal{L}_{>\hat{0}}$ we define a poset $(\mathrm{Bl}_X \mathcal{L}, \prec)$ on the set of elements

$$\mathrm{Bl}_X \mathcal{L} = \{\, Y \mid Y \in \mathcal{L}, Y \not\geq X \,\} \cup \{\, Y' \mid Y \in \mathcal{L}, Y \not\geq X, \text{ and } Y \vee X \text{ exists in } \mathcal{L} \,\}.$$

The order relation $<$ in $\mathcal{L}$ determines the order relation $\prec$ within the two parts of $\mathrm{Bl}_X \mathcal{L}$ described above,

$$Y \prec Z, \quad \text{for } Y < Z \text{ in } \mathcal{L},$$
$$Y' \prec Z', \quad \text{for } Y < Z \text{ in } \mathcal{L},$$

and additional order relations between elements of these two parts are defined by

$$Y \prec Z', \quad \text{for } Y < Z \text{ in } \mathcal{L},$$

where in all three cases it is assumed that $Y, Z \not\geq X$ in $\mathcal{L}$. We call $\mathrm{Bl}_X \mathcal{L}$ the *combinatorial blowup* of $\mathcal{L}$ in X.

In fact, the poset $\mathrm{Bl}_X \mathcal{L}$ is again a semilattice. Figure 7 explains what is going on.

The construction does the following: it removes the closed upper interval on top of X from $\mathcal{L}$, and then marks the set of elements in $\mathcal{L}$ that are not larger or equal X, but have a join with X in $\mathcal{L}$. This subset of $\mathcal{L}$ (in fact, a lower ideal in the sense of order theory) is doubled and any new element Y' in the copy is

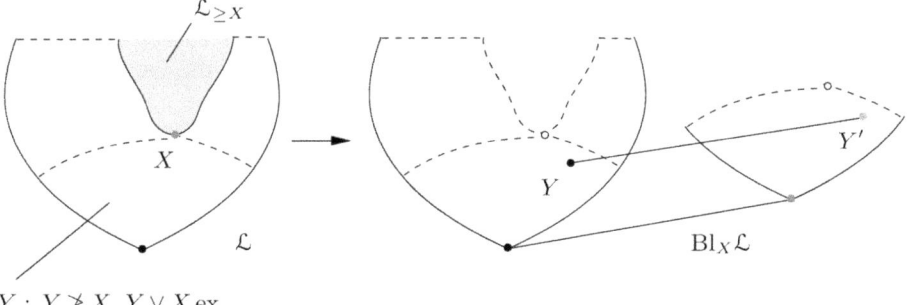

$Y : Y \not\geq X,\ Y \vee X$ ex.

Figure 7. A combinatorial blowup.

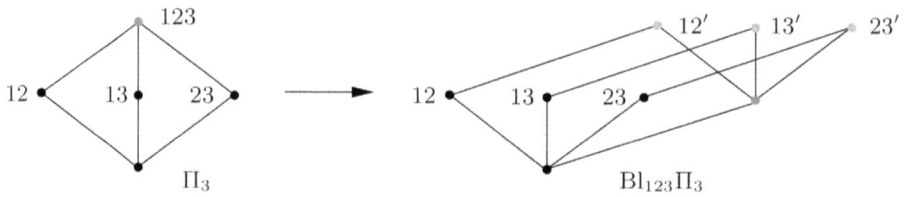

Figure 8. The combinatorial blowup of Π_3 in 123.

defined to be covering the original element Y in $\mathcal{L}$. The order relations in the remaining, respectively the doubled, part of $\mathcal{L}$ stay the same as before.

In Figure 8 we give a concrete example: the combinatorial blowup of the maximal element 123 in Π_3, $\mathrm{Bl}_{123}\Pi_3$. The result should be compared with Figure 2. In fact, $\mathrm{Bl}_{123}\Pi_3$ is the face poset of the divisor stratification in $Y_{A_2} = \mathrm{Bl}_{\{0\}}V$.

The following theorem shows that the three concepts introduced above — combinatorial building sets, nested sets, and combinatorial blowups — fit together so as to provide a combinatorial analogue of the De Concini–Procesi model construction.

THEOREM 3.6 [Feichtner and Kozlov 2004, Theorem 3.4]. *Let $\mathcal{L}$ be a semilattice, $\mathcal{G}$ a combinatorial building set in $\mathcal{L}$, and $G_1, \ldots, G_t$ a linear order on $\mathcal{G}$ that is nonincreasing with respect to the partial order on $\mathcal{L}$. Then, consecutive combinatorial blowups in $G_1, \ldots, G_t$ result in the face poset of the nested set complex $\mathcal{N}(\mathcal{L}, \mathcal{G})$:*

$$\mathrm{Bl}_{G_t}(\ldots(\mathrm{Bl}_{G_2}(\mathrm{Bl}_{G_1}\mathcal{L}))\ldots) = \mathcal{F}(\mathcal{N}(\mathcal{L}, \mathcal{G})).$$

3.2. An algebra defined for atomic lattices. For any atomic lattice, we define a family of graded commutative algebras based on the notions of building sets and nested sets given above. Our exposition here and in Section 4.2 follows [Feichtner and Yuzvinsky 2004]. Restricting our attention to atomic lattices is not essential for the definition. However, for various algebraic considerations and for geometric interpretations (compare Section 4.2) it is convenient to assume that the lattice is atomic.

DEFINITION 3.7. Let $\mathcal{L}$ be a finite atomic lattice, $\mathfrak{A}(\mathcal{L})$ its set of atoms, and $\mathcal{G}$ a building set in $\mathcal{L}$. We define the algebra $D(\mathcal{L}, \mathcal{G})$ of $\mathcal{L}$ with respect to $\mathcal{G}$ as

$$D(\mathcal{L}, \mathcal{G}) := \mathbb{Z}\left[\{x_G\}_{G \in \mathcal{G}}\right] \Big/ \mathcal{J},$$

where the ideal of relations $\mathcal{J}$ is generated by

$$\prod_{i=1}^{t} x_{G_i} \qquad \text{for } \{G_1, \ldots, G_t\} \notin \mathcal{N}(\mathcal{L}, \mathcal{G}),$$

$$\sum_{G \geq H} x_G \qquad \text{for } H \in \mathfrak{A}(\mathcal{L}).$$

Observe that this algebra is a quotient of the face ring of the nested set complex $\mathcal{N}(\mathcal{L}, \mathcal{G})$.

EXAMPLE 3.8 (ALGEBRAS ASSOCIATED TO Π_3 AND Π_4). For Π_3 and its only building set $\mathcal{G}_{\max} = \Pi_3 \setminus \{\hat{0}\}$, the algebra reads as follows:

$$D(\Pi_3, \mathcal{G}_{\max}) = \mathbb{Z}\,[x_{12}, x_{13}, x_{23}, x_{123}] \Big/ \left\langle \begin{array}{l} x_{12}x_{13},\ x_{12}x_{23},\ x_{13}x_{23} \\ x_{12} + x_{123},\ x_{13} + x_{123},\ x_{23} + x_{123} \end{array} \right\rangle$$

$$\cong \mathbb{Z}\,[x_{123}]/\langle x_{123}^2 \rangle.$$

For Π_4 and its minimal building set $\mathcal{G}_{\min}$, we obtain the following algebra after simplifying slightly the presentation:

$$D(\Pi_4, \mathcal{G}_{\min}) \cong \mathbb{Z}\,[x_{123}, x_{124}, x_{134}, x_{234}, x_{1234}] \Big/$$

$$\left\langle \begin{array}{ll} x_{ijk}\,x_{1234} & \text{for all } 1 \le i < j < k \le 4 \\ x_{ijk}\,x_{i'j'k'} & \text{for all } ijk \ne i'j'k' \\ x_{ijk}^2 + x_{1234}^2 & \text{for all } 1 \le i < j < k \le 4 \end{array} \right\rangle.$$

There is an explicit description for a Gröbner basis of the ideal $\mathcal{I}$, which in particular yields an explicit description for a monomial basis of the graded algebra $D(\mathcal{L}, \mathcal{G})$.

THEOREM 3.9. (1) [Feichtner and Yuzvinsky 2004, Theorem 2] *The following polynomials form a Gröbner basis of the ideal* $\mathcal{I}$:

$$\prod_{G \in \mathcal{S}} x_G \quad \text{for } \mathcal{S} \notin \mathcal{N}(\mathcal{L}, \mathcal{G}),$$

$$\prod_{i=1}^{k} x_{A_i} \left(\sum_{G \ge B} x_G \right)^{d(A,B)},$$

where $A_1, \ldots, A_k$ *are maximal elements in a nested set* $\mathcal{H} \in \mathcal{N}(\mathcal{L}, \mathcal{G})$, $B \in \mathcal{G}$ *with* $B > A = \bigvee_{i=1}^{k} A_i$, *and* $d(A, B)$ *is the minimal number of atoms needed to generate* B *from* A *by taking joins.*

(2) [Feichtner and Yuzvinsky 2004, Corollary 1] *The resulting linear basis for the graded algebra* $D(\mathcal{L}, \mathcal{G})$ *is given by the following set of monomials:*

$$\prod_{A \in \mathcal{S}} x_A^{m(A)},$$

where $\mathcal{S}$ *is running over all nested subsets of* $\mathcal{G}$, $m(A) < d(A', A)$, *and* A' *is the join of* $\mathcal{S} \cap \mathcal{L}_{<A}$.

Part (2) of Theorem 3.9 generalizes a basis description by Yuzvinsky [1997] for $D(\mathcal{L}, \mathcal{G})$ in the case of $\mathcal{G}$ being the minimal building set in an intersection lattice $\mathcal{L}$ of a complex hyperplane arrangement. Yuzvinsky's basis description has also been generalized in a somewhat different direction by Gaiffi [1997], namely for closely related algebras associated with complex subspace arrangements.

We will return to the algebra $D(\mathcal{L}, \mathcal{G})$ and discuss its geometric significance in Section 4.2.

4. Returning to Geometry

4.1. Understanding stratifications in wonderful models. We first relate the combinatorial setting of building sets and nested sets developed in Section 3.1 to its origin, the De Concini–Procesi model construction. Here is how to recover the original notion of building sets [De Concini and Procesi 1995a, Definition in §2.3], we call them *geometric building sets*, from our definitions:

DEFINITION 4.1 (GEOMETRIC BUILDING SETS). Let $\mathcal{L}$ be the intersection lattice of an arrangement of subspaces in real or complex vector space V and cd : $\mathcal{L} \to \mathbb{N}$ a function on $\mathcal{L}$ assigning the codimension of the corresponding subspace to each lattice element. A subset $\mathcal{G}$ in $\mathcal{L}$ is a *geometric building set* if it is a building set in the sense of 3.1, and for any $X \in \mathcal{L}$ the codimension of X is equal to the sum of codimensions of its factors, $F_{\mathcal{G}}(X)$:

$$\operatorname{cd}(X) = \sum_{Y \in F_{\mathcal{G}}(X)} \operatorname{cd}(Y).$$

An easy example shows that the notion of geometric building sets indeed is more restrictive than the notion of combinatorial building sets. For arrangements of hyperplanes, however, the notions coincide [Feichtner and Kozlov 2004, Proposition 4.5.(2)].

EXAMPLE 4.2 (GEOMETRIC VERSUS COMBINATORIAL BUILDING SETS). Let $\mathcal{A}$ denote the following arrangement of 3 subspaces in $\mathbb{R}^4$:

$$A_1 : \ x_4 = 0, \quad A_2 : \ x_1 = x_2 = 0, \quad A_3 : \ x_1 = x_3 = 0.$$

The intersection lattice $\mathcal{L}(\mathcal{A})$ is a boolean algebra on 3 elements; we depict the lattice with its codimension labelling in Figure 9. The set of atoms obviously is a combinatorial building set. However, any geometric building set must contain the intersection $A_2 \cap A_3$: its codimension is *not* the sum of codimensions of its (combinatorial) factors A_2 and A_3.

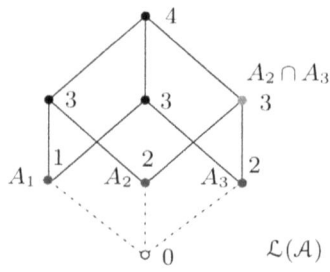

Figure 9. Geometric versus combinatorial building sets.

As we mentioned before, there are wonderful model constructions for arrangement complements $\mathcal{M}(\mathcal{A})$ that start from an arbitrary geometric building set $\mathcal{G}$ of the intersection lattice $\mathcal{L}(\mathcal{A})$ [De Concini and Procesi 1995a, 3.1]: In Definition 2.2, replace the product on the right hand side of (2–1) by a product over building set elements in $\mathcal{L}$, and obtain the wonderful model $Y_{\mathcal{A},\mathcal{G}}$ by again taking the closure of the image of $\mathcal{M}(\mathcal{A})$ under Ψ. In Definition 2.3, replace the linear extension of $\mathcal{L}^{\mathrm{op}}_{>\hat{0}}$ by a nonincreasing linear order on the elements in $\mathcal{G}$, and obtain the wonderful model $Y_{\mathcal{A},\mathcal{G}}$ by successive blowups of subspaces in $\mathcal{G}$, and of proper transforms of such.

The key properties of these models are analogous to those listed in Theorem 2.5, where in part (2), lattice elements are replaced by building set elements, and in part (3), chains in $\mathcal{L}$ as indexing sets of nonempty intersections of irreducible components of divisors are replaced by nested sets. Hence, the face poset of the stratification of $Y_{\mathcal{A},\mathcal{G}}$ given by irreducible components of divisors and their intersections coincides with the face poset of the nested set complex $\mathcal{N}(\mathcal{L},\mathcal{G})$. Compare Examples 2.6 and 3.3, where we found that nested sets with respect to the minimal building set $\mathcal{G}_{\min}$ in Π_4 index nonempty intersections of irreducible divisor components in the arrangement model $Y_{\mathcal{A}_3,\mathcal{G}_{\min}}$.

While the intersection lattice $\mathcal{L}(\mathcal{A})$ captures the combinatorics of the stratification of V given by subspaces of $\mathcal{A}$ and their intersections, the nested set complex $\mathcal{N}(\mathcal{L},\mathcal{G})$ captures the combinatorics of the divisor stratification of the wonderful model $Y_{\mathcal{A},\mathcal{G}}$. More than that: combinatorial blowups turn out to be the right concept to describe the incidence change of strata during the construction of wonderful arrangement models by successive blowups:

THEOREM 4.3 [Feichtner and Kozlov 2004, Proposition 4.7(1)]. *Let $\mathcal{A}$ be a complex subspace arrangement, $\mathcal{G}$ a geometric building set in $\mathcal{L}(\mathcal{A})$, and $G_1, \ldots, G_t$ a nonincreasing linear order on $\mathcal{G}$. Let $\mathrm{Bl}_i(\mathcal{A})$ denote the result of blowing up strata $G_1, \ldots, G_i$, for $i \leq t$, and denote by $\mathcal{L}_i$ the face poset of the stratification of $\mathrm{Bl}_i(\mathcal{A})$ by proper transforms of subspaces in $\mathcal{A}$ and the exceptional divisors. Then the poset $\mathcal{L}_i$ coincides with the successive combinatorial blowups of $\mathcal{L}$ in $G_1, \ldots G_i$:*

$$\mathcal{L}_i = \mathrm{Bl}_{G_i}(\ldots (\mathrm{Bl}_{G_2}(\mathrm{Bl}_{G_1}\mathcal{L}))\ldots).$$

Combinatorial building sets, nested sets and combinatorial blowups occur in other situations and prove to be the right concept for describing stratifications in more general model constructions. This applies to the *wonderful conical compactifications* of MacPherson and Procesi [1998] as well as to models for mixed subspace and halfspace arrangements and for stratified real manifolds by Gaiffi [2003].

Also, combinatorial blowups describe the effect which stellar subdivisions in polyhedral fans have on the face poset of the fans. In fact, combinatorial blowups describe the incidence change of torus orbits for resolutions of toric varieties by consecutive blowups in closed torus orbits. This implies, in particular, that for any toric variety and for any choice of a combinatorial building set in the face poset of its defining fan, we obtain a resolution of the variety with torus orbit structure prescribed by the nested set complex associated to the chosen building set. We believe that such combinatorially prescribed resolutions can prove useful in various concrete situations (see [Feichtner and Kozlov 2004, Section 4.2] for further details).

There is one more issue about nested set stratifications of maximal wonderful arrangement models that we want to discuss here, mostly in perspective of applications in Section 5. According to Definition 2.2, any point in the model Y_A can be written as a collection of a point in V and lines in V, one line for each element in $\mathcal{L}(A)$. There is a lot of redundant information in this rendering, e.g., points on the open stratum $\pi^{-1}(\mathcal{M}(A))$ are fully determined by their first "coordinate entry", the point in $\mathcal{M}(A) \subseteq V$.

Here is a more economic encoding of a point ω on Y_A [Feichtner and Kozlov 2003, Section 4.1]: we find that ω can be uniquely written as

$$\omega = (x, H_1, \ell_1, H_2, \ell_2, \ldots, H_t, \ell_t) = (x, \ell_1, \ell_2, \ldots, \ell_t), \qquad (4\text{-}1)$$

where x is a point in V, the $H_1, \ldots, H_t$ form a descending chain of subspaces in $\mathcal{L}_{>\hat{0}}$, and the ℓ_i are lines in V. More specifically, $x = \pi(\omega)$, and H_1 is the maximal lattice element that, as a subspace of V, contains x. The line ℓ_1 is orthogonal to H_1 and corresponds to the coordinate entry of ω indexed by H_1 in $\mathbb{P}(V/H_1)$. The lattice element H_2, in turn, is the maximal lattice element that contains both H_1 and ℓ_1. The specification of lines ℓ_i, i.e., lines that correspond to coordinates of ω in $\mathbb{P}(V/H_i)$, and the construction of lattice elements H_{i+1}, continues analogously for $i \geq 2$ until a last line ℓ_t is reached whose span with H_t is not contained in any lattice element other than the full ambient space V. Observe that the H_i are determined by x and the sequence of lines ℓ_i; we choose to include the H_i in order to keep the notation more transparent.

The full coordinate information on ω can be recovered from (4-1) by setting $H_0 = \bigcap A$, $\ell_0 = \langle x \rangle$, and retrieving the coordinate ω_H indexed by $H \in \mathcal{L}_{>\hat{0}}$ as

$$\omega_H = \langle \ell_j, H \rangle / H \in \mathbb{P}(V/H),$$

where j is chosen from $\{1, \ldots, t\}$ such that $H \leq H_j$, but $H \not\leq H_{j+1}$.

A nice feature of this encoding is that for a given point ω in Y_A we can tell the open stratum in the nested set stratification which contains it:

PROPOSITION 4.4 [Feichtner and Kozlov 2003, Proposition 4.5]. *A point ω in a maximal arrangement model Y_A is contained in the open stratum indexed by the chain $H_1 > H_2 > \ldots > H_t$ in $\mathcal{L}_{>\hat{0}}$ if and only if its point/line description (4-1) reads $\omega = (x, H_1, \ell_1, H_2, \ell_2, \ldots, H_t, \ell_t)$.*

4.2. A wealth of geometric meaning for $D(\mathcal{L}, \mathcal{G})$. We turn to the algebra $D(\mathcal{L}, \mathcal{G})$ that we defined for any atomic lattice $\mathcal{L}$ and combinatorial building set $\mathcal{G}$ in $\mathcal{L}$ in Section 3.2. We give two geometric interpretations for this algebra; one is restricted to $\mathcal{L}$ being the intersection lattice of a complex hyperplane arrangement and originally motivated the definition of $D(\mathcal{L}, \mathcal{G})$, the other applies to any atomic lattice and provides for a somewhat unexpected connection to toric varieties.

We comment briefly on the projective version of wonderful arrangement models that we need in this context (see [De Concini and Procesi 1995a, § 4] for details). For any arrangement of linear subspaces $\mathcal{A}$ in V, a model for its projectivization $\mathbb{P}\mathcal{A} = \{\mathbb{P}A \mid A \in \mathcal{A}\}$ in $\mathbb{P}V$, i.e., for $\mathcal{M}(\mathbb{P}\mathcal{A}) = \mathbb{P}V \setminus \bigcup \mathbb{P}\mathcal{A}$, can be obtained by replacing the ambient space V by its projectivization $\mathbb{P}V$ in the model constructions 2.2 and 2.3. The constructions result in a smooth projective variety that we denote by $Y_{\mathcal{A}}^{\mathbb{P}}$. A model $Y_{\mathcal{A},\mathcal{G}}^{\mathbb{P}}$ for a specific geometric building set $\mathcal{G}$ in $\mathcal{L}$ can be obtained analogously. In fact, under the assumption that $\mathbb{P}(\bigcap \mathcal{A})$ is contained in the building set $\mathcal{G}$, the affine model $Y_{\mathcal{A},\mathcal{G}}$ is the total space of a (real or complex) line bundle over the projective model $Y_{\mathcal{A},\mathcal{G}}^{\mathbb{P}}$ which is isomorphic to the divisor component in $Y_{\mathcal{A},\mathcal{G}}$ indexed with $\bigcap \mathcal{A}$.

The most prominent example of a projective arrangement model is the minimal wonderful model for the complex braid arrangement, $Y_{\mathcal{A}_{n-2}^{\mathbb{C}}, \mathcal{G}_{\min}}$. It is isomorphic to the Deligne–Knudson–Mumford compactification $\overline{M}_{0,n}$ of the moduli space of n-punctured complex projective lines [De Concini and Procesi 1995a, 4.3].

Here is the first geometric interpretation of $D(\mathcal{L}, \mathcal{G})$ in the case of $\mathcal{L}$ being the intersection lattice of a complex hyperplane arrangement.

THEOREM 4.5 [De Concini and Procesi 1995b; Feichtner and Yuzvinsky 2004]. *Let $\mathcal{L} = \mathcal{L}(\mathcal{A})$ be the intersection lattice of an essential arrangement of complex hyperplanes $\mathcal{A}$ and $\mathcal{G}$ a building set in $\mathcal{L}$ which contains the total intersection of $\mathcal{A}$. Then, $D(\mathcal{L}, \mathcal{G})$ is isomorphic to the integral cohomology algebra of the projective arrangement model $Y_{\mathcal{A},\mathcal{G}}^{\mathbb{P}}$:*

$$D(\mathcal{L}, \mathcal{G}) \;\cong\; H^*(Y_{\mathcal{A},\mathcal{G}}^{\mathbb{P}}, \mathbb{Z}).$$

EXAMPLE 4.6 (COHOMOLOGY OF BRAID ARRANGEMENT MODELS). The projective arrangement model $Y_{\mathcal{A}_2}^{\mathbb{P}}$ is homeomorphic to the exceptional divisor in $Y_{\mathcal{A}_2} = \mathrm{Bl}_{\{0\}} \mathbb{C}^2$, hence to $\mathbb{CP}^1$. Its cohomology is free of rank 1 in degrees 0 and 2 and zero otherwise. Compare with $D(\Pi_3, \mathcal{G}_{\max})$ in Example 3.8.

The projective arrangement model $Y_{\mathcal{A}_3, \mathcal{G}_{\min}}^{\mathbb{P}}$ is homeomorphic to $\overline{M}_{0,5}$, whose cohomology is known to be free of rank 1 in degrees 0 and 4, free of rank 5 in degree 2, and zero otherwise. At least the coincidence of ranks is easy to verify in comparison with $D(\Pi_4, \mathcal{G}_{\min})$ in Example 3.8.

Theorem 4.5 in fact gives an elegant presentation for the integral cohomology of $\overline{M}_{0,n} \cong Y_{\mathcal{A}_{n-2},\mathcal{G}_{\min}}^{\mathbb{P}}$ in terms of generators and relations:

$$H^*(\overline{M}_{0,n}) \cong D(\Pi_{n-1}, \mathcal{G}_{\min})$$

$$\cong \mathbb{Z}\left[\{x_S\}_{S \subseteq [n-1], |S| \geq 2}\right] \Big/ \left\langle \begin{array}{ll} x_S\,x_T & \text{for } S \cap T \neq \varnothing,\, S \not\subseteq T,\, T \not\subseteq S, \\ \displaystyle\sum_{\{i,j\} \subseteq S} x_S \text{ for } 1 \leq i < j \leq n-1 \end{array} \right\rangle.$$

A lot of effort has been spent on describing the cohomology of $\overline{M}_{0,n}$ (see [Keel 1992]), but none of the presentations comes close to the simplicity of the one stated above.

A nice expression for the Hilbert function of $H^*(\overline{M}_{0,n})$ has been derived in [Yuzvinsky 1997] as a consequence of a monomial linear basis for minimal projective arrangement models presented there.

To propose a more general geometric interpretation for $D(\mathcal{L}, \mathcal{G})$, we start by describing a polyhedral fan $\Sigma(\mathcal{L}, \mathcal{G})$ for any atomic lattice $\mathcal{L}$ and any combinatorial building set $\mathcal{G}$ in $\mathcal{L}$.

DEFINITION 4.7 (A SIMPLICIAL FAN REALIZING $\mathcal{N}(\mathcal{L}, \mathcal{G})$). Let $\mathcal{L}$ be an atomic lattice with set of atoms $\mathfrak{A} = \{A_1, \ldots, A_n\}$, $\mathcal{G}$ a combinatorial building set in $\mathcal{L}$. For any $G \in \mathcal{G}$ define the characteristic vector v_G in $\mathbb{R}^n$ by

$$(v_G)_i := \begin{cases} 1 \text{ if } G \geq A_i, \\ 0 \text{ otherwise,} \end{cases} \qquad \text{for } i = 1, \ldots, n.$$

The simplicial fan $\Sigma(\mathcal{L}, \mathcal{G})$ in $\mathbb{R}^n$ is the collection of cones

$$V_\mathcal{S} := \mathrm{cone}\{v_G \mid G \in \mathcal{S}\}$$

for $\mathcal{S}$ nested in $\mathcal{G}$.

By construction, $\Sigma(\mathcal{L}, \mathcal{G})$ is a rational, simplicial fan that realizes the nested set complex $\mathcal{N}(\mathcal{L}, \mathcal{G})$. The fan gives rise to a (noncompact) smooth toric variety $X_{\Sigma(\mathcal{L}, \mathcal{G})}$ [Feichtner and Yuzvinsky 2004, Proposition 2].

EXAMPLE 4.8 (THE FAN $\Sigma(\Pi_3, \mathcal{G}_{\max})$ AND ITS TORIC VARIETY). We depict $\Sigma(\Pi_3, \mathcal{G}_{\max})$ in Figure 10. The associated toric variety is the blowup of $\mathbb{C}^3$ in $\{0\}$ with the proper transforms of coordinate axes removed.

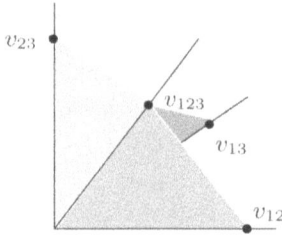

Figure 10. The simplicial fan $\Sigma(\Pi_3, \mathcal{G}_{\max})$.

The algebra $D(\mathcal{L}, \mathcal{G})$ here gains another geometric meaning, this time for *any* atomic lattice $\mathcal{L}$. The abstract algebraic detour of considering $D(\mathcal{L}, \mathcal{G})$ in this general setting is rewarded by a somewhat unexpected return to geometry:

THEOREM 4.9 [Feichtner and Yuzvinsky 2004, Theorem 3]. *For an atomic lattice* $\mathcal{L}$ *and a combinatorial building set* $\mathcal{G}$ *in* $\mathcal{L}$, $D(\mathcal{L}, \mathcal{G})$ *is isomorphic to the Chow ring of the toric variety* $X_{\Sigma(\mathcal{L}, \mathcal{G})}$,

$$D(\mathcal{L}, \mathcal{G}) \cong \mathrm{Ch}^*(X_{\Sigma(\mathcal{L}, \mathcal{G})}).$$

5. Adding Arrangement Models to the Geometer's Toolbox

Let a differentiable action of a finite group Γ on a smooth manifold M be given. The goal is to modify the manifold by blowups so as to have the group act on the resolution $\widetilde{M}$ with abelian stabilizers — the quotient $\widetilde{M}/\Gamma$ then has much more manageable singularities than the original quotient. Such modifications for the sake of simplifying quotients have been of crucial importance at various places. One instance is Batyrev's work [1999] on stringy Euler numbers, which in particular implies a conjecture of Reid [1992], and constitutes substantial progress towards higher dimensional MacKay correspondence.

There are two observations that point to wonderful arrangement models as a possible tool in this context. First, the model construction is equivariant if the initial setting carries a group action: if a finite group Γ acts on a real or complex vector space V, and the arrangement $\mathcal{A}$ is Γ-invariant, then the arrangement model $Y_{\mathcal{A}, \mathcal{G}}$ carries a natural Γ-action for any Γ-invariant building set $\mathcal{G} \subseteq \mathcal{L}(\mathcal{A})$. Second, the model construction is not bound to arrangements. In fact, locally finite stratifications of manifolds which are local subspace arrangements, i.e., locally diffeomorphic to arrangements of linear subspaces, can be treated in a fully analogous way. In the complex case, the construction has been pushed to so-called *conical stratifications* in [MacPherson and Procesi 1998] with a real analogue in [Gaiffi 2003].

The significance of De Concini–Procesi model constructions for abelianizing group actions on complex varieties has been recognized by Borisov and Gunnells [2002], following work of Batyrev [1999; 2000]. Here we focus on the real setting.

5.1. Learning from examples: permutation actions in low dimension.
Consider the action of the symmetric group $\mathfrak{S}_n$ on real n-dimensional space by permuting coordinates:

$$\sigma(x_1, \ldots, x_n) = (x_{\sigma(1)}, \ldots, x_{\sigma(n)}) \qquad \text{for } \sigma \in \mathfrak{S}_n, \ x \in \mathbb{R}^n.$$

Needless to say, we find a wealth of nonabelian stabilizers: For a point $x \in \mathbb{R}^n$ that induces the set partition $\pi = (B_1 | \ldots | B_t)$ of $\{1, \ldots, n\}$ by pairwise coinciding coordinate entries, the stabilizer of x with respect to the permutation action is the Young subgroup $\mathfrak{S}_\pi = \mathfrak{S}_{B_1} \times \ldots \times \mathfrak{S}_{B_t}$ of $\mathfrak{S}_n$, where $\mathfrak{S}_{B_i}$ denotes the symmetric subgroup of $\mathfrak{S}_n$ permuting the coordinates in B_i for $i = 1, \ldots, t$.

The locus of nontrivial stabilizers for the permutation action of $\mathfrak{S}_n$, in fact, is a familiar object: it is the rank $(n-1)$ braid arrangement $\mathcal{A}_{n-1}$. A natural idea that occurs when trying to abelianize a group action by blowups is to resolve the locus of *nonabelian stabilizers* in a systematic way. We look at some low-dimensional examples.

EXAMPLE 5.1 (THE PERMUTATION ACTION OF $\mathfrak{S}_3$). We consider $\mathfrak{S}_3$ acting on real 2-space $V \cong \mathbb{R}^3/\Delta$. The locus of nontrivial stabilizers consists of the 3 hyperplanes in $\mathcal{A}_2$: for $x \in H_{ij} \setminus \{0\}$, $\operatorname{stab} x = \langle (ij) \rangle \cong \mathbb{Z}_2$; in fact, 0 is the only point having a nonabelian stabilizer, namely it is fixed by all of $\mathfrak{S}_3$.

Blowing up $\{0\}$ in V according to the general idea outlined above, we recognize the maximal wonderful model for $\mathcal{A}_2$ that we discussed in Example 2.4.

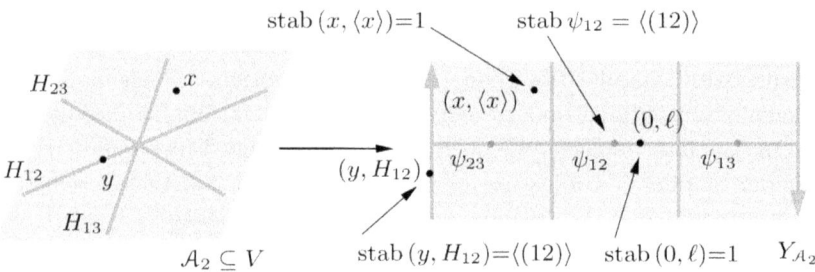

Figure 11. $\mathfrak{S}_3$ acting on $Y_{\mathcal{A}_2}$.

By construction, $\mathfrak{S}_3$ acts coordinate-wise on $Y_{\mathcal{A}_2}$. For points on proper transforms of hyperplanes $(y, H_{ij}) \in D_{ij}$, $1 \le i < j \le 3$, stabilizers are of order two: $\operatorname{stab}(y, H_{ij}) = \langle (ij) \rangle \cong \mathbb{Z}_2$. Otherwise, stabilizers are trivial, unless we are looking at one of the three points ψ_{ij} marked in Figure 11. E.g., for $\psi_{12}=(0, \langle (1, -1, 0) \rangle)$, $\operatorname{stab}\psi_{12} = \langle (12) \rangle \cong \mathbb{Z}_2$. Although the transposition (12) does not fix the line $\langle (1, -1, 0) \rangle$ point-wise, it fixes ψ_{12} as a point in $Y_{\mathcal{A}_2}$! We see that transpositions $(ij) \in \mathfrak{S}_3$ act on the open Möbius band $Y_{\mathcal{A}_2}$ by central symmetries in ψ_{ij}.

Observe that the nested set stratification is not fine enough to distinguish stabilizers: as the points ψ_{ij} show, stabilizers are not isomorphic on open strata.

EXAMPLE 5.2 (THE PERMUTATION ACTION OF $\mathfrak{S}_4$). We now consider $\mathfrak{S}_4$ acting on real 3-space $V \cong \mathbb{R}^4/\Delta$. The locus of nonabelian stabilizers consists of the triple intersections of hyperplanes in $\mathcal{A}_3$, i.e., the subspaces contained in the minimal building set $\mathcal{G}_{\min}$ in $\mathcal{L}(\mathcal{A}_3)=\Pi_4$. Our general strategy suggests to look at the arrangement model $Y_{\mathcal{A}, \mathcal{G}_{\min}}$.

We consider a situation familiar to us from Example 2.6. In Figure 12, we illustrate the situation after the first blowup step in the construction of $Y_{\mathcal{A}, \mathcal{G}_{\min}}$, i.e., the exceptional divisor after blowing up $\{0\}$ in V with the stratification induced by the hyperplanes of $\mathcal{A}_3$. To complete the construction of $Y_{\mathcal{A}, \mathcal{G}_{\min}}$, another 4 blowups in the triple intersections of hyperplanes are necessary, the result of

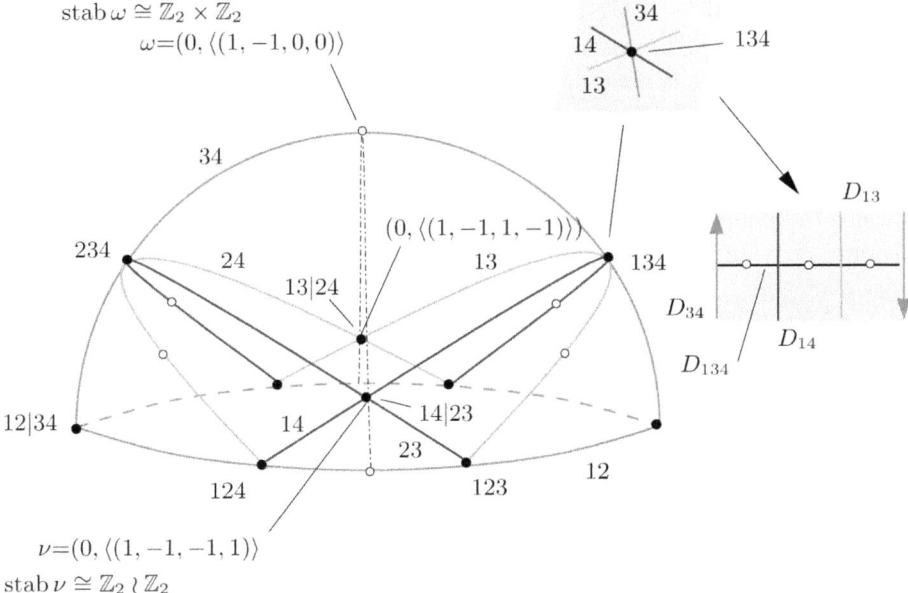

Figure 12. $\mathfrak{S}_4$ acting on $\mathrm{Bl}_{\{0\}}V$, where $V = \mathbb{R}^4/\Delta$.

which we illustrate locally for the triple intersection corresponding to 134. Triple intersections of hyperplanes in $\mathrm{Bl}_{\{0\}}V$ have stabilizers isomorphic to $\mathfrak{S}_3$ — the further blowups in triple intersections are indeed necessary to obtain an abelianization of the permutation action.

Again, we observe that the nested set stratification on $Y_{A,\mathcal{G}_{\min}}$ does not distinguish stabilizers: we indicate subdivisions of nested set strata resulting from nonisomorphic stabilizers by dotted lines, respectively unfilled points in Figure 12.

We now look at stabilizers of points on the model $Y_{A,\mathcal{G}_{\min}}$. We find points with stabilizers isomorphic to $\mathbb{Z}_2$ — any generic point on a divisor D_{ij} will be such. We also find points with stabilizers isomorphic to $\mathbb{Z}_2 \times \mathbb{Z}_2$, e.g., the point ω on D_{1234} corresponding to the line $\langle(1,-1,0,0)\rangle$.

But, on $Y_{A,\mathcal{G}_{\min}}$ we also find points with *nonabelian* stabilizers! For example, the intersection of D_{14} and D_{23} on D_{1234} corresponding to the line $\langle(1,-1,-1,1)\rangle$ is stabilized by both (14) and $(12)(34)$ in $\mathfrak{S}_4$, which do not commute. In fact, the stabilizer is isomorphic to $\mathbb{Z}_2 \wr \mathbb{Z}_2$.

This observation shows that blowing up the locus of nonabelian stabilizers is not enough to abelianize the action! Further blowups in double intersections of hyperplanes are necessary, which suggests, contrary to our first assumption, the *maximal* arrangement model Y_{A_3} as an abelianization of the permutation action.

Some last remarks on this example: observe that stabilizers of points on Y_{A_3} all are elementary abelian 2-groups. We will later see that the strategy of resolving finite group actions on real vector spaces and even manifolds by constructing a suitable maximal De Concini–Procesi model does not only abelianize the action, but yields stabilizers isomorphic to elementary abelian 2-groups.

Also, it seems we cannot do any better than that within the framework of blowups, i.e., we neither can get rid of nontrivial stabilizers, nor can we reduce the rank of nontrivial stabilizers any further. The divisors D_{ij} are stabilized by transpositions (ij) which supports our first claim. For the second claim, consider the point $\omega = (0, \ell_1)$ in Y_{A_3} with $\ell_1 = \langle (1, -1, 0, 0) \rangle$ (here we use the encoding of points on arrangement models proposed in (4–1)). We have seen above that $\operatorname{stab}\omega \cong \mathbb{Z}_2 \times \mathbb{Z}_2$, in fact $\operatorname{stab}\omega = \langle (12) \rangle \times \langle (34) \rangle$. Blowing up Y_{A_3} in ω means to again glue in an open Möbius band. Points on the new exceptional divisor $D_\omega \cong \mathbb{RP}^1$ will be parameterized by tupels $(0, \ell_1, \ell_2)$, where ℓ_2 is a line orthogonal to ℓ_1 in V. A generic point on this stratum will be stabilized only by the transposition (12), specific points however, e.g., $(0, \ell_1, \langle (0, 0, 1, -1) \rangle)$ will still be stabilized by all of $\operatorname{stab}\omega \cong \mathbb{Z}_2 \times \mathbb{Z}_2$.

5.2. Abelianizing a finite linear action. Following the basic idea of proposing De Concini–Procesi arrangement models as abelianizations of finite group actions and drawing from our experiences with the permutation action on low-dimensional real space in Section 5.1 we here treat the case of finite linear actions.

Let a finite group Γ act linearly and effectively on real n-space $\mathbb{R}^n$. Without loss of generality, we can assume that the action is orthogonal [Vinberg 1989, Section 2.3, Theorem 1]; we fix the appropriate scalar product throughout.

Our strategy is to construct an arrangement of subspaces $\mathcal{A}(\Gamma)$ in real n-space, and to propose the maximal wonderful model $Y_{\mathcal{A}(\Gamma)}$ as an abelianization of the given action.

CONSTRUCTION 5.3 (THE ARRANGEMENT $\mathcal{A}(\Gamma)$). For any subgroup H in Γ, define a linear subspace

$$L(H) := \operatorname{span}\{ \ell \mid \ell \text{ line in } \mathbb{R}^n \text{ with } H \circ \ell = \ell \}, \qquad (5\text{--}1)$$

the linear span of all lines in V that are invariant under the action of H.

Denote by $\mathcal{A}(\Gamma) = \mathcal{A}(\Gamma \circlearrowleft \mathbb{R}^n)$ the arrangement of proper subspaces in $\mathbb{R}^n$ that are of the form $L(H)$ for some subgroup H in Γ.

Observe that the arrangement $\mathcal{A}(\Gamma)$ never contains any hyperplane: if $L(H)$ were a hyperplane for some subgroup H in Γ, then also its orthogonal line ℓ would be invariant under the action of H. By definition of $L(H)$, however, ℓ would then be contained in $L(H)$ which in turn would be the full ambient space.

THEOREM 5.4 [Feichtner and Kozlov 2005, Thm. 3.1]. *For any effective linear action of a finite group Γ on n-dimensional real space, the maximal wonderful arrangement model $Y_{\mathcal{A}(\Gamma)}$ abelianizes the action. Moreover, stabilizers of points on the arrangement model are isomorphic to elementary abelian 2-groups.*

The first example coming to mind is the permutation action of $\mathfrak{S}_n$ on real n-space. We find that $\mathcal{A}(\mathfrak{S}_n)$ is the rank 2 truncation of the braid arrangement,

$\mathcal{A}_{n-1}^{\mathrm{rk}\geq 2}$, i.e., the arrangement consisting of subspaces in $\mathcal{A}_{n-1}$ of codimension at least 2. For details, see [Feichtner and Kozlov 2005, Section 4.1]. In earlier work [Feichtner and Kozlov 2003], we had already proposed the maximal arrangement model of the braid arrangement as an abelianization of the permutation action. We proved that stabilizers on $Y_{\mathcal{A}_{n-1}}$ are isomorphic to elementary abelian 2-groups by providing explicit descriptions of stabilizers based on an algebraic-combinatorial set-up for studying these groups.

5.3. Abelianizing finite differentiable actions on manifolds. We now look at a generalization of the abelianization presented in Section 5.2. Assume that Γ is a finite group that acts differentiably and effectively on a smooth real manifold M. We first observe that such an action induces a linear action of the stabilizer stab x on the tangent space $T_x M$ at any point x in M. Hence, locally we are back to the setting that we discussed before: For any subgroup H in stab x, we can define a linear subspace $L(x, H) := L(H)$ of the tangent space $T_x M$ as in (5–1), and we can combine the nontrivial subspaces to form an arrangement $\mathcal{A}_x := \mathcal{A}(\text{stab} \circlearrowleft T_x M)$ in $T_x M$.

Combined with the information that a model construction in the spirit of De Concini–Procesi arrangement models exists also for local subspace arrangements, we need to stratify the manifold so as to locally reproduce the arrangement $\mathcal{A}_x$ in any tangent space $T_x M$. Here is how to do that:

CONSTRUCTION 5.5 (THE STRATIFICATION $\mathfrak{L}$). For any $x \in M$, and any subgroup H in stab x, define a normal (!) subgroup $F(x, H)$ in H by

$$F(x, H) = \{h \in H \mid h \circ y = y \text{ for any } y \in L(x, H)\};$$

$F(x, H)$ is the subgroup of elements in H that fix all of $L(x, H)$ point-wise. Define $\mathfrak{L}(x, H)$ to be the connected component of the fixed point set of $F(x, H)$ in M that contains x. Now combine these submanifolds so as to form a locally finite stratification

$$\mathfrak{L} = (\mathfrak{L}(x, H))_{x \in M, H \leq \text{stab } x}.$$

Observe that, as we tacitly did for stratifications induced by arrangements or by irreducible components of divisors, we only specify strata of proper codimension.

The stratification $\mathcal{L}$ locally coincides with the tangent space stratifications coming from our linear setting. Technically speaking: for any $x \in M$, there exists an open neighborhood U of x in M, and a stab x-equivariant diffeomorphism $\Phi_x : U \to T_x M$ such that

$$\Phi_x(\mathfrak{L}(x, H)) = L(x, H) \tag{5–2}$$

for any subgroup H in stab x. In particular, (5–2) shows that the stratification $\mathfrak{L}$ of M is a local subspace arrangement.

THEOREM 5.6 [Feichtner and Kozlov 2005, Theorem 3.4]. *Let a finite group Γ act differentiably and effectively on a smooth real manifold M. Then the wonderful* *model $Y_{\mathfrak{L}}$ induced by the locally finite stratification $\mathfrak{L}$ of M abelianizes the action. Moreover, stabilizers of points on the model $Y_{\mathfrak{L}}$ are isomorphic to elementary abelian 2-groups.*

EXAMPLE 5.7 (ABELIANIZING THE PERMUTATION ACTION ON $\mathbb{RP}^2$). We take a small nonlinear example: the permutation action of $\mathfrak{S}_3$ on the real projective plane induced by $\mathfrak{S}_3$ permuting coordinates in $\mathbb{R}^3$.

We picture $\mathbb{RP}^2$ by its upper hemisphere model in Figure 13, where we agree to place the projectivization of $\Delta^\perp$ on the equator. The locus of nontrivial stabilizers of the $\mathfrak{S}_3$ permutation action consists of the projectivizations of hyperplanes $H_{ij}: x_i = x_j$, for $1 \leq i < j \leq 3$, and three additional points Ψ_{ij} on $\mathbb{P}\Delta^\perp$ indicated in Figure 13. The $\mathfrak{S}_3$ action can be visualized by observing that transpositions $(ij) \in \mathfrak{S}_3$ act as reflections in the lines $\mathbb{P}H_{ij}$, respectively.

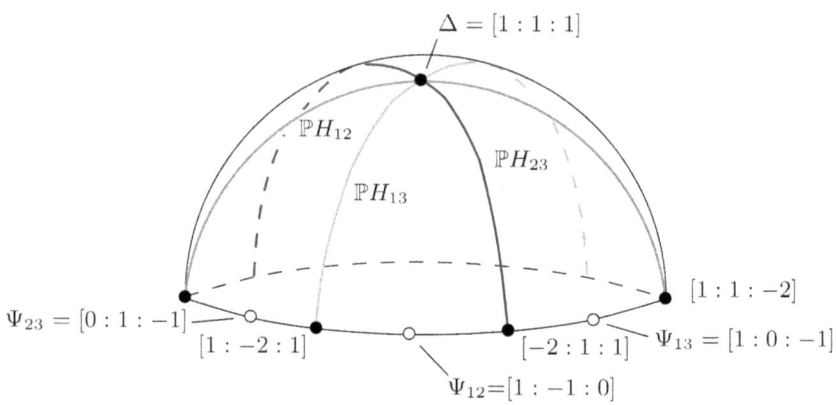

Figure 13. $\mathfrak{S}_3$ acting on $\mathbb{RP}^2$: the stabilizer stratification.

We find that the arrangements $\mathcal{A}_\ell$ in the tangent spaces $T_\ell\mathbb{RP}^2$ are empty, unless $\ell = [1{:}1{:}1]$. Hence, (5–2) allows us to conclude that the $\mathfrak{L}$-stratification of $\mathbb{RP}^2$ consists of a single point, $[1{:}1{:}1]$. Observe that the $\mathfrak{S}_3$-action on $T_{[1{:}1{:}1]}\mathbb{RP}^2$ coincides with the permutation action of $\mathfrak{S}_3$ on $\mathbb{R}^3/\Delta$.

The wonderful model $Y_{\mathfrak{L}}$ hence is a Klein bottle, the result of blowing up $\mathbb{RP}^2$ in $[1{:}1{:}1]$, i.e., glueing a Möbius band into the punctured projective plane.

Observe that the $\mathfrak{L}$-stratification is coarser than the codimension 2 truncation of the stabilizer stratification: The isolated points Ψ_{ij} on $\mathbb{P}\Delta^\perp$ have nontrivial stabilizers, but do not occur as strata in the $\mathfrak{L}$-stratification.

Acknowledgments

I thank the organizers of the Combinatorial and Discrete Geometry workshop held at MSRI in November 2003, for creating an event of utmost breadth, a true kaleidoscope of topics unified by the ubiquity of geometry *and* combinatorics.

References

[Batyrev 1999] V. V. Batyrev, "Non-Archimedean integrals and stringy Euler numbers of log-terminal pairs", *J. Eur. Math. Soc.* **1**:1 (1999), 5–33.

[Batyrev 2000] V. V. Batyrev, "Canonical abelianization of finite group actions", preprint, 2000. Available at math.AG/0009043.

[Borisov and Gunnells 2002] L. A. Borisov and P. E. Gunnells, "Wonderful blowups associated to group actions", *Selecta Math. (N.S.)* **8**:3 (2002), 373–379.

[De Concini and Procesi 1995a] C. De Concini and C. Procesi, "Wonderful models of subspace arrangements", *Selecta Math. (N.S.)* **1**:3 (1995), 459–494.

[De Concini and Procesi 1995b] C. De Concini and C. Procesi, "Hyperplane arrangements and holonomy equations", *Selecta Math. (N.S.)* **1**:3 (1995), 495–535.

[Deligne et al. 2000] P. Deligne, M. Goresky, and R. MacPherson, "L'algèbre de cohomologie du complément, dans un espace affine, d'une famille finie de sous-espaces affines", *Michigan Math. J.* **48** (2000), 121–136.

[Feichtner and Kozlov 2003] E. M. Feichtner and D. N. Kozlov, "Abelianizing the real permutation action via blowups", *Int. Math. Res. Not.* no. 32 (2003), 1755–1784.

[Feichtner and Kozlov 2004] E. M. Feichtner and D. N. Kozlov, "Incidence combinatorics of resolutions", *Selecta Math. (N.S.)* **10**:1 (2004), 37–60.

[Feichtner and Kozlov 2005] E. M. Feichtner and D. N. Kozlov, "A desingularization of real differentiable actions of finite groups", *Int. Math. Res. Not.* no. 15 (2005).

[Feichtner and Müller 2005] E. M. Feichtner and I. Müller, "On the topology of nested set complexes", *Proc. Amer. Math. Soc.* **133** (2005), 999–1006.

[Feichtner and Yuzvinsky 2004] E. M. Feichtner and S. Yuzvinsky, "Chow rings of toric varieties defined by atomic lattices", *Invent. Math.* **155**:3 (2004), 515–536.

[Feichtner and Ziegler 2000] E. M. Feichtner and G. M. Ziegler, "On cohomology algebras of complex subspace arrangements", *Trans. Amer. Math. Soc.* **352**:8 (2000), 3523–3555.

[Fulton and MacPherson 1994] W. Fulton and R. MacPherson, "A compactification of configuration spaces", *Ann. of Math. (2)* **139**:1 (1994), 183–225.

[Gaiffi 1997] G. Gaiffi, "Blowups and cohomology bases for De Concini-Procesi models of subspace arrangements", *Selecta Math. (N.S.)* **3**:3 (1997), 315–333.

[Gaiffi 2003] G. Gaiffi, "Models for real subspace arrangements and stratified manifolds", *Int. Math. Res. Not.* no. 12 (2003), 627–656.

[Goresky and MacPherson 1988] M. Goresky and R. MacPherson, *Stratified Morse theory*, vol. 14, Ergebnisse der Mathematik und ihrer Grenzgebiete (3) [Results in Mathematics and Related Areas (3)], Springer, Berlin, 1988.

[Keel 1992] S. Keel, "Intersection theory of moduli space of stable n-pointed curves of genus zero", *Trans. Amer. Math. Soc.* **330**:2 (1992), 545–574.

[Longueville 2000] M. de Longueville, "The ring structure on the cohomology of coordinate subspace arrangements", *Math. Z.* **233**:3 (2000), 553–577.

[Longueville and Schultz 2001] M. de Longueville and C. A. Schultz, "The cohomology rings of complements of subspace arrangements", *Math. Ann.* **319**:4 (2001), 625–646.

[MacPherson and Procesi 1998] R. MacPherson and C. Procesi, "Making conical compactifications wonderful", *Selecta Math. (N.S.)* **4**:1 (1998), 125–139.

[Morgan 1978] J. W. Morgan, "The algebraic topology of smooth algebraic varieties", *Inst. Hautes Études Sci. Publ. Math.* no. 48 (1978), 137–204.

[Orlik and Solomon 1980] P. Orlik and L. Solomon, "Combinatorics and topology of complements of hyperplanes", *Invent. Math.* **56**:2 (1980), 167–189.

[Reid 1992] M. Reid, "The MacKay correspondence and the physicists' Euler number", lecture notes, University of Utah and MSRI, 1992.

[Stanley 1997] R. P. Stanley, *Enumerative combinatorics*, vol. 1, Cambridge Studies in Advanced Mathematics **49**, Cambridge University Press, Cambridge, 1997.

[Vinberg 1989] E. B. Vinberg, *Linear representations of groups*, Basler Lehrbücher **2**, Birkhäuser, Basel, 1989.

[Yuzvinsky 1997] S. Yuzvinsky, "Cohomology bases for the De Concini–Procesi models of hyperplane arrangements and sums over trees", *Invent. Math.* **127**:2 (1997), 319–335.

[Yuzvinsky 1999] S. Yuzvinsky, "Rational model of subspace complement on atomic complex", *Publ. Inst. Math. (Beograd) (N.S.)* **66(80)** (1999), 157–164.

[Yuzvinsky 2002] S. Yuzvinsky, "Small rational model of subspace complement", *Trans. Amer. Math. Soc.* **354**:5 (2002), 1921–1945.

EVA MARIA FEICHTNER
DEPARTMENT OF MATHEMATICS
ETH ZURICH
8092 ZURICH
SWITZERLAND
feichtne@math.ethz.ch

Combinatorial and Computational Geometry
MSRI Publications
Volume **52**, 2005

Thinnest Covering of a Circle by Eight, Nine, or Ten Congruent Circles

GÁBOR FEJES TÓTH

ABSTRACT. Let r_n be the maximum radius of a circular disc that can be covered by n closed unit circles. We show that $r_n = 1 + 2\cos(2\pi/(n-1))$ for $n = 8$, $n = 9$, and $n = 10$.

1. Introduction

What is the maximum radius r_n of a circular disk which can be covered by n closed unit circles? The determination of r_n for $n \leq 4$ is an easy task: we have $r_1 = r_2 = 1$, $r_3 = 2/\sqrt{3}$ and $r_4 = \sqrt{2}$. The problem of finding r_5 has been motivated by a game popular on fairs around the turn of the twentieth century [Neville 1915; Ball and Coxeter 1987, pages 97–99]. The goal of the game was to cover a circular space painted on a cloth by five smaller circles equal to each other. The difficulty consisted in the restriction that an "on-line algorithm" had to be used, that is no circle was allowed to be moved once it had been placed. Neville [1915] conjectured that $r_5 = 1.64100446\ldots$ and this has been verified by K. Bezdek [1979; 1983] who also determined the value of $r_6 = 1.7988\ldots$. The proofs of these cases are complicated. The case $n = 7$ is again easy. We have $r_7 = 2$ and if 7 unit circles cover a circle C_7 of radius 2, then one of them is concentric with C_7 while the centers of the other circles lie in the vertices of a regular hexagon of side $\sqrt{3}$ concentric with C_7. In his thesis Dénes Nagy [1975] claimed without proof that $r_n = 1 + 2\cos\big(2\pi/(n-1)\big)$ for $n = 8$ and $n = 9$ and that, as for $n = 7$, the best arrangement has $(n-1)$-fold rotational symmetry. He conjectured the same for $n = 10$. Krotoszyński [1993] claimed to have proved this even for $n \leq 11$. Unfortunately, his proof contains some errors. In fact, Melissen and Schuur (see [Melissen 1997]) gave a counter example for $n = 11$.

Part of this work was done at the Mathematical Sciences Research Institute at Berkeley, CA, where the author was participating in a program on Discrete and Computational Geometry. The research was also supported by OTKA grants T 030012, T 038397, and T 043520.

In this note we settle the cases $n = 8$, $n = 9$, and $n = 10$.

THEOREM. *Let r_n be the maximum radius of a circular disk which can be covered by n unit circles. Then we have for $n = 8$, 9, and 10*

$$r_n = 1 + 2 \cos \frac{2\pi}{n-1}.$$

Moreover, if for $n = 8$, $n = 9$, or $n = 10$, n unit circles cover a circle C_n of radius r_n, then one of them is concentric with C_n and the centers of the other circles are situated in the vertices of a regular $(n-1)$-gon at a distance $2 \sin(\pi/(n-1))$ from the center of C_n.

The analogous problems of the thinnest covering of a square and an equilateral triangle with a given number of equal circles, as well as the dual problem concerning the densest packing of a given number of equal circles in a circle, a square or an equilateral triangle have been investigated intensively. A comprehensive account can be found in [Melissen 1997].

Generally, given a compact set C in a metric space, one can consider the problems of the densest packing of n balls in C and the thinnest covering of C with n balls. In lack of similarity the problems are formulated in a dual form. Let $r_C(n)$ be the maximum number with the property that n balls of radius $r_C(n)$ can be packed in C and let $R_C(n)$ be the minimum number with the property that n balls of radius $r_C(n)$ can cover C. The basic task is, of course, to design effective algorithms determining the values of $r_C(n)$ and $R_C(n)$, as well as the corresponding arrangements. So far only the case of $r_C(n)$ for C a square has been solved, by an algorithm devised by Peikert [1994]; see also [Peikert et al. 1992]. Exact solutions are generally known only for small values of n. The only exception is the problem of densest packing of circles in an equilateral triangle. When C is an equilateral triangle, $r_C(n)$ is known for all n of the form $k(k+1)/2$, the triangular numbers; see [Groemer 1960; Oler 1961]. If C is an equilateral triangle with side-length 1, we have for such triangular numbers $r_C(n) = 1/2(k + \sqrt{3} - 1)$. The optimal arrangement is given by the regular triangular lattice.

Many conjectured best arrangements of circles, both for packing and for covering, have been constructed using different heuristic algorithms. The examples show that optimal arrangements quite often contain freely movable circles. This raises the following questions.

Does there exist a compact set C for which for infinitely many n an optimal packing of (covering with) n congruent circles contains a freely movable circle? Does there exist a C for which there is no n at all such that an optimal packing of (covering with) n congruent circles contains a freely movable circle? Is there a constant c, possibly depending on C but independent of n such that the number of freely movable circles in an optimal arrangement is at most c?

The densest packing of $n = k(k+1)/2$ circles in an equilateral triangle shows another interesting phenomenon. According a conjecture of Erdős and Oler [Croft et al. 1991, page 248] if n is a triangular number, then $r_C(n) = r_C(n-1)$, that is, the optimal arrangement for $n-1$ circles is obtained by removing one circle from the optimal arrangement of n circles. The conjecture has been confirmed for $n = 6$ and $n = 10$ [Melissen 1997]. There is a similar situation on the sphere: it is known (see [Rankin 1955], for example) that if $C = S^d$, the d-dimensional sphere, then $r_C(d+3) = r_C(d+4) = \ldots = r_C(2d+2)$. This suggests the following question.

For a given compact set C, are there natural numbers $k = k(C)$ and $K = K(C)$ such that $r_C(n) > r_C(n+k)$ and $R_C(n) > R_C(n+K)$ for every n?

I conjecture that the answer is yes if $C = S^d$ and also if C is a compact convex set in Euclidean or spherical space, but I would not be surprised if the answer turned out to be no for general compact sets, or even for convex bodies in hyperbolic geometry.

2. Proof of the Theorem

For the proof we modify the argument used by Schütte [1955] for the determination of the thinnest covering of the sphere by 5 and 7 congruent caps. Clearly, it suffices to show the second statement of the theorem, from that it follows immediately that no circle of radius greater than r_n can be covered by n unit circles ($n = 8$, $n = 9$, or $n = 10$). The proof of the three cases are similar, however the case $n = 10$ is more complicated. We shall leave two of the more involved discussions for $n = 10$ to Section 3. In the treatment of all three cases the functions $f_r(\alpha)$ and $F_r(\alpha)$ defined for $0 \le \alpha \le \pi$ by

$$f_r(\alpha) = 2 \arcsin \frac{\sin(\alpha/2)}{r}$$

and

$$F_r(\alpha) = r^2 \left(\arcsin \frac{\sin(\alpha/2)}{r} - \frac{1}{2} \sin \left(2 \arcsin \frac{\sin(\alpha/2)}{r} \right) \right) + \frac{\sin \alpha}{2}$$

play an important role. Here $r > 2$ is not a variable but a parameter.

The geometric meaning of $f_r(\alpha)$ and $F_r(\alpha)$ is the following: Let C be a circle of radius r centered at o and let $\tilde{C}$ be a unit circle with center $\tilde{o} \in C$ such that $\operatorname{bd} C$ and $\operatorname{bd} \tilde{C}$ intersect, say in a and b. If $\angle a \tilde{o} b = \alpha$, then $f(\alpha) = \angle aob$ and $F(\alpha)$ is the area of the domain bounded by the segments $\tilde{o}a$, $\tilde{o}b$ and the arc ab of $\operatorname{bd} C$.

We have

$$f_r'(\alpha) = \frac{\cos(\alpha/2)}{\left(r^2 - \sin^2(\alpha/2)\right)^{1/2}}, \qquad f_r''(\alpha) = -\frac{(r^2 - 1)\sin(\alpha/2)}{2\left(r^2 - \sin^2(\alpha/2)\right)^{3/2}},$$

$$F_r'(\alpha) = \frac{\sin\alpha\sin(\alpha/2)}{2(r^2 - \sin^2(\alpha/2))^{1/2}} + \frac{\cos\alpha}{2},$$

$$F_r''(\alpha) = -\left(\sin\frac{\alpha}{2}\right)\left(\frac{\cos\alpha\left(\sin^2(\alpha/2) - r^2\right) - r^2\cos^2(\alpha/2)}{2(r^2 - \sin^2(\alpha/2))^{3/2}} + \cos\frac{\alpha}{2}\right).$$

Hence it is easily seen that $f_r(\alpha)$ is concave and strictly increasing for $0 \le \alpha \le \pi$. The concavity of $F_r(\alpha)$ needs some calculation. To check it, we have to show that

$$\cos\alpha\left(\sin^2\frac{\alpha}{2} - r^2\right) - r^2\cos^2\frac{\alpha}{2} + 2\cos\frac{\alpha}{2}\left(r^2 - \sin^2\frac{\alpha}{2}\right)^{3/2} > 0.$$

Introducing the abbreviations $s = \sin(\alpha/2)$ and $c = \cos(\alpha/2)$, we have

$$\cos\alpha\left(s^2 - r^2\right) - r^2c^2 + 2c\left(r^2 - s^2\right)^{3/2}$$
$$> \cos\alpha\left(s^2 - r^2\right) - r^2c^2 + 2c\left(r^2 - s^2\right)$$
$$= \left(2c^2 - 1\right)\left(1 - c^2 - r^2\right) - r^2c^2 + 2c\left(r^2 - 1 + c^2\right)$$
$$= \left(r^2 - 1\right)\left(1 - c\right)\left(1 + 3c\right) + 2c^3\left(1 - c\right) > 0.$$

Let C_n be a circle of radius r_n centered at o and let $C_0, \ldots, C_{n-1}$ be closed unit circles with centers $o_0, \ldots, o_{n-1}$ covering C_n. We assume that for a circle C_i, $i = 0, \ldots, n-1$, for which $C_i \cap \operatorname{bd} C_n \ne \varnothing$ the centers of C_i and C_n lie on the same side of the radical axis of the circles C_i and C_n. Otherwise we reflect C_i in this radical axis and still get a covering of C_n. Let $\bar{C}_0, \ldots, \bar{C}_{n-1}$ be unit circles in the position described in the theorem, that is so that $\bar{C}_0$ is concentric with C_n and the centers $\bar{o}_1, \ldots, \bar{o}_{n-1}$ of $\bar{C}_1, \ldots, \bar{C}_{n-1}$ are situated in the vertices of a regular $(n-1)$-gon at distance $2\sin(\pi/(n-1))$ from o. We are going to show that the two arrangements of circles are congruent.

The following lemma claims that the two arrangements of circles $\{C_i\}_{i=0}^n$ and $\{\bar{C}_i\}_{i=0}^n$ have the same topological structure.

LEMMA. *Exactly one of the circles $\{C_i\}_{i=0}^{n-1}$ is contained in* $\operatorname{int} C_n$. *Moreover, no three of the circles intersecting* $\operatorname{bd} C_n$ *can have a common point.*

Since $r_n > 2$, there is a circle, say C_0, which is contained in $\operatorname{int} C_n$. Observe that an arc of $\operatorname{bd} C_n$ which is covered by a unit circle spans at o an angle not greater than $2\arcsin(1/r_n)$. Since

$$6\arcsin\frac{1}{r_8} = 2.76326081\ldots < \pi \quad \text{and} \quad 7\arcsin\frac{1}{r_9} = 2.989550105\ldots < \pi,$$

it follows that if $n = 8$ or $n = 9$, then $C_i \cap \operatorname{bd} C_n \ne \varnothing$ for $i = 1, \ldots, n-1$. We also observe that three unit circles with a common point cannot cover a part of $\operatorname{bd} C_n$ whose angle spanned at o exceeds $2\arcsin(2/r_n)$. Since

$$4\arcsin\frac{1}{r_8} + \arcsin\frac{2}{r_8} = 2.942412903\ldots < \pi$$

and

$$5 \arcsin \frac{1}{r_9} + \arcsin \frac{2}{r_9} = 3.111686536\ldots < \pi,$$

it follows that for $n = 8$ and 9 no three of the circles C_i, $i = 1, \ldots, n-1$, can have a common point.

This argument breaks down when $n = 10$. We have

$$7 \arcsin \frac{1}{r_{10}} = 2.841948021 \ldots < \pi,$$

showing that at most two of the circles $\{C_i\}_{i=0}^{9}$ are contained in $\operatorname{int} C_{10}$; however

$$8 \arcsin \frac{1}{r_{10}} = 3.24794059 \ldots > \pi,$$

so we cannot exclude in this way that two of the circles $\{C_i\}_{i=0}^{9}$ are contained in $\operatorname{int} C_{10}$. Also the proof that no three of the circles intersecting $\operatorname{bd} C_n$ can have a common point requires a different argument. Melissen [1997, pp. 108–111] proved the Lemma for $n = 10$ using an argument based on the investigation of distances. In Section 3 we repeat Melissen's argument for the proof of the first part of the Lemma and give an alternative proof for the second statement, using estimations of areas.

Let D_i, $i = 0, \ldots, n-1$, be the Dirichlet cell of C_i with respect to C_n. From the considerations above it follows that each vertex in the cell complex of the Dirichlet cells is trihedral, D_0 is an $(n-1)$-gon, while for $i = 1, \ldots, n-1$, D_i is a curved quadrilateral bounded by three line segments and an arc of $\operatorname{bd} C_n$. Thus the cell complex of the cells D_i is isomorphic to the cell complex of the Dirichlet cells $\bar{D}_i$ of the circles $\bar{C}_i$.

We introduce some notations. We describe them for the circles C_i. The same symbols with a bar will be used for the corresponding objects and quantities for the circles $\bar{C}_i$ (see Figure 1).

Let the vertices of D_0 be $p_1, \ldots, p_{n-1}$ and let the vertices of Dirichlet cells on $\operatorname{bd} C_n$ be $q_1, \ldots, q_{n-1}$. We write $p_n = p_1$, $q_n = q_1$ and assume, as we may without loss of generality, that the notation is chosen so that the vertices of D_i are p_i, p_{i+1}, q_{i+1}, q_i for $i = 1, \ldots, n-1$. We write

$$\alpha_i = \measuredangle q_i o_i q_{i+1}, \quad \beta_i = \measuredangle p_i o_i q_i, \quad \gamma_i = \measuredangle p_{i+1} o_i q_{i+1},$$
$$\delta_i = \measuredangle p_i o_i p_{i+1}, \quad \varepsilon_i = \measuredangle p_i o_0 p_{i+1}.$$

We note that the assumption that o and o_i lie on the same side of the radical axis of C_n and C_i implies that

$$\alpha_i \leq \pi$$

for $i = 1, \ldots, n-1$. It is easy to check that

$$\bar{\alpha}_i = \frac{6\pi}{n-1}, \quad \bar{\beta}_i = \bar{\gamma}_i = \frac{(n-5)\pi}{n-1}, \quad \text{and} \quad \bar{\delta}_i = \bar{\varepsilon}_i = \frac{2\pi}{n-1}.$$

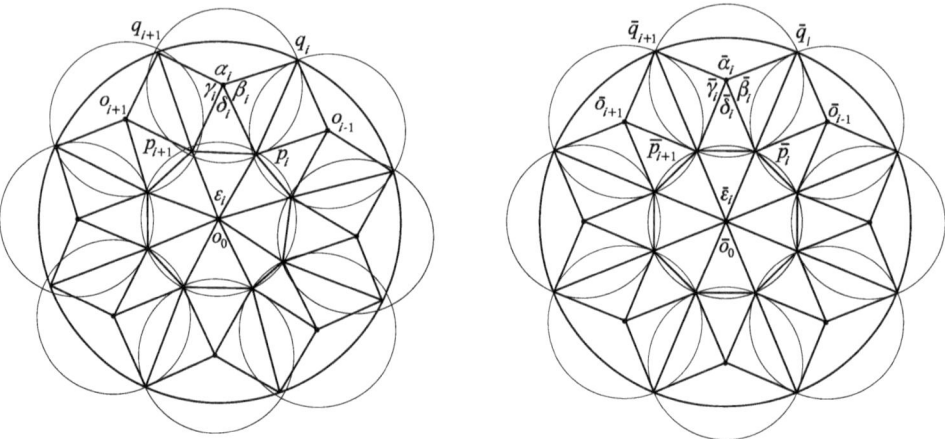

Figure 1.

Using the relation $\bar{\alpha}_i = 6\pi/(n-1)$ one can verify that

$$f\left(\frac{6\pi}{n-1}\right) = \frac{2\pi}{n-1}.$$

We dissect C_n into the triangles $T_i = p_i o_0 p_{i+1}$, $T_i^* = p_i o_i p_{i+1}$, $T_i^- = p_i o_i q_i$, $T_i^+ = p_{i+1} o_i q_{i+1}$ and into the regions R_i bounded by the segments $o_i q_i$, $o_i q_{i+1}$ and the arc $q_i q_{i+1}$ of bd C_n, for $i = 1, \ldots, n-1$. We shall estimate the total area of these domains and show that it is less than the area of C_n unless the arrangement of the circles $C_0, \ldots, C_{n-1}$ is congruent to that of the circles $\bar{C}_0, \ldots, \bar{C}_{n-1}$.

Observe that the triangles T_i and T_i^* are congruent, so that

$$\sum_{i=1}^{n-1} \delta_i = \sum_{i=1}^{n-1} \varepsilon_i = 2\pi = \sum_{i=1}^{n-1} \bar{\delta}_i = \sum_{i=1}^{n-1} \bar{\varepsilon}_i \tag{1}$$

and

$$\sum_{i=1}^{n-1} (\beta_i + \gamma_i) = (n-1)2\pi - \sum_{i=1}^{n-1} (\alpha_i + \delta_i) = (n-2)2\pi - \sum_{i=1}^{n-1} \alpha_i. \tag{2}$$

We have

$$2\pi = \sum_{i=1}^{n-1} \angle q_i o q_{i+1} \le \sum_{i=1}^{n-1} f_{r_n}(\alpha_i) \le (n-1) f_{r_n}\left(\frac{\sum_{i=1}^{n-1} \alpha_i}{n-1}\right).$$

Hence we get

$$\sum_{i=1}^{n-1} \alpha_i \ge (n-1) f_{r_n}^{-1}\left(\frac{2\pi}{n-1}\right) = (n-1)\frac{6\pi}{n-1} = \sum_{i=1}^{n-1} \bar{\alpha}_i. \tag{3}$$

Now we are in the position to estimate the total area of the parts of C_n. Using Jensen's inequality we get

$$\sum_{i=1}^{n-1}(a(T_i)+a(T_i^*)) \le \sum_{i=1}^{n-1}\sin\varepsilon_i \le (n-1)\sin\frac{\sum_{i=1}^{n-1}\varepsilon_i}{n-1}, \qquad (4)$$

$$\sum_{i=1}^{n-1}(a(T_i^-)+a(T_i^+)) \le \frac{1}{2}\sum_{i=1}^{n-1}(\sin\beta_i+\sin\gamma_i) \le (n-1)\sin\frac{\sum_{i=1}^{n-1}(\beta_i+\gamma_i)}{n-1}, \quad (5)$$

$$\sum_{i=1}^{n-1}a(R_i) \le \sum_{i=1}^{n-1}F_{r_n}(\alpha_i) \le (n-1)F_{r_n}\left(\frac{\sum_{i=1}^{n-1}\alpha_i}{n-1}\right). \qquad (6)$$

In view of (1) we have

$$(n-1)\sin\frac{\sum_{i=1}^{n-1}\varepsilon_i}{n-1} = \sum_{i=1}^{n-1}\sin\bar{\varepsilon}_i = \sum_{i=1}^{n-1}(a(\bar{T}_i)+a(\bar{T}_i^*)).$$

Write

$$\alpha = \sum_{i=1}^{n-1}\frac{\alpha_i}{n-1}.$$

Then we have, in view of (2),

$$\frac{\sum_{i=1}^{n-1}(\beta_i+\gamma_i)}{n-1} = \frac{2(n-2)\pi}{n-1} - \alpha;$$

hence, by (5) and (6),

$$\sum_{i=1}^{n-1}(a(T_i^-)+a(T_i^+)+a(R_i)) \le (n-1)\left(\sin\left(\frac{2(n-2)\pi}{n-1}-\alpha\right)+F_{r_n}(\alpha)\right).$$

The function

$$\sin\left(\frac{2(n-2)\pi}{n-1}-\alpha\right)+F_{r_n}(\alpha)$$

is concave for $0 \le \alpha \le \pi$ and, as it can be checked numerically, decreasing for $\alpha = 6\pi/(n-1)$. Therefore it is decreasing for $6\pi/(n-1) \le \alpha \le \pi$. Observing that

$$\frac{6\pi}{n-1} = \bar{\alpha}_i \quad \text{and} \quad \frac{2(n-2)\pi}{n-1} - \frac{6\pi}{n-1} = \frac{2(n-10)\pi}{n-1} = \bar{\beta}_i = \bar{\gamma}_i,$$

we deduce that

$$\sum_{i=1}^{n-1}(a(T_i^-)+a(T_i^+)+a(R_i)) \le (n-1)\left(\sin\frac{2(n-10)\pi}{n-1}+F_{r_n}\left(\frac{6\pi}{n-1}\right)\right)$$

$$= \frac{1}{2}\sum_{i=1}^{n-1}(\sin\bar{\beta}_i+\sin\bar{\gamma}_i)+\sum_{i=1}^{n-1}F_{r_n}(\bar{\alpha}_i)$$

$$= \sum_{i=1}^{n-1}(a(\bar{T}_i^-)+a(\bar{T}_i^+)+a(\bar{R}_i)). \qquad (7)$$

Adding inequalities (4) and (7) we get

$$a(C_n) = \sum_{i=1}^{n-1}(a(T_i) + a(T_i^*) + a(T_i^-) + a(T_i^+) + a(R_i))$$

$$\leq \sum_{i=1}^{n-1}(a(\bar{T}_i) + a(\bar{T}_i^*) + a(\bar{T}_i^-) + a(\bar{T}_i^+) + a(\bar{R}_i)) = a(\bar{C}_n).$$

Therefore we have equality in all of the inequalities (5)–(7). This can only occur if the arrangements of the circles $C_0, \ldots, C_{n-1}$ and $\bar{C}_0, \ldots, \bar{C}_{n-1}$ are congruent.

3. Proof of the Lemma for $n = 10$

Let $C_0, \ldots, C_9$ be closed unit circles covering the circle C_{10} of radius r_{10}. As in the previous section, we denote the center of C_i, $i = 0, \ldots, 9$ by o_i and the center of C_{10} by o. We shall follow the argument of Melissen to show that no eight of the circles can cover $\operatorname{bd} C_{10}$. Suppose that $\operatorname{bd} C_{10} \subset \bigcup_{i=0}^{7} C_i$. Since the angular measure of an arc of $\operatorname{bd} C_{10}$ covered by a unit circle is at most $2\arcsin(1/r_{10})$ and

$$7\arcsin \frac{1}{r_{10}} = 2.841948021 \ldots < \pi,$$

no proper subset of the circles C_i, $i = 0, \ldots, 7$ covers $\operatorname{bd} C_{10}$, hence no three of the arcs $C_i \cap \operatorname{bd} C_{10}$, $0 \leq i \leq 7$ intersect. This property defines a cyclic order of the arcs $C_i \cap \operatorname{bd} C_{10}$. We assume that the notation is chosen so that this cyclic order coincides with the order of the indices, that is $C_0 \cap C_1 \cap \operatorname{bd} C_{10} \neq \emptyset, \ldots, C_6 \cap C_7 \cap \operatorname{bd} C_{10} \neq \emptyset, C_7 \cap C_0 \cap \operatorname{bd} C_{10} \neq \emptyset$. We choose points $q_1, \ldots, q_8$ from the sets $C_0 \cap C_1 \cap \operatorname{bd} C_{10}, \ldots, C_7 \cap C_0 \cap \operatorname{bd} C_{10}$, respectively.

Recall that the maximum angular measure of an arc of $\operatorname{bd} C_{10}$ covered by three unit circles with a common point is $2\arcsin(2/r_{10})$. Since

$$\arcsin \frac{2}{r_{10}} + 5\arcsin \frac{1}{r_{10}} = 2.940546309 \ldots < \pi,$$

no three of the circles C_i, $i = 0, \ldots, 7$ have a common point. Let

$$p_i = \operatorname{bd} C_{i-1} \cap C_i \cap \operatorname{int} C_{10}$$

for $i = 1, \ldots, 7$, and $p_8 = C_7 \cap C_0 \cap \operatorname{int} C_{10}$ (see Figure 2).

The main observation of Melissen is that the points p_i, $i = 1, \ldots, 8$, cannot be covered by two circles. This follows easily from the following result:

PROPOSITION. *We have*

$$|p_i p_{i+3}| > 2 \quad and \quad |p_i p_{i+4}| > 2$$

for $i = 1, \ldots, 8$, *with* $p_i = p_{i+8}$.

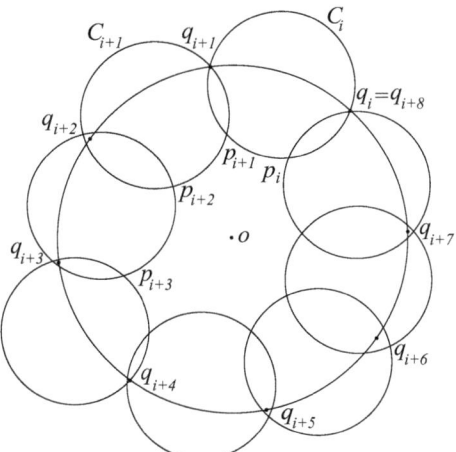

Figure 2.

Indeed, the first inequality readily implies that one of the circles C_8 and C_9 contains all points p_i with an odd subscript and the other contains all points with an even subscript. This, however, contradicts the second inequality.

In order to estimate the distances between the points p_i we need a lower bound for the distance $|op_i|$. Let $h(\vartheta)$ be the minimum distance between o and a point of intersection of the boundaries of two unit circles that cover an arc of angular measure ϑ from $\operatorname{bd} C_{10}$, with $0 \leq \vartheta \leq 4\arcsin(1/r_{10})$. It is easy to see that this minimum distance is achieved in the symmetric position when each of the unit circles cover an arc of angular measure $\frac{\vartheta}{2}$ from $\operatorname{bd} C_{10}$. Using some trigonometry we calculate that

$$h(\vartheta) = r_{10}\cos\frac{\vartheta}{2} - 2\sqrt{1 - r_{10}^2\sin^2\frac{\vartheta}{4}}\cos\frac{\vartheta}{4}.$$

Writing $s = \sin^2\frac{\vartheta}{4}$ we have

$$h'(\vartheta) = \frac{\sqrt{s}\left(\sqrt{1 - r_{10}^2 s^2} - r_{10}\cos\frac{\vartheta}{4}\right)^2}{\sqrt{1 - r_{10}^2 s^2}}$$

and

$$h''(\vartheta) = \frac{-2r_{10}(1 - r_{10}^2 s)^{3/2}\cos\frac{\vartheta}{2} + (1 + r_{10}^2 - 6r_{10}^2 s + 4r_{10}^4 s^2)\cos\frac{\vartheta}{4}}{8(1 - r_{10}^2 s^2)^{3/2}}.$$

It immediately follows that $h(\vartheta)$ is increasing. Observing that

$$1 + r_{10}^2 - 6r_{10}^2 s + 4r_{10}^4 s^2 > r_{10}^2 - \frac{5}{4}$$

we get

$$h''(\vartheta) > \frac{-2r_{10}\cos\frac{\vartheta}{2} + (r_{10}^2 - \frac{5}{4})\cos\frac{\vartheta}{4}}{8} = \frac{-4r_{10}\cos^2\frac{\vartheta}{4} + (r_{10}^2 - \frac{5}{4})\cos\frac{\vartheta}{4} + 2r_{10}}{8}.$$

The minimum of the right side is $\frac{1}{8}\left(r_{10}^2 - 2r_{10} - \frac{5}{4}\right) > 0$, showing that $h(\vartheta)$ is convex.

Now we are in the position to estimate the distances between the points p_i. Write $\psi = \angle p_i o p_{i+3}$ and $\xi = \angle p_i o p_{i+4}$. Then we have, on the one hand,

$$\psi = 2\pi - \angle p_{i+3} o q_{i+4} - \sum_{j=4}^{6} \angle q_{i+j} o q_{i+j+1} - \angle q_{i+7} o p_i$$

$$\geq 2\pi - 2\arcsin\frac{2}{r_{10}} - 6\arcsin\frac{1}{r_{10}} = \psi_{\min} = 2.026062\ldots > \frac{\pi}{2}$$

and

$$\xi = 2\pi - \angle p_{i+4} o q_{i+5} - \sum_{j=45}^{6} \angle q_{i+j} o q_{i+j+1} - \angle q_{i+7} o p_i$$

$$\geq 2\pi - 2\arcsin\frac{2}{r_{10}} - 4\arcsin\frac{1}{r_{10}} = \xi_{\min} = 2.838048136\ldots > \frac{\pi}{2},$$

and on the other hand,

$$\psi = \angle p_i o q_{i+1} + \angle q_{i+1} o q_{i+2} + \angle q_{i+2} o p_{i+3}$$

$$\leq 2\arcsin\frac{2}{r_{10}} + 2\arcsin\frac{1}{r_{10}} = 2.633152\ldots < \pi$$

and

$$\xi = \angle p_i o q_{i+1} + \angle q_{i+1} o q_{i+2} + \angle q_{i+2} o p_{i+3} + \angle p_{i+3} o p_{i+4}$$

$$\leq 2\arcsin\frac{2}{r_{10}} + 4\arcsin\frac{1}{r_{10}} = 2\pi - \xi_{\min}.$$

By the law of cosines we get

$$|p_i p_{i+3}| = \sqrt{|op_i|^2 + |op_{i+3}|^2 - 2|op_i||op_{i+3}|\cos\psi},$$

$$|p_i p_{i+4}| = \sqrt{|op_i|^2 + |op_{i+4}|^2 - 2|op_i||op_{i+4}|\cos\xi}.$$

Let ϑ_1, ϑ_2, and ϑ_3 be the angular measure of the arc of $\operatorname{bd} C_{10}$ covered by the pair of circles C_{i-1}, C_i, C_{i+2}, C_{i+3}, and C_{i+3}, C_{i+4}, respectively. As the triangles $p_i o p_{i+3}$ and $p_i o p_{i+4}$ are obtuse, we get lower bounds for $|p_i p_{i+3}|$ and $|p_i p_{i+4}|$ if we substitute for $|op_i|$, $|op_{i+3}|$, and $|op_{i+4}|$ their minimum values and for $\cos\psi$ and $\cos\xi$ their maximum values:

$$|p_i p_{i+3}| \geq \sqrt{h^2(\vartheta_1) + h^2(\vartheta_2) - 2h(\vartheta_1)h(\vartheta_2)\cos\psi_{\min}},$$

$$|p_i p_{i+4}| \geq \sqrt{h^2(\vartheta_1) + h^2(\vartheta_3) - 2h(\vartheta_1)h(\vartheta_3)\cos\xi_{\min}}.$$

We have, for $j = 2, 3$,

$$\vartheta_1 + \vartheta_j \leq 2\pi - 8\arcsin\frac{1}{r_{10}}.$$

Since $h(\vartheta)$ is increasing and convex, therefore,

$$h(\vartheta_1) + h(\vartheta_j) \geq 2h\left(\pi - 4\arcsin\frac{1}{r_{10}}\right).$$

The functions $\sqrt{h_1^2 + h_2^2 - 2h_1 h_2 \cos \psi_{\min}}$ and $\sqrt{h_1^2 + h_2^2 - 2h_1 h_2 \cos \xi_{\min}}$ are homogeneous of degree one in the variables h_1 and h_2, thus, they are convex. They are also increasing in both variables. Therefore

$$|p_i p_{i+3}| \geq \sqrt{2h^2(\pi - 4 \arcsin(1/r_{10}))(1 - \cos \psi_{\min})} =$$

$$= 2h(\pi - 4 \arcsin(1/r_{10})) \sin \frac{\psi_{\min}}{2} = 2.02349 \ldots > 2$$

and

$$|p_i p_{i+4}| \geq \sqrt{2h^2(\pi - 4 \arcsin(1/r_{10}))(1 - \cos \xi_{\min})} =$$

$$= 2h(\pi - 4 \arcsin(1/r_{10})) \sin \frac{\xi_{\min}}{2} = 2.357538 \ldots > 2.$$

This completes the proof of the Proposition and at the same time the proof of the first part of the Lemma.

It remains to show the second part of the Lemma, namely that if nine of the circles $C_0, \ldots, C_9$ intersect $\operatorname{bd} C_{10}$, then no three of them can have a common point. This part of the Lemma can be settled by estimating areas in a similar way as we did in the previous section.

Suppose that $C_0 \cap \operatorname{bd} C_{10} = \varnothing$ and $C_i \cap \operatorname{bd} C_{10} \neq \varnothing$ for $i = 1, \ldots, 9$. We shall scrutinize the cell complex formed by the Dirichlet cells D_i of the circles C_i, $0 \leq i \leq 9$, with respect to C_{10}. We may assume that $D_i \cap \operatorname{bd} C_{10} \neq \varnothing$ for $i = 1, \ldots, 9$, otherwise $\operatorname{bd} C_{10}$ is covered by eight circles, which we already excluded. Without loss of generality we may suppose that the arcs $D_i \cap \operatorname{bd} C_{10}$, $i = 1, \ldots, 9$, are situated on $\operatorname{bd} C_{10}$ in their natural cyclic order.

We shall exclude the possibility that three of the Dirichlet cells $D_1, \ldots, D_9$ intersect. We note that three circles can intersect without their corresponding Dirichlet cells having a common point, however the case when no three of the cells $D_1, \ldots, D_9$ intersect has been already discussed in the previous section. Observe that

$$D_i \cap D_{i \pm j} = \varnothing \quad \text{for} \quad i = 1, \ldots, 9, \quad j = 3, 4. \tag{8}$$

Otherwise the circles $C_i, C_{i \pm 1} \ldots, C_{i \pm j}$ cover from $\operatorname{bd} C_{10}$ an arc whose angular measure is at most $2 \arcsin(2/r_{10})$, while the angular measure of the arc covered by the other $9 - j - 1 \leq 5$ circles cannot exceed $10 \arcsin(1/r_{10})$. Since $\arcsin(2/r_{10}) + 5 \arcsin(2/r_{10}) = 2.940546309 \ldots < \pi$, this is impossible.

Suppose that three of the cells $D_1, \ldots, D_9$ intersect. In view of (8) they must belong to consecutive indices. Assume, say, that $D_1 \cap D_2 \cap D_3 \neq \varnothing$. If no further triple of the cells $D_1, \ldots, D_9$ intersect, then the cells are arranged as depicted on the left side of Figure 3, where the Dirchlet cells are drawn by broken lines. We shall refer to this situation as Case 1.

If there is another intersecting triple, say D_i, D_{i+1}, and D_{i+2}, among the cells $D_1, \ldots, D_9$, then $\{1, 2, 3\} \cap \{i, i+1, i+2\} \neq \varnothing$. Otherwise the total angular

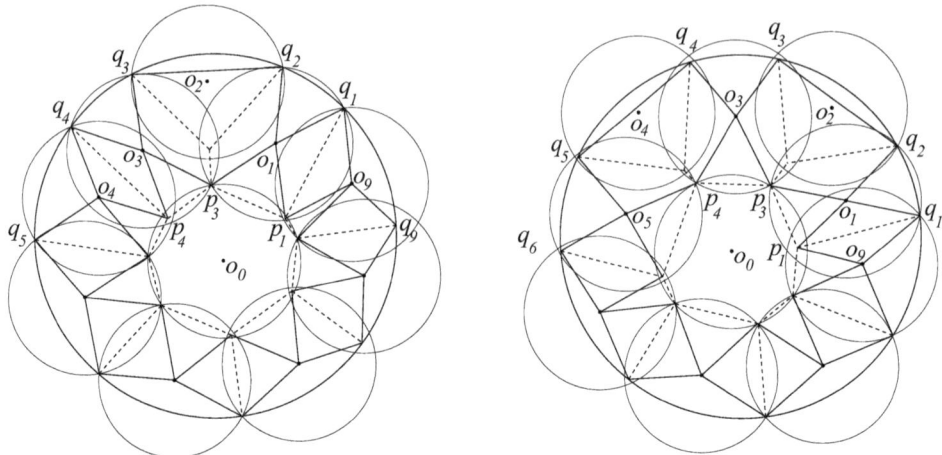

Figure 3.

measure of the arcs of bd C_{10} covered by the circles is at most $4\arcsin(2/r_{10}) + 6\arcsin(1/r_{10}) = 6.078289194\ldots < 2\pi$. D_2 cannot be the member of another intersecting triple of cells. For, if D_2, D_3, and D_4 or D_9, D_1, and D_2 intersect, then $D_1 \cap D_4 \neq \varnothing$ or $D_9 \cap D_3 \neq \varnothing$, which is impossible by (8). Thus, the only candidates for another triple of intersecting cells are $\{D_3, D_4, D_5\}$ and $\{D_8, D_9, D_1\}$. As these triples are disjoint, only one of them can have a nonempty intersection. Hence, the other case we have to investigate is that besides D_1, D_2, and D_3, say D_3, D_4, and D_5 have a common point. This is Case 2 which is represented on the right side of Figure 3.

As before, we denote by o_i, $i = 1, \ldots, 10$, the center of C_i, and by q_i, $i = 1, \ldots, 9$, the vertices of Dirichlet cells on bd C_{10} choosing the notation so that q_i is common to D_{i-1} and D_i. We denote the vertices of D_0 in their consecutive order for the two cases by $p_1, p_3, p_4, \ldots, p_9$ and $p_1, p_3, p_5, \ldots, p_9$, respectively, so that $p_1 q_1$ is the side common to D_1 and D_9 (see Figure 3). We divide C_{10} into the following regions:

(i) the 16-gon $P_1 = p_1 o_1 p_3 o_3 p_4 o_4 p_5 o_5 p_6 o_6 p_7 o_7 p_8 o_8 p_9 o_9$, the pentagon $P_2 = q_2 o_1 p_3 o_3 q_3$, the segment S_1 cut off from C_{10} by the chord $q_2 q_3$, the quadrilaterals $Q_i = o_i q_{i+1} o_{i+1} p_{i+1}$, $3 \leq i \leq 9$, and the regions R_i, $1 \leq i \leq 9$, $i \neq 2$, bounded by the segments $o_i q_i$, $o_i q_{i+1}$ and the arc $q_i q_{i+1}$ of bd C_n in Case 1;

(ii) the 14-gon $P_1 = p_1 o_1 p_3 o_3 p_5 o_5 p_6 o_6 p_7 o_7 p_8 o_8 p_9 o_9$, the pentagons

$$P_2 = q_2 o_1 p_3 o_3 q_3 \quad \text{and} \quad P_3 = q_4 o_3 p_5 o_5 q_5,$$

the two segments S_1 and S_2 cut off from C_{10} by the chords $q_2 q_3$ and $q_4 q_5$, respectively, the quadrilaterals $Q_i = o_i q_{i+1} o_{i+1} p_{i+1}$, $5 \leq i \leq 9$, and the regions R_i, $1 \leq i \leq 9$, $i \neq 2, 4$, bounded by the segments $o_i q_i$, $o_i q_{i+1}$ and the arc $q_i q_{i+1}$ of bd C_n in Case 2.

As we saw in the previous section, P_1 can be dissected into pairs of congruent triangles one half of the triangles making up the cell D_0. Hence we get

$$a(P_1) \leq 8 \sin \frac{\pi}{4} = 5.656854249\ldots \tag{9}$$

in Case 1 and

$$a(P_1) \leq 7 \sin \frac{2\pi}{7} = 5.472820377\ldots \tag{9'}$$

in Case 2.

The length of four sides of the pentagon P_2 (P_3) is bounded above by 1, while the length of the fifth side is at most 2. The area of such a pentagon cannot exceed the area of a pentagon with four sides of length 1 and one side of length 2 inscribed into a circle. The radius r of the circle is determined implicitly by the equation

$$4 \arcsin \frac{1}{2r} + \arcsin \frac{1}{r} = \pi.$$

$r_0 = 1.07326$ is an upper bound for r and

$$r_0^2 \left(\frac{1}{2} \sin 2 \arcsin \frac{1}{r_0} + 2 \sin \frac{\pi - \arcsin \frac{1}{r.}}{2} \right) = 2.284572282\ldots$$

is an upper bound for the area of the pentagon. Thus, we have

$$a(P_2) \leq 2.284572282\ldots \tag{10}$$

and, in Case 2,

$$a(P_2) + a(P_3) \leq 4.5691944564\ldots. \tag{10'}$$

The area of S_1 (S_2) cannot exceed

$$r_{10}^2 \left(\arcsin \frac{1}{r_{10}} - \frac{1}{2} \sin 2 \arcsin \frac{1}{r_{10}} \right) = 0.27675335\ldots,$$

the area of a segment of C_{10} cut off by a chord of length 2. Hence

$$a(S_1) \leq 0.27675335\ldots \tag{11}$$

and

$$a(S_1) + a(S_2) \leq 0.533506699\ldots. \tag{11'}$$

Using the rough estimate $a(Q_i) \leq 1$ we get

$$\sum_{i=3}^{9} a(Q_i) \leq 7 \tag{12}$$

and

$$\sum_{i=5}^{9} a(Q_i) \leq 5, \tag{12'}$$

respectively.

We estimate the total area of the regions R_i using the method developed in the previous section. Let $x = \angle q_2 o q_3$ in Case 1 and $x = \angle q_2 o q_3 + \angle q_4 o q_5$ in

Case 2. Then we have $x \leq 2 \arcsin(1/r_{10})$ and $x \leq 4 \arcsin(1/r_{10})$, respectively. Writing $\alpha_i = \angle q_i o_i q_{i+1}$ we have

$$2\pi - x = \sum_{1 \leq i \leq 9,\, i \neq 2} \angle q_i o q_{i+1} \leq \sum_{1 \leq i \leq 9,\, i \neq 2} f_{r_{10}}(\alpha_i) \leq 8 f_{r_{10}} \left(\frac{1}{8} \sum_{1 \leq i \leq 9,\, i \neq 2} \alpha_i \right)$$

and

$$2\pi - x = \sum_{1 \leq i \leq 9,\, i \neq 2,4} \angle q_i o q_{i+1} \leq \sum_{1 \leq i \leq 9,\, i \neq 2,4} f_{r_{10}}(\alpha_i) \leq 8 f_{r_{10}} \left(\frac{1}{8} \sum_{1 \leq i \leq 9,\, i \neq 2,4} \alpha_i \right),$$

hence

$$\frac{1}{8} \sum_{1 \leq i \leq 9,\, i \neq 2} \alpha_i \geq f_{r_{10}}^{-1} \left(\frac{2\pi - x}{8} \right) \geq f_{r_{10}}^{-1} \left(\frac{\pi - \arcsin \frac{1}{r_{10}}}{4} \right) = 2.028453422\ldots$$

and

$$\frac{1}{7} \sum_{1 \leq i \leq 9,\, i \neq 2,\, 4} \alpha_i \geq f_{r_{10}}^{-1} \left(\frac{\pi - x}{7} \right) \geq f_{r_{10}}^{-1} \left(\frac{2\pi - 4 \arcsin \frac{1}{r_{10}}}{7} \right) = 1.948256547\ldots,$$

respectively.

Observing that $F_{r_{10}}(\alpha)$ is decreasing for $\alpha \geq 1.9$ we get for the total area of the regions R_i the estimate

$$\sum_{1 \leq i \leq 9,\, i \neq 2} a(R_i) \leq \sum_{1 \leq i \leq 9,\, i \neq 2} F_{r_{10}}(\alpha_i) \leq 8 F_{r_{10}} \left(\frac{1}{8} \sum_{1 \leq i \leq 9,\, i \neq 2} \alpha_i \right)$$

$$\leq 8 F_{r_{10}} \left(f_{r_{10}}^{-1} \left(\frac{\pi - \arcsin \frac{1}{r_{10}}}{4} \right) \right) = 4.92397937\ldots \qquad (13)$$

in Case 1 and

$$\sum_{1 \leq i \leq 9,\, i \neq 2,4} a(R_i) \leq \sum_{1 \leq i \leq 9,\, i \neq 2,4} F_{r_{10}}(\alpha_i) \leq 7 F_{r_{10}} \left(\frac{1}{7} \sum_{1 \leq i \leq 9,\, i \neq 2,4} \alpha_i \right)$$

$$\leq 7 F_{r_{10}} \left(f_{r_{10}}^{-1} \left(\frac{2\pi - 4 \arcsin \frac{1}{r_{10}}}{7} \right) \right) = 4.332295377\ldots \qquad (13')$$

in Case 2.

From inequalities (9)–(13) we conclude that

$$a(C_{10}) = a(P_1) + a(P_2) + a(S_1) + \sum_{i=3}^{9} a(Q_i) + \sum_{1 \leq i \leq 9,\, i \neq 2} a(R_i) \leq$$

$$\leq 20.14216 < 20.1422 < a(C_{10}).$$

in Case 1, and it follows from $(9')$–$(13')$ that

$$a(C_{10}) = a(P_1) + a(P_2) + a(P_3) + a(S_1) + a(S_2) + \sum_{i=5}^{9} a(Q_i) + \sum_{1 \le i \le 9, \, i \ne 2,4} a(R_i)$$

$$\le 20 < 20.1422 < a(C_{10})$$

in Case 2, yielding in both cases a contradiction.

This completes the proof of the Lemma.

References

[Ball and Coxeter 1987] W. W. R. Ball and H. S. M. Coxeter, *Mathematical recreations and essays*, Dover Publications Inc., New York, 1987. Original edition, 1892.

[Bezdek 1979] K. Bezdek, *Optimal covering of circles*, Thesis, Budapest, 1979. In Hungarian.

[Bezdek 1983] K. Bezdek, "Über einige Kreisüberdeckungen", *Beiträge Algebra Geom.* no. 14 (1983), 7–13.

[Croft et al. 1991] H. T. Croft, K. J. Falconer, and R. K. Guy, *Unsolved problems in geometry*, Problem Books in Mathematics, Springer, New York, 1991.

[Groemer 1960] H. Groemer, "Über die Einlagerung von Kreisen in einen konvexen Bereich", *Math. Z* **73** (1960), 285–294.

[Krotoszyński 1993] S. Krotoszyński, "Covering a disk with smaller disks", *Studia Sci. Math. Hungar.* **28**:3-4 (1993), 277–283.

[Melissen 1997] H. Melissen, *Packing and covering with circles*, Ph.D. Thesis, 1997.

[Nagy 1975] D. Nagy, *Coverings and their applications*, Thesis, Budapest, 1975. In Hungarian.

[Neville 1915] E. H. Neville, "On the solutions of numerical functional equations, illustrated by an account of a popular puzzle and its solution", *Proc. London Math. Soc.* (2) **14** (1915), 308–326.

[Oler 1961] N. Oler, "A finite packing problem", *Canad. Math. Bull.* **4** (1961), 153–155.

[Peikert 1994] R. Peikert, "Dichteste Packungen von gleichen Kreisen in einem Quadrat", *Elem. Math.* **49**:1 (1994), 16–26.

[Peikert et al. 1992] R. Peikert, D. Würtz, M. Monagan, and C. de Groot, "Packing circles in a square: a review and new results", pp. 45–54 in *System modelling and optimization* (Zurich, 1991), Lecture Notes in Control and Inform. Sci. **180**, Springer, Berlin, 1992.

[Rankin 1955] R. A. Rankin, "The closest packing of spherical caps in n dimensions", *Proc. Glasgow Math. Assoc.* **2** (1955), 139–144.

[Schütte 1955] K. Schütte, "Überdeckungen der Kugel mit höchstens acht Kreisen", *Math. Ann.* **129** (1955), 181–186.

GÁBOR FEJES TÓTH
ALFRÉD RÉNYI INSTITUTE OF MATHEMATICS
HUNGARIAN ACADEMY OF SCIENCES
P.O.BOX 127
H-1364 BUDAPEST
HUNGARY
  gfejes@renyi.hu

Combinatorial and Computational Geometry
MSRI Publications
Volume **52**, 2005

On the Complexity of Visibility Problems
with Moving Viewpoints

PETER GRITZMANN AND THORSTEN THEOBALD

ABSTRACT. We investigate visibility problems with moving viewpoints in
n-dimensional space. We show that these problems are NP-hard if the underlying bodies are balls, $\mathcal{H}$-polytopes, or $\mathcal{V}$-polytopes. This is contrasted
by polynomial time solvability results for fixed dimension. We relate the
computational complexity to existing algebraic-geometric aspects of the
visibility problems, to the theory of packing and covering, and to the view
obstruction problem from diophantine approximation.

1. Introduction

Computer graphics and visualization deal with preparing data in order to show
("visualize") these data on a (two-dimensional) computer screen. In computer
graphics, the original data typically stem from the three-dimensional Euclidean
space $\mathbb{R}^3$, whereas in scientific visualization the data might originate from spaces
of much higher dimension (e.g., in information visualization or high-dimensional
sphere models in statistical mechanics) [Swayne et al. 1998].

In these scenarios, *visibility computations* play a central role [O'Rourke 1997].
In the simplest case, we are given a fixed viewpoint $v \in \mathbb{R}^n$, and the scene consists
of a set $\mathcal{B}$ of bodies. Now the task is to compute a suitable two-dimensional
projection of the scene ("to render the scene") that reflects which part of the
scene is *visible* from the viewpoint v. In a more dynamic setting, the viewpoint
can be moved interactively (see [Bern et al. 1994; Lenhof and Smid 1995], for
example). However, in general, after each movement of the viewpoint a new
rendering process is necessary. In order to speed up this process, commercial
renderers apply caching techniques [Wernecke 1994].

Mathematics Subject Classification: 68Q17, 68U05, 52C45, 11J13, 52A37.

Keywords: Visibility, computational complexity, computational geometry, simultaneous homogeneous diophantine approximation, view obstruction, computational convexity.

From the algorithmic and geometric point of view it is desirable to establish a more global view of the scene in advance and answer questions like: Which of these bodies can be seen (at least partially) from *some* viewpoint within a given viewpoint area? The bodies which are not even partially visible from any of these viewpoints can be removed from the whole visualization process in advance. In the case of dense scenes (like in the visualization of dense crystals, consisting of many atoms) this can reduce the time consumption of the rendering processes significantly. In n-dimensional space invisibility of a body is a sufficient criterion for its invisibility in any low-dimensional projection.

As yet, algorithmic treatment of visibility computations with moving viewpoints in dimension at least three still bears many challenges (see the recent papers [Devillers et al. 2003; Durand 2002; Durand et al. 1997; Wang and Zhu 2000]). A main reason for this can be seen in various intrinsic difficulties in the underlying complexity-theoretical, geometric and algebraic questions.

In the present paper, we analyze the *binary Turing machine complexity* of visibility computations in spaces of variable dimension. The classes of geometric bodies under consideration are that of balls, that of polytopes represented as the convex hull of finitely many points ("$\mathcal{V}$-polytopes"), and that of polytopes represented by an intersection of finitely many halfspaces ("$\mathcal{H}$-polytopes"). Roughly speaking, we show the following results that characterize the borderline between tractable and hard. If the dimension of the space is part of the input, then checking visibility of a given body B in the scene is NP-hard for all three classes. Moreover, these hardness results persist even for very restricted classes of polytopes. In the case where the given body B degenerates to a single point, we can prove also membership in NP for the two classes of polytopes. If however, the dimension is fixed then the visibility problem becomes solvable in polynomial time for all three classes. (For precise statements of the results see Theorems 2 and 3.)

Moreover, we relate these complexity results to existing results from several other perspectives. From the *algebraic-geometric point of view*, visibility computations with moving viewpoints require the study of the interaction of the geometric bodies with lines. In particular, it is essential to characterize certain extreme situations which correspond to common tangent lines to a given set of bodies. In dimension 2, the resulting geometric questions typically remain rather elementary (see [O'Rourke 1997; Pocchiola and Vegter 1996]). However, in dimension 3 already, and even for simple types of bodies, such as balls, the underlying geometric problems have a high algebraic degree and hence give rise to difficult questions of real algebraic geometry [Macdonald et al. 2001; Theobald 2002]. See Section 4 for details.

We also relate our complexity results to Hornich and Fejes Tóth configurations from the theory of packing and covering. Our results imply that already the test whether a given visibility configuration is a Hornich or Fejes Tóth configuration is an NP-hard problem.

Finally, we establish a link between our hardness results and the *view obstruction* or *lonely runner* conjecture from diophantine approximation [Wills 1968; Cusick 1973; Bienia et al. 1998]. Let $\|x\|_I$ denote the distance of a real number x to a nearest integer. Then, for each positive integer n, let

$$\kappa(n) = \inf_{v_1,\ldots,v_n \in \mathbb{N}} \ \sup_{\tau \in [0,1]} \ \min_{1 \leq i \leq n} \|\tau v_i\|_I,$$

a measure for simultaneous homogeneous diophantine approximation. Wills [1968] and later Cusick [1973] conjectured that $\kappa(n) = 1/(n+1)$. Although this conjecture has been investigated in a series of papers in the last 30 years (see the list of references in [Chen and Cusick 1999]), the exact value of $\kappa(n)$ is known only for values up to 5. Our hardness results can be seen as a complexity-theoretical indication why the number-theoretical view obstruction problem is hard.

The present paper is organized as follows. In Section 2, we introduce the necessary notation and review known algorithmic results in dimension 3. In Section 3, we determine the computational complexity of the considered visibility problems. Finally, in Section 4, we establish connections between our complexity-theoretical results and the other mentioned fields.

2. Preliminaries and Known Results

Throughout this paper $\mathbb{R}^n$ denotes n-dimensional Euclidean space; $\langle \cdot, \cdot \rangle$ and $\|\cdot\|$ denote the Euclidean scalar product and norm; and $\mathbb{B}^n = \{x \in \mathbb{R}^n : \|x\| \leq 1\}$ and $\mathbb{S}^{n-1} = \{x \in \mathbb{R}^n : \|x\| = 1\}$ denote the unit ball und unit sphere.

2.1. Geometric objects and the model of computation. The geometric objects under consideration are particular convex bodies. A *convex body* (or simply *body*) is a bounded, closed, and convex set which contains interior points. Our model of computation is the binary Turing machine model: all relevant convex bodies are presented by certain rational numbers, and the size of the input is defined as the length of the binary encoding of the input data (see, e.g., [Garey and Johnson 1979; Gritzmann and Klee 1994; Grötschel et al. 1993]).

Specifically, a $\mathcal{B}$-*ball* B is a ball that is represented by a triple $(n; c, \rho)$ with $n \in \mathbb{N}$, $c \in \mathbb{Q}^n$, and $\rho^2 \in (0, \infty) \cap \mathbb{Q}$ such that $B = c + \rho \mathbb{B}^n$. Let $\mathcal{B}^n$ denote the class of all $\mathcal{B}$-balls in $\mathbb{R}^n$, and set $\mathcal{B} = \bigcup_{n \in \mathbb{N}} \mathcal{B}^n$.

For rational polytopes we distinguish between $\mathcal{H}$- and $\mathcal{V}$-presentations [Gritzmann and Klee 1994]. A $\mathcal{V}$-*polytope* is a polytope P which is represented by a tuple $(n; k; v_1, \ldots, v_k)$ with $n, k \in \mathbb{N}$, and $v_1, \ldots, v_k \in \mathbb{Q}^n$ such that $P = \operatorname{conv}\{v_1, \ldots, v_k\}$, i.e., P is the convex hull of $v_1, \ldots, v_k$. An $\mathcal{H}$-*polytope* is a polytope P represented by a tuple $(n; k; A; b)$ with $n, k \in \mathbb{N}$, a rational $k \times n$-matrix A, and $b \in \mathbb{Q}^k$ such that $P = \{x \in \mathbb{R}^n : Ax \leq b\}$. Let $\mathcal{P}_{\mathcal{H}}^n$ and $\mathcal{P}_{\mathcal{V}}^n$ denote the classes of $\mathcal{H}$- and $\mathcal{V}$-polytopes in $\mathbb{R}^n$, respectively, and set $\mathcal{P}_{\mathcal{H}} = \bigcup_{n \in \mathbb{N}} \mathcal{P}_{\mathcal{H}}^n$, $\mathcal{P}_{\mathcal{V}} = \bigcup_{n \in \mathbb{N}} \mathcal{P}_{\mathcal{V}}^n$.

For fixed dimension $\mathcal{H}$- and $\mathcal{V}$-presentations of a polytope can be transformed into each other in polynomial time. If, however, the dimension is part of the input then the size of one presentation may be exponential in the size of the other [McMullen 1970].

2.2. Visibility problems. A *ray issuing from* x is a set of the form $x + [0, \infty)w$ with some $w \in \mathbb{R}^n \setminus \{0\}$. If a ray issues from the origin then it is also called a *0-ray.* For $m + 1$ bodies $B_0, B_1, \ldots, B_m$ from a class $\mathcal{X}$ we call B_0 *visible (with respect to $B_1, \ldots, B_m$)* if there exists a visibility ray r for B_0, i.e., a ray issuing from some point $p \in B_0$ satisfying $r \cap B_i = \varnothing$ for all $1 \leq i \leq m$.

Our definition of algorithmic visibility problems depends on the class $\mathcal{X}$ of geometric bodies. Note that the dimension of the ambient space is part of the input.

Problem VISIBILITY$_{\mathcal{X}}$:

Instance: m, n, bodies $B_0, B_1, \ldots, B_m \subset \mathbb{R}^n$ from the class $\mathcal{X}$.

Question: Decide whether B_0 is visible with respect to $B_1, \ldots, B_m$.

A visibility problem is called *anchored* if B_0 is a single point located at the origin. With regard to a more restricted viewing region, we call B_0 *visible from the positive orthant* (with respect to $B_1, \ldots, B_m$) if there exists a visibility ray for B_0 contained in the (closed) positive orthant.

Problem QUADRANT VISIBILITY$_{\mathcal{X}}$:

Instance: m, n, bodies $B_0, B_1, \ldots, B_m \subset \mathbb{R}^n$ from the class $\mathcal{X}$.

Question: Decide whether B_0 is visible from the positive orthant with respect to $B_1, \ldots, B_m$.

In the basic form we do not require the bodies to be disjoint. We add the index $\varnothing$ if the input bodies $B_0, \ldots, B_m$ are required to be disjoint (e.g., VISIBILITY$_{\mathcal{B}, \varnothing}$). If $\mathcal{X} = \mathcal{P}_{\mathcal{H}}$ or $\mathcal{X} = \mathcal{P}_{\mathcal{V}}$, we will usually denote the bodies by $P_0, \ldots, P_m$.

REMARK 1. Using the techniques presented in the treatment of QUADRANT VISIBILITY, it is also possible to prove hardness results for many other classes of restricted viewing regions. For the sake of simplicity, we restrict ourselves to the one example of that type that is relevant for the view obstruction problem.

Let e_i be the i-th unit vector in $\mathbb{R}^n$. For $c \in \mathbb{R}^n$ and $\rho_1, \ldots, \rho_n > 0$, conv$\{c \pm \rho_i e_i : 1 \leq i \leq n\}$ is called a *cross polytope* in $\mathbb{R}^n$. A *parallelotope* is a polytope $c + \sum_{i=1}^n [-1, 1]z_i$ with $c \in \mathbb{R}^n$ and linearly independent $z_1, \ldots, z_n \in \mathbb{R}^n$.

For a set $A \subset \mathbb{R}^n$ let

$$\text{pos } A = \left\{ \sum_{i=1}^k \lambda_i x_i : k \in \mathbb{N}, x_1, \ldots, x_k \in A, \lambda_1, \ldots, \lambda_k \geq 0 \right\}$$

denote the *positive hull* of A.

For $c \in \mathbb{R}^n$ and a j-flat F, $d(c, F)$ denotes the Euclidean distance of c from F.

3. Complexity Results for Variable Dimension

3.1. Main results. We analyze the binary Turing machine complexity of the visibility problems for the case of variable dimension. Our main intractability results are summarized in the following theorem.

THEOREM 2. (a) For $\mathcal{X} \in \{\mathcal{B}, \mathcal{P_H}, \mathcal{P_V}\}$ the problems VISIBILITY$_{\mathcal{X}}$ and QUADRANT VISIBILITY$_{\mathcal{X}}$ are NP-hard. The hardness persists even if the instances are restricted to those for which the bodies are disjoint. Moreover, in case of $\mathcal{H}$-polytopes the hardness persists if all bodies are axis-aligned cubes, and in case of $\mathcal{V}$-polytopes if all bodies are axis-aligned cross polytopes.
(b) For $\mathcal{X} \in \{\mathcal{P_H}, \mathcal{P_V}\}$ the anchored versions of VISIBILITY$_{\mathcal{X}}$ and QUADRANT VISIBILITY$_{\mathcal{X}}$ are NP-complete.

These hardness results are contrasted by the following positive results for *fixed* dimension.

THEOREM 3. Let $\mathcal{X} \in \{\mathcal{B}, \mathcal{P_H}, \mathcal{P_V}\}$, and the dimension n be fixed. Then VISIBILITY$_{\mathcal{X}}$ and QUADRANT VISIBILITY$_{\mathcal{X}}$ can be solved in polynomial time.

3.2. Informal description of the hardness proofs. Let us consider an anchored visibility problem.

In order to show NP-hardness, we provide reductions from the well-known NP-complete 3-satisfiability (3-SAT) problem [Garey and Johnson 1979]. Let $\mathcal{C} = \mathcal{C}_1 \wedge \ldots \wedge \mathcal{C}_k$ denote a 3-SAT formula with clauses $\mathcal{C}_1, \ldots, \mathcal{C}_k$ in the variables $\eta_1, \ldots, \eta_n$. Further, let $\bar{\eta}_i$ denote the complement of a variable η_i, and let the literals η_i^1 and η_i^{-1} be defined by $\eta_i^1 = \eta_i$, $\eta_i^{-1} = \bar{\eta}_i$. Let the clause $\mathcal{C}_i$ be of the form

$$\mathcal{C}_i = \eta_{i_1}^{\tau_{i_1}} \vee \eta_{i_2}^{\tau_{i_2}} \vee \eta_{i_3}^{\tau_{i_3}}, \qquad (3\text{-}1)$$

where $\tau_{i_1}, \tau_{i_2}, \tau_{i_3} \in \{-1, 1\}$ and $1 \le i_1, i_2, i_3 \le n$ are pairwise different indices.

In our reduction, we construct an anchored visibility problem in $\mathbb{R}^n$. The reduction consists of two ingredients. First we enforce that any visibility 0-ray has a direction which is close to a direction in the set $\{-1, 1\}^n$. For this purpose, consider the cube $[-1, 1]^n$. For each of the $2n$ facets of the cube we construct a suitable body (a ball or a polytope) whose positive hull covers the whole facet with the exception of "regions near the vertices". We call these bodies *structural bodies*. Figure 1, left shows this situation for the 3-dimensional case of a ball. Any visibility 0-ray can then be naturally associated with a 0-ray in one of the directions $\{-1, 1\}^n$; this imposes a discrete structure on the problem. The $2n$ structural bodies are always part of the construction, independent of the specific 3-SAT formula. The positions of each of these $2n$ bodies will depend linearly on some positive parameter γ. In fact, all bodies can be moved radially and their size be appropriately adjusted so that the crucial covering properties persist. The parameters will be used later to make the bodies disjoint. In order to define the "region near a vertex" we consider Figure 1, right. For every vertex v of $[-1, 1]^n$

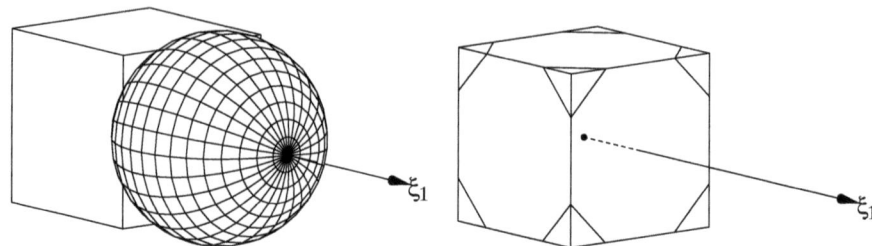

Figure 1. Imposing discrete structure. Left: Placing structural bodies. Right: Vertex simplices.

let the *vertex simplex of v* be defined as the convex hull of v and those n points which result by dividing exactly one component of v by 2. The construction will be such that any point in the boundary of $[-1, 1]^n$ that is not contained in the positive hull of a structural body will be contained in some vertex simplex.

In the second step, we relate satisfying assignments of a clause (3–1) to certain visibility 0-rays. Let $t : \{\text{TRUE}, \text{FALSE}\} \to \{-1, 1\}$ be defined by $t(\text{TRUE}) = 1$ and $t(\text{FALSE}) = -1$. We utilize the correspondence between a truth assignment $a = (\alpha_1, \ldots, \alpha_n)^T \in \{\text{TRUE}, \text{FALSE}\}^n$ to the variables $\eta_1, \ldots, \eta_n$ and the 0-ray with direction $(t(\alpha_1), \ldots, t(\alpha_n))^T \in \{-1, 1\}^n$.

For each clause (3–1), we construct a body whose positive hull contains the set

$$\{x = (\xi_1, \ldots, \xi_n)^T \in \{-1, 1\}^n : \xi_{i_1} = -t(\tau_{i_1}) \wedge \xi_{i_2} = -t(\tau_{i_2}) \wedge \xi_{i_3} = -t(\tau_{i_3})\}$$

as well as the corresponding vertex simplices, but which does not contain the set

$$\{x \in \{-1, 1\}^n : \xi_{i_1} = t(\tau_{i_1}) \vee \xi_{i_2} = t(\tau_{i_2}) \vee \xi_{i_3} = t(\tau_{i_3})\}.$$

Again, the position of each body depends linearly on some positive parameter δ, which will be used to achieve disjointness of the bodies.

The construction will guarantee that a truth assignment a satisfies the given 3-SAT formula $\mathcal{C}$ if and only if there exists a visibility 0-ray.

3.3. The case of balls. The following simple and well-known distance formula is needed in the subsequent constructions. Here, for $x \in \mathbb{R}^n$ let $x^2 := \langle x, x \rangle$.

REMARK 4. Let $c \in \mathbb{R}^n$, $p \in \mathbb{R}^n$ and $q \in \mathbb{R}^n \setminus \{0\}$. Then the distance from c to the line $p + \mathbb{R}q$ is given by

$$d(c, p + \mathbb{R}q)^2 = (p - c)^2 - \frac{\langle q, (p - c) \rangle^2}{q^2}.$$

PROOF. The line $p + \mathbb{R}q$ has distance ρ from c if and only if the quadratic equation $(p + \lambda q - c)^2 = \rho^2$ in λ has a solution of multiplicity two. This gives the equation

$$\frac{\langle q, (p - c) \rangle^2}{q^2} - (p - c)^2 + \rho^2 = 0. \qquad \square$$

LEMMA 5. ANCHORED VISIBILITY$_{\mathcal{B},\varnothing}$ *is* NP-*hard. Also,* ANCHORED VISIBILITY$_{\mathcal{B}}$
is NP-*hard even if the instances are restricted to* (*not necessarily disjoint*) *balls*
of the same radius.

PROOF. We complete the construction outlined in Section 3.2 so as to provide a
polynomial time reduction from 3-SAT to ANCHORED VISIBILITY$_{\mathcal{B},\varnothing}$. Without
loss of generality let $n \geq 4$.

Let us consider the $2n$ structural balls $S_i(\gamma_i) = (n; s_i(\gamma_i), \sigma_i(\gamma_i))$, $1 \leq i \leq 2n$,
where γ_i is the scaling parameter of S_i as described above. Naturally, we place
these balls symmetrically so that their centers lie on coordinate axes, i.e., let

$$s_i(\gamma_i) = \gamma_i e_i \text{ and } s_{n+i}(\gamma_{n+i}) = -\gamma_{n+i} e_i, \quad 1 \leq i \leq n.$$

In order to specify the radii $\sigma_i(\gamma_i)$ of the structural balls, let us consider $S_1(\gamma_1)$;
see Figure 1, left. (The construction of the other balls is done analogously.) For
convenience of notation, we shortly write $S = (n; s, \sigma)$.

In order to impose the discrete structure we will satisfy the following two
conditions. Firstly, pos(S) must not contain the vertices $\{1\} \times \{-1, 1\}^{n-1}$. Secondly, pos($S$) must contain those points which result from the vertices of the
facet $\{1\} \times [-1, 1]^{n-1}$ after dividing exactly one of the last $n - 1$ components
by 2. The two conditions will yield an upper and a lower bound for σ.

We start with the first condition. Since any of the 0-rays $\{1\} \times \{-1, 1\}^{n-1}$
has the same distance from the center s, it suffices to consider $[0, \infty)q$ with
$q = (1, 1, \ldots, 1)^T$. By Remark 4,

$$d(s, [0, \infty)q)^2 = \gamma^2 \frac{n-1}{n}.$$

Consequently, we have to choose $\sigma^2 < \gamma^2(n-1)/n$. For the second condition,
consider the point $q = (1, \ldots, 1, 1/2)^T$. Then Remark 4 yields

$$d(s, [0, \infty)q)^2 = \gamma^2 \frac{4n-7}{4n-3}.$$

Therefore, a ball centered in s with radius σ satisfying

$$\gamma^2 \frac{4n-7}{4n-3} < \sigma^2 < \gamma^2 \frac{n-1}{n} \tag{3-2}$$

guarantees the two conditions. The construction of structural balls for all $2n$
facets guarantees that any point in a facet of $[-1, 1]^n$ that is not contained in
the positive hull of a structural ball is contained in a facet of some vertex simplex.

Now we turn to the balls $C_i(\delta_i) = (c_i(\delta_i), \rho_i(\delta_i))$, $1 \leq i \leq k$, representing the
k clauses. For notational convenience we describe the construction for the clause

$$\mathcal{C} = \eta_1^{-1} \vee \eta_2^1 \vee \eta_3^{-1}.$$

It should of course be clear that the construction works just as well for other
clauses. We abbreviate the ball for the clause $\mathcal{C}$ by $C = (n; c, \rho)$ (without referring explicitly to the dependence on the parameter $\delta := \delta_i$). We set $c =$

$\delta(1, -1, 1, 0, \ldots, 0)^T$, hence all the Boolean variables $\eta_4, \ldots, \eta_n$ are treated similarly.

In order to represent the given clause by the ball C we guarantee the following two properties. First, none of the 0-rays spanned by $\{-1, 1\}^n \setminus (1, -1, 1) \times \{-1, 1\}^{n-3}$ must be contained in pos(C). Within this set of rays, the ray $[0, \infty)q$ with $q = (1, -1, -1, 1, \ldots, 1)^T$ (among others) leads to the smallest distance from C. Remark 4 implies

$$d(c, [0, \infty)q)^2 = \delta^2 \frac{3n-1}{n},$$

which yields the condition $\rho^2 < \delta^2 (3n-1)/n$.

Moreover, we guarantee the following second property. The positive hull of C must contain all the points in $(1, -1, 1) \times \{-1, 1\}^{n-3}$ as well as their vertex simplices. Among all these points and among the vertices of the vertex simplices, the vector $q = (1, -1, 1/2, 1, \ldots, 1)^T$ leads to the ray with the largest distance from c. Remark 4 implies

$$d(c, [0, \infty)q)^2 = \delta^2 \frac{12n-34}{4n-3}.$$

Hence, a ball centered in c with radius ρ satisfying

$$\delta^2 \frac{12n-34}{4n-3} < \rho^2 < \delta^2 \frac{3n-1}{n}$$

guarantees the two conditions for the clause ball. Note that the upper bound implies that the origin is not contained in the ball.

As yet, the definitions of the $2n$ structural balls and the k clause balls depend on the positive parameters $\gamma_1, \ldots, \gamma_{2n}$ and $\delta_1, \ldots, \delta_k$, respectively. By choosing these parameters appropriately, we make the balls disjoint. Since $\sigma_i < \gamma_i \sqrt{(n-1)/n}$ for the structural balls, we choose the parameter γ_i of the i-th structural ball successively so that

$$\gamma_i - \gamma_i \sqrt{\frac{n-1}{n}} > \gamma_{i-1} + \gamma_{i-1} \sqrt{\frac{n-1}{n}}.$$

Then

$$(\gamma_i e_i - \gamma_j e_j)^2 > (\sigma_i + \sigma_j)^2 \quad \text{for } i > j.$$

Setting $\gamma_0 = 1$, this leads to the condition

$$\gamma_i > \left(\frac{1 + \sqrt{(n-1)/n}}{1 - \sqrt{(n-1)/n}} \right)^i = \left(2n - 1 + 2\sqrt{n \cdot (n-1)} \right)^i.$$

Hence, choosing $\gamma_i = (4n-1)^i$ for $1 \leq i \leq 2n$ guarantees that the structural balls are pairwise disjoint. Note that the binary logarithm of these numbers grows only polynomially in the number of balls, i.e., we can choose rational centers and (squares of) radii of the structural balls of polynomial size. Similarly, the parameters $\delta_1, \ldots, \delta_k$ of the clause balls can be chosen. In particular, when also choosing δ_1 sufficiently large, then the clause balls are disjoint from the structural balls.

Now it is easy to show that the given 3-SAT formula $\mathcal{C}$ can be satisfied if and only if the single point B_0 is visible. Let $(\alpha_1, \ldots, \alpha_n)^T$ be a satisfying assignment of $\mathcal{C}$. Then there does not exist any ball B in the construction whose positive hull contains the 0-ray in direction $(t(\alpha_1), \ldots, t(\alpha_n))^T$. Hence, B_0 is visible. Conversely, let q be a visibility ray for B_0. Due to the structural balls the ray q intersects with the vertex simplex of some vector $v = (\nu_1, \ldots, \nu_n)^T \in \{-1, 1\}^n$. Consequently, the truth assignment $(t^{-1}(\nu_1), \ldots, t^{-1}(\nu_n))^T$ is a satisfying assignment because otherwise the positive hull of some clause ball would contain the vertex simplex of v. Hence, $\mathcal{C}$ can be satisfied.

In order to achieve the result for balls of the same size, the role of σ and γ (respectively, ρ and δ) in (3–2) is interchanged in the sense that the radius σ is now given and a condition on γ is imposed. Clearly, these conditions for $\gamma_1, \ldots, \gamma_{2n}$ can be satisfied in the same way as the conditions on the radius before. $\qquad\square$

COROLLARY 6. VISIBILITY$_{\mathcal{B}, \varnothing}$ is NP-*hard*.

PROOF. It is obvious that the proof for the case that B_0 is a single point generalizes to the case of a nondegenerated ball centered in 0 with some sufficiently small radius $\sigma_0 > 0$. In the following we will outline the precise argument.

Let $S_i = (n; s_i, \sigma_i)$, $1 \le i \le 2n$, and $C_j = (n; c_j, \rho_j)$, $1 \le j \le k$, be the structural balls and the clause balls in the proof of Lemma 5, and set $\hat{B}_0 = (n; 0, \sigma_0)$, where σ_0 is such that the inequalities given in the proofs of Lemma 5 hold for both σ_i, ρ_j and for $\sigma_i' := \sigma_i - \sigma_0$, $\rho_j' := \rho_j - \sigma_0$, $1 \le i \le 2n$, $1 \le j \le k$. Further, let $S_i' = (n; s_i, \sigma_i - \sigma_0)$, $1 \le i \le 2n$, and $C_j' = (n; c_j, \rho_j - \sigma_0)$, $1 \le j \le k$. Then it follows from the fact that $(B_0, S_1, \ldots, S_{2n}, C_1, \ldots, C_k)$ and $(B_0, S_1', \ldots, S_{2n}', C_1', \ldots, C_k')$ are Yes-instances of the visibility problem if and only if the given Boolean expression is satisfiable that the same holds for $(\hat{B}_0, S_1, \ldots, S_{2n}, C_1, \ldots, C_k)$. $\qquad\square$

3.4. The case of $\mathcal{V}$-polytopes

LEMMA 7. ANCHORED VISIBILITY$_{\mathcal{P}_\mathcal{V}, \varnothing}$ is NP-*hard. This result persists if the instance are restricted to axes-aligned cross polytopes.*

PROOF. We establish a polynomial time reduction from 3-SAT to ANCHORED VISIBILITY$_{\mathcal{P}_\mathcal{V}, \varnothing}$ based on the framework in Section 3.2. Again we assume $n \ge 4$.

This time, we choose the $2n$ structural bodies as cross polytopes of the form $S_i(\gamma_i) = \mathrm{conv}\{s_i(\gamma_i) + \sigma_{ij}(\gamma_i)e_j : 1 \le j \le n\}$ with rational coefficients $s_i(\gamma_i)$, $\sigma_{ij}(\gamma_i)$ depending on the scaling parameter γ_i. The centers of the cross polytopes are defined by

$$s_i(\gamma_i) = \gamma_i e_i \text{ and } s_{n+i}(\gamma_{n+i}) = -\gamma_{n+i}e_i, \quad 1 \le i \le 2n.$$

Now we specify the coefficients σ_{ij}. We describe the construction of $S_1(\gamma_1)$ which for simplicity will be abbreviated by $S = \text{conv}\{s + \sigma_j e_j : 1 \leq j \leq n\}$. The construction of the other structural bodies is then similar.

For any choice of the parameters $\sigma_2, \ldots, \sigma_n > 0$, the $(n-1)$-dimensional cross polytope $S' = \text{conv}\{s + \sigma_j e_j : 2 \leq j \leq n\}$ is contained in the hyperplane $\xi_1 = \gamma$. Similar to the case of the balls, two conditions are imposed on the choice of $\sigma_2, \ldots, \sigma_n$. Firstly, the positive hull of S' must not contain the vertices $\{1\} \times \{-1, 1\}^n$. Secondly, the positive hull of S' must contain those points resulting from the vertices of the facet $\{1\} \times [-1, 1]^{n-1}$ by dividing exactly one of the last $n - 1$ components by 2.

We choose $\sigma_2 = \ldots = \sigma_n$. The necessary upper and lower bounds for σ_2 result as follows. Without loss of generality we consider the 0-ray $[0, \infty)(1, \ldots, 1)^T$. The vertex $\gamma(1, \ldots, 1)^T$ of $\gamma[-1, 1]^n$ is contained in a facet of the $(n-1)$-dimensional cross polytope $\text{conv}\{s \pm \gamma(n-1)e_j : 2 \leq j \leq n\}$. On the other hand, the point $\gamma(1, 1, 1, \ldots, 1, 1/2)^T$ is contained in a facet of the $(n-1)$-dimensional cross polytope with vertices $\text{conv}\{s \pm \gamma(n - 3/2)e_j\}$, $2 \leq j \leq n$. Hence, if σ_2 satisfies

$$\gamma\left(n - \frac{3}{2}\right) < \sigma_2 < \gamma(n-1),$$

the two conditions enforcing the discrete structure are satisfied.

In order to make the $(n-1)$-dimensional polytope S' full-dimensional we consider some ε with $0 < \varepsilon < \gamma$. Then $s - \varepsilon e_1 \in \text{pos } S'$. Hence, by adding the vertices $s \pm \varepsilon e_1$ we obtain an n-dimensional cross polytope S with $\text{pos}(S) = \text{pos}(S')$.

Now we show how to represent a clause by a cross polytope. Again, we describe the construction for the clause $\eta_1^{-1} \vee \eta_2^1 \vee \eta_3^{-1}$. The associated cross polytope will be of the form $C = \text{conv}\{c \pm \rho_j e_j : 2 \leq j \leq n\}$ with $c = \delta(1, -1, 1, 0, \ldots, 0)^T$ and coefficients ρ_j (also depending on the parameter δ). For any choice of parameters $\rho_4, \ldots, \rho_n > 0$, the $(n-3)$-dimensional cross polytope $C' = \text{conv}\{c \pm \rho_j e_j : 4 \leq j \leq n\}$ is contained in the $(n-3)$-dimensional flat $\xi_1 = \delta, \xi_2 = -\delta, \xi_3 = \delta$. We choose $\rho_4 = \ldots = \rho_n$. As before, we add the vertices $c \pm \varepsilon e_j$, $1 \leq j \leq 3$, for some parameter $0 < \varepsilon < \delta$ to obtain a full-dimensional cross polytope. If $\rho_4 = 2(n-3)$ then the point $\delta(1, -1, 1/2, 1, \ldots, 1)^T$ is contained in the n-dimensional cross polytope. Hence, by choosing $\rho_4 > 2(n-3)$ the positive hull of C contains all the points in $(1, -1, 1) \times \{-1, 1\}^{n-3}$ as well as their vertex simplices. Moreover, since $\text{pos}(C)$ is contained in the cone defined by $\xi_1 \geq 0, \xi_2 \leq 0, \xi_3 \geq 0$, none of the vectors in $\{-1, 1\}^n \setminus (1, -1, 1) \times \{-1, 1\}^{n-3}$ is contained in the positive hull of the cross polytope.

Similarly to the proof of Lemma 5, we can choose the parameters $\gamma_1, \ldots, \gamma_{2n}$, $\delta_1, \ldots, \delta_k$, and ε (for making the bodies n-dimensional) in such a way that the bodies are pairwise disjoint and that their encoding lengths remain polynomially bounded. Hence, the polynomial time reduction from 3-SAT follows in the same way as in the proof of Theorem 5.   □

Using an inclusion technique like in Corollary 6 we readily obtain the following corollary.

COROLLARY 8. VISIBILITY$_{\mathcal{P}_\mathcal{V}, \varnothing}$ is NP-hard even for axis-aligned cross polytopes.

LEMMA 9. ANCHORED VISIBILITY$_{\mathcal{P}_\mathcal{V}}$ is contained in NP.

PROOF. Let $(m; n; P_0, \ldots, P_m)$ be an instance of ANCHORED VISIBILITY$_{\mathcal{P}_\mathcal{V}}$ with $P_0 = \{0\}$ and $\mathcal{V}$-polytopes $P_1, \ldots, P_m$, and let $\mathcal{F}_{n-2}(P_i)$ denote the set of all $(n-2)$-dimensional faces of P_i, $1 \le i \le m$. The set of all linear subspaces $\text{lin } F$, $F \in \mathcal{F}_{n-2}(P_i)$, naturally decomposes the unit sphere $\mathbb{S}^{n-1}$ into $(n-1)$-dimensional sectors. For two 0-rays belonging to the (relative) interior (w.r.t. $\mathbb{S}^{n-1}$) of the same sector either both of them are visibility rays or none of them is. Each 0-ray through a vertex of a sector can be computed in polynomial time. In particular, two such vertices have a distance that is bounded below by a polynomial in the input. Hence for each sector there does indeed exist a polynomial size vector specifying a ray that meets the sector in its (relative) interior (w.r.t. $\mathbb{S}^{n-1}$). Hence there exists a polynomial size certificates for candidates for visibility rays.

It remains to show that it can be verified in polynomial time that a given witness ray does not intersect any of the polytopes P_i. Since the number of polytopes is bounded by the input length of the instance, it suffices to explain this polynomial verification method for a single polytope $P \in \{P_1, \ldots, P_m\}$. Let the $\mathcal{V}$-presentation of P be $P = \text{conv}\{v_1, \ldots, v_k\}$. P does not intersect the ray $[0, \infty)q$ if and only if the system

$$\sum_{i=1}^{k} \mu_i v_i = \lambda q,$$
$$\sum_{i=1}^{k} \mu_i = 1,$$
$$\mu_i \ge 0 \quad \text{for } 1 \le i \le k,$$
$$\lambda \ge 0$$

does not have a solution. This can be checked in polynomial time by linear programming.   □

3.5. The case of $\mathcal{H}$-polytopes

LEMMA 10. ANCHORED VISIBILITY$_{\mathcal{P}_\mathcal{H}}$ and VISIBILITY$_{\mathcal{P}_\mathcal{H}}$ are NP-hard. These statements persist if we restrict the polytopes to be axis-aligned n-dimensional unit cubes. The hardness also persists if we restrict the polytopes to be disjoint axis-aligned n-dimensional cubes.

PROOF. We give a polynomial time reduction from 3-SAT, but this time the proof differs from the framework in Section 3.2. We begin with the anchored version, in which P_0 is a single point located in the origin.

Let $\mathcal{C} = \mathcal{C}_1 \wedge \ldots \wedge \mathcal{C}_k$ be an instance of 3-SAT with clauses $\mathcal{C}_1, \ldots, \mathcal{C}_k$ in the variables $\eta_1, \ldots, \eta_n$. Let

$$\mathcal{C}_i = \eta_{i_1}^{\tau_{i1}} \vee \eta_{i_2}^{\tau_{i2}} \vee \eta_{i_3}^{\tau_{i3}}.$$

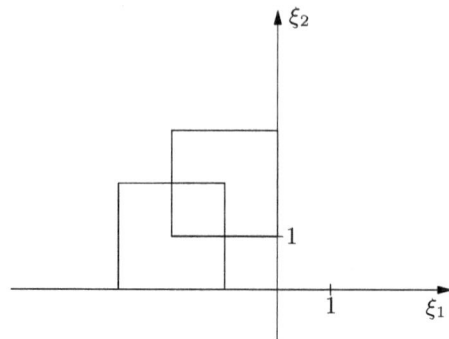

Figure 2. In order to represent the 2-clause $y_1^1 \vee y_2^{-1}$, all visibility rays in the orthant $\xi_1 \leq 0$, $\xi_2 \geq 0$ have to be blocked. This can be achieved by placing two unit squares centered at $(-2,1)^T$ and $(-1,2)^T$.

We construct a set of axis-aligned unit cubes ensuring that 0-rays spanned by any (nonzero) vector $b = (\beta_1, \ldots, \beta_n)^T$ with $\mathrm{sgn}(\beta_{i_1}) \in \{-\tau_{i_1}, 0\}$, $\mathrm{sgn}(\beta_{i_2}) \in \{-\tau_{i_2}, 0\}$, $\mathrm{sgn}(\beta_{i_3}) \in \{-\tau_{i_3}, 0\}$ cannot be visibility rays. Figure 2 depicts the idea of the construction for two variables y_1 and y_2 and the 2-clause $y_1^1 \vee y_2^{-1}$. Define the $2n - 3$ axis-aligned unit cubes

$$P_1 = -2\tau_{i_1} e_{i_1} - \tau_{i_2} e_{i_2} - \tau_{i_3} e_{i_3} + [-1,1]^n \,,$$
$$P_2 = -\tau_{i_1} e_{i_1} - 2\tau_{i_2} e_{i_2} - \tau_{i_3} e_{i_3} + [-1,1]^n \,,$$
$$P_3 = -\tau_{i_1} e_{i_1} - \tau_{i_2} e_{i_2} - 2\tau_{i_3} e_{i_3} + [-1,1]^n \,.$$
$$P_j' = -\tau_{i_1} e_{i_1} - \tau_{i_2} e_{i_2} - \tau_{i_3} e_{i_3} + 2e_j + [-1,1]^n \,, \quad j \in \{1, \ldots, n\} \setminus \{i_1, i_2, i_3\}\,,$$
$$P_j'' = -\tau_{i_1} e_{i_1} - \tau_{i_2} e_{i_2} - \tau_{i_3} e_{i_3} - 2e_j + [-1,1]^n \,, \quad j \in \{1, \ldots, n\} \setminus \{i_1, i_2, i_3\}\,.$$

All these cubes are contained in the set $\{x = (\xi_1, \ldots, \xi_n)^T \in \mathbb{R}^n : \mathrm{sgn}(\xi_{i_1}) \in \{-\tau_{i_1}, 0\}$, $\mathrm{sgn}(\xi_{i_2}) \in \{-\tau_{i_2}, 0\}$, $\mathrm{sgn}(\xi_{i_3}) \in \{-\tau_{i_3}, 0\}\}$, and none of the cubes contains the origin. The union of the $2n - 3$ cubes contains all facets of the cube

$$-\tau_{i_1} e_{i_1} - \tau_{i_2} e_{i_2} - \tau_{i_3} e_{i_3} + [-1,1]^n$$

except the three facets which are contained in one of the hyperplanes $\xi_{i_1} = 0$, $\xi_{i_2} = 0$, or $\xi_{i_3} = 0$. Namely, P_1, P_2, and P_3 contain the facets in the hyperplanes $\xi_{i_1} = -2\tau_{i_1}$, $\xi_{i_2} = -2\tau_{i_2}$, and $\xi_{i_3} = -2\tau_{i_3}$, respectively, and for $j \in \{1, \ldots, n\} \setminus \{i_1, i_2, i_3\}$ the cubes P_j and P_j' contain the facets in the hyperplanes $\xi_j = 1$ and $\xi_j = -1$. Hence, a ray $[0, \infty)b$ intersects one of the $2n - 3$ cubes if and only if $\mathrm{sgn}(\beta_{i_1}) \in \{-\tau_{i_1}, 0\}$, $\mathrm{sgn}(\beta_{i_2}) \in \{-\tau_{i_2}, 0\}$, and $\mathrm{sgn}(\beta_{i_3}) \in \{-\tau_{i_3}, 0\}$.

Altogether, a ray $[0, \infty)b$ is a visibility ray for P_0 if and only if C can be satisfied. Hence, ANCHORED VISIBILITY$_{\mathcal{P}_\mathcal{H}}$ is NP-hard even if we restrict the polytopes to be axis-aligned n-dimensional unit cubes. Note that if the instance cannot be satisfied then the union of the polytopes in our construction contains the boundary of the cube $[-1,1]^n$. Hence, the single point P_0 can be replaced by the cube $[-1, 1]^n$, which shows that VISIBILITY$_{\mathcal{P_H}}$ is NP-hard even if the polytopes are restricted to be axis-aligned n-dimensional unit cubes.

In order to show that ANCHORED VISIBILITY$_{\mathcal{P_H}, \varnothing}$ and VISIBILITY$_{\mathcal{P_H}, \varnothing}$ are NP-hard, we can scale the cubes as in the earlier proofs.   □

LEMMA 11. ANCHORED VISIBILITY$_{\mathcal{P_H}}$ *is contained in* NP.

PROOF. The proof is analogous to that of Lemma 9.   □

3.6. Polynomial solvability results for fixed dimension.

In order to prove the polynomial solvability results for fixed dimension, we use the fact that the theory of real closed fields can be decided in polynomial time [Ben-Or et al. 1986; Collins 1975]. More precisely, for rational polynomials $p_1(\xi_1, \ldots, \xi_n)$, $\ldots$, $p_l(\xi_1, \ldots, \xi_n)$ in the variables $\xi_1, \ldots, \xi_n$, a *Boolean formula over* $p_1, \ldots, p_l$ is defined as a Boolean combination (allowing the operators $\wedge$, $\vee$, $\neg$) of polynomial equations and inequalities of the type $p_i(\xi_1, \ldots, \xi_n) = 0$ or $p_i(\xi_1, \ldots, \xi_n) \leq 0$. We consider the following decision problem for quantified Boolean formulas over the real numbers.

Problem REAL QUANTIFIER ELIMINATION:

Instance: n, l, rational polynomials $p_1(\xi_1, \ldots, \xi_n), \ldots, p_l(\xi_1, \ldots, \xi_n)$, a Boolean formula $\varphi(\xi_1, \ldots, \xi_n)$ over $p_1, \ldots, p_l$, and quantifiers $Q_1, \ldots, Q_n \in \{\forall, \exists\}$.

Question: Decide the truth of the statement

$$Q_1(\xi_1 \in \mathbb{R}) \, \ldots \, Q_n(\xi_n \in \mathbb{R}) \quad \varphi(\xi_1, \ldots, \xi_n).$$

In [Ben-Or et al. 1986; Collins 1975] it was shown:

PROPOSITION 12. *For fixed dimension* n, REAL QUANTIFIER ELIMINATION *can be decided in polynomial time.*

REMARK 13. Despite of this polynomial solvability result for fixed dimension, current implementations are only capable of dealing with very small dimensions. Generally, there are two approaches towards practical solutions of decision problems over the reals. One is based on Collins' cylindrical algebraic decomposition (CAD) [Collins 1975], and the other is the critical point method ([Grigor'ev and Vorobjov 1988]; for the state of the art see [Aubry et al. 2002]).

In order to prove polynomial solvability of VISIBILITY$_{\mathcal{B}}$ for fixed dimension, we formulate the problem algebraically. We represent a ray $p + \lambda q$, $\lambda \geq 0$, by its initial vector $p \in \mathbb{R}^n$ and a direction vector $q \in \mathbb{R}^n$ with $\|q\| = 1$. B_0 is visible with respect to $B_1 = (n; c_1, \rho_1), \ldots, B_m = (n; c_m, \rho_m)$ if and only if there exist $p, q \in \mathbb{R}^n$ such that for all $\lambda \in \mathbb{R}$ the following formula holds:

$$\|q\|^2 = 1$$

and $$\|p - c_0\|^2 \leq \rho_0^2$$

and $(\lambda < 0 \text{ or } \|p + \lambda q - c_i\|^2 \geq \rho_i^2)$ for $1 \leq i \leq m$.

Hence, we have to decide the truth of the following statement:

$$\exists p \in \mathbb{R}^n \; \exists q \in \mathbb{R}^n \; \forall \lambda \in \mathbb{R}$$

$$\|q\|^2 = 1 \;\wedge\; \|p - c_0\|^2 \le \rho_0^2 \;\wedge\; \left((\lambda < 0 \;\vee\; \|p + \lambda q - c_i\|^2 \ge \rho_i^2)\; \text{for}\; 1 \le i \le m\right) .$$

After expanding the Euclidean norm and applying some trivial transformations (such as establishing the mentioned normal form $p_i(\xi_1, \dots, \xi_n) \le 0$ for the polynomial inequalities), this is a quantified Boolean formula of the required form. Hence, Proposition 12 implies the following statement.

LEMMA 14. *For fixed dimension n, VISIBILITY$_\mathcal{B}$ can be solved in polynomial time.*

For the case of $\mathcal{H}$-polytopes, let $P_i = \{x \in \mathbb{R}^n : A_i x \le b_i\}$ with $A_i \in \mathbb{Q}^{k_i \times n}$, $b_i \in \mathbb{Q}^{k_i}$, $0 \le i \le m$. P_0 is visible if and only if there exist $p, q \in \mathbb{R}^n$ such that for all $\lambda \in \mathbb{R}$ we have

$$\|q\|^2 = 1$$

$$\text{and} \qquad A_0 p \le b_0$$

$$\text{and} \quad \neg\left(A_i(p + \lambda q) \le b_i\right) \quad \text{for}\; 1 \le i \le m.$$

Applying Proposition 12 on this formulation we can conclude:

LEMMA 15. *For fixed dimension n, VISIBILITY$_{\mathcal{P}_\mathcal{H}}$ can be solved in polynomial time.*

Since for fixed dimension n, a $\mathcal{V}$-polytope can be transformed into a $\mathcal{H}$-polytope in polynomial time [Dyer 1983], this also implies

COROLLARY 16. *For fixed dimension n, VISIBILITY$_{\mathcal{P}_\mathcal{V}}$ can be solved in polynomial time.*

Similarly, by small modifications of the proofs, the polynomial time solvability results for VISIBILITY can also be transferred to QUADRANT VISIBILITY.

4. On the Frontiers of the Results and Their Relations with Our Other Fields

4.1. Relations with algebraic geometry. Theorems 2 and 3 do not guarantee membership of VISIBILITY$_\mathcal{B}$ in NP. Let us illuminate this situation from the algebraic point of view. First note that even though quantifier elimination methods can decide ANCHORED VISIBILITY$_\mathcal{B}$ or VISIBILITY$_\mathcal{B}$ for fixed dimension in polynomial time (see Lemma 14), it is not known how to compute a short witness of a positive solution with these methods (see [Ben-Or et al. 1986]).

For "Yes" instances of ANCHORED VISIBILITY$_\mathcal{B}$ or VISIBILITY$_\mathcal{B}$ there always exists a ray in the closure of all visibility rays whose underlying line is simultaneously tangent to several balls. Hence, the question of membership in NP is tightly connected to the algebraic characterization of the lines simultaneously tangent to a given set of balls in $\mathbb{R}^n$. In particular, it is essential to characterize the lines tangent to $2n - 2$ balls, since the Grassmannian of lines in n-space has dimension $2n - 2$ (i.e., a line in $\mathbb{R}^n$ has $2n - 2$ degrees of freedom). In [Sottile and Theobald 2002] it was shown that for $n \geq 3$, $2n - 2$ balls in general position in $\mathbb{R}^n$ have $3 \cdot 2^{n-1}$ (complex) common tangent lines. Hence, the visibility problem in dimension n is tightly connected to an algebraic problem of degree $3 \cdot 2^{n-1}$.

Similarly, Theorems 2 and 3 do not guarantee membership of VISIBILITY$_{\mathcal{P}_\mathcal{H}}$ or VISIBILITY$_{\mathcal{P}_\mathcal{V}}$ in NP. These questions are tightly connected to the common transversals to $2n - 2$ given $(n-2)$-dimensional flats in $\mathbb{R}^n$. The generic number of (complex) transversals to $2n - 2$ given $(n-2)$-flats in $\mathbb{R}^n$ is

$$\frac{1}{n}\binom{2n-2}{n-1};$$

(see, e.g., [Kleiman and Laksov 1972; Sottile 1997]).

In both cases (balls and polytopes), the algebraic degree is reflected by our hardness results in the Turing machine model.

4.2. Relations with the theory of packing and covering.

Concerning NP-hardness, Theorem 2 does not include a result for ANCHORED VISIBILITY$_{\mathcal{B},\varnothing}$ or VISIBILITY$_{\mathcal{B},\varnothing}$ if the balls are unit balls. However, the following statement shows that in "No"-instances of VISIBILITY$_{\mathcal{B},\varnothing}$ the number of balls necessarily grows exponentially in the input dimension n. Even if this does not rule out the existence of a polynomial time algorithm (since the running time of the algorithm is not measured in terms of the dimension n but in the overall length of the input size which in this case is exponential in n), it might give a useful sufficient criterion for large input dimensions.

LEMMA 17. *Let $n \geq 6$, $m \in \mathbb{N}$, and let $B_0, B_1, \ldots, B_m$ be disjoint unit balls in $\mathbb{R}^n$. If $m < \sqrt{3n}\, e^{(3/8)(n-1)}$ then B_0 is visible with respect to $B_1, \ldots, B_m$.*

PROOF. Without loss of generality we can assume that B_0 is the unit ball centered at the origin. Let $0 < r < 1$ and H be a hyperplane in $\mathbb{R}^n$ at distance r from the origin. Then the set of points on the unit sphere separated from the origin by H is called an r-cap. Since any ball B_i, $1 \leq i \leq m$, is disjoint from B_0, an elementary geometric computation shows that pos(B_i) intersects the unit sphere in an r-cap with $\sqrt{3}/2 < r < 1$. A necessary condition for B_0 being invisible is that these r-caps cover the unit sphere. Let $\tau(n, r)$ denote the minimum number of r-caps covering the unit sphere. By Lemma 5.2 in [Brieden et al. 1998], we have for $r > 2/\sqrt{n}$

$$\tau(n, r) \geq 2r\sqrt{n}\, e^{r^2(n-1)/2}.$$

Substituting the value $r = \sqrt{3}/2$ into this formula yields the desired estimation.
$\square$

Moreover, the problem VISIBILITY is closely related to difficult problems in the theory of packing and covering (see [Tóth 1971] or [Zong 1999, Chapter 12]). A *Hornich configuration* in $\mathbb{R}^n$ is a set $\{B_1, \ldots, B_m\}$ of disjoint unit balls with $\{B_1, \ldots, B_m\} \cap \mathbb{B}^n = \varnothing$ such that the origin is not visible with respect to $B_1, \ldots, B_m$. The *Hornich number* h_n is the smallest number m of disjoint unit balls $B_1, \ldots, B_m$ such that $\{B_1, \ldots, B_m\}$ is a Hornich configuration. Hence, for the class of unit spheres, ANCHORED VISIBILITY$_{\mathcal{B}, \varnothing}$ asks whether a given configuration is a Hornich configuration. Similarly, a *Fejes Tóth configuration* in $\mathbb{R}^n$ is a set $\{B_0, \ldots, B_m\}$ of disjoint unit balls such that B_0 is not visible with respect to $B_1, \ldots, B_m$. The *Fejes Tóth number* ℓ_n in $\mathbb{R}^n$ is the smallest number m of disjoint unit balls $B_0, \ldots, B_m$ such that there exists a Fejes Tóth configuration with m balls.

Even in dimension 3, the Hornich number h_3 is not known, and the best known bounds are $30 \le h_3 \le 42$. Lower and upper bounds for general dimensions n can be found in [Zong 1999]. Concerning the Fejes Tóth number, Zong gave the upper bound $\ell_n \le (8e)^n (n+1)^{n-1} n^{(n^2+n-2)/2}$ [Zong 1997].

If the balls are allowed to be of different radius then Theorem 2 implies that already the test whether a given configuration is a (generalized) Hornich or Fejes Tóth configuration is NP-hard.

4.3. Quadrant visibility and view obstruction. In Sections 3.2–3.5 our hardness results for VISIBILITY were based on reductions from 3-SAT in which any assignment $a \in \{\text{TRUE}, \text{FALSE}\}^n$ was identified with one of the 2^n quadrants in $\mathbb{R}^n$. For that reason, the question arises whether the hardness results still hold for more restricted viewing areas, say, for those viewing areas which are contained in a single quadrant.

In the following we prove the correponding part of Theorem 2.

LEMMA 18. ANCHORED QUADRANT VISIBILITY$_{\mathcal{B}, \varnothing}$ *is NP-hard.* ANCHORED QUADRANT VISIBILITY$_{\mathcal{B}}$ *is NP-hard even if all balls are restricted to (not necessarily disjoint) balls of the same radius.*

PROOF. Once more, we provide a reduction from 3-SAT, and therefore consider a 3-SAT formula in the variables $\eta_1, \ldots, \eta_n$. The essential idea of the reduction is to construct an instance of QUADRANT VISIBILITY in $(n+1)$-dimensional space $\mathbb{R}^{n+1}$. The 0-ray with direction $v := (1, \ldots, 1)^T$ is contained in the positive orthant Q of $\mathbb{R}^{n+1}$. By considering a hyperplane which is orthogonal to v and which intersects $(0, \infty)v$, we transfer the proof ideas of ANCHORED VISIBILITY to ANCHORED QUADRANT VISIBILITY.

In order to simplify the notation, we apply an orthogonal transformation to transform the diagonal ray $[0, \infty)v$ into $[0, \infty)e_{n+1}$, the nonnegative part of the ξ_{n+1}-axis. By this operation, Q is transformed into a cone Q'. As in the proof of Lemma 5, we impose a discrete structure on the visibility problem. Namely, for some positive parameter $\tau > 0$ specified below, we associate the

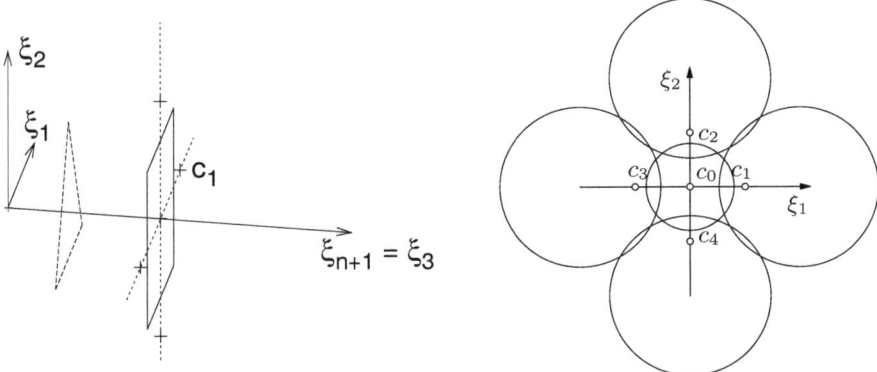

Figure 3. The figure shows how to impose discrete structure on ANCHORED QUADRANT VISIBILITY in case $n = 2$ and $\gamma_0 = \ldots = \gamma_{2n} =: \gamma$ (so all the centers of the structural balls are contained in the hyperplane $\xi_{n+1} = \gamma$). The positive hull of the triangle on the left represents Q', the positive orthant after the orthogonal transformation. The right figure shows the section of the balls through the hyperplane $\xi_{n+1} = \gamma$.

2^n truth assignments $\{\text{TRUE}, \text{FALSE}\}^n$ with the 0-rays spanned by the vectors $\{1\} \times \{-\tau, \tau\}^n$. Note that the set $\{1\} \times [-\tau, \tau]^n$ is an n-dimensional cube in $\mathbb{R}^{n+1}$.

In order to achieve this discrete structure, we place $2n + 1$ structural balls $S_i(\gamma_i, \tau) = (n; s_i(\gamma_i, \tau), \sigma_i(\gamma_i, \tau))$, $0 \le i \le 2n$, at the centers $c_0 = \gamma_0 e_{n+1}$, $c_i = \gamma_i(e_{n+1} + \tau e_i)$, $c_{n+i} = \gamma_{n+i}(e_{n+1} - \tau e_i)$, $1 \le i \le n$. In contrast to the proofs for ANCHORED VISIBILITY, the centers of the structural balls do not only depend on positive parameters γ_i, but also on the global positive parameter τ. Figure 3 shows this situation for the case $n = 2$. The parameter τ is chosen so that the n-dimensional cube $\{1\} \times [-\tau, \tau]^n$ is contained in Q'. The radii $s_i(\gamma_i, \tau)$, $1 \le i \le n$, of the structural balls can be chosen such that any visibility ray must be close to a vertex of the n-dimensional cube; this establishes the discrete structure. In a second step, the parameters γ_i can be used to scale the balls in order to make them disjoint.

Then, similarly to the proof of Lemma 5, we can construct balls representing the clauses of the 3-SAT formula in order to complete the desired polynomial time reduction.   □

It is easy to see that the hardness result can be extended to the case of QUADRANT VISIBILITY$_{\mathcal{B}, \varnothing}$, where B_0 is a proper ball. Moreover, by combining the proofs in Sections 3.4 and 3.5 with a lifting into $\mathbb{R}^{n+1}$, the hardness results can also be established for the case of $\mathcal{V}$- and $\mathcal{H}$-polytopes. (For $\mathcal{H}$-polytopes, the construction from the proof of Lemma 10 is carried out in the hyperplane given

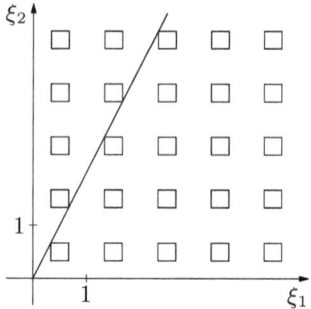

Figure 4. The situation of the view obstruction problem in $\mathbb{R}^2$. Here $\lambda(2) = \frac{1}{3}$.

by $\xi_{n+1} = \gamma$; and – as in that lemma – the construction manages without any structural bodies.)

Note that the proof technique of Lemma 18 can also be generalized to establish hardness results for other classes of viewing areas.

The problem ANCHORED QUADRANT VISIBILITY is related to the problem of diophantine approximation introduced by Wills [Wills 1968] of determining

$$\kappa(n) = \inf_{v_1,\dots,v_n \in \mathbb{N}} \sup_{\tau \in [0,1]} \min_{1 \le i \le n} \|\tau v_i\|_I .$$

Based on the pigeonhole principle, Wills showed $\frac{1}{2n} \le \kappa(n) \le \frac{1}{n+1}$ and conjectured $\kappa(n) = \frac{1}{n+1}$. This conjecture was later restated by Cusick [Cusick 1973] who interpreted it as a visibility problem called *view obstruction*. Let $C = [-\frac{1}{2}, \frac{1}{2}]^n$. For some factor $\alpha > 0$, consider the infinite set of cubes

$$\left\{ \left(\gamma_1 + \tfrac{1}{2}, \dots, \gamma_n + \tfrac{1}{2}\right)^T + \alpha C \; : \; \gamma_1, \dots, \gamma_n \in \mathbb{N}_0 \right\}. \tag{4–1}$$

Now the problem is to determine the supremum of $\alpha > 0$ such that there exists a visibility ray in the strictly positive orthant (see Figure 4). This supremum, called $\lambda(n)$, can be written as

$$\lambda(n) = 2 \sup_{\omega_1,\dots,\omega_n \in (0,\infty)} \inf_{\xi \in (0,\infty)} \max_{1 \le i \le n} \left\|\omega_i \xi - \tfrac{1}{2}\right\|_I .$$

The connection between Wills' problem and the view obstruction problem is established by the statement that for $n \ge 2$ we have $\lambda(n) = 1 - 2\kappa(n)$ (see [Wills 1968; Cusick 1973]).

Yet another approach to the same core problem called *lonely runner* has been given in [Bienia et al. 1998]. In spite of many research efforts during the last 30 years, the exact value of $\kappa(n)$ is known only for values up to 5 ([Bohman et al. 2001]). For $n \ge 5$, only upper and lower bounds have been determined. If one considers balls instead of cubes [Cusick and Pomerance 1984], then also the exact values for the view obstruction problem are known only up to dimension 5 [Dumir et al. 1996].

Although, of course, the view obstruction problem involves an infinite number of bodies, our complexity results for finite instances can be seen as a certain complexity-theoretical indication for the hardness of the computation of $\lambda(n)$ for larger n. Namely, by Theorem 3, for fixed dimension ANCHORED VISIBILITY or ANCHORED QUADRANT VISIBILITY can be solved in polynomial time. However, if the dimension is part of the input, then the problem becomes NP-hard by Theorem 2. In a nonrigorous sense, this can be seen as a quantification of the strong influence of the dimension compared to the other input parameters.

References

[Aubry et al. 2002] P. Aubry, F. Rouillier, and M. S. E. Din, "Real solving for positive dimensional systems", *J. Symb. Comp.* **34** (2002), 543–560.

[Ben-Or et al. 1986] M. Ben-Or, D. Kozen, and J. Reif, "The complexity of elementary algebra and geometry", *J. Comput. System Sci.* **32** (1986), 251–264.

[Bern et al. 1994] M. Bern, D. Dobkin, D. Eppstein, and R. Grossman, "Visibility with a moving point of view", *Algorithmica* **11** (1994), 360–378.

[Bienia et al. 1998] W. Bienia, L. Goddyn, P. Gvozdjak, A. Sebö, and M. Tarsi, "Flows, view obstructions, and the lonely runner", *J. Comb. Theory, Ser. B* **72** (1998), 1–9.

[Bohman et al. 2001] T. Bohman, R. Holzman, and D. Kleitman, "Six lonely runners", *Electron. J. Combin.* **8**:2 (2001), R3.

[Brieden et al. 1998] A. Brieden, P. Gritzmann, R. Kannan, V. Klee, L. Lovász, and M. Simonovits, "Approximation of diameters: randomization doesn't help", pp. 244–251 in *Proc. 39th Annual Symposium on Foundations of Computer Science* (Palo Alto, CA), 1998.

[Chen and Cusick 1999] Y.-G. Chen and T. W. Cusick, "The view-obstruction problem for n-dimensional cubes", *J. Number Theory* **74** (1999), 126–133.

[Collins 1975] G. E. Collins, "Quantifier eliminations for real closed fields by cylindrical algebraic decomposition", pp. 134–183 in *Automata theory and formal languages* (Second GI Conference, Kaiserslautern, 1975), edited by H. Brakhage, Lecture Notes in Computer Science **33**, Springer, Berlin, 1975.

[Cusick 1973] T. W. Cusick, "View-obstruction problems", *Aequationes Math.* **9** (1973), 165–170.

[Cusick and Pomerance 1984] T. W. Cusick and J. Pomerance, "View-obstruction problems III", *J. Number Theory* **19** (1984), 131–139.

[Devillers et al. 2003] O. Devillers, V. Dujmovic, H. Everett, X. Goaoc, S. Lazard, H.-S. Na, and S. Petitjean, "The expected number of 3D visibility events is linear", *SIAM J. Computing* **32** (2003), 1586–1620.

[Dumir et al. 1996] V. C. Dumir, R. J. Hans-Gill, and J. B. Wilker, "The view-obstruction problem for spheres in R^n", *Monatshefte Math.* **122** (1996), 21–34.

[Durand 2002] F. Durand, "The 3D visibility complex", *ACM Transactions on Graphics* **21** (2002), 176–206.

[Durand et al. 1997] F. Durand, G. Drettakis, and C. Puech, "The 3D visibility complex: a unified data structure for global visibility of scenes of polygons and smooth objects", in *Proc. 9th Canadian Conf. on Comput. Geom.* (Kingston, ON), 1997.

[Dyer 1983] M. E. Dyer, "The complexity of vertex enumeration methods", *Math. Oper. Res.* **8** (1983), 381–402.

[Garey and Johnson 1979] M. R. Garey and D. S. Johnson, *Computers and intractability: A guide to the theory of NP–completeness*, Freeman, 1979.

[Grigor'ev and Vorobjov 1988] D. Y. Grigor'ev and N. N. Vorobjov, Jr., "Solving systems of polynomial inequalities in subexponential time", *J. Symb. Comp.* **5** (1988), 37–64.

[Gritzmann and Klee 1994] P. Gritzmann and V. Klee, "On the complexity of some basic problems in computational convexity I: Containment problems", *Discrete Math.* **136** (1994), 129–174.

[Grötschel et al. 1993] M. Grötschel, L. Lovász, and A. Schrijver, *Geometric algorithms and combinatorial optimization*, Algorithms and combinatorics **2**, Springer, Berlin, 1993.

[Kleiman and Laksov 1972] S. L. Kleiman and D. Laksov, "Schubert calculus", *Amer. Math. Monthly* **79** (1972), 1061–1082.

[Lenhof and Smid 1995] H.-P. Lenhof and M. Smid, "Maintaining the visibility map of spheres while moving the viewpoint on a circle at infinity", *Algorithmica* **13** (1995), 301–312.

[Macdonald et al. 2001] I. G. Macdonald, J. Pach, and T. Theobald, "Common tangents to four unit balls in $\mathbb{R}^3$", *Discrete Comput. Geom.* **26** (2001), 1–17.

[McMullen 1970] P. McMullen, "The maximum numbers of faces of a convex polytope", *Mathematika* **17** (1970), 179–184.

[O'Rourke 1997] J. O'Rourke, "Visibility", pp. 467–479 in *Handbook of discrete and computational geometry*, edited by J. E. Goodman and J. O'Rourke, CRC Press, Boca Raton (FL), 1997.

[Pocchiola and Vegter 1996] M. Pocchiola and G. Vegter, "Topologically sweeping visibility complexes via pseudotriangulations", *Discrete Comput. Geom.* **16** (1996), 419–453.

[Sottile 1997] F. Sottile, "Enumerative geometry for the real Grassmannian of lines in projective space", *Duke Math. J.* **87** (1997), 59–85.

[Sottile and Theobald 2002] F. Sottile and T. Theobald, "Lines tangent to $2n-2$ spheres in $\mathbb{R}^n$", *Trans. Amer. Math. Soc.* **354** (2002), 4815–4829.

[Swayne et al. 1998] D. F. Swayne, D. Cook, and A. Buja, "XGobi: interactive dynamic data visualization in the X window system", *J. Comput. Graphical Statistics* **7** (1998), 113–130.

[Theobald 2002] T. Theobald, "An enumerative geometry framework for algorithmic line problems in $\mathbb{R}^3$", *SIAM J. Computing* **31** (2002), 1212–1228.

[Tóth 1971] L. F. Tóth, *Lagerungen in der Ebene, auf der Kugel und im Raum*, Grundlehren der math. Wissenschaften **65**, Springer, Berlin, 1971.

[Wang and Zhu 2000] C. A. Wang and B. Zhu, "Three-dimensional weak visibility: Complexity and applications", *Theor. Comp. Sci.* **234** (2000), 219–232.

[Wernecke 1994] J. Wernecke, *The Inventor Mentor: Programming object-oriented 3D Graphics with Open Inventor*, Addison-Wesley, Reading (MA), 1994.

[Wills 1968] J. M. Wills, "Zur simultanen homogenen diophantischen Approximation I", *Monatshefte Math.* **72** (1968), 254–263.

[Zong 1997] C. Zong, "A problem of blocking light rays", *Geom. Dedicata* **67** (1997), 117–128.

[Zong 1999] C. Zong, *Sphere packings*, Universitext, Springer, New York, 1999.

PETER GRITZMANN
ZENTRUM MATHEMATIK
TECHNISCHE UNIVERSITÄT MÜNCHEN
BOLTZMANNSTR. 3
D–85747 GARCHING BEI MÜNCHEN
GERMANY
  gritzman@ma.tum.de

THORSTEN THEOBALD
INSTITUT FÜR MATHEMATIK, MA 6-2
TECHNISCHE UNIVERSITÄT BERLIN
STRASSE DES 17. JUNI 136
D–10623 BERLIN
GERMANY
  theobald@math.tu-berlin.de

Combinatorial and Computational Geometry
MSRI Publications
Volume **52**, 2005

Cylindrical Partitions of Convex Bodies

ALADÁR HEPPES AND WŁODZIMIERZ KUPERBERG

ABSTRACT. A cylindrical partition of a convex body in $\mathbb{R}^n$ is a partition of the body into subsets of smaller diameter, obtained by intersecting the body with a collection of mutually parallel convex-base cylinders. Convex bodies of constant width are characterized as those that do not admit a cylindrical partition. The main result is a finite upper bound, exponential in n, on the minimum number $b_c(n)$ of pieces needed in a cylindrical partition of every convex body of nonconstant width in $\mathbb{R}^n$. (A lower bound on $b_c(n)$, exponential in $\sqrt{n}$, is a consequence of the construction of Kalai and Kahn for counterexamples to Borsuk's conjecture.) We also consider cylindrical partitions of centrally symmetric bodies and of bodies with smooth boundaries.

1. Introduction and Preliminaries

Throughout this article, M denotes a compact subset of $\mathbb{R}^n$ containing at least two points. By diam M we denote the maximum distance between points of M, but *diameter of M* also means the line segment connecting any pair of points of M that realize this distance (ambiguity is always avoided by the context). A *Borsuk partition* of M is a family of subsets of M, each of diameter smaller than diam M, whose union contains M. The *Borsuk partition number of M*, denoted by $b(M)$, is the minimum number of sets needed in a Borsuk partition of M. It is obvious that $b(M)$ is finite. It is also obvious that the maximum of $b(M)$ over all bounded sets M in $\mathbb{R}^n$ exists and is bounded above exponentially in n, since every set of diameter d is contained in a ball of radius d. Therefore the *n-th Borsuk partition number*, denoted by $b(n)$, and defined as the minimum number of sets needed for a Borsuk partition of any bounded set in $\mathbb{R}^n$, is finite. Since a Borsuk

Mathematics Subject Classification: 52A20, 52A37.

Keywords: convex body, diameter, constant width, Borsuk's conjecture, cylindrical partition.

Heppes acknowledges, with thanks, support from the Hungarian Research Foundation OTKA, Grant Numbers T 037752 and T 038397. Kuperberg gratefully acknowledges support received from the Mathematical Sciences Research Institute in Berkeley, CA during the week-long workshop on Combinatorial and Discrete Geometry, November 17-21, 2003.

partition of a ball in $\mathbb{R}^n$ requires at least $n+1$ sets, it follows that $b(n) \geq n+1$. K. Borsuk [1933] conjectured that $b(n) = n+1$, to which a counterexample was found by G. Kalai and J. Kahn [1993] in dimension $n = 1325$. In fact, Kalai and Kahn proved that, for large n, $b(n)$ is bounded below exponentially in $1.2^{\sqrt{n}}$. (At the time of writing of this paper, the lowest dimension known for which Borsuk's conjecture fails is 298; see [Hinrichs and Richter 2003].)

In this paper we consider a special kind of Borsuk partitions, which we will call cylindrical partitions; we ask related questions and provide some answers.

DEFINITION. By a *cylinder* in the direction of a line l we understand a closed set that contains every line parallel to l that intersects the set. A cylinder's cross-section perpendicular to its direction is called the *base* of the cylinder. Let M be a compact set, let l be a line, and let $\{M_i\}$ be a Borsuk partition of M. We say that the partition is *cylindrical* and that l defines its direction, provided that each of the sets M_i is the intersection of M with a cylinder parallel to l. For brevity, we say "cylindrical partition" instead of "cylindrical Borsuk partition," assuming automatically that the pieces of M are of diameter smaller than $\operatorname{diam} M$.

For the purpose of studying the problem of existence and minimum cardinality of cylindrical partitions of M (over variable M) one can always replace M with its convex hull $\operatorname{Conv} M$. Therefore it will be assumed from now on that M is a convex body, unless otherwise specified.

The *width of M in the direction of line l* is the distance between the pair of hyperplanes of support of M that are perpendicular to l and enclose M between them. If M has the same width in every direction, then M is said to be a *body of constant width* (see [Heil and Martini 1993, p. 363]). It is easy to see that if segment d is of maximum length among all chords of M parallel to d, then there exist two parallel hyperplanes of support of M, each containing an end of d. It follows that a body of constant width has a diameter in every direction. Hence:

PROPOSITION 1. *If M is a body of constant width, then M does not admit a cylindrical partition in any direction (not even into an infinite number of pieces).*

The converse to the above is true as well:

PROPOSITION 2. *If M is a bounded set of nonconstant width, then there exists a direction in which M admits a finite cylindrical partition.*

PROOF. Denote the diameter of M by d. Let P_1 and P_2 be a pair of parallel hyperplanes supporting M from opposite sides and determining a width $d_1 < d$. Let M_1 be the perpendicular projection of M to P_1. There exists a finite partition of M_1 (in P_1) into sets of diameter smaller than $\sqrt{d^2 - d_1^2}$. This partition gives rise (perpendicularly to P_1) to a finite cylindrical partition of M. $\square$

COROLLARY 3. *Bodies of constant width are characterized among convex bodies as those that do not admit a finite cylindrical partition.*

The above characterization of bodies of constant width and the subsequent investigations were inspired by a previous result of A. Heppes [1959] characterizing curves of constant width in the plane.

DEFINITION. If M is a set of nonconstant width, let $b_c(M)$, the *Borsuk cylindrical partition number of M*, or the *cylindrical partition number of M* for short, denote the smallest number of pieces into which M can be cylindrically partitioned.

The notion defined below is analogous to the n-th Borsuk number:

DEFINITION. The maximum of $b_c(M)$ over all sets M of nonconstant width in $\mathbb{R}^n$ is called the *n-th cylindrical partition number* and is denoted by $b_c(n)$.

The above proof of Proposition 2 may seem to indicate that already in the plane the cylindrical partition number of a set of nonconstant width may be arbitrarily large. But in Section 3 we show that there is an upper bound for $b_c(M)$ depending on the dimension of the ambient space only, which justifies the definition of $b_c(n)$. Specifically, we show that $b_c(n)$ is bounded above exponentially in n. In our proof we apply the following classical result concerning bodies of constant width (see [Bonnesen and Fenchel 1974, p. 129]), obtained by E. Meissner [1911] for $n = 2$ and $n = 3$, and generalized to all n by B. Jessen [1928]:

THEOREM 4 (MEISSNER–JESSEN). *Every convex body in $\mathbb{R}^n$ can be enlarged, without increasing its diameter, to a body of constant width.*

Henceforth, the distance from point A to point B, the segment with ends A and B, and the line containing them are denoted by AB, $\overline{AB}$, and $\overleftrightarrow{AB}$, respectively.

DEFINITION. Let s be a segment and let l be a line containing neither of the two ends of s. The *angle at which s is seen from l* is the measure of the smallest dihedral angle with its edge on l and containing s. We denote this angle by $\angle(l, s)$.

Observe that the above definition is meaningful in every dimension $n \geq 2$, although for $n = 2$ the angle is always $0°$ or $180°$. In the next section we present a lemma concerning the minimum angle at which a diameter of a bounded set can be seen from the line of another diagonal.

2. Diameters of a Bounded Set

Here we relax the standing assumption that M is a convex body. We do not even require M to be compact, only to be bounded. In what follows, α_0 denotes the measure of the dihedral angle in a regular tetrahedron, $\alpha_0 = \arccos \frac{1}{3} = 70.52\ldots°$.

LEMMA 5. *Let M be a bounded set and let d_1 and d_2 be diameters of M that do not have a common end point. Then*

$$\angle(\overleftrightarrow{d_1}, d_2) \geq \alpha_0,$$

equality being attained if and only if the convex hull of $d_1 \cup d_2$ is a regular tetrahedron.

PROOF. (F. Santos, private communication, 2003). Assume for simplicity that diam $M = 1$ and let l denote the line of d_1. Observe that d_2 cannot have an end on l, hence $\angle(l, d_2)$ is well defined. If the segments d_1 and d_2 intersected, then we would have $\angle(l, d_2) = 180°$, hence we can assume otherwise. Then the lines of d_1 and d_2 cannot intersect at all, and, obviously, they cannot be parallel. Since any pair of skew lines in $\mathbb{R}^n$ with $n > 3$ determine a 3-dimensional flat containing them, we can now assume that M is a subset of $\mathbb{R}^3$. Hence all we need to prove is the following:

ASSERTION. Among all tetrahedra of diameter 1 and with two nonadjacent edges of length 1, the minimum of the dihedral angle at either one of the two edges is attained on, and only on, the regular tetrahedron.

Let $T = ABCD$ be a tetrahedron with $AB = CD = 1$, and with all four of its remaining edges of length at most 1. Obviously, T lies in a lens-like set L that is in the common part of two unit balls, one centered at A and the other at B. Let p be the plane containing $\overleftrightarrow{AB}$ and parallel to $\overleftrightarrow{CD}$, let h denote the distance between p and $\overleftrightarrow{CD}$, and let p_h be the plane parallel to p and containing $\overleftrightarrow{CD}$. The set $L_h = L \cap p_h$ is the common part of two circular disks (in p_h) of radius $\sqrt{1 - h^2}$ each, their centers one unit apart. The boundary of L_h is the union of two circular arcs: the A-arc, consisting of points in L_h one unit away from A, and the B-arc, consisting of points one unit away from B. Clearly, edge $\overline{CD}$ of T lies in L_h; we will vary the position of that edge within L_h in order to minimize the dihedral angle at edge $\overline{AB}$.

Rotating $\overline{CD}$ (within L_h) about either of its ends to a position "more parallel to" $\overleftrightarrow{AB}$ (that is, to decrease the angle between $\overleftrightarrow{CD}$ and $\overleftrightarrow{AB}$) decreases the dihedral angle at $\overline{AB}$. By combining at most two such rotations we can bring $\overline{CD}$ to a position in which one of its ends, say C, lies on the B-arc, and the other one, D, on the A-arc of L_h. Then we have $AD = BC = 1$. Observe that unless the equality $AD = BC = 1$ held already before, this change requires at least one rotation, thus it actually decreases the dihedral angle at $\overline{AB}$.

Finally, if $AC < 1$, then increasing the length of $\overline{AC}$ while keeping the length of the remaining edges fixed results in a decrease of the dihedral angle at $\overleftrightarrow{AB}$ (and the same holds about lengthening the edge $\overline{BD}$). To prove this fact, consider a sufficiently small sphere centered at point B. The intersection of the sphere with T is an isosceles (spherical) triangle $t^* = A^* D^* C^*$ (labeling to reflect the correspondence to points A, D, and C) with legs $A^* D^*$ and $C^* D^*$ whose length remains constant, since triangles ABD and CBD remain rigid, hinged on their common edge $\overline{BD}$. Observe that the angle of t^* at A^* is the dihedral angle of T at the edge $\overline{AB}$. Now, as AC increases, the constant-length legs of the isosceles triangle t^* open wider, and the two equal angles at its base decrease.

This argument proves that the dihedral angle at $\overline{AB}$ attains its minimum when, and only when, T is regular, which completes the proof of the lemma. $\square$

3. An Exponential Upper Bound for $b_c(n)$

In this section we prove our main result:

THEOREM 6. *There is a constant k such that $b_c(n) \leq k^n$ for every n.*

PROOF. Let K be a convex body of nonconstant width in $\mathbb{R}^n$ and let $\widehat{K}$ be a body of constant width containing K and of the same diameter as K, whose existence is provided by Theorem 4. Assume $\operatorname{diam} K = \operatorname{diam} \widehat{K} = 1$. Then there is a line l such that $d = l \cap \widehat{K}$ is of length 1, while the segment $l \cap K$ is shorter than 1, hence is a proper subset of d. Let H be the hyperplane perpendicular to l and let O denote the intersection point $l \cap H$. There exists a round cylinder C about l of radius r small enough so that $\operatorname{diam}(C \cap K) < 1$. Let B_r denote the base of C in H, which is an $(n-1)$-dimensional ball of radius r, centered at O. Let S denote the boundary of B_r, which is a sphere of dimension $n-2$.

Consider a covering of S with the smallest number s_{n-2} of congruent spherical caps C_i of angular diameter α slightly smaller than $\alpha_0 = 70.5\ldots^\circ$ (α_0 is the dihedral angle in a regular tetrahedron, as in Lemma 5). By a simple argument involving a saturated packing with caps and a rough volume estimate, s_{n-2} is bounded above exponentially in n. (C. A. Rogers [1963] gives a very good, specific upper bound obtained by a refined analysis.) Let V_i denote the cone (in H) composed of rays emanating from O and passing through C_i, and let W_i be the subset of V_i obtained by chopping off a small tip of V_i, say W_i is the convex hull of the closure of the set $V_i \setminus B_r$, a truncated cone. The family of $1 + s_{n-2}$ subsets of H consisting of B_r and the truncated cones W_i ($i = 1, 2, \ldots, s_{n-2}$) covers H. This covering gives rise to a family of cylinders in the direction of l whose intersections with K form a cylindrical partition of K. Indeed: every diagonal of K is a diagonal of $\widehat{K}$, and every diagonal of $\widehat{K}$ other than d, either:

(i) has a common end with d, in which case that end lies outside the union of the cylinders over the truncated cones W_i, or

(ii) has no common end with d and therefore is seen from l at an angle greater than or equal to α_0, which implies that it cannot be contained in any one of the cylinders over the sets W_i.

Since C does not contain any diameter of K by design and none of the cylinders over the truncated cones W_i does either, and since each of these cylindrical pieces is convex, we have a cylindrical partition of K, with an exponential (in n) upper bound on the number of pieces. $\square$

REMARK. The construction in the proof of Theorem 6 actually demonstrates that $b_c(n) \leq 1 + s_{n-2}$. However, in case K is smooth, *i.e.*, at every point of the boundary of K the support hyperplane is unique, no diagonal of K has a common end with another diagonal of $\widehat{K}$. Then the cylinders over the cones V_i, $1 \leq i \leq s_{n-2}$, form a cylindrical partition of K already (the "central piece" C is not needed). Thus, $b_c(K) \leq s_{n-2}$ for every n-dimensional smooth convex body K of nonconstant width.

The construction described above in the proof of Theorem 6, combined with the fact that the necessary number of pieces in a Borsuk partition of the n-dimensional regular simplex is $n + 1$, yield:

COROLLARY 7. *A constant k exists such that the inequalities $n + 1 \leq b_c(n) \leq k^n$ hold. In dimensions up to 4, we have, more specifically, $b_c(2) = 3$, $4 \leq b_c(3) \leq 7$, and $5 \leq b_c(4) \leq 15$. Moreover, by virtue of the remark above, if K is a smooth convex body of nonconstant width in $\mathbb{R}^n$, then $b_c(K) = 2$ for $n = 2$, $b_c(K) \leq 6$ for $n = 3$, and $b_c(K) \leq 14$ for $n = 4$.*

The inequality $b_c(3) \leq 7$ follows from $5\alpha_0 < 360° < 6\alpha_0$, i.e., $s_1 = 6$. The inequality $b_c(4) \leq 15$ is obtained by the fact that the 2-sphere (the boundary of the 3-dimensional ball) can be covered with 14 congruent spherical caps of spherical diameter smaller than α_0, i.e., $s_2 = 14$. Indeed, the smallest diameter of 14 congruent spherical caps that can cover the 2-sphere is approximately 69.875° (see [Fejes Tóth 1969]), just a little bit less than α_0.

REMARK. Let A_n denote the counterexample of Kalai and Kahn [1993] to Borsuk's conjecture in $\mathbb{R}^n$, whose Borsuk partition number is bounded below exponentially by $\sqrt{n}$. Since each of the sets A_n is finite, the set $K_n = \text{Conv} A_n$, being a polyhedron, is of nonconstant width. It follows that $b_c(K_n) \geq b(K_n)$.

Consequently, we have:

COROLLARY 8. *There exist constants $k_1 > 1$ and k_2 such that, for large n,*

$$k_1^{\sqrt{n}} \leq b_c(n) \leq k_2^n.$$

4. Cylindrical Partitions of Bodies With Central Symmetry

Under the assumption of central symmetry of the body to be cylindrically partitioned, the upper bound on the number of pieces needed is much lower than the bound obtained in the previous section:

THEOREM 9. *Let K be a centrally symmetric convex body in $\mathbb{R}^n$ other than a ball. Then $b_c(K) \leq n$, and the inequality is sharp.*

PROOF. Assume that O is the symmetry center of K and that $\text{diam}\, K = 1$. Then the ball B of radius $1/2$ and centered at O contains K as a proper subset. Therefore one of the diameters of B, say d, has both of its ends outside K. Let H be a hyperplane perpendicular to d and passing through O. The set $H \cap B$ is an $(n-1)$-dimensional ball (in H) of diameter 1, hence its boundary S, a sphere of dimension $n-2$, can be covered by n congruent caps c_i of diameter smaller than

$180°$, with centers placed at the vertices of a regular $(n-1)$-simplex inscribed in S. Let C_i be the cone composed of rays that emanate from O and pass through c_i. By erecting a cylinder D_i parallel to d with base C_i $(i = 1, 2, \ldots, n)$, we obtain a cylindrical partition of K into n pieces, since neither of the convex sets $D_i \cap K$ contains a pair of antipodes of B.

The inequality is sharp, since any Borsuk partition of a ball in $\mathbb{R}^n$ with a pair of small antipodal congruent caps cut off requires n pieces.   □

5. Final Remarks and Some Open Problems

Our Proposition 1, Proposition 2, and Corollary 3 can be generalized to n-dimensional Minkowski spaces by methods described in [Averkov and Martini 2002]. The classical Meissner–Jessen theorem (Theorem 4 here) has been generalized to convex bodies in n-dimensional Minkowski spaces by G. D. Chakerian and H. Groemer [1983]. Therefore it is perhaps possible that Theorem 6, our main result, can be so generalized as well, although the magnitude of the upper bound may depend on the unit ball in the Minkowski space.

One could generalize the concept of cylindrical partitions in $\mathbb{R}^n$ by considering "k-cylinders" obtained as a Cartesian product of a set lying in an $(n-k)$-dimensional flat with a k-dimensional flat (a 1-cylinder would then be a "usual" cylinder, *i.e.*, a product with a line). But, because of their connection to bodies of constant width, we decided to deal with cylindrical partitions based on the usual cylinders only.

M. Lassak [1982] proved that $b(n) \le 2^{n-1}+1$, and from a result of O. Schramm [1988] on covering a body of constant width with its smaller homothetic copies it follows that $b(n) \le 5n\sqrt{n}(4 + \log n) \left(\frac{3}{2}\right)^{n/2}$, presenting an upper bound of order of magnitude $\left(\sqrt{1.5}\right)^n$. The precise asymptotic behavior of $b(n)$ remains unknown.

The problem of determining the precise asymptotic behavior of $b_c(n)$ as $n \to \infty$ (let alone the exact values), appears to be extremely difficult, just as, or perhaps even more so than, the similar problem for $b(n)$. But it seems reasonable to expect some improvements on the bounds given in Corollary 7 and in Theorem 8. In particular, we feel that the upper bound in $4 \le b_c(3) \le 7$ can be lowered, perhaps all the way down to 4. Also, one should be able to narrow the gap between the lower and upper bounds in $5 \le b_c(4) \le 15$.

And, finally, it seems strange that the seemingly natural inequality $b(n) \le b_c(n)$ is not obvious at all; perhaps it may even be false for some n. It is *a priori* conceivable that in some dimension n, the value of $b(n)$ is attained on a body (or bodies) of constant width only, and that in such dimension, $b_c(n)$, being defined by bodies of *nonconstant* width, is smaller than $b(n)$. Paradoxical as it may seem, thus far such possibility has not been excluded. However, it is quite obvious that $b(n) \le b_c(n) + 1$, because every convex body of constant width can be reduced to a convex body of nonconstant width by separating from it one small piece.

Personal Acknowledgements

We thank Gábor Fejes Tóth and Greg Kuperberg for their valuable comments and suggestions for improvements, and we give special thanks to Francisco Santos for bringing Lemma 5 to its present "best possible" form, with an elementary, short proof.

References

[Averkov and Martini 2002] G. Averkov and H. Martini, "A characterization of constant width in Minkowski planes", preprint, Technische Universität Chemnitz, 2002.

[Bonnesen and Fenchel 1974] T. Bonnesen and W. Fenchel, *Theorie der konvexen Körper*, Springer, Berlin, 1974.

[Borsuk 1933] K. Borsuk, "Drei Sätze über die n dimensionale Euklidische Sphäre", *Fund. Math.* **20** (1933), 177–190.

[Chakerian and Groemer 1983] G. D. Chakerian and H. Groemer, "Convex bodies of constant width", pp. 49–96 in *Convexity and its applications*, edited by P. M. Gruber and J. M. Wills, Birkhäuser, Basel, 1983.

[Fejes Tóth 1969] G. Fejes Tóth, "Kreisüberdeckungen der Sphäre", *Studia Sci. Math. Hungar.* **4** (1969), 225–247.

[Heil and Martini 1993] E. Heil and H. Martini, "Special convex bodies", pp. 347–385 in *Handbook of convex geometry*, vol. A, North-Holland, Amsterdam, 1993.

[Heppes 1959] A. Heppes, "On characterisation of curves of constant width", *Mat. Lapok* **10** (1959), 133–135.

[Hinrichs and Richter 2003] A. Hinrichs and C. Richter, "New sets with large Borsuk numbers", *Discrete Math.* **270**:1-3 (2003), 137–147.

[Jessen 1928] B. Jessen, "Über konvexe Punktmengen konstanten Breite", *Math. Z.* **29** (1928), 378–380.

[Kahn and Kalai 1993] J. Kahn and G. Kalai, "A counterexample to Borsuk's conjecture", *Bull. Amer. Math. Soc.* (*N.S.*) **29**:1 (1993), 60–62.

[Lassak 1982] M. Lassak, "An estimate concerning Borsuk partition problem", *Bull. Acad. Polon. Sci. Sér. Sci. Math.* **30**:9-10 (1982), 449–451 (1983).

[Meissner 1911] E. Meissner, "Über Punktmengen konstanten Breite", *Vjschr. naturforsch. Ges. Zürich* **56** (1911), 42–50.

[Rogers 1963] C. A. Rogers, "Covering a sphere with spheres", *Mathematika* **10** (1963), 157–164.

[Schramm 1988] O. Schramm, "Illuminating sets of constant width", *Mathematika* **35**:2 (1988), 180–189.

ALADÁR HEPPES
VÉRCSE U. 24/A
H-1124 BUDAPEST
HUNGARY
  hep9202@helka.iif.hu

WŁODZIMIERZ KUPERBERG
DEPARTMENT OF MATHEMATICS
AUBURN UNIVERSITY
AUBURN, AL 36849
UNITED STATES
  kuperwl@auburn.edu

Combinatorial and Computational Geometry
MSRI Publications
Volume **52**, 2005

Tropical Halfspaces

MICHAEL JOSWIG

ABSTRACT. As a new concept tropical halfspaces are introduced to the
(linear algebraic) geometry of the tropical semiring $(\mathbb{R}, \min, +)$. This yields
exterior descriptions of the tropical polytopes that were recently studied
by Develin and Sturmfels [2004] in a variety of contexts. The key tool
to the understanding is a newly defined sign of the tropical determinant,
which shares remarkably many properties with the ordinary sign of the
determinant of a matrix. The methods are used to obtain an optimal
tropical convex hull algorithm in two dimensions.

1. Introduction

The set $\mathbb{R}$ of real numbers carries the structure of a semiring if equipped
with the *tropical addition* $\lambda \oplus \mu = \min\{\lambda, \mu\}$ and the *tropical multiplication*
$\lambda \odot \mu = \lambda + \mu$, where $+$ is the ordinary addition. We call the triplet $(\mathbb{R}, \oplus, \odot)$ the
tropical semiring[1]. It is an equally simple and important fact that the operations
$\oplus, \odot : \mathbb{R} \times \mathbb{R} \to \mathbb{R}$ are continuous with respect to the standard topology of $\mathbb{R}$. So
the tropical semiring is, in fact, a topological semiring. Considering the tropical
scalar multiplication

$$\lambda \odot (\mu_0, \dots, \mu_d) = (\lambda + \mu_0, \dots, \lambda + \mu_d)$$

(and componentwise tropical addition) turns the set $\mathbb{R}^{d+1}$ into a semimodule.

The study of the linear algebra of the tropical semiring and, more generally,
of idempotent semirings, has a long tradition. Applications to combinatorial
optimization, discrete event systems, functional analysis etc. abound. For an
introduction see [Baccelli et al. 1992]. A recent contribution in the same vein,
with many more references, is [Cohen et al. 2004].

This work has been carried out while visiting the Mathematical Sciences Research Institute in
Berkeley for the special semester on Discrete and Computational Geometry.

[1]Other authors reserve the name *tropical semiring* for $(\mathbb{N} \cup \{+\infty\}, \min, +)$ and call $(\mathbb{R} \cup \{+\infty\}, \min, +)$ the min-plus-semiring.

Convexity in the tropical world (and even in a more general setting) was first studied by Zimmermann [1977]. Following the approach of Develin and Sturmfels [2004] here we stress the point of view of discrete geometry. We recall some of the key definitions. A subset $S \subset \mathbb{R}^{d+1}$ is *tropically convex* if for any two points $x, y \in S$ the *tropical line segment*

$$[x, y] = \{\lambda \odot x \oplus \mu \odot y \mid \lambda, \mu \in \mathbb{R}\}$$

is contained in S. The *tropical convex hull* of a set $S \subset \mathbb{R}^{d+1}$ is the smallest tropically convex set containing S; it is denoted by $\operatorname{tconv} S$. It is easy to see [Develin and Sturmfels 2004, Proposition 4] that

$$\operatorname{tconv} S = \{\lambda_1 \odot x_1 \oplus \cdots \oplus \lambda_n \odot x_n \mid \lambda_i \in \mathbb{R}, \ x_i \in S\}.$$

A *tropical polytope* is the tropical convex hull of finitely many points. Since any convex set in $\mathbb{R}^{d+1}$ is closed under tropical multiplication with an arbitrary scalar, it is common to identify tropically convex sets with their respective images under the canonical projection onto the d-dimensional *tropical projective space*

$$\mathbb{TP}^d = \{\mathbb{R} \odot x \mid x \in \mathbb{R}^{d+1}\} = \mathbb{R}^{d+1}/\mathbb{R}(1, \dots, 1).$$

In explicit computations we often choose *canonical coordinates* for a point $x \in \mathbb{TP}^d$, meaning the unique nonnegative vector in the class $\mathbb{R} \odot x$ which has at least one zero coordinate. For visualization purposes, however, we usually normalize the coordinates by choosing the first one to be zero (which can then be omitted): This identification $(\xi_0, \dots, \xi_d) \mapsto (\xi_1 - \xi_0, \dots, \xi_d - \xi_0) : \mathbb{TP}^d \to \mathbb{R}^d$ is a homeomorphism.

Develin and Sturmfels observed that *tropical simplices*, that is tropical convex hulls of $d+1$ points in $\mathbb{TP}^d$ (in sufficiently general position), are related to Isbell's [1964] *injective envelope* of a finite metric space; see [Develin and Sturmfels 2004, Theorem 29] and the Erratum. Isbell's injective envelope in turn coincides with the *tight span* of a finite metric space that arose in the work of Dress and others; see [Dress et al. 2002] and its list of references. In a way, tropical simplices may be understood as nonsymmetric analogues of injective hulls or tight spans.

Additionally, tropical polytopes are interesting also from a purely combinatorial point of view: They bijectively correspond to the regular polyhedral subdivisions of products of simplices; see [Develin and Sturmfels 2004, Theorem 1].

The present paper studies tropical polytopes as geometric objects in their own right. It is shown that, at least to some extent, it is possible to develop a theory of tropical polytopes in a fashion similar to the theory of ordinary convex polytopes. The key concept introduced to this end is the notion of a tropical halfspace. One of our main results, Theorem 4.7, gives a characterization of tropical halfspaces in terms of the tropical determinant, which is the same as the min-plus-permanent already studied by Yoeli [1961] and others; see also [Richter-Gebert et al. 2005]. The proof leads to the definition of the *faces* of a tropical polytope in a natural way. In the investigation, in particular, we prove that the faces form a distributive lattice; see Theorem 3.7. Moreover, as one would expect by analogy to ordinary convex polytopes, the tropical polytopes are precisely the bounded intersections of finitely many tropical halfspaces; see Theorem 3.6.

It is a further consequence of our results on tropical polytopes that some concepts and ideas from computational geometry can be carried over from ordinary convex polytopes to tropical polytopes. In Section 5 this leads us to a comprehensive solution of the convex hull problem in $\mathbb{TP}^2$. The general tropical convex hull problem in arbitrary dimension is certainly interesting, but this is beyond our current scope.

The paper closes with a selection of open questions.

2. Hyperplanes and Halfspaces

We start this section with some observations concerning the topological aspects of tropical convexity. As already mentioned the tropical projective space $\mathbb{TP}^d$ is homeomorphic to $\mathbb{R}^d$ with the usual topology. Moreover, the space $\mathbb{TP}^d$ carries a natural metric: For a point $x \in \mathbb{TP}^d$ with canonical coordinates $(\xi_0, \ldots, \xi_d)$ let

$$\|x\| = \max\{\xi_0, \ldots, \xi_d\}$$

be the *tropical norm* of x. Equivalently, for arbitrary coordinates $(\xi'_0, \ldots, \xi'_d) \in \mathbb{R} \odot x$ we have that $\|x\| = \max\{|\xi'_i - \xi'_j| \mid i \neq j\}$. We prove a special case of [Cohen et al. 2004, Theorem 17]:

LEMMA 2.1. *The map*

$$\mathbb{TP}^d \times \mathbb{TP}^d \to \mathbb{R} : (x, z) \mapsto \|x - z\|$$

is a metric.

PROOF. By definition the map is nonnegative. Moreover, it is clearly definite and symmetric. We prove the triangle inequality: Assume that $x = (\xi_0, \ldots, \xi_d)$, $z = (\zeta_0, \ldots, \zeta_d)$, and that $y = (\eta_0, \ldots, \eta_d)$ be a third point. Then

$$
\begin{aligned}
\|x - z\| &= \max\{|(\xi_i - \zeta_i) - (\xi_j - \zeta_j)| \mid i \neq j\} \\
&= \max\{|(\xi_i - \xi_j) - (\eta_i - \eta_j) + (\eta_i - \eta_j) - (\zeta_i - \zeta_j)| \mid i \neq j\} \\
&\leq \max\{|(\xi_i - \xi_j) - (\eta_i - \eta_j)| + |(\eta_i - \eta_j) - (\zeta_i - \zeta_j)| \mid i \neq j\} \\
&\leq \max\{|(\xi_i - \eta_i) - (\xi_j - \eta_j)| \mid i \neq j\} + \max\{|(\eta_i - \zeta_i) - (\eta_j - \zeta_j)| \mid i \neq j\} \\
&= \|x - y\| + \|y - z\|. \qquad \square
\end{aligned}
$$

The topology induced by this metric coincides with the quotient topology on $\mathbb{TP}^d$ (and thus with the natural topology of $\mathbb{R}^d$). In particular, $\mathbb{TP}^d$ is locally compact and a set $C \subset \mathbb{TP}^d$ is compact if and only if it is closed and bounded. Tacitly we will always assume that $d \geq 2$.

PROPOSITION 2.2. *The topological closure of a tropically convex set is tropically convex.*

PROOF. Let S be a tropically convex set with closure $\bar{S}$. Then, by [Develin and Sturmfels 2004, Proposition 4], $\mathrm{tconv}(\bar{S})$ is the set of points in $\mathbb{TP}^d$ which can be obtained as tropical linear combinations of points in $\bar{S}$. Now the claim follows from the fact that tropical addition and multiplication are continuous. □

From the named paper by Develin and Sturmfels we quote a few results which will be useful in our investigation.

THEOREM 2.3 [Develin and Sturmfels 2004, Theorem 15]. *A tropical polytope has a canonical decomposition as a finite ordinary polytopal complex, where the cells are both ordinary and tropical polytopes.*

PROPOSITION 2.4 [Develin and Sturmfels 2004, Proposition 20]. *The intersection of two tropical polytopes is again a tropical polytope.*

PROPOSITION 2.5 [Develin and Sturmfels 2004, Proposition 21] (see also [Helbig 1988]). *For each tropical polytope P there is a unique minimal set $\mathrm{Vert}(P) \subset P$ with $\mathrm{tconv}(\mathrm{Vert}(P)) = P$.*

The elements of $\mathrm{Vert}(P)$ are called the *vertices* of P. The following is implied by Theorem 2.3. There is also an easy direct proof which we omit, however.

PROPOSITION 2.6. *A tropical polytope is compact.*

The *tropical hyperplane* defined by the *tropical linear form* $a = (\alpha_0, \dots, \alpha_d) \in \mathbb{R}^{d+1}$ is the set of points $(\xi_0, \dots, \xi_d) \in \mathbb{TP}^d$ such that the minimum

$$\min\{\alpha_0 + \xi_0, \dots, \alpha_d + \xi_d\} = \alpha_0 \odot \xi_0 \oplus \cdots \oplus \alpha_d \odot \xi_d$$

is attained at least twice. The point $-a$ is contained in the tropical hyperplane defined by a, and it is called its *apex*. Note that any two tropical hyperplanes only differ by a translation.

PROPOSITION 2.7 [Develin and Sturmfels 2004, Proposition 6]. *Tropical hyperplanes are tropically convex.*

The complement of a tropical hyperplane $\mathcal{H}$ in $\mathbb{TP}^d$ has $d+1$ connected components corresponding to the facets of an ordinary d-simplex. We call each such connected component an *open sector* of $\mathcal{H}$. The topological closure of an open sector is a *closed sector*. It is easy to prove that each (open or closed) sector is convex both in the ordinary and in the tropical sense.

EXAMPLE 2.8. Consider the zero tropical linear form $0 \in \mathbb{R}^{d+1}$. The open sectors of the corresponding tropical hyperplane $\mathcal{Z}$ are the sets $S_0, \dots, S_d$, where

$$S_i = \{(\xi_0, \dots, \xi_d) \mid \xi_i < \xi_j \text{ for all } j \neq i\}.$$

The closed sectors are the sets $\bar{S}_0, \ldots, \bar{S}_d$, where

$$\bar{S}_i = \{(\xi_0, \ldots, \xi_d) \mid \xi_i \le \xi_j \text{ for all } j \ne i\}.$$

In canonical coordinates this can be expressed as follows:

$$S_i = \{(\xi_0, \ldots, \xi_d) \mid \xi_i = 0 \text{ and } \xi_j > 0, \text{ for } j \ne i\}$$

and

$$\bar{S}_i = \{(\xi_0, \ldots, \xi_d) \mid \xi_i = 0 \text{ and } \xi_j \ge 0, \text{ for } j \ne i\}.$$

Just as any two tropical hyperplanes are related by a translation, each translation of a sector is again a sector. We call such sectors *parallel*.

The following simple observation is one of the keys to the structural results on tropical polytopes in the subsequent sections. It characterizes the solvability of one tropical linear equation. For related results see [Akian et al. 2005].

PROPOSITION 2.9. *Let $x_1, \ldots, x_n \in \mathbb{TP}^d$. Then $0 \in \mathrm{tconv}\{x_1, \ldots, x_n\}$ if and only if each closed sector $\bar{S}_k$ of the zero tropical linear form contains at least one x_i.*

PROOF. We write ξ_{ij} for the canonical coordinates of the x_i in $\mathbb{R}^{d+1}$. Then all the $n(d+1)$ entries in the matrix

$$\begin{pmatrix} \xi_{10} & \cdots & \xi_{1d} \\ \vdots & \ddots & \vdots \\ \xi_{n0} & \cdots & \xi_{nd} \end{pmatrix}$$

are nonnegative. Hence

$$0 = \lambda_1 \odot x_1 \oplus \cdots \oplus \lambda_n \odot x_n$$

(with $\lambda_i \ge 0$, as we may assume without loss of generality) if and only if $\min\{\lambda_1 + \xi_{1k}, \ldots, \lambda_n + \xi_{nk}\} = 0$ for all k. We conclude that zero is in the tropical convex hull of $x_1, \ldots, x_n$ if and only if for all k there is an i such that $\xi_{ik} = 0$ or, equivalently, $x_i \in \bar{S}_k$. $\square$

Throughout the following we abbreviate $[d+1] = \{0, \ldots, d\}$, and we write $\mathrm{Sym}(d+1)$ for the symmetric group of degree $d+1$ acting on the set $[d+1]$. Let e_i be the i-th unit vector of $\mathbb{R}^{d+1}$. Observe that under the natural action of $\mathrm{Sym}(d+1)$ on $\mathbb{TP}^d$ by permuting the unit vectors tropically convex sets get mapped to tropically convex sets. The set of all k-element subsets of a set Ω is denoted by $\binom{\Omega}{k}$.

We continue our investigation with the construction of a two-parameter family of tropical polytopes.

EXAMPLE 2.10. We define the k-th *tropical hypersimplex* in $\mathbb{TP}^d$ as

$$\Delta_k^d = \mathrm{tconv}\left\{ \sum_{i \in J} -e_i \;\middle|\; J \in \binom{[d+1]}{k} \right\} \subset \mathbb{TP}^d.$$

It is essential that

$$\mathrm{Vert}(\Delta_k^d) = \left\{ \sum_{i \in J} -e_i \ \middle|\ J \in \binom{[d+1]}{k} \right\},$$

for all $k > 0$: This has to do with the fact that the symmetric group $\mathrm{Sym}(d+1)$ acts on the set, due to which either all or none of the points $\sum_{i \in J} -e_i$ is a vertex. But from Proposition 2.5 we know that $\varnothing \neq \mathrm{Vert}(\Delta_k^d) \subseteq \{\sum_{i \in J} -e_i \mid J \in \binom{[d+1]}{k}\}$, and hence the claim follows. Develin, Santos, and Sturmfels [Develin et al. 2005] construct tropical polytopes from matroids. The tropical hypersimplices arise as the special case of uniform matroids.

It is worth-while to look at two special cases of the previous construction.

EXAMPLE 2.11. The first tropical hypersimplex in $\mathbb{TP}^d$ is the d-dimensional *tropical standard simplex* $\Delta^d = \Delta_1^d = \mathrm{tconv}\{-e_0, \ldots, -e_d\}$. Note that Δ^d is a tropical polytope which at the same time is an ordinary polytope.

EXAMPLE 2.12. The second tropical hypersimplex $\Delta_2^d \subset \mathbb{TP}^d$ is the tropical convex hull of the $\binom{d+1}{2}$ vectors $-e_i - e_j$ for all pairs $i \neq j$. The tropical polytope Δ_2^d is not convex in the ordinary sense. It is contained in the tropical hyperplane $\mathcal{Z}$ corresponding to the zero tropical linear form. For $d = 2$ see Figure 1 below.

PROPOSITION 2.13. *The second tropical hypersimplex $\Delta_2^d \subset \mathbb{TP}^d$ is the intersection of the tropical hyperplane $\mathcal{Z}$ corresponding to the zero tropical linear form with the set of points whose tropical norm is bounded by 1.*

PROOF. Clearly, $-e_i - e_j \in \mathcal{Z}$ for $i \neq j$. We have to show that a point x with canonical coordinates $(\xi_0, \ldots, \xi_d)$ and $\|x\| \leq 1$ such that, e.g., $\xi_0 = \xi_1 = 0$, is a tropical linear combination of the $\binom{d+1}{2}$ vertices of Δ_2^d. We compute

$$x = (0, 0, 1, \ldots, 1) \oplus \xi_2 \odot (0, 1, 0, 1, \ldots, 1) \oplus \cdots \oplus \xi_d \odot (0, 1, \ldots, 1, 0),$$

and hence the claim.   □

In particular, this implies that Δ_2^d contains Δ_k^d, for all $k > 2$. A similar computation further shows that $\Delta_k^d \supsetneq \Delta_{k+1}^d$, for all k.

EXAMPLE 2.14. The ordinary d-dimensional ± 1-cube

$$C^d = \left\{ (0, \xi_1, \ldots, \xi_d) \ \middle|\ -1 \leq \xi_i \leq 1 \right\}$$

is a tropical polytope: $C^d = \mathrm{tconv}\{-e_0 - 2e_1, \ldots, -e_0 - 2e_d, e_1 + \ldots + e_d\}$.

One way of reading Proposition 2.13 is that the intersection of the tropical hyperplane corresponding to the zero tropical linear form with the ordinary ± 1-cube is a tropical polytope. An important consequence is the following.

COROLLARY 2.15. *The (nonempty) intersection of a tropical polytope with a tropical hyperplane is again a tropical polytope.*

PROOF. Let $P \subset \mathbb{TP}^d$ be a tropical polytope and $\mathcal{H}$ a tropical hyperplane. As usual, up to a translation we can assume that $\mathcal{H} = \mathcal{Z}$ corresponds to the zero tropical linear form. By Proposition 2.13 the intersection $P \cap \mathcal{Z}$ is contained in a suitably scaled copy of the second tropical hypersimplex Δ_2^d. Now the claim follows from Proposition 2.4. $\qquad\square$

A *closed tropical halfspace* in $\mathbb{TP}^d$ is the union of at least one and at most d closed sectors of a fixed tropical hyperplane. Hence it makes sense to talk about the *apex* of a tropical halfspace. An *open tropical halfspace* is the complement of a closed one. Clearly, the topological closure of an open tropical halfspace is a closed tropical halfspace. To each (open or closed) tropical halfspace $\mathcal{H}^+$ there is an *opposite* (open or closed) tropical halfspace $\mathcal{H}^-$ formed by the sectors of the corresponding tropical hyperplane which are not contained in $\mathcal{H}^+$. Two halfspaces are *parallel* if they are formed of parallel sectors.

LEMMA 2.16. *Let $a + \bar{S}_k \subset \mathbb{TP}^d$ be a closed sector, for some $k \in [d+1]$, and $b \in a + \bar{S}_k$ a point inside. Then the parallel sector $b + \bar{S}_k$ is contained in $a + \bar{S}_k$.*

Note that this includes the case where b is a point in the boundary $a + (\bar{S}_k \setminus S_k)$. The proof of the lemma is omitted.

PROPOSITION 2.17. *Each closed tropical halfspace is tropically convex.*

PROOF. Let $\mathcal{H}^+$ be a closed tropical halfspace. Without loss of generality, we can assume that $\mathcal{H}^+$ is the union the of closed sectors $\bar{S}_{i_1}, \ldots, \bar{S}_{i_l}$ of the tropical hyperplane $\mathcal{Z}$ corresponding to the zero tropical linear form. Since we already know that each $\bar{S}_{i_k}$ is tropically convex, it suffices to consider, e.g., $x \in \bar{S}_{i_1}$ and $y \in \bar{S}_{i_2}$ and to prove that $[x, y] \subset \mathcal{H}^+$. Let $(\xi_0, \ldots, \xi_d)$ and $(\eta_0, \ldots, \eta_d)$ be the canonical coordinates of $x, y \in \mathbb{TP}^d$, respectively. Since $x \in \bar{S}_{i_1}$ and $y \in \bar{S}_{i_2}$ we have that $\xi_{i_1} = 0$ and $\eta_{i_2} = 0$. Then the minimum

$$\min\{\lambda + \xi_0, \ldots, \lambda + \xi_d, \mu + \eta_0, \ldots, \mu + \eta_d\}$$

is $\lambda = \lambda + \xi_{i_1}$ or $\mu = \mu + \eta_{i_2}$, for arbitrary $\lambda, \mu \in \mathbb{R}$. This is equivalent to $\lambda \odot x + \mu \odot y \in \bar{S}_{i_1} \cup \bar{S}_{i_2}$, which implies the claim. $\qquad\square$

A similar argument shows that open tropical halfspaces are tropically convex.

COROLLARY 2.18. *The boundary of a tropical halfspace is tropically convex.*

PROOF. The boundary of a closed tropical halfspace $\mathcal{H}^+$ is the intersection of $\mathcal{H}^+$ with its opposite closed tropical halfspace $\mathcal{H}^-$. $\qquad\square$

TROPICAL SEPARATION THEOREM 2.19. *Let P be tropical polytope, and $x \notin P$ a point outside. Then there is a closed tropical halfspace containing P but not x.*

PROOF. From Proposition 2.9 we infer that there is a closed sector $x + \bar{S}_k$ of the tropical hyperplane with apex x which is disjoint from P. Now e_k is the unique coordinate vector such that $e_k \notin \bar{S}_k$. Since P is compact and $\bar{S}_k$ is closed there is some $\varepsilon > 0$ such that the closed sector $x + \varepsilon e_k + \bar{S}_k$ is disjoint from P. The complement of the open sector $x + \varepsilon e_k + S_k$ is a closed tropical halfspace of the desired kind.   □

Tropical halfspaces implicitly occur in [Cohen et al. 2004]. In particular, their results imply the Tropical Separation Theorem. In fact, a variation of this result already occurs in [Zimmermann 1977]. Another variant of the same is the Tropical Farkas Lemma [Develin and Sturmfels 2004, Proposition 9].

3. Exterior Descriptions of Tropical Polytopes

Throughout this section let $P \subset \mathbb{TP}^d$ be a tropical polytope. Like their ordinary counterparts tropical polytopes also have an exterior description.

LEMMA 3.1. *The tropical polytope P is the intersection of the closed tropical halfspaces that it is contained in.*

PROOF. Let P' be the intersection of all the tropical halfspaces which contain P. Clearly, P' contains P. Suppose that there is a point $x \in P' \setminus P$. Then, by the Tropical Separation Theorem, there is a closed tropical halfspace which contains P but not x. This contradicts the assumption that P' is the intersection of all such tropical halfspaces.   □

Of course, the set of closed tropical halfspaces that contain the given tropical polytope P is partially ordered by inclusion. A closed tropical halfspace is said to be *minimal* with respect to P if it is a minimal element in this partial order.

A key observation in what follows is that the minimal tropical halfspaces come from a small set of candidates only. For a given finite set of points $p_1, \ldots, p_n \in \mathbb{TP}^d$ let the *standard affine hyperplane arrangement* be generated by the ordinary affine hyperplanes

$$p_i + \big\{ (0, \xi_1, \ldots, \xi_d) \in \mathbb{R}^{d+1} \mid \xi_j = 0 \big\} \text{ and } p_i + \big\{ (0, \xi_1, \ldots, \xi_d) \in \mathbb{R}^{d+1} \mid \xi_j = \xi_k \big\}.$$

For an example illustration see Figure 3. A *pseudovertex* of P is a vertex of the standard affine hyperplane arrangement with respect to $\text{Vert}(P)$ which is contained in the boundary ∂P. In [Develin and Sturmfels 2004] our pseudovertices are called the *vertices*.

Here is a special case of [Develin and Sturmfels 2004, Proposition 18].

PROPOSITION 3.2. *The bounded cells of the standard affine hyperplane arrangement are tropical polytopes which are at the same time ordinary convex polytopes.*

PROPOSITION 3.3. *The apex of a closed tropical halfspace that is minimal with respect to P is a pseudovertex of P.*

PROOF. Let $\mathcal{H}^+$ be a closed tropical halfspace, with apex a, which minimally contains P. Suppose that a is not a vertex of the standard affine hyperplane arrangement $\mathfrak{A}$ generated by $\text{Vert}(P)$, but rather a is contained in the relative interior of some cell C of $\mathfrak{A}$ of dimension at least one. Now there is some $\varepsilon > 0$ such that for each point a' in C with $\|a'-a\| < \varepsilon$ the closed tropical halfplane with apex a' and parallel to $\mathcal{H}^+$ still contains P. For each $a' \in \mathcal{H}^+$ the corresponding translate is contained in $\mathcal{H}^+$ and hence $\mathcal{H}^+$ is not minimal. Contradiction.

It remains to show that $a \in P$. Again suppose the contrary. Then, by the Tropical Separation Theorem 2.19, there is a closed halfspace $\mathcal{H}_1^+$ containing P but not a. Now, since $\mathcal{H}^+$ is minimal, $\mathcal{H}_1^+$ is not contained in $\mathcal{H}^+$ and, in particular, $\mathcal{H}_1^+$ is not parallel to $\mathcal{H}^+$. As $a \notin \mathcal{H}_1^+$ the closed tropical halfspace $\mathcal{H}_2^+$ with apex a which is parallel to $\mathcal{H}_1^+$ contains P. We infer that $\mathcal{H}^+ \cap \mathcal{H}_2^+ \subsetneq \mathcal{H}^+$ is a closed tropical halfspace (with apex a) which contains P. This contradicts the minimality of $\mathcal{H}^+$. $\qquad\square$

COROLLARY 3.4. *There are only finitely many closed tropical halfspace which are minimal with respect to P.*

PROOF. The standard affine hyperplane arrangement generated by $\mathrm{Vert}(P)$ is finite, and thus there are only finitely many pseudovertices. Since there are only $2^{d+1} - 2$ closed affine halfspaces with a given apex,[2] the claim now follows from Proposition 3.3. $\qquad\square$

COROLLARY 3.5. *The tropical polytope P is the intersection of the (finitely many) minimal closed tropical halfspaces that it is contained in.*

We can now prove our first main result.

THEOREM 3.6. *The tropical polytopes are exactly the bounded intersections of finitely many tropical halfspaces.*

PROOF. Let P be the bounded intersection of finitely many tropical halfspaces $\mathcal{H}_1^+, \dots, \mathcal{H}_m^+$. Then P is the union of (finitely many) bounded cells of the standard affine hyperplane arrangement corresponding to the apices of $\mathcal{H}_1^+, \dots, \mathcal{H}_m^+$. By Proposition 3.2 each of those cells is the tropical convex hull of its pseudovertices. Since P is tropically convex, this property extends to P, and P is a tropical polytope.

The converse follows from Corollary 3.5. $\qquad\square$

Ordinary polytope theory is combinatorial to a large extent. This is due to the fact that many important properties of an ordinary polytope are encoded into its face lattice. While it is tempting to start a combinatorial theory of tropical polytopes from the results that we obtained so far, this turns out to be quite intricate. Here we give a brief sketch, while a more detailed discussion will be picked up in a forthcoming second paper.

A *boundary slice* of the tropical polytope P is the tropical convex hull of $\mathrm{Vert}(P) \cap \partial\mathcal{H}^+$ where $\mathcal{H}^+$ is a closed tropical halfspace containing P. The

---

[2]The Example 3.9 shows that there may indeed be more than one minimal halfspace with a given apex.

boundary slices are partially ordered by inclusion; we call a maximal element of this finite partially ordered set a *facet* of P. Let $F_1, \ldots, F_m$ be facets of P. Then the set

$$F_1 \sqcap \ldots \sqcap F_m = \mathrm{tconv}(\mathrm{Vert}(P) \cap F_1 \ldots \cap F_m)$$

is called a *proper face* of P provided that $F_1 \sqcap \ldots \sqcap F_m \neq \varnothing$. The sets $\varnothing$ and P are the *nonproper faces*. The faces of a tropical polytope are partially ordered by inclusion, the maximal proper faces being the facets. Note that, by definition, faces of tropical polytopes are again tropical polytopes.

THEOREM 3.7. *The face poset of a tropical polytope is a finite distributive lattice.*

PROOF. We can extend the definition of $\sqcap$ to arbitrary faces, this gives the *meet* operation. There is no choice for the *join* operation then: $G \sqcup H$ is the meet of all facets containing G and H, for arbitrary faces G and H. Denote the set of facets containing the face G by $\Phi(G)$, that is, $G = \sqcap \Phi(G)$. It is immediate from the definitions that $\Phi(G \sqcup H) = \Phi(G) \cap \Phi(H)$ and $\Phi(G \sqcap H) = \Phi(G) \cup \Phi(H)$. Hence the absorption and distributive laws are inherited from the boolean lattice of subsets of the set of all facets. $\square$

In order to simplify some of the discussion below we shall introduce a certain nondegeneracy condition: A set $S \subset \mathbb{TP}^d$ is called *full*, if it is not contained in the boundary of any tropical halfspace. If S is not contained in any tropical hyperplane, then, clearly, S is full. As the Example 3.9 below shows, however, the converse does not hold.

EXAMPLE 3.8. The tropical standard simplex

$$\Delta^2 = \mathrm{tconv}\{(0,1,1),(1,0,1),(1,1,0)\}$$

is not contained in a tropical hyperplane, and hence it is full; see Figure 1, left. It is the intersection of the three minimal closed tropical halfspaces $(1,0,0) + \bar{S}_0$, $(0,1,0) + \bar{S}_1$, and $(0,0,1) + \bar{S}_2$. The tropical line segments $[(0,1,1),(1,0,1)]$, $[(1,0,1),(1,1,0)]$, and $[(1,1,0),(0,1,1)]$ form the facets. The three vertices form the only other proper faces.

EXAMPLE 3.9. The second tropical hypersimplex

$$\Delta_2^2 = \mathrm{tconv}\{(1,0,0),(0,1,0),(0,0,1)\}$$

is a full tropical triangle in the tropical plane $\mathbb{TP}^2$, which is contained in the tropical line $\mathcal{Z}$ corresponding to the zero tropical linear form. There are *six* minimal closed tropical halfspaces: $\bar{S}_0 \cup \bar{S}_1$, $\bar{S}_1 \cup \bar{S}_2$, $\bar{S}_0 \cup \bar{S}_2$, $(1,0,0) + \bar{S}_0$, $(0,1,0) + \bar{S}_1$, and $(0,0,1) + \bar{S}_2$. Note that the three closed tropical halfspaces $\bar{S}_0 \cup \bar{S}_1$, $\bar{S}_1 \cup \bar{S}_2$, and $\bar{S}_0 \cup \bar{S}_2$ share the origin as their apex. The tropical line segments $[(1,0,0),(0,1,0)]$, $[(0,1,0),(0,0,1)]$, and $[(0,0,1),(1,0,0)]$ form the facets. As in the example above, the three vertices form the only other proper faces. The triangle is depicted in Figure 1, right.

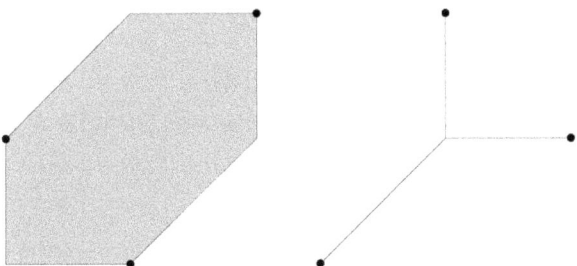

Figure 1. Two full tropical triangles in $\mathbb{TP}^2$: the tropical standard simplex Δ^2 (left) and the second tropical hypersimplex $\Delta_2^2 \subset \mathbb{TP}^2$, which has collinear vertices. Both have the same face lattice as an ordinary triangle.

REMARK 3.10. It is a consequence of Proposition 2.4 and Corollary 2.15 that boundary slices of tropical polytopes are tropical polytopes, but they need not be faces: E.g., the intersection of Δ_2^2 with the boundary of the halfspace

$$(0,0,1/2) + (\bar{S}_1 \cup \bar{S}_2)$$

is the tropical (and at the same time ordinary) line segment $[(0,0,1/2),(0,0,1)]$, which contains the face $(0,0,1)$ and is properly contained in the intersection $[(1,0,0),(0,0,1)] \cap [(0,1,0),(0,0,1)]$ of two facets.

REMARK 3.11. It is easy to see that the face lattice of a tropical n-gon in $\mathbb{TP}^2$ (which is necessarily full) is always the same as that of an ordinary n-gon. This will play a role in the investigation of the algorithmic point of view in Section 5.

Minimal closed tropical halfspaces can be recognized by the intersection of their boundaries with P, provided that P is full.

PROPOSITION 3.12. *Assuming that P is full, let $\mathcal{H}_1^+$ and $\mathcal{H}_2^+$ be closed tropical halfspaces which are minimal with respect to P. If $\partial\mathcal{H}_1^+ \cap P = \partial\mathcal{H}_2^+ \cap P$ then $\mathcal{H}_1^+ = \mathcal{H}_2^+$.*

PROOF. If P is full then it is impossible that for any closed tropical halfspace $\mathcal{H}^+$ containing P the opposite closed halfspace $\mathcal{H}^-$ also contains P.

Let a_1 and a_2 be the respective apices of $\mathcal{H}_1^+$ and $\mathcal{H}_2^+$. By Lemma 3.3 the points a_1 and a_2 are contained in P and hence $a_1 \in \partial\mathcal{H}_2^+$ and $a_2 \in \partial\mathcal{H}_1^+$. In particular, $a_1 \in \mathcal{H}_2^-$. Therefore, the closed tropical halfspace $(a_1 - a_2) + \mathcal{H}_2^+$ with apex a_1 which is parallel to $\mathcal{H}_2^+$ is a closed tropical halfspace containing P. Since $\mathcal{H}_1^+$ is minimal, $\mathcal{H}_1^+ \subseteq (a_1 - a_2) + \mathcal{H}_2^+$ and hence $\mathcal{H}_1^+ \subseteq \mathcal{H}_2^+$. Symmetrically, $\mathcal{H}_2^+ \subseteq \mathcal{H}_1^+$, and the claim follows.   □

REMARK 3.13. The familiarity of the names for the objects defined could inspire the question whether tropical polytopes and, more generally, point configurations in the tropical projective space can be studied in the framework of oriented matroids. However, as the example in Figure 1 shows, the boundaries of tropical halfspaces spanned by a given set of points do *not* form a pseudo-hyperplane arrangement, in general.

4. Tropical Determinants and Their Signs

For algorithmic approaches to ordinary polytopes it is crucial that the incidence of a point with an affine hyperplane can be characterized by the vanishing of a certain determinant expression. Moreover, by evaluating the sign of that same determinant, it is possible to distinguish between the two open affine halfspaces which jointly form the complement of the given affine hyperplane. This section is about the tropical analog.

Let $M = (m_{ij}) \in \mathbb{R}^{(d+1) \times (d+1)}$ be a matrix. Then the *tropical determinant* is defined as

$$\text{tdet}\, M = \bigoplus_{\sigma \in \text{Sym}(d+1)} \bigodot_{i=0}^{d} m_{i,\sigma(i)} = \min\{m_{0,\sigma(0)} + \ldots + m_{d,\sigma(d)} \mid \sigma \in \text{Sym}(d+1)\}.$$

Now M is *tropically singular* if the minimum is attained at least twice, otherwise it is *tropically regular*. Tropical regularity coincides with the strong regularity of a matrix studied by Butkovič [1994]; see also [Burkard and Butkovič 2003].

The following is proved in [Richter-Gebert et al. 2005, Lemma 5.1].

PROPOSITION 4.1. *The matrix M is tropically singular if and only if the $d + 1$ points in $\mathbb{TP}^d$ corresponding to the rows of M are contained in a tropical hyperplane.*

From the definition of tropical singularity it is immediate that M is tropically regular if and only if its transpose M^{tr} is. Hence the above proposition also applies to the columns of M.

The *tropical sign* of tdet M, denoted as tsgn M, is either 0 or ± 1, and it is defined as follows. If M is singular, then tsgn $M = 0$. If M is regular, then there is a unique $\sigma \in \text{Sym}(d + 1)$ such that $m_{0,\sigma(0)} + \ldots + m_{d,\sigma(d)} = \text{tdet}\, M$. We let the tropical sign of M be the sign of this permutation σ. See also [Baccelli et al. 1992, §3.5.1] and Remark 4.9 below.

As it turns out the tropical sign shares some key properties with the (sign of the) ordinary determinant.

PROPOSITION 4.2. *Let $M \in \mathbb{R}^{(d+1) \times (d+1)}$.*

(1) *If M contains twice the same row (column), then tsgn $M = 0$.*
(2) *If the matrix M' is obtained from M by exchanging two rows (columns), we have tsgn $M' = -\text{tsgn}\, M$.*
(3) *tsgn $M^{\text{tr}} = \text{tsgn}\, M$.*

PROOF. The first property follows from Proposition 4.1. The second one is immediate from the definition of the tropical sign. And since permuting the

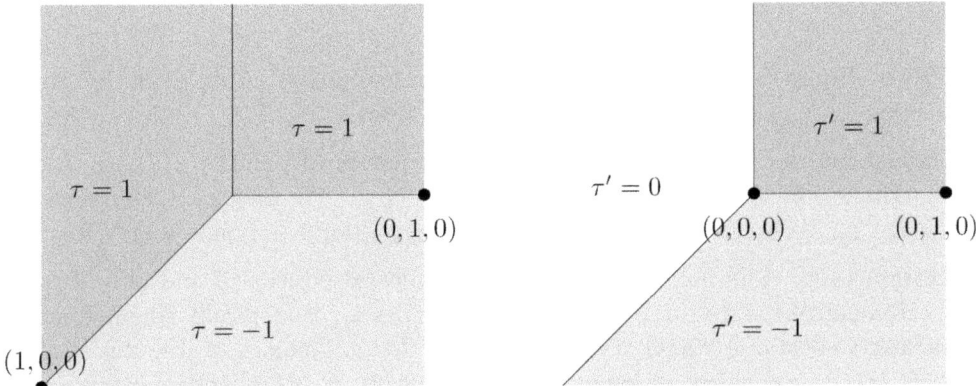

Figure 2. Values of $\tau_{p,q}$ in $\mathbb{TP}^2$ for two pairs of points: nondegenerate case for $\tau = \tau_{(1,0,0),(0,1,0)}$ (left) and degenerate case for $\tau' = \tau_{(0,0,0),(0,1,0)}$ (right). On the tropical line spanned by the two black points the values are zero in both cases.

rows of a matrix is the same as permuting the columns with the inverse, we conclude that $\operatorname{tsgn} M^{\mathrm{tr}} = \operatorname{tsgn} M$. □

While the behavior of the sign of the ordinary determinant under scaling a row (column) by $\lambda \in \mathbb{R}$ depends on the sign of λ, the tropical sign is invariant under this operation. Given $v_0, \ldots, v_d \in \mathbb{R}^{d+1}$ we write $(v_0, \ldots, v_d)$ for the $(d+1) \times (d+1)$-matrix with row vectors $v_0, \ldots, v_d$.

LEMMA 4.3. *For $v_0, \ldots, v_d \in \mathbb{R}^{d+1}$ and $\lambda_0, \ldots, \lambda_d \in \mathbb{R}$ we have $\operatorname{tsgn}(\lambda_0 \odot v_0, \ldots, \lambda_d \odot v_d) = \operatorname{tsgn}(v_0, \ldots, v_d)$.*

In fact, tsgn is a function on $(d+1)$-tuples of points in the tropical projective space $\mathbb{TP}^d$. For given $p_1, \ldots, p_d$, consider the function

$$\tau_{p_1, \ldots, p_d} : \mathbb{TP}^d \to \{-1, 0, 1\} : x \mapsto \operatorname{tsgn}(x, p_1, \ldots, p_d).$$

Note that we do allow the case where the points $p_1, \ldots, p_d$ are not in *general position*, that is, they may be contained in more than one tropical hyperplane; see the example in Figure 4, right.

EXAMPLE 4.4. Consider the real $(d+1) \times (d+1)$-matrix formed of the vertices $-e_0, \ldots, -e_d$ of the tropical standard simplex Δ^d. Then we have

$$\operatorname{tdet}(-e_0, \ldots, -e_d) = -d,$$

and the matrix is regular: The unique minimum is attained for the identity permutation, hence

$$\operatorname{tsgn}(-e_0, \ldots, -e_d) = 1,$$

or equivalently, $\tau_{-e_1, \ldots, -e_d}(-e_0) = 1$.

PROPOSITION 4.5. *The function $\tau_{p_{\bullet},\ldots,p_d}$ is constant on each connected component of the set $\mathbb{TP} \setminus \tau_{p_{\bullet},\ldots,p_d}^{-1}(0)$.*

PROOF. Equip the set $\{-1, 0, 1\}$ with the discrete topology. Away from zero the function $\tau_{p_{\bullet},\ldots,p_d}$ is continuous, and the result follows.   □

Throughout the following we keep a fixed sequence of points $p_1, \ldots, p_d$, and we write π_{ij} for the j-th canonical coordinate of p_i. We frequently abbreviate $\tau = \tau_{p_{\bullet},\ldots,p_d}$ as well as $p(\sigma) = \pi_{1,\sigma(1)} + \ldots + \pi_{d,\sigma(d)}$ for $\sigma \in \mathrm{Sym}(d+1)$.

REMARK 4.6. The points $p_1, \ldots, p_d$ are in general position if and only if no $d \times d$-minor of the $d \times (d+1)$-matrix with entries π_{ij} is tropically singular; see [Richter-Gebert et al. 2005, Theorem 5.3]. In the terminology of [Develin et al. 2005] this is equivalent to saying that the matrix (π_{ij}) has maximal tropical rank d.

THEOREM 4.7. *The set $\{x \in \mathbb{TP}^d \mid \tau(x) = 1\}$ is either empty or the union of at most d open sectors of a fixed tropical hyperplane. Conversely, each such union of open sectors arises in this way.*

PROOF. We can assume that $\tau(x) = 1$ for some $x \in \mathbb{TP}^d$, since otherwise there is nothing left to prove. From Proposition 4.1 we know that the $d + 1$ points $x, p_1, \ldots, p_d$ are not contained in a tropical hyperplane, and hence they are the vertices of a full tropical d-simplex $\Delta = \mathrm{tconv}\{x, p_1, \ldots, p_d\}$. Consider the facet $F = \mathrm{tconv}\{p_1, \ldots, p_d\}$, and let $\mathcal{H}^+$ be the unique corresponding closed tropical halfspace which is minimal with respect to Δ and for which we have $\partial\mathcal{H}^+ \cap \Delta = F$. Let a be the apex of $\mathcal{H}^+$, and let $a + S_k$ be the open sector containing x. By construction $a + S_k \subset \mathcal{H}^+$.

Assume that $\tau(y) \neq \tau(x)$ for some $y \in a + S_k$. Then there exists a point $z \in [x, y]$ with $\tau(z) = 0$. Since sectors are tropically convex, $z \in a + S_k$. By Proposition 4.1 there exists a tropical hyperplane $\mathcal{K}$ which contains the points $z, p_1, \ldots, p_d$. Let $\mathcal{K}^+$ be the minimal closed tropical halfspace of the tropical hyperplane $\mathcal{K}$ containing $x, p_1, \ldots, p_d$. As $\mathcal{H}^+$ and $\mathcal{K}^+$ are both minimal with respect to the tropical simplex Δ, the Proposition 3.12 says that $\mathcal{H}^+ = \mathcal{K}^+$, and in particular, $a + S_k \ni z$ is an open sector of $\mathcal{K}$. The latter contradicts, however, $z \in \mathcal{K}$.

For the converse, it surely suffices to consider the tropical hyperplane $\mathcal{Z}$ corresponding to the zero tropical linear form, since otherwise we can translate. We have to prove that for each set $K \subset [d+1]$ with $1 \leq \#K \leq d$ there is a set of points $u_1, \ldots, u_d \in \mathcal{Z}$ such that

$$\{x \in \mathbb{TP}^d \mid \tau_{u_{\bullet},\ldots,u_d}(x) = 1\} = \bigcup \{S_k \mid k \in K\}.$$

More specifically, we will even show that, for arbitrary K, those d points can be chosen among the $\binom{d+1}{2}$ vertices of the second tropical hypersimplex Δ_2^d; see Example 2.12. Since the symmetric group $\mathrm{Sym}(d+1)$ acts on the set of open sectors of $\mathcal{Z}$ as well as on the set $\mathrm{Vert}(\Delta_2^d)$, it suffices to consider one set of sectors for each possible cardinality $1, \ldots, d$. Let us first consider the case where d is odd and $K = \{0, 2, 4, \ldots, d-1\}$ is the set of even indices, which has cardinality $(d+1)/2$. We set

$$q_i = \begin{cases} -e_i - e_{i+1} & \text{for } i < d \\ -e_0 - e_d & \text{for } i = d. \end{cases}$$

If we want to evaluate $\tau_{q_1, \ldots, q_d}(x)$ for some point $x \in \mathbb{TP}^d \setminus \mathcal{Z}$ with canonical coordinates $(\xi_0, \ldots, \xi_d)$, we compute the tropical determinant and the tropical sign of $\mathrm{tdet}(x, q_1, \ldots, q_d)$, which in canonical row coordinates looks as follows:

$$Q_d = \begin{pmatrix} \xi_0 & \xi_1 & \xi_2 & \xi_3 & \xi_4 & \cdots & \xi_d \\ 1 & 0 & 0 & 1 & 1 & \cdots & 1 \\ 1 & 1 & 0 & 0 & 1 & \cdots & 1 \\ \vdots & \vdots & \ddots & \ddots & \ddots & \ddots & \vdots \\ 1 & 1 & \cdots & 1 & 0 & 0 & 1 \\ 1 & 1 & \cdots & 1 & 1 & 0 & 0 \\ 0 & 1 & \cdots & 1 & 1 & 1 & 0 \end{pmatrix}.$$

Since $x \notin \mathcal{Z}$, there is a unique permutation $\sigma_x \in \mathrm{Sym}(d+1)$ such that $\mathrm{tdet}(Q) = \xi_{\sigma_x(0)} + q(\sigma_x)$. We can verify that $\mathrm{tdet}(Q) = 0$ in all cases and that

$$\sigma_x = \begin{cases} (0) & \text{if } x \in S_0, \\ (0 \ k \ k+1 \ \cdots \ d) & \text{if } x \in S_k \text{ for } k > 0. \end{cases}$$

Here we make use of the common cycle notation for permutations, and (0) denotes the identity. For $k > 0$ this means that σ_x is a $(d+2-k)$-cycle, which is an even permutation if and only if k is even, since d is odd. We infer that $\tau_{q_1, \ldots, q_d}(x) = 1$ if and only if $x \in S_k$ for k even.

We now discuss the case where $\#K \geq (d+1)/2$ and d is arbitrary. As in the case above, by symmetry, we can assume that $K = \{0, 2, 4, \ldots, 2(l-1), 2l-1, 2l, \ldots, d\}$ for some $l < \lfloor d/2 \rfloor$. We define

$$q_i' = -e_0 - e_i,$$

for all $i \geq 2l+1$, and we are concerned with the matrix $(x, q_1, \ldots, q_l, q_{l+1}', \ldots, q_d')$, which, in canonical row coordinates, looks like this:

$$Q_d^l = \begin{pmatrix} \xi_0 & \xi_1 & \xi_2 & \xi_3 & \xi_4 & \cdots & \xi_{2l-1} & \xi_{2l} & \cdots & \xi_d \\ 1 & 0 & 0 & 1 & 1 & \cdots & 1 & 1 & \cdots & 1 \\ 1 & 1 & 0 & 0 & 1 & \cdots & 1 & 1 & \cdots & 1 \\ \vdots & \vdots & \ddots & \ddots & \ddots & \ddots & \vdots & \vdots & \ddots & \vdots \\ 1 & 1 & \cdots & 1 & 0 & 0 & 1 & 1 & \cdots & 1 \\ 1 & 1 & \cdots & 1 & 1 & 0 & 0 & 1 & \cdots & 1 \\ 0 & 1 & \cdots & 1 & 1 & 1 & 0 & 1 & \cdots & 1 \\ \vdots & \vdots & \ddots & \vdots & \vdots & \vdots & \ddots & \ddots & \ddots & \vdots \\ 0 & 1 & \cdots & 1 & 1 & 1 & \cdots & 1 & 0 & 1 \\ 0 & 1 & \cdots & 1 & 1 & 1 & \cdots & 1 & 1 & 0 \end{pmatrix}.$$

Note that the upper left $2l \times 2l$-submatrix is exactly Q_l. Hence the same reasoning now yields

$$\sigma_x = \begin{cases} (0) & \text{if } x \in S_0, \\ (0 \; k \; k+1 \; \cdots \; 2l-1) & \text{if } x \in S_k \text{ for } k > 0, \end{cases}$$

and σ_x is an even permutation if and only if k is even or $k > 2l$.

Scrutinizing the matrices Q_d and Q_d^l yields that none of their $d \times (d+1)$-submatrices consisting of all rows but the first contains a tropically singular minor. Equivalently, the points $q_1, \ldots, q_d$ as well as the points $q_1, \ldots, q_l, q'_{l+1}, \ldots, q'_d$ are in general position. But then the set

$$\left\{ x \in \mathbb{TP}^d \mid \tau_{q_1, \ldots, q_l, q'_{l+1}, \ldots, q'_d}(x) = -1 \right\}$$

is just the union of the sectors in the complement $[d+1] \setminus K$, and since further, according to Proposition 4.2, $\tau_{q_1, \ldots, q_d} = -\tau_{q_2, q_1, q_3, \ldots, q_l, q'_{l+1}, \ldots, q'_d}$, the argument given so far already covers the remaining case of $\#K < (d+1)/2$. This completes the proof. $\square$

Now, for the fixed set of points $p_1, \ldots, p_d$, we can glue together the connected components of $\mathbb{TP}^d \setminus \tau_{p_1, \ldots, p_d}^{-1}(0)$ into two (if $\tau_{p_1, \ldots, p_d} \not\equiv 0$) large chunks according to their tropical sign: To this end we define the *closure* of the function $\tau_{p_1, \ldots, p_d}$ as follows. Let $\bar{\tau}_{p_1, \ldots, p_d}(x) = \varepsilon$ if there is a neighborhood U of x such that $\tau_{p_1, \ldots, p_d}$ restricted to $U \setminus \tau_{p_1, \ldots, p_d}^{-1}(0)$ is identically ε; otherwise let $\bar{\tau}_{p_1, \ldots, p_d}(x) = 0$. Clearly, if $\tau_{p_1, \ldots, p_d}(x) \neq 0$ then $\bar{\tau}_{p_1, \ldots, p_d}(x) = \tau_{p_1, \ldots, p_d}(x)$.

Theorem 4.7 then implies the following.

COROLLARY 4.8. *The set $\left\{ x \in \mathbb{TP}^d \mid \bar{\tau}(x) = 1 \right\}$ is empty or a closed tropical halfspace. Conversely, each closed tropical halfspace arises in this way.*

REMARK 4.9. One can show that $\bar{\tau}_{p_1, \ldots, p_d}(x) = 1$ if and only if all optimal permutations, that is, all $\sigma \in \mathrm{Sym}(d+1)$ with $\mathrm{tdet}(x, p_1, \ldots, p_d) = \xi_{\sigma(0)} + p(\sigma)$ are even. In this sense our function $\bar{\tau}$ captures the sign of the determinant in the symmetrized min-plus-algebra as defined in [Baccelli et al. 1992, §3.5.1].

COROLLARY 4.10. *For each point $x = (\xi_0, \ldots, \xi_d) \in \mathbb{TP}^d$ with $\bar{\tau}_{p_1, \ldots, p_d}(x) = 0$ there are two permutations σ and σ' of opposite sign such that*

$$\mathrm{tdet}(x, p_1, \ldots, p_d) = \xi_{\sigma(0)} + p(\sigma) = \xi_{\sigma'(0)} + p(\sigma').$$

5. Convex Hull Algorithms in $\mathbb{TP}^2$

For points in the ordinary Euclidean plane the known algorithms can be phrased easily in terms of sign evaluations of certain determinants. It turns out that the results of the previous sections can be used to "tropify" many ordinary convex hull algorithms.

In this section we do not use canonical coordinates for points in the tropical projective space, but rather we normalize by setting the first coordinate to zero.

This way the description of the algorithms can be made in the ordinary affine geometry language more easily. In particular, a point in $(0, \xi_1, \xi_2) \in \mathbb{TP}^2$ is determined by its x-*coordinate* ξ_1 and its y-*coordinate* ξ_2. We hope that this helps to see the strong relationship between the ordinary convex hull problem in $\mathbb{R}^2$ and the tropical convex hull problem in $\mathbb{TP}^2$. Moreover, this way it may be easier to interpret the illustrations.

Consider a set $S = \{p_1, \dots, p_n\} \subset \mathbb{TP}^2$. Let bottom$(S)$ be the lowest point (least y-coordinate) of S with ties broken by taking the rightmost (highest x-coordinate) one. Similarly, let right(S) be the rightmost one with ties broken by taking the highest, top(S) the highest with ties broken by taking the rightmost, and left(S) the leftmost one with ties broken by taking the highest. Clearly, some of the four points defined may coincide. If a set of points is in *general position*, that is, for any two points of the input their three rays are pairwise distinct, then there are unique points with minimum and maximum x- and y-coordinate respectively. In this case there are no ties.

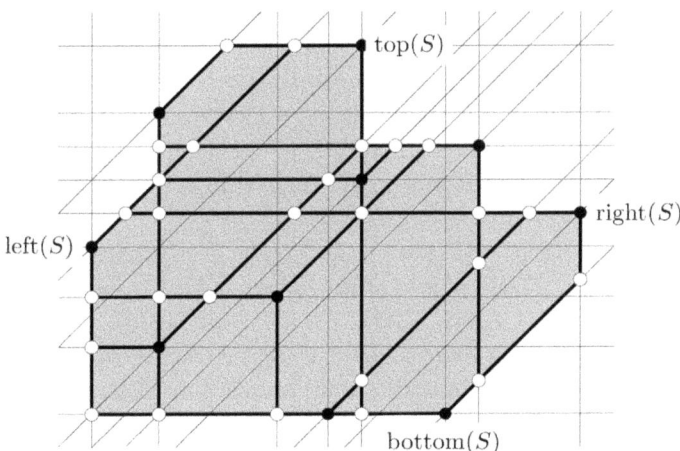

Figure 3. Standard affine line arrangement generated by a set of points $S \subset \mathbb{TP}^2$, displayed in black. The white points are the pseudovertices on tropical line segments between any two points. Additionally, the tropical convex hull is marked.

LEMMA 5.1. *The points* bottom(S), right(S), top(S), left(S) *are vertices of the tropical polygon* tconv(S). *Moreover,* [bottom(S), left(S)] *is a facet provided that* bottom$(S) \neq$ left(S).

PROOF. By definition, the closed sector bottom$(S) + \bar{S}_1$ does not contain any point of S other than bottom(S). This certifies that, indeed, the point bottom(S) is a vertex because of Propositions 2.5 and 2.9. We omit the proofs of the remaining statements, which are similar. $\square$

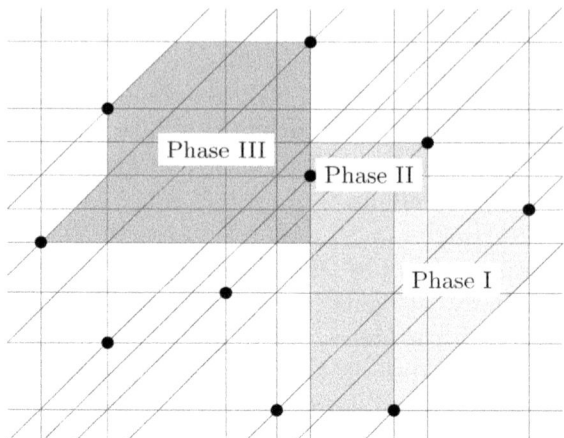

Figure 4. Three phases of Algorithm A.

Note that due to the special shape of the tropical lines, it is important how to break the ties. For instance, if two points have the same lowest y-coordinate, then the left one is also on the boundary, but not necessarily a vertex; see Figure 3.

A key difference between tropical versus ordinary polytopes is that the former only have few directions for their (half-)edges. This can be exploited to produce convex hull algorithms which do not have a directly corresponding ordinary version.

Through each point $p = (0, \xi, \eta) \in \mathbb{TP}^2$ there is a unique tropical line consisting of three ordinary rays emanating from p: We respectively call the sets $\{(0, \xi + \lambda, \eta) \mid \lambda \geq 0\}$, $\{(0, \xi, \eta + \lambda) \mid \lambda \geq 0\}$, and $\{(0, \xi - \lambda, \eta - \lambda) \mid \lambda \geq 0\}$ the *horizontal*, *vertical*, and *skew* ray through x. If we have a second point $p' = (0, \xi', \eta')$ then we can compare them according to the relative positions of their three rays. This way there is a natural notion of left and right, but there are two notions of above and below, which we wish to distinguish carefully: p' is *y-above* p if $\eta' > \eta$, and it is *skew-above* if $\eta' - \xi' > \eta - \xi$.

The introduction of the sign of the tropical determinant now clearly allows to take most ordinary 2-dimensional convex hull algorithms and produce a "tropified" version with little effort. For instance a suitable tropical version of Graham's scan provides a worst-case optimal $O(n \log n)$-algorithm. We omit the details since we describe a different algorithm with the same complexity. The commonly expected output of an ordinary convex hull algorithm in two dimensions is the list of vertices in counter-clockwise order. As the results in the previous section imply that the combinatorics of tropical polygons in $\mathbb{TP}^2$ does not differ from the ordinary, we adopt this output strategy.

The data structures for the Algorithm A below is are three doubly-linked lists L, Y, B such that each input point occurs exactly once in each list. It is important that all three lists can be accessed at their front and back with constant cost.

In order to obtain a concise description we assume that the input set S is in general position. For input not in general position the notions of left, right, above, and below have to be adapted as above. The complexity of the algorithm remains the same.

Input : $S \subset \mathbb{TP}^2$ finite
Output: list of vertices of tconv(S) in counter-clockwise order

sort S from left to right and store the result in list L
sort S from y-below to y-above and store the result in list Y
sort S from skew-below to skew-above and store the result in list B

$H \leftarrow$ empty list; $v \leftarrow$ front(Y); $w \leftarrow$ next(v, Y)
while w *y-below* back(L) **do**
   if w *skew-below* v **then**
      $v \leftarrow w$; append v to H
   $w \leftarrow$ next(w, Y)
$v \leftarrow$ back(L); append v to H

$w \leftarrow$ previous(v, L)
while w *to the right of* back(Y) **do**
   if w *y-above* v **then**
      $v \leftarrow w$; append v to H
   $w \leftarrow$ previous(w, L)
$v \leftarrow$ back(Y); append v to H;

$w \leftarrow$ back(B)
while w *skew-above* front(L) **do**
   if w *to the left of* v **then**
      $v \leftarrow w$; append v to H
   $w \leftarrow$ previous(w, B)
$v \leftarrow$ front(L); append v to H

if $v \neq$ front(Y) **then**
   append front(Y) to H

return H

Algorithm A: Triple sorting algorithm.

PROPOSITION 5.2. *The Algorithm* A *correctly computes the vertices of the tropical convex hull in counter-clockwise order.*

PROOF. The algorithm has an initialization and three phases, where each phase corresponds to one of the three while-loops; for an illustration see Figure 4. In the first phase all the vertices between bottom(S) and right(S) are enumerated, in the second phase the vertices between right(S) and top(S), and in the third phase the vertices between top(S) and left(S).

By Lemma 5.1 the point front(Y) = bottom(S) is a vertex. Throughout the algorithm the following invariant in maintained: v is the last vertex found, and w is an input point not yet processed, which is a candidate for the next vertex in counter-clockwise order. We have a closer look at Phase I, the remaining being similar. If w is a vertex between bottom(S) and right(S) then it will be y-above of v, hence we process the points according to their order in the sorted list Y. However, none of those vertices can be y-above back(L) = right(S), therefore the stop condition. Under these conditions w is a vertex if and only if w is skew-below the tropical line segment $[v, \text{right}(S)]$.   □

The worst-case complexity of the algorithm based on triple sorting is $O(n \log n)$. If, however, the points are uniformly distributed at random, say, in the unit square, then by applying Bucket Sort, we can sort the input in an expected number of $O(n)$ steps; see [Cormen et al. 2001].

If only few of the input points are actually vertices of the convex hull, then it is easy to beat an $O(n \log n)$ algorithm. For ordinary planar convex hulls the Jarvis' march algorithm is known as an easy-to-describe method which is output-sensitive in this sense. We sketch a "tropified" version, we will be instrumental later. Its complexity is $O(nh)$, where h is the number of vertices.

Input : $S \subset \mathbb{TP}^2$ finite
Output: list of vertices of tconv(S) in counter-clockwise order

$v_0 \leftarrow$ bottom(S); $v \leftarrow v_0$; $H \leftarrow$ empty list
repeat
    $w \leftarrow$ some point in S
    for $p \in S \setminus \{w\}$ **do**
        if $\bar{\tau}_{v,w}(p) = -1$ *or* $(\bar{\tau}_{v,w}(p) = 0$ *and* $\|p - v\| > \|w - v\|)$ **then**
            ∟ $w \leftarrow p$
    $v \leftarrow w$; $S \leftarrow S \setminus \{v\}$; append v to H
until $v = v_0$
return H

Algorithm B: Tropical Jarvis' march algorithm.

In the ordinary case, Chan [1996] has given a worst-case optimal $O(n \log h)$ algorithm, based on a combination of Jarvis' march and a divide-and-conquer approach. We sketch how the same ideas can be used to obtain an $O(n \log h)$ convex hull algorithm in $\mathbb{TP}^2$. If we split the input into $\lceil n/m \rceil$ parts of size at most m, then we can use our $O(n \log n)$ algorithm based on triple sorting to compute the $\lceil n/m \rceil$ hulls in $O((n/m)(m \log m)) = O(n \log m)$ time. Now we use Jarvis' march to combine the $\lceil n/m \rceil$ tropical convex hulls into one. The crucial observation is that each vertex of the big tropical polygon is also a vertex of one of the $\lceil n/m \rceil$ small tropical polygons. Therefore, in order to compute the next vertex of the big tropical polygon in the counter clockwise order, we first compute the tropical tangent through the current vertex to each of the small tropical polygons. In each small tropical polygon this can be done by binary searching the vertices in their cyclic order; this requires $O(\log m)$ steps per small tropical polygon and per vertex of the big tropical polygon. Summing up this gives a total of $O(n \log m + h((n/m) \log m)) = O(n(1 + h/m) \log m)$ operations. That is to say, if we could know the number of vertices of the big tropical polygon beforehand, then we could split the input into portions of size at most h, thus arriving at a complexity bound of $O(n \log h)$. But this can be achieved by repeated guessing as has been suggested by Chazelle and Matoušek [1995].

We summarize our findings in the following result, which is identical to the ordinary case. Note that, as in the ordinary case, one has an $\Omega(n \log n)$ worst case lower bound for the complexity of the two-dimensional tropical convex hull problem which comes from sorting. In this sense our output-sensitive algorithm is optimal.

THEOREM 5.3. *The complexity of the problem to compute the tropical convex hull of n points in $\mathbb{TP}^2$ with h vertices is as follows*:

(1) *There is an output-sensitive $O(n \log h)$-algorithm.*
(2) *There is an algorithm which requires expected linear time for random input.*

6. Concluding Remarks

One of the main messages of this paper is that, with suitably chosen definitions, it is possible to build up a theory of tropical polytopes quite similar to the one for ordinary convex polytopes. But, of course, very many items are still missing. We list a few open questions, and the reader will easily find more.

(1) How are the face lattices of tropical polytopes related to the face lattices of ordinary convex polytopes? In particular, how do the face lattices of tropical polytopes in $\mathbb{TP}^3$ look alike?

(2) It is known [Develin and Sturmfels 2004, Lemma 22] that the tropical convex hull of n points in $\mathbb{TP}^d$ is the bounded subcomplex of some $(n + d)$-dimensional unbounded ordinary convex polyhedron (defined in terms in inequalities). Hence the tropical convex hull problem can be reduced to solving a (dual) ordinary convex hull problem, followed by a search of the bounded faces in the face lattice. The question is: How does an intrinsic tropical convex hull algorithm look alike that works in arbitrary dimensions? Here *intrinsic* means that the algorithm should not take a detour via that face lattice computation in the realm of ordinary convex polytopes. While the complexity status of the ordinary convex

hull problem is notoriously open (output-sensitive with varying dimension) it is well conceivable that the tropical analog is, in fact, easier. An indication may be the easy to check certificate in Proposition 2.9 which leads to a simple and fast algorithm for discarding the nonvertices among the input points, a task which is polynomially solvable in the ordinary case, but which requires an LP-type oracle.

(3) What is the proper definition of a tropical triangulation? Such a definition would say that a tropical triangulation should be a subdivision into tropical simplices which meet properly. The precise notion of meeting is subtle, however. While it is obvious that the standard intersection as sets does not do any good, the more refined way by extending the ⊓ operation also leads to surprising examples. A meaningful definition of a tropical triangulation should lead to one solution of the previous problem.

(4) Can the dimension of an arbitrary tropical polytope, which is not necessarily full, computed in polynomial time? Here *dimension* means the same as *tropical rank*. In fact, this is Question (1) at the end of the paper [Develin et al. 2005].

(5) As mentioned in Remark 3.13 point configurations in the tropical projective space do not generate an oriented matroid in the usual way. But does there exist a more general notion than an oriented matroid which encompasses the tropical case?

Acknowledgments

I am grateful to Julian Pfeifle and Francisco Santos for many helpful conversations on the contents of this paper and, especially, for suggesting better names for some of the notions defined than I could have come up with. I am indebted to Bernd Sturmfels for requiring the correction of a few minor errors. Finally, I would like to thank the anonymous referee who helped alleviate my ignorance of previously published results on max-plus-algebras.

References

[Akian et al. 2005] M. Akian, S. Gaubert, and V. Kolokoltsov, "Set coverings and invertibility of functional Galois connections", in *Idempotent mathematics and mathematical physics*, edited by G. L. Litvinov and V. P. Maslov, Contemp. Math. **377**, Amer. Math. Soc., Providence, RI, 2005. Available at arxiv.org/math.FA/0403441. to appear.

[Baccelli et al. 1992] F. L. Baccelli, G. Cohen, G. J. Olsder, and J.-P. Quadrat, *Synchronization and linearity: An algebra for discrete event systems*, Wiley, Chichester, 1992.

[Burkard and Butkovič 2003] R. E. Burkard and P. Butkovič, "Max algebra and the linear assignment problem", *Math. Program.* **98**:1-3, Ser. B (2003), 415–429. Integer programming (Pittsburgh, PA, 2002).

[Butkovič 1994] P. Butkovič, "Strong regularity of matrices—a survey of results", *Discrete Appl. Math.* **48**:1 (1994), 45–68.

[Chan 1996] T. M. Chan, "Optimal output-sensitive convex hull algorithms in two and three dimensions", *Discrete Comput. Geom.* **16**:4 (1996), 361–368. Eleventh Annual Symposium on Computational Geometry (Vancouver, BC, 1995).

[Chazelle and Matoušek 1995] B. Chazelle and J. Matoušek, "Derandomizing an output-sensitive convex hull algorithm in three dimensions", *Comput. Geom.* **5**:1 (1995), 27–32.

[Cohen et al. 2004] G. Cohen, S. Gaubert, and J.-P. Quadrat, "Duality and separation theorems in idempotent semimodules", *Linear Algebra Appl.* **379** (2004), 395–422. Tenth Conference of the International Linear Algebra Society.

[Cormen et al. 2001] T. H. Cormen, C. E. Leiserson, R. L. Rivest, and C. Stein, *Introduction to algorithms*, Second ed., MIT Press, Cambridge, MA, 2001.

[Develin and Sturmfels 2004] M. Develin and B. Sturmfels, "Tropical convexity", *Doc. Math.* **9** (2004), 1–27. Erratum, ibid., pp. 205–206.

[Develin et al. 2005] M. Develin, B. Sturmfels, and F. Santos, "On the rank of a tropical matrix", pp. 213–242 in *Combinatorial and computational geometry*, edited by J. E. Goodman et al., Math. Sci. Res. Inst. Publ. **52**, Cambridge U. Press, New York, 2005.

[Dress et al. 2002] A. Dress, K. T. Huber, and V. Moulton, "An explicit computation of the injective hull of certain finite metric spaces in terms of their associated Buneman complex", *Adv. Math.* **168**:1 (2002), 1–28.

[Helbig 1988] S. Helbig, "On Carathéodory's and Kreĭn-Milman's theorems in fully ordered groups", *Comment. Math. Univ. Carolin.* **29**:1 (1988), 157–167.

[Isbell 1964] J. R. Isbell, "Six theorems about injective metric spaces", *Comment. Math. Helv.* **39** (1964), 65–76.

[Richter-Gebert et al. 2005] J. Richter-Gebert, B. Sturmfels, and T. Theobald, "First steps in tropical geometry", in *Idempotent mathematics and mathematical physics*, edited by G. L. Litvinov and V. P. Maslov, Contemp. Math. **377**, Amer. Math. Soc., Providence, RI, 2005. Available at arxiv.org/math.FA/0403441. To appear.

[Yoeli 1961] M. Yoeli, "A note on a generalization of Boolean matrix theory", *Amer. Math. Monthly* **68** (1961), 552–557.

[Zimmermann 1977] K. Zimmermann, "A general separation theorem in extremal algebras", *Ekonom.-Mat. Obzor* **13**:2 (1977), 179–201.

MICHAEL JOSWIG
INSTITUT FÜR MATHEMATIK, MA 6-2
TECHNISCHE UNIVERSITÄT BERLIN
10623 BERLIN, GERMANY
joswig@math.tu-berlin.de

Combinatorial and Computational Geometry
MSRI Publications
Volume **52**, 2005

Two Proofs for Sylvester's Problem
Using an Allowable Sequence of Permutations

HAGIT LAST

ABSTRACT. The famous Sylvester's problem is: Given finitely many non-collinear points in the plane, do they always span a line that contains precisely two of the points? The answer is yes, as was first shown by Gallai in 1944. Since then, many other proofs and generalizations of the problem appeared. We present two new proofs of Gallai's result, using the powerful method of allowable sequences.

1. Introduction

Sylvester [1893] raised the following problem: Given finitely many noncollinear points in the plane, do they always span a simple line (that is, a line that contains precisely two of the points)? The answer is yes, as was first shown by Gallai [1944].

By duality, the former question is equivalent to the question: Given finitely many straight lines in the plane, not all passing through the same point, do they always determine a simple intersection point (a point that lies on precisely two of the lines)?

A natural generalization is to find a lower bound on the number of simple lines (or simple points, in the dual version). The dual version of this question can be generalized to pseudolines. The best lower bound [Csima and Sawyer 1993] states that an arrangement of n pseudolines in the plane determines at least $6n/13$ simple points. The conjecture [Borwein and Moser 1990] is that there are at least $n/2$ simple points for $n \neq 7, 13$. For the history of Sylvester's problem, with its many proofs and generalizations, see [Borwein and Moser 1990; Nilakantan 2005].

This paper presents two new proofs of Gallai's result using allowable sequences. A proof of Gallai's result using allowable sequences was given recently by Nilakantan [2005], but it differs from the two given here.

The notion of *allowable sequences* was introduced by Goodman and Pollack [1980]. It has proved to be a very effective tool in discrete and computational geometry; for a broad discussion see [Goodman and Pollack 1993]. Here is a short description of the notion.

Let S be a set of n points in the plane, let L be the set of the lines spanned by S, and let $\{k_1, k_2, \ldots, k_m\}$ be the m different slopes of the lines according to a fixed coordinate system. We choose a directed line l in the plane with a point P on it, such that l does not contain any point of S and is not orthogonal to any line in L.

Here is the construction of $\mathcal{A}_{l,P}(S)$, the allowable sequence of permutations of a point set S, according to the directed line l and the point P: We label the points of S according to their orthogonal projection on l and we get the first permutation $\pi_0 = 1, \ldots, n$. Let l rotate counterclockwise around P by $180°$ and look at the orthogonal projections of the labeled points of S on l as it rotates. A new permutation arises whenever l passes through a direction orthogonal to one of the slopes $k_1, k_2, \ldots, k_m$. It follows that along the course of this rotation, beside π_0, we will get m different permutations: $\pi_1, \ldots, \pi_m$. Define $\mathcal{A}_{l,P}(S) = \{\pi_0, \pi_1, \pi_2, \ldots, \pi_m\}$.

For each $1 \le i \le m$, whenever l passes through a direction orthogonal to k_i, the new permutation that arises differs from the previous one by reversing the order of the consecutive elements whose corresponding points of S lie on a line of slope k_i. Such reversed consecutive elements are called *a reversed substring*. If t lines in L have a slope equal to k_i, the permutation that corresponds to k_i has t disjoint reversed substrings. A reversed substring of length 2 is called *a simple switch*. A simple switch corresponds to a simple line.

Three important properties of $\mathcal{A}_{l,P}(S)$ are:

1. $\mathcal{A}_{l,P}(S)$ is a sequence of permutations of the elements $\{1, 2, \ldots, n\}$, where n is the cardinality of S.

2. The first permutation is $\pi_0 = 1, \ldots, n-1, n$, and the last is $\pi_m = n, n - 1, \ldots, 1$. Here m is the number of different slopes of the lines spanned by S. If the points of S are not collinear, then $m > 1$ (actually $m \ge n - 1$, as was proved in [Scott 1970]).

3. In the course of the sequence of permutations, every pair $i < j$ switches exactly once and so each permutation differs from the previous one by reversing at least one increasing substring. Only increasing substrings are reversed.

For example, if $\pi_i = 1, 7, 2, 4, 6, 3, 5$, then $N_1 = 1, 7$, $N_2 = 2, 4$, $N_3 = 2, 4, 6$, and $N_4 = 3, 5$ are its increasing substrings, and so π_{i+1} is obtained from π_i by reversing the order of one or more of these substrings.

For the convenience of writing the proofs in Section 2, we would like to assume that in each step only one increasing substring is reversed. We can arrange this by replacing each permutation that contains t reversed substrings by t permutations, as we reverse a single substring at a time. The length of the new sequence of permutations, $\mathcal{A}$, is the cardinality of L and satisfies the condition that each permutation differs from the previous one by reversing a single increasing substring.

2. The Proofs

Let S be a set of n noncollinear points in the plane. We will show the existence of a simple spanned line by proving that $\mathcal{A}$ contains a permutation with a simple switch. Assume, for a contradiction, that each reversed substring has length at least 3.

Since S is a set of noncollinear points, then $\mathcal{A}$ has length greater than 2, with $\pi_0 = 1, 2, \ldots, n$ and $\pi_m = n, n-1, \ldots, 1$ $(m > 1)$.

For $1 \leq r \leq m$, denote by J_r the reversed substring of π_r and denote by I_r the increasing substring of π_{r-1} which is reversed at π_r. J_r and I_r consist of the same set of elements, in J_r the elements are in decreasing order and in I_r they are in increasing order. For example, if $\pi_1 = 1, 2, 5, 4, 3$, $\pi_2 = 5, 2, 1, 4, 3$, then, $I_2 = 1, 2, 5$, $J_2 = 5, 2, 1$.

For $J_r = a_1, a_2, \ldots, a_{k-1}, a_k$, we will refer to $a_2, \ldots, a_{k-1}$ as its *internal elements*. By our assumption, every I_r as well as every J_r has an internal element.

PROOF 1. We show that an internal element of a reversed substring cannot change its location before a simple switch occurs.

For every $0 \leq k \leq m$ and every element $1 \leq a \leq n$, denote by $T_k(a)$ the location of the element a in π_k. For example, $T_m(n-1) = 2$.

If $T_k(a) \neq T_{k-1}(a)$, we say that J_K *changed the location of the element a*. If $T_k(a) > T_{k-1}(a)$, we say that *a moves to the right at π_k*.

A reversed substring, J_r, is *centrally symmetric*, if it is symmetric around the middle of the permutation. For example, If $\pi_1 = 1, 2, 3, \underline{6, 5, 4}, 7, 8, 9$, then $J_1 = 6, 5, 4$ is centrally symmetric.

Let s be the smallest number such that J_s changes the location of an element which was an internal element in J_t for $t < s$. Such s must exist, otherwise, all internal elements of J_1 are already on their final positions at π_m. This means that J_1 is centrally symmetric. But then J_2 cannot be centrally symmetric and so its internal elements must later change their locations in order to be on their final positions at π_m.

Let a be an internal element of J_t with $t < s$, such that J_s changes the location of a. Without loss of generality, $T_s(a) > T_t(a)$. Since a moves to the right, there exist b, c such that a, b, c are consecutive elements of π_{s-1} and $a < b < c$. Since a is an internal element of J_t, there are d, e such that d, a, e are consecutive elements of π_t and $d > a > e$.

Let π_l, $t < l < s-1$, be the first permutation in which b is the right neighbor of a. Then there exist f, g such that a, b, f, g are consecutive elements of π_l and $a < b > f > g$. Since $T_{s-1}(a) = T_l(a)$, it follows that $T_{s-1}(c) = T_l(f)$. That means that before a moves to the right at π_s, f needs to change its location. But

f is an internal element in J_l and so, no J_d, $l < d < s$, can change the location of f (otherwise, it contradicts the definition of s). We conclude that such s cannot exist, which leads a contradiction. □

SECOND PROOF. A substring of three consecutive elements x, y, z in a permutation is called *a bad triplet* if $x < z$ but x, y, z are not in an increasing order.

Let π_l be the last permutation that contains a bad triplet x, y, z. Such π_l exists because π_1 has a bad triplet but π_m does not. For example, if π_1, π_m are $\pi_1 = 1, 4, 3, 2, 5, 6$, $\pi_m = 6, 5, 4, 3, 2, 1$, then π_1 has two bad triplets $1, 4, 3$ and $3, 2, 5$. π_m is in decreasing order, so it contains no bad triplet.

To get a contradiction, we show here that at least one of the permutations that follows π_l contains a bad triplet.

Suppose that none of the permutations that follows π_l contains a bad triplet. Then either x or z (but not both) are elements of J_{l+1}. Assume that $x \in J_{l+1}$ (similar arguments can be used for the case $z \in J_{l+1}$).

We define the *closed interval* $[a, b]_d$ to be the part of the permutation π_d that contains the consecutive elements between a and b including a and b. Example, for $\pi_d : 6, 3, 2, 1, 5, 4$ $[3, 5]_d = 3, 2, 1, 5$.

We now consider two cases:

Case 1: $x, y \in J_{l+1}$.

Then x is the right neighbor of y in J_{l+1}, and J_{l+1} contains at least one more element to the right of x. Let a be the rightmost element of J_{l+1} and b its left neighbor. Then $z > x \geq b > a$, from which follows that b, a, z are consecutive elements of π_{l+1} satisfying $b > a < z$ and $b < z$, which means that b, a, z is a bad triplet.

Case 2: $x \in J_{l+1}$, $y \notin J_{l+1}$.

Let $s = \min\{k \mid k > l + 1 \text{ and } x \in J_k \text{ is not the leftmost element in } J_k\}$. Such s exists since $z > x$ and z, x are not yet reversed at π_{l+1}. Denote by c the left neighbor of x in J_s. Then x, c are consecutive elements of π_{s-1} and $x < c$.

Let $t = \max\{k \mid k < s \text{ and } x \in J_k\}$. Note that since x is an element of J_{l+1} and $l + 1 < s$, such t exists and satisfies $l + 1 \leq t < s$. Also, note that since x is the leftmost element of J_{l+1}, x is the leftmost element in J_t.

Let a, b be the two right neighbors of x in J_t. Then x, a, b are three consecutive elements of π_t and $x > a > b$.

Since $x \notin J_r$ for $t < r < s$, it follows that in order for c to be the right neighbor of x in π_{s-1}, c must switch with b first, and then with a, in permutations between t and s. So there exists r, $t < r < s$, such that $c, b \in J_r$ and there exists q, $r < q < s$, such that $c, a \in J_q$.

We claim that for every j satisfying $t \leq j < s$, $[x, b]_j$ contains no increasing substring of length greater than 2. Also, the three rightmost elements in $[x, b]_j$ are in decreasing order.

We will prove it by induction. For $j = t$ the claim holds. By the induction hypothesis, the three rightmost elements in $[x, b]_{j-1}$ are in decreasing order and

$I_j \not\subset [x,b]_{j-1}$. Since, in addition, $x \notin I_j$, it follows that if I_j contains elements of $[x,b]_{j-1}$, it must contain b only. If it does, the three rightmost elements of J_j are the three rightmost elements of $[x,b]_j$ and are in decreasing order.

Any increasing substring in $[x,b]_j$ can consist of only two elements, each of which belongs to a different reversed substring involving b. This completes the proof of the claim. By the definition of r, for every j satisfying $r \le j < s$ we have $c \in [x,b]_j$, but by the above claim, $I_j \not\subset [x,b]_{j-1}$, which implies that c cannot switch with a in a permutation that precedes π_s. So q as defined above cannot exist: a contradiction. $\qquad\Box$

Acknowledgments

We thank Gil Kalai and Rom Pinchasi for useful discussions.

References

[Borwein and Moser 1990] P. Borwein and W. O. J. Moser, "A survey of Sylvester's problem and its generalizations", *Aequationes Math.* **40**:2-3 (1990), 111–135.

[Csima and Sawyer 1993] J. Csima and E. T. Sawyer, "There exist $6n/13$ ordinary points", *Discrete Comput. Geom.* **9**:2 (1993), 187–202.

[Goodman and Pollack 1980] J. E. Goodman and R. Pollack, "On the combinatorial classification of nondegenerate configurations in the plane", *J. Combin. Theory Ser. A* **29**:2 (1980), 220–235.

[Goodman and Pollack 1993] J. E. Goodman and R. Pollack, "Allowable sequences and order types in discrete and computational geometry", pp. 103–134 in *New trends in discrete and computational geometry*, edited by J. Pach, Algorithms and Combinatorics **10**, Springer-Verlag, Berlin, 1993.

[Nilakantan 2005] N. Nilakantan, "On some extremal problems in combinatorial geometry", pp. 477–492 in *Combinatorial and computational geometry*, edited by J. E. Goodman et al., Math. Sci. Res. Inst. Publ. **52**, Cambridge U. Press, New York, 2005.

[Scott 1944] P. R. Scott, "Solution to Problem 4065", *Amer. Math. Monthly* **51** (1944), 169–171.

[Scott 1970] P. R. Scott, "On the sets of directions determined by n points", *Amer. Math. Monthly* **77** (1970), 502–505.

[Sylvester 1893] J. J. Sylvester, "Mathematical question 11851", *Educational Times* **46** (1893), 156.

HAGIT LAST
INSTITUTE OF MATHEMATICS
THE HEBREW UNIVERSITY
91904 JERUSALEM
ISRAEL
hagitl@math.huji.ac.il

Combinatorial and Computational Geometry
MSRI Publications
Volume **52**, 2005

A Comparison of Five Implementations of 3D Delaunay Tessellation

YUANXIN LIU AND JACK SNOEYINK

ABSTRACT. When implementing Delaunay tessellation in 3D, a number of engineering decisions must be made about update and location algorithms, arithmetics, perturbations, and representations. We compare five codes for computing 3D Delaunay tessellation: qhull, hull, CGAL, pyramid, and our own tess3, and explore experimentally how these decisions affect the correctness and speed of computation, particularly for input points that represent atoms coordinates in proteins.

1. Introduction

The Delaunay tessellation is a useful canonical decomposition of the space around a given set of points in a Euclidean space E^3, frequently used for surface reconstruction, molecular modelling and tessellating solid shapes [Delaunay 1934; Boissonnat and Yvinec 1998; Okabe et al. 1992]. The Delaunay tessellation is often used to compute its dual Voronoi diagram, which captures proximity. In its turn, it is often computed as a convex hull of points lifted to the paraboloid of revolution in one dimension higher [Brown 1979; Brown 1980]. As we sketch in this paper, there are a number of engineering decisions that must be made by implementors, including the type of arithmetic, degeneracy handling, data structure representation, and low-level algorithms.

We wanted to know what algorithm would be fastest for a particular application: computing the Delaunay tessellation of points that represent atoms coordinates in proteins, as represented in the PDB (Protein Data Bank) format [Berman et al. 2000]. Atoms in proteins are well-packed, so points from PDB files tend to be evenly distributed, with physically-enforced minimum separation distances. Coordinates in PDB files have a limit on precision: because they have an 8.3f field specification in units of ångstroms, they may have three decimal digits before the decimal place (four if the number is positive), and three digits

This research has been partially supported by NSF grant 0076984.

after. Thus, positions need at most 24 bits, with differences between neighboring atoms usually needing 12 bits. Since the experimental techniques do not give accuracies of thousandths or even hundredths of ångstroms, we may even reduce these limits.

We therefore decided to see whether we could stretch the use of standard IEEE 754 double precision floating point arithmetic [IEEE 1985] to perform Delaunay computation for this special case. We implemented a program, tess3 [Liu and Snoeyink n.d.], which we sketch, and compared it with four popular codes that are available for testing: qhull [Barber et al. 1996], the CGAL geometry library's Delaunay hierarchy [Boissonnat et al. 2002; Devillers 1998], pyramid [Shewchuk 1998] and hull [Clarkson 1992]. Our program, designed to handle limited precision, uniformly-spaced input using only double precision floating point arithmetic, was fastest on both points from PDB files and randomly generated input points, although it did compute incorrect tetrahedra for one of the 20,393 PDB files that did not satisfy the input assumptions.

The performance of Delaunay code is affected by a number of algorithmic and implementation choices. We compare these choices made by all five programs in an attempt to better understand what makes a Delaunay program work well in practice. In Section 2, we review the problem of computing the Delaunay tessellation and describe the main algorithmic approaches and implementation issues. In Section 3, we compare the programs for computing the Delaunay tessellation. In section 4, we show experiments that compare all five programs in speed, and some experiments that look at the performance of tess3 in detail.

There are many other programs that can compute the Delaunay tessellation. These include nnsort [Watson 1981; Watson 1992], detri [Edelsbrunner and Mücke 1994], Proshape [Koehl et al. n.d.] and Ciel [Ban et al. 2004] — the last two are targeted particularly at computations on proteins. The candidate programs are selected because they are the fastest programs we are able to find for our test input sets — PDB files and randomly generated points of size up to a million. Other work [Boissonnat et al. 2002] tests Delaunay programs on other input distributions including scanned surfaces, which do not satisfy our input assumptions.

2. Delaunay Tessellation

There are several common elements in the five programs that we survey.

Definition. The *Delaunay diagram* in E^3 can be defined for a finite set of point sites P: Given a set of sites $P' \subseteq P$, if we can find a sphere that touches every point of P' and is empty of sites of P, then the relative interior of the convex hull of P' is in the Delaunay diagram. The Delaunay diagram is dual to the *Voronoi diagram* of P, which is defined as the partition of E^3 into maximally-connected regions that have the same set of closest sites of P. The Delaunay diagram completely partitions the convex hull of P.

If the sites are in *general position*, in the sense that no more than four points are co-spherical and no more than three are co-planar, then the convex hulls in the Delaunay diagram become simplices. The programs we survey have different approaches to enforce or simulate general position, so that the Delaunay tessellation can always be represented by a simplicial complex.

Representation. A simplicial complex can be represented by its full facial lattice: its vertices, edges, triangles and tetrahedra and their incident relationships. A programmer will usually choose to store only a subset of the simplices and the incidence relationships, deriving the rest as needed.

All five programs store the set of tetrahedra, and for each tetrahedron t, references to its vertices and *neighbors*—a neighbor is another tetrahedron that shares a common triangle with t. A *corner* is a vertex reference in a tetrahedron. Two corners are opposite if their tetrahedra are neighbors, but neither is involved in the shared triangle.

It is common to include a point at infinity, ∞, so that for every triangle $\{a, b, c\}$ on the convex hull, there is a tetrahedron $\{\infty, a, b, c\}$. Thus, each tetrahedron in the tessellation has exactly 4 neighbors.

Incremental construction. Each of the five programs compute the Delaunay tessellation incrementally, adding one point at a time. A new point p is added in two steps: First, a point location routine finds the tetrahedron (or some sphere) that was formerly empty, but that now contains the new point p. Second, an update routine removes tetrahedra that no longer have an empty sphere after adding p and fills in the hole with tetrahedra emanating from p. The running time of an incremental algorithm is proportional to the number of tetrahedra considered in point location, plus the total number of tetrahedra created.

The worst-case number of tetrahedra created in adding a vertex is linear, so the total number of tetrahedra is at most quadratic. This is also the worst-case number in any one tessellation, and simple examples, such as $n/2$ points on each of two skew lines or curves, give a matching lower bound. Nevertheless, linear-size Delaunay tessellations are most commonly observed—the practice is better than the theory predicts. Some theoretical works explain this under assumptions on the input such as random points or uniform samples from surfaces [Attali et al. 2003; Dwyer 1991; Erickson 2002].

For the linear-sized Delaunay tessellations observed in practice, point location can actually become the bottleneck in 3D, as it is in 2d, because the number of new tetrahedra from adding a new vertex is so small. There are a wide variety of point location algorithms in the programs we survey, so we will discuss this primarily in Section 3.

Numerical computations. The geometric tests in Delaunay code are performed by doing numerical computations. The most important is the InSphere test. Let p be a point whose Cartesian coordinates are p_x, p_y and p_z. We can represent p by a tuple $(p_1, p_x, p_y, p_z, p_q)$, where $p_1 = 1$ is a homogenizing coordinate and $p_q = p_x^2 + p_y^2 + p_z^2$. Mathematically, any positive scalar multiple of p can be taken to represent the same point, but for computation, we prefer the computer graphics convention that $p_1 = 1$, and assume that the Cartesian coordinates are b-bit integers. The special point $\infty = (0, 0, \ldots, 0, 1)$, representing the *point at infinity*, is the sole exception. Four noncoplanar points a, b, c and d define an oriented sphere and point p lies inside, on, or outside of the sphere depending on whether the sign of $\texttt{InSphere}(a, b, c, d; p)$ in equation 2–1 is negative, zero, or positive.

$$\texttt{InSphere}(a, b, c, d; p) = \begin{vmatrix} a_1 & a_x & a_y & a_z & a_q \\ b_1 & b_x & b_y & b_z & b_q \\ c_1 & c_x & c_y & c_z & c_q \\ d_1 & d_x & d_y & d_z & d_q \\ p_1 & p_x & p_y & p_z & p_q \end{vmatrix} \qquad (2\text{--}1)$$

Note that if one of the four points on the sphere is ∞, the determinant is equal to an orientation determinant that tests a point against a plane. Therefore, when a tetrahedron includes the ∞ vertex, we can still use this determinant to perform the InSphere test on its sphere and a chosen point; the test will return the position of the point with respect to an "infinite sphere" that is an oriented convex hull plane.

Computers store numbers with limited precision and perform floating point operations that could result in round-off errors. In the Delaunay algorithms, round-off errors change the sign of a determinant and produce the wrong answer for an InSphere test. Therefore, we look at the *bit complexity* of the numerical operations: Assuming that the input numbers are b-bit integers, how large can the results of an algebraic evaluation be as a function of b?

The InSphere determinant can be expanded into an alternating sum of multiplicative terms, each of degree five. Therefore, if we use the determinant directly, we need at least $5b$ bits to compute each multiplicative term correctly. The determinant itself can take no longer than $5b$ bits, since the InSphere determinant gives the volume of a parallelepiped in $\mathbb{R}^4$, where the thickness of the parallelepiped along the x, y, z and the lifted dimension take no more than b, b, b and $2b$ bits, respectively.

Knowing that, e.g., a is a finite point, and that the homogenizing coordinate for points is unity, we can rewrite the determinant to depend on the differences in coordinates, rather than absolute coordinates by just subtracting the row a from all finite points, and then evaluate the determinant. The last coordinate can also be made smaller by lifting after subtraction, although it adds extra squaring operations that must be done within each InSphere determinant.

When an InSphere determinant is zero, then the five points being tested lie on a sphere, and are not in general position. (Subjecting the points to a random perturbation will make them no longer co-spherical, except for a set of measure zero.) Edelsbrunner and Mücke [1990] showed how to simulate general position for determinant computations by infinitesimal perturbations of the input points, and there have been many approaches since. We describe the approaches taken by the different programs in Section 3.

3. Comparison of Delaunay Codes

With this background, we elaborate on the engineering choices made in the five programs for representation, arithmetic, perturbation, update and point location. A summary table is provided at the end of the section.

Implementation goals. The five programs that we survey were implemented with different goals in mind.

CGAL is a `C++` geometric algorithm library that includes a `Delaunay_trian-gulation_3` class that encapsulates functions for Delaunay tessellation. It also supports vertex removal [Devillers and Teillaud 2003]. It uses traits classes to support various types of arithmetic and point representations; we tested `Simple_cartesian<double>`, which uses floating point arithmetic only, and

$$\text{Static\_filters<Filtered\_kernel<Simple\_cartesian<double>>>},$$

which guarantees that the signs returned by geometric tests are computed exactly by using exact arithmetic whenever its floating pointer filter "sees" that, before a geometry test, floating point computation might produce erroneous signs.

Clarkson's hull [1992] computes convex hull of dimension 2, 3 and 4 by an incremental construction that can either shuffle the input points or take them as is. It uses a low bit-complexity algorithm to evaluate signs of determinants in double-precision floating point.

Qhull [Barber et al. 1996], initially developed at the geometry center of University of Minnesota, is a popular program for computing convex hulls in general dimensions. It supports many geometric queries over the convex hull and connects to geomview for display.

Shewchuk's pyramid [1998] was developed primarily to generate quality tessellation of a solid shape. In addition to taking points and producing the Delaunay tessellation, it can take lines and triangles and compute a conforming Delaunay, adding points on these features until the final tessellation contains, for each input feature, a set of edges or triangles is a partition of that feature.

Our program, tess3, specializes in the Delaunay tessellation of near-uniformly spaced points with limited precision, of the sort found in the crystallographic structures deposited in the PDB [Berman et al. 2000]. We have been pleased to find that it also works with NMR structures, which often have several variants of the same structure in the same file, and therefore violate the separation assumptions under which our code was developed.

Representation. Each program stores pointers from tetrahedra to their neighbors. Pyramid and tess3 have special ways to indicate which corners in a pair of neighboring tetrahedra correspond: Pyramid stores four bits with each neighbor pointer to indicate the orientation of the neighboring tetrahedron and location of the vertices of the shared triangle. Tess3 uses a corner-based representation that is a refinement of the structure of [Paoluzzi et al. 1993] or [Kettnet et al. 2003]. An array stores all the corners so that each subsequent block of four corners is one tetrahedron. Each corner points to its vertex and its *opposite* corner — the corner in the neighboring tetrahedron across the shared triangle. Each block is stored with vertices in increasing order, except that the first two may be swapped to keep the orientation positive. The correspondence between vertices in neighboring tetrahedra, where vertex $0 \le i < 4$ is replaced by vertex at position $0 \le j < 4$, can be recorded in a table indexed by i, j. This supports operations such as walking through tetrahedra, or cycling around an edge without requiring conditional tests.

Since a tetrahedron's sphere can be used repeatedly for InSphere tests, the minors of the determinant expanded along the last row can be pre-computed and stored in a vector S so that the test becomes a simple dot product:

$$\text{InSphere}(a, b, c, d; p) = S \cdot p.$$

Hull, pyramid, and tess3 store these sphere vectors.

Incremental computation. Each of the programs must update the data structures as tetrahedra are destroyed and created. One of the biggest decisions is whether an algorithm uses flipping [Edelsbrunner and Shah 1992] to always maintain a tessellation of the convex hull, or uses the Bowyer–Watson approach [Bowyer 1981; Watson 1981] of removing all destroyed tetrahedra, then filling in with new. We have observed in our experiments that flipping assigns neighbor pointers to twice as many tetrahedra, since many tetrahedra created by flips with a new vertex p are almost immediately destroyed by other flips with p.

Amenta, Choi and Rote [2003] pointed out that the number of tetrahedra is not the only consideration. Since modern memory architecture is hierarchical, and the paging policies favor programs that observe locality of reference, a major concern is *cache coherence:* a sequence of recent memory references should be clustered locally rather than randomly in the address space. A program implementing a randomized algorithm does not observe this rule and can be dramatically slowed down when its address space no longer fits in main memory. Their Biased Randomized Insertion Order (BRIO) preserves enough randomness in the input points so that the performance of a randomized incremental algorithm is unchanged but orders the points by spatial locality to improve cache coherence. More specifically, they first partition the input points into $O(\log n)$

sets as follows: Randomly sample half of the input points and put them into the first set; repeatedly make the next set by randomly sampling half of the previous set. Order the sets in the reverse order they are created. Finally, the points within each set are ordered by first bucketing them with an octree and traverse the buckets in a depth-first order.

To partition the points, tess3 uses a deterministic approach that we call *bit-levelling*. Bit-levelling group the points whose three coordinates share i trailing zeros (or any other convenient, popular, bit pattern) in the ith level. Levels are inserted in increasing order, and points within each level are ordered along a space-filling curve. With experimentally-determined data, the least-significant bits tend to be random, so bit-levelling generates a sample without the overhead of generating random bits. The real aim for bit-levelling, however, is to reduce the bit-complexity of the InSphere computation. Recall that when we evaluate the determinant for the InSphere test, one point can be used for the local origin and subtracted from all finite points. Using floating point, the effective number of coordinate bits in the mantissa is reduced if some of the most- and/or least-significant bits agree. Since the points are assumed to be evenly distributed (the next section describes how tess3 adds all the points ordered along a Hilbert curve), in the final levels the points used for InSphere tests tend to be close and share some most-significant bits. Since bit-levelling forms the ith level by grouping points with the same i least-significant bits, giving cancellation in the early, sparse levels as well.

Point location. In theory, point location is not the bottleneck for devising optimal 3D Delaunay algorithms. In practice, however, the size of the neighborhood updated by inserting a new point is close to constant, and point location to find the tetrahedron containing a new point p can be more costly than updating the tessellation if not done carefully.

Hull and qhull implement the two standard ways to perform point location in randomized incremental constructions of the convex hull: Hull maintains the history of all simplices, and searches the history dag to insert a new point. Qhull maintains a conflict list for each facet of the convex hull in the form of an *outside set*, which is the set of points yet to be processed that can "see" the facet. These are equivalent in the amount of work done, although the history dag is larger, and the conflict list requires that all points be known in advance.

The other programs invoke some form of walk through the tetrahedra during the point location. The simplest kind of walk visits one tetrahedron at each step, choose a triangle face f (randomly out of at most three) so that p and the tetrahedron are on the opposite side of the plane through f and walk to the neighboring tetrahedron across f. The walk always terminates by the acyclic theorem from Edelsbrunner [1989]. We will refer to this walk as *remembering stochastic walk* following Devillers et al. [2002], who also provides a comparison with other possible walking schemes.

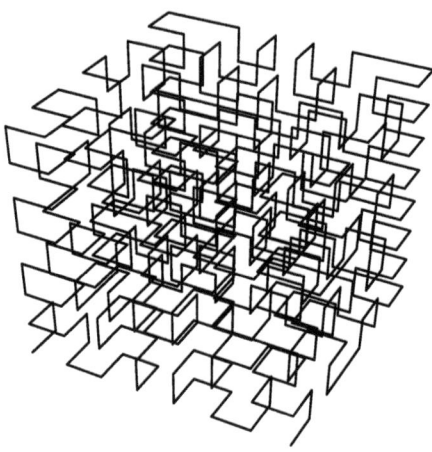

Figure 1. Hilbert curve for an $8 \times 8 \times 8$ grid.

Tess3 uses the remembering stochastic walk for point location. It locates a sphere, rather than a tetrahedron, containing the new point p from the last tetrahedron created. Tess3 uses sphere equations to perform the plane test. Suppose neighboring tetrahedra t_1 and t_2 share a triangle in plane P_{12}, have vertices q_1 and q_2 that are not on P_{12}, and have circumspheres $S_1 \neq S_2$. Tess3 can determine the side of plane P_{12} that contains p by the sign of $P_{12} \cdot p = q_1(S_2 \cdot p) - q_2(S_1 \cdot p)$. Note that this reuses the InSphere tests already performed with p, and reduces the orientation determinant to the difference of two dot products. When $S_1 = S_2$, a degenerate configuration, we would have to test the plane, but this happens rarely enough that tess3 simply chooses the side randomly.

To make the walks short, tess3 initially places all input points into a grid of $N \times N \times N$ bins, which it visits in Hilbert curve order so that nearby points in space have nearby indices [Moon et al. 2001]. To order a set of points with a Hilbert curve, tess3 subdivides a bounding cube into $(2^i)^3$ boxes and reorders the points using counting sort on the index of the box on the Hilbert curve that contains each point. Points in a box can be reordered recursively until the number of points in each subbox is small. Parameter i is chosen large enough so that few recursive steps are needed, and small enough that the permutation can be done in a cache-coherent manner. We find that having $(2^3)^3 = 512$ boxes works well; ordering 1 million points takes between 1–2 seconds on common desktop machines.

CGAL implements the Delaunay hierarchy scheme from [2002]. It combines a hierarchical point location data structure with the remembering stochastic walk.

The Delaunay hierarchy first creates a sequence of levels so that the 0th level is P, and each subsequent level is produced by random sampling a constant fraction of the points from the previous level. Next, Delaunay tessellation is created for each level, and the tetrahedra that share vertices between levels are linked. To locate p, at each step, a walk is performed within a level to find the vertex closest to p. This vertex is then used as the starting point for the next step. The hierachical tessellation makes the asymptopic point location time to be $O(() \log(n))$, which is optimal, while the walk, along with appropiately chosen parameter for the sizes of the levels, allow the space used the data structure to be small.

Pyramid uses the *jump-and-walk* introduced in [Mücke et al. 1996]. To locate p in a mesh of m tetrahedra, it measures the distance from p to a random sample of $m^{1/4}$ tetrahedra, then walk from the closest of these to the tetrahedron containing p. Each step of the walk visits a tetrahedron t, shoots a ray from the centroid of t towards p, and go to the neighboring tetrahedron intersected by the ray. In the worst case, this walk may visit almost all tetrahedra, but under some uniformity assumptions the walk takes $O(n^{1/4})$ steps, which is an improvement over $O(n^{1/3})$ steps that a walk would have required without the initial sampling.

Contrasting the asymptotic behavior of the Delaunay hierarchy and the jump-and-walk, we should point out that the difference between $(n^1/4)$ and $\log(n)$ is small for practical value of n; the Delaunay hierarchy, however, makes no assumption about the point distribution.

Numerical computations. Each of the programs takes a different approach to reducing or eliminating errors in numerical computation.

Qhull and tess3 use floating point operations exclusively, and are written so that they do not crash if the arithmetic is faulty, but they may compute incorrect structures. Qhull checks for structural errors, and can apply heuristics to repair them in postprocessing. Tess3 assumes that input points have limited precision and are well distributed, and uses bit-levelling and Hilbert curve orders to try to ensure that the low-order or high-order (or both) bits agree, and that the bit differences take even fewer bits of mantissa. We explore this in more detail in the experiments.

CGAL has many options for evaluating geometric tests exactly. It can use interval arithmetic [Pion 1999], without or with static filtering [Devillers and Pion 2003] or an adapted filtering that guarantees correctness for integers of no more than 24 bits. We list these options in increasing speed, though static filtering is usually recommended because it makes no assumption about the input and is still quite competitive in speed.

Hull uses a low bit-complexity algorithm for evaluating the sign of an orientation determinant that is based on Graham–Schmidt orthogonalization. The idea is that since we care only about the sign of the determinant, we can manipulate the determinant so far as its sign does not change. The implementation uses only double precision floating operations and is able to compute the signs of InSphere determinants exactly for input whose coordinates have less than 26 bits.

Pyramid uses multilevel filtering [Shewchuk 1996] and an exact arithmetic to implement its geometric tests.

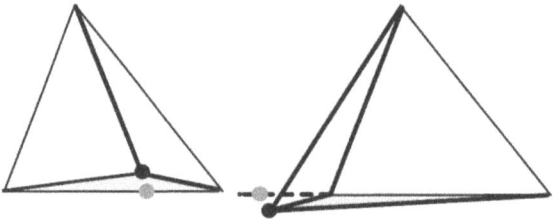

Figure 2. Perturbing point inside (left) or outside (right) produces flat triangles (shaded).

Perturbation to handle degeneracies. In Delaunay computation, InSphere determinants equal to zero are degeneracies — violations of the general position assumption that affect the running of the algorithm. They occur when a point is incident either on the Delaunay sphere of some tetrahedron or on the plane of the convex hull, which can be considered a sphere through the point at infinity.

Qhull allows the user to select a policy when the input contains degeneracies or the output contains errors: either it perturbs the input numerically and tries again, or it attempts to repair the outputs with some heuristics.

Edelsbrunner and Mücke [1990] showed how to simulate general position for the Delaunay computation directly, but advocated "perturbing in the lifted space" as easier. For lifted points, perturbation can be handled by simple policies: either treat all 0s as positive or treat them all as negative. These are consistent with perturbing a point outside or inside the convex hull in 4D, respectively. These perturbation schemes have three short-comings; usually only the third has impact on practice.

(i) The output from the perturbation is dependent on the insertion order of the points.

(ii) Perturbing the lifted points in 4D may produce a tessellation that is not the Delaunay tessellation of any actual set of points.

(iii) The perturbation (either the "in" or the "out" version) may produce "flat" tetrahedra near the convex hull. Figure 2 illustrates the 2D analog.

Hull and pyramid perturb points inside.

Tess3 first perturbs a point p down in the lifted dimension so that it is not on any finite sphere; next, if p is on an infinite sphere S, p is perturbed either into or away from the convex hull in 3D depending on these two cases: If q is inside the finite neighbor of S, q is perturbed into the convex hull; otherwise, q is perturbed away. This perturbation guarantees that there are no flat tetrahedra (handling 3), yet is still simple to implement.

CGAL perturbs the point on an infinite sphere the same way as tess3 but uses a more involved scheme for perturbing the point on a finite sphere [Devillers and Teillaud 2003]. It has the advantage that the perturbation of a point is determined by its index, which is independent from the insertion order; their scheme also guarantees that there are no flat tetrahedra (handling 1, 3).

Program	F	point location	E	C	degeneracy	L
CGAL	N	Delaunay hierarchy	Y	N	Perturbing E^3	C++
Hull	N	history dag	Y	Y	Perturb points into hull in E^4	C
Pyramid	Y	jump-and-walk	Y	N	Perturb points into hull in E^4. Remove flat tetrahedra by post-processing	C
QHull	N	outside set	N	N	Perturb points into hull in E^4. Remove flat tetrahedra by post-processing.	C
Tess3	N	Hilbert ordering, zig-zag walk	N	Y	Perturbation in E^4 with no flat tetrahedra.	C

Table 1. Program comparison summary. Column abbreviations: F = uses flips; E = exact; C = uses caching spheres; L = programming language. Versions and dates: CGAL version 2.4; hull and pyramid obtained in March 2004; qhull version 2003.1; Tess3 last revised in 9/2003.

A note on the weighted Delaunay tessellation. The definition of the Delaunay tessellation can be generalized easily to a weighted version, which associates each site p with a real number p_w. Recall that a point p in the Delaunay tessellation is represented by a tuple $(p_1, p_x, p_y, p_z, p_q)$, where p_q is the lifted coordinate. In the weighted version, we let $p_q = p_x^2 + p_y^2 + p_z^2 - p_w$, and the tessellation is the projection of the lower convex hull of the lifted points, as in the unweighted version. Note that a weighted site can be *redundant*: If it is in the interior of the convex hull, then it is not a new vertex of the tessellation. The weighted Delaunay tessellation has a number of applications in computational biology, such as computing the alpha shape [Edelsbrunner and Mücke 1994] and the skin surface [Cheng et al. 2001]. For these applications, the weight for a point site is the squared radius of the atom, and no redundant site occurs because of the physically-enforced minimum separation distances between atoms.

Each of the programs we study in this paper has been extended or can be easily modified to handle weighted points. For the programs that compute the tessellation via convex hull, namely hull and qhull, the weights are handled simply by changing the lifting computation. Pyramid uses flipping to maintain the tessellation and has to take extra care to insure that the flipping does not get stuck, which can happen [Edelsbrunner and Shah 1996]. Tess3 tries to locate a sphere which has the new point inside, so if the point is redundant, the location

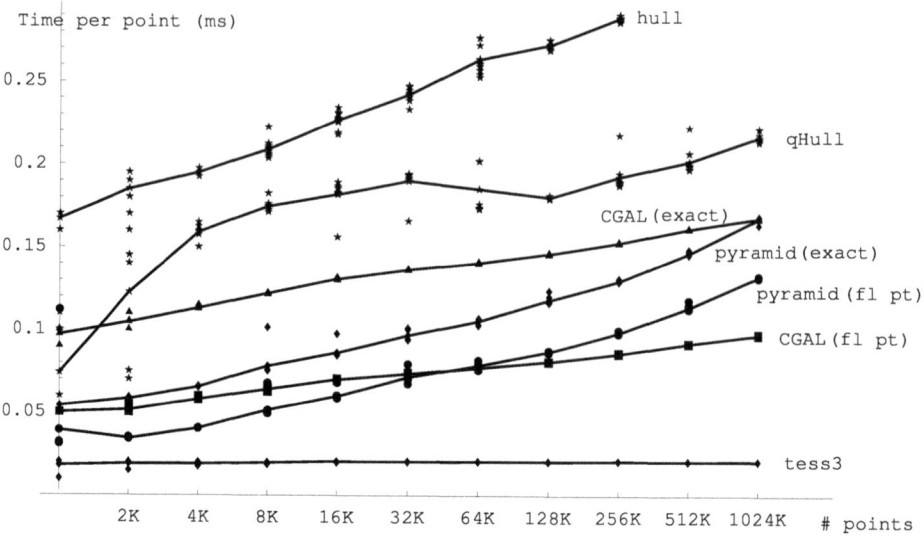

Figure 3. Running time of the programs with 10 bit random points.

routine does not terminate. However, as discussed before, if the input come from protein atoms, this does not happen. CGAL's Delaunay hierarchy currently does not handle weighted points, though its regular triangulation code, which is developed separately and is slower, does.

We chose not to study the performance of the programs with the weighted Delaunay tessellation, partly to make the comparisons easier and mostly because in our application, the weights come from radii of the atoms that differ very little, which implies that resulting tessellation will be similar. For input from PDB files, we have never observed any performance difference between the weighted and the unweighted version.

4. Experiments

In this section we report on experiments running the five programs on randomly generated points and on PDB files. We first report on running time. Then, because tess3 uses only standard floating point arithmetic, we report on the (small number of) errors that it makes.

We have tried to use the latest available codes of these programs. Hull and pyramid codes were given to us by the authors. CGAL and Qhull codes were downloaded from their web sites. The latest version of CGAL in April, 2004 is 3.0.1; however, we found that it is more than two times slower than CGAL 2.4 due to compiler issues. (Sylvain Pion, an author of the CGAL code, has found a regression in the numerical computation code generated by gcc that probably explains the slow-down.) We therefore proceed to use CGAL 2.4. Qhull 2003.1 we used is the latest version.

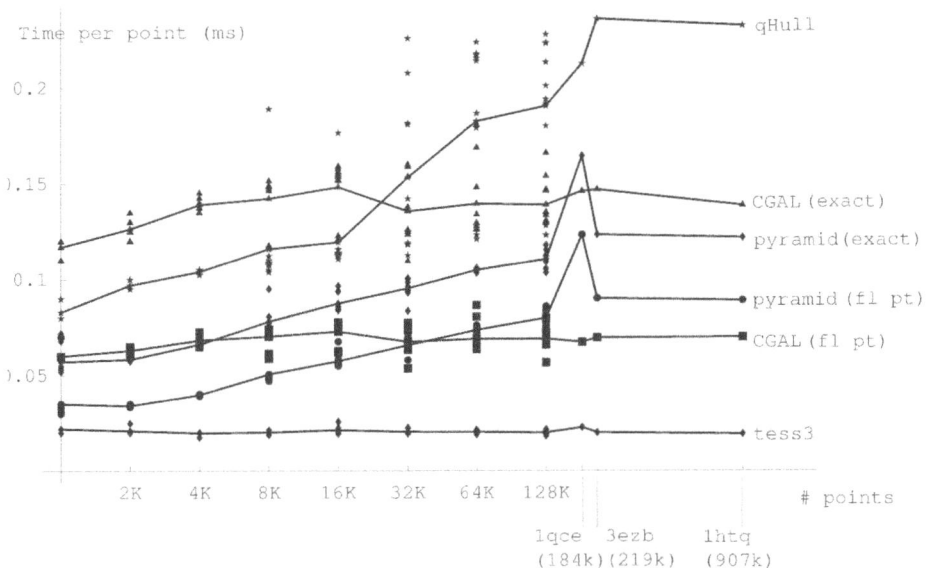

Figure 4. Running time of the programs with PDB files.

The plots in Figures 3 and 4 show the running time comparisons using random data and PDB data as input, respectively, using a logarithmic scale on the x axis and the running time per point in micro-seconds on the y axis. Hull's running time is much slower than the rest of the programs, with time per point between 0.4–0.6 ms. In Figure 4, we omitted it so other plots can be compared more easily. The timings are performed on a single processor of an AMD Athlon 1.4GHZ machine with 2GB of memory, running Red Hat Linux 7.3. Using time per point removes the expected linear trend and allows easier comparison across the entire x-coordinate range. Lines indicate the averages of ten runs; individual runs are plotted with markers. We should also mention that CGAL's running time seems to be affected most by compiler changes, with the slowest as much as 2.5 times slower than the fastest.[1]

We generated random data by choosing coordinates uniformly from 10-bit nonnegative integers. This ensures that the floating point computations of both Qhull and tess3 are correct. For the PDB data, for each input size n that is indicated on the x-axis, we try to find 10 files whose number of atoms are closest to n, though there is only one (with the indicated name) for each of the three largest sizes. We have posted online [Liu and Snoeyink n.d.] the names of these PDB files, as well as the program used to generate the random data.

---

[1] The timing plots are produced with a version that is roughly 1.5 times slower than the fastest we have been able to compile. The reason for this is that the public machine that compiled the fastest binary had a Linux upgrade, and, for unknown reasons, we could not since reproduce the speed on that machine (or other machines we have tried).

	total created sph./tetra	MkSph. (μs)	InSph. (μs) fl.pt. exct.	Update (μs)	Point location fl.pt. exct.	Mem. (MB)
CGAL	2,760,890	–	0.06^p 18.5^p 0.24^t 1.72^t	0.1^p 16.1^t	$21.8\%^p$ $25.3\%^p$ $22.1\%^t$ $27.9\%^t$	39
Hull	2,316,338	10.02	0.14 –	2.40	– 73.1%	401
Pyramid	$5,327,541^f$ $2,662,496^n$	–	0.21 0.72	2.44	50.2% 38.1%	57
QHull	2,583,320	0.65	0.12	>4.39	9.0% –	172
Tess3	2,784,736	0.13	0.04 –	2.42	$3.88\%^h$ – $0.43\%^w$	77

Table 2. Summary of timings and memory usage, running the programs against the same 100k randomly generated points with 10 bit coordinates. Dates and versions as in Table 1. Notes: For pyramid tetrahedra creation, numbers marked f include all initialized by flipping and marked n include only those for which new memory is allocated — equivalently, only those not immediately destroyed by a flip involving the same new point. For CGAL timings, p indicates profiler and t direct timing. For tess3 point location, h includes the preprocessing to order the points along a Hilbert curve; w is walk only.

There are a few immediate conclusions: The ordering of programs, tess3 < CGAL (fp) < pyramid (fp) < pyramid(ex) & CGAL < Qhull < hull, is consistent, although hull is particularly slow with the PDB files in comparison and is therefore not shown. In Figure 3 and 4, we can see a clear penalty for exact arithmetic, because even when an exact arithmetic package is able to correctly evaluate a predicate with a floating point filter, it must still evaluate and test an error bound to know that it was correct. Time per point shows some increase for everything but CGAL and tess3, which we believe is due to point location.

To further explain the difference in these programs' running time, we used the gcc profiler to determine the time-consuming routines. There are caveats to doing so; function level profiling turns off optimizations such as inlining, and adds overhead to each function call, which is supposed to be factored out, but may not be. (This affects CGAL the most, with its templated C++ functions we could not get reasonable profiler numbers, so we also tried to time its optimized code, but this has problems with clock resolution.) The table shows some of our findings for running the programs against the same 100k randomly generated points with 10 bit coordinates.

The "total created spheres/tetra" column shows that flipping must initialize many more tetrahedra. The MakeSphere and InSphere columns, which record time to make sphere equations and test points against them, indicate that there are speed advantages to using native floating point arithmetic for numerical computations. Even simple floating point filters must check error bounds for computations. Note that for the programs that do not cache spheres, the InSphere test

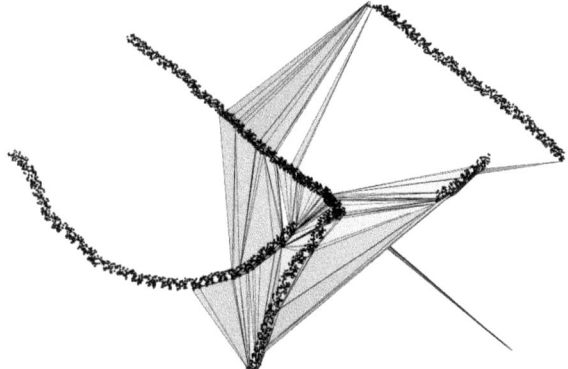

Figure 5. 1H1K points and bad tetrahedra.

is a determinant computation. The Update column indicates the time to update the tetrahedral complex and does not include any numerical computation time. The Point Location column indicates the percentage of time a program spends in point location (for tess3, this number includes the time for sorting the points along the Hilbert curve). The Memory column indicates the total amount of memory the programs occupy in the end.

As we can see from the table, tess3 benefited particulary from its fast point location. Caching sphere equations also helped speed up the numerical computation. A version of tess3 that does not cache sphere equations is about 20 percent slower. We observed some bottlenecks of the other programs: Qhull's data structure is expensive to update and the code contains debugging and option tests; Hull's exact arithmetic incurs a significant overhead even when running on points with few bits; Pyramid was bogged down mainly by its point location, which samples many tetrahedra.

Point ordering. Since tess3 does not use exact arithmetic, we did additional runs using audit routines to check the correctness of the output. We first check the topological correctness — that is, whether our data structure indeed represents a simplicial complex (it always has) — we then check the geometric correctness by testing (with exact arithmetic) for each tetrahedron if any neighboring tetrahedron vertex is inside its sphere. We also did some runs checking every InSphere test.

For the random data with 10 bits there are no errors, although we do find geometric errors for larger numbers of bits. For 20,393 PDB files, our program computes topologically correct output on all files and geometrically correct output on all except one.

Figure 5 displays the 266 incorrect tetrahedra, and shows that the assumptions of uniform distribution are egregiously violated. The comments to 1H1K state: "This entry corresponds to only the RNA model which was built into the blue tongue virus structure data. In order to view the whole virus in conjunction

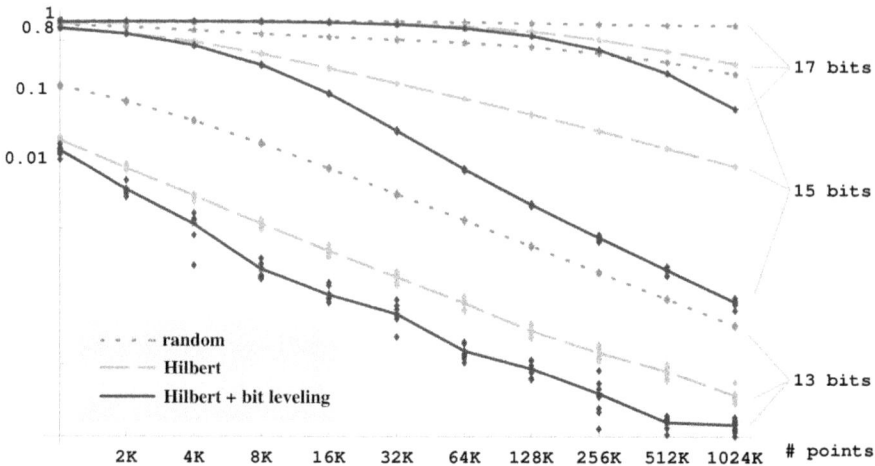

Figure 6. Semilog plot showing percentage of InSphere tests with round-off errors by number of points n and number of coordinate bits, for three orderings. We plot a dot for each of 10 runs for given n and bit number, and draw lines through the averages of 10 runs.

with the nucleic acid template, this entry must be seen together with PDB entry 2BTV."

We investigate how much ordering points along a Hilbert curve and bit-levelling helps speed up tess3 and make it more resistent to numerical problems. Figure 6 shows a log-log plot of the percentage of InSphere tests that contain round-off errors with three different orderings: random, Hilbert ordering only, and Hilbert ordering combined with bit-levelling. The percentages of errors are affected by both the number of coordinate bits and the number of points in the input; the plot illustrates variations in both of these controls. Given an input with a certain number of coordinate bits, we can see that the combined ordering has the lowest amount of numerical errors — and the difference becomes more dramatic as the number of input points increases. We should emphasize that the InSphere errors here are observed during the incremental construction and the final output always contains much fewer errors. For example, for the combined ordering, no output contains an error until the number of coordinate bits reaches 17.

Since BRIO [Amenta et al. 2003] also uses a spatial-locality preserving ordering to speed up point location, we close by comparing BRIO insertion order with a Hilbert curve order. Figure 7 compares the running times of CGAL, which uses a randomized point location data structure, under the BRIO and Hilbert insertion orders. The Hilbert curve is faster on average and has a smaller deviation. This suggests that for input points that are uniformly distributed, adding randomness into the insertion ordering perhaps will only slow down the program.

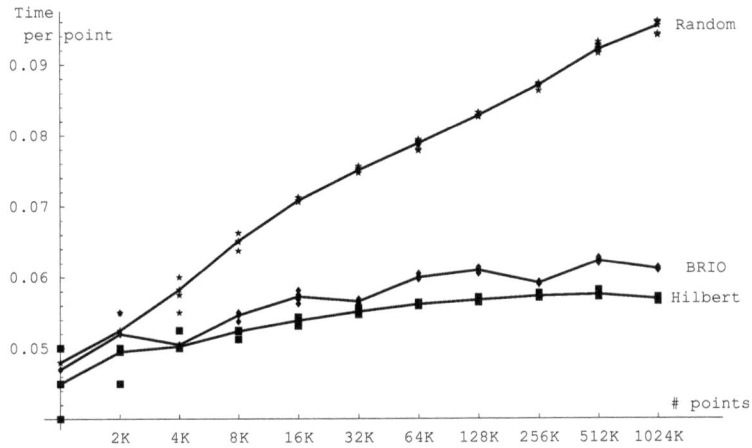

Figure 7. Running time of the CGAL Delaunay hierarchy using random, BRIO and Hilbert point orders.

5. Conclusions

We have surveyed five implementations of 3D Delaunay tessellation and compared their speed on PDB files and randomly generated data. The experiments show that Hull and QHull, the two programs that solve the more general problem of convex hull construction in 4D, are slower, penalized by not doing point location in 3D. Amongst the other three programs, tess3 is the fastest because its point location is carefully engineered for input points that are uniformly distributed in space. Exact arithmetic with filtering is quite efficient, as demonstrated by CGAL and Pyramid, but still incurs an overhead. We show that it is possible to have an implementation that works well even when straightforward bit-complexity analysis suggests otherwise.

Acknowledgments

We thank Jonathan Shewchuk, Kenneth Clarkson and Sunghee Choi for providing us with their codes and Sylvain Pion for the quick response to our questions about CGAL and many helpful suggestions. We also thank the anoymous referees who have made many helpful suggestions to improve the paper.

References

[Amenta et al. 2003] N. Amenta, S. Choi, and G. Rote. Incremental constructions con BRIO. In *Proceedings of the Nineteenth ACM Symposium on Computational Geometry*, pages 211–219, 2003.

[Attali et al. 2003] D. Attali, J.-D. Boissonnat, and A. Lieutier. Complexity of the Delaunay triangulation of points on surfaces: the smooth case. In *Proceedings of the Nineteenth ACM Symposium on Computational Geometry*, pages 201–210, 2003.

[Ban et al. 2004] Y.-E. A. Ban, H. Edelsbrunner, and J. Rudolph. Interface surfaces for protein-protein complexes. In *Proceedings of the Eighth Annual International Conference on Computational Molecular Biology*, pages 205–212, 2004.

[Barber et al. 1996] C. B. Barber, D. P. Dobkin, and H. Huhdanpaa. The quickhull algorithm for convex hulls. *ACM Trans. Math. Softw.*, 22(4):469–483, Dec. 1996. http://www.qhull.org/.

[Berman et al. 2000] H. M. Berman, J. Westbrook, Z. Feng, G. Gilliland, T. N. Bhat, H. Weissig, I. N. Shindyalov, and P. E. Bourne. The protein data bank. *Nucleic Acids Research*, 28(1):235–242, 2000. http://www.rcsb.org.

[Boissonnat and Yvinec 1998] J.-D. Boissonnat and M. Yvinec. *Algorithmic geometry*. Cambridge University Press, UK, 1998. Translated by Hervé Brönnimann.

[Boissonnat et al. 2002] J.-D. Boissonnat, O. Devillers, S. Pion, M. Teillaud, and M. Yvinec. Triangulations in CGAL. *Comput. Geom. Theory Appl.*, 22:5–19, 2002.

[Bowyer 1981] A. Bowyer. Computing Dirichlet tesselations. *Comput. J.*, 24:162–166, 1981.

[Brown 1979] K. Q. Brown. Voronoi diagrams from convex hulls. *Inform. Process. Lett.*, 9(5):223–228, 1979.

[Brown 1980] K. Q. Brown. *Geometric transforms for fast geometric algorithms*. Ph.D. thesis, Dept. Comput. Sci., Carnegie-Mellon Univ., Pittsburgh, PA, 1980. Report CMU-CS-80-101.

[Cheng et al. 2001] H.-L. Cheng, T. K. Dey, H. Edelsbrunner, and J. Sullivan. Dynamic skin triangulation. *Discrete and Computational Geometry*, 25:525–568, 2001.

[Clarkson 1992] K. L. Clarkson. Safe and effective determinant evaluation. In *Proc. 31st IEEE Symposium on Foundations of Computer Science*, pages 387–395, 1992.

[Delaunay 1934] B. Delaunay. Sur la sphère vide. A la memoire de Georges Voronoi. *Izv. Akad. Nauk SSSR, Otdelenie Matematicheskih i Estestvennyh Nauk*, 7:793–800, 1934.

[Devillers 1998] O. Devillers. Improved incremental randomized Delaunay triangulation. In *Proc. 14th Annu. ACM Sympos. Comput. Geom.*, pages 106–115, 1998.

[Devillers 2002] O. Devillers. The Delaunay hierarchy. *International Journal of Foundations of Computer Science*, 13:163–180, 2002.

[Devillers and Pion 2003] O. Devillers and S. Pion. Efficient exact geometric predicates for Delaunay triangulations. In *Proc. 5th Workshop Algorithm Eng. Exper.*, pages 37–44, 2003.

[Devillers and Teillaud 2003] O. Devillers and M. Teillaud. Perturbations and vertex removal in a 3D Delaunay triangulation. In *Proc. 14th ACM-SIAM Sympos. Discrete Algorithms (SODA)*, pages 313–319, 2003.

[Devillers et al. 2002] O. Devillers, S. Pion, and M. Teillaud. Walking in a triangulation. *Internat. J. Found. Comput. Sci.*, 13:181–199, 2002.

[Dwyer 1991] R. A. Dwyer. Higher-dimensional Voronoi diagrams in linear expected time. *Discrete Comput. Geom.*, 6:343–367, 1991.

[Edelsbrunner 1989] H. Edelsbrunner. An acyclicity theorem for cell complexes in d dimensions. In *Proc. 5th Annu. ACM Sympos. Comput. Geom.*, pages 145–151, 1989.

[Edelsbrunner and Mücke 1990] H. Edelsbrunner and E. P. Mücke. Simulation of simplicity: A technique to cope with degenerate cases in geometric algorithms. *ACM Trans. Graph.*, 9(1):66–104, 1990.

[Edelsbrunner and Mücke 1994] H. Edelsbrunner and E. P. Mücke. Three-dimensional alpha shapes. *ACM Trans. Graph.*, 13(1):43–72, Jan. 1994.

[Edelsbrunner and Shah 1992] H. Edelsbrunner and N. R. Shah. Incremental topological flipping works for regular triangulations. In *Proc. 8th Annu. ACM Sympos. Comput. Geom.*, pages 43–52, 1992.

[Edelsbrunner and Shah 1996] H. Edelsbrunner and N. R. Shah. Incremental topological flipping works for regular triangulations. *Algorithmica*, 15:223–241, 1996.

[Erickson 2002] J. Erickson. Dense point sets have sparse Delaunay triangulations. In *Proceedings of the Thirteenth Annual ACM-SIAM Symposium on Discrete Algorithms*, pages 125–134. Society for Industrial and Applied Mathematics, 2002.

[IEEE 1985] *IEEE Standard for binary floating point arithmetic, ANSI/IEEE Std 754 − 1985*. IEEE Computer Society, New York, NY, 1985. Reprinted in SIGPLAN Notices, 22(2):9–25, 1987.

[Kettnet et al. 2003] L. Kettner, J. Rossignac, and J. Snoeyink. The Safari interface for visualizing time-dependent volume data using iso-surfaces and a control plane. *Comp. Geom. Theory Appl.*, 25(1-2):97–116, 2003.

[Koehl et al. n.d.] P. Koehl, M. Levitt, and H. Edelsbrunner. Proshape. http://biogeometry.duke.edu/software/proshape/.

[Liu and Snoeyink n.d.] Y. Liu and J. Snoeyink. Sphere-based computation of Delaunay diagrams in 3d and 4d. In preparation. http://www.cs.unc.edu/~liuy/tess3.

[Moon et al. 2001] B. Moon, H. V. Jagadish, C. Faloutsos, and J. H. Saltz. Analysis of the clustering properties of the Hilbert space-filling curve. *Knowledge and Data Engineering*, 13(1):124–141, 2001.

[Mücke et al. 1996] E. P. Mücke, I. Saias, and B. Zhu. Fast randomized point location without preprocessing in two- and three-dimensional Delaunay triangulations. In *Proc. 12th Annu. ACM Sympos. Comput. Geom.*, pages 274–283, 1996.

[Okabe et al. 1992] A. Okabe, B. Boots, and K. Sugihara. *Spatial tessellations: Concepts and applications of Voronoi diagrams*. John Wiley & Sons, Chichester, UK, 1992.

[Paoluzzi et al. 1993] A. Paoluzzi, F. Bernardini, C. Cattani, and V. Ferrucci. Dimension-independent modeling with simplicial complexes. *ACM Trans. Graph.*, 12(1):56–102, Jan. 1993.

[Pion 1999] S. Pion. Interval arithmetic: An efficient implementation and an application to computational geometry. In *Workshop on Applications of Interval Analysis to Systems and Control*, pages 99–110, 1999.

[Shewchuk 1996] J. R. Shewchuk. Robust adaptive floating-point geometric predicates. In *Proc. 12th Annu. ACM Sympos. Comput. Geom.*, pages 141–150, 1996.

[Shewchuk 1998] J. R. Shewchuk. Tetrahedral mesh generation by Delaunay refinement. In *Proc. 14th Annu. ACM Sympos. Comput. Geom.*, pages 86–95, 1998.

[Watson 1981] D. F. Watson. Computing the n-dimensional Delaunay tesselation with applications to Voronoi polytopes. *Comput. J.*, 24(2):167–172, 1981.

[Watson 1992] D. F. Watson. *Contouring: A guide to the analysis and display of spatial data*. Pergamon, 1992.

YUANXIN LIU
DEPARTMENT OF COMPUTER SCIENCE
CAMPUS BOX 3175, SITTERSON HALL
UNC-CHAPEL HILL
CHAPEL HILL, NC 27599-3175
  liuy@cs.unc.edu

JACK SNOEYINK
DEPARTMENT OF COMPUTER SCIENCE
CAMPUS BOX 3175, SITTERSON HALL
UNC-CHAPEL HILL
CHAPEL HILL, NC 27599-3175
  snoeyink@cs.unc.edu

Combinatorial and Computational Geometry
MSRI Publications
Volume **52**, 2005

The Bernstein Basis and Real Root Isolation

BERNARD MOURRAIN, FABRICE ROUILLIER,
AND MARIE-FRANÇOISE ROY

ABSTRACT. In this mostly expository paper we explain how the Bernstein basis, widely used in computer-aided geometric design, provides an efficient method for real root isolation, using de Casteljau's algorithm. We discuss the link between this approach and more classical methods for real root isolation. We also present a new improved method for isolating real roots in the Bernstein basis inspired by Roullier and Zimmerman.

Introduction

Real root isolation is an important subroutine in many algorithms of real algebraic geometry [Basu et al. 2003] as well as in exact geometric computations, and is also interesting in its own right.

Our approach to real root isolation is based on properties of the Bernstein basis. We first recall Descartes' Law of Signs and give a useful partial reciprocal to it. Section 2 contains the definition and main properties of the Bernstein basis. In the third section, several variants of real root isolation based on the Bernstein basis are given. In the fourth section, the link with more classical real root isolation methods [Uspensky 1948] is established. We end the paper with a few remarks on the computational efficiency of the algorithms described.

1. Descartes' Law of Signs

The *number of sign changes*, $\mathrm{V}(a)$, in a sequence $a = a_0, \ldots, a_p$ of elements in $\mathbb{R} \setminus \{0\}$ is defined by induction on p by

$$\mathrm{V}(a_0) = 0,$$
$$\mathrm{V}(a_0, \ldots, a_p) = \begin{cases} \mathrm{V}(a_1, \ldots, a_p) + 1 & \text{if } a_0 a_1 < 0, \\ \mathrm{V}(a_1, \ldots, a_p) & \text{if } a_0 a_1 > 0. \end{cases}$$

This definition extends to any finite sequence a of elements in $\mathbb{R}$ by considering the finite sequence b obtained by dropping the zeros in a and defining $V(a) = V(b)$, with the convention $V(\varnothing) = 0$.

Let $P = a_p X^p + \cdots + a_0$ be a univariate polynomial in $\mathbb{R}[X]$. We write $V(P)$ for the number of sign changes in $a_0, \ldots, a_p$ and $\mathrm{pos}(P)$ for the number of positive real roots of P, counted with multiplicity.

We state the famous Descartes' law of signs, of 1636. (Descartes' text appears in [Struik 1969, pp. 90–91]. See also [Basu et al. 2003], for example, for a proof.)

THEOREM 1.1 (DESCARTES' LAW OF SIGNS).

(i) $\mathrm{pos}(P) \leq V(P)$.
(ii) $V(P) - \mathrm{pos}(P)$ *is even.*

In general, it is not possible to conclude much about the number of roots on an interval using only Theorem 1.1.

An instance where Descartes' law of signs permits a sharp conclusion is the following.

THEOREM 1.2. *Let*

$$\mathcal{D} = \{(x+iy) \in \mathbb{R}[i] \mid x < -\tfrac{1}{2}, (x+1)^2 + y^2 < 1\}$$

be the part of the open disk with center $(-1,0)$ and radius 1 which is to the left of the line $x = -\tfrac{1}{2}$ in $\mathbb{R}^2 = \mathbb{R}[i]$. If $P \in \mathbb{R}[X]$ is square-free and has either no roots or exactly one simple root in $(0, +\infty)$, and all its complex roots in $\mathcal{D}$, then $V(P) = 0$ or $V(P) = 1$ and

(i) *P has one root in $(0, +\infty)$ if and only if $V(P) = 1$,*
(ii) *P has no root in $(0, +\infty)$ if and only if $V(P) = 0$.*

The proof of the theorem relies on the following lemmas.

LEMMA 1.3. *For $A, B \in \mathbb{R}[X]$*

$$V(A) = 0, \; V(B) = 0 \implies V(AB) = 0.$$

PROOF. Obvious.   □

LEMMA 1.4. *For $A, B \in \mathbb{R}[X]$*

$$V(A) = 1, \; B = X + b, \; b \geq 0 \implies V(AB) = 1.$$

PROOF. If $b = 0$, $V(AB) = V(A) = 1$. Now, let $b > 0$. Let

$$A = a_d X^d + a_{d-1} X^{d-1} + \cdots + a_0,$$

and suppose, without loss of generality, that $a_d = 1$. Since $V(A) = 1$ and $a_d = 1$, there exists k such that

$$a_i \begin{cases} \geq 0 & \text{if } i > k, \\ < 0 & \text{if } i = k, \\ \leq 0 & \text{if } i < k. \end{cases} \tag{1-1}$$

Letting c_i be the coefficient of X^i in AB and making the convention that $a_{d+1} = a_{-1} = 0$, we have

$$c_i = \begin{cases} a_{i-1} + a_i b \geq 0 & \text{if } k+1 < i \leq d, \\ a_{k-1} + a_k b < 0 & \text{if } i = k, \\ a_{i-1} + a_i b \leq 0 & \text{if } i < k, \end{cases}$$

and $c_{d+1} = a_d > 0$. So, whatever the sign of c_{k+1}, $\mathrm{V}(AB) = 1$. □

LEMMA 1.5. *If* $\mathrm{V}(A) = 1$, $B = X^2 + bX + c$ *with* $b > 1$, $b > c > 0$, *then* $\mathrm{V}(AB) = 1$.

PROOF. Let $A = a_d X^d + a_{d-1} X^{d-1} + \cdots + a_0$ and suppose without loss of generality that $a_d = 1$. Since $\mathrm{V}(A) = 1$ and $a_d = 1$, there exists k such that (1–1) is satisfied. Letting c_i be the coefficient of X^i in AB and making the convention that $a_{d+2} = a_{d+1} = a_{-1} = a_{-2} = 0$, we have

$$c_i = \begin{cases} a_{i-2} + a_{i-1}b + a_i c \geq 0 & \text{for } k+2 < i \leq d+2, \\ a_{k-2} + a_{k-1}b + a_k c < 0 & \text{for } i = k, \\ a_{i-2} + a_{i-1}b + a_i c \leq 0 & \text{for } i < k. \end{cases}$$

The only way to have $\mathrm{V}(AB) > 1$ would be to have $c_{k+1} > 0, c_{k+2} < 0$, but this is impossible since

$$c_{k+2} - c_{k+1} = a_{k+2}c + a_{k+1}(b-c) + a_k(1-b) - a_{k-1} > 0.$$ □

PROOF OF THEOREM 1.2. Notice first that by Theorem 1.1, $\mathrm{V}(P) = 1$ implies that P has one root in $(0, +\infty)$, and $\mathrm{V}(P) = 0$ implies that P has no root in $(0, +\infty)$. Note also that

- if $X + a$ has its root in $(0, +\infty)$, then $a < 0$ and $\mathrm{V}(X+a) = 1$,
- if $X + b$ has its root in $(-\infty, 0]$, then $b \geq 0$ and $\mathrm{V}(X+b) = 0$,
- if $X^2 + bX + c$ has its roots in $\mathcal{D}$, then $b > 1, b > c > 0$ and $\mathrm{V}(X^2 + bX + c) = 0$.

Now decompose P into irreducible factors of degree 1 and 2 over $\mathbb{R}$. If P has one root a in $(0, +\infty)$, $\mathrm{V}(X + a) = 1$. Starting from $X + a$ and multiplying successively by the other irreducible factors of P, we get polynomials with sign variations equal to 1, using Lemma 1.4 and Lemma 1.5. Finally, $\mathrm{V}(P) = 1$.

If P has no root in $(0, +\infty)$, starting from 1 and multiplying successively by the irreducible factors of P, we get polynomials with sign variations equal to 0, using Lemma 1.3. Finally, $\mathrm{V}(P) = 0$. □

2. The Bernstein Basis

The Bernstein basis is widely used in computer-aided design [Farin 1990]. We recall some of its main properties, in order to use them for real root isolation in the next section.

NOTATION 2.1. Let P be a polynomial of degree $\leq p$. The *Bernstein polynomials* of degree p for c, d are the

$$B_{p,i}(c,d) = \binom{p}{i} \frac{(X-c)^i(d-X)^{p-i}}{(d-c)^p},$$

for $i = 0, \ldots, p$.

REMARK 2.2. Note that $B_{p,i}(c,d) = B_{p,p-i}(d,c)$ and that

$$B_{p,i}(c,d) = \frac{(X-c)}{d-c} \frac{p}{i} B_{p-1,i-1}(c,d).$$

Since the multiplicity of the polynomial $B_{p,i}(c,d)$ at $x = c$ is i and $B_{p,i}(c,d)$ is a polynomial of degree p, we immediately deduce that the polynomials $B_{p,i}(c,d)$, $i = 0, \ldots, p$ are linearly independent and form a basis of the vector space of polynomials of degree $\leq p$.

Here are some simple transformations, useful to understand the connection between the Bernstein basis and the monomial basis.

Reciprocal polynomial in degree p: $\mathrm{Rec}_p(P(X)) = X^p P(1/X)$. The nonzero roots of P are the inverses of the nonzero roots of $\mathrm{Rec}(P)$.

Contraction by ratio λ: for every nonzero λ, $C_\lambda(P(X)) = P(\lambda X)$. The roots of $C_\lambda(P)$ are of the form x/λ, where x is a root of P.

Translation by c: for every c, $T_c(P(X)) = P(X-c)$. The roots of $T_c(P(X))$ are of the form $x+c$ where x is a root of P.

PROPOSITION 2.3. *Let* $P = \sum_{i=0}^p b_i B_{p,i}(d,c) \in \mathbb{R}[X]$ *be of degree* $\leq p$. *Let*

$$T_{-1}\left(\mathrm{Rec}_p(C_{d-c}(T_{-c}(P)))\right) = \sum_{i=0}^p c_i X^i.$$

Then

$$\binom{p}{i} b_i = c_{p-i}.$$

PROOF. Performing the contraction of ratio $d-c$ after translating by $-c$ transforms

$$\binom{p}{i} \frac{(X-c)^i(d-X)^{p-i}}{(d-c)^p} \quad \text{into} \quad \binom{p}{i} X^i(1-X)^{p-i}.$$

Translating by -1 after taking the reciprocal polynomial in degree p transforms

$$\binom{p}{i} X^i(1-X)^{p-i} \quad \text{into} \quad \binom{p}{i} X^{p-i}. \qquad \square$$

Let P be of degree p. We denote by $b = b_0, \ldots, b_p$ the coefficients of P in the Bernstein basis of c, d. Let $n(P; (c,d))$ be the number of roots of P in (c, d) counted with multiplicities.

PROPOSITION 2.4. (i) $V(b) \geq n(P; (c,d))$.
(ii) $V(b) - n(P; (c,d))$ *is even.*

PROOF. This follows immediately from Descartes' law of signs (Theorem 1.1), using Proposition 2.3. Indeed, the image of (c, d) under the translation by $-c$ followed by the contraction of ratio $d-c$ is $(0, 1)$. The image of $(0, 1)$ under the inversion $z \mapsto 1/z$ is $(1, +\infty)$. Finally, translating by -1 gives $(0, +\infty)$. $\quad\square$

We now describe a special case where the number $V(b)$ coincides with the number of roots of P on (c, d). Let $d > c$, and $\mathcal{C}(c, d)_0$ be the closed disk with center $(c, 0)$ and radius $d-c$, and let $\mathcal{C}(c, d)_1$ be the closed disk with center $(d, 0)$ and radius $d-c$.

THEOREM 2.5 (THEOREM OF 2 CIRCLES). *If P is square-free and has either no root or exactly one simple root in (c, d) and P has no complex root in $\mathcal{C}(c, d)_0 \cup \mathcal{C}(c, d)_1$, then*

(i) *P has one root in (c, d) if and only if $V(b) = 1$,*
(ii) *P has no root in (c, d) if and only if $V(b) = 0$.*

PROOF. We identify $\mathbb{R}^2$ with $\mathbb{C} = \mathbb{R}[i]$. The image of the complement of $\mathcal{C}(c, d)_0$ (resp. $\mathcal{C}(c, d)_1$) under the translation by $-c$ followed by the contraction of ratio $d-c$ is the complement of $\mathcal{C}(0, 1)_0$ (resp. $\mathcal{C}(0, 1)_1$). The image of the complement of $\mathcal{C}(0, 1)_0$ under the inversion $z \mapsto 1/z$ is

$$\{(x+iy) \in \mathbb{R}[i] \mid 0 < x^2 + y^2 < 1\}.$$

The image of the complement of $\mathcal{C}(0, 1)_1$ under the inversion $z \mapsto 1/z$ is

$$\{(x+iy) \in \mathbb{R}[i] \mid x < \tfrac{1}{2}\}.$$

The image of the complement of $\mathcal{C}(0, 1)_0 \cup \mathcal{C}(0, 1)_1$ under $z \mapsto 1/z$ is

$$\{(x+iy) \in \mathbb{R}[i] \mid 0 < x^2 + y^2 < 1,\ x < \tfrac{1}{2}\}.$$

Translating this region by -1, we get the region

$$\mathcal{D} = \{(x+iy) \mid x < -\tfrac{1}{2},\ (x+1)^2 + y^2 < 1\}$$

defined in Theorem 1.2.

The statement then follows from Theorem 1.2 and Proposition 2.3. $\quad\square$

Notice that this result which is a weaker version of the two-circles theorem presented in [Mehlhorn 2001], and related to [Ostrowski 1950], is given for the sake of simplicity. Indeed, one can use instead the two-circles $D\left(\tfrac{1}{2} \pm \tfrac{i}{2\sqrt{3}}, \tfrac{1}{\sqrt{3}}\right)$, as proved in the works cited.

The coefficients $b = b_0, \ldots, b_p$ of P in the Bernstein basis of c, d give a rough idea of the shape of the polynomial P on the interval c, d. The *control line of P on $[c, d]$* is the union of the segments $[M_i, M_{i+1}]$ for $i = 0, \ldots, p-1$, with

$$M_i = \left(\frac{i\,d + (p-i)\,c}{p}, b_i\right).$$

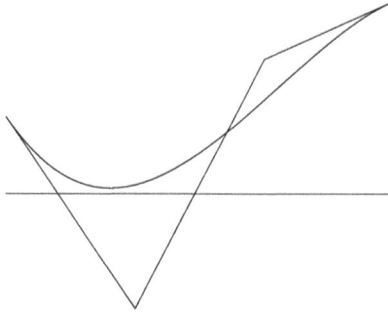

Figure 1. Graph of P on $[0, 1]$ and the control line.

It is clear from the definitions that the graph of P goes through M_0 and M_p and that the line M_0, M_1 (resp. M_{p-1}, M_p) is tangent to the graph of P at M_0 (resp. M_p).

EXAMPLE 2.6. We take $p = 3$, and consider the polynomial P with coefficients $(4, -6, 7, 10)$ in the Bernstein basis for $0, 1$

$$(1 - X)^3, \ 3(1 - X)^2 X, \ 3(1 - X)X^2, \ X^3.$$

We draw the graph of P on $[0, 1]$, the control line, and the X-axis in Figure 1.

The *control polygon of P on $[c, d]$* is the convex hull of the points M_i for $i = 1, \ldots, p$.

EXAMPLE 2.7. Continuing Example 2.6, we draw the graph of P on $[0, 1]$ and the control polygon in Figure 2.

An important and well-known property of the Bernstein polynomials is the following:

PROPOSITION 2.8. *The graph of P on $[c, d]$ is contained in the control polygon of P on $[c, d]$.*

PROOF. In order to prove the proposition, it is enough to prove that any line L above (respectively under) all the points in the control polygon of P on $[c, d]$ is

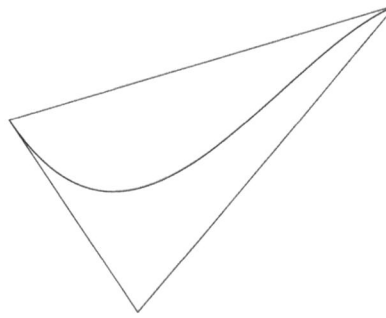

Figure 2. Graph of P on $[0, 1]$ and the control polygon.

above (respectively under) the graph of P on $[c,d]$. If L is defined by $Y = aX + b$, let us express the polynomial $aX + b$ in the Bernstein basis. Since

$$1 = \left(\frac{X-c}{d-c} + \frac{d-X}{d-c}\right)^p,$$

the binomial formula gives

$$1 = \sum_{i=0}^{p} \binom{p}{i} \left(\frac{X-c}{d-c}\right)^i \left(\frac{d-X}{d-c}\right)^{p-i} = \sum_{i=0}^{p} B_{p,i}(c,d).$$

Since

$$X = \left(d\left(\frac{X-c}{d-c}\right) + c\left(\frac{d-X}{d-c}\right)\right)\left(\frac{X-c}{d-c} + \frac{d-X}{d-c}\right)^{p-1},$$

the binomial formula together with Remark 2.2 gives

$$X = \sum_{i=0}^{p-1} \left(d\left(\frac{X-c}{d-c}\right) + c\left(\frac{d-X}{d-c}\right)\right) B_{p-1,i}(c,d)$$

$$= \sum_{i=0}^{p} \left(\frac{id+(p-i)c}{p}\right) B_{p,i}(c,d).$$

Thus,

$$aX + b = \sum_{i=0}^{p} \left(a\left(\frac{id+(p-i)c}{p}\right) + b\right) B_{p,i}(c,d).$$

It follows immediately that if L is above every M_i, that is, if

$$a\left(\frac{id+(p-i)c}{p}\right) + b \geq b_i$$

for every i, then L is above the graph of P on $[c,d]$, since $P = \sum_{i=0}^{p} b_i B_{p,i}(c,d)$ and the Bernstein polynomials of c,d are nonnegative on $[c,d]$. A similar argument holds for L under every M_i. $\qquad\square$

The following algorithm computes the coefficients of P in the Bernstein bases of c,e and e,d from the coefficients of P in the Bernstein basis of c,d.

ALGORITHM 2.9 (DE CASTELJAU).
INPUT: a list $b = b_0, \ldots, b_p$ representing a polynomial P of degree $\leq p$ in the Bernstein basis of c,d, and a number $e \in \mathbb{R}$.
OUTPUT: the list $b' = b'_0, \ldots, b'_p$ representing P in the Bernstein basis of c,e and the list $b'' = b''_0, \ldots, b''_p$ representing P in the Bernstein basis of e,d.
1. Define $\alpha = (d-e)/(d-c)$ and $\beta = (e-c)/(d-c)$.
2. Initialization: $b_j^{(0)} := b_j$, $j = 0, \ldots, p$.
3. For $i = 1, \ldots, p$
      For $j = 0, \ldots, p-i$
         compute $b_j^{(i)} := \alpha b_j^{(i-1)} + \beta b_{j+1}^{(i-1)}$.
4. Output $b' = b_0^{(0)}, \ldots, b_0^{(j)}, \ldots, b_0^{(p)}$ and $b'' = b_0^{(p)}, \ldots, b_j^{(p-j)}, \ldots, b_p^{(0)}$.

De Casteljau's algorithm can be visualized by means of the triangle

$$
\begin{array}{ccccccccc}
b_0^{(0)} & & b_1^{(0)} & & \cdots & & \cdots & & b_{p-1}^{(0)} & & b_p^{(0)} \\
& b_0^{(1)} & & & \cdots & & & \cdots & & b_{p-1}^{(1)} \\
& & \cdots & & \cdots & & \cdots & & \cdots \\
& & & \cdots & & \cdots & & \cdots & & \cdots \\
& & & & b_0^{(p-1)} & & b_1^{(p-1)} & & \cdots \\
& & & & & b_0^{(p)}
\end{array}
$$

where $b_j^{(i)} := \alpha b_j^{(i-1)} + \beta b_{j+1}^{(i-1)}$, $\alpha = (d-e)/(d-c)$ and $\beta = (e-c)/(d-c)$.

The coefficients of P in the Bernstein basis of c,d appear in the top side of the triangle and the coefficients of P in the Bernstein basis of c,e and e,d appear in the two other sides of the triangle.

NOTATION 2.10. We denote by $\tilde{a}$ the list obtained by reversing the list a.

PROOF OF CORRECTNESS OF DE CASTELJAU'S ALGORITHM. It is enough to prove the part of the claim concerning c,e. Indeed, by Remark 2.2, $\tilde{b}$ represents P in the Bernstein basis of d,c, and the claim is obtained by applying de Casteljau's Algorithm to $\tilde{b}$ at e. The output is $\tilde{b}''$ and $\tilde{b}'$ and the conclusion follows using again Remark 2.2.

Let $\delta_{p,i}$ be the list of length $p+1$ consisting of zeros except a 1 at the $i+1$-th place. Note that $\delta_{p,i}$ is the list of coefficients of $B_{p,i}(c,d)$ in the Bernstein basis of c,d. We will prove that the coefficients of $B_{p,i}(c,d)$ in the Bernstein basis of c,e coincide with the result of de Casteljau's Algorithm 2.9 performed with input $\delta_{p,i}$. The correctness of de Casteljau's Algorithm 2.9 for c,e then follows by linearity.

First notice that, since $\alpha = (d-e)/(d-c)$ and $\beta = (e-c)/(d-c)$,

$$
\frac{d-X}{d-c} = \alpha \frac{X-c}{e-c} + \frac{e-X}{e-c},
$$

$$
\frac{X-c}{d-c} = \beta \frac{X-c}{e-c}.
$$

Thus

$$
\left(\frac{d-X}{d-c}\right)^{p-i} = \sum_{k=0}^{p-i} \binom{p-i}{k} \alpha^k \left(\frac{X-c}{e-c}\right)^k \left(\frac{e-X}{e-c}\right)^{p-i-k},
$$

$$
\left(\frac{X-c}{d-c}\right)^i = \beta^i \left(\frac{X-c}{e-c}\right)^i.
$$

It follows that

$$
B_{p,i}(c,d) = \binom{p}{i} \sum_{j=i}^{p} \binom{p-i}{j-i} \alpha^{j-i} \beta^i \left(\frac{X-c}{e-c}\right)^j \left(\frac{e-X}{e-c}\right)^{p-j}.
$$

Since

$$\binom{p}{i}\binom{p-i}{j-i} = \binom{j}{i}\binom{p}{j},$$

$$B_{p,i}(c,d) = \sum_{j=i}^{p} \binom{j}{i} \alpha^{j-i} \beta^i \binom{p}{j} \left(\frac{X-c}{e-c}\right)^j \left(\frac{e-X}{e-c}\right)^{p-j}.$$

Finally,

$$B_{p,i}(c,d) = \sum_{j=i}^{p} \binom{j}{i} \alpha^{j-i} \beta^i B_{p,j}(c,e).$$

On the other hand, we prove by induction on p that de Casteljau's Algorithm with input $\delta_{p,i}$ outputs the list $\delta'_{p,i}$ starting with i zeros and with $(j+1)$-th element $\binom{j}{i}\alpha^{j-i}\beta^i$ for $j = i, \dots, p$.

The result is clear for $p = i = 0$. If de Casteljau's Algorithm applied to $\delta_{p-1,i-1}$ outputs $\delta'_{p-1,i-1}$, the equality

$$\binom{j}{i}\alpha^{j-i}\beta^i = \alpha\binom{j-1}{i}\alpha^{j-i-1}\beta^i + \beta\binom{j-1}{i-1}\alpha^{j-i}\beta^{i-1}$$

proves by induction on j that the algorithm applied to $\delta_{p,i}$ outputs $\delta'_{p,i}$. So the coefficients of $B_{p,i}(c,d)$ in the Bernstein basis of c, e coincide with the output of the algorithm with input $\delta_{p,i}$. $\square$

de Casteljau' Algorithm works both ways.

COROLLARY 2.11. *Let b, b' and b'' be the lists of coefficients of P in the Bernstein basis of (c,d), (c,e) and (e,d) respectively.*

(i) *De Casteljau's Algorithm applied to b with weights $\alpha = (d-e)/(d-c)$ and $\beta = (e-c)/(d-c)$ outputs b' and b''.*

(ii) *De Casteljau's Algorithm applied to b' with weights $\alpha = (e-d)/(e-c)$ and $\beta = (d-c)/(e-c)$ outputs b and $\tilde{b}''$.*

(iii) *De Casteljau's Algorithm applied to b'' with weights $\alpha = (d-c)/(d-e)$ and $\beta = (c-e)/(d-e)$ outputs $\tilde{b}'$ and b.*

De Casteljau's Algorithm gives a geometric construction of the control polygon of P on $[c,e]$ and on $[e,d]$ from the control polygon of P on $[c,d]$. The points of the new control polygons are constructed by taking iterated barycenters with weights α and β.

EXAMPLE 2.12. Continuing Example 2.7, de Casteljau's Algorithm gives

$$
\begin{array}{ccccccc}
4 & & -6 & & 7 & & 10 \\[4pt]
& -1 & & \frac{1}{2} & & \frac{17}{2} & \\[4pt]
& & -\frac{1}{4} & & \frac{9}{2} & & \\[4pt]
& & & \frac{17}{8} & & &
\end{array}
$$

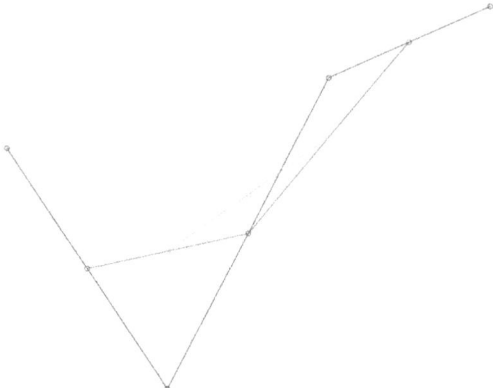

Figure 3. Control line of P on $[0, \frac{1}{2}]$.

We construct the control line of P on $[0, \frac{1}{2}]$ from the control line of P on $[0, 1]$ as explained in Figure 3.

We then draw the graph of P on $[0, 1]$ and the control line on $[0, \frac{1}{2}]$ in Figure 4.

3. Real Root Isolation in the Bernstein Basis

Let P be a polynomial of degree p in $\mathbb{R}[X]$. We are going to explain how to characterize the roots of P in $\mathbb{R}$, performing exact computations. The roots of P in $\mathbb{R}$ will be described by intervals with rational end points. Our method will be based on Descartes' law of signs (Theorem 1.1) and the properties of the Bernstein basis studied in the preceding section.

PROPOSITION 3.1. *Let b, b' and b'' be the lists of coefficients of P in the Bernstein basis of c, d; c, e; and e, d. If $c < e < d$, then*

$$V(b') + V(b'') \leq V(b).$$

Moreover if $P(e) \neq 0$, $V(b) - V(b') - V(b'')$ is even.

PROOF. The proof of the proposition is based on the following easy observations:

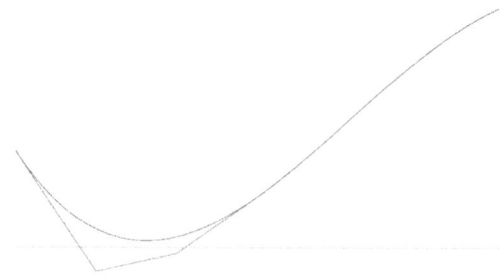

Figure 4. Graph of P on $[0, 1]$ and control line on $[0, \frac{1}{2}]$.

(i) Inserting in a list $a = a_0, \ldots, a_n$ a value x in $[a_i, a_{i+1}]$ if $a_{i+1} \geq a_i$ (resp. in $[a_{i+1}, a_i]$ if $a_{i+1} < a_i$) between a_i and a_{i+1} does not modify the number of sign variations.

(ii) Removing from a list $a = a_0, \ldots, a_n$ with first nonzero $a_k, k \geq 0$, and last nonzero $a_\ell, k \leq \ell \leq n$, an element $a_i, i \neq k, i \neq \ell$ decreases the number of sign variation by an even (possibly zero) natural number.

Indeed the lists $b^{(j)}$ defined from the values $b_j^{(i)}$ (see de Casteljau's algorithm), as follows:

$$b^{(0)} = b_0^{(0)}, \ldots, \ldots, \ldots, \ldots, b_p^{(0)}$$
$$b^{(1)} = b_0^{(0)}, b_0^{(1)}, \ldots, \ldots, \ldots, b_{p-1}^{(1)}, b_p^{(0)}$$
$$\cdots$$
$$b^{(i)} = b_0^{(0)}, \ldots, \ldots, b_0^{(i)}, \ldots, \ldots, b_{p-i}^{(i)}, \ldots, \ldots, b_p^{(0)}$$
$$\cdots$$
$$b^{(p-1)} = b_0^{(0)}, \ldots, \ldots, \ldots, b_0^{(p-1)}, b_1^{(p-1)}, \ldots, \ldots, \ldots, b_p^{(0)}$$
$$b^{(p)} = b_0^{(0)}, \ldots, \ldots, \ldots, \ldots, \quad b_0^{(p)} \quad, \ldots, \ldots, \ldots, \ldots, b_p^{(0)}$$

are successively obtained by inserting intermediate values and removing elements that are not end points, since when $c < e < d$, $b_j^{(i)}$ is between $b_j^{(i-1)}$ and $b_{j+1}^{(i-1)}$, for $i = 1, \ldots, p, j = 0, \ldots, p-i-1$. Thus $V(b^{(p)}) \leq V(b)$ and the difference is even. Since

$$b' = b_0^{(0)}, \ldots, \ldots, \ldots, \ldots, \quad b_0^{(p)},$$
$$b'' = b_0^{(p)} \quad, \ldots, \ldots, \ldots, \ldots, b_p^{(0)},$$

$V(b') + V(b'') \leq V(b^{(p)})$, and $V(b') + V(b'') \leq V(b)$. If $P(e) \neq 0$, it is clear that $V(b^{(p)}) = V(b') + V(b'')$, since $b_0^{(p)} = P(e) \neq 0$. □

EXAMPLE 3.2. Continuing Example 2.12, we observe, denoting by b, b' and b'', the lists of coefficients of P in the Bernstein basis of $0, 1$, $0, \frac{1}{2}$, and $\frac{1}{2}, 1$, that $V(b) = 2$. This is visible on Figure 1: the control line for $[0, 1]$ cuts twice the X-axis. Similarly, $V(b') = 2$. This is visible on Figure 4: the control line for $[0, \frac{1}{2}]$ also cuts twice the X-axis. Similarly, it is easy to check that $V(b'') = 0$.

We cannot decide from this information whether P has two roots on $(0, \frac{1}{2})$ or no root on $(0, \frac{1}{2})$.

Suppose that $P \in \mathbb{R}[X]$ is a polynomial of degree p with all its real zeros in $(-2^\ell, 2^\ell)$ and is square-free. Consider natural numbers k and c such that $0 \leq c \leq 2^k$ and define

$$a_{c,k} = \frac{-2^{\ell+k} + c2^{\ell+1}}{2^k}.$$

It is clear that, for k big enough, the polynomial P has at most one root in $(a_{c,k}, a_{c+1,k})$ and has no other complex root in $\mathcal{C}(a_{c,k}, a_{c+1,k})_0 \cup \mathcal{C}(a_{c,k}, a_{c+1,k})_1$.

Let $b(P, c, k)$ be the list of coefficients of P in the Bernstein basis of $(a_{c,k}, a_{c+1,k})$. Note that $b(P, 0, 0)$, the list of coefficients of P in the Bernstein basis of $(-2^\ell, 2^\ell)$, can easily be computed from P, using Proposition 2.3.

Using Theorem 2.5, it is possible to decide, for k big enough, whether P has exactly one root in $(a_{c,k}, a_{c+1,k})$ or has no root on $(a_{c,k}, a_{c+1,k})$ by testing whether $V(b(P, c, k))$ is zero or one.

EXAMPLE 3.3. Continuing Example 3.2, let us study the roots of P on $(0, 1)$, as a preparation to a more formal description of Algorithm 3.4 (B1 Real Root Isolation).

The Bernstein coefficients of P for $(0, 1)$ are $4, -6, 7, 10$. There may be roots of P on $(0, 1)$ as there are sign variations in its Bernstein coefficients.

As seen in Example 3.2, a first application of de Casteljau's Algorithm with weights $\alpha = \beta = \frac{1}{2}$ gives

$$
\begin{array}{ccccccc}
4 & & -6 & & 7 & & 10 \\
& -1 & & \frac{1}{2} & & \frac{17}{2} & \\
& & -\frac{1}{4} & & \frac{9}{2} & & \\
& & & \frac{17}{8} & & &
\end{array}
$$

There may be roots of P on $(0, \frac{1}{2})$ as there are sign variations in the Bernstein coefficients of P which are $32, -8, -2, 17$. There are no roots of P on $(\frac{1}{2}, 1)$.

We apply once more de Casteljau's Algorithm with weights $\frac{1}{2}, \frac{1}{2}$:

$$
\begin{array}{ccccccc}
4 & & -1 & & -\frac{1}{4} & & \frac{17}{8} \\
& \frac{3}{2} & & -\frac{5}{8} & & \frac{15}{16} & \\
& & \frac{7}{16} & & \frac{5}{32} & & \\
& & & \frac{19}{64} & & &
\end{array}
$$

There are no sign variations on the sides of the triangle so there are no roots of P on $(0, \frac{1}{4})$ and on $(\frac{1}{4}, \frac{1}{2})$.

An *isolating list for* P is a finite list L of rational points and disjoint open intervals with rational end points of $\mathbb{R}$ such that each point or interval of L contains exactly one root of P in $\mathbb{R}$ and every root of P in $\mathbb{R}$ belongs to an element of L.

ALGORITHM 3.4 (B1 REAL ROOT ISOLATION).

INPUT: a square-free nonzero polynomial $P \in \mathbb{R}[X]$, an interval $(-2^\ell, 2^\ell)$ containing the roots of P in $\mathbb{R}$, the list $b(P, 0, 0)$ of the Bernstein coefficients of P for $(-2^\ell, 2^\ell)$.

OUTPUT: a list $L(P)$ isolating for P.

1. Initialization: Define Pos $:= \{b(P, 0, 0)\}$ and $L(P) := \varnothing$.
2. While Pos is nonempty:

   - Remove $b(P, c, k)$ from Pos.
   - If $V(b(P, c, k)) = 1$ then insert $(a_{c,k}, a_{c+1,k})$ in $L(P)$.

- If $V(b(P, c, k)) > 1$ then
  - Compute $b(P, 2c, k+1)$ and $b(P, 2c+1, k+1)$ using de Casteljau's Algorithm with weights $(\frac{1}{2}, \frac{1}{2})$ and insert them in Pos.
  - If $P(a_{2c+1,k+1}) = 0$ then insert $a_{2c+1,k+1}$ in $L(P)$.

3. Output $L(P)$.

The hypotheses are not a real loss of generality since, given any polynomial Q, a square-free polynomial P having the same roots at Q can be computed using the gcd of Q and Q' (see for example [Basu et al. 2003]).

Moreover, setting

$$Q = c_p X^p + \ldots + c_0,$$

$$C(Q) = \sum_{0 \le i \le p} \left| \frac{c_i}{c_p} \right|,$$

the absolute value of any root of Q in $\mathbb{R}$ is smaller than $C(Q)$ [Mignotte and Ștefănescu 1999; Basu et al. 2003], so that it is easy, knowing Q, to compute ℓ such that $(-2^\ell, 2^\ell)$ contains the roots of Q in $\mathbb{R}$.

Since each subdivision yields after a scaling and a shift, new polynomials on $[0, 1]$ for which the distance between the roots if doubled, by the two-circles theorem, the maximal number h of the subdivisions is bounded by

$$h \le \lceil \log_2(2/s) \rceil,$$

where s is the minimal distance between the complex roots of Q. Using classical bounds on this minimal distance between the roots of a polynomial Q with integer coefficients [Mignotte and Ștefănescu 1999; Basu et al. 2003], one can prove that

$$h \le (p-1) \log_2 \|Q\|_2 + \tfrac{1}{2}(p+2) \log_2 p + 1$$

(where $\|Q\|_2$ is the 2-norm of the coefficient vector of Q) and that the binary complexity of computing the square-free part P of Q, computing ℓ such that $(-2^\ell, 2^\ell)$ contains the roots of Q in $\mathbb{R}$, and performing Algorithm 3.4 (B1 Real Root Isolation) for P, is $O(p^6(\tau + \log_2 p)^2)$, where p is a bound on the degree of Q and τ a bound on the bitsize of the coefficients of Q [Basu et al. 2003]. The coefficients of the elements of the $b(P, c, k)$ computed in the algorithm are rational numbers of bitsize $O(p^2(\tau + \log_2 p))$ [Basu et al. 2003]. Since there are at most $2p$ values of $b(P, c, k)$ in Pos throughout the computation, and there are $p+1$ coefficients in each $b(P, c, k)$, the workspace of the algorithm is $O(p^4(\tau + \log_2 p))$.

An improved version of Algorithm 3.4 (B1 Real Root Isolation) is based on the following idea, inspired from [Rouillier and Zimmermann 2004]: since every $b(P, c, k)$ computed in the algorithm carries the whole information about P, it is not necessary to store the value of $b(P, c, k)$ at all the nodes, and the workspace of the algorithm can be improved.

It will be necessary to convert the Bernstein coefficients of P on an interval $(a_{d,m}, a_{d+1,m})$ into the Bernstein coefficients of P on an interval $(a_{c,k}, a_{c+1,k})$.

ALGORITHM 3.5 (CONVERT).
INPUT: (c, k), (d, m), and $b(P, d, m)$, Bernstein coefficients of P on $(a_{d,m}, a_{d+1,m})$.
OUTPUT: the Bernstein coefficients $b(P, c, k)$ of P on $(a_{c,k}, a_{c+1,k})$.

1. Initialize $b := b(P, d, m)$.
2. Let $c = c_0 + \cdots + c_{n-1} 2^{n-1} + c_n 2^n + \cdots + c_{k-1} 2^{k-1}$ and $d = d_0 + \cdots + d_{n-1} 2^{n-1} + d_n 2^n + \cdots + d_{m-1} 2^{m-1}$, with $c_i \in \{0, 1\}$, $c_n \neq d_n$, $c_i = d_i$ for every $i < n$.
3. For i in $m-1, \ldots, n$:

   - If $d_i = 0$ then apply de Casteljau's Algorithm to b, weights $(-1, 2)$ and output b', b''. Update $b := b'$.

   - If $d_i = 1$ then apply de Casteljau's Algorithm to b with weights $(2, -1)$ and output b', b''. Update $b := b''$.

4. For i in $n, \ldots, k-1$:

   - If $c_i = 0$ then apply de Casteljau's Algorithm to b with weights $(\frac{1}{2}, \frac{1}{2})$ and output b', b''. Update $b := b'$.

   - If $c_i = 1$ then apply de Casteljau's Algorithm to b with weights $(\frac{1}{2}, \frac{1}{2})$ and output b', b''. Update $b := b''$.

5. Output b.

The correctness of this algorithm clearly follows from that of de Casteljau's Algorithm.

It is now easy to describe the improved real root isolation method.

ALGORITHM 3.6 (B2 REAL ROOT ISOLATION).
INPUT: a square-free nonzero polynomial $P \in \mathbb{R}[X]$, an interval $(-2^\ell, 2^\ell)$ containing the roots of P in $\mathbb{R}$, the list $b(P, 0, 0)$ of the Bernstein coefficients of P for $(-2^\ell, 2^\ell)$.
OUTPUT: a list $L(P)$ isolating for P.

1. Initialization: Set Pos $:= \{(0, 0)\}$, $L(P) := \varnothing$, $d := 0$, $m := 0$.
2. While Pos is nonempty:

   - Remove the first element (c, k) of Pos.
   - Compute $b(P, c, k)$ from $b(P, d, m)$ using Algorithm 3.5 (Convert).
   - If $V(b(P, c, k)) = 1$ then insert $(a_{c,k}, a_{c+1,k})$ in $L(P)$.
   - If $V(b(P, c, k)) > 1$ then:
     - Insert $(2c, k+1), (2c+1, k+1)$ at the beginning of Pos.
     - If $P(a_{2c+1,k+1}) = 0$ then insert $a_{2c+1,k+1}$ in $L(P)$.
   - Update $d := c$, $m := k$.

3. Output $L(P)$.

The next lemma is the key result for analyzing the complexity of this algorithm. (Note that the set of (c, k) such that $b(P, c, k)$ is computed in Algorithm 3.4 (B1 Real Root Isolation) is naturally equipped with a binary tree structure, denoted by T: (d, m) is a child of (c, k) if $d = 2c$ or $d = 2c + 1$, and $m = k + 1$.)

LEMMA 3.7. *In Algorithm 3.6 (B2 Real Root Isolation), the leaves of T are visited once, the nodes of T with one child are visited twice and the nodes of T with two children are visited three times*

PROOF. Easy by induction on the depth of T, noting that if (c, k) and (d, m) are nodes of T, $a_{c,k} < a_{d,m}$ if and only if every visit of (c, k) takes place before (d, m) is visited. $\square$

The binary complexity of Algorithm 3.6 (B2 Real Root Isolation) for P, is $O(p^6(\tau + \log_2 p)^2)$, similarly to Algorithm 3.4 (B1 Real Root Isolation), since every node in the tree T is visited at most three times in by Lemma 3.7. However Algorithm 3.6 (B2 Real Root Isolation) uses only $O(p^3(\tau + \log_2 p))$ workspace, since only one vector of Bernstein coefficients is stored throughout the computation, rather than $O(p^4(\tau + \log_2 p))$ workspace in Algorithm 3.4 (B1 Real Root Isolation).

We can also perform the computation using interval arithmetic. The basic idea of interval arithmetic is that real numbers are represented by intervals with rational bounds encoded as floating point numbers with a fixed precision. The advantages of interval arithmetic is that it is much quicker than exact arithmetic, and it allows us to compute with polynomials known approximately.

The interval arithmetic we consider is indexed by two natural numbers u, n, defining the precision. The u, n-intervals are of the form $[\frac{i}{2^u}, \frac{j}{2^u}]$, with i and j being integers between -2^u and 2^u, $i \le j$, and I and J being integers between -2^n and 2^n. A consistent interval arithmetic is compatible with the arithmetic operations: if α and β are two real numbers represented respectively by two intervals A and B, the result $A \odot B$ of any arithmetic operation $\odot$ will contain the real number $\alpha \odot \beta$. In the next paragraph, we assume working with a multi-precision interval arithmetic such as in [Revol and Rouillier 2002] (where u can be arbitrary fixed by the user). In order to perform Algorithm 3.6 (B2 Real Root Isolation) in this arithmetic, we only need to double an interval, subtract two intervals and compute the average of two intervals.

The sign of an interval $[a, b]$, where $a \le b$, is defined as follows:

$$
\mathrm{sign}[a, b] = \begin{cases} 0 & \text{if } a = b = 0, \\ 1 & \text{if } a > 0, \\ -1 & \text{if } b < 0, \\ ? & \text{if } a \le 0 \le b, \ a \ne 0, \ b \ne 0. \end{cases}
$$

The number of sign variations in a list $A = [a_0, b_0], \ldots, [a_p, b_p]$ of intervals with rational end points is defined as follows:

- If $\text{sign}[a_i, b_i] \neq ?$ for all $i = 0, \dots, p$, set

$$V([a_0, b_0], \dots, [a_p, b_p]) = V(\text{sign}[a_0, b_0], \dots, \text{sign}[a_p, b_p]).$$

- If, for every $i = 1, \dots, p$ such that $\text{sign}[a_i, b_i] = ?$, both $\text{sign}[a_{i-1}, b_{i-1}]$ and $\text{sign}[a_{i+1}, b_{i+1}]$ are defined and moreover $\text{sign}[a_{i-1}, b_{i-1}]\,\text{sign}[a_{i+1}, b_{i+1}] < 0$, then set

$$V(A) = V(B),$$

where B is obtained by removing from A all the $[a_i, b_i]$ such that $\text{sign}[a_i, b_i] = ?$.
- Otherwise set $V(A) = ?$.

EXAMPLE 3.8. If $A = [1, 2], [-2, -1]$, $V(A) = 1$. If $A = [1, 2], [-1, 1]$, $V(A) = ?$. If $A = [1, 2], [-1, 1], [-2, -1]$, $V(A) = 1$.

ALGORITHM 3.9 (B3 REAL ROOT ISOLATION).
INPUT: an integer ℓ, a precision u, n, a list $\bar{b}(0, 0)$ with $p+1$ elements which are u, n-intervals, and whose first and last element do not contain 0.
OUTPUT: a list L and a list N of intervals such that for every polynomial P such that $(-2^\ell, 2^\ell)$ contains the roots of P in $\mathbb{R}$, and whose Bernstein coefficients for $(-2^\ell, 2^\ell)$ belong to $\bar{b}(0, 0)$, there exists one and only one root of P in each interval of L and all the other roots of P in $(-2^\ell, 2^\ell)$, belong to an interval of N.

1. Initialization: Compute $V(\bar{b}(0, 0))$, using Proposition 2.3 and u, n-arithmetic define $\text{Pos} := \{(0, 0)\}$, $L := \varnothing$, $N := \varnothing$, $d := 0$, $m := 0$.
2. While Pos is nonempty:

   - Remove the first element (c, k) of Pos.
   - Compute $\bar{b}(c, k)$ from $\bar{b}(d, m)$ by Algorithm 3.5 (Convert), using u, n-arithmetic.
   - If $V(\bar{b}(c, k)) = 1$ then insert $(a_{c,k}, a_{c+1,k})$ in L.
   - If $V(\bar{b}(c, k)) > 1$ then insert $(2c, k+1), (2c+1, k+1)$ at the beginning of Pos.
   - If $V(\bar{b}(c, k)) = ?$ then insert $(a_{c,k}, a_{c+1,k})$ in N.
   - Update $d := c$, $m := k$.

3. Output L, N.

Interval arithmetic can be used as well when the polynomial P is known exactly. In this case we can compute the square-free part of P and it is easy to design a variant of Algorithm 3.9 and output a list of isolating intervals by augmenting precision, examining again the intervals where no decision has been taken yet.

ALGORITHM 3.10 (B4 REAL ROOT ISOLATION).
INPUT: a square-free $P \in \mathbb{R}[X]$, and the list $b(P, 0, 0)$ of Bernstein coefficients of P for $(-2^\ell, 2^\ell)$, where $(-2^\ell, 2^\ell)$ contains the roots of P in $\mathbb{R}$.
OUTPUT: a list $L(P)$ isolating for P.

1. Initialization: u such that the elements of $b(P,0,0)$ belong to $(-2^u, 2^u)$, $n :=$ 1. Compute $\mathrm{V}(b(P,0,0))$, define Pos $:= \{(0,0)\}$, $L(P) := \varnothing$, $N(P) := \varnothing$, $d := 0$, $m := 0$.
2. While Pos is nonempty:
   - Remove the first element (c,k) of Pos.
   - Compute $b(P,c,k)$ from $b(P,d,m)$ by Algorithm 3.5 (Convert) using u, n-arithmetic.
   - If $\mathrm{V}(b(P,c,k)) = 1$ then insert $(a_{c,k}, a_{c+1,k})$ in $L(P)$.
   - If $\mathrm{V}(b(P,c,k)) > 1$ then:
     - Insert $(2c, k+1), (2c+1, k+1)$ at the beginning of Pos.
     - If $P(a_{2c+1,k+1}) = 0$ then insert $a_{2c+1,k+1}$ to $L(P)$.
   - If $\mathrm{V}(b(P,c,k)) = ?$ then insert $(a_{c,k}, a_{c+1,k})$ in $N(P)$.
   - Update $d := c$, $m := k$.
3. If $N(P) \neq \varnothing$ then update $n := n+1$, Pos $= N(P)$ and go to step 2.
4. Output $L(P)$.

4. Real Root Isolation in the Monomial Basis

The preceding methods for real root isolation are adapted to polynomials given in the Bernstein basis. However in many cases, the polynomials are given in the monomial basis, and the conversion to the Bernstein basis is computationally expensive. It is thus natural to look for real root isolation algorithms adapted to the case where the polynomials are expressed in the monomial basis.

Such algorithms for real root isolation in the monomial basis are very classical, and have been studied extensively, starting from [Uspensky 1948] (see [Rouillier and Zimmermann 2004] for a bibliography). We prove here that their correctness is an immediate consequence of the correctness of the corresponding algorithms in the Bernstein basis.

Rather than looking at the Bernstein coefficients of the same polynomial P on varying intervals, we are going to consider different polynomials closely related to P on each interval. We need some notation. Suppose as before that $P \in \mathbb{R}[X]$ is a polynomial of degree p with all its real zeros in $(-2^\ell, 2^\ell)$ and is square-free, consider natural numbers k and c such that $0 \leq c \leq 2^k$ and define

$$a_{c,k} = \frac{-2^{\ell+k} + c2^{\ell+1}}{2^k}.$$

We define

$$P_{c,k} := \mathrm{C}_{2^{\ell+1-k}}(\mathrm{T}_{-a_{c,k}}(P)).$$

$P_{c,k}$ is simply the result of the transformation operated on P when the segment $(a_{c,k}, a_{c+1,k})$ is sent to $(0,1)$ by a translation followed by a contraction.

The following lemma is the key result making the connection between the real root isolation in the monomial basis and the Bernstein basis.

LEMMA 4.1. *Let $Q_{c,k} := \mathrm{T}_{-1}\left(\mathrm{Rec}_p(P_{c,k})\right)$. Then $\mathrm{V}(Q_{c,k}) = \mathrm{V}(b(P,c,k))$.*

PROOF. Immediate by Proposition 2.3.   □

The four algorithms of the preceding section have analogous versions in the monomial basis [Rouillier and Zimmermann 2004]. We describe only the algorithms corresponding to the conversion from one interval to another and the improved root isolation algorithm.

ALGORITHM 4.2 (M1 CHANGE INTERVAL).
INPUT: (c,k), (d,m) and the polynomial $P_{d,m}$.
OUTPUT: the polynomial $P_{c,k}$.

1. Let $c = c_0 + \cdots + c_{n-1}2^{n-1} + c_n 2^n + \cdots + c_{k-1}2^{k-1}$ and $d = d_0 + \cdots + d_{n-1}2^{n-1} + d_n 2^n + \cdots + d_{m-1}2^{m-1}$, with $c_i \in \{0,1\}$, $c_n \neq d_n$, $c_i = d_i$ for every $i < n$. and $R := P_{d,m}$.
2. For i from $m-1$ to n:
   - If $d_i = 0$, then $R := \mathrm{C}_2(R)$.
   - If $d_i = 1$, then $R := \mathrm{C}_2(\mathrm{T}_{-1}(R))$.
3. For i from n to $k-1$:
   - If $c_i = 0$, then $R := \mathrm{C}_{1/2}(R)$.
   - If $c_i = 1$, then $R := \mathrm{C}_{1/2}(\mathrm{T}_{-1/2}(R))$.
4. Output R.

The correctness of the algorithm follows clearly from the definition of $P_{c,k}$.

It is now easy to describe the improved real root isolation method in the monomial basis.

ALGORITHM 4.3 (M2 REAL ROOT ISOLATION).
INPUT: a square-free nonzero polynomial $P \in \mathbb{R}[X]$, and an interval $(-2^\ell, 2^\ell)$ containing the roots of P in $\mathbb{R}$.
OUTPUT: a list $L(P)$ isolating for P.

1. Initialization: Define $\mathrm{Pos} := \{(0,0)\}$, $L(P) := \varnothing$, $d := 0$, $m := 0$.
2. While Pos is nonempty:
   - Remove the first element (c,k) of Pos.
   - Compute $P_{c,k}$ from $P_{d,m}$ using Algorithm 3.5 (Change Interval). Take $Q_{c,k} := \mathrm{T}_{-1}\left(\mathrm{Rec}_p(P_{c,k})\right)$.
   - If $\mathrm{V}(Q_{c,k}) = 1$, then insert $(a_{c,k}, a_{c+1,k})$ in $L(P)$.
   - If $\mathrm{V}(Q_{c,k}) > 1$ then:
     - Insert $(2c, k+1), (2c+1, k+1)$ at the beginning of Pos.
     - If $P(a_{2c+1,k+1}) = 0$ then insert $a_{2c+1,k+1}$ in $L(P)$.
   - Update $d := c$, $m := k$.
3. Output $L(P)$.

The correctness of Algorithm 4.3 follows from the correctness of Algorithm 3.6 and Lemma 4.1.

The complexity analysis of the real root isolation method in the monomial basis and in the Bernstein basis are quite similar.

5. Efficiency of the Methods

The experimental behavior of Algorithms 4.3 [M2 Real Root Isolation in monomial basis], more precisely of its interval arithmetic variants is excellent, and real root isolation can be performed by this method for polynomials of degree several thousands and with coefficients of bit size several hundred (see [Rouillier and Zimmermann 2004] for details on these experimental results).

The experimental behavior in the case of the Bernstein basis has not been studied fully yet, but the first experiments indicate that the algorithms presented here are as efficient as the corresponding ones in the monomial basis, if the polynomial is initially given in the Bernstein basis.

Implementations of these algorithms are available in the libraries RS (see http://fgbrs.lip6.fr/salsa/Software/) and SYNAPS (see http://www-sop.inria.fr/galaad/software/synaps/).

References

[Basu et al. 2003] S. Basu, R. Pollack, and M.-F. Roy, *Algorithms in real algebraic geometry*, Algorithms and Computation in Mathematics **10**, Springer, Berlin, 2003.

[Farin 1990] G. Farin, *Curves and surfaces for computer aided geometric design*, Academic Press, Boston, 1990.

[Mehlhorn 2001] K. Mehlhorn, "A remark on the Sign Variation Method for real root isolation", preprint, 2001. Available at http://www.mpi-sb.mpg.de/~mehlhorn/ftp/Descartes.ps.

[Mignotte and Ştefănescu 1999] M. Mignotte and D. Ştefănescu, *Polynomials*, Series in Disc. Math. and Theoret. Computer Science, Springer-Verlag, Singapore, 1999.

[Ostrowski 1950] A. M. Ostrowski, "Note on Vincent's theorem", *Ann. of Math.* (2) **52** (1950), 702–707. Reprinted as pp. 728–733 in *Collected Mathematical Papers*, vol. 1, Birkhäuser, Basel, 1983.

[Revol and Rouillier 2002] N. Revol and F. Rouillier, "Motivations for an arbitrary precision interval arithmetic and the MPFI library", pp. 155–161 in *Proceedings of the Workshop on Validated Computing* (Toronto, 2002), 2002. Revised version to appear in *Reliable Computing*.

[Rouillier and Zimmermann 2004] F. Rouillier and P. Zimmermann, "Efficient isolation of polynomial's real roots", *J. Comput. Appl. Math.* **162**:1 (2004), 33–50.

[Struik 1969] D. J. Struik (editor), *A source book in mathematics, 1200–1800*, edited by D. J. Struik, Harvard University Press, Cambridge, MA, 1969.

[Uspensky 1948] J. V. Uspensky, *Theory of equations*, MacGraw-Hill, New York, 1948.

BERNARD MOURRAIN
GALAAD, INRIA
BP 93
06902 SOPHIA ANTIPOLIS
FRANCE
    mourrain@sophia.inria.fr

FABRICE ROUILLIER
SALSA, INRIA/LIP6
8, RUE DU CAPITAINE SCOTT
75015 PARIS
FRANCE
    Fabrice.Rouillier@inria.fr

MARIE-FRANÇOISE ROY
IRMAR, CNRS
UNIVERSITÉ DE RENNES I
CAMPUS DE BEAULIEU
35042 RENNES CEDEX
FRANCE
    marie-francoise.roy@univ-rennes1.fr

Combinatorial and Computational Geometry
MSRI Publications
Volume **52**, 2005

Extremal Problems Related to the Sylvester–Gallai Theorem

NIRANJAN NILAKANTAN

ABSTRACT. We discuss certain extremal problems in combinatorial geometry, including Sylvester's problem and its generalizations.

1. Introduction

Many interesting problems in combinatorial geometry have remained unsolved or only partially solved for a long time. From time to time breakthroughs are made. In this survey, we shall discuss the known results about some metric and nonmetric problems. In particular, we shall discuss the Sylvester–Gallai problem and the Dirac–Motzkin conjecture on the existence and number of ordinary lines, the Dirac conjecture on the number of connecting lines, and the problem of distinct and repeated distances. The main focus will be on versions of these problems in the Euclidean and real projective plane.

The method of allowable sequences will be described as a tool to give purely combinatorial solutions to extremal problems in combinatorial geometry.

2. Sylvester's Problem

Sylvester [1893] posed a question in the *Educational Times* that was to remain unsolved for 40 years until it was raised again by Erdős [1943]. Then it was soon solved by Gallai [1944], who gave an affine proof. More followed: Steinberg's proof in the projective plane and others by Buck, Grünwald and Steenrod, all collected in [Steinberg et al. 1944]; Kelly's Euclidean proof [1948], and others, including [Motzkin 1951; Lang 1955; Williams 1968].

We give the following definitions before we state the problem and its solutions.

Let P be a finite set of 3 or more noncollinear points in the plane. Let F be a finite collection of simple closed curves in the real projective plane which do not separate the plane, every two of which have exactly one point in common, where they cross. F is known as a *pseudoline arrangement*.

CONNECTING LINE: a line containing two or more points of P.
ORDINARY LINE: a connecting line which has exactly two points of P on it.
VERTEX: an intersection of two or more lines of a straight line arrangement or pseudolines of a pseudoline arrangement.
ORDINARY POINT: a vertex which is the intersection of exactly two lines or two pseudolines.

Sylvester asked for a proof of the statement that every set P of noncollinear points always determines an ordinary line. In the dual, one has to show that any straight line arrangement in which not all lines are concurrent has an ordinary point. By the principle of duality, proofs for point configurations carry over trivially into proofs for line arrangements and vice versa. The canonical correspondence maps the point (a, b) to the line $y = -ax + b$.

Levi [1926] introduced the notion of a pseudoline defined above. A natural question to ask is whether every pseudoline arrangement in which not all pseudolines are concurrent has an ordinary vertex. This is more general than the question of whether every straight line arrangement has an ordinary vertex, since every straight-line arrangement has an equivalent pseudoline arrangement, but there exist unstretchable pseudoline arrangements [Grünbaum 1970; Goodman and Pollack 1980b].

Solutions to Sylvester's problem. We now show some of the techniques used to solve Sylvester's problem in the both the primal and dual versions, and in the Euclidean as well as the projective plane.

Gallai's proof (affine). Choose any point $p_1 \in P$. If p_1 lies on an ordinary line, we are done, so we may *assume that p_1 does not lie on any ordinary line*. Project p_1 to infinity and consider the set of lines containing p_1. These lines are parallel, and there are at least two such lines. Let s be a connecting line not through p_1 which forms the smallest angle with the parallel lines:

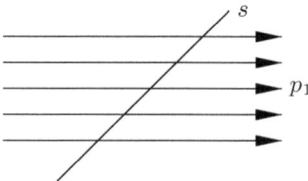

We assert that s is ordinary. If not, it must have at least 3 points p_2, p_3, p_4, as in the figure at the top of the next page. The connecting line through p_1 and p_3 has another point p_5, since it is not ordinary (this point is shown in two possible positions in the figure). Then, either $p_5 p_2$ or $p_5 p_4$ forms a smaller angle with the parallel lines than s, contradicting the hypothesis that s forms the smallest angle.

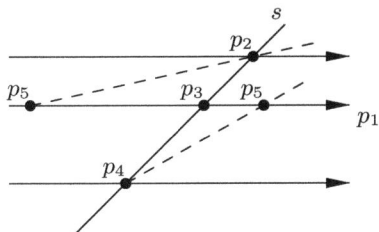

Kelly's proof (Euclidean). We have the set P of points not all collinear and the set S of connecting lines determined by P. Any point in P and any connecting line not through the point determine a perpendicular distance from the point to the line. The collection of all these distances is finite, because P and S are finite, so there is a smallest such distance. Let $p^* \in P$ and $s^* \in S$ be a nonincident pair realizing this smallest distance, and let q be the foot of the perpendicular line from p^* to s^*:

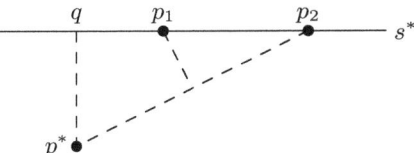

Then s^* is ordinary; otherwise it would contain three points of P, at least two of them lying on the same side of q. Let these two points be p_1 and p_2, with p_1 between q and p_2. Now the distance from p_1 to the connecting line p^*p_2 would be less than the distance from p^* to s^*, giving a contradiction.

Steinberg's proof (projective). With S and P as above, take any p in P. If p lies on an ordinary line we are done, so we may *assume that p lies on no ordinary line.* Let l be a line through p that is not a connecting line, that is, one that contains no point of P apart from p. Let Q be the set of intersections of l with lines in S, and take $q \in Q$ next to p (meaning that one of the open segments determined by p and q on the projective line l contains no element of Q). Let s be a line of S through q; then s must be ordinary. Otherwise, there would be three points of P on s, say p_1, p_2, p_3 (arranged in that order in $s \setminus \{q\}$; note that q is not in P, by our choice of l):

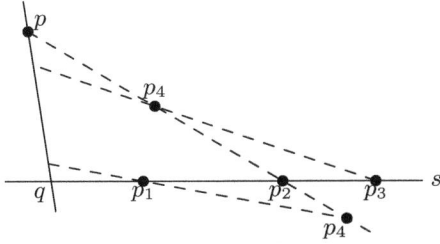

The line through p and p_2 would then contain another point of P, say p_4, since p lies on no ordinary line; then p_1p_4 or p_3p_4 would meet the forbidden segment pq (see the figure where two possibilities for p_4 are shown).

The Dirac–Motzkin conjecture. Having determined the existence of an ordinary line (or point, in the dual problem), attention was turned to the problem of establishing the number of ordinary lines (or points). For P an *allowable* set of points — one not all of whose elements are collinear — let $m(P)$ denote the number of ordinary lines determined by P. Define

$$m(n) = \min_{|P|=n} m(P),$$

where P ranges over all allowable sets of points of cardinality $|P| = n$.

De Bruijn and Erdős [1948] proved that $m(n) \geq 3$, and this was proved again by Dirac [1951], who conjectured that there were at least $\lfloor n/2 \rfloor$ ordinary lines. In a different context, Melchior [1941] proved again the $m(n) \geq 3$ bound. Motzkin [1951] improved this to $m(n) > \sqrt{2n} - 2$. Kelly and Moser [1958] improved the lower bound to $3n/7$. Kelly and Rottenberg [1972] proved the same result for pseudoline arrangements. In 1980, Hansen gave a lengthy "proof" of Dirac's $\lfloor n/2 \rfloor$ conjecture, but it was found to be incorrect by Csima and Sawyer [1993], who nonetheless proved that there exist at least $6n/13$ ordinary lines.

Creating point configurations with few ordinary lines is hard. When n is odd, we know of configurations where the conjecture is tight only when $n = 7$ and $n = 13$. The former is shown by the Kelly–Moser configuration [1958]:

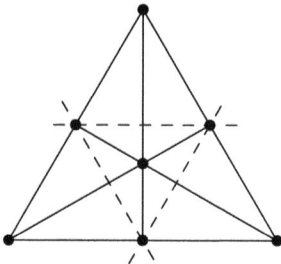

and the latter by the Crowe–McKee configuration [1968]. The Böröczky configurations [Crowe and McKee 1968] are valid for all even n; they are most easily visualized dually — here is the case $n = 12$, with dots marking ordinary vertices:

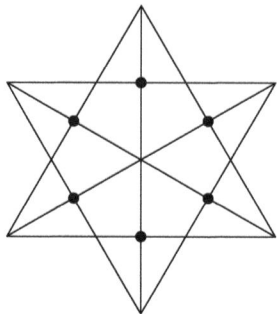

Solutions to the generalized problem. We now outline the techniques used in the progress towards settling the conjecture.

PENCIL: a collection of lines all of which intersect at a single point.
NEAR-PENCIL: a collection of lines all but one of which intersect at a single point.

According to the moment's convenience, we assume given either some arrangement L of lines not forming a pencil or near-pencil, or a configuration of points not all collinear. We seek to prove a lower bound for the number of ordinary points in the first case, and ordinary lines in the second.

Melchior's proof of the existence of 3 ordinary points. The lines of L partition the real projective plane into polygonal regions. Let V, E and F denote the number of vertices, edges and faces in the partition. By Euler's formula,

$$V - E + F = 1.$$

Let f_i denote the number of faces with exactly i sides and v_i the number of vertices incident with exactly i lines. Since the lines are not all concurrent, every face has at least three sides, so $f_2 = 0$. Then,

$$V = \sum_{i \geq 2} v_i, \quad F = \sum_{i \geq 3} f_i, \quad 2E = \sum_{i \geq 3} i f_i = 2 \sum_{i \geq 2} i v_i.$$

This implies that

$$3 = 3V - E + 3F - 2E = 3 \sum_{i \geq 2} v_i - \sum_{i \geq 2} i v_i + 3 \sum_{i \geq 3} f_i - \sum_{i \geq 3} i f_i$$

$$= \sum_{i \geq 2} (3 - i) v_i + \sum_{i \geq 3} (3 - i) f_i,$$

and hence that

$$v_2 = 3 + \sum_{i \geq 4} (i - 3) v_i + \sum_{i \geq 4} (i - 3) f_i \geq 3 + \sum_{i \geq 4} (i - 3) v_i.$$

Thus, any finite set of nonconcurrent lines has at least 3 ordinary points.

Motzkin's proof of the existence of $O(\sqrt{n})$ ordinary lines. Consider a point $p \in P$ not lying on any ordinary line. (If there is no such point, there are at least $n/2$ ordinary lines and we are done.) Consider the set of connecting lines not passing through p. These partition the plane into regions, and p lies in one of these, which is called its *cell* C. If p has at least 3 lines on the boundary of its cell, then all the lines in the boundary of the region containing p must be ordinary.

It is easy to see that no point of P can lie on the edges of the cell C. Suppose one of the lines l on the boundary of the cell is not ordinary, that is, l has 3 points p_1, p_2, p_3 labeled so that p_1, x separate p_2, p_3, where x is a point on l not in P on the boundary of C (see figure on the next page). The line pp_1 is not ordinary by hypothesis, and therefore contains a point q of P. But then either

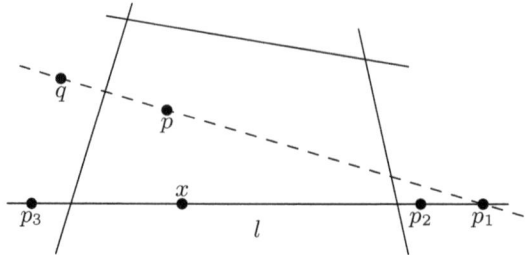

qp_2 or qp_3 cuts the cell C, contradicting the fact that C is the polygonal region containing p.

Thus, the ordinary lines partition the plane into polygonal regions, and all the points which do not lie an any ordinary line lie in one of these regions. It is easy to see that no region can have more than one point.

Now, m ordinary lines determine at most $\binom{m}{2} + 1$ regions, and can have at most $2m$ points of P on them. Since every point is on an ordinary line or in a cell, we have $\binom{m}{2} + 1 + 2m \geq n$, implying that $m \geq \sqrt{2n} - 2$.

Kelly and Moser's proof of the existence of $3n/7$ ordinary lines. Let P be the set of points and S the set of connecting lines. We denote a generic point by p and a generic line by s. The set of lines of S which do not go through p subdivide the plane into polygonal regions. p is contained in one of these polygonal regions, which is called its *residence*.

NEIGHBOR OF p: a line of S containing the edges of the residence of p.
ORDER OF p: the number of ordinary lines passing through p.
RANK OF p: the number of neighbors of p which are ordinary lines.
INDEX OF p: the sum of its order and rank.

THEOREM 1. *If a point q has precisely one neighbor, then S is a near-pencil.*

This is because the neighboring line is the only line which does not pass through q, and all the other lines pass through q:

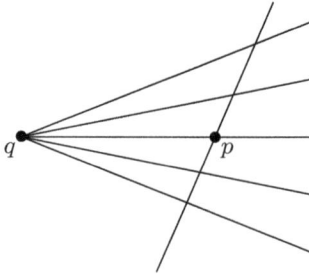

THEOREM 2. *If a point p has precisely two neighbors, then S is a near-pencil.*

The lines of S that do not pass through p form a pencil, or else p would have at least three neighbors. Let q be the vertex of the pencil. Let s_i and s_j be two lines through q and p_i and p_j be points on s_i and s_j respectively, different from q. The connecting line through p_i and p_j does not pass through q and therefore passes through p. Thus, only one line of S passes through p, and all the rest pass through q, as in the previous figure.

As a consequence of the previous two theorems, we have:

THEOREM 3. *If S is not a near-pencil, each point of P has at least three neighbors.*

THEOREM 4. *If the order of p is zero, every neighbor of p is an ordinary line.*

This was proved in [Motzkin 1951]; we gave the proof on page 483.

THEOREM 5. *Any point of P not of order two has index at least three.*

If the order is zero and S is not a near-pencil, the rank is at least three. If the order is at least three, there is nothing more to prove. If the order is one, the rank is at least two and the correct proof of this was given by Dirac in his review of Kelly and Moser's article [Dirac 1959].

THEOREM 6. *If a line s of S is a neighbor of three points p_1, p_2, p_3, then the points of P which lie on s are on the connecting lines determined by p_1, p_2, p_3.*

Three points that have a common neighbor cannot be collinear: if p_1, p_2 separate x, p_3, where x is the intersection of s with p_1p_2, then s cannot lie on the boundary of p_3's cell. Let the intersections of p_1p_2, p_2p_3, p_3p_1 with s be x_3, x_1, x_2 respectively. If p is a point of P on s such that x_ix_j separate x_kp, then pp_i and pp_j separate s from p_k. Here, i, j, k is some permutation of $1, 2, 3$.

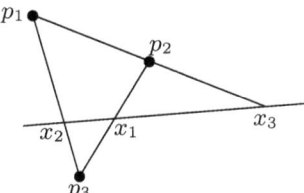

This implies the following.

THEOREM 7. *A line l of S is a neighbor of at most four points.*

Suppose l was the neighbor of five points $p_1, \ldots, p_5$. Looking at p_1, p_2, p_3, we see that at least 2 of x_1, x_2, x_3 must be elements of P. Assume that x_2, x_3 are elements of P. However, neither x_2 nor x_3 can be on the lines p_1p_4 or p_1p_5. This means that one of the points of P on l is not on the connecting lines of the set p_1, p_4, p_5, implying that l is not a neighbor of one of the three points.

THEOREM 8. *If I_i is the index of the point p_i, then*

$$m \geq \frac{1}{6} \sum_{i=1}^{n} I_i$$

Since each ordinary line can be counted at most six times — four times as a neighbor and twice as being incident with each of its points — the sum of the index over all the points is greater than six times the number of ordinary lines.

THEOREM 9. $m \geq 3n/7$.

Let k be the number of points of order 2. Then

$$m \geq \frac{3(n-k) + 2k}{6} = \frac{3n - k}{6},$$

which leads to $6m \geq 3n - k \geq 3n - m$ since $m \geq k$ (trivially). Hence $m \geq 3n/7$.

Proof by Csima and Sawyer. Csima and Sawyer improved upon Kelly and Moser by showing that except for pencils and the Kelly–Moser configuration the number s of ordinary points in a configuration of n lines is at least $\frac{6}{13}n$, with eqaulity occuring for the McKee configuration. They generalize the Kelly–Moser proof in the following way. In the Kelly–Moser proof, the sum of the indices of each point was compared to the six times the number of ordinary lines to get the desired bound. In the Csima–Sawyer result, the index is a weighted sum of the order and the rank. The following is a sketch of their proof for an arrangement of lines, and works for arrangements of pseudolines as well.

ATTACHED: An ordinary point which not on a line but associated to it, by proximity. For instance, in the proof of Kelly and Moser, the ordinary lines on the boundary of the cell of a point are attached to it.

TYPE of a line l: The pair $T(l) = (\mu, \nu)$, if there are exactly μ ordinary points on l and ν ordinary points attached to l.

α-WEIGHT of a line l of type (μ, ν): : the number $w_\alpha(l) = \alpha\mu + \nu$.

THEOREM 10. *Suppose Γ is a finite configuration of lines in the real projective plane having two lines of type $(2, 0)$ that intersect in an ordinary point. Then Γ is the Kelly–Moser configuration.*

THEOREM 11. *Apart from pencils, if $T(l) \neq (2, 0)$, then $w_1(l) \geq 3$.*

This is a restatement of a theorem of Kelly and Moser, which asserts that the index of a point which is not of order two is at least three.

THEOREM 12. *If l_1 and l_2 have an ordinary intersection in any configuration other than pencils, then $w_1(l_1) + w_1(l_2) \geq 5$.*

THEOREM 13. *Except for pencils and the Kelly–Moser configuration, $s \geq \frac{6}{13}n$.*

Partition the ordinary points into the sets

σ = ordinary points that lie on a line of type $(2, 0)$,

τ = ordinary points that do not lie on a line of type $(2, 0)$.

and the lines into sets of bad, good and fair lines:

$$\mathcal{B} = \text{lines } l \text{ of type } (2,0),$$

$$\mathcal{G} = \text{lines } l \text{ that contain a point in } \sigma \text{ but } l \notin \mathcal{B},$$

$$\mathcal{F} = \text{lines } l \text{ that do not contain a point in } \sigma.$$

The set $\mathcal{G}$ is further partitioned into sets

$$\mathcal{G}_j = \text{lines } l \text{ in } \mathcal{G} \text{ which contain exactly } j \text{ points of } \sigma$$

Consider two lines l and m. If their intersection is in σ, we can assume without loss of generality that $l \in \mathcal{B}$. Then m has a 1-weight of at least three, and lies in $\mathcal{G}$. Thus, each point in σ appears on exactly one line from $\mathcal{B}$ and one line from $\mathcal{G}$. If $B = |\mathcal{B}|$, $G = |\mathcal{G}|$, $F = |\mathcal{F}|$, and $G_j = |\mathcal{G}_j|$, we have

$$G = \sum_j G_j \sum_{j \geq 1} G_j = |\sigma| = 2B.$$

If $l \in \mathcal{G}_1$, then $T(l) = (\mu, \nu) \geq (1,0)$, and $w_1(l) = \mu + \nu \geq 3$, and since $\alpha \geq 1$, we have $w_\alpha(l) = \alpha\mu + \nu \geq \alpha + 2$. If $l \in \mathcal{G}_2$, then $w_\alpha(l) \geq 2\alpha + 1$. If $l \in \mathcal{G}_j$ for $j \geq 3$, then $w_\alpha(l) \geq j\alpha$. If $l \in \mathcal{B}$, then $w_\alpha(l) = 2\alpha$, and if $l \in \mathcal{F}$, then $w_\alpha(l) \geq 3$.
Thus,

$$\sum_{l \in \Gamma} w_\alpha(l) = \sum_{l \in \mathcal{B}} w_\alpha(l) + \sum_j \sum_{m \in \mathcal{G}_j} + \sum_{l \in \mathcal{F}} w_\alpha(l)$$

$$\geq 2\alpha B + (\alpha + 2)G_1 + (2\alpha + 1)G_2 + \sum_{j \geq 3} j\alpha G_j + 3F$$

$$= 2\alpha B + \alpha\left(\sum_{j \geq 1} jG_j\right) + 2G_1 + G_2 + 3F$$

$$= (4\alpha - 2)B + 3G_1 + 3G_2 + \sum_{j \geq 3} jG_j + 3F$$

$$\geq (4\alpha - 2)B + 3G + 3F$$

Choosing $\alpha = \frac{5}{4}$ we get,

$$\sum_{l \in \Gamma} w_{5/4}(l) \geq 3B + 3G + 3F = 3n.$$

Consider a matrix with rows labeled by the lines l and columns labeled by the ordinary points. If the ith line is incident with the j-th ordinary point, the (i,j)-th entry of the matrix is $\frac{5}{4}$. If the j-th point is attached to the i-th line, the (i,j)-th entry is 1. All other entries are zero.

An ordinary point P is attached to at most four lines. Therefore, the column sum is at most $2(\frac{5}{4}) + 4 = \frac{13}{2}$. The sum over all the rows is exactly $\sum_{l \in \Gamma} w_{5/4}(l) \geq 3n$. Consequently,

$$3n \leq \sum_{l \in \Gamma} w_{5/4}(l) \leq \frac{13}{2}s.$$

3. Allowable Sequences

The notion of allowable sequences has proved very effective in determining the combinatorial classification of configurations of the plane.

A configuration of n points is an ordered n-tuple of distinct points in the plane. The points are labeled $1, 2, \ldots, n$. Given a configuration C and a directed line l which is not orthogonal to any line determined by two points of C, the orthogonal projection of C on l determines a permutation of $1, 2, \ldots, n$. As the line l rotates in a counterclockwise direction about a fixed point, we obtain a periodic sequence of permutations which is called the circular sequence of the configuration.

Allowable sequences are circular sequences constrained by the following properties:

1. Succesive permutations differ only by having the order of two or more adjacent numbers switched.
2. If a move results in the reversal of a pair ij then every other pair is reversed subsequently before i and j switch again.

Allowable sequences and the Sylvester problem. The point configurations encountered in the Sylvester problem must take into account highly degenerate cases. Since many points may be collinear, the corresponding circular sequence will have switches in which more than two adjacent numbers are reversed. The problem of showing the existence of an ordinary line is equivalent to the problem of determining whether a simple switch occurs.

History of the use of allowable sequences. Though the concept was introduced by Goodman and Pollack [1980a] to study the Erdős-Szekeres conjecture, it has been very useful in solving a range of problems which depend mainly on the order types of the point configuration. In particular, it has been used to show that

- every pseudoline arrangement of less than nine lines is stretchable [Goodman and Pollack 1980b];
- the number of directions determined by $2n$ points is at least $2n$ [Ungar 1982];
- the number of k-sets among a set of n points is $O(nk^{1/2})$ [Edelsbrunner and Welzl 1985];
- the maximum number of at most k-sets is $O(nk)$ [Welzl 1986];
- pseudoline arrangements are semispace equivalent if and only if they have the same allowable sequence modulo local equivalence [Goodman and Pollack 1984].

Properties determined by allowable sequences.

- $i_1, i_2, \ldots, i_k$ are collinear if and only if they switch simultaneously
- i is in the convex hull of $i_1, i_2, \ldots, i_k$ if and only if in every permutation in the sequence, i is preceded by one of $i_1, i_2, \ldots, i_k$.

- i is an extreme point if and only if some permutation begins with i
- ij is parallel to kl if and only if they both switch simultaneously
- ijk turn counterclockwise if and only if ij precedes ik, written as $ij \prec ik$.
- ij separates k from m if and only if when ij switches, k and m are on opposite sides of the substring ij in the permutation.

Using allowable sequences, Edelsbrunner and Welzl [1985] were able to derive improved upper bounds for the k−set problem viz. that the number of k−sets is $O(n\sqrt{k})$. Welzl [1986] generalized this result to bound the number of at most k−sets in a configuration of n points. Ungar [1982] was able to settle the conjecture regarding the number of directions determined by a configuration of points.

As an example of the power of allowable sequences, we give the following proof by Ungar.

Ungar's proof for the number of directions determined by $2n$ points.
We pay special attention to switches which straddle the midpoint of a permutation. A switch in which some indices cross the midpoint is called a crossing move. The ith crossing move causes an increasing string straddling the midpoint to be reversed. If d_i denotes the distance from the midpoint to the nearest end of the string, then, at the ith crossing move, exactly $2d_i$ indices cross the midpoint.

Since every index must cross the midpoint, if there are t crossing moves in all, then
$$2d_1 + 2d_2 + \cdots + 2d_t \geq 2n$$
since some indices can cross more than once.

Between two crossing moves, there must be at least $d_i + d_{i+1} - 1$ noncrossing moves, since we must first tear down a decreasing string of length d_i and build an increasing string of length d_{i+1}, and a decreasing string can be shortened by at most one in a switch (an increasing string can be increased by at most one in a switch).

Thus, the total number of switches between the first crossing move and when this same crossing move occurs in reverse corresponds to a half period and has $\sum(d_i + d_{i+1} - 1 + 1) = \sum(2d_i) \geq 2n$.

This is a tight lower bound, since the regular $2n$-gon determines exactly $2n$ directions, as in the Böröczky configuration of page 482.

When the number of points is odd, say $2n + 1$, the number of directions can be shown to be at least $2n$, since all but one point must cross the position $n + 1$ in the permutation.

4. Colored Extensions of Sylvester's Problem

Let $\{P_i\}$ be a collection of sets of points, and let all points in the same set be assigned a color. A line is monochromatic if it passes through at least two points of the same color and no points of any other color. The following problem

is attributed to Graham and Newman: Given a finite set of points in the plane
colored either red or blue, and not all collinear, must there exist a monochromatic
line? Motzkin [1967] solved the problem in the dual, showing there must exist a
monochromatic point in an arrangement of colored lines. The proof is sketched
in Section 4 (page 490). Chakerian [1970] and Stein gave additional proofs.

MONOCHROMATIC POINT: an intersection point in an arrangement of colored
    lines where all the lines intersecting at that point have the same color.

Consider the following question: Does there exist for every k a set of points in
the plane so that if one colors the points by two colors in an arbitrary way, there
should always be at least one line which contains at least k points, all of whose
points have the same color? This is known to be true for $k = 3$, but nothing is
known for larger values of k.

Various generalizations of this problem to higher dimensions have been proposed and solved [Chakerian 1970; Borwein 1982; Borwein and Edelstein 1983;
Tingley 1975; Baston and Bostock 1978].

Clearly, we cannot insist that the monochromatic line be ordinary without
additional restrictions. In the search for ordinary lines in the colored setting,
Fukuda [1996] raised the following question. Let R be a set of red points and
B be a set of blue points in the plane, not all on the same line. If R and B
are separated by a line and their sizes differ by at most one, then there exists
an ordinary bichromatic line, that is, a line with exactly one red point and one
blue point. This conjecture is shown not to be true for small n in [Finschi and
Fukuda 2003].

Pach and Pinchasi [2000] have shown that there exist bichromatic lines with
few points.

Motzkin's solution of the existence of a bichromatic point.

THEOREM 14. *Let S and T be two sets of nonconcurrent lines in the real projective plane colored red and blue respectively. At least one of the intersection
points in $S \cup T$ is monochromatic.*

Suppose S and T do not define any monochromatic vertex. Then, every intersection point w of two red lines has a blue line passing through it. These lines
can be ordered so that the blue line lies in between the red lines. Since not all
the blue lines are concurrent, there is some other blue line that does not pass
through this intersection point. The new blue line forms a triangle wxy with the
two red lines, as shown here (blue = gray):

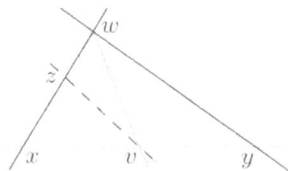

Consider such a triangle that is minimal in the sense that it does not completely contain another such triangle. This must exist because there only finitely many triangles in the arrangement.

The intersection point v of the two blue lines must be monochromatic. If not, there must exist a red line through v producing a triangle vzx of the original type (two red lines and a blue line) which is contained in the minimal triangle, contradicting the assumption that wxy is minimal.

5. Connecting Lines and Dirac's Conjecture

Another interesting problem concerns the connecting lines of a set of P points. Define an $i-$line to be a connecting line containing exactly i points of P and let $t_i(P)$ denote the number of $i-$lines determined by P. Also, let $t(P) = \sum_{i \geq 2} t_i(P)$. Let $r(n)$ be the minimum over all configurations of n points of the maximum number of connecting lines from a single point. i.e.

$$r(n) = \min_{P \subset R^n} \max_{p \in P} t(p)$$

Dirac [1951] asked whether one of the n points must always be incident with at least $\lfloor \frac{n}{2} \rfloor$ of the connecting lines. He showed that this was the best possible by placing all the points evenly on two intersecting lines. He also proved a trivial lower bound of $\sqrt{n}$. In [Grünbaum 1972] a list of exceptions to this formulation is enumerated.

Erdős relaxed the problem by asking whether it could be shown that $r(n) \geq cn$. The more general question he raised was the following. Is it true that there exists an absolute constant c independent of k and n such that if $0 \leq k \leq 2$ and $t_i(P) = 0$ for $i > n - k$ then

$$ckn < t(P) < 1 + kn$$

The upper bound is trivial, and the lower bound was shown by Beck [1983] and Szemerédi and Trotter [1983], but with very small constants. Clarkson et al. [1990] improved the constant significantly.

The question of whether $t(P) \geq n$ was raised by Erdős [1943] and proved by various people including Erdős and Hanani [Hanani 1951]. Kelly and Moser [1958] were able to prove that

$$t(P) \geq kn - \tfrac{1}{2}(3k + 2)(k - 1)$$

if k is small compared to n and any connecting line contains at most $n - k$ points.

6. A Solution for Sylvester's Problem Using Allowable Sequences

We now look at a simple application of allowable sequences to solve Sylvester's problem.

Consider an allowable sequence of permutations of $1, \ldots, n$. In the first half-period, each permutation is obtained from the previous one by the switch of a substring that is monotonically increasing. We shall pay special attention to the switches involving 1 or n. We claim that the first switch involving a substring to the right of n or a substring to the left on 1 in the permutation is simple, thus proving the theorem.

Assume that n makes a switch before 1 makes a switch. Similar arguments hold for 1 if this not the case. This assumption implies that the first switch involving n does not involve 1.

Every substring switch involving n has n at the end of the substring before switch. After the switch, the right of n in the permutation consists of a concatenation of substrings, each of which is monotonically decreasing, since a switch turns an increasing substring into a decreasing one. Note that either n is involved in a simple switch, in which case there is nothing further to prove, or else each switch involving n has length at least three.

If there have been no switches to the right of n, the length of the longest monotonically increasing substring to the right of n is at most two, which can happen only at the end of one substring and the beginning of another formed by the switches involving n. Thus, the first switch involving elements to the right of n in the permutation has a length of exactly two.

There must be at least one such switch, since:

(i) n must switch at least twice as there is no switch of length n, which corresponds to the case when all the points are collinear, an excluded case.

(ii) We have assumed that n switches before 1, implying that 1 is not involved in the first switch involving n, which in turn implies that the elements to the right n are not always monotonically decreasing.

References

[Baston and Bostock 1978] V. J. Baston and F. A. Bostock, "A Gallai-type problem", *J. Combinatorial Theory Ser. A* **24**:1 (1978), 122–125.

[Beck 1983] J. Beck, "On the lattice property of the plane and some problems of Dirac, Motzkin and Erdős in combinatorial geometry", *Combinatorica* **3** (1983), 281–297.

[Borwein 1982] P. Borwein, "On monochrome lines and hyperplanes", *J. Combin. Theory Ser. A* **33**:1 (1982), 76–81.

[Borwein and Edelstein 1983] P. Borwein and M. Edelstein, "A conjecture related to Sylvester's problem", *Amer. Math. Monthly* **90** (1983), 389–390.

[de Bruijn and Erdős 1948] N. G. de Bruijn and P. Erdős, "On a combinatorial problem", *Nederl. Akad. Wetensch., Proc.* **51** (1948), 1277–1279 = *Indagationes Math.* **10** (1948), 421–423.

[Chakerian 1970] G. D. Chakerian, "Sylvester's problem on collinear points and a relative", *Amer. Math. Monthly* **77** (1970), 164–167.

[Clarkson et al. 1990] K. L. Clarkson, H. Edelsbrunner, L. J. Guibas, M. Sharir, and E. Welzl, "Combinatorial complexity bounds for arrangements of curves and spheres", *Discrete Comput. Geom.* **5**:2 (1990), 99–160.

[Coxeter 1948] H. S. M. Coxeter, "A problem of collinear points", *Amer. Math. Monthly* **55** (1948), 26–28.

[Crowe and McKee 1968] D. W. Crowe and T. A. McKee, "Sylvester's problem on collinear points", *Math. Mag.* **41** (1968), 30–34.

[Csima and Sawyer 1993] J. Csima and E. T. Sawyer, "There exist $6n/13$ ordinary points", *Discrete Comput. Geom.* **9**:2 (1993), 187–202.

[Dirac 1951] G. A. Dirac, "Collinearity properties of sets of points", *Quart. J. Math. Oxford* (2) **2** (1951), 221–227.

[Dirac 1959] G. A. Dirac, review of [Kelly and Moser 1958] in *Mathematical Reviews* (MR20:3494), 1959.

[Edelsbrunner and Welzl 1985] H. Edelsbrunner and E. Welzl, "On the number of line separations of a finite set in the plane", *J. Combin. Theory Ser. A* **38**:1 (1985), 15–29.

[Erdős 1943] P. Erdős, "Problem 4065", *Amer. Math. Monthly* **50** (1943), 169–171.

[Finschi and Fukuda 2003] L. Finschi and K. Fukuda, "Combinatorial generation of small point configurations and hyperplane arrangements", pp. 425–440 in *Discrete and computational geometry*, edited by B. Aronov et al., Algorithms Combin. **25**, Springer, Berlin, 2003.

[Fukuda 1996] K. Fukuda, question raised at the problem session at the AMS–IMS–SIAM Joint Summer Research Conference on Discrete and Computational Geometry (South Hadley, MA), 1996.

[Goodman and Pollack 1980a] J. E. Goodman and R. Pollack, "On the combinatorial classification of nondegenerate configurations in the plane", *J. Combin. Theory Ser. A* **29**:2 (1980), 220–235.

[Goodman and Pollack 1980b] J. E. Goodman and R. Pollack, "Proof of Grünbaum's conjecture on the stretchability of certain arrangements of pseudolines", *J. Combin. Theory Ser. A* **29**:3 (1980), 385–390.

[Goodman and Pollack 1984] J. E. Goodman and R. Pollack, "Semispaces of configurations, cell complexes of arrangements", *J. Combin. Theory Ser. A* **37**:3 (1984), 257–293.

[Grünbaum 1970] B. Grünbaum, "The importance of being straight", pp. 243–254 in *Proc. Twelfth Biennial Sem. Canad. Math. Congr. on Time Series and Stochastic Processes, Convexity and Combinatorics* (Vancouver, B.C., 1969), Canad. Math. Congr., Montreal, Que., 1970.

[Grünbaum 1972] B. Grünbaum, *Arrangements and spreads*, CBMS Regional Conf. Series in Math. **10**, Amer. Math. Soc., Providence, RI, 1972.

[Hanani 1951] H. Hanani, "On the number of straight lines determined by n points", *Riveon Lematematika* **5** (1951), 10–11.

[Kelly and Moser 1958] L. M. Kelly and W. O. J. Moser, "On the number of ordinary lines determined by n points", *Canad. J. Math.* **1** (1958), 210–219.

[Kelly and Rottenberg 1972] L. M. Kelly and R. Rottenberg, "Simple points in pseudoline arrangements", *Pacific J. Math.* **40** (1972), 617–622.

[Lang 1955] G. D. W. Lang, "The dual of a well-known theorem", *The Mathematical gazette* **39** (1955), 124–135.

[Levi 1926] F. Levi, "Die Teilung der projektiven Ebene durch Gerade oder Pseudogerade", *Berichte über die Verhandlungen der Königlich Sächsischen Gesellschaft der Wissenschaften zu Leipzig. Math-Physik Classe* **78** (1926), 256–267.

[Melchior 1941] E. Melchior, "Über Vielseite der projektiven Ebene", *Deutsche Math.* **5** (1941), 461–475.

[Motzkin 1951] T. Motzkin, "The lines and planes connecting the points of a finite set", *Trans. Amer. Math. Soc.* **70** (1951), 451–464.

[Motzkin 1967] T. S. Motzkin, "Nonmixed connecting lines", *Notices of the American Mathematical Society* **14** (1967), 837–837.

[Pach and Pinchasi 2000] J. Pach and R. Pinchasi, "Bichromatic Lines with Few Points", *J. Combin. Theory Ser. A* **53** (2000), 326–335.

[Steinberg et al. 1944] R. Steinberg, R. C. Buck, T. Grünwald, and R. E. Steenrod, "Solution to Problem 4065", *The American Mathematical Monthly* **51** (1944), 169–171.

[Sylvester 1893] J. J. Sylvester, "Mathematical Question 11851", *Educational Times* **59** (1893), 385–394.

[Szemerédi and Trotter 1983] E. Szemerédi and W. T. Trotter, Jr., "Extremal problems in discrete geometry", *Combinatorica* **3**:3-4 (1983), 381–392.

[Tingley 1975] D. Tingley, "Monochromatic lines in the plane", *Math. Mag.* **48**:5 (1975), 271–274.

[Ungar 1982] P. Ungar, "$2N$ noncollinear points determine at least $2N$ directions", *J. Combin. Theory Ser. A* **33**:3 (1982), 343–347.

[Welzl 1986] E. Welzl, "More on k-sets of finite sets in the plane", *Discrete Comput. Geom.* **1**:1 (1986), 95–100.

[Williams 1968] V. C. Williams, "A proof of Sylvester's theorem on collinear points", *Amer. Math. Monthly* **75** (1968), 980–982.

NIRANJAN NILAKANTAN
COMPUTER SCIENCE DEPARTMENT
COURANT INSTITUTE
NEW YORK UNIVERSITY
421 WARREN WEAVER HALL
251 MERCER STREET
NEW YORK, NY 10012
UNITED STATES
nilakant@cs.nyu.edu

Combinatorial and Computational Geometry
MSRI Publications
Volume **52**, 2005

A Long Noncrossing Path Among Disjoint Segments in the Plane

JÁNOS PACH AND ROM PINCHASI

ABSTRACT. Let $\mathcal{L}$ be a collection of n pairwise disjoint segments in general position in the plane. We show that one can find a subcollection of $\Omega(n^{1/3})$ segments that can be completed to a noncrossing simple path by adding rectilinear edges between endpoints of pairs of segments. On the other hand, there is a set $\mathcal{L}$ of n segments for which no subset of size $(2n)^{1/2}$ or more can be completed to such a path.

1. Introduction

Since the publication of the seminal paper of Erdős and Szekeres [1935], many similar results have been discovered, establishing the existence of various regular subconfigurations in large geometric arrangements. The classical tool for proving such theorems is Ramsey theory [Graham et al. 1990]. However, the size of the regular substructures guaranteed by Ramsey's theorem are usually very small (at most logarithmic) in terms of the size n of the underlying arrangement. In most cases, the results are far from optimal. One can obtain better bounds (n^{ε} for some $\varepsilon > 0$) by introducing some linear orders on the elements of the arrangement and applying some Dilworth-type theorems [1950] for partially ordered sets [Pach and Törőcsik 1994; Larman et al. 1994; Pach and Tardos 2000]. A simple one-dimensional prototype of such a statement is the Erdős-Szekeres lemma: any sequence of n real numbers has a monotone increasing or monotone decreasing subsequence of length $\lceil \sqrt{n} \rceil$. In this note, we give a new application of this idea.

A collection $\mathcal{L}$ of segments in the plane is in *general position* if no two elements of $\mathcal{L}$ are parallel, all of their endpoints are distinct, and no three endpoints are collinear. A polygonal path $P = p_1 p_2 \ldots p_n$ is called *simple* if no pair of its vertices coincide, i.e., $p_i \neq p_j$ whenever $i \neq j$. It is called *noncrossing* if no

Research supported by NSF Grant CCR-00-98246. János Pach has also been supported by a PSC-CUNY Research Award and by grants from NSA and the Hungarian Research Foundation OTKA.

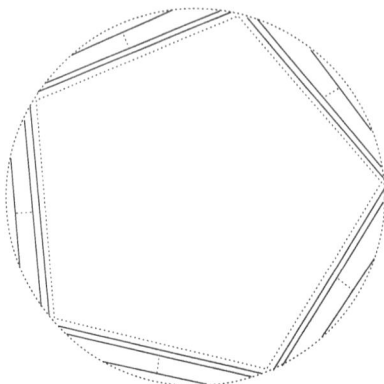

Figure 1. An arrangement of segments showing that $f(n) \leq 2\sqrt{2n}$.

two edges share an interior point. A polygonal path P is called *alternating* with respect to $\mathcal{L}$ if every other edge of P belongs to $\mathcal{L}$.

We consider the following old problem of unknown origin: what is the maximum length $f(n)$ of an alternating path that can be found in any collection of n pairwise disjoint segments in the plane in general position? This question appears in a list of open problems in computational geometry collected and annotated by Urrutia [2002]. The easy construction described there can be slightly improved to show that $f(n) \leq 2\sqrt{2}\sqrt{n}$ for $n = 2k^2$. Consider a $2k$-gon inscribed in a circle C and replace each of its edges e with k pairwise disjoint chords of C, almost parallel to e, that are farther away from the center of C than e is. (See Figure 1.) It seems likely that the order of magnitude of this bound is not far from optimal. For some similar problems, see [Hoffmann and Tóth 2003; Mirzaian 1992; Pach and Rivera-Campo 1998; Rappaport et al. 1990].

First we consider the special case when all segments cross the same line.

THEOREM 1. *Let $\mathcal{L}$ be a collection of n pairwise disjoint segments in general position in the plane, all of whose members cross a given line. Then one can select $\Omega(n^{1/2})$ segments from $\mathcal{L}$ that can be completed to a noncrossing simple alternating path.*

THEOREM 2. *The maximum length $f(n)$ of an alternating path that can be found in any collection of n pairwise disjoint segments in the plane satisfies $f(n) = \Omega(n^{1/3})$.*

PROOF OF THEOREM 2 ASSUMING THEOREM 1. By the Dilworth theorem (for example), any collection $\mathcal{L}$ of n pairwise disjoint segments has a subcollection $\mathcal{L}_1$ consisting of least $n^{1/3}$ segments whose projections to the x-axis are pairwise disjoint, or a subcollection $\mathcal{L}_2$ consisting of at least $n^{2/3}$ segments, all of which can be crossed by a line parallel to the y-axis. In the first case, the elements of $\mathcal{L}_1$ can be connected to form an alternating path. In the second case, we can apply Theorem 1. □

2. Proof of Theorem 1

Assume without loss of generality that all segments cross the y-axis, no two of them are parallel, and all $2n$ coordinates of their endpoints are distinct. The above-below relation between the crossings of the segments with the y-axis induces a natural linear order on the elements of $\mathcal{L}$. We apply the Erdős-Szekeres lemma to find a subsequence of $\mathcal{L}$ consisting of $\lceil \sqrt{n} \rceil$ segments with increasing or decreasing slopes with respect to this order. Since we can always flip the plane about the y-axis, we may assume that the slopes of the elements of this subsequence are monotone increasing. In what follows, for convenience we assume that $\sqrt{n}$ and all other numbers that appear in the argument (except the coordinates of the endpoints) are integers satisfying the necessary divisibility conditions so that we do not have to use "floor" and "ceiling" operations. This will not effect the asymptotic results obtained in this paper.

To be more precise, we find a sequence of at least $\sqrt{n}$ segments $s_1, \ldots, s_m$ ($m = \sqrt{n}$) of $\mathcal{L}$ such that if $i < j$, then s_i is above s_j and the slope of s_i is smaller than that of s_j (see Figure 2).

Partition $s_1, \ldots, s_m$ into $k = m/5$ groups, each consisting of 5 consecutive segments. That is, let $G_i = \{s_{5(i-1)+1}, \ldots, s_{5(i-1)+5}\}$ for every $1 \leq i \leq k$. For each G_i, apply again the Erdős-Szekeres lemma and find a subsequence of 3 segments such that the x-coordinates of their right endpoints form an increasing or a decreasing sequence. By flipping the plane about the x-axis, if necessary, we can also assume that for at least half of the G_is, these sequences are decreasing. From now on, we disregard all other segments. Summarizing: we now have $k/2$ groups $L_1, \ldots, L_{k/2}$, each consisting of 3 elements of $\mathcal{L}$. For each $1 \leq i \leq k/2$, let $L_i = \{\ell_1^i, \ell_2^i, \ell_3^i\}$, where ℓ_b^a is above $\ell_{b'}^{a'}$ and its slope is smaller, whenever $a < a'$, or if $a = a'$ and $b < b'$. Moreover, for a fixed a and any $b < b'$, the x-coordinate of the right endpoint of ℓ_b^a is larger than that of $\ell_{b'}^a$. Let $\mathcal{S} := L_1 \cup \cdots \cup L_{k/2}$.

Denote by p_b^a and q_b^a the left endpoint and the right endpoint of ℓ_b^a, respectively. For any two points r, s, let $[r, s]$ stand for the segment connecting r and s.

Define a set of auxiliary segments as follows. For $1 \leq a \leq k/2$ and $b = 1, 2$, let $z_b^a = [q_b^a, q_{b+1}^a]$. We say that z_b^a is *bad*, if there is a segment in $\mathcal{S}$ that meets the interior of z_b^a. For any segment $\ell_j^t \in \mathcal{S}$ meeting the interior of z_b^a, we have $t > a$, because all elements of $\cup_{t<a} L_t$ lie strictly above z_b^a, otherwise they would cross ℓ_b^a. Define the *witness index* of a bad segment z_b^a as the smallest index $t > a$ with the property that there exists an ℓ_j^t meeting the interior of z_b^a.

LEMMA 2.1. *If the witness index of a bad segment z_b^a is t, then ℓ_1^t meets z_b^a. Moreover, q_1^t must belong to the interior of the region enclosed by the y-axis, ℓ_b^a, ℓ_{b+1}^a, and z_b^a.*

PROOF. We know that $t > a$ and that for some j the segment ℓ_j^t crosses z_b^a. Assume that $j > 1$. Let W denote the region bounded by the y-axis, ℓ_{b+1}^a, z_b^a, and ℓ_j^t. The segment ℓ_1^t lies above ℓ_j^t, and the x-coordinate of its right endpoint

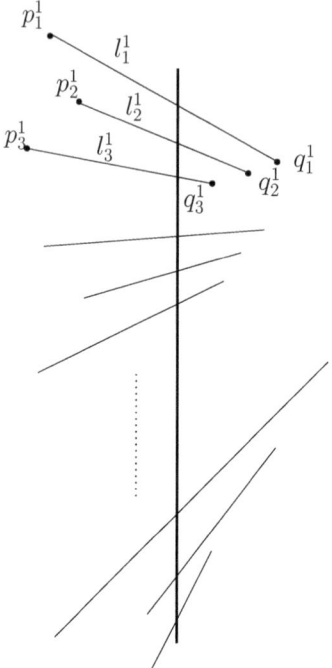

Figure 2. The segments l_1^i, l_2^i, l_3^i.

q_1^t is larger than the x-coordinate of q_j^t. Clearly, the intersection point r of ℓ_j^t and z_b^a is the rightmost corner of the boundary of W. There is a point on ℓ_1^t whose x-coordinate is the same as that of r. This point must lie above r and outside the region W. Since ℓ_1^t crosses the y-axis above ℓ_j^t and below ℓ_{b+1}^a, at a boundary point of R, and it has a point outside W, it must have another crossing with the boundary of W. Using the fact that the elements of S are pairwise disjoint, this second crossing must belong to z_b^a.

As for the second part of the lemma, let R denote the region bounded by the y-axis, ℓ_b^a, ℓ_{b+1}^a, and z_b^a. We have seen that ℓ_1^t meets the boundary of R (at a point of z_b^a). Since ℓ_1^t is disjoint from both ℓ_b^a and ℓ_{b+1}^a, and it intersects the y-axis below ℓ_{b+1}^a, it follows that ℓ_1^t cannot cross the boundary of R a second time. Therefore, q_1^t must belong to the interior of R.   □

LEMMA 2.2. *No two different bad segments can have the same witness index.*

PROOF. Assume to the contrary that t is the witness index of two bad segments, z_b^a and $z_{b'}^{a'}$. Suppose without loss of generality that ℓ_b^a lies above $\ell_{b'}^{a'}$. We know that both of them lie above ℓ_1^t. As in the proof of Lemma 2.1, let R denote the region bounded by the y-axis, ℓ_b^a, ℓ_{b+1}^a, and z_b^a. Similarly, let R' denote the region bounded by the y-axis, $\ell_{b'}^{a'}$, $\ell_{b'+1}^{a'}$, and $z_{b'}^{a'}$. R and R' do not overlap. Indeed, since the elements of S are pairwise disjoint, R and R' could overlap only if $\ell_{b'}^{a'}$ crossed z_b^a. However, this would contradict the minimality of t.

On the other hand, by Lemma 2.1, ℓ_1^t must intersect both z_b^a and $z_{b'}^{a'}$, and its right endpoint q_1^t must belong to the interiors of both R and R'. We thus obtained the desired contradiction.   □

Now we are in a position to prove Theorem 1.

By Lemma 2.2, the number of bad segments is at most $k/2$. We say that an index i $(1 \leq i \leq k/2)$ is *good* if at least one of the segments z_1^i, z_2^i is not bad. Obviously, at least $k/2 - \frac{1}{2}(k/2) = k/4$ indices between 1 and $k/2$ are good. Assume without loss of generality that the first $k/4$ indices are good. To complete the proof it is sufficient to show how to draw a noncrossing simple alternating path P that uses the segments ℓ_2^i, ℓ_3^i (and perhaps even ℓ_1^i) for $1 \leq i \leq k/4 = \Omega(\sqrt{n})$.

Let the first points of P be $q_1^1, p_1^1, q_2^1, p_2^1, q_3^1, p_3^1$, in this order. That is, so far we have built a "zigzag" path that uses the segments $\ell_1^1, \ell_2^1, \ell_3^1$. Since 2 is a good index, there exists a segment z_j^2 $(j = 1$ or $2)$ which is not bad. Let us extend P by adding the vertices p_j^2, q_j^2, q_{j+1}^2, and hence adding the edges ℓ_j^2 (from left to right) and z_j^2. Next we can add the point p_{j+1}^2 and, if $j = 1$, also the points q_3^2, p_3^2, zigzagging just like before. Continuing in the same manner, we build a path P using at least two edges from each group L_i $(i \leq k/4)$. It is easy to check that P is a *noncrossing* path, because (1) its edges belonging to $\mathcal{L} \subset \mathcal{S}$ are pairwise disjoint; (2) its edges to the left of the y-axis do not cross any other edge, by the assumption that the slopes of the elements of $\mathcal{S}$ form an increasing sequence; (3) its edges to the right of the y-axis are not bad, therefore they do not cross any other edge of P. This completes the proof of Theorem 1.

References

[Dilworth 1950] R. P. Dilworth, "A decomposition theorem for partially ordered sets", *Ann. of Math.* (2) **51** (1950), 161–166.

[Erdős and Szekeres 1935] P. Erdős and G. Szekeres, "A combinatorial problem in geometry", *Compositio Math.* **2** (1935), 463–470.

[Graham et al. 1990] R. L. Graham, B. L. Rothschild, and J. H. Spencer, *Ramsey theory*, Wiley, New York, 1990.

[Hoffmann and Tóth 2003] M. Hoffmann and C. D. Tóth, "Alternating paths through disjoint line segments", *Inform. Process. Lett.* **87**:6 (2003), 287–294.

[Larman et al. 1994] D. Larman, J. Matoušek, J. Pach, and J. Töröcsik, "A Ramsey-type result for convex sets", *Bull. London Math. Soc.* **26**:2 (1994), 132–136.

[Mirzaian 1992] A. Mirzaian, "Hamiltonian triangulations and circumscribing polygons of disjoint line segments", *Comput. Geom.* **2**:1 (1992), 15–30.

[Pach and Rivera-Campo 1998] J. Pach and E. Rivera-Campo, "On circumscribing polygons for line segments", *Comput. Geom.* **10**:2 (1998), 121–124.

[Pach and Tardos 2000] J. Pach and G. Tardos, "Cutting glass", *Discrete Comput. Geom.* **24**:2-3 (2000), 481–495.

500   JÁNOS PACH AND ROM PINCHASI

[Pach and Törőcsik 1994] J. Pach and J. Törőcsik, "Some geometric applications of Dilworth's theorem", *Discrete Comput. Geom.* **12**:1 (1994), 1–7.

[Rappaport et al. 1990] D. Rappaport, H. Imai, and G. T. Toussaint, "Computing simple circuits from a set of line segments", *Discrete Comput. Geom.* **5**:3 (1990), 289–304.

[Urrutia 2002] J. Urrutia, "Open problems in computational geometry", pp. 4–11 in *LATIN 2002: Theoretical informatics* (Cancun), edited by S. Rajsbaum, Lecture Notes in Comput. Sci. **2286**, Springer, Berlin, 2002.

JÁNOS PACH
CITY COLLEGE, CUNY
and
COURANT INSTITUTE OF MATHEMATICAL SCIENCES
NEW YORK UNIVERSITY
NEW YORK, NY 10012
UNITED STATES
    pach@cims.nyu.edu

ROM PINCHASI
DEPARTMENT OF MATHEMATICS
TECHNION – ISRAEL INSTITUTE OF TECHNOLOGY
HAIFA 32000
ISRAEL
    room@math.technion.ac.il

Combinatorial and Computational Geometry
MSRI Publications
Volume **52**, 2005

On a Generalization of Schönhardt's Polyhedron

JÖRG RAMBAU

ABSTRACT. We show that the nonconvex twisted prism over an n-gon cannot be triangulated without new vertices. For this, it does not matter what
the coordinates of the n-gon are as long as the top and the bottom n-gon
are congruent and the twist is not too large. This generalizes Schönhardt's
polyhedron, which is the nonconvex twisted prism over a triangle.

1. The Background

Lennes [1911] was the first to present a simple three-dimensional nonconvex
polyhedron whose interior cannot be triangulated without new vertices. The
more famous example, however, was given by Schönhardt [1927]: he observed
that in the nonconvex twisted triangular prism (subsequently called "Schönhardt's polyhedron") every diagonal that is not a boundary edge lies completely
in the exterior. This implies that there can be no triangulation of it without new
vertices because there is simply no interior tetrahedron: all possible tetrahedra
spanned by four of its six vertices would introduce new edges. Moreover, he
proved that every simple polyhedron with the same properties must have at
least six vertices. Later, further such nonconvex, nontriangulable polyhedra with
an arbitrary number of points have been presented. Among them, Bagemihl's
polyhedron [1948] also has the feature that every nonfacial diagonal is in the
exterior.

The nonconvex twisted prism over an arbitrary n-gon would arguably be the
most natural generalization of Schönhardt's polyhedron. Surprisingly enough,
there has been no proof so far that it cannot be triangulated without new vertices. One of the reasons seems to be that — in contrast to Schönhardt's and
Bagemihl's polyhedra — not every nonfacial diagonal lies completely outside the
polygonal prism. Yet, the nonconvex twisted polygonal prism indeed cannot be

Supported by a general membership at the Mathematical Sciences Research Institute Berkeley
(MSRI) in the special semester "Discrete and Computational Geometry" (2003); research at
MSRI is supported in part by NSF grant DMS-9810361.

triangulated without new vertices, as we will show below. For this, it does not matter what the coordinates of the n-gon are as long as the top and the bottom n-gon are congruent and the twist is just a perturbation by rotation, i.e., it is not too large.

There is also a convex variant of Schönhardt's polyhedron, the untwisted triangular prism. Consider the two possible cyclically symmetric triangulation of its boundary quadrilaterals. They appear if we untwist the Schönhardt polyhedron and keep the diagonals on the boundary quadrilaterals. Neither such boundary triangulation can be extended to the interior without new vertices. The reason is analogous to the Schönhardt case: every possible tetrahedron would induce at least one diagonal that intersects one of the prescribed diagonals. We will show below the corresponding generalization to the polygonal prism: there is no a triangulation of the general (untwisted) polygonal prism extending a cyclically symmetric triangulation of the boundary quadrilaterals.

Besides the fact that the (frequently asked) question about the existence of triangulations of the nonconvex twisted polygonal prism deserves a conclusive answer at last, we mention one other motivation for studying problems like this. Deciding the existence of a triangulation without new vertices for a given polyhedron is NP-hard [Ruppert and Seidel 1992]. In studying the twisted polygonal prism we surprisingly hit the borderline between existence and nonexistence of triangulations without new vertices in a single type of point configurations, and this could make the twisted or untwisted polygonal prism a handy gadget for NP-hardness proofs. A similar pattern appears, e.g., in a proof that finding *minimal* triangulations of polytopes is NP-hard [Below et al. 2000].

2. The Objects

Consider a two-dimensional point configuration $C_n := \{v_0, v_1, \ldots, v_{n-1}\}$ in strictly convex position labeled counterclockwise. Fix a point o in the interior of C_n in $\mathbb{R}^2$. For $\alpha \in [0, 2\pi)$, let $C_n(\alpha)$ be a copy of C_n rotated by α around the point o (rotation by an angle in $(0, 2\pi)$ means counterclockwise rotation). We call the corresponding points $w_0, w_1, \ldots, w_{n-1}$. The *Cayley embedding* of C_n and $C_n(\alpha)$ is defined by

$$P_n(\alpha) := \mathrm{conv}\big((C_n \times \{0\}) \cup (C_n(\alpha) \times \{1\})\big).$$

A *triangulation* of a three-dimensional polyhedron P is a dissection into finitely many tetrahedra such that any two intersect in a common face (possibly empty). For a triangulation of P and a simplex F of arbitrary dimension we say T *uses* F if F is a face of some tetrahedron in T. Faces are denoted by their sets of vertices. A *triangulation without new vertices* or a *v-triangulation* of P is a triangulation all of whose vertices are vertices of P.

$P_n := P_n(0)$ is known as a *prism* over C_n. The *cyclic set of diagonals*

$$D_c := \big\{ \{v_i, w_{i+1}\} \ : \ i = 0, 1, \dots, n-1 \big\}$$

induces a triangulation of the quadrilateral facets of $P_n(0)$ into the triangles $\{v_i, w_i, w_{i+1}\}$ and $\{v_i, w_{i+1}, v_{i+1}\}$, $i = 0, 1, \dots, n-1$ (indices taken modulo n).

The continuity of the determinant function ensures that there is an $\alpha > 0$ such that no full-dimensional tetrahedron in $P_n(0)$ has a reversed orientation (sign of determinant of the points in homogeneous coordinates) in $P_n(\alpha)$. In that case, the *vertical edges* $\{v_i, w_i\}$ and the *reverse cyclic edges* $\{w_i, v_{i+1}\}$ are among the boundary edges of $P_n(\alpha)$, for all $i = 0, 1, \dots, n-1$. For such an α, we call $P_n(\alpha)$ a *convex twisted prism over* C_n. ($P_n(\alpha)$ is a convex twisted prism over C_n if and only if the map sending $v_i, w_i \in P_n(\alpha)$ to the corresponding $v_i, w_i \in P_n(0)$ induces a weak map of oriented matroids [Björner et al. 1993].)

For a convex twisted prism over C_n, the *cyclic set of tetrahedra* is the set of tetrahedra

$$T_c := \big\{ \{v_i, v_{i+1}, w_i, w_{i+1}\} \ : \ i = 0, 1, \dots, n-1 \big\}.$$

Any pair of consecutive such tetrahedra intersects in a common edge.

3. The Results

THEOREM 3.1. *For all $n \geq 3$, no prism $P_n(0)$ over an n-gon admits a triangulation without new vertices that uses the cyclic set D_c of diagonals.*

THEOREM 3.2. *For all $n \geq 3$ and all sufficiently small $\alpha > 0$, no convex twisted prism $P_n(\alpha)$ admits a triangulation that contains the cyclic set T_c of tetrahedra.*

We define the *nonconvex twisted prism* $\check{P}_n(\alpha)$ to be the topological closure of $P_n(\alpha) \backslash T_c$. Since the twist is not too large, this is a nonconvex simple polyhedron. Here is now the generalization of Schönhard's polyhedron:

COROLLARY 3.3. *For all $n \geq 3$ and all sufficiently small $\alpha > 0$, the nonconvex twisted prism $\check{P}_n(\alpha)$ cannot be triangulated without new vertices.*

REMARK 3.4. When C_n is a regular triangle and $\alpha \in (0, 2\pi/3)$, the twisted prism $P_3(\alpha)$ coincides with Schönhardt's twisted prism.

4. The Tools

For a more detailed background about the following consult [Huber et al. 2000] and the references therein.

Minkowski sums and mixed subdivisions. Let P and Q be point configurations in $\mathbb{R}^2$. Then the *Minkowski sum of P and Q scaled by $\lambda \in (0, 1)$* is the point configuration

$$(1-\lambda)P + \lambda Q := \{(1-\lambda)p + \lambda q \ : \ p \in P, q \in Q\} \subset \mathbb{R}^2.$$

We make the following simplifying assumption: we consider only generic $\lambda \in (0, 1)$, for which $(1-\lambda)p + \lambda q = (1-\lambda)p' + \lambda q'$ implies that $p = p'$ and $q = q'$. A *mixed cell* in $(1-\lambda)P + \lambda Q$ is the Minkowski sum $(1-\lambda)\sigma + \lambda\tau$ of subsets $\sigma \subseteq P$ and $\tau \subseteq Q$. A *mixed subdivision* of $(1-\lambda)P + \lambda Q$ is a dissection of $(1-\lambda)P + \lambda Q$ into finitely many mixed cells such that any two intersect in common faces (possibly empty).

A two-dimensional mixed cell is *fine* if it is the Minkowski sum of either two edges or of a point and a triangle. In the first case, the cell is a parallelogram, in the second case the cell is a triangle. A mixed subdivision is *fine* if it contains only fine mixed cells.

Cayley embeddings. Let P and Q as above. Then the *Cayley embedding* of P and Q is the point configuration

$$\mathcal{C}(P, Q) := \{(p, 0) \ : \ p \in P\} \cup \{(q, 1) \ : \ q \in Q\} \subset \mathbb{R}^3.$$

For example, $P_n(\alpha)$ from above is a Cayley embedding for all α.

The Cayley trick. The Cayley trick states that for all P and Q as above, triangulations of $\mathcal{C}(P, Q)$ are in one-to-one correspondence with fine mixed subdivisions of $(1-\lambda)P + \lambda Q$ for all $\lambda \in (0, 1)$. We will only need the fact that every triangulation of $\mathcal{C}(P, Q)$ induces a fine mixed subdivision of $(1-\lambda)P + \lambda Q$ for all $\lambda \in (0, 1)$.

The correspondence is given by intersecting $\mathcal{C}(P, Q)$ with a horizontal hyperplane H_λ at height λ. The intersection of any tetrahedron in a triangulation of $\mathcal{C}(P, Q)$ with H_λ is a fine mixed cell in $((1-\lambda)P + \lambda Q) \times \{\lambda\} \subset \mathbb{R}^3$. Since intersection with affine hyperplanes preserves face relations, the set of all fine mixed cells so obtained yields a fine mixed subdivision of $(1-\lambda)P + \lambda Q$.

Applied to $P_n(\alpha)$ this means: each triangulation of $P_n(\alpha)$ induces a fine mixed subdivision of $S_n(\alpha, \lambda) := (1-\lambda)C_n + \lambda C_n(\alpha)$ for every $\lambda \in (0, 1)$. In summary, we have the following correspondences between objects in the Cayley embedding and the Minkowski sum:

$P_n(\alpha)$	$S_n(\alpha, \lambda)$
tetrahedra	fine mixed polygons
tetrahedra with a triangle on the top or the bottom	fine mixed triangles
tetrahedra with edges on both top and bottom	fine mixed parallelograms
nonhorizontal triangles	fine mixed edges
nonhorizontal edges	fine mixed points
orientation	counterclockwise orientation

Since the Minkowski sum lives in one dimension less than the Cayley embedding, we rather work with $S_n(\alpha, \lambda)$.

5. The Proofs

Let $\alpha \geq 0$ be small enough such that $P_n(\alpha)$ is a prism or a twisted prism. Fix a (small) $\varepsilon \in (0,1)$ such that $|\varepsilon(v_j - v_i)| < |(1-\varepsilon)(w_j - w_i)|$ for all $i, j = 0, 1, \ldots, n-1$. (All following considerations are also true for arbitrary $\varepsilon \in (0,1)$; the choice of a small ε makes some arguments more transparent) In particular, the scaled Minkowski sum $S_n(\alpha) := S_n(\alpha, 1-\varepsilon) = \varepsilon P_n + (1-\varepsilon)P_n(\alpha)$ does not contain multiple points. (We use ε here instead of λ as in the Cayley trick of the previous page to generate the impression of a small scaling factor.) For brevity, we will use the notation (i, j) for the point $\varepsilon v_i + (1-\varepsilon)w_j$, $i, j = 0, 1, \ldots, n-1$.

Some notions and notation. In all what follows, we use the term "edges" not only for boundary edges but also for interior edges, sometimes called "diagonals". Consider mixed edges. All mixed edges are, by definition, Minkowski sums of either a point and an edge or of an edge and a point. In our notation, they are of the form $(e, i) := \{(k, i), (l, i)\}$ or of the form $(j, e) := \{(j, k), (j, l)\}$ for some edge (or diagonal, see above) $e = \{k, l\}$ in C_n or $C_n(\alpha)$, resp.

The following notions are motivated by regarding ε as being small. We highlight the most important one as a definition.

DEFINITION 5.1 (SHORT AND LONG EDGES). Call a mixed edge *short* if it is of the form (e, i), call it *long* otherwise. The short mixed edge $e_i := \{(i, i), (i+1, i)\}$ is called *special*.

The special edges are interesting in S_n because — via the Cayley trick — they correspond to triangles that are incompatible with the cyclic set of diagonals D_c in P_n. Moreover, they are interesting in $S_n(\alpha)$ for $\alpha > 0$ because the cyclic set of tetrahedra T_c covers the corresponding triangles in $P_n(\alpha)$ so that in any triangulation containing T_c no other cell can use them.

For $i = 0, 1, \ldots, n-1$, there are the convex sub-n-gons $(C_n, i) := \varepsilon C_n + (1-\varepsilon)w_i$ and $(i, C_n(\alpha)) := \varepsilon v_i + (1-\varepsilon)C_n(\alpha)$ in S_n. By construction, all (C_n, i) are scaled translates of C_n, and all $(i, C_n(\alpha))$ are scaled translates of $C_n(\alpha)$, which itself is an angle-preserving image of C_n under a (small) rotation that we call $r(\alpha)$. The *long* translation that shifts (C_n, i) to (C_n, j) along the long edge $\{(i, i), (i, j)\}$ is denoted by T_{ij}; the *short* translation that moves $(i, C_n(\alpha))$ to $(j, C_n(\alpha))$ along the short edge $\{(i, i), (j, i)\}$ is denoted by t_{ij}. Note that we regard C_n, (C_n, i), and $(j, C_n(\alpha))$ as point configurations in convex position, not as two-dimensional polytopes. The corresponding polytopes will be denoted by $\mathrm{conv}(C_n)$, $\mathrm{conv}(C_n, i)$, and $\mathrm{conv}(j, C_n(\alpha))$, resp.

Call the n-gons (C_n, i) *small* and the n-gons (j, C_n) *large*. Similarly, we call a mixed triangle with only short edges *small*; we call a mixed triangle with only long edges *large*. By definition of the Minkowski sum, each mixed triangle is either small or large. We can regard short mixed edges as edges that have both end points in the same small sub-n-gon. The special short mixed edge e_i lies in the boundary of $S_n(\alpha)$. Figure 1 illustrates the setup.

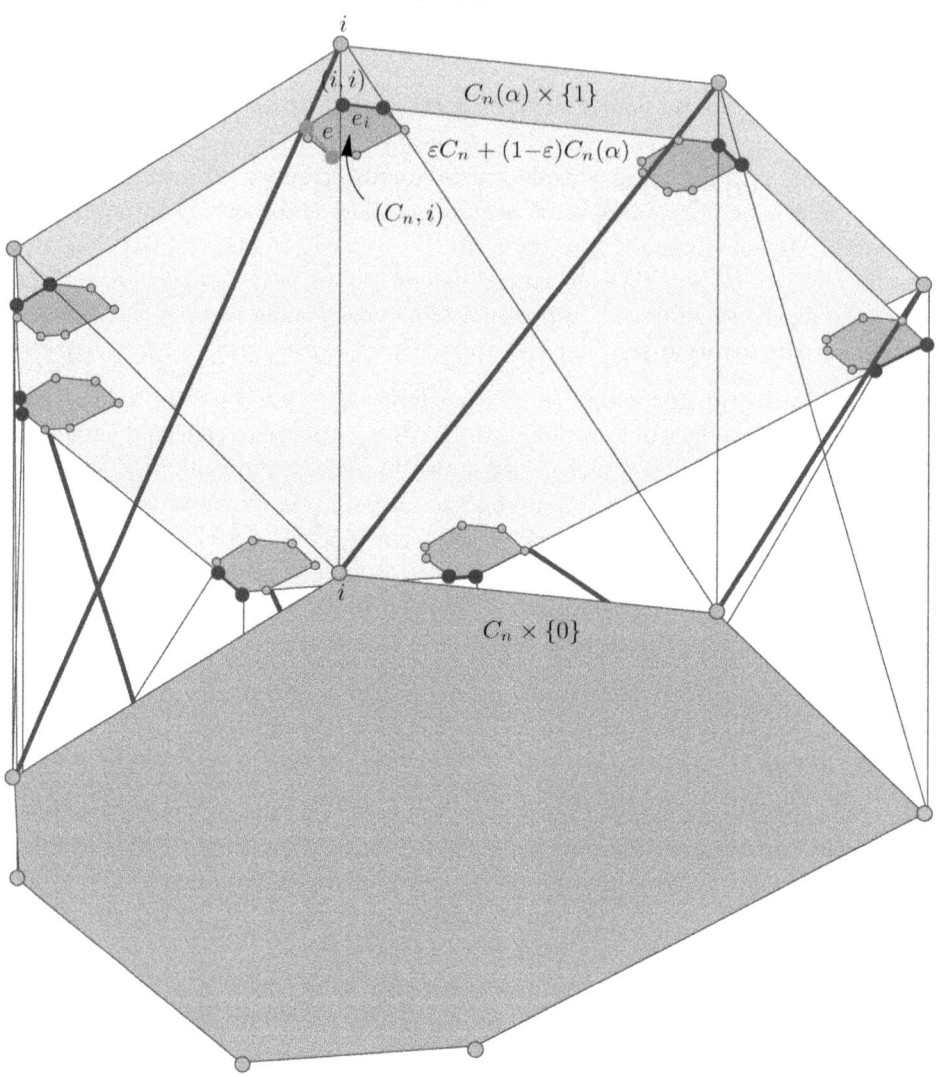

i

(i, i)

e e_i

$C_n(\alpha) \times \{1\}$

$\varepsilon C_n + (1-\varepsilon)C_n(\alpha)$

(C_n, i)

i

$C_n \times \{0\}$

Figure 1. Cutting the Cayley embedding of two n-gons with a horizontal hyperplane close to the top yields their Minkowski sum scaled as in $S_n(\alpha)$; the cyclic set of diagonals and the special edges are drawn thicker.

Road-map of the proofs. Note that any triangulation of P_n that uses the cyclic set of diagonals induces a mixed subdivision M of S_n in which no special edge e_i is used. Consider any nonspecial short edge e in M in some small n-gon (C_n, i). Then the "region" between e and e_i must be covered by M somehow. We want to show that this cannot be accomplished unless at least one special edge is used. We even show that at least one special edge must be used as an edge of some mixed triangle (Theorem 5.10).

How can the region between e and e_i be subdivided? There must be a cell adjacent to e on the same side as e_i. If we use a mixed triangle, i.e., a small triangle, then we harvest new short edges in the same small n-gon. One of these new short edges is "closer" to e_i in a sense to be defined precisely below, and we can proceed. If we use a mixed parallelogram then there is another short edge e' opposite to e in some other small n-gon (C_n, j) at a "partner vertex" j of e. But the "regions" containing potential partner vertices for e' towards e_j will turn out to be strictly smaller than for e.

But what happens if we use a mixture of mixed triangles and parallelograms? It fact, both ideas from above can be merged by using a certain lexicographic partial order on short edges, in which the short edges that are hit by "chasing the mixed subdivision M towards special short edges" are strictly decreasing. This shows that not all special short edges can be avoided by M.

We can make this idea precise for both the prism and the twisted prism. In the latter case, it is no surprise that even all special edges must be used, since they are boundary edges of $S_n(\alpha)$. However, using the cyclic set of tetrahedra means covering all special short edges by parallelograms, and we will show that at least one of them must be in a small triangle.

In the sequel, we will formalize these arguments in order to obtain rigorous proofs of Theorems 3.1 and 3.2.

Ordering short mixed edges. For the following, let e be a short edge in (C_n, i). We want to give an orientation to the halfplanes separated by the line $l(e)$ spanned by e. If $e = e_i$, then we make use of the fact that e_i is in the boundary of S_n, thus $l(e)$ is a supporting hyperplane for S_n. Therefore, we can define the positive side $l(e)^+$ of e to be the halfplane not containing S_n. If $e \neq e_i$, we define the positive side $l(e)^+$ of e to be the halfplane containing e_i. This idea of investigating the subdivision between e and e_i can now be formulated as looking at cells on the positive side of $l(e)$.

The following is a simple observation.

LEMMA 5.2. *Let σ be a mixed parallelogram in $S_n(\alpha)$ with short edges e and e'. Then:*

(i) *If σ is on the positive sides or on the negative sides of both of its short edges then $l(e)$ and $l(e')$ have opposite orientations.*

(ii) *If σ is on the positive side of e and on the negative side of e', or vice versa, then $l(e)$ and $l(e')$ have parallel orientations.*

One of the cases mentioned in Lemma 5.2 can actually never occur. This will allow us to keep on finding new cells on the positive sides of short edges.

LEMMA 5.3 (ORIENTATION LEMMA). *There is no fine mixed 2-cell σ in S_n on the positive side of all of its short edges.*

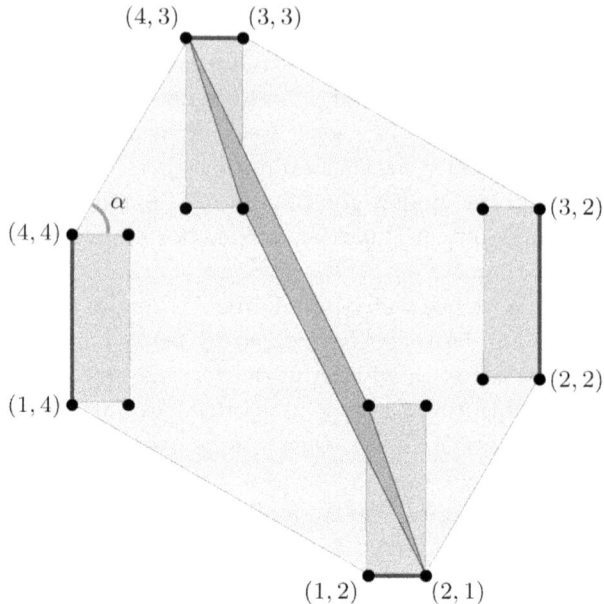

Figure 2. Parallelograms which are on the positive sides of both of their short edges exist when α is too large; in the picture $\alpha = \pi/3$. However, it can be seen that the bad parallelogram flips its orientation when $P_4(\alpha)$ is untwisted.

REMARK 5.4. The correctness of the Orientation Lemma heavily depends on the congruence of the top/bottom polygons of $P_n(\alpha)$ and on the restriction of α. That the lemma is false in even slightly more general situations can be seen in the example in Figure 2.

PROOF. Assume, for the sake of contradiction, that σ is a mixed 2-cell in S_n lying on the positive side of all of its short edges. Since σ contains the short edge e, it must be either a small triangle or a parallelogram.

Consider the case where σ is a small triangle on the positive side of all of its edges. The special edge e_i cannot be an edge of σ, since σ is contained in conv S_n, and $l(e_i)^+$ was defined to be the side of $l(e_i)$ not containing S_n. By definition of the orientations of short edges other than e_i, we conclude that e_i must be contained in σ. Since (C_n, i) is convex, this can only be the case if e_i is an edge of σ: contradiction.

Therefore, σ must be a parallelogram lying on the positive sides of both of its short edges e in (C_n, i) and e' in (C_n, j) for some $i, j \in \{0, 1, \dots, n-1\}$. We first consider this in the case of the prism, i.e., when $\alpha = 0$. We will also include the degenerate case, i.e., where σ is a line segment, into our considerations. Since $\sigma \subset l(e)^+ \cap l(e')^+$, the orientations of e and e' must be opposite (Lemma 5.2). In terms of translations, $T_{ij}(l(e)^+) = l(e')^-$ and $T_{ji}(l(e')^+ = l(e)^-$. By definition of the orientation, e_i is on the positive side of e, and hence $(i, i) \in l(e)^+$. Similarly,

$(j, j) \in l(e')^+$. This implies

$$(i, i) \in l(e)^+,$$
$$(j, i) = T_{ji}(j, j) \in T_{ji}(l(e')^+) = l(e)^-,$$
$$(i, j) = T_{ij}(i, i) \in T_{ij}(l(e)^+) = l(e')^-,$$
$$(j, j) \in l(e')^+.$$

$(5\text{--}1)$

These are necessary conditions for a nondegenerate σ being on the positive side of both of its short edges. While being on the positive side of short edges does not make sense for degenerate σ, Conditions (5–1) have a meaning in the degenerate case as well. For further reference, we call these necessary conditions the *orientation conditions*.

Since $\alpha = 0$, the points (i, i), (j, i), (i, j), and (j, j) lie on a straight line ℓ. Since ε is very small, the points appear on ℓ in the order (i, i), (j, i), (i, j), and (j, j). This tells us that ℓ starts in $l(e)^+$, enters $l(e)^-$, and then returns into $l(e)^+$. This implies that $\ell = l(e)$. By the symmetric argument, also $\ell = l(e')$. Therefore, σ is a segment. Moreover, its short edges are actually $e = \{(i, i), (j, i)\}$ and $e' = \{(i, j), (j, j)\}$ because the points in (C_n, i) are in strictly convex position.

This shows that a nondegenerate σ cannot exist in the prism. Moreover, we have learned the following useful fact: if the points (i, i), (j, i), (i, j), and (j, j) satisfy the orientation conditions (5–1) for the short edges e and e' of some (possibly degenerate) parallelogram σ then $\sigma = \{(i, i), (j, i), (i, j), (j, j)\}$.

Since σ cannot exist in the prism, consider the case where $\alpha > 0$ so that $P_n(\alpha)$ is still a twisted prism. That means, no full-dimensional tetrahedron in P_n switches orientation during the twisting towards $P_n(\alpha)$. That implies that no full-dimensional parallelogram in $S_n(0)$ changes its orientation with respect to its short edges (by the Cayley trick correspondence, page 504; easy exercise in linear algebra).

Now, untwist $P_n(\alpha)$, and hence σ. Then, σ must degenerate to a segment in P_n. During the untwist, for all $\alpha > 0$ the points (i, i), (j, i), (i, j), and (j, j) must always satisfy the orientation conditions. Since the conditions define a closed space and untwisting changes all data continuously in α, they must also hold in the degenerate position $\alpha = 0$. Hence, σ must be of the form $\{(i, i), (j, i), (i, j), (j, j)\}$ for some $i, j \in \{0, 1, \dots, n-1\}$. In particular, $e = \{(i, i), (j, i)\}$.

We finally show that during the twist, σ folds up in the "wrong" direction. Consider the order of the short edges incident to (i, i) counterclockwise starting at an edge of S_n. In this order e_i is the first edge, by definition. Twisting P_n again counterclockwise by α will turn the slope of the short edge $e = \{(i, i), (j, i)\}$ counterclockwise into the slope of the long edge $\{(i, i), (i, j)\}$. Therefore, the long edge $\{(i, i), (i, j)\}$ and the special short edge e_i are on different sides of e. This means, σ lies on the negative side of e: contradiction. $\square$

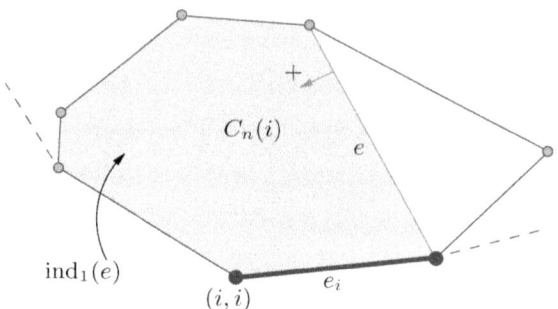

Figure 3. Primary index $\mathrm{ind}_1(e)$ of a short edge e.

The following quantity defines how close a short edge is to the corresponding special short edge. See Figure 3 for an illustration.

DEFINITION 5.5 (PRIMARY INDEX). We define the *primary index* $\mathrm{ind}_1(e)$ of any short edge e in $S_n(\alpha)$ by

$$\mathrm{ind}_1(e) := \mathrm{vol}\big(\mathrm{conv}(C_n, i) \cap l(e)^+\big).$$

We now turn our attention to measuring how many short partner edges a short edge can find to build a parallelogram on its positive side. Consider the unique line $l(e, i)$ parallel to e through (i, i). Let $l(e, i, \alpha)$ be the line that is obtained from $l(e, i)$ by a rotation by $-\alpha$ around (i, i). Its orientation is obtained by rotating the orientation of $l(e)$ by $-\alpha$ as well. The resulting positive halfplane defined by $l(e, i, \alpha)$ is called $l(e, i, \alpha)^+$.

LEMMA 5.6 (PARTNER LEMMA). *Let σ be a mixed parallelogram with short edges e and e' so that σ lies on the positive side of e. Assume, e lies in the small polygon (C_n, i) and e' lies in the small polygon (C_n, j). Then (j, i) lies in the interior of $l(e, i, \alpha)^+$.*

PROOF. Assume, for the sake of contradiction, that (j, i) lies in $l(e, i, \alpha)^-$. By definition, e_i is inside $l(e)^+$. Since e_i is a boundary edge of $S_n(\alpha)$, one of the long edges E of σ must separate e_i from σ. Let $(k, i) := E \cap e$, where $k = i$ is possible.

Let β be the angle from e to E around (k, i). This angle is the same as the angle from $l(e, i)$ to $\{(i, i), (i, j)\}$ around (i, i): the short translation t_{ki} moves (k, i) to (i, i), E onto $\{(i, i), (i, j)\}$, and e into $l(e, i) \cap \mathrm{conv}\, S_n(\alpha)$. There are two cases: either $0 < \beta < \pi$ or $-\pi < \beta < 0$.

If $0 < \beta < \pi$ then the slope of e turns counterclockwise around (k, i) into the slope of E. Since σ, and hence E, are in $l(e)^+$, the interior of the positive side $l(e)^+$ of $l(e)$ can be characterized as follows: a point $x \in \mathbb{R}^2$ is in the interior of $l(e)^+$ if and only if the angle from e to $\{(k, i), x\}$ around (k, i) is in the interval $(0, \pi)$. Since the orientation of $l(e, i)$ is parallel to this, the analogous characterization holds for the interior of $l(e, i)^+$. The characterization of the interior of the positive side $l(e, i, \alpha)^+$ of $l(e, i, \alpha)$ is analogous.

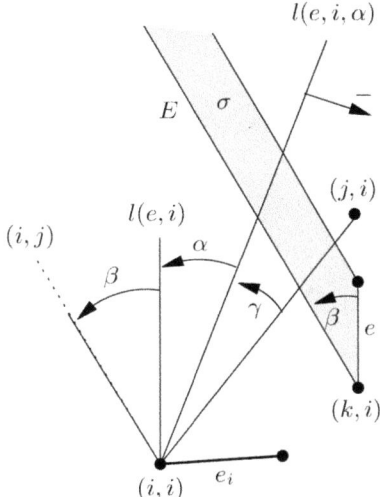

Figure 4. The case $0 < \beta < \pi$ in the proof of the Partner Lemma.

Let γ be the angle from $\{(i,i),(j,i)\}$ to $l(e,i,\alpha)$ around (i,i). The assumption that (j,i) lies in $l(e,i,\alpha)^-$ can now be expressed as $-\gamma \in [-\pi,0] \iff \gamma \in [0,\pi]$. The angle from $\{(i,i),(j,i)\}$ to $\{(i,i),(i,j)\}$ around (i,i) equals α, by construction of $P_n(\alpha)$. (See Figure 4 for an illustration.) Therefore:

$$\alpha = \angle(\{(i,i),(j,i)\},\{(i,i),(i,j)\})$$
$$= \angle(\{(i,i),(j,i)\},l(e,i,\alpha))) + \angle(l(e,i,\alpha),l(e,i))) + \angle(l(e,i),\{(i,i),(j,i)\})$$
$$= \underbrace{\gamma}_{\in[0,\pi]} + \alpha + \underbrace{\beta}_{\in(0,\pi)}$$
$$\in (\alpha, \alpha + 2\pi).$$

This is a contradiction.

If $-\pi < \beta < 0$ then we get analogously $\gamma \in [-\pi,0]$. (See Figure 5 for an illustration.) Thus:

$$\alpha = \angle(\{(i,i),(j,i)\},\{(i,i),(i,j)\})$$
$$= \angle(\{(i,i),(j,i)\},l(e,i,\alpha))) + \angle(l(e,i,\alpha),l(e,i))) + \angle(l(e,i),\{(i,i),(j,i)\})$$
$$= \underbrace{\gamma}_{\in[-\pi,0]} + \alpha + \underbrace{\beta}_{\in(-\pi,0)}$$
$$\in (\alpha - 2\pi, \alpha).$$

Contradiction again, and we are done.   □

The following secondary index measures for any short edge the size of the region in which partner edges for a parallelogram can be found. See Figure 6 for a sketch.

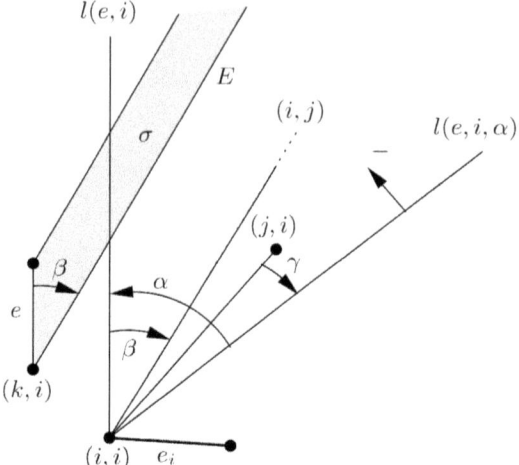

Figure 5. The case $-\pi < \beta < 0$ in the proof of the Partner Lemma.

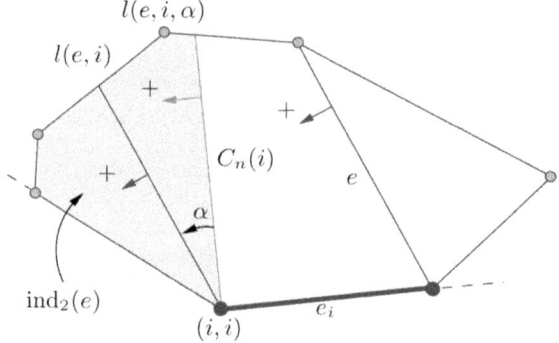

Figure 6. Secondary index $\mathrm{ind}_2(e)$ of a short edge e.

DEFINITION 5.7 (SECONDARY INDEX). The *secondary index* of a short edge e is defined as

$$\mathrm{ind}_2(e) := \mathrm{vol}\big(\mathrm{conv}(C_n, i) \cap l(e, i, \alpha)^+\big).$$

We can now define a lexicographic partial order induced by primary and secondary index. This will turn out to be the crucial relation among short edges in M. It is the partial order that will always decrease when we "chase M along short edges towards special short edges".

DEFINITION 5.8. Let e and e' be short edges in M'. Then

$$e \prec e' : \Longleftrightarrow \begin{cases} \text{either} & \mathrm{ind}_1(e) < \mathrm{ind}_1(e') \\ \text{or} & \mathrm{ind}_1(e) = \mathrm{ind}_1(e') \text{ and } \mathrm{ind}_2(e) < \mathrm{ind}_2(e'). \end{cases}$$

The following lemma is the formalization of "chasing the mixed subdivision towards special short edges".

LEMMA 5.9 (ORDER LEMMA). *Let e be a short edge in a mixed subdivision M of $S_n(\alpha)$. Then the following hold:*

(i) $\mathrm{ind}_1(e) \geq 0$ *and* $\mathrm{ind}_2(e) \geq 0$.

(ii) $\mathrm{ind}_1(e) = 0$ *if and only if* $e = e_i$ *for some* $i = 0, 1, \ldots, n-1$.

(iii) *If* $e \neq e_i$ *for all* $i = 0, 1, \ldots, n-1$, *then there exists another short edge* e' *in M with* $e' \prec e$; *moreover, there exists a 2-cell σ such that both e and e' are short edges of σ, and σ is on the positive side of e and on the negative side of* e'.

PROOF. Assertions (a) and (b) are true by definition.

In order to prove (c), consider a short edge e in M. Assume that e is in (C_n, i) and that $e \neq e_i$. Then the mixed subdivision M must contain cells that subdivide the convex hull of e and e_i. In particular, there must be a cell σ on the positive side of e. There are two cases: Either σ is a simplex containing only short edges inside (C_n, i), or σ is a parallelogram containing two short and two long edges.

Case 1: The cell σ is a simplex with short edges. By construction, $l(e)^+$ contains σ. By Lemma 5.3, σ lies on the negative side of one of its short edges, say e'. Then $l(e')^+$ does not contain σ. Moreover, since (C_n, i) is convex, $l(e)$ and $l(e')$ do not cross inside $\mathrm{conv}(C_n, i)$. Thus, $l(e')^+ \cap \mathrm{conv}(C_n, i) \subseteq l(e)^+ \cap \mathrm{conv}(C_n, i) \setminus \sigma$. Therefore, $\mathrm{ind}_1(e') \leq \mathrm{ind}_1(e) - \mathrm{vol}(\sigma) < \mathrm{ind}_1(e)$, whence $e' \prec e$.

Case 2: The cell σ is a parallelogram containing two short and two long edges. Consider the short edge e' in σ opposite to e. It lies in (C_n, j) for some $j = 0, 1, \ldots, n-1$ with $j \neq i$.

We first prove that e and e' have the same primary index. By Lemma 5.3, σ lies on the negative side of e'. By construction, σ lies on the positive side of e. Therefore, by Lemma 5.2, the parallel lines $l(e)$ and $l(e')$ have parallel orientations. That means, $T_{ij}(l(e)^+) = l(e')^+$. Because $T_{ij}(\mathrm{conv}(C_n, i)) = \mathrm{conv}(C_n, j)$, we conclude $\mathrm{ind}_1(e') = \mathrm{ind}_1(e)$.

Next, we show that the secondary index of e' is strictly smaller than that of e. By Lemma 5.6, (j, i) lies in the interior of $l(e, i, \alpha)^+$. This implies that $(j, j) = T_{ij}(j, i)$ lies in the interior of $T_{ij}(l(e, i, \alpha)^+)$. Since the parallel lines $l(e)$ and $l(e')$ have parallel orientations, the parallel lines $l(e, i, \alpha)$ and $l(e', j, \alpha)$ also have parallel orientations. Thus, $l(e', j, \alpha)^+$ is strictly contained in $T_{ij}(l(e, i, \alpha)^+)$. Therefore,

$$\mathrm{ind}_2(e') = \mathrm{vol}\big(\mathrm{conv}(C_n, j) \cap l(e', j, \alpha)^+\big)$$
$$= \mathrm{vol}\big(\mathrm{conv}\, T_{ij}(C_n, i) \cap l(e', j, \alpha)^+\big)$$
$$< \mathrm{vol}\big(\mathrm{conv}\, T_{ij}(C_n, i) \cap T_{ij}(l(e, i, \alpha)^+)\big)$$
$$= \mathrm{vol}\big(\mathrm{conv}(C_n, i) \cap (l(e, i, \alpha)^+)\big)$$
$$= \mathrm{ind}_2(e').$$

This proves that $e' \prec e$, and (iii) is proven as well. $\square$

The neighborhood of special short edges. We are now in a position to prove the main property of mixed subdivisions of $S_n(\alpha)$.

THEOREM 5.10. *Let $\alpha \geq 0$ such that $P_n(\alpha)$ is a prism or a twisted prism. Then every mixed subdivision M of $S_n(\alpha)$ contains at least one triangle one of whose edges is some special short edge.*

REMARK 5.11. If α is too large then not only the Order Lemma is false but also Theorem 5.10, which can be seen in Figure 7. Theorem 3.2, however, might still be true for large α because the cyclic set of tetrahedra defines parallelograms that are incompatible with the parallelogram that is on the positive sides of both of its short edges in Figure 7. One could consider all $\alpha \geq 0$ for which the face lattice of $P_n(\alpha)$ equals the one of the twisted prism in our sense. Since the existence of triangulations depends on the orientations of tetrahedra (the oriented matroid) rather than on the face lattice, we decided not to investigate this any further. If the top and the bottom n-gons are not congruent, Theorem 5.10 — and even Theorem 3.1 — do not hold either, as can be seen in Figure 8.

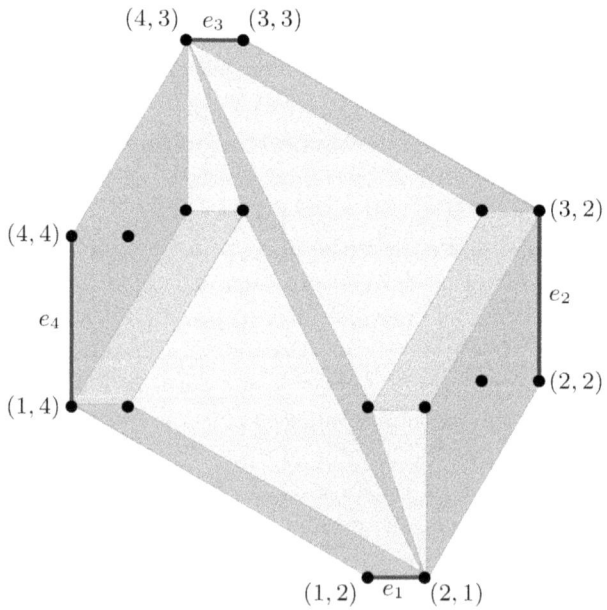

Figure 7. When α is too large (here $\alpha = \pi/3$), there exists a mixed subdivision where no special edge is covered by a mixed triangle; the parallelogram of Figure 2 serves as kind of an adapter between two part of the subdivision that would be incompatible otherwise. This mixed subdivision disappears when $P_4(\alpha)$ is untwisted. The indicated mixed subdivision does, however, not contradict the statement in Theorem 3.2 for larger α, since it does not use the parallelograms corresponding to the cyclic set of tetrahedra.

Figure 8. Congruence of top and bottom n-gons is important: even if the top and the bottom n-gon of a Cayley embedding of two n-gons are normally equivalent, there may be triangulations using the cyclic set of diagonals of the resulting combinatorial polygonal prism; the figure shows the corresponding mixed subdivision; note that indeed no special edge is used.

PROOF. Since every triangulation of $P_n(\alpha)$ induces a triangulation of its top and its bottom polygon, at least one short triangle must be used. Not all of its short edges can be edges of $S_n(\alpha)$. Therefore, there is a short edge having cells on both of its sides. Hence, there is at least on 2-cell that is on the positive side of some short edge. By Lemma 5.3, every such cell lies on the negative side of one of its other short edges.

Let σ be a cell on the positive side of its short edge e and on the negative side of its short edge e' such that e' is minimal with respect to "$\prec$". Then, by Lemma 5.3(iii), e' is a special edge.

Every parallelogram σ with a special short edge e_i must lie on the negative side of e_i, since the positive side of e_i is outside $S_n(\alpha)$. Therefore, the parallelogram σ lies on the same side of e_i as (C_n, i). Assume the opposite edge e of σ lies in (C_n, j) for some $j \in \{0, 1, \ldots, n-1\}$. Then, by Lemma 5.2, σ lies on the opposite side of e as (C_n, j). In particular, σ lies on the opposite side of e as e_j, which means, σ lies also on the negative side of e. $\qquad\square$

PROOF OF THEOREM 3.1 (PRISM). For the sake of contradiction, assume that there is a triangulation T of P_n that uses the cyclic set D_c of diagonals. Using the Cayley trick, T induces a fine mixed subdivision M of S_n that uses, among others, the set of points $(i, i+1)$ for all $i = 0, 1, \ldots, n-1$, corresponding to the cyclic set of diagonals (labels again regarded modulo n). The triangles in the quadrilateral facets of P_n induce the mixed edges $\{(i, i), (i, i+1)\}$ in the boundary of S_n. They already cover the whole boundary of S_n. Thus, the special edges $e_i := \{(i, i), (i+1, i)\}$ in the boundary of S_n, which correspond to the reverse cyclic set of diagonals in the quadrilateral facets of P_n, are not used in M. However, by Theorem 5.10, at least one e_i must be in M: contradiction. $\qquad\square$

PROOF OF THEOREM 3.2 (TWISTED PRISM). For the sake of contradiction, assume that there is a triangulation T of P_n that uses the cyclic set T_c of tetrahedra. Construct the corresponding mixed subdivision M of $S_n(\alpha)$. The set M_c of mixed cells corresponding to T_c are parallelograms that cover all the special edges e_i. Therefore, there can be no other cell that contains a special edge. By Theorem 5.10, there must be at least one mixed triangle containing a special edge e_i: contradiction.   □

Acknowledgements

I thank Yuri Romanovsky for bringing up this problem and Günter Ziegler for communicating it to me.

References

[Bagemihl 1948] F. Bagemihl, "On indecomposable polyhedra", *Amer. Math. Monthly* **55** (1948), 411–413.

[Below et al. 2000] A. Below, J. A. De Loera, and J. Richter-Gebert, "Finding minimal triangulations of convex 3-polytopes is NP-hard", pp. 65–66 in *Proceedings of the Eleventh Annual ACM-SIAM Symposium on Discrete Algorithms* (San Francisco, 2000), ACM, New York, 2000.

[Björner et al. 1993] A. Björner, M. Las Vergnas, B. Sturmfels, N. White, and G. M. Ziegler, *Oriented matroids*, Encyclopedia of Mathematics and its Applications **46**, Cambridge University Press, Cambridge, 1993.

[Huber et al. 2000] B. Huber, J. Rambau, and F. Santos, "The Cayley trick, lifting subdivisions and the Bohne-Dress theorem on zonotopal tilings", *J. Eur. Math. Soc.* **2**:2 (2000), 179–198.

[Lennes 1911] N. J. Lennes, "Theorems on the simple finite polygon and polyhedron", *Amer. J. Math.* **33** (1911), 37–62.

[Ruppert and Seidel 1992] J. Ruppert and R. Seidel, "On the difficulty of triangulating three-dimensional nonconvex polyhedra", *Discrete Comput. Geom.* **7**:3 (1992), 227–253.

[Schönhardt 1927] E. Schönhardt, "Über die Zerlegung von Dreieckspolyedern in Tetraeder", *Math. Annalen* **89** (1927), 309–312.

JÖRG RAMBAU
ZUSE-INSTITUTE BERLIN
TAKUSTR. 7
14195 BERLIN
GERMANY
  rambau@zib.de

Combinatorial and Computational Geometry
MSRI Publications
Volume **52**, 2005

On Hadwiger Numbers of Direct Products of Convex Bodies

ISTVÁN TALATA

ABSTRACT. The Hadwiger number $H(K)$ of a d-dimensional convex body K is the maximum number of mutually nonoverlapping translates of K that can touch K. We define $H^*(K)$ analogously, with the restriction that all touching translates of K are pairwise disjoint. In this paper, we verify a conjecture of Zong [1997] by showing that for any $d_1, d_2 \geq 3$ there exist convex bodies K_1 and K_2 such that K_i is d_i-dimensional, $i = 1, 2$, and $H(K_1 \times K_2) > (H(K_1) + 1)(H(K_2) + 1) - 1$ holds, where $K_1 \times K_2$ denotes the direct product of K_1 and K_2. To obtain the inequality, we prove that if K is the direct product of n convex discs in the plane and there are exactly k parallelograms among its factors, then $H^*(K) = 4^k(4 \cdot 6^{n-k} + 1)/5$. Based on this formula, we also establish that for every $d \geq 3$ there exists a strictly convex d-dimensional body K fulfilling $H(K) \geq \frac{16}{35}(\sqrt{7})^{d-1}$.

1. Introduction and Main Results

The *Hadwiger number* $H(K)$ of a d-dimensional convex body K is the maximum number of mutually nonoverlapping translates of K that can be arranged so that all touch K. Often $H(K)$ is called the *translative kissing number of* K as well. $H^*(K)$ is defined analogously with the restriction that all touching translates of K are pairwise disjoint. Trivially, $H^*(K) \leq H(K)$. It is known that $H(K) \leq 3^d - 1$ [Hadwiger 1957], with equality attained only for parallelotopes [Groemer 1961].

Let $A_i \subseteq \mathbb{R}^{d_i}$, $i = 1, 2 \ldots, n$, for some positive integer n. We denote by $A_1 \times A_2 \times \ldots \times A_n$ the direct product of the A_i's in their given order, which is the collection of the ordered n-tuples $\{(x_1, x_2, \ldots x_n) \mid x_i \in A_i, 1 \leq i \leq n\}$, and

Mathematics Subject Classification: Primary 52C17, 52A21; Secondary 52C07.

Keywords: Hadwiger number, translative kissing number, finite packing, strictly convex body, Minkowski metric, Cartesian product, direct product.

The work is in part supported by grant no. T038397 of the Hungarian National Science Foundation (OTKA) and by a Magyary fellowship of the Foundation for the Hungarian Higher Education and Research.

it is identified with a subset of $\mathbb{R}^d$, where $d = \sum_{i=1}^{n} d_i$, by writing the coordinates of the x_i's consecutively in one d-tuple. It is also called the Cartesian product of the A_i's, and sometimes it is denoted by $\prod_{i=1}^{n} A_i$ as well. Clearly, the direct product of convex bodies is also a convex body. If $A \subseteq \mathbb{R}^d$, then A^n stands for the direct product of n copies of A.

Let K_i be a d_i-dimensional convex body, for $i = 1, 2$. Observe that if $\mathcal{C}_i$ is a packing with translates of K_i, $i = 1, 2$, then $\mathcal{C}_1(\times)\mathcal{C}_2 = \{A \times B \mid A \in \mathcal{C}_1, B \in \mathcal{C}_2\}$ is a packing with translates of $K_1 \times K_2$. By looking at which translates of $K_1 \times K_2$ touch each other in $\mathcal{C}_1(\times)\mathcal{C}_2$, we get the general inequality

$$H(K_1 \times K_2) \geq (H(K_1) + 1)(H(K_2) + 1) - 1. \qquad (1\text{--}1)$$

Zong [1997] proved that there is equality in (1–1) when $\min(d_1, d_2) \leq 2$. He also conjectured that there are some large integers d_1, d_2 for which inequality (1–1) is strict for suitable d_i-dimensional convex bodies K_i, $i = 1, 2$. In the following theorem, we verify Zong's conjecture, and we even show that more is true: We provide examples for a strict inequality in (1–1) for every $d_1, d_2 \geq 3$.

THEOREM 1.1. *For every $d_1, d_2 \geq 3$, there is a d_1-dimensional convex body K_1 and a d_2-dimensional convex body K_2 such that*

$$H(K_1 \times K_2) \geq (H(K_1) + 1)(H(K_2) + 1) + 16 \cdot 3^{d_1+d_2-6} - 1 \qquad (1\text{--}2)$$

holds.

To prove Theorem 1.1, we rely on the value of $H^*(K)$ when K is the direct product of two circles. In the following proposition, we determine that quantity in a more general setting when the convex body is the direct product of finitely many arbitrary convex discs. (By a convex disc we always mean a two-dimensional convex body.)

PROPOSITION 1.2. *Let $D_1, D_2, \ldots D_n$ be convex discs, $n \geq 1$. If there are exactly k parallelograms among the discs, then*

$$H^*(D_1 \times D_2 \times \ldots \times D_n) = 4^k \left(\frac{4(6^{n-k}) + 1}{5} \right) \qquad (1\text{--}3)$$

holds.

Note that one can prove

$$H^*(K_1 \times K_2) \geq H^*(K_1)H^*(K_2) \qquad (1\text{--}4)$$

the same way as (1–1). Proposition 1.2 shows that there can be strict inequality in (1–4), e.g., that is the case when K_1 and K_2 are convex discs that are different from parallelograms.

Since $H(K) < 3^d - 1$ for any convex body K different from a parallelotope [Groemer 1961], one may ask how large $H(K)$ can be, when K belongs to some specific class of convex bodies, in which the shapes of the bodies are very different from a parallelotope, for example, when the bodies are strictly convex, i.e., their boundaries contain no segment. Based on Proposition 1.2, for every $d \geq 3$ we are able to show the existence of a strictly convex d-dimensional body, having relatively large Hadwiger number, as $d \to \infty$.

THEOREM 1.3. *For every $d \geq 3$ there exists a strictly convex d-dimensional body S such that*

$$H(S) \geq \frac{16}{35}(\sqrt{7})^d \approx 2.6457^{d-o(d)} \tag{1-5}$$

holds.

The *lattice kissing number* $H_L(K)$ is the maximum number of those translates that touch K in a lattice packing of K. Trivially, $H_L(K) \leq H(K)$. Although in several cases $H(K) = H_L(K)$ holds (for example, it holds for every convex disc by [Grünbaum 1961]), it can happen that $H(K) - H_L(K) > 0$ [Zong 1994]. In fact, $H(K) - H_L(K) \geq 2^{d-1}$ holds for some d-dimensional convex body K, for every $d \geq 3$ [Talata 1998a], showing that there can be exponentially large gap between $H(K)$ and $H_L(K)$. Minkowski [1896/1910] (see also [Cassels 1959]) showed that $H_L(K) \leq 2^{d+1} - 2$ holds for strictly convex d-dimensional bodies; thus Theorem 1.3 implies new asymptotic bounds showing that both the gap and the ratio between $H(K)$ and $H_L(K)$ can be relatively large.

COROLLARY 1.4. *For every integer $d \geq 1$, denote by $\mathcal{K}_d$ the collection of all d-dimensional convex bodies. Then*

$$\max_{K \in \mathcal{K}_d} (H(K) - H_L(K)) \geq (\sqrt{7})^{d-1} - 2^{d+1} + 2 \approx 2.6457^{d-o(d)} \tag{1-6}$$

and

$$\max_{K \in \mathcal{K}_d} (H(K)/H_L(K)) \geq \tfrac{8}{35}(\sqrt{7}/2)^d \approx 1.3228^{d-o(d)} \tag{1-7}$$

hold.

We would like to note that the proofs of Theorems 1.1 and 1.3 (and the proof of the lower bound part for $H^*(\prod_{i=1}^{n} D_i)$ in Proposition 1.2) are constructive: thus, general methods are given to construct bodies of different shapes, based on some initially given convex discs and some parameters, implying for example, that with respect to the Hausdorff metric, in every neighbourhood of any d-dimensional parallelotope there are convex bodies which possess the properties described in the Theorems 1.1 and 1.3. Furthermore, when the initial discs are unit circles, we can calculate the actual values for those parameters to make the definitions of those bodies.

We conclude this section with two conjectures. First, we suggest that there may not be any kind of analogue of Zong's formula [1997] for $H(K_1 \times K_2)$ when K_1 and K_2 are sufficiently high dimensional convex bodies. That is, we conjecture that $H(K_1 \times K_2)$ can not be expressed as a function of $H(K_1)$ and $H(K_2)$ in general.

CONJECTURE 1.1. *For some $d_1, d_2 \geq 3$, there exist a pair (K_1, K_1') of d_1-dimensional convex bodies and a pair (K_2, K_2') of d_2-dimensional convex bodies such that $H(K_1) = H(K_1')$ and $H(K_2) = H(K_2')$, but $H(K_1 \times K_2) \neq H(K_1' \times K_2')$.*

Second, we consider a quantity similar to $H(K)$: The touching number $t(K)$ of a d-dimensional convex body K is defined as the maximum number of mutually touching translates of K. We have $t(K) \leq 2^d$, with equality exactly for parallelotopes [Danzer and Grünbaum 1962]. It is conjectured in [Füredi et al. 1991] that for strictly convex bodies $t(K) \leq (2 - \varepsilon)^d$ holds for some $\varepsilon > 0$. We conjecture the analogous inequality for $H(K)$ in case of strictly convex bodies.

CONJECTURE 1.2. *There exists an absolute constant $\varepsilon > 0$ such that*

$$H(K) \leq (3 - \varepsilon)^d \tag{1–8}$$

holds whenever K is a strictly convex d-dimensional body.

We organize the remaining part of the paper in the following way: In Section 2 we introduce notation and recall some facts. Then we prove Proposition 1.2, Theorem 1.1 and Theorem 1.3 in Sections 3, 4 and 5, respectively. In those sections we usually prove various statements organized in lemmas and propositions so that we can combine them to get the desired theorem or proposition.

2. Preliminaries

For arbitrary $A, B \subseteq \mathbb{R}^d$ and $\alpha, \beta \in \mathbb{R}$, let $\alpha A + \beta B = \{\alpha a + \beta b \mid a \in A, b \in B\}$. We write $A + v$ instead of $A + \{v\}$, and further, we write $A - B$ instead of $A + (-1)B$. The notation $\mathrm{conv}(\cdot)$ stands for the convex hull and $[a, b]$ stands for the segment whose endpoints are $a, b \in \mathbb{R}^d$. If $c \in \mathbb{R}$, then $\{c\}$ denotes the fractional part of c, that is, $\{c\} = c - [c]$, where $[c]$ is the largest integer which does not exceed c. In the text, we always avoid confusion with the similar notation for a one-element set by using fractional parts only in inequalities. We denote by $|S|$ the cardinality of a set S. We use the notation o_d for the origin of $\mathbb{R}^d$.

We denote by ∂K the boundary of a convex body K. For an o_d-symmetric convex body K, let dist_K be the distance function of the Minkowski metric whose unit ball is K. Note that we denote the usual Euclidean distance simply by dist. Recall that $\mathrm{dist}_{K_1 \times K_2 \times \ldots \times K_n} = \max_{1 \leq i \leq n}(\mathrm{dist}_{K_i})$. In a metric space, a set S is called r-*discrete* for some $r > 0$ if the distance between any two points of S is at least r in the given metric. If the distance is larger than r we say S is r^+-*discrete*.

A *Hadwiger configuration* of a convex body K is a collection of mutually nonoverlapping translates of K which all touch K. It is easy to see that any collection $\{K + v_i\}_{i=1}^n$ of translates of a convex body K is a Hadwiger configuration of K if and only if $v_i \in \partial(K - K)$ and $\mathrm{dist}_{K-K}(v_i, v_j) \geq 1$, for every $i \neq j$ [Talata 1998b]. Clearly, $K + v_i$ and $K + v_j$ are touching if and only if

$\text{dist}_{K-K}(v_i, v_j) = 1$. Thus $H(K)$ is the maximum cardinality of a 1-discrete subset of $\partial(K - K)$ in the metric dist_{K-K}. Furthermore, it is not difficult to see that $H(K) + 1$ is the maximum cardinality of a 1-discrete subset $S \subseteq K - K$ in the metric dist_{K-K}, and $|S| = H(K) + 1$ holds only if $o_d \in S$. (To see this, observe that if $S \subseteq K - K$ is 1-discrete with respect to dist_{K-K}, then replacing each $p \in S \setminus \{o_d\}$ with that $q \in \partial(K - K)$ for which $p \in [o_d, q]$, we get a 1-discrete set with respect to dist_{K-K}.) Similarly, one can obtain that $H^*(K)$ is the maximum cardinality of a 1^+-discrete subset of $K - K$ in the metric dist_{K-K}. Note if K is o_d-symmetric, then $K - K$ can be replaced by K in the preceding characterizations of $H(K)$ and $H^*(K)$, since then $K - K = 2K$.

3. Determining $H^*(K)$ for Direct Products of Convex Discs

In this section, we prove Proposition 1.2. First we prove several lemmas, then we combine those to get the actual proof of the proposition. Note that in some cases we even allow 0-dimensional convex bodies to appear as factors in direct products for sake of completeness. Observe that $K_1 \times K_2 \cong K_2$ when K_1 is a 0-dimensional convex body (i.e., K_1 is a point).

LEMMA 3.1. *Let K be a d-dimensional convex body, $d \geq 0$, and let I be a segment. Then $H^*(K \times I) = 2H^*(K)$.*

PROOF. We may assume that K is o_d-symmetric and $I = [-1, 1]$. If $S \subseteq K$ is 1^+-discrete in the metric dist_K, then $S \times \{-1, 1\}$ is 1^+-discrete in the metric $\text{dist}_{K \times I}$, implying $H^*(K \times I) \geq 2H^*(K)$. On the other hand, if $S \subseteq K \times I$ is 1^+-discrete in the metric $\text{dist}_{K \times I}$, then let $S_1 = S \cap (K \times [-1, 0])$, and $S_2 = S \cap (K \times [0, 1])$. Now, let $\pi : K \times I \to K$ be the projection of the direct product body to the first factor. Then both $\pi(S_1)$ and $\pi(S_2)$ are 1^+-discrete subsets of K in the metric dist_K, implying $H^*(K \times I) \leq 2H^*(K)$. $\square$

Observe that an immediate consequence of Lemma 3.1 is that $H^*(K \times P) = 2^n H(K)$ holds if P is an n-dimensional parallelotope, $n \geq 1$.

LEMMA 3.2. *Let K be a d-dimensional convex body, $d \geq 0$, and let D be a convex disc. Then $H^*(K \times D) \leq 6H^*(K) - 1$.*

PROOF. We may assume that both K and D are symmetric about the origin. Let $S \subseteq K \times D$ be 1^+-discrete in the metric $\text{dist}_{K \times D}$, and let $s_0 \in S$. Define $\pi_1 : K \times D \to K$ and $\pi_2 : K \times D \to D$ as projections of the direct product body to its first and second factor, respectively. Consider an affine regular hexagon H inscribed to D, having vertices $v_1, v_2, \ldots, v_6$. We may even assume that H is chosen in a way that $\pi_2(s_0) \in [o_2, v_1]$; see [Fejes Tóth 1972]. Then the segments $[o_2 v_i]$, $1 \leq i \leq 6$, divide H into six regions U_i, $1 \leq i \leq 6$, such that their diameters are equal to 1 in the metric dist_D, and $\bigcup_{i=1}^{6} U_i = D$. Let $S_i = S \cap (K \times U_i)$. Then $\pi_1(S_i) \subseteq K$ is 1^+-discrete in the metric dist_K, implying $|S_i| \leq H^*(K)$ for each i. But s_0 is contained in two S_i's, implying $H^*(K \times I) \leq 6H^*(K) - 1$. $\square$

LEMMA 3.3. *Let C be a centrally symmetric convex disc, different from a parallelogram, and let k be a positive integer. Let $m = 6k - 1$. Then there exists a sequence $S = \langle s_i \rangle_{i=0}^{m-1}$ of points of ∂C such that for every $i \geq j$, $\operatorname{dist}_C(s_i, s_j) > 1$ holds if and only if $\frac{1}{6} < \{ \frac{i-j}{m} \} < \frac{5}{6}$ is satisfied.*

PROOF. If C is a circle, then it is easy to check that $\langle s_i \rangle_{i=0}^{m-1}$ can be chosen as consecutive vertices of a regular m-gon. For general C, we describe a little bit more sophisticated construction for S. Pick an affine regular hexagon H that is inscribed to C. Since C is not a parallelogram, it can be seen that we can choose H in a way that no side of H is longer than any segment in ∂C which is parallel to that side. Fix a positive constant $\varepsilon < 1$. If s_i' is already defined, let v_{i+1} be the first point chosen on ∂C in counterclockwise direction for which $\operatorname{dist}_C(v_i, v_{i+1}) = 1 + \varepsilon$. Let $V = \langle v_i \rangle_{i=0}^{m-1}$. It is easy to check that if ε is small enough, then every point of V lies in a small neighbourhood of some vertex of H, and $v_0, v_6, v_{12}, \ldots v_1, v_7 \ldots$ etc. are consecutive points on ∂C. Now, to get S, order the points of V consecutively in counterclockwise direction, starting with v_0, so $s_0 = v_0, s_1 = v_6, s_2 = v_{12} \ldots$ etc. It is easy to check that for any $i \geq j$, $\operatorname{dist}_C(s_i, s_j) > 1$ holds if and only if $\min(|i - j|, |m + j - i|) \geq k$, from which one can get that S has the property required in the lemma.   □

LEMMA 3.4. *Let n and q be integers, $n \geq 1$, $q \geq 3$. Let*

$$m = \frac{(q - 2) \cdot q^n + 1}{q - 1}.$$

Then for every positive integer $j \leq m - 1$, there is an integer i, $1 \leq i \leq n$, such that

$$\frac{1}{q} < \left\{ \frac{q^{n-i} j}{m} \right\} < \frac{q - 1}{q}.$$

PROOF. Define a sequence by $a_0 = 1$ and $a_i = qa_{i-1} - 1$, for every $i \geq 1$. It is easy to check that $a_n = q^n - \sum_{i=0}^{n-1} q^i = ((q - 2)(q^n) + 1)/(q - 1)$, for every $n \geq 0$. That is, $m = a_n$.

Let $1 \leq j \leq a_n - 1$. We claim that there are integers $t, z, j_1 \geq 0$ such that $j = (q - 2)q^t z + j_1$, $1 \leq t \leq n - 1$, $a_{t-1} \leq j_1 \leq a_t - a_{t-1}$ and $z \leq \sum_{i=0}^{n-t-1} q^i$ hold. To see this, we express j in the number system of base q as $(b_{n-1}, b_{n-2}, \ldots, b_1, b_0)_q$. That is, $j = \sum_{i=0}^{n-1} b_i q^i$, where $b_i \in \{0, 1, \ldots, q - 1\}$ for every i. Note $a_n - 1 = \sum_{i=0}^{n-1} (q - 2)q^i$. If $k \leq n - 1$, then $a_k - a_{k-1} = (q - 2)q^{k-1}$ and $a_{k-1} = (q - 1) + \sum_{i=1}^{k-2} (q - 2)q^i$ also hold. We distinguish three cases. First, if $b_i \in \{0, q - 2\}$ for every i, then let $t = 1 + \min\{k \mid b_k = q - 2\}$. Second, if $k_0 \neq q - 1$, where $k_0 = \max\{k \mid b_k \notin \{0, q - 2\}\}$, then let $t = k_0 + 1$. Third, if $k_0 = q - 1$, then let $t = \min\{k \mid k > k_0, b_k = 0\}$. In all cases, let $j_1 = \sum_{i=0}^{t-1} b_i q^i$ and $z = \sum_{i=0}^{n-t-1} c_i q^i$, where $c_i = b_{i+t}/(q - 2) \in \{0, 1\}$ for $1 \leq i \leq n - t - 1$. It is easy to check that the defined t, z and j_1 all have the claimed properties.

We show that the lemma holds for $i = t$. Clearly, $q^{n-t} j = (q - 2)q^n z + q^{n-t} j_1$ holds, thus $(q - 2)q^n z = ((q - 1)a_n - 1)z$ implies the equality $\{ q^{n-t} j / a_n \} =$

$\{(q^{n-t}j_1 - z)/a_n\}$. Now, on one hand, $q^{n-t}j_1/a_n < (q-1)/q$ by $j_1 \leq (q-2)q^{t-1}$. On the other hand, $(q^{n-t}j_1 - z)/a_n \geq (q^{n-t}a_{t-1} - \sum_{i=0}^{n-t-1} q^i)/a_n$. Observe that $a_k = q^{k-i}a_i - \sum_{i=0}^{k-i-1} q^i$, for every $k > i$. This can be proved by induction on $k-i$. Consequently, $(q^{n-t}j_1 - z)/a_n \geq a_{n-1}/a_n > 1/q$. This completes the proof of the lemma. □

PROOF OF PROPOSITION 1.2. From Lemma 3.1 follows that $H^*(K \times P) = 4H^*(K)$ for any parallelogram P, implying that in the following it is enough to consider (1–3) for $k = 0$. Assume that $K = \prod_{i=1}^n D_i$ is a direct product of convex discs all different from a parallelogram. We may also assume that all the discs D_i are symmetric about o_2. On the one hand, for $d = 0$, Lemma 3.2 implies $H^*(D_1) \leq 5$, thus repeated applications of Lemma 3.2 yield $H^*(K) \leq c_n$, where c_n is defined as $c_0 = 1$, $c_i = 6c_{i-1} - 1$ for every $i \geq 1$. Since $c_n = (4(6^n) + 1)/5$, consequently $H^*(K) \leq (4(6^n) + 1)/5$. On the other hand, applying Lemma 3.3 for $k = c_{n-1}$, $m = c_n$ and $C = D_i$, for any $1 \leq i \leq n$, we obtain a sequence $\langle s_i(j) \rangle_{j=0}^{m-1}$ of points of ∂D_i such that for every j and j_0, $\mathrm{dist}_{D_i}(s_i(j), s_i(j_0)) > 1$ is equivalent with $1/6 < \{(j - j_0)/m\} < 5/6$. Now, define a point $p_j = \prod_{i=1}^n s_i(b(i,j))$, for every $0 \leq j \leq m - 1$, where $0 \leq b(i,j) \leq m - 1$, $b(i,j) \equiv 6^{n-i}j \pmod{m}$. Then $S = \{p_j\}_{j=1}^{m-1} \subseteq \partial K$, and S is 1^+-discrete in the metric $\mathrm{dist}_K = \max_{1 \leq i \leq n}(\mathrm{dist}_{D_i})$, since for every $j_1 \neq j_2$, by applying Lemma 3.4 for $q = 6$ and $j = j_1 - j_2$, there is an index i such that $1/6 < \{6^{n-i}(j_1 - j_2)/m\} < 5/6$, that is equivalent with $\mathrm{dist}_{D_i}\big(s_i(b(i,j_1)), s_i(b(i,j_2))\big) > 1$. Consequently, $H^*(K) \geq c_n$. □

4. Verifying a Conjecture of Zong

In this section, we prove Theorem 1.1. For a set $S \subseteq \mathbb{R}^d$ and any $x \in \mathbb{R}$ we denote by (S, x) the set $S \times \{x\} \subseteq \mathbb{R}^{d+1}$. Further, if $C \subseteq \mathbb{R}^2$, then let $C(r) = r \cdot C$. From now on, I stands for the interval $[-1, 1]$ in the paper. Let C be a centrally symmetric convex disc, $0 < \varepsilon < 1$, $0 < \delta \leq 1$. We define a three-dimensional convex body as the convex hull of four suitable homothetic copies of C placed in $\mathbb{R}^3$: Let $B(C, \varepsilon, \delta) = \mathrm{conv}(C_1, C_2, -C_2, -C_1)$, where $C_1 = ((1 - \delta)C) \times \{1\}$ and $C_2 = C \times \{1 - \varepsilon\}$. Next we prove two lemmas. Combining these, first we get Theorem 1.1 for $d_1 = d_2 = 3$ in Proposition 4.3, then we prove it in general.

LEMMA 4.1. *Let C be an arbitrary centrally symmetric convex disc that is different from a parallelogram, $0 < \varepsilon \leq 1/3$, and $0 < \delta < \delta_0$, where $\delta_0 < 1$ is a positive constant that depends on C only (when C is a circle, one can choose $\delta_0 = 1 - (2\sin(\frac{\pi}{5}))^{-1} \approx 0.1493$). If $B = B(C, \varepsilon, \delta)$, then $H(B) = 16$ holds.*

PROOF. We may assume that C is symmetric about the origin. Let α_n be the largest possible value for the minimum distance occuring in a set of n points of ∂C with respect to the metric dist_C, for any $n \geq 1$. If C is a circle, then α_n is the side length of a regular n-gon insribed into C. Observe that $H(C) = 6$

implies $\alpha_6 \geq 1$ and $\alpha_8 < 1$. By Proposition 1.2 we have $H^*(C) = 5$, from which $\alpha_5 > 1$ follows. Let $\delta_0 = \min(\alpha_8, 1 - (1/\alpha_5))$.

First we show $H(B) \geq 16$. Let $V_i \subseteq \partial C$ be a set of i points that is α_i-discrete in the metric dist_C, for $i = 5, 6$. Let $V = ((1-\delta)V_5, 1) \cup (V_6, 0) \cup (-(1-\delta)V_5, -1)$. It is easy to check that V is a 1-discrete subset of ∂B in the metric dist_B. Thus $H(B) \geq 16$.

Next we prove $H(B) \leq 16$. First we introduce further notation. We define the projection functions $h : \mathbb{R}^3 \to \mathbb{R}$ and $\pi : \mathbb{R}^3 \to \mathbb{R}^2$ by $h(x) = x_3$ and $\pi(x) = (x_1, x_2)$ if $x = (x_1, x_2, x_3) \in \mathbb{R}^3$. For $c \in \mathbb{R}$, let $P(c)$ be the plane $\{x \in \mathbb{R}^3 \mid h(x) = c\}$. Denote by $P^+(c)$ the open halfspace $\{x \in \mathbb{R}^3 \mid h(x) > c\}$.

Let $\mathcal{C}$ be a Hadwiger configuration of B. We can partition $\mathcal{C}$ into $\{\mathcal{C}_1, \mathcal{C}_2, \mathcal{C}_3\}$ in a way that for $B + v \in \mathcal{C}$ we have $B + v \in \mathcal{C}_1$ if $h(v) \geq 2\varepsilon$, $B + v \in \mathcal{C}_2$ if $h(v) \leq -2\varepsilon$, and $B + v \in \mathcal{C}_3$ otherwise. Let $n_i = |\mathcal{C}_i|$, $i = 1, 2, 3$. We may assume $n_1 \geq n_2$. It is clear that for every $B' \in \mathcal{C}_1$, $B' \cap P(1+\varepsilon)$ is a translate of $(C, 1+\varepsilon)$, and $B' \cap P^+(1+\varepsilon) \neq \varnothing$, so $\mathcal{U} = \{\pi(B') \mid B' \in \mathcal{C}_1\}$ is a packing of n_1 translates of C, each having common point with the C. This immediately gives $n_1 \leq 7$.

Now we consider when $n_1 \geq 6$. Let $\mathcal{C}' = \{B' \in \mathcal{C} \mid B' \cap P \neq \varnothing, B' \cap P^+ \neq \varnothing\}$, where $P = P(1 - 2\varepsilon)$ and $P^+ = P(1 - 2\varepsilon)^+$. Observe that $\{\pi(B' \cap P) \mid B' \in \mathcal{C}'\}$ is a packing of sets all containing a translate of $C(1 - \delta)$, where $\delta < \alpha_8$, and having centers of symmetry on $\partial(2C)$. This implies $|\mathcal{C}'| \leq 7$. If $C \in \mathcal{U}$, then at least five other members of $\mathcal{U}$ touch C. But if $B' = B + v \in \mathcal{C}_1$, and $\pi(B')$ is such a touching disc, then B' touches B at a point $p = (1/2)v$ for which $\varepsilon \leq h(p) \leq 1 - \varepsilon$, and thus $h(v) \leq 2 - 2\varepsilon$. Therefore $|\mathcal{C}' \cap \mathcal{C}_1| \geq 5$. By $\varepsilon \leq 1/3$ we have $\mathcal{C}_3 \subseteq \mathcal{C}'$. Thus $n_3 + 5 \leq |\mathcal{C}'|$, implying $n_3 \leq 2$. Therefore $|\mathcal{C}| \leq 7 + 2 + 7 = 16$. If $C \notin \mathcal{U}$, then $n_1 = 6$, and there are at least three members of $\mathcal{U}$ that touch C. (To see this, one can replace $\mathcal{U}$ by a Hadwiger configuration of C similarly as we did in Section 2 by "pushing out" the translates, and then use the description of all possible Hadwiger configurations of six translates by [Swanepoel 2000] to observe that at most three translates can be "pushed back". Note that for circles the claim be easily shown directly, using angles determined by the translation vectors). Similarly to the case $C \in \mathcal{U}$, we get $|\mathcal{C}' \cap \mathcal{C}_1| \geq 3$, implying $n_3 + 3 \leq |\mathcal{C}'|$ and thus $n_3 \leq 4$. Then $|\mathcal{C}| \leq 6 + 4 + 6 = 16$.

Finally, we consider when $n_1 \leq 5$. Since for $\{\pi(B' \cap P(0)) \mid B' \in \mathcal{C}_3\}$ one can easily show by $\varepsilon \leq 1/3$ that it is a packing of translates of C, all touching C, we have $n_3 \leq 6$. Combining the upper bounds, we get $|\mathcal{C}| = 5 + 6 + 5 \leq 16$. $\square$

LEMMA 4.2. *Let $\{C_i\}_{i=1}^n$ be a collection of n centrally symmetric convex discs that are different from parallelograms, $n \geq 1$. If $0 < \varepsilon < 1$, $0 < \delta \leq \gamma$, where $\gamma < 1$ is a positive constant that depends on $\{C_i\}_{i=1}^n$ only, and $B_i = B(C_i, \varepsilon, \delta)$, $1 \leq i \leq n$, then*

$$H\left(\prod_{i=1}^n B_i\right) \geq \frac{4(19)^n + 9^n}{5} - 1. \tag{4-1}$$

If every C_i is a circle, one can choose $\gamma = \delta_n = 1 - \left(2\sin\left(\frac{\pi}{6} + \frac{5\pi}{4(6^{n+1})+6}\right)\right)^{-1}$. In particular, if $\delta \leq \gamma$, then $H(B_1 \times B_2) \geq 304$. (Note if C_1, C_2 are circles, then one can choose $\gamma = \delta_2 = 1 - \left(2\sin\left(\frac{5\pi}{29}\right)\right)^{-1} \approx 0.030169$.)

PROOF. Let $K = \prod_{i=1}^{n} B_i$. Let $A_i(j) = (C_i(1-\delta), j)$, for $j = -1, 1$, and let $A_i(0) = (C_i, j)$, $1 \leq i \leq n$. Clearly, $A_i(j) \subseteq B_i$. Moreover, if $p \in A_i(j)$ and $q \in A_i(k)$, $j \neq k$, then $\operatorname{dist}_{B_i}(p, q) \geq 1$. Let $D = \prod_{i=1}^{n} M_i$, where $M_i \in \{A_i(j)\}_{j=-1,0,1}$, and M_i is chosen in an arbitrary way. Then, there is a permutation π of the $3n$ coordinates so that $\pi(D) = U \times W \times Z$, where $U = \prod_{M_i \neq A_i(0)} C_i(1-\delta)$, $W = \prod_{M_i = A_i(0)} C_i$ and Z is a single vector having coordinates from the set $\{-1, 0, 1\}$. Denote by $2m$ the dimension of U. By Proposition 1.2, for some $\gamma_0 > 0$ there is a $(1/(1-\gamma_0))$-discrete set $S_1 \subseteq U$ in the metric dist_U having cardinality $c_n = (4(6^m)+1)/5$, and by $H(C_i) = 6$, there is a 1-discrete set $S_2 \subseteq W$ in the metric dist_W having cardinality 7^{n-m}. Let $X = \pi^{-1}(S_1 \times S_2 \times Z)$. Let Y be the union of such sets X when M_i's are choosen all possible ways, and let γ be the minimum of all occuring γ_0's. Clearly, $Y \subseteq K$ and Y is 1-discrete in the metric dist_K if $(1-\delta)/(1-\gamma) \geq 1$, that is, $\delta \leq \gamma$. Thus $H(K) + 1 \geq |Y|$. If every C_i is a unit circle, then S_1 is a subset of the direct products of inscribed regular c_n-gons G_i, and $1/(1-\gamma)$ can be chosen as the minimum distance that is larger than 1 and occurs among the vertices of G_i. Corresponding to the choices of the sets M_i, we can count the cardinality of Y:

$$|Y| = \sum_{m=0}^{n} \binom{n}{m} \frac{4(6^{n-m})+1)}{5}(7^m)(2^{n-m}) = \frac{4}{5}(19^n) + \frac{1}{5}(9^n), \qquad (4\text{--}2)$$

Finally, based on $H(K) \geq |Y| - 1$ and (4–2), we get (4–1). □

Combining Lemma 4.1 and Lemma 4.2 for $n = 2$, it readily implies the following.

PROPOSITION 4.3. *Let C_1, C_2 be arbitrary convex discs that are different from parallelograms, $0 < \varepsilon \leq 1/3$, and $0 < \delta < \mu$, where $\mu < 1$ is a positive constant that depends on C_1, C_2 only. Then $B_i = B(C_i, \varepsilon, \delta)$, $i = 1, 2$, satisfies*

$$H(B_1 \times B_2) \geq (H(B_1) + 1)(H(B_2) + 1) + 15. \qquad (4\text{--}3)$$

If C_1, C_2 are circles, then one can choose $\mu = 0.03$.

PROOF OF THEOREM 1.1. For any $d_1, d_2 \geq 3$, let $K_i = B \times I^{d_i - 3}$, $i = 1, 2$, where I^{d-3} is a $(d-3)$-dimensional cube. Since $H(K \times I^n) + 1 = 3^n(H(K) + 1)$ holds for every convex body K and positive integer n by Zong [1997], from Proposition 4.3 one can immediately deduce (1–2). □

5. Hadwiger Numbers of Strictly Convex Bodies

In this section, we prove Theorem 1.3. First we show that for every odd integer $d \geq 3$ there exists a d-dimensional convex body for which $H^*(K)$ is relatively large. After that we prove a similar statement for arbitrary $d \geq 3$, which will imply Theorem 1.3.

PROPOSITION 5.1. *For every odd integer $d \geq 3$ there exists a d-dimensional convex body K such that*

$$H^*(K) \geq \frac{8(\sqrt{7})^{d-1} + 2(\sqrt{2})^{d-1}}{5} \tag{5-1}$$

holds.

PROOF. Let $n = (d-1)/2$. Consider $K_0 = \prod_{i=1}^{n} D_i$ where every D_i is a strictly convex disc that is symmetric about the origin. Let $\pi_i : K_0 \to D_i$ be the projection to the ith factor of the direct product. Denote by J an arbitrary subset of $N = \{1, 2, \ldots, n\}$. Let $m = |J|$, $P_J = \prod_{i \in J} D_i$, and let $Q_J = \prod_{i \in N \smallsetminus J} D_i$. Then $g_J(K_0) = P_J \times Q_J$ for some permutation g_J of the coordinates. By Proposition 1.2, there is a set $S_J \subseteq \partial P_J$ of cardinality $(4(6^m) + 1)/5$ which is a 1^+-discrete set in the metric dist_{P_J}. Let $T_i(J) = \pi_i(S)$, for $i \in J$, and let $V_i = \bigcup_{\{J : i \in J\}} T_i(J)$, for every $1 \leq i \leq n$. We may assume that $\pi_i(S_J) \subseteq \partial D_i$ holds for every i and J, by moving out the points of $\pi_i(S_J)$ towards ∂D_i on a ray emanating from the center o_2 if necessary. We can even perturb the elements of every occuring set S_J if necessary so that $o_2 \notin (p+q)/2$ holds for every $p, q \in V_i$, $p \notin q$, and S_J still remains 1^+-discrete in the metric dist_{K_1} and $\pi_i(S_J) \subseteq \partial D_i$ is still holds for every i. Let $W_i = \mathrm{conv}(V_i)$, $1 \leq i \leq n$, $W = \prod_{i=1}^{n} W_i$, and let $K = \mathrm{conv}((W, 1), (-W, -1))$. Denote by X_J the set $g_J^{-1}(S_J \times \{o_{n-m}\})$, and let $X = \bigcup_{J \subseteq N} X_J$. Observe that if $p, q \in X$, $p \neq q$, then either $p, q \in X_J$ for some J, or $p \in X_J$, $q \in X_M$ for some $J, M \subseteq N$, $J \neq M$. In the first case, there is an index $i \in J$ for which $\mathrm{dist}_{D_i}(\pi_i(p), \pi_i(q)) > 1$. In the second case, there is an index $i \in (J \setminus M) \cup (M \setminus J)$, for which either $\pi_i(p) = o_2$ and $\pi_i(q) \in \partial D_i$, or $\pi_i(q) = o_2$ and $\pi_i(p) \in \partial D_i$ holds, implying $\mathrm{dist}_{(W_i - W_i)/2}(\pi_i(p), \pi_i(q)) > 1$. Let $Y = (X, 1) \cup (-X, -1)$. It is easy to see that $Y \subseteq K$ and Y is 1^+-discrete in the metric dist_K. Counting the cardinality of Y by the corresponding choices of J, we get

$$|Y| = 2 \sum_{m=0}^{n} \binom{n}{m} \frac{4(6^m) + 1)}{5} = \frac{8(7^n) + 2^{n+1}}{5}. \tag{5-2}$$

By $H^*(K) \geq |Y|$, we obtain (5–1).   □

If $d \geq 4$ is even, then one can apply Proposition 5.1 in dimension $d-1$ and combine that with the cylindrical construction of Lemma 3.1 to get a d-dimensional convex body K with $H^*(K) \geq (16(\sqrt{7})^{d-2} + 4(\sqrt{2})^{d-2})/5$. Comparing this formula with (5–1), we get the following.

COROLLARY 5.2. *For every integer $d \geq 3$ there exists a d-dimensional convex body K such that*

$$H^*(K) \geq \frac{16(\sqrt{7})^{d-2} + 4(\sqrt{2})^{d-2}}{5} \geq \frac{16}{35}(\sqrt{7})^d \approx 2.6457^{d-o(d)}. \tag{5-3}$$

PROOF OF THEOREM 1.3. Consider the collection $\mathcal{K}_d$ of all d-dimensional convex bodies equipped with the Hausdorff metric [Schneider 1993]. Note that $H^*(K)$ is not decreasing in a sufficiently small neighbourhood of K (this latter is obvious by the description of $H^*(K)$ in terms of 1^+-discrete subsets, see Section 2), and the strictly convex bodies form a dense set in $\mathcal{K}_d$. Therefore we can apply Corollary 5.2 to get a convex body $K \in \mathcal{K}_d$ for which (5–3) holds, and we can pick a strictly convex body S sufficiently close to it in the Hausdorff metric so that $H(S) \geq H^*(S) \geq H^*(K)$. □

REMARK 1. Instead of proving only existence, one can also construct strictly convex bodies of various shapes having the properties of Theorem 1.3: By the proof of Proposition 5.1 and the paragraph following that we have a description of an o_d-symmetric convex polytope K that fulfils (5–3), for every $d \geq 3$. We also have a description of a 1^+-discrete set $Y \subseteq \partial K$ in the metric dist_K whose cardinality is at least the lower bound appearing in (5–3). Denote by τ the minimum distance occuring in Y with respect to the metric dist_K. Then $K \subseteq \text{int}(\tau K)$, therefore to each facet F of τK we can find a Euclidean ball $B(F)$ which touches F at a point $p \in \text{relint}F$ and contains K, just the radius of the ball needs to be sufficiently large. Let $S = \bigcap\{B(F) \mid F \text{ is a facet of } \tau K\}$. Then S is strictly convex, and (1–5) holds.

REMARK 2. In particular, when K is constructed in the proof of Proposition 5.1 by applying Proposition 1.2 for direct products of unit circles, then one can explicitely define a strictly convex body S fulfilling (1–5): If d is odd, then K is chosen as $\text{conv}((W, 1), (-W, -1))$ where W is the direct product of n copies of a regular c_n-gon, inscribed into a unit circle. One can check $\tau = 2/(1+\cos(\pi/c_n))$, and for every facet F of τK, the body K is contained in a ball $B(F)$ that touches F at its baricenter (that is, at $(1/|\text{vert}(F)|) \sum_{v \in \text{vert}(F)} v$) and has radius $(n+4)\tau^2/(\tau^2-1)$, so S can be chosen as the intersection of such balls. To make the definiton more expicit, one may even calculate the centers of the balls $B(F)$ in terms of the vertices of G. The case when d is even can be treated similarly.

REMARK 3. Finally, we note that similarly to the proof of Theorem 1.3, one can show that every dense subcollection of the space of all d-dimensional convex bodies contains a member S for which (1–5) holds.

References

[Cassels 1959] J. W. S. Cassels, *An introduction to the geometry of numbers*, Grundlehren der math. Wissenschaften **99**, Springer, Berlin, 1959. Second edition, 1971.

[Danzer and Grünbaum 1962] L. Danzer and B. Grünbaum, "Über zwei Probleme bezüglich konvexer Körper von P. Erdős und von V. L. Klee", *Math. Z.* **79** (1962), 95–99.

[Fejes Tóth 1972] L. Fejes Tóth, *Lagerungen in der Ebene auf der Kugel und im Raum*, Springer, Berlin, 1972.

[Füredi et al. 1991] Z. Füredi, J. C. Lagarias, and F. Morgan, "Singularities of minimal surfaces and networks and related extremal problems in Minkowski space", pp. 95–109 in *Discrete and computational geometry* (New Brunswick, NJ, 1989/1990), edited by J. E. Goodman et al., DIMACS Ser. Discrete Math. Theoret. Comput. Sci. **6**, Amer. Math. Soc., Providence, RI, 1991.

[Groemer 1961] H. Groemer, "Abschätzungen für die Anzahl der konvexen Körper, die einen konvexen Körper berühren", *Monatsh. Math.* **65** (1961), 74–81.

[Grünbaum 1961] B. Grünbaum, "On a conjecture of H. Hadwiger", *Pacific J. Math.* **11** (1961), 215–219.

[Hadwiger 1957] H. Hadwiger, "Über Treffanzahlen bei translationsgleichen Eikörpern", *Arch. Math.* **8** (1957), 212–213.

[Minkowski 1896/1910] H. Minkowski, *Geometrie der Zahlen* (2 vol.), Teubner, Leipzig, 1896/1910.

[Schneider 1993] R. Schneider, *Convex bodies: the Brunn–Minkowski theory*, Encyclopedia of Mathematics and its Applications **44**, Cambridge University Press, Cambridge, 1993.

[Swanepoel 2000] K. J. Swanepoel, "Gaps in convex disc packings with an application to 1-Steiner minimum trees", *Monatsh. Math.* **129**:3 (2000), 217–226.

[Talata 1998a] I. Talata, "Exponential lower bound for the translative kissing numbers of *d*-dimensional convex bodies", *Discrete Comput. Geom.* **19**:3, Special Issue (1998), 447–455.

[Talata 1998b] I. Talata, "On a lemma of Minkowski", *Period. Math. Hungar.* **36**:2-3 (1998), 199–207.

[Zong 1994] C. M. Zong, "An example concerning the translative kissing number of a convex body", *Discrete Comput. Geom.* **12**:2 (1994), 183–188.

[Zong 1997] C. M. Zong, "The translative kissing number of the Cartesian product of two convex bodies, one of which is two-dimensional", *Geom. Dedicata* **65**:2 (1997), 135–145.

ISTVÁN TALATA
DEPARTMENT OF GEOMETRY
EÖTVÖS UNIVERSITY
PÁZMÁNY PÉTER SÉTÁNY 1/C
BUDAPEST, H-1117
HUNGARY
talata@cs.elte.hu

Combinatorial and Computational Geometry
MSRI Publications
Volume **52**, 2005

Binary Space Partitions: Recent Developments

CSABA D. TÓTH

ABSTRACT. A binary space partition tree is a data structure for the representation of a set of objects in space. It found an increasing number of applications over the last decades. In recent years, intensifying research focused on its combinatorial properties, which affect directly the efficiency of applications. Important advances were made on binary space partitions for disjoint line segments in the plane and for axis-aligned objects in higher dimensions. New research directions were also initiated on some realistic polygonal scenes and on kinetic binary space partitions. This paper attempts to give an overview of these results and reiterates some of the most pressing open problems.

1. Introduction

The *binary space partition tree* is a geometric data structure obtained by a recursive partitioning scheme, called *binary space partition* (for short, *BSP*) over a set of input objects: The space is partitioned along a hyperplane into two half-spaces, then either half-space is partitioned recursively until every subproblem contains only a trivial fraction of the input objects. The concept of BSP has emerged from the computer graphics community in the seventies. It was originally designed to assist efficient hidden-surface removal algorithms for moving viewpoints, but it has later found widespread applications in many areas of computational and combinatorial geometry.

In many of the applications, the bottle neck of the space complexity is the size of the BSP tree they rely on. Combinatorial research focused on determining the worst case complexity of BSPs for certain classes of inputs. Despite the simplicity of the BSP algorithm, it is often challenging to determine the so-called *partition complexity* even for simple object classes such as disjoint line segments in the plane, or axis-aligned boxes in higher dimensions. Ideally, the partition hyperplanes do not split the input objects, and the size of the BSP tree is linear in terms of the input size. In many cases, however, it is inevitable that input objects are also fragmented, and the size of the data structure becomes superlinear.

This paper surveys results on the combinatorial properties of BSPs from the last four-five years. Inevitably, we include less recent results as well because early ideas and observations are often used, sometimes in an enhanced form, to obtain new results. Before we move on to the latest developments, let us define the binary space partition and the partition complexity, and recall a few applications and early results.

Definitions. A binary space partition tree is a recursive partition scheme for an input set of pairwise interior disjoint objects in $\mathbb{R}^d$, $d \in \mathbb{N}$. If the input contains two full-dimensional objects or a lower-dimensional object, we partition the space by a hyperplane h and recursively apply two binary space partitions for the objects clipped in each of the two open half-spaces of h. If the input is at most one full-dimensional object (and no lower-dimensional object), we stop.

The partition algorithm naturally corresponds to a binary tree: Every node corresponds to the input of a recursive call of the BSP: The root corresponds to the initial input set, the two children of a non-leaf node correspond to the inputs of its two subproblems. The BSP tree *data structure* is based on this binary tree: Every leaf stores at most one full-dimensional object which is the input of the corresponding subproblem; and every non-leaf node stores the splitting hyperplane and the (lower-dimensional) objects of the corresponding subproblem that lie on the splitting hyperplane. As a convention, the non-leaf nodes store only k-dimensional fragments of k-dimensional objects lying on the splitting hyperplane in $\mathbb{R}^d$, $0 \leq k \leq d$. For example, if a splitting hyperplane h crosses an input segment s then the point $h \cap s$ is never stored. Figure 1 depicts an example for a recursive partitioning and corresponding BSP tree for four input objects.

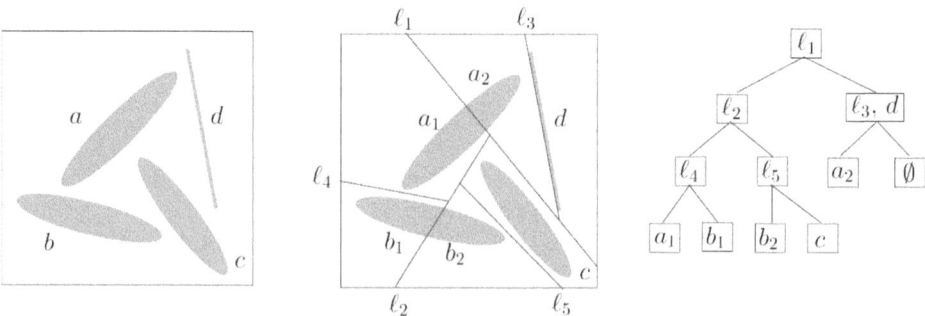

Figure 1. Three disjoint ellipses and a line segment (left), binary space partition with five partition lines (middle), and the corresponding BSP tree (right).

The binary space partition algorithm also defines a convex space partition: Every node of the BSP tree corresponds to a convex polytope, which is the

intersection of the halfspaces bounded by the splitting hyperplanes of its ances-
tors. The root corresponds to the space $\mathbb{R}^d$. A splitting hyperplane stored at a
non-leaf node splits the corresponding convex polytope into the two polytopes
corresponding to the two child nodes. The polytope corresponding to a node is
also called the *cell* of the (sub)problem because it contains every input object of
the (sub)problem. Observe that the set of all leaves of a BSP tree correspond to
a convex partition of the space.

An important special BSP, called *autopartition*, is defined for at most $(d-1)$-
dimensional input objects in $\mathbb{R}^d$. An autopartition is a BSP where every splitting
hyperplane contains an input object of the corresponding (sub)problem.

A BSP has two important parameters: Its *size* is the total number of fragments
of objects stored at the nodes of the BSP tree; its *height* is the height of the BSP
tree. The BSP in Figure 1 has size 6 and height 3. An easy way to compute
the size of the BSP is to sum the number of input objects and the number of
cuts, the events when a splitting hyperplane partitions (a fragment of) an input
object. If the BSP does not make "useless cuts", that is, if every splitting plane
partitions the convex hull of the input objects, then the BSP size is an upper
bound on the number of non-leaf nodes of the BSP tree.

There are many BSPs for every input set depending on the choice of the
splitting hyperplane at each subproblem. The *partition complexity* of a set S of
objects is the size of the smallest BSP for S. The minimal size BSP does not
necessarily comes with minimal heights, though. The *autopartition complexity* of
a set S is the size of the smallest autopartition for S. Clearly, the autopartition
complexity is never smaller than the partition complexity.

The convention that objects lying in a partition hyperplane are not partitioned
further leads to somewhat counterintuitive phenomenon: A set of objects in $\mathbb{R}^d$
with superlinear partition complexity will have linear partition complexity when
embedded into $\mathbb{R}^{d+1}$ because a single splitting hyperplane contains them all. One
may define another binary partitioning scheme which would partition recursively
the fragments of objects lying in each splitting hyperplane. The minimal number
of fragments under such an alternative partition scheme may be much higher than
the partition complexity. In this survey, we focus on combinatorial properties of
the partition complexity.

1.1. Applications. The initial and most prominent applications, where the
BSP tree itself is stemming from, lie in computer graphics: BSPs support fast
hidden-surface removal [Schumacker et al. 1969; Fuchs et al. 1980; Murali 1998]
and *ray tracing* [Naylor and Thibault 1986] for moving viewpoints. Rendering
is used for visualizing spatial opaque surfaces on the screen. A common and
efficient rendering technique is the so-called *painter's method*. It draws every
object sequentially according to the *depth order* or *back-to-front* order, starting
with the deepest object and proceeding with the objects closer to the viewer.

When all the objects have been drawn, every pixel represents the color of the object closest to the viewpoint.

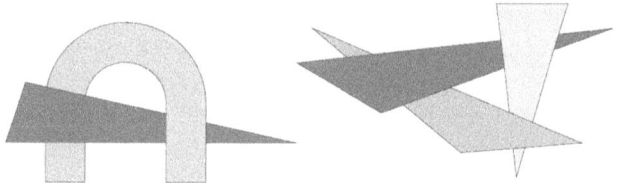

Figure 2. Objects may have cyclic overlaps.

Unfortunately, some sets of objects have cyclic overlaps, where no valid depth order exists (Figure 2). A BSP partitions the input objects into fragments for which the depth order always exists for every viewpoint. It is easy to extract the depth order of the fragments by a simple traversal of the BSP tree. An additional benefit of BSPs is that the depth order can easily be updated as the viewpoint moves continuously: It is enough to swap the two children of a node of the BSP tree when the viewpoint crosses the corresponding splitting hyperplane. (Computing the depth order, if it exists, for a *fixed* viewpoint may be easier than constructing a BSP, however; see [de Berg 1993].)

The first applications of BSPs were soon followed by many others, e.g., *constructive solid geometry* [Thibault and Naylor 1987; Naylor et al. 1990; Buchele 1999] and *shadow generation* [Chin and Feiner 1989; Batagelo and Jr. 1999]. The BSP for the faces of a full dimensional polyhedral solid can represent the solid and its boundary with Boolean set operations. This, in turn, can be used for efficient real-time shadow generation by computing the union of shadow volumes obtained from traversals of the BSP tree in depth order for each light source position. BSPs were also used in *surface simplification* [Agarwal and Suri 1998; Shaffer and Garland 2001; Pauly et al. 2002] for clustering sample points.

Other applications of BSP trees include *range counting* [Agarwal and Matoušek 1992], *point location* [Arya et al. 2000; de Berg 2000], *collision detection* [Ar et al. 2000; Ar et al. 2002], *robotics* [Ballieux 1993], *graph drawing* [Asano et al. 2003], and *network design* [Mata and Mitchell 1995]. BSP trees generalize some of the most commonly used geometric search structures such as quadtrees/octrees, Kd-trees, and BAR-trees [Duncan et al. 2001], each of which has numerous applications on its own right.

Early research. Research on combinatorial properties of BSPs began with two influential papers of Paterson and Yao [1990; 1992]. They considered BSPs for disjoint $(d-1)$-dimensional hyper-rectangles in $\mathbb{R}^d$ (including line segments in the plane). Many of their initial observations (about free cuts, anchored segments, and round-robin algorithms) are key elements of the most recent partition algorithms, as well.

Paterson and Yao [1990] gave a randomized and a deterministic algorithm to construct a BSP of size $O(n \log n)$ and height $O(\log n)$ for n disjoint line segments in the plane. The height bound is optimal apart from a constant factor, but the size bound is not known to be tight. A lower bound construction, where the partition complexity of n disjoint line segments is $\Omega(n \log n / \log \log n)$, is discussed in Section 3.

They also proved that the partition complexity of n disjoint $(d-1)$-dimensional simplices in $\mathbb{R}^d$, $d \geq 3$, is $O(n^{d-1})$, which is best possible for $d = 3$ and tight for autopartitions in every dimension [Paterson and Yao 1990]. In three-space, there are n disjoint line segments, lying along a hyperbolic paraboloid, whose partition complexity is $\Theta(n^2)$. The same lower bound carries over to hyper-rectangles in $\mathbb{R}^d$ with the same $\Theta(n^2)$ partition complexity. The gap between the upper and lower bounds in higher dimensions has resisted research efforts so far.

OPEN PROBLEM. *What is the partition complexity of n disjoint $(d-1)$-dimensional simplices in $\mathbb{R}^d$, for $d \geq 4$?*

Every BSP computes a convex partition of the free space around the input objects. The size of the convex partition is the number of leaves of the BSP tree. The partition complexity of $(d-1)$-dimensional simplices is, therefore, cannot be smaller than the minimum convex partition of a polyhedron with n faces in $\mathbb{R}^d$. It would be tempting to derive lower bounds on the BSP size from the convex partitioning, unfortunately, no super-quadratic lower bound is known for this problem [Chazelle 1984] for any $d \in \mathbb{N}$.

OPEN PROBLEM. *What is the size of a minimum convex partition of a polyhedron with n faces in $\mathbb{R}^d$, $d > 3$?*

In this survey we focus on the worst case (maximum) partition complexity of a set of objects of a certain type. Computation or approximation of the *optimal* size BSP for *specific* instances seems to be an elusive open problem (see e.g., [Agarwal et al. 2000c]).

OPEN PROBLEM. *Is it NP-hard or is there a polynomial algorithm to compute the partition complexity of n given disjoint line segments in the plane?*

OPEN PROBLEM. *Is it APX-hard or is there a polynomial time approximation scheme to compute the partition complexity of n given disjoint line segments in the plane?*

Road map. In Section 2, we continue with recalling earlier ideas and observations which proved to be omnipresent in later results on BSPs. This section can be considered a short warm up course on the basics on binary space partitioning, which is essential in understanding the sometimes intricate partition schemes and lower bound constructions. Section 3 sketches the proof for two closely related results on disjoint line segments in the plane: An $\Omega(n \log n / \log \log n)$ lower bound construction and a constructive upper bound on segments with k

distinct orientations, $k \in \mathbb{N}$. Section 4 gathers a series of results about BSPs for axis-aligned input objects, where all splitting hyperplanes are also axis-aligned. We conclude the paper with two short sections related to other fields of computational geometry: Section 5 reviews results on realistic object classes which sometimes allow for linear partition complexity. Section 6 deals with kinetic BSPs for continuously moving input objects.

2. Preliminaries

Many of the resent results on BSPs are based on or are extending basic concepts known for decades. In this section, we present the most important ideas with references the their recent extensions or enhancements.

Free cuts. A hyperplane h that does not split any input object but partitions the *set* of objects into smaller sets (i.e., at least two of $\text{int}(h^-)$, h, and $\text{int}(h^+)$ intersect input objects), is called a *free cut*. A minimum size BSP should use free cuts whenever they are available [Paterson and Yao 1990]. A BSP that uses free cuts only is called *perfect*. De Berg, de Groot, and Overmars [de Berg et al. 1997a] designed an $O(n^2 \log n)$ time algorithm to detect if a perfect BSP exists for n disjoint lien segments in the plane.

Cutting along a $(d-1)$-dimensional object whose (relative) boundary lies on the boundary of the cell is clearly a free cut (Figure 3). In the plane, this implies, for instance, that if segment s is cut already at points p and q, then the subproblem containing the middle fragment $pq \subset s$ can be partitioned along pq without cutting any input segment. Therefore, if we track the cuts of a single input segment through an optimal binary space partitioning, then only the two *endings* (the portion between the segment endpoint and the first or last cut) can be further fragmented. This observation leads to an $O(n \log n)$ upper bound on the partition complexity of n disjoint line segments [Paterson and Yao 1990], but it is a basic element in all BSP algorithms.

Free cuts along objects. Consider a $(d-1)$-dimensional object s in $\mathbb{R}^d$. If a fragment $\hat{s}$ of s is bounded by cuts (previous splitting planes), then a hyperplane along $\hat{s}$ is a free cut for the subproblem containing $\hat{s}$. As a consequence, we can assume that whenever a splitting hyperplane cuts a fragment $\hat{s}$ of s then $\hat{s}$ is adjacent to the (relative) boundary of s. If a splitting hyperplane h partitions $\hat{s}$ but does not partition the common (relative) boundary of s and $\hat{s} \subset s$ (e.g. in Figure 3), then this commmon boundary is entirely on one side of h and the fragment of $\hat{s}$ on the other side of h is bounded by cuts and so it is a free cut. For each final fragment of s adjacent to the boundary of s, there is one such cut on every level of the BSP tree. Therefore the partition complexity of n disjoint $(d-1)$-dimensional simplices in $\mathbb{R}^d$ is upper bounded by the product of the size and the height of a BSP for their $(d-2)$-dimensional boundary simplices. For example, we obtain an $O(n \log n)$ size BSP for n disjoint line segments in the plane from a $2n$ size and $\lceil \log 2n \rceil$ height Kd-tree of the $2n$ segment endpoints.

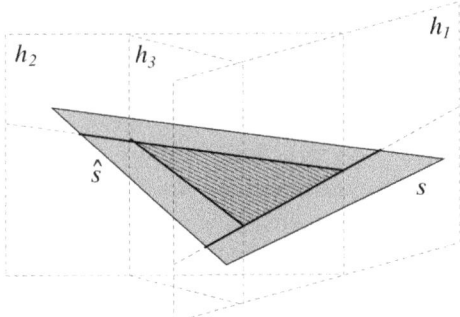

Figure 3. An input triangle s in $\mathbb{R}^3$ is cut by three vertical splitting planes h_1, h_2, and h_3. The striped triangle fragment is a free cut for the cell bounded by the three splitting planes.

Paterson and Yao [1990] showed that cutting along all input objects in a random order and applying free cuts whenever possible results in an $O(n^{d-1})$ expected size BSP for n disjoint $(d-1)$-dimensional simplices in $\mathbb{R}^d$, $d \geq 3$. This BSP, however, can have linear expected height in the worst case, e.g., if the input is the face set of a convex polytope. In three-space, this gives a randomized BSP of $O(n^2)$ size and $O(n)$ height for n disjoint triangles.

We can construct deterministically a BSP of $O(\log^2 n)$ height and $O(n^2 \log^2 n)$ size: Project all triangle edges to the xy-plane, and cover them with $O(n^2)$ pairwise non-crossing segments. By [Paterson and Yao 1990], there is a $O(n^2 \log(n^2))$ size and $O(\log(n^2))$ height BSP for these segments. The lifting of the planar splitting lines to vertical splitting planes in $\mathbb{R}^3$ along with all possible free cuts gives a BSP for the triangles. Refining this idea, Agarwal et al. [2000c] designed a BSP of size $O(n^2 \log^2 n)$ size and $O(\log n)$ height for n disjoint triangles; while Agarwal, Erickson, and Guibas [Agarwal et al. 1998] reported a randomized BSP of expected $O(n^2)$ size and $O(\log n)$ height.

Cycles. Assume that we are given disjoint $(d-1)$-dimensional objects in R^d. If the hyperplane spanned by any object is disjoint from all other objects, then every hyperplane through an object is a free cut and we obtain a perfect BSP of size n by partitioning along them in an arbitrary order. We can define a binary relation between two objects a and b saying that $a \succ b$ if and only if the hyperplane through a splits b. A BSP of size n using free cuts only still exists if $\succ$ is an acyclic relation: No object is cut if we always split the space along a minimum element with respect to $\succ$. Sets with large partition complexity ought to have many *cycles* w.r.t. $\succ$. Figure 4 depicts examples of cycles in the plane and in three space. Cycles are vital in the analysis of BSPs for disjoint line segments in the plane.

Anchored segments. A line segment is *anchored* if one of its endpoints lies on the boundary of the cell. An anchored segment has only a uni-directional extension that might split other segments. Paterson and Yao [1992] found a small *BSP for axis-aligned anchored segments* in the plane that cuts non-anchored segments at most four times. Using this BSP as a subroutine inductively, they obtain an $O(n)$ size BSP for n disjoint axis-parallel line segments: If a segment is partitioned into c pieces during a BSP for anchored segments, then $c-2$ pieces are free-cuts in their proper subproblems and are not fragmented any further, the remaining two pieces are anchored and are taken care of by the next call of the subroutine for anchored segments. This idea was extended and turned out to be extremely fruitful in a number of recent results [Tóth 2003b; Tóth 2003a]

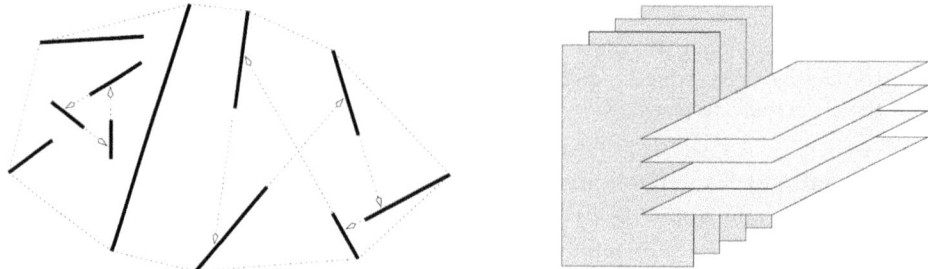

Figure 4. A set of line segments with a free cut and two cycles, one of which is an anchored cycle (left). Axis-aligned rectangles where $\succ$ defines a complete bipartite graph (right).

De Berg et al. [1997b] noticed a useful property of *cycles of anchored segments*. Consider a cycle $(a_1, a_2, \ldots, a_k)$ of anchored segments, where $a_i \succ a_{i+1}$ for $i = 1, 2, \ldots, k-1$ and $a_k \succ a_1$. Assume, furthermore, that the first anchored segment hit by the extension of a_i is a_{i+1} for $i = 1, 2, \ldots k-1$, and a_1 for $i = k$ (Figure 4). A *cycle cut* for $(a_1, a_2, \ldots, a_k)$ is a partition subroutine where we cut sequentially along the segments $a_k, a_{k-1}, \ldots, a_1$ in this order. Note that only the first splitting line can cut anchored segments. Indeed, $a_{k-1} \succ a_k$, so the extension of a_{k-1} hits a_k. The split along a_{k-1} does not extend beyond a_k, along which we have made a previous cut. Similarly, partitions along $a_{k-2}, \ldots, a_2, a_1$ are all free cuts w.r.t. anchored segments of the initial problem. This property of cycles of anchored segments in the plane unfortunately does not generalize to higher dimensions.

One possible higher dimensional generalization of anchored segments was introduced by Dumitrescu et al. [2004] in the context of BSPs for axis-aligned k-flats in $\mathbb{R}^d$. An axis-aligned k-flat b is called an ℓ-*stabber*, $0 \le \ell \le k$, if two opposite $(k-\ell)$-faces of b lie on the boundary of the convex hull. Tóth [2003a] defined another generalization: A 1-stabber b is a *shelf* with respect to a cell C if $C \setminus b$ is simply connected. He showed that the partition complexity of shelves is $O(n \log n)$ in $\mathbb{R}^3$.

Round-robin partition. Researchers tried to construct small size BSPs by computing, for each subproblem, a splitting plane that optimizes some objective function [Paterson and Yao 1992; Nguyen 1996; Nguyen and Widmayer 1995]. Intuitively, an "optimal" cut should cut few objects and partition the input into almost equal size subproblems. Local optimization rarely lead to near optimal BSPs, one notable exception is the case of axis-parallel segments in $\mathbb{R}^d$, $d \geq 3$. Paterson and Yao [1992] considered the product of the number of segments for each axis-parallel direction as an objective function. They showed that locally optimal axis-aligned cuts (i.e., cuts decreasing maximally this value) build up to an $O(n^{d/(d-1)})$ size and $O(\log n)$ height BSP, which is best possible. The orientation of optimal axis-aligned cuts in recursive calls varies in an irregular order.

A simple but efficient *round-robin BSP* scheme for axis-aligned objects in $\mathbb{R}^d$ works in rounds: Each round partitions the space into 2^d pieces in d recursive steps along axis-aligned hyperplanes of all d orientations. The round-robin BSP scheme for points in $\mathbb{R}^d$ is the well known Kd-tree. For axis-aligned line segments in $\mathbb{R}^d$, a round-robin BSP makes a cut recursively along the median hyperplanes of the segment endpoints. This BSP has $O(n^{(d-1)/d})$ size and $O(\log n)$ height, which is best possible for n axis-aligned segments [Paterson and Yao 1992]. Recently, round-robin schemes [Dumitrescu et al. 2004; Hershberger et al. 2004] lead to non-trivial bounds on the axis-aligned partition complexity of (disjoint) k-flats in $\mathbb{R}^d$ for certain values of k.

3. Line Segments in the Plane

The best known upper bound on the partition complexity of n disjoint line segments in the plane is $O(n \log n)$. Paterson and Yao [1990] found an astonishingly simple randomized BSP of expected $O(n \log n)$ size. The trapezoid decomposition method of Preparata [1981] (see also [Preparata and Shamos 1985]) gives a deterministic BSP of this size as well. It was widely believed that disjoint segments in the plane have linear partition complexity, this was supported by experiments and by linear upper bounds for certain input classes such as axis-parallel segments [Paterson and Yao 1992], anchored segments, and segments of similar length [de Berg et al. 1997b]. Tóth [2003c] gave a family of n disjoint line segments for every n, $n \in \mathbb{N}$, whose partition complexity is $\Theta(n \log n/ \log \log n)$. An extension of these ideas lead to an $O(n \log k)$ bound on the partition complexity of n disjoint line segments with k distinct orientations.

Lower bound construction. We show how the idea of cycles over segments in the plane can be developed into a lower bound construction whose partition complexity is $\Theta(n \log n/ \log \log n)$. For the sake of simplicity, we focus on autopartitions, where every cut is made along an input segment. We start with the simple observation that a BSP cuts at least one segment of every cycle. If every segment appears in a unique cycle of size 3, then this already implies that the partition complexity of n segments is at least $n + n/3$. To guarantee a higher rate of fragmentation we design a recursive construction on k levels: The construction S_i on level i, $1 < i \leq k$, consists of copies of the construction S_{i-1} of level $i - 1$. On the lowest level, S_0 consists of a single line segment.

We squeeze S_{i-1} into a long and skinny rectangle, and build congruent copies of this deformed S_{i-1} into S_i. Note that the partition complexity is invariant under affine transformations. The copies of S_{i-1} are so skinny that their position in S_i can be described by disjoint line segments, which we call *sticks*. The lower bound on the partition complexity depends on three elements: *(a)* If S_i consists of x copies of S_{i-1} arranged in cycles then we can guarantee that any BSP makes at least $\Omega(x)$ cuts on sticks. *(b)* If a cut of a stick (that is, a cut through a copy of S_{i-1}) implies that $\Omega(|S_{i-1}|)$ true segments of S_{i-1} are cut, then the arrangement of sticks at level i actually guarantees $\Omega(n)$ cuts on the input segments. *(c)* Once we make sure that cycles of sticks at distinct levels induce distinct cuts on each input segment, then the entire construction gives a lower bound of $\Omega(kn)$ on the partition complexity, where k is the number of levels. It remains to show how to find a construction satisfying all three conditions with $k = \Omega(\log n / \log \log n)$ levels.

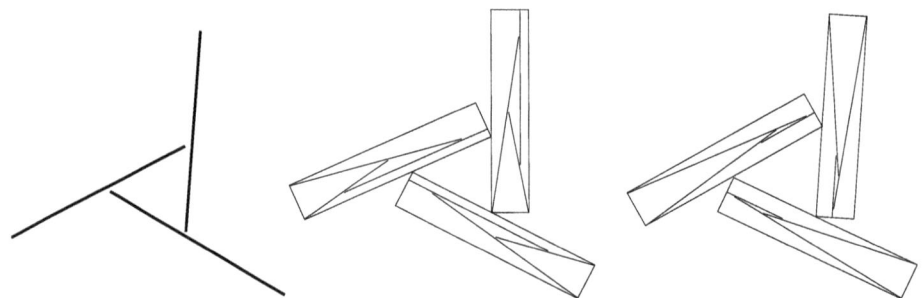

Figure 5. A cycle (left) and two cycles composed of cycles enclosed in long and skinny boxes (middle and right). A line spanned by segments in one rectangle either pierces a cycle in another rectangle (middle) or not (right).

A splitting line ℓ intersecting a copy of S_{i-1} does not necessarily cut *all* the true segments in S_{i-1}. If ℓ cuts all sub-constructions S_{i-2} within S_{i-1}, then it destroys the cycle arrangement of the sticks in S_{i-1} (Figure 5, middle) and so this cycle arrangement does not induce any cuts on segments of S_{i-1}. Therefore we assume that the sticks of S_{i-1} are arranged so that no line ℓ spanned by an input segment in another copy of S_{i-1} destroys their cycle (Figure 5, right). If, however, a line ℓ cuts only a fixed fraction c_{i-1}, $0 < c_{i-1} < 1$, of the sticks of S_{i-1}, then it cuts only $\left(\prod_{j=1}^{i} c_j \right) \cdot n$ true segments. Since we want $\prod_{i=1}^{k} c_i$ to be a constant independent of $k = \Omega(\log n / \log \log n)$, a cut through S_{i-1} must actually cut the vast majority of the sticks of S_{i-1}, which requires an asymmetric cyclic arrangement of sticks.

Consider an asymmetric cycle in S_i with three sticks such that two sticks correspond to i^2 copies of S_{i-1} (lying in disjoint skinny rectangles along the same segment), and the third stick corresponds to only one copy of S_{i-1} (Figure 6, left). If every line ℓ cutting S_i cuts through both heavy sticks, then ℓ cuts $\prod_{i=1}^{k} c_i = \prod_{i=1}^{k} 2i^2/(i^2+1) = O(1)$ fraction of the true segments. Recall that s cycle cut may split only one sticks in the cycle. This means that an acyclic cycle could be partitioned efficiently by cutting the light stick only, and none of the $2i^2$ copies of S_{i-1} along the heavy sticks (Figure 6, left).

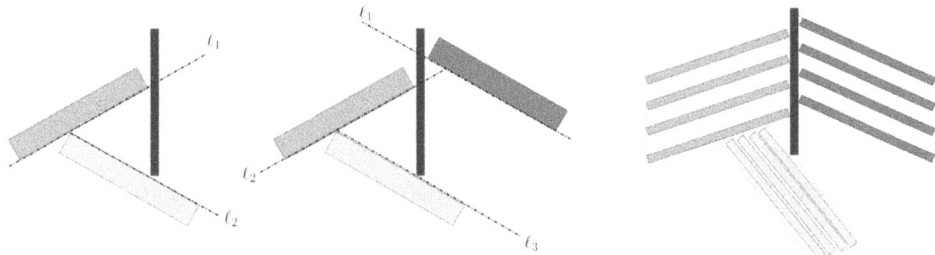

Figure 6. An asymmetric cycle (left), an asymmetric cycle with a one-step staircase structure on either side of the smallest rectangle (middle), and with a four step staircase (right).

We can overcome this difficulty with one additional idea: We want that if neither of the heavy sticks is cut by a line ℓ, then the light stick should be cut *many times*. Specifically, we want that any BSP of the stick arrangement at level i either cuts a heavy stick once or it cuts the light stick i^2 times. Both scenarios incur the same number of cuts on sticks, and on true segments, too. We can force the light stick to be cut many times by arranging i^2 copies of S_{i-1} in a staircase fashion on either side of the light stick. (Figure 6, middle and right). Technically, we break each heavy stick into i^2 distinct sticks in a staircase-like structure as depicted in Figure 6, right).

Tóth [2003c] shows that both the autopartition and partition complexity of this construction is $\Omega(n \log n / \log \log n)$. The partition complexity of n disjoint line segments in the plane is, thus, $O(n \log n)$ and $\Omega(n \log n / \log \log n)$.

OPEN PROBLEM. *What is the partition complexity of n disjoint line segments in the plane?*

Convex cycles. The importance of cycles of anchored segments was already noted by de Berg, de Groot, and Overmars [de Berg et al. 1997b]. The segments of a cycle $(a_1, a_2, \ldots, a_k)$, however, can be quite irregular: A cycle cut can partition a non-anchored segment into k fragments (Figure 4). Tóth [2003b] defined a *convex cycle*, which is a cycle of anchored segments lying along sides of a con-

vex polygon (Figure 7). Cutting sequentially along segments of a convex cycle
partitions any non-anchored segment into at most four pieces. To make sure
that every set of segments in a cycle contains a subset forming a convex cycle,
we *drop* the condition that $a_k \succ a_1$ in the definition of convex cycles (Figure 7,
right).

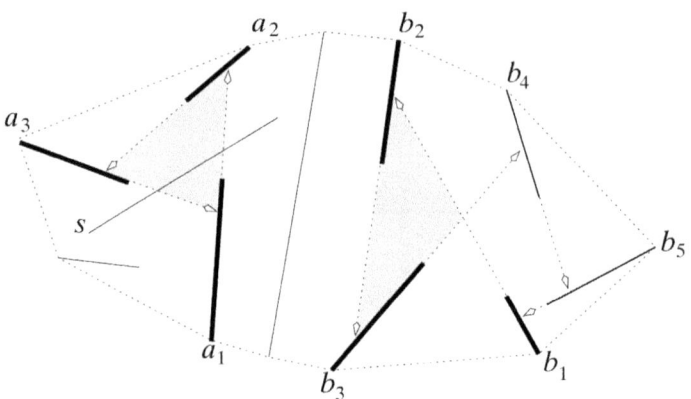

Figure 7. A convex cycle (a_1, a_2, a_3) and a cycle $(b_1, b_2, \ldots, b_5)$ that contains
a convex cycle (b_1, b_2, b_3) with $b_3 \not\succ b_1$.

Limited number of line directions. Paterson and Yao [1992] proved, using anchored segments, that the partition complexity of n disjoint axis-parallel
segments in the plane is at most $3n$. This was further improved to $2n - 1$ [Du-
mitrescu et al. 2004] (see Section 4). The proof techniques, however, do not
generalize for cases where the segments have three or more distinct orientations.
Tóth [2003b] proved recently that for any k, $k \in \mathbb{N}$, the autopartition complexity of n disjoint line segments with k distinct orientations is $O(nk)$, while their
partition complexity is $O(n \log k)$. The latter bound is asymptotically tight for
$k = O(1)$ and also for the lower bound construction described above, that is,
where n disjoint segments have $k = 2^{O(\log n / \log \log n)}$ distinct orientations.

The partitioning algorithm for a set S of disjoint segments works in *phases*:
Every phase i performs a BSP for a fixed set A_i of anchored segments such that
no segment of S is cut more than $O(\log k)$ times (respectively, $O(k)$ times in
the case of autopartitions). At an odd phase $2i - 1$, we let A_{2i-1} be the set
of *lower anchored segments* at the current subproblem, that is, all (fragments
of) segments whose lower endpoint is on the boundary of a cell. Similarly, at
an even phase $2i$, A_{2i} is the set of all upper anchored segments. The result
follows immediately, since if a segment $s \in S$ is cut in phase i then it is split into
$c \log k$ fragments, $c \log k - 2$ of which are free cuts in their subproblem and 2 are
anchored fragments. The two endings of s may be cut again $O(\log k)$ times in

phase $i+1$ and $i+2$ during a BSP for upper and lower anchored segments, but s is no longer cut in later phases.

A phase i calls a BSP subroutine P for the set A_i in each cell. $P(A_i)$ makes two types of cuts: (1) A *segment-cut* splits a cell along a line ℓ spanned by a segment $s \in A_i$. It is applied if ℓ does not cut any anchored segment in the cell. (2) A *cycle-cut* partitions along a convex cycle of anchored segments (as described in Section 2). Every cycle-cut is followed by a *cleanup step*, where we apply the subroutine P for subsets $B_i \subset A_i$ of those anchored segments of A_i which are split by the cycle cut. One can ensure that B_i has at most half as many distinct orientations as A_i in each resulting subcell (in case of autopartitions, B_i has at least one fewer distinct orientations than A_i in each subcell).

For every phase, the segment-cut and cycle-cut steps can be assigned into different levels according to recursive calls where subroutine P applied them. Since the number of distinct orientations decreases by a factor of two in each level, there are at most $\log k$ levels (at most k levels for autopartitions). One can also ensure by carefully isolating the region in which a cycle-cut splits anchored segments of A_i that every input segment is cut $O(1)$ times in each level. This implies that every segment is cut at most $O(\log k)$ times in total ($O(k)$ times for autopartitions).

4. Axis-aligned Binary Space Partitions

A k-dimensional object is *axis-aligned*, if each of its ℓ-dimensional faces is parallel to ℓ coordinate axes, $1 \leq \ell \leq k$. In an orthogonal coordinate system, axis-aligned objects are also called orthogonal, rectilinear, or isothetic. Note that BSPs are invariant under affine transformations of the Euclidean space, and so the partition complexity of axis-aligned objects is independent of the angles among the coordinate axes. BSPs for axis-aligned objects are important, because in many applications input objects are axis-aligned by nature or are modeled by their axis-aligned bounding boxes.

The simplest axis-aligned objects are the boxes, defined as a cross product $\prod_{i=1}^{d}[a_i, b_i]$ of d intervals in $\mathbb{R}^d$. The *extent dimensions* of a box B are the coordinates i where $a_i < b_i$, while in every non-extent dimension, $a_i = b_i$. An axis-aligned k-flat is a box with k extent dimensions. An axis-parallel line segment, for example, is a 1-flat.

Every partition algorithm in this section uses axis-aligned splitting hyperplanes only, such BSPs are said to be *axis-aligned*. Note that every cell in an axis-aligned BPS is an axis-aligned box. For most axis-aligned input classes, the axis-aligned BSPs are best possible *among all BSPs* ignoring constant or logarithmic factors, because they match the partition complexity known for the class. In some cases, lower bounds known for *axis-aligned BSPs* are higher by a constant or logarithmic factor than those for the general BSPs. Of course, it is

easy to construct instances where the smallest BSP and the smallest axis-aligned
BSP have significantly different sizes.

Segments and rectangles in the plane. After Paterson and Yao [1992]
showed that the partition complexity of n disjoint axis-parallel line segments in
the plane is $O(n)$, it remained to determine the exact value of the coefficient
hidden in the asymptotic notation. A partition algorithm due to d'Amore and
Franciosa [1992], originally designed for axis-aligned boxes, always computes an
axis-aligned BSP of size at most $2n - 1$. Dumitrescu, Mitchell, and Sharir [2004]
almost matched this bound with a construction (Figure 8, middle) whose axis-
aligned partition complexity is $2n - o(n)$.

For disjoint axis-aligned rectangles, Berman, DasGupta, and Muthukrishnan
[Berman et al. 2002] showed that the partition complexity is at most $3n-1$. Their
partition algorithm is a blend of previous algorithms [Paterson and Yao 1992;
d'Amore and Franciosa 1992], combined with a charging scheme. In [Dumitrescu
et al. 2004] a construction is given where the axis-aligned partition complexity
is at least $\frac{7}{3}n - o(n)$. The authors point out, however, that their construction
cannot grant a lower bound that would match $3n - 1$, since it does have an
axis-aligned BSP of size $2.444n + o(n)$.

OPEN PROBLEM. *What is the coefficient of the linear term in the (axis-aligned)
partition complexity of disjoint axis-aligned boxes in the plane?*

A tradeoff between size and height. Arya [2002] noticed that the height of a
linear size *axis-aligned* BSP for disjoint axis-parallel segments cannot always be
logarithmic. Examining the construction on Figure 8, left, he showed that there
is a tradeoff between the size and the height of axis-aligned BSP trees. If the
height of such a tree is h then its size is $\Omega(n \log n / \log h)$. A more complicated,
but stronger formulation of his result says that the BSP-tree of height h must
have size $\Omega(nr)$ in the worst case where $r \in \mathbb{N}$, $\sum_{i=0}^{r} \binom{h}{i} \le n/8$.

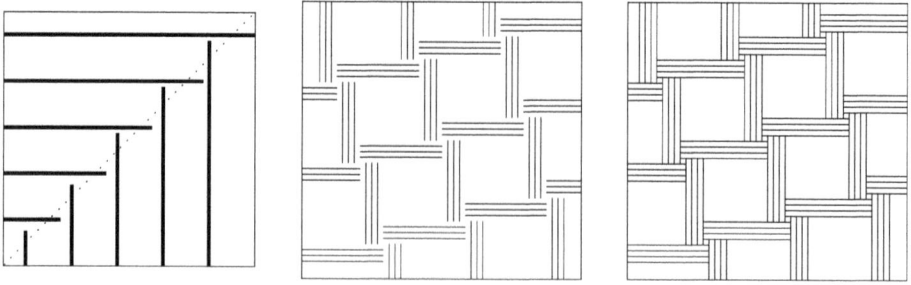

Figure 8. Arya's construction (left), Dumitrescu's construction with $2 \cdot 3 \cdot 3 \cdot 4 = 72$
disjoint axis-parallel line segments (middle), and the corresponding axis-aligned
tiling of size $72 + 3^3 + (3 + 1)^2$ (right).

On the two extremities, this implies that an axis-aligned BSP-tree of size $O(n)$ has height $n^{\Omega(1)}$; and one of height $O(\log n)$ has size $\Omega(n \log n)$. The set of segments in Arya's construction, however, has linear partition complexity: All segments become anchored after a diagonal cut (dotted line on Figure 8, left). No tradeoff is known so far for the size and height of general BSP trees.

OPEN PROBLEM. *Is there a tradeoff between the size and the height of the BSP trees for disjoint line segments (with arbitrary orientations)?*

k-flats in d-dimensions. Paterson and Yao [1992] proved that the axis-aligned partition complexity of n disjoint line segments in $\mathbb{R}^d$, $d > 2$, lies between $(n/d)^{d/(d-1)} + n$ and $d \cdot (n/d)^{d/(d-1)} + n$. There has been no improvement on the coefficients on the main terms so far. Their lower bound construction is a cubic grid $\prod_{i=1}^{d} \{1, 2, \ldots, (n/d)^{1/(d-1)}\}$ where d lines of distinct orientation stab each grid cell. Their upper bound can be obtained by a round-robin partition algorithm applying median cuts in all d directions.

Dumitrescu, Mitchell, and Sharir obtained an $O(n^{d/(d-k)})$ upper bound for the partition complexity of n axis-aligned k-flats in $\mathbb{R}^d$ [Dumitrescu et al. 2004]. It is not difficult to see that a round-robin cutting scheme delivers this bound, too: Let every round consist of d recursive cuts in all d directions such that every splitting hyperplane cuts through the median of the vertex set of the fragments clipped to the cell. The total number of fragments may increase by a factor of 2^k in each round, because every (fragment of a) k-flat can be split along each of its k extents. At the same time, the number of fragments in one subproblem decreases by a factor of 2^{d-k}. So there are at most $(\log_2 n)/(d-k)$ rounds, and the BSP size is $O(n \cdot n^{k/(d-k)})$. Dumitrescu et al. [2004], in fact, applied a somewhat different partition scheme where all cuts for each hyperplane direction are made simultaneously. They also give a matching lower bound of $\Omega(n^{d/(d-k)})$ for the case where $k < d/2$.

The formula $\Theta(n^{d/(d-k)})$ is no longer valid for disjoint objects if $k \geq d/2$ (the axis-aligned partition complexity is $\Theta(n^{5/3})$ for $k = 2$ and $d = 4$ [Dumitrescu et al. 2004]). The reason for this is that *disjointness* plays no role if $k < d/2$, but it is restrictive for $k \geq d/2$. Indeed, if two objects whose extent dimensions together do not contain all d dimensions intersect, then they have a common non-extent dimension where their coordinates coincide. Any system of k-flats in $\mathbb{R}^d$, $k < d/2$, can be perturbed into a pairwise disjoint set by translating each object independently by a small random vector. If the random vectors are sufficiently small, then this perturbation does not decrease the partition complexity (every BSP for the perturbed set translates into a BSP of less than or equal size for the original input). Two flats whose extent dimensions contain all d dimensions may not always be separated by an arbitrarily small perturbation. The simplest examples are a set of axis-parallel segments in the plane and in $\mathbb{R}^3$, respectively: A small random perturbations removes all segment-segment intersection in three-space, but intersections may prevail in the plane.

The disjointness requirement is more apparent for the following generalization of free cut segments in higher dimensions: An axis-aligned box is a *stabber* in dimension i if its extent contains the extent of the bounding box of the input in dimension i. It is an M-*stabber* for a set $M \subseteq \{1, 2, \ldots, d\}$, if it is a stabber in every dimension $i \in M$; we can also say that it is a k-stabber if $k = |M|$. An M_1-stabber and an M_2-stabber intersect if $M_1 \cup M_2 = \{1, 2, \ldots, d\}$. Stabbers were defined by Dumitrescu, Mitchell, and Sharir [Dumitrescu et al. 2004]. Later Hershberger, Suri, and Tóth [Hershberger et al. 2004] gave an alternative definition saying that a box is k-stabber, if all its j-faces, $j < k$, lie on the boundary of the bounding box.

The concept of stabbers is crucial in any axis-aligned BSP algorithm. Let us mention just two of their useful properties: (i) A $(d-1)$-stabber is a free-cut in $\mathbb{R}^d$. (ii) Apart from at most 2^k fragments incident to the vertices of b, every fragment of a k-flat b is a ℓ-stabber for some $1 \leq \ell \leq k$. A set of disjoint stabbers corresponds to a pairwise intersecting set system: Let us map every stabber b to the set $\varphi(b) \subset \{1, 2, \ldots, d\}$ of dimensions in which b is *not* a stabber. $\{\varphi(b) : b\}$ is a pairwise intersecting set system; and vice versa: Given a pairwise intersecting set system D on $\{1, 2, \ldots, d\}$, there is a set of stabbers in $\mathbb{R}^d$ whose image under φ is D.

Tilings. In every lower bound construction where the partition complexity is high, the free space surrounding the input objects is also complex in the sense that its smallest convex partition is often superlinear. Researchers expected that if there is no space among the (convex) input objects (or, if the convex partition of the free space has the same order of magnitude as the input), then the partition complexity would be significantly smaller. A *tiling* in $\mathbb{R}^d$ is a set of d-dimensional objects that partition the space. An *axis-aligned tiling* is a set of full-dimensional axis-aligned boxes that partition the space.

Already in the plane, the worst case partition complexity of axis-aligned tilings is smaller than that of disjoint boxes. Berman, DasGupta, and Muthukrishnan [Berman et al. 2002] showed that every axis-aligned tiling of size n has an axis-aligned BSP of size at most $2n$ (recall that the axis-aligned partition complexity of disjoint rectangles is at least $\frac{7}{3}n - o(n)$ [Dumitrescu et al. 2004]). This is asymptotically optimal, since a construction from [Dumitrescu et al. 2004], originally designed for disjoint line segments, can be converted to an axis-aligned tiling. Their lower bound construction for axis-parallel line segments consists of $2k(k+1)$ bundles of k parallel lines (Figure 8, middle). These $2k^3 + 2k^2)$ segments can be blown up to skinny axis-aligned boxes such that each bundle fills an axis-aligned rectangle, and the free space between the rectangles can be covered by $k^2 + (k+1)^2 = O(k^2)$ interior-disjoint axis-aligned boxes (Figure 8, right). A total of $n = 2k^3 + O(k^2)$ boxes tile the plane, and their axis-aligned BSP cannot be smaller than that of the original k^3 segments we have started with, that is, at least $4k^3 - O(k^2)$.

The difference in the partition complexity of disjoint and plane filling rectangles was only the coefficient of the linear term. In three- and higher dimensions the upper bounds on the partition complexity of tilings is smaller in exponent than that of disjoint boxes. Hershberger and Suri [2003] observed that every tiling where all $(d-3)$-dimensional faces of every tile lie on the convex hull has a linear size axis-aligned BSP. (This is not true for disjoint axis-aligned boxes, in general: In the known worst-case constructions, all $(d-3)$-dimensional faces lie on the convex hull.) This implies that it is enough to partition an axis-aligned tiling until every cell is empty of $(d-3)$-dimensional faces, and then a linear size BSP in each cell results in a BSP for the input tiling.

Hershberger, Suri, and Tóth [Hershberger et al. 2004] apply a round-robin BSP for the $(d-3)$-dimensional faces of an axis-aligned tiling in $\mathbb{R}^d$. One can show that the number of $(d-3)$-faces in each subproblem decreases by a factor of 2^3. Therefore the round-robin scheme terminates in $\frac{1}{3}\log n + O(1)$ rounds. In each round of the round-robin scheme, every $(d-2)$-face is split into at most 2^{d-2} fragments. In the course of $\frac{1}{3}\log n + O(1)$ rounds, every $(d-2)$-face is partitioned into $(2^{d-2})^{\div \log n + O(1)} = O(n^{(d-2)/3})$ fragments. Since the $(d-1)$-faces that become free-cut in their subproblem are not fragmented any further, every $(d-1)$-face and every tile is partitioned into $O(n^{(d-2)/3})$ fragments, as well. The round-robin scheme for $(d-3)$-faces partitions the axis-aligned tiling in $\mathbb{R}^d$ into $n \cdot O(n^{(d+1)/3}) = O(n^{(d+1)/3})$ fragments such that no $(d-3)$-face lies in the interior of a cell. By applying a linear size BSP for the tiling clipped in each resulting cell, we obtain a BSP of size $O(n^{(d+1)/3})$ for an axis-parallel tiling with n tiles in $\mathbb{R}^d$. By contrast, the best known upper bound on the partition complexity of n disjoint full-dimensional boxes in $\mathbb{R}^d$ is $O(n^{d/2})$ only.

In dimensions $d = 3$, the partition complexity of axis-aligned tilings of size n is $O(n^{4/3})$, which is tight by a construction of Hershberger and Suri [2003]. They start with the lower bound construction of Paterson and Yao [1992] for axis-parallel line segments in three space, which consists of $3k^2$ axis-parallel line segments such that three segments of distinct directions stab every unit cube in $[0, k] \times [0, k] \times [0, k]$ grid. Then they replace every segment by an air-tight bundle of k axis-aligned skinny boxes, and fill the free space between the bundles by $2k^3$ pairwise disjoint boxes. They obtain a tiling with $n = 5k^3$ boxes. By the argument of Paterson and Yao, each grid cell corresponds to a cut through a bundle of k skinny boxes, which gives a total of $k \cdot k^3 = \Theta(n^{4/3})$ cuts of tiles. Generalizing this idea and ideas of Dumitrescu et al. [2004], Hershberger, Suri, and Tóth [Hershberger et al. 2004] gave a construction in d-dimensions for which the axis-aligned partition complexity is $\Omega(n^{\beta(d)})$, where $\lim_{d \to \infty} \beta(d) = (1 + \sqrt{5})/2 = 1.618$. Apart from the staggering gap between the upper and lower bounds for $d \geq 4$, many problems remain open on BSPs of axis-aligned and of general tilings.

OPEN PROBLEM. *What is the (axis-aligned) partition complexity of n space filling axis-aligned boxes in $\mathbb{R}^d$, $d \geq 4$?*

OPEN PROBLEM. *What is the partition complexity of n space filling simplices in $\mathbb{R}^d$, $d \geq 3$?*

OPEN PROBLEM. *What is the partition complexity of n tetrahedra that form the triangulation of a convex polytope in d-space, $d \geq 3$?*

Fat rectangles. Another characteristics of the lower bound constructions for n axis-aligned rectangles in $\mathbb{R}^3$ is that the rectangles are long and skinny (they are axis-parallel line segments fattened by a small $\varepsilon > 0$ in all extent dimensions). It was reasonable to expect that disjoint fat objects have small partition complexity. An axis-aligned k-flat is *fat* if the ratio of a largest inscribed and a smallest circumscribed cubes[1] in the k-dimensional space spanned by the object is bounded by a constant.

De Berg [1995] proved that the partition complexity of *full dimensional* disjoint fat objects is linear in every dimension. This is true for any set of disjoint fat objects even if they are not axis-aligned (Section 5 deals with further extensions of this result). On the other hand, n disjoint fat but not axis-aligned rectangles may have $\Theta(n^2)$ partition complexity [Paterson and Yao 1990].

Agarwal et al. [2000b; 1998] considered n disjoint axis-aligned fat 2-flats in $\mathbb{R}^3$ and proved that their partition complexity is $n \cdot 2^{O(\sqrt{\log n})}$. Tóth [2003a] improved this upper bound and gave an algorithm to compute a BSP of $O(n \log^8 n)$ size and $O(\log^4 n)$ height. He also gave a lower bound construction for which the axis-aligned partition complexity is $\Omega(n \log n)$.

The main difficulty in dealing with a fat rectangle r in three-space is that the fragments of r clipped to a cell of a subproblem is not necessarily fat anymore. Every fragment, however, belongs to one of the following three types, each of which has some useful properties that help our partitioning algorithm.

For an input rectangle r and a cell C, we say that the fragment $r' = r \cap C$ is
- a *corner* if r' is incident to a vertex of the original axis-parallel rectangle r,
- *bridge* if one extent of r' is the same as the corresponding extent of C (r' is a 1-stabber) and the other lies in the interior of the corresponding extent of C (see Figure 9), or
- a *shelf* if one extent of r' is the same as the corresponding extent of C and the other is incident to one endpoint of the other extent of C (see Figure 9).

Every rectangle has only four vertices, therefore in every level of the BSP tree, at most four sub-problems contain corners of an input rectangle r. This implies that the number of cuts on corners of an input r is bounded by (four times) the height of the BSP. A *bridge* fragment r' of a fat rectangle r is not necessarily

---

[1]Traditionally, an object is called *fat* if the ratio of the radii of a circumscribed and an inscribed *ball* is bounded by a constant. Our definition, in terms of cubes, is equivalent to the standard definition (with a different constant threshold).

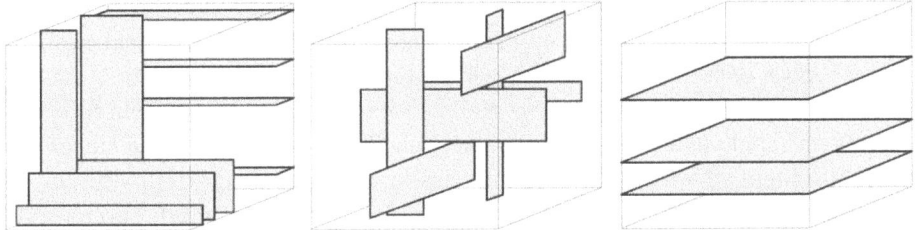

Figure 9. Shelves, bridges, and free-cuts clipped to a cell C in $\mathbb{R}^3$.

fat. It still preserves part of the fatness of r, a property we call *semi-fatness*: If r is α-fat then the extent of r' in the interior of that of C is at most α times longer than the extent containing the corresponding extent of C. The semi-fatness allows us to separate bridges of different orientations (see subroutine P_3 below). Finally, there is an $O(n \log n)$ size BSP for n disjoint shelves such that it partitions *every* rectangle into at most $O(\log^2 n)$ fragments.

We outline the main ideas to construct a $O(\log^4 n)$ height BSP for disjoint axis-aligned fat rectangles. The partition algorithm is composed of a four level hierarchy of BSPs: We give a glimpse of each level and explain the lowest level (a BSP for shelves) in detail below.

The main algorithm, P_1, makes cuts recursively along the median xy-plane of the vertices of the input rectangles. Since the number of rectangle vertices in a cell is halved in each step, P_1 terminates in $O(\log n)$ rounds. After every median cut of P_1, a *cleanup step* is applied in each cell C, which is a BSP *for the stabbers* w.r.t. C and partitions *every rectangle* into $O(\log^6 n)$ fragments. If the cleanup step P_2 makes a cut along a stabber $r \cap C$, then subsequent median cuts of P_1 cannot partition $r \cap C$ anymore. This ensures that the median cuts performed by P_1 may split any rectangle $O(\log n)$ times only.

The BSP P_2 for a set S of stabbers makes cuts recursively along the median xz-plane of the vertex set of S. Each median cut of P_2 is followed by another cleanup subroutine, P_3, which in each cell separates the stabbers of S from other stabbers (i.e., fragments that became stabbers due to the last median cut by P_2). Finally, P_3 makes cuts along yz-planes, and calls a BSP P_4 for all shelves after every such step. We describe the BSP P_4 for a set S of n shelves in a cell. P_4 has $O(n \log n)$ size and it partitions every rectangle into $O(\log n)$ fragments.

Consider a set $S_\perp$ of $n_\perp$ disjoint shelves in a cell C. Assume for simplicity that they are all adjacent to the bottom side of C. The subroutine BSP for $S_\perp$ is a simple recursive algorithm: We split C by a horizontal plane h through the highest upper edge of a shelf in $S_\perp$, then we split the subcell below h along the highest shelf (a free cut) and along the median shelf, and recursively call this subroutine for each resulting subproblem (Figure 10 shows an example). The algorithm terminates in $\log n_\perp$ rounds because the number of shelves in a subproblem is halved in each round. None of the shelves is split, and we show that every axis-parallel segment disjoint from these shelves is cut $O(\log n_{\perp})$ times: A segment parallel to stabbing extent of the shelves is never cut, a vertical segment is cut at most once in every round, which amounts to $O(\log n_{\perp})$ cuts. Lastly, consider a segment s orthogonal to the shelves. In subproblems where s is a stabbing segment, it lies above the highest shelf and so it cannot be cut at all. In a subproblem where an endpoint of s lies in the interior of a cell, s be cut by the splitting planes along the median shelf, which totals to at most $2 \log n_{\perp}$ cuts during the entire subroutine.

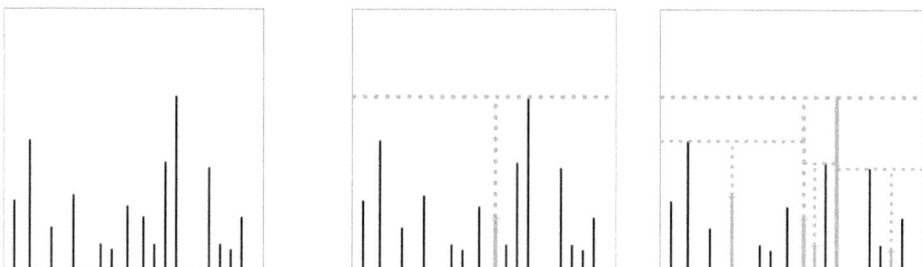

Figure 10. First two rounds of the BSP for shelves.

This was a BSP for shelves along the one side of a cell C. We still need to integrate six BSPs for each of the six sides of C: If we naïvely call the subroutine six times sequentially for the subproblems and for each side then the number of cuts on axis-parallel segments might be $(O(\log n))^{6}$. We use, instead, the concept of *overlay of BSPs* to keep the number of cuts on every segment $O(\log n)$. We compute independently six BSPs for the shelves adjacent to each side, then we "overlay" them one after another: A cut along a splitting plane made in one BSP may incur splits in several cells resulting from previous BSPs. The combination is still a valid BSP because every subproblem is split by a hyperplane. The number of cuts on each segment is the sum of the cuts made by each of the six BSPs rather than their product. The overlay of BSPs is a key concept to keep the number of cuts low throughout this algorithm.

In general we would like know the partition complexity of n disjoint axis-aligned fat k-flats in $\mathbb{R}^{d}$, $1 < k < d$. It is easy to show a lower bound of $\Omega(n^{(d-k+1)/(d-k)})$. Indeed, consider a worst case construction for n axis-parallel line segments in $\mathbb{R}^{d-k+1}$, whose partition complexity is $\Omega(n^{(d-k+1)/(d-k)})$ by Paterson and Yao [1992]. Embed it into a $(d-k+1)$-dimensional affine subspace H of $\mathbb{R}^{d}$ and expand it to a k-dimensional square orthogonally to H. We obtain n pairwise disjoint k-dimensional squares in $\mathbb{R}^{d}$. The size of any BSP for this set is $\Omega(n^{(d-k+1)/(d-k)})$. We expect that this bound can be attained apart from a polylogarithmic factor, that is, out of the k extents of a k-dimensional fat rectangle, $k - 1$ extents are essentially redundant with respect to the partition complexity.

OPEN PROBLEM. *What is the (axis-aligned) partition complexity of n disjoint fat axis-aligned k-flats in $\mathbb{R}^d$, $1 < k < d$?*

5. From Fatness to Guardable Scenes

The worst case constructions for the partition complexity of general and axis-aligned disjoint objects in 3-space use long and skinny input objects. This observation suggests that disjoint fat objects must have small partition complexity. A full dimensional object is fat if the ratio of an circumscribed and inscribed ball is bounded by a constant. Fat objects are often easier to handle. Among other things, the union of fat objects have small combinatorial complexity in many cases [Efrat and Sharir 2000; Pach and Tardos 2002; Pach et al. 2003]. A result that every set of disjoint fat objects has linear partition complexity [de Berg 1995] initiated a flurry of work on extensions defining new object classes which also have linear partition complexity by analogous arguments.

Fat full dimensional objects De Berg, de Groot, and Overmars [1997b] have proved that the partition complexity of n disjoint convex fat objects in the plane is $O(n)$. De Berg [1995] has later found an elegant proof that establishes linear partition complexity for disjoint full dimensional fat polyhedra with constant number of vertices (e.g., fat tetrahedra) in $\mathbb{R}^d$, $d \in \mathbb{N}$, as well. His partition algorithm generates the BSP tree in $O(n \log n)$ time for any $d \in \mathbb{N}$, where the constant of proportionality depends on the dimension. We briefly describe his method below.

Consider an orthogonal coordinate system and then compute the axis-aligned bounding box of every fat object. Generate a BSP *for the set V of vertices of all bounding boxes* by repeating one of the following two steps starting with a bounding cube C of V: (*i*) Partition the cubic cell C into 2^d congruent cubes if at least two sub-cubes have non-empty intersection with V. (*ii*) Otherwise every point of $V \cap C$ lies in one of the subcubes C'. In this case, partition C along the sides of the bounding cube of $(V \cap C) \cup c'$ and (subsets of) where c' is the common vertex of C and C'. Every subproblem that still contains a vertex of V is a cube (allowing recursion). The subcells of the complement of the bounding box of $(V \cap C) \cup c'$ in C are not necessarily fat but they can be expressed as the union of $O(1)$ cubes. Every cell of the resulting partition is empty of bounding box vertices of the fat input objects. This implies that every resulting cell intersects only a constant number of input objects. In each subproblem, one can complete the BSP with a constant number of cuts if every fat object is either convex or polyhedral with constant combinatorial complexity.

Note that the argument above used only that the input objects are fat, and that they are pairwise disjoint. The extensions of disjoint fat scenes define properties of the whole input scene rather than properties of individual objects.

Relaxations. De Berg [2000] noticed that the above argument goes through verbatim if, instead of disjoint fat objects, he requires a much weaker property: unclutteredness. A set of objects is *uncluttered* if every axis-aligned cube intersecting more than a constant number of input objects contains a vertex of the bounding box of an input object. It is easy to see that every set of disjoint fat objects is uncluttered. The converse is false and so the class of uncluttered sets is strictly larger than that of fat objects.

De Berg et al. [2003] further explored possible relaxations of fatness and unclutteredness that still grants linear partition complexity. A set of n objects is *guardable* if there are $O(n)$ *guard* points such that every axis-aligned box that intersects more than a constant number of objects must contain a guard point. Every uncluttered set is guardable with the obvious choice for guard points being the bounding box vertices. On the other hand, there are guardable sets that are cluttered (Figure 11, right). Unlike fatness and unclutteredness [de Berg et al. 2002], however, there is no polynomial time algorithm known to decide if a set of objects is guardable or not, let alone finding a guard points for a given guardable set [de Berg et al. 2003].

SCC
∪
guardable
∪
uncluttered
∪
disjoint fat

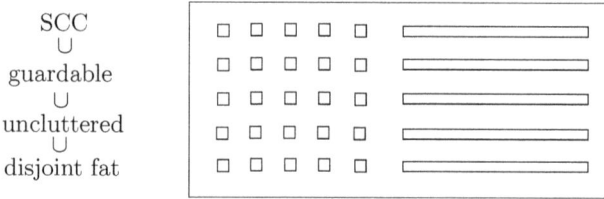

Figure 11. The relations between the classes of polygonal scenes (left), and a guardable but cluttered set of boxes (right).

Another property likely to imply low partition complexity is the *small simple-cover complexity* (SSC), defined originally by Mitchell, Mount, and Suri [Mitchell et al. 1997]. We cite the definition from [de Berg et al. 2002]: A set of n objects has *SSC*, if their convex hull can be covered by $O(n)$ balls such that the number of objects intersecting each ball is bounded by a small constant. De Berg et al. [2003] proved that in the plane a set of objects has SSC if and only if it is guardable. In higher dimensions, however, SSC is a strictly broader property than guardability. It is not known if there is a small size BSP for an SSC set.

OPEN PROBLEM. *What is the partition complexity of a set of objects with small simple-cover complexity in $\mathbb{R}^d$, $d \geq 3$?*

Tobola and Nechvile [2003] extended the definition of unclutteredness to sets of axis-aligned objects where the neighborhood of every object is uncluttered after a linear transformation. A set S of axis-aligned objects is *locally uncluttered* if for every object $s \in S$ the fragments of objects clipped in a neighborhood $N(s)$

of s are uncluttered after a linear transformation T_s. The neighborhood $N(s)$ is the axis-aligned box obtained from s by fattening each extent with a multiple of the longest extent of s. Tobola and Nechvile [2003] show that every locally uncluttered set of n objects in $\mathbb{R}^d$ has an $O(n)$ size axis-aligned BSP.

6. Kinetic Binary Space Partitions

The first motivation for binary space partitions was efficient hidden-surface removal from a *moving* viewpoint. The input objects of the BSP were, however, assumed to be static. Recent research on BSPs for moving objects was set in the model of *kinetic data structures* (KDS) of [Basch et al. 1999]. In this model, objects move continuously along a given trajectory (flight plan), typically along a line or a low degree algebraic curve. The splitting hyperplanes are defined by faces of the input objects, and so they move continuously, too. The BSP is updated only at discrete *events*, though, when the combinatorial structure of the BSP changes.

In the KDS paradigm, the goal is to maintain at all times a BSP whose size and height is below a reasonable threshold, which is usually close to the partition complexity available for static input. Efficiency of the algorithms are measured by additional parameters: the total *space* required during the algorithm, *total time*, which is broken down into the *total number of events* and the maximum *update time* for an event.

First Agarwal et al. [2000c] studied the BSP in the KDS setting. They consider n moving line segments in the plane such that the trajectory of every segment endpoint is a constant degree polynomial and the segments are pairwise disjoint at all times. They design a randomized algorithm to maintain a BSP of expected size $O(n \log n)$ in $O(n \log n)$ space, the number of events is $O(n^2)$ and each event requires an expected $O(\log n)$ update time.

Instead of adapting a previously known static BSP algorithm [Paterson and Yao 1990] to KDS paradigm, they start out from a randomized incremental BSP, which is an adaptation of the *vertical (trapezoid) decomposition* of Mulmuley [1990] and Seidel [1991]. They fix a random permutation of the segments, which is the only random choice in the algorithm. In their BSP, every splitting line is either vertical or a free cut (i.e., lies along an input segment); they call such BSPs *cylindrical*. By restricting the possible splitting planes, they reduce the number of possible event types: In fact, the only possible combinatorial event is that a segment endpoint starts or stops crossing a vertical splitting line. A careful analysis of the effects of these events on the BSP tree establishes the bounds on the update time and the total space requirement.

All known kinetic BSPs are based on vertical decomposition. This choice keeps the *types* of the possible combinatorial events under control, but the *number* of events can be suboptimal. If the x coordinates of every two segment endpoints are swapped during the motion, then the number of events can be $\Theta(n^2)$ for any *cylindric* kinetic BSP. For arbitrary kinetic BSPs, no such lower bound is known. Agarwal et al. [2000a] established lower bounds on the combinatorial changes every kinetic BSP has to go through: They give a set P of n points in the plane all moving along axis-parallel lines with constant velocity such that any kinetic BSP for P experiences $\Omega(n^{3/2})$ combinatorial changes.

OPEN PROBLEM. *How many combinatorial changes occur in the kinetic BSP of n points moving with constant velocity in the plane?*

Agarwal, Erickson, and Guibas [Agarwal et al. 1998] adapted the cylindrical BSP method to potentially intersecting line segments. They maintain a BSP of size $O(n \log n + k)$ and height $O(\log n)$ in total time $O(n \log^2 n + k \log n)$, all in expectation, where k is the number of intersecting segment pairs. They apply this algorithm to derive a kinetic BSP for n disjoint triangles in $\mathbb{R}^3$, for which they maintain a randomized BSP of expected $O(n^2)$ size and $O(\log n)$ height in $O(n^2 \log n)$ total time.

De Berg, Comba, and Guibas [de Berg et al. 2001] give a *deterministic* kinetic BSP for disjoint segments in the plane. Their algorithm is based on a *static* BSP algorithm due to Paterson and Yao [1990] using segment trees. They maintain a BSP of size $O(n \log n)$ and height $O(\log^2 n)$. The segment tree changes every time when the x-coordinates of two segment endpoints swap. So if every endpoint moves along a constant-degree polynomial trajectory, then the number of events is $O(n^2)$, each requiring $O(\log^2 n)$ update time. The average update time is $O(\log n)$, though, similarly to the randomized algorithm of [Agarwal et al. 2000c].

OPEN PROBLEM. *Is there a kinetic BSP of size $O(n \operatorname{polylog} n)$ and height $O(\operatorname{polylog} n)$ with $o(n^2)$ events for n disjoint line segments whose endpoints are moving along straight line trajectories?*

Acknowledgments

I thank the volume editors, Eli Goodman, János Pach, and Emo Welzl, for their encouragement to compile this survey paper. I am also indebted to the anonymous referees whose critical comments helped tremendously to improve the presentation of this paper.

Work on this survey paper was done while at the Department of Computer Science, University of California at Santa Barbara.

References

[Agarwal and Matoušek 1992] P. K. Agarwal and J. Matoušek, "On range searching with semialgebraic sets", pp. 1–13 in *Mathematical foundations of computer science* (Prague, 1992), Lecture Notes in Comput. Sci. **629**, Springer, Berlin, 1992.

[Agarwal and Suri 1998] P. K. Agarwal and S. Suri, "Surface approximation and geometric partitions", *SIAM J. Comput.* **27**:4 (1998), 1016–1035.

[Agarwal et al. 1998] P. K. Agarwal, J. Erickson, and L. J. Guibas, "Kinetic binary space partitions for intersecting segments and disjoint triangles (extended abstract)", pp. 107–116 in *Proceedings of the Ninth Annual ACM-SIAM Symposium on Discrete Algorithms* (San Francisco, 1998), ACM, New York, 1998.

[Agarwal et al. 2000a] P. K. Agarwal, J. Basch, M. de Berg, L. J. Guibas, and J. Hershberger, "Lower bounds for kinetic planar subdivisions", *Discrete Comput. Geom.* **24**:4 (2000), 721–733.

[Agarwal et al. 2000b] P. K. Agarwal, E. F. Grove, T. M. Murali, and J. S. Vitter, "Binary space partitions for fat rectangles", *SIAM J. Comput.* **29**:5 (2000), 1422–1448.

[Agarwal et al. 2000c] P. K. Agarwal, L. J. Guibas, T. M. Murali, and J. S. Vitter, "Cylindrical static and kinetic binary space partitions", *Comput. Geom.* **16**:2 (2000), 103–127.

[Ar et al. 2000] S. Ar, B. Chazelle, and A. Tal, "Self-customized BSP trees for collision detection", *Comput. Geom. Theory Appl.* **15**:1-3 (2000), 91–102.

[Ar et al. 2002] S. Ar, G. Montag, and A. Tal, "Deferred, self-organizing BSP trees", *Comput. Graph. Forum* **21**:3 (2002), 269–278.

[Arya 2002] S. Arya, "Binary space partitions for axis-parallel line segments: size-height tradeoffs", *Inform. Process. Lett.* **84**:4 (2002), 201–206.

[Arya et al. 2000] S. Arya, T. Malamatos, and D. M. Mount, "Nearly optimal expected-case planar point location", pp. 208–218 in *41st Annual Symposium on Foundations of Computer Science* (Redondo Beach, CA, 2000), IEEE Comput. Soc. Press, Los Alamitos, CA, 2000.

[Asano et al. 2003] T. Asano, M. de Berg, O. Cheong, L. J. Guibas, J. Snoeyink, and H. Tamaki, "Spanning trees crossing few barriers", *Discrete Comput. Geom.* **30**:4 (2003), 591–606.

[Ballieux 1993] C. Ballieux, "Motion planning using binary space partitions", Tech. Rep. Inf/src/93-25, Utrecht University, 1993.

[Basch et al. 1999] J. Basch, L. J. Guibas, and J. Hershberger, "Data structures for mobile data", *J. Algorithms* **31**:1 (1999), 1–28.

[Batagelo and Jr. 1999] H. C. Batagelo and I. C. Jr., "Real-time shadow generation using BSP trees and stencil buffers", pp. 93–102 in *Proc. 12th Brazilian Sympos. Comp. Graph. and Image Processing* (Campinas, SP), edited by J. Stolfi and C. L. Tozzi, IEEE, Los Alamitos, CA, 1999.

[de Berg 1993] M. de Berg, *Ray shooting, depth orders and hidden surface removal*, Lecture notes in computer science **703**, Springer, Berlin, 1993.

[de Berg 1995] M. de Berg, "Linear size binary space partitions for fat objects", pp. 252–263 in *Algorithms—ESA '95* (Corfu, 1995), Lecture Notes in Comput. Sci. **979**, Springer, Berlin, 1995.

[de Berg 2000] M. de Berg, "Linear size binary space partitions for uncluttered scenes", *Algorithmica* **28**:3 (2000), 353–366.

[de Berg et al. 1997a] M. de Berg, M. M. de Groot, and M. H. Overmars, "Perfect binary space partitions", *Comput. Geom.* **7**:1-2 (1997), 81–91.

[de Berg et al. 1997b] M. de Berg, M. de Groot, and M. Overmars, "New results on binary space partitions in the plane", *Comput. Geom.* **8**:6 (1997), 317–333.

[de Berg et al. 2001] M. de Berg, J. Comba, and L. J. Guibas, "A segment-tree based kinetic BSP", pp. 134–140 in Proc. 17th ACM Sympos. on Comput. Geom. (Somerville, MA, 2001), ACM Press, 2001.

[de Berg et al. 2002] M. de Berg, M. J. Katz, A. F. van der Stappen, and J. Vleugels, "Realistic input models for geometric algorithms", *Algorithmica* **34**:1 (2002), 81–97.

[de Berg et al. 2003] M. de Berg, H. David, M. J. Katz, M. Overmars, A. F. van der Stappen, and J. Vleugels, "Guarding scenes against invasive hypercubes", *Comput. Geom.* **26**:2 (2003), 99–117.

[Berman et al. 2002] P. Berman, B. Dasgupta, and S. Muthukrishnan, "Exact size of binary space partitionings and improved rectangle tiling algorithms", *SIAM J. Discrete Math.* **15**:2 (2002), 252–267.

[Buchele 1999] S. F. Buchele, *Three-dimensional binary space partitioning tree and constructive solid geometry tree construction from algebraic boundary representations*, Ph.D. thesis, University of Texas, Austin, 1999.

[Chazelle 1984] B. Chazelle, "Convex partitions of polyhedra: a lower bound and worst-case optimal algorithm", *SIAM J. Comput.* **13**:3 (1984), 488–507.

[Chin and Feiner 1989] N. Chin and S. Feiner, "Near real-time shadow generation using BSP trees", *Computer Graphics* **23**:3 (1989), 99–106.

[d'Amore and Franciosa 1992] F. d'Amore and P. G. Franciosa, "On the optimal binary plane partition for sets of isothetic rectangles", *Inform. Process. Lett.* **44**:5 (1992), 255–259.

[Dumitrescu et al. 2004] A. Dumitrescu, J. S. B. Mitchell, and M. Sharir, "Binary space partitions for axis-parallel segments, rectangles, and hyperrectangles", *Discrete Comput. Geom.* **31**:2 (2004), 207–227.

[Duncan et al. 2001] C. A. Duncan, M. T. Goodrich, and S. Kobourov, "Balanced aspect ratio trees: combining the advantages of k-d trees and octrees", *J. Algorithms* **38**:1 (2001), 303–333.

[Efrat and Sharir 2000] A. Efrat and M. Sharir, "On the complexity of the union of fat convex objects in the plane", *Discrete Comput. Geom.* **23**:2 (2000), 171–189.

[Fuchs et al. 1980] H. Fuchs, Z. M. Kedem, and B. Naylor, "On visible surface generation by a priori tree structures", *Comput. Graph.* **14**:3 (1980), 124–133.

[Hershberger and Suri 2003] J. Hershberger and S. Suri, "Binary space partitions for 3D subdivisions", pp. 100–108 in *Proceedings of the Fourteenth Annual ACM-SIAM Symposium on Discrete Algorithms* (Baltimore, MD, 2003), ACM, New York, 2003.

[Hershberger et al. 2004] J. Hershberger, S. Suri, and C. D. Tóth, "Binary space partitions of orthogonal subdivisions", pp. 230–238 in *Proc. 20th Sympos. Comput. Geom.* (Brooklyn, NY, 2004), ACM Press, 2004.

[Mata and Mitchell 1995] C. S. Mata and J. S. B. Mitchell, "Approximation algorithms for geometric tour and network design problems", pp. 360–369 in *Proc. 11th Sympos. Comput. Geom.* (Vancouver, 1995), ACM Press, 1995.

[Mitchell et al. 1997] J. S. B. Mitchell, D. M. Mount, and S. Suri, "Query-sensitive ray shooting", *Internat. J. Comput. Geom. Appl.* **7**:4 (1997), 317–347.

[Mulmuley 1990] K. Mulmuley, "A fast planar partition algorithm, I", *J. Symbolic Comput.* **10**:3-4 (1990), 253–280.

[Murali 1998] T. M. Murali, *Efficient hidden-surface removal in theory and in practice*, Ph.D. thesis, D. Thesis, Department of Computer Science, Brown University, Providence, RI, 1998.

[Naylor and Thibault 1986] B. Naylor and W. Thibault, "Application of BSP trees to ray-tracing and CSG evaluation", Tech. Rep. GIT-ICS 86/03, Georgia Institute of Tech., 1986.

[Naylor et al. 1990] B. Naylor, J. A. Amanatides, and W. Thibault, "Merging BSP trees yields polyhedral set operations", *Comput. Graph.* **24**:4 (1990), 115–124.

[Nguyen 1996] V. H. Nguyen, *Optimal binary space partitions for orthogonal objects*, Diss. 11818, ETH Zürich, 1996.

[Nguyen and Widmayer 1995] V. H. Nguyen and P. Widmayer, "Binary space partitions for sets of hyperrectangles", pp. 59–72 in *Algorithms, concurrency and knowledge (Pathumthani, 1995)*, Lecture Notes in Comput. Sci. **1023**, Springer, Berlin, 1995.

[Pach and Tardos 2002] J. Pach and G. Tardos, "On the boundary complexity of the union of fat triangles", *SIAM J. Comput.* **31**:6 (2002), 1745–1760.

[Pach et al. 2003] J. Pach, I. Safruti, and M. Sharir, "The union of congruent cubes in three dimensions", *Discrete Comput. Geom.* **30**:1 (2003), 133–160.

[Paterson and Yao 1990] M. S. Paterson and F. F. Yao, "Efficient binary space partitions for hidden-surface removal and solid modeling", *Discrete Comput. Geom.* **5**:5 (1990), 485–503.

[Paterson and Yao 1992] M. S. Paterson and F. F. Yao, "Optimal binary space partitions for orthogonal objects", *J. Algorithms* **13**:1 (1992), 99–113.

[Pauly et al. 2002] M. Pauly, M. H. Gross, and L. Kobbelt, "Efficient simplification of point-sampled surfaces", pp. 163–170 in *Proc. 13th IEEE Visualization Conference*, 2002.

[Preparata 1981] F. P. Preparata, "A new approach to planar point location", *SIAM J. Comput.* **10**:3 (1981), 473–482.

[Preparata and Shamos 1985] F. P. Preparata and M. I. Shamos, *Computational geometry*, Springer, New York, 1985.

[Schumacker et al. 1969] R. A. Schumacker, R. Brand, M. Gilliland, and W. Sharp, "Study for applying computer-generated images to visual simulation", Tech. Rep. AFHRL–TR–69–14, U.S. Air Force Human Resources Laboratory, 1969.

[Seidel 1991] R. Seidel, "A simple and fast incremental randomized algorithm for computing trapezoidal decompositions and for triangulating polygons", *Comput. Geom.* **1**:1 (1991), 51–64.

[Shaffer and Garland 2001] E. Shaffer and M. Garland, "Efficient adaptive simplification of massive meshes", pp. 127–134 in *Proc. 12th IEEE Visualization Conference*, 2001.

[Thibault and Naylor 1987] W. C. Thibault and B. F. Naylor, "Set operations on polyhedra using binary space partitioning trees", *Comput. Graphics* **21**:4 (1987), 153–162.

[Tobola and Nechvile 2003] P. Tobola and K. Nechvile, "Linear binary space partitions and hierarchy of object classes", pp. 64–67 in *Proc. 15th Canadian Conf. Comput. Geom.* (Halifax, NS, 2003), 2003.

[Tóth 2003a] C. D. Tóth, "Binary space partition for orthogonal fat rectangles", pp. 494–505 in *Algorithms—ESA 2003*, Lecture Notes in Comput. Sci. **2832**, Springer, Berlin, 2003.

[Tóth 2003b] C. D. Tóth, "Binary space partitions for line segments with a limited number of directions", *SIAM J. Comput.* **32**:2 (2003), 307–325.

[Tóth 2003c] C. D. Tóth, "A note on binary plane partitions", *Discrete Comput. Geom.* **30**:1 (2003), 3–16.

CSABA D. TÓTH
DEPARTMENT OF MATHEMATICS
MASSACHUSETTS INSTITUTE OF TECHNOLOGY
77 MASSACHUSETTS AVE., ROOM 2-336
CAMBRIDGE, MA 02139
UNITED STATES OF AMERICA
toth@math.mit.edu

Combinatorial and Computational Geometry
MSRI Publications
Volume **52**, 2005

The Erdős–Szekeres Theorem: Upper Bounds and Related Results

GÉZA TÓTH AND PAVEL VALTR

ABSTRACT. Let $\mathrm{ES}(n)$ denote the least integer such that among any $\mathrm{ES}(n)$ points in general position in the plane there are always n in convex position. In 1935, P. Erdős and G. Szekeres showed that $\mathrm{ES}(n)$ exists and $\mathrm{ES}(n) \leq \binom{2n-4}{n-2} + 1$. Six decades later, the upper bound was slightly improved by Chung and Graham, a few months later it was further improved by Kleitman and Pachter, and another few months later it was further improved by the present authors. Here we review the original proof of Erdős and Szekeres, the improvements, and finally we combine the methods of the first and third improvements to obtain yet another tiny improvement.

We also briefly review some of the numerous results and problems related to the Erdős–Szekeres theorem.

1. Introduction

In 1933, Esther Klein raised the following question. Is it true that for every n there is a least number — which we will denote by $\mathrm{ES}(n)$ — such that among any $\mathrm{ES}(n)$ points in general position in the plane there are always n in convex position?

This question was answered in the affirmative in a classical paper of Erdős and Szekeres [1935]. In fact, they showed (see also [Erdős and Szekeres 1960/1961]) that

$$2^{n-2} + 1 \leq \mathrm{ES}(n) \leq \binom{2n-4}{n-2} + 1.$$

The lower bound, $2^{n-2}+1$, is sharp for $n = 2, 3, 4, 5$ and has been conjectured to be sharp for all n. However, the upper bound,

$$\binom{2n-4}{n-2} + 1 \approx c\frac{4^n}{\sqrt{n}},$$

Tóth was supported by grant OTKA-T-038397.
Valtr was supported by projects LN00A056 and 1M0021620808 of the Ministry of Education of the Czech Republic.

was not improved for 60 years. Recently, Chung and Graham [1998] managed to improve it by 1. Shortly after, Kleitman and Pachter [1998] showed that $ES(n) \le \binom{2n-4}{n-2} + 7 - 2n$. A few months later the present authors [Tóth and Valtr 1998] proved that $ES(n) \le \binom{2n-5}{n-2} + 2$, which is a further improvement, roughly by a factor of 2.

In this note we review the original proof of Erdős and Szekeres, all three improvements, and then we combine the ideas of the first and third improvements to obtain the following result, which is a further improvement by 1.

THEOREM 1. For $n \ge 5$, any set of $\binom{2n-5}{n-2} + 1$ points in general position in the plane contains n points in convex position. That is, $ES(n) \le \binom{2n-5}{n-2} + 1$.

Next section contains a brief review of some of the numerous results and problems related to the Erdős–Szekeres theorem.

2. Some Related Results

Many researchers have been motivated by the Erdős–Szekeres theorem. Here we mention only a small part of the research related to the Erdős–Szekeres theorem. See [Morris and Soltan 2000; Bárány and Károlyi 2001; Braß et al. 2005] for the latest survey.

Empty polygons. A famous open problem related to the Erdős–Szekeres theorem is the *empty–hexagon problem*. Let P be a finite set of points in general position in the plane. A subset $Q \subset P, |Q| = n$, is called an *n-hole* (or an *empty convex n-gon*) *in* P, if it is in convex position and its convex hull contains no further points of P. Let $g(n)$ be the smallest positive integer such that any P, $|P| \ge g(n)$, in general position contains an n-hole. It is easy to see that $g(3) = 3$, $g(4) = 5$. Harborth [Harborth 1978] proved $g(5) = 10$. Horton [Horton 1983] gave a construction showing that no finite $g(7)$ exists.

The empty–hexagon problem: *Is there a finite $g(6)$?*

Using a computer search, Overmars [Overmars 2003] found a set of 29 points in general position having no empty hexagon. Thus, if $g(6)$ exists then $g(6) \ge 30$.

Let $X_k(P)$ be the number of empty k-gons in an n-element point set P in general position, for $k \ge 0$; as special cases $X_0(P) = 1$, $X_1(P) = n$, $X_2(P) = \binom{n}{2}$, since every subset of P of size up to 2 is considered an empty polygon. There are several equalities and inequalities involving these parameters. Ahrens et al. [1999] proved general results giving the following interesting equalities on the numbers $X_k(P)$:

$$\sum_{k \ge 0}(-1)^k X_k(P) = 0, \qquad \sum_{k \ge 1}(-1)^k k X_k(P) = -|P \cap \mathrm{Int}(P)|,$$

where $|P \cap \mathrm{Int}(P)|$ is the number of interior points of P. Pinchasi et al. [$\geq$ 2005] proved these two equalities by a simple argument (the "continuous motion" method) and gave also some other equalities and inequalities, such as

$$X_4(P) \geq X_3(P) - \tfrac{1}{2}n^2 - O(n), \qquad X_5(P) \geq X_3(P) - n^2 - O(n).$$

Let $Y_k(n) = \min_{|P|=n} X_k(P)$ be the minimum number of empty convex k-gons in a set of n points. By the construction of Horton, $Y_k(n) = 0$ for $k \geq 7$. For $k \leq 6$, the best known bounds are

$$n^2 - 5n + 10 \leq Y_3(n) \leq 1.6195...n^2 + o(n^2),$$

$$\binom{n-3}{2} + 6 \leq Y_4(n) \leq 1.9396...n^2 + o(n^2),$$

$$3\left\lfloor \frac{n}{12} \right\rfloor \leq Y_5(n) \leq 1.0206...n^2 + o(n^2),$$

$$0 \leq Y_6(n) \leq 0.2005...n^2 + o(n^2).$$

The lower bounds are given in [Dehnhardt 1987], the upper bounds in [Bárány and Valtr 2004].

Convex bodies. Several authors [Bisztriczky and Fejes Tóth 1989; 1990; Pach and Tóth 1998; Tóth 2000] have extended the Erdős–Szekeres theorem to families of pairwise disjoint convex sets, instead of points.

A family of pairwise disjoint convex sets is said to be in *convex position* if none of its members is contained in the convex hull of the union of the others.

It is easy to construct an arbitrarily large family of pairwise disjoint convex sets such that no three or more of them are in convex position. So, without any additional condition on the family, we cannot generalize the Erdős–Szekeres theorem.

For points we had the condition "no three points are on a line", that is, "any three points are in convex position". Therefore, the most natural condition to try for families of convex sets is "any three convex sets are in convex position".

Bisztriczky and Fejes Tóth [1989] proved that there exists a function $P_3(n)$ such that if a family $\mathcal{F}$ of pairwise disjoint convex sets has more than $P_3(n)$ members, and any *three* members of $\mathcal{F}$ are in convex position, then $\mathcal{F}$ has n members in convex position. In [Bisztriczky and Fejes Tóth 1990] they showed that this statement is true with a function $P_3(n)$, triply exponential in n. Pach and Tóth [1998] further improved the upper bound on $P_3(n)$ to a simply exponential function. The best known lower bound for $P_3(n)$ is the classical lower bound for the original Erdős–Szekeres theorem, $2^{n-2} \leq P_3(n)$.

In the case of points, if we have a stronger condition that every *four* points are in convex position, then the problem becomes uninteresting; in this case all points are in convex position.

In case of convex sets, the condition "every four are in convex position" does not make the problem uninteresting, but it still turns out to be a rather strong condition. Let $\mathcal{F}$ be a family of pairwise disjoint convex sets. If any k members of $\mathcal{F}$ are in convex position, then we say that $\mathcal{F}$ satisfies *property P_k*. If no n members of $\mathcal{F}$ are in convex position, then we say that $\mathcal{F}$ satisfies *property P^n*. Property P_k^n means that both P_k and P^n are satisfied. Using these notions, the above cited result of Pach and Tóth states that if a family $\mathcal{F}$ satisfies property P_3^n, then $|\mathcal{F}| \leq \binom{2n-4}{n-2}^2$.

Bisztriczky and Fejes Tóth [1990] raised the following more general question. What is the maximum size $P_k(n)$ of a family $\mathcal{F}$ satisfying property P_k^n? Some of their bounds were later improved in [Pach and Tóth 1998] and [Tóth 2000]. The best known bounds are

$$2^{n-2} \leq P_3(n) \leq \binom{2n-4}{n-2}^2,$$

$$2 \left\lfloor \frac{n+1}{4} \right\rfloor^2 \leq P_4(n) \leq n^3,$$

$$n - 1 + \left\lfloor \frac{n-1}{k-2} \right\rfloor \leq P_5(n) \leq 6n - 12,$$

$$n - 1 + \left\lfloor \frac{n-1}{k-2} \right\rfloor \leq P_k(n) \leq n + \frac{1}{k-5}n \quad \text{for } k \geq 6.$$

See [Erdős and Szekeres 1960/1961] for the first line, [Pach and Tóth 1998] for the first and second, and [Bisztriczky and Fejes Tóth 1990; Tóth 2000] for the last two.

Pach and Tóth [2000] investigated the case when the sets are not necessarily disjoint.

The partitioned version. It follows from the exponential upper bound on the number $\mathrm{ES}(n)$ by a simple counting argument that for a given n every "huge" set of points in general position in the plane contains "many" n-point subsets in convex position. However, geometric arguments yield much stronger results.

A *convex n-clustering* is defined as a finite planar point set in general position which can be partitioned into n finite sets $X_1, X_2, \ldots, X_n$ of *equal* size such that $x_1 x_2 \ldots x_n$ is a convex n-gon for each choice $x_1 \in X_1$, $x_2 \in X_2$, $\ldots$, $x_n \in X_n$.

The positive fraction Erdős–Szekeres theorem [Bárány and Valtr 1998] states that for any n any sufficiently large finite set X of points in general position contains a convex n-clustering of size $\geq \varepsilon_n |X|$, where $\varepsilon_n > 0$ is independent of X. Pór [2003] and Pór and Valtr [2002], answering a question of Bárány, proved a partitioned version of the Erdős–Szekeres theorem: for any n there are two positive constants c_n, c_n' such that any finite X in general position can be partitioned into at most c_n convex n-clusterings and a remaining set of at most c_n' points. The optimal constants $1/\varepsilon_n, c_n'$ are exponential in n, while c_n is at least exponential in n and at most of order $n^{O(n^2)}$. For details see [Pór and Valtr 2002].

The positive fraction Erdős–Szekeres theorem for collections of convex sets can be found in [Pach and Solymosi 1998], and the partitioned Erdős–Szekeres theorem for collections of convex sets can be found in [Pór and Valtr ≥ 2005].

3. The Upper Bound of Erdős and Szekeres

Definition. The points $(x_1, y_1), (x_2, y_2), \ldots, (x_n, y_n)$, $x_1 < x_2 < \ldots < x_n$, form an n-*cap* if

$$\frac{y_2 - y_1}{x_2 - x_1} > \frac{y_3 - y_2}{x_3 - x_2} > \ldots > \frac{y_n - y_{n-1}}{x_n - x_{n-1}}.$$

Similarly, they form an n-*cup* if

$$\frac{y_2 - y_1}{x_2 - x_1} < \frac{y_3 - y_2}{x_3 - x_2} < \ldots < \frac{y_n - y_{n-1}}{x_n - x_{n-1}}.$$

THEOREM 2 [Erdős and Szekeres 1935]. *Let $f(n, m)$ be the least integer such that any set of $f(n, m)$ points in general position in the plane contains either an n-cap or an m-cup. Then*

$$f(n, m) = \binom{n + m - 4}{n - 2} + 1.$$

The following observation has a key role in the proof of the Erdős–Szekeres theorem.

OBSERVATION 1. *If a point v is the rightmost point of a cap and also the leftmost point of a cup then the cap or the cup can be extended to a larger cap or cup, respectively.*

PROOF. Let u be the second point of the cap from the right, and let w be the second point of the cup from the left. Now, depending on the angle uvw, either the cap can be extended by w, or the cup can be extended by u. See Figure 1. □

Figure 1. Either the cap can or the cup can be extended.

PROOF THAT $f(n, m) \leq \binom{n+m-4}{n-2} + 1$. We use double induction on n and m. The statement trivially holds for $n = 2$ and any m, and for $m = 2$ and any n. Let $n, m \geq 3$ and suppose that the statement holds for $(n, m-1)$ and for $(n-1, m)$. Take $\binom{n+m-4}{n-2} + 1$ points in general position. By induction we know that any subset of at least $\binom{n+m-5}{n-3} + 1$ points contains either an $(n-1)$-cap or an m-cup. In the latter case we are done, so we can assume that any subset of at least $\binom{n+m-5}{n-3} + 1$ points contains an $(n-1)$-cap. Take an $(n-1)$-cap and remove its right endpoint from the point set. Since we still have at least $\binom{n+m-5}{n-3}+1$ points, we have another $(n-1)$-cap, remove its right endpoint again, and continue until we have $\binom{n+m-5}{n-3}$ points left. We have removed $\binom{n+m-4}{n-2}+1-\binom{n+m-5}{n-3}=\binom{n+m-5}{m-3}+1$ points, each of them a right endpoint of some $(n-1)$-cap. But the set of these points, by induction, contains either an n-cap or an $(m-)1$-cup. In the first case we are done. In the second case we have an $(m-1)$-cup whose left endpoint v is the right endpoint of some $(n-1)$-cap. Observation 1 then finishes the induction step. $\qquad\square$

PROOF OF THE ERDŐS–SZEKERES THEOREM. Since $\mathrm{ES}(n)\leq f(n,n)$, we have $\mathrm{ES}(n)\leq\binom{2n-4}{n-2}+1$. $\qquad\square$

Erdős and Szekeres [1935] also proved that the bound $f(n,m)\leq\binom{n+m-4}{n-2}+1$ is tight for any n,m. But it does not imply that the bound for $\mathrm{ES}(n)$ is tight as well. The best known lower bound is $2^{n-2}+1\leq\mathrm{ES}(n)$ [Erdős and Szekeres 1960/1961] and in fact it is conjectured to be tight.

4. Three Improvements

THEOREM 3 [Chung and Graham 1998]. For $n\geq 4$,
$$\mathrm{ES}(n)\leq\binom{2n-4}{n-2}.$$

PROOF. Take $\binom{2n-4}{n-2}$ points in general position. Let A be the set of those points which are right endpoints of some $(n-1)$-cap. Just as above, we can argue that $|A|\geq\binom{2n-4}{n-2}-\binom{2n-5}{n-3}=\binom{2n-5}{n-3}$. If $|A|>\binom{2n-5}{n-3}$, then A contains either an n-cap or an $(n-1)$-cup. In the first case we are done immediately, in the second we have an $(n-1)$-cup whose left endpoint is also a right endpoint of some $(n-1)$-cap and we are done as in the previous proof. So we can assume that $|A|=\binom{2n-5}{n-3}$. Let B be the set of the other points, clearly $|B|=\binom{2n-5}{n-3}$. Let $b\in B$. The set $\{b\}\cup A$ has size $\binom{2n-5}{n-3}+1$ so again it contains either an n-cap or an $(n-1)$-cup. In the case of n-cap we are done, so we can assume that it contains an $(n-1)$-cup for any choice of b. If the left endpoint of this $(n-1)$-cup is an element of A, we are done by Observation 1, since we have an $(n-1)$-cup whose left endpoint is also a right endpoint of some $(n-1)$-cap. So, the left endpoint of this $(n-1)$-cup is b. Therefore, any $b\in B$ is the left endpoint of an $(n-1)$-cup whose right endpoint is in A. We can argue analogously, that any $a\in A$ is the right endpoint of an $(n-1)$-cap whose left endpoint is in B. Let S be the set of all segments ab, where $a\in A$, $b\in B$, and there is an $(n-1)$-cup or $(n-1)$-cap whose right endpoint is a and left endpoint is b. Let ab be the element of S with the largest slope. Suppose that ab represents an $(n-1)$-cup, the other case is analogous. We know that there is an $(n-1)$-cap whose right endpoint is a and left endpoint is b'. Now it is easy to see that either the $(n-1)$-cup and b', or the $(n-1)$-cap and b determine a convex n-gon. This concludes the proof; see Figure 2. $\qquad\square$

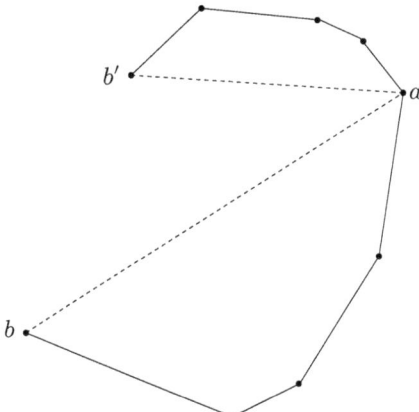

Figure 2. Either b' can be added to the cup, or b to the cap.

THEOREM 4 [Kleitman and Pachter 1998]. For $n \geq 4$,

$$\mathrm{ES}(n) \leq \binom{2n-4}{n-2} - 2n + 7.$$

PROOF. We say that a point set is *vertical* if its two leftmost points have the same x-coordinate. Observe, that any point set can be made vertical by an appropriate rotation. We define caps and cups for vertical sets just like for any set of points, the only difference is that now the vertical edge determined by the two leftmost points is allowed to be the leftmost edge of a cup or a cap; see Figure 3.

Let $f_v(n, m)$ be the least integer such that any *vertical* set of $f_v(n, m)$ points in general position contains either an n-cap or an m-cup. Take $f_v(n, m) - 1$ points in a vertical point set with no n-caps and m-cups. Let a and b be the two leftmost points such that a is above b. Let A be the set of those points which are right endpoints of some $(n-1)$-cap, and B be the set of the other points. Since the two leftmost points do not belong to A, B is a vertical point set. If $|B| \geq f_v(n-1, m)$ then B has an $(n-1)$-cap or an m-cup. The first case contradicts the definition of A, the second case contradicts the assumption that we do not have an m-cup. So, $|B| \leq f_v(n-1, m) - 1$. Now consider the set $A' = A \cup \{b\}$ and suppose that $|A'| \geq f(n, m-1)$. Then A' has an n-cap or an $(m-1)$-cup. The first case is a contradiction immediately, in the second case consider the left endpoint of that $(m-1)$-cup. If it is b, then it can be extended to an m-cup by a, a contradiction. If it is in A, then the usual argument works, we have an $(n-1)$-cup whose left endpoint is also a right endpoint of some $(n-1)$-cap and one of them can be extended by Observation 1. So $|A| = |A'| - 1 \leq f(n, m-1) - 2$. Combining the two inequalities we get that

$$f_v(n, m) \leq f_v(n-1, m) + f(n, m-1) - 2,$$

and an analogous argument shows that

$$f_v(n, m) \leq f_v(n, m-1) + f(n-1, m) - 2.$$

Using the known values of $f(n, m)$, and that $f_v(n, 3) = f_v(3, n) = n$, we get that $f_v(n, m) \leq \binom{n+m-4}{n-2} + 7 - n - m$, and the result follows. In fact, the inequality obtained for $f_v(n, m)$ is sharp [Kleitman and Pachter 1998]. □

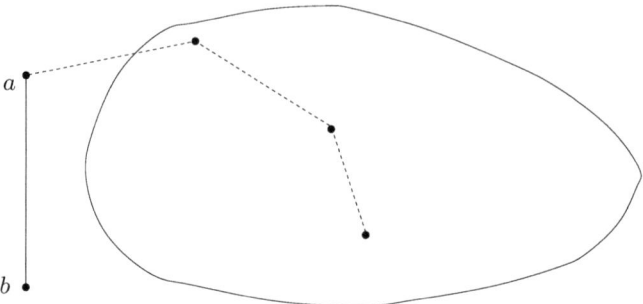

Figure 3. A vertical point set with a 5-cap.

THEOREM 5 [Tóth and Valtr 1998]. For $n \geq 3$,

$$\mathrm{ES}(n) \leq \binom{2n-5}{n-2} + 2.$$

PROOF. Take $\binom{2n-5}{n-3}$ points in general position. Suppose that the set P does not contain n points in convex position. Let x be a vertex of the convex hull of P. Let y be a point outside the convex hull of P such that none of the lines determined by the points of $P \setminus \{x\}$ intersects the segment $\overline{xy}$. Finally, let ℓ be a line through y which avoids the convex hull of P.

Consider a projective transformation T which maps the line ℓ to the line at infinity, and maps the segment $\overline{xy}$ to the vertical half-line $v^-(x')$, emanating downwards from $x' = T(x)$. We get a point set $P' = T(P)$ from P. Since ℓ avoided the convex hull of P, the transformation T does not change convexity on the points of P, that is, any subset of P is in convex position if and only if the corresponding points of P' are in convex position. So the assumption holds also for P', no n points of P' are in convex position. By the choice of the point y, none of the lines determined by the points of $P' \setminus \{x'\}$ intersects $v^-(x')$. Therefore, any m-cap in the set $Q' = P' \setminus \{x'\}$ can be extended by x' to a convex $(m+1)$-gon.

Since no n points of P' are in convex position, Q' cannot contain any n-cup or $(n-1)$-cap. Therefore, by the Lemma,

$$|Q'| \leq f(n-1, n) - 1 = \binom{2n-5}{n-2}, \quad |P| \leq \binom{2n-5}{n-2} + 1,$$

and the theorem follows. □

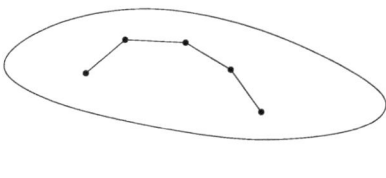

Figure 4. Any $(n-1)$-cap can be extended by x' to a convex n-gon.

5. A Combination of Two Methods

PROOF OF THEOREM 1. Suppose that the set P does not contain n points in convex position and $|P| = \binom{2n-5}{n-2} + 1$. Let x be a vertex of the convex hull of P and y be a point outside the convex hull of P so close to x that none of the lines determined by the points of $P \setminus \{x\}$ intersects the segment $\overline{xy}$. Finally, let ℓ be a line through y which avoids the convex hull of P.

Consider a projective transformation T which maps the line ℓ to the line at infinity, and maps the segment $\overline{xy}$ to the vertical half-line $v^-(x')$, emanating downwards from $x' = T(x)$. We get a point set P' from P. Just like in the previous proof, T does not change convexity on the points of P. Let $P'' = P' \setminus \{x'\}$. By the assumption, P'' does not contain any $(n-1)$-cap or n-cup.

Let A be the set of those points of P'' which are right endpoints of some $(n-2)$-cap, and let $B = P'' \setminus A$. If $|A| > \binom{2n-6}{n-3}$ then A contains either an $(n-1)$-cap or an $(n-1)$-cup. The first case contradicts the assumption, in the second case we have an $(n-1)$-cup whose left endpoint is also a right endpoint of some $(n-2)$-cap, so, either the $(n-1)$-cup or the $(n-2)$-cap can be extended by one point and we get a contradiction. So, $|A| \le \binom{2n-6}{n-3}$. If $|B| > \binom{2n-6}{n-2}$, then B contains either an $(n-2)$-cap or an n-cup. The first case contradicts the definition of A, since we find a right endpoint of some $(n-2)$-cap in B, the second case contradicts the assumption. So $|B| \le \binom{2n-6}{n-2}$. But then $|P''| = |A| + |B| \le \binom{2n-6}{n-3} + \binom{2n-6}{n-2} = \binom{2n-5}{n-2} = |P''|$, therefore, $|A| = \binom{2n-6}{n-3}$ and $|B| = \binom{2n-6}{n-2}$.

Let $b \in B$. The set $\{b\} \cup A$ has size $\binom{2n-6}{n-3} + 1$ so again it contains either an $(n-1)$-cap or an $(n-1)$-cup. In the case of $(n-1)$-cap we are done, so we can assume that it is an $(n-1)$-cup for any choice of b. If the left endpoint of this $(n-1)$-cup is an element of A, we have an $(n-1)$-cup whose left endpoint is also a right endpoint of some $(n-2)$-cap, so, either the $(n-1)$-cup or the $(n-2)$-cap can be extended by one point and we get a contradiction again. Hence the left endpoint of the $(n-1)$-cup is b. Therefore, any $b \in B$ is the left endpoint of an

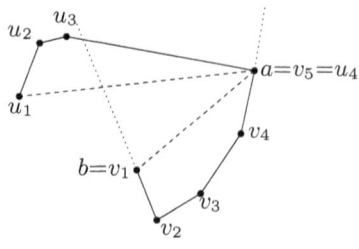

Figure 5. $x', u_1, u_2, u_3, v_1, v_2$ determine a convex hexagon.

$(n-1)$-cup whose right endpoint is in A. We can argue analogously, considering the sets $\{a\} \cup B$, that any $a \in A$ is the right endpoint of an $(n-2)$-cap whose left endpoint is in B.

Let S be the set of all segments ab, where $a \in A$, $b \in B$, and there is either an $(n-1)$-cup or $(n-2)$-cap whose right endpoint is a and left endpoint is b. Let ab be the element of S with the largest slope. Suppose that ab represents an $(n-1)$-cup. The argument in the other case is analogous. Let $b = v_1, v_2, \ldots, v_{n-1} = a$ be the points of the $(n-1)$-cup from left to right. We know that there is also an $(n-2)$-cap whose right endpoint is a and left endpoint in B. Let $u_1, u_2, \ldots, u_{n-2} = a$ be its points from left to right. If u_{n-3} lies above the line $v_1 v_2$, then $u_j, v_1, v_2, \ldots, v_{n-1}$ determine a convex n-gon and we are done. Otherwise $x', u_1, u_2, \ldots, u_{n-3}, v_1, v_2$ determine a convex n-gon; see Figure 5. This concludes the proof. $\qquad\square$

References

[Ahrens et al. 1999] C. Ahrens, G. Gordon, and E. W. McMahon, "Convexity and the beta invariant", *Discrete Comput. Geom.* **22**:3 (1999), 411–424.

[Bárány and Károlyi 2001] I. Bárány and G. Károlyi, "Problems and results around the Erdős–Szekeres convex polygon theorem", pp. 91–105 in *Discrete and computational geometry* (Tokyo, 2000), edited by J. Akiyama et al., Lecture Notes in Comput. Sci. **2098**, Springer, Berlin, 2001.

[Bárány and Valtr 1998] I. Bárány and P. Valtr, "A positive fraction Erdős–Szekeres theorem", *Discrete Comput. Geom.* **19**:3 (1998), 335–342.

[Bárány and Valtr 2004] I. Bárány and P. Valtr, "Planar point sets with a small number of empty convex polygons", 2004.

[Bisztriczky and Fejes Tóth 1989] T. Bisztriczky and G. Fejes Tóth, "A generalization of the Erdős–Szekeres convex n-gon theorem", *J. Reine Angew. Math.* **395** (1989), 167–170.

[Bisztriczky and Fejes Tóth 1990] T. Bisztriczky and G. Fejes Tóth, "Convexly independent sets", *Combinatorica* **10**:2 (1990), 195–202.

[Braß et al. 2005] P. Braß, W. Moser, and J. Pach, "Convex polygons and the Erdős–Szekeres problem", pp. 326–341 in *Research problems in discrete geometry*, Springer, Berlin, 2005.

[Chung and Graham 1998] F. R. K. Chung and R. L. Graham, "Forced convex n-gons in the plane", *Discrete Comput. Geom.* **19**:3 (1998), 367–371.

[Dehnhardt 1987] K. Dehnhardt, *Leere konvexe Vielecke in ebenen Punktmengen*, dissertation, Technische Univ. Braunschweig, 1987.

[Erdős and Szekeres 1935] P. Erdős and G. Szekeres, "A combinatorial problem in geometry", *Compositio Math.* **2** (1935), 463–470.

[Erdős and Szekeres 1960/1961] P. Erdős and G. Szekeres, "On some extremum problems in elementary geometry", *Ann. Univ. Sci. Budapest. Eötvös Sect. Math.* **3–4** (1960/1961), 53–62.

[Harborth 1978] H. Harborth, "Konvexe Fünfecke in ebenen Punktmengen", *Elem. Math.* **33**:5 (1978), 116–118.

[Horton 1983] J. D. Horton, "Sets with no empty convex 7-gons", *Canad. Math. Bull.* **26**:4 (1983), 482–484.

[Kleitman and Pachter 1998] D. Kleitman and L. Pachter, "Finding convex sets among points in the plane", *Discrete Comput. Geom.* **19**:3, Special Issue (1998), 405–410.

[Morris and Soltan 2000] W. Morris and V. Soltan, "The Erdős–Szekeres problem on points in convex position—a survey", *Bull. Amer. Math. Soc. (N.S.)* **37**:4 (2000), 437–458.

[Overmars 2003] M. Overmars, "Finding sets of points without empty convex 6-gons", *Discrete Comput. Geom.* **29**:1 (2003), 153–158.

[Pach and Solymosi 1998] J. Pach and J. Solymosi, "Canonical theorems for convex sets", *Discrete Comput. Geom.* **19**:3 (1998), 427–435.

[Pach and Tóth 1998] J. Pach and G. Tóth, "A generalization of the Erdős–Szekeres theorem to disjoint convex sets", *Discrete Comput. Geom.* **19**:3 (1998), 437–445.

[Pach and Tóth 2000] J. Pach and G. Tóth, "Erdős–Szekeres-type theorems for segments and noncrossing convex sets", *Geom. Dedicata* **81**:1-3 (2000), 1–12.

[Pinchasi et al. ≥ 2005] R. Pinchasi, R. Radoičić, and M. Sharir, "On empty convex polygons in a planar point set". Manuscript.

[Pór 2003] A. Pór, "A partitioned version of the Erdős–Szekeres theorem for quadrilaterals", *Discrete Comput. Geom.* **30**:2 (2003), 321–336.

[Pór and Valtr 2002] A. Pór and P. Valtr, "The partitioned version of the Erdős–Szekeres theorem", *Discrete Comput. Geom.* **28**:4 (2002), 625–637.

[Pór and Valtr ≥ 2005] A. Pór and P. Valtr, "The partitioned Erdős–Szekeres theorem for convex sets". In preparation.

[Tóth 2000] G. Tóth, "Finding convex sets in convex position", *Combinatorica* **20**:4 (2000), 589–596.

[Tóth and Valtr 1998] G. Tóth and P. Valtr, "Note on the Erdős–Szekeres theorem", *Discrete Comput. Geom.* **19**:3, Special Issue (1998), 457–459.

GÉZA TÓTH
RÉNYI INSTITUTE
HUNGARIAN ACADEMY OF SCIENCES
REÁLTANODA U. 13-15
BUDAPEST, 1053
HUNGARY

PAVEL VALTR
DEPARTMENT OF APPLIED MATHEMATICS
and
INSTITUTE FOR THEORETICAL COMPUTER SCIENCE (ITI)
CHARLES UNIVERSITY
MALOSTRANSKÉ NÁM. 25, 118 00
PRAHA 1
CZECH REPUBLIC

Combinatorial and Computational Geometry
MSRI Publications
Volume **52**, 2005

On the Pair-Crossing Number

PAVEL VALTR

ABSTRACT. By a *drawing* of a graph G, we mean a drawing in the plane such that vertices are represented by distinct points and edges by arcs. The *crossing number* $\mathrm{cr}(G)$ of a graph G is the minimum possible number of crossings in a drawing of G. The *pair-crossing number* pair-$\mathrm{cr}(G)$ of G is the minimum possible number of (unordered) crossing pairs in a drawing of G. Clearly, pair-$\mathrm{cr}(G) \leq \mathrm{cr}(G)$ holds for any graph G. Let $f(k)$ be the maximum $\mathrm{cr}(G)$, taken over all graphs G with pair-$\mathrm{cr}(G) = k$. Obviously, $f(k) \geq k$. Pach and Tóth [2000] proved that $f(k) \leq 2k^2$. Here we give a slightly better asymptotic upper bound $f(k) = O(k^2/\log k)$. In case of x-monotone drawings (where each vertical line intersects any edge at most once) we get a better upper bound $f^{\mathrm{mon}}(k) \leq 4k^{4/3}$.

1. Introduction

By a *drawing* of a graph G, we mean a drawing in the plane such that vertices are represented by distinct points and edges by arcs. The arcs are allowed to cross, but they may not pass through vertices (except for their endpoints) and no point is an internal point of three or more arcs. Two arcs may have only finitely many common points. A *crossing* is a common internal point of two arcs. A *crossing pair* is a pair of edges which cross each other at least once. A drawing is *planar*, if there are no crossings in it. A *subdrawing* of a drawing is defined analogously as a subgraph of a graph.

The *crossing number* $\mathrm{cr}(G)$ of a graph G is the minimum possible number of crossings in a drawing of G. The *pair-crossing number* pair-$\mathrm{cr}(G)$ of G is the minimum possible number of (unordered) crossing pairs in a drawing of G.

In this paper we investigate the relation between the crossing number and the pair-crossing number. Clearly, pair-$\mathrm{cr}(G) \leq \mathrm{cr}(G)$ holds for any graph G. The problem of deciding whether $\mathrm{cr}(G) = $ pair-$\mathrm{cr}(G)$ holds for every G appears quite challenging. Let $f(k)$ be the maximum $\mathrm{cr}(G)$, taken over all graphs G with pair-$\mathrm{cr}(G) = k$. Obviously, $f(k) \geq k$. Pach and Tóth [2000] proved that

This work was supported by project 1M0021620808 of The Ministry of Education of the Czech Republic.

$f(k) \leq 2k^2$. In fact, they proved this bound in a stronger version when the pair-crossing number is replaced by the so-called odd-crossing number, which is the minimum number of pairs of edges in a drawing that cross each other an odd number of times. Here we find a slightly better asymptotic upper bound on $f(k)$:

THEOREM 1. $f(k) = O(k^2/\log k)$.

The improvement is small but its proof gives some insight to the structure of possible counterexamples to $f(k) = k$.

We get a significantly subquadratic upper bound in the case of *(x-)monotone* drawings. A drawing D is *monotone* if every edge is drawn as an x-monotone curve, meaning that no vertical line intersects it more than once. The *monotone crossing number* $\mathrm{cr}^{\mathrm{mon}}(G)$ is the minimum possible number of crossings in a monotone drawing of G. The *monotone pair-crossing number* pair-$\mathrm{cr}^{\mathrm{mon}}(G)$ is defined analogously — it is the minimum possible number of (unordered) crossing pairs in a monotone drawing of G. Let $f^{\mathrm{mon}}(k)$ be the maximum $\mathrm{cr}^{\mathrm{mon}}(G)$, taken over all graphs G with pair-$\mathrm{cr}^{\mathrm{mon}}(G) = k$. Obviously, $f^{\mathrm{mon}}(k) \geq k$.

THEOREM 2. $f^{\mathrm{mon}}(k) \leq 4k^{4/3}$.

Theorem 1 is proved in Section 2 and Theorem 2 in Section 3.

Remarks. **1.** It is possible that our results hold also if the (monotone) pair-crossing number is replaced by the so-called (monotone) odd-crossing number (see [Pach and Tóth 2000] for the definition of the odd-crossing number and for a similar result). We did not investigate this question.

2. Some related results can be found in [Kolman and Matoušek 2004]. In particular, these authors prove that

$$\mathrm{cr}(G) = O\left(\log^3 |V| \left(\text{pair-cr}(G) + \sum_{v \in V} (\deg v)^2\right)\right)$$

for any graph $G = (V, E)$.

3. One could hope to prove $f(k) = k$ by a contradiction, considering local modifications of a drawing witnessing $f(k) > k$. We tried this approach but it does not seem to work in some straightforward way. Our difficulties with this approach might have an explanation in an example [Kratochvíl and Matoušek 1994] of a drawing in which it is not possible to eliminate multiple crossings of edge pairs without introducing new crossing pairs.

2. A Logarithmic Improvement over the Quadratic Bound

Here we give a simple proof of $f(k) \leq 2k^2$ and then refine the method, thereby obtaining $f(k) = O(k^2/\log k)$.

A simple proof of the quadratic bound. The bound $f(k) \leq 2k^2$ can be proved easily; this was probably known to experts but, as far as we know, hasn't appeared in print. Let G be a graph with pair-cr$(G) = k$. Consider a drawing D_0 of G witnessing pair-cr$(G) = k$. At most $2k$ edges, the *bad edges*, are involved in at least one crossing in D_0. The remaining edges, the *good edges*, form a planar subdrawing D_{pl} of D_0. Each of the bad edges is drawn in a single face of D_{pl}. Let us choose a drawing D of G that extends D_{pl} such that each bad edge is drawn within a single face of D_{pl}, and the number of crossings is minimized among all such drawings.

We now show that every two edges cross at most once in the drawing D. Suppose on the contrary that x_1, x_2 are common crossings of two edges e, f. We swap the portions of e and f between x_1 and x_2, thereby eliminating x_1, x_2 and introducing no new crossings (see Figure 1). If the swap creates selfintersections of e or f, we easily eliminate them without introducing any new crossings (see Figure 2). We get a contradiction with the minimum number of crossings in D.

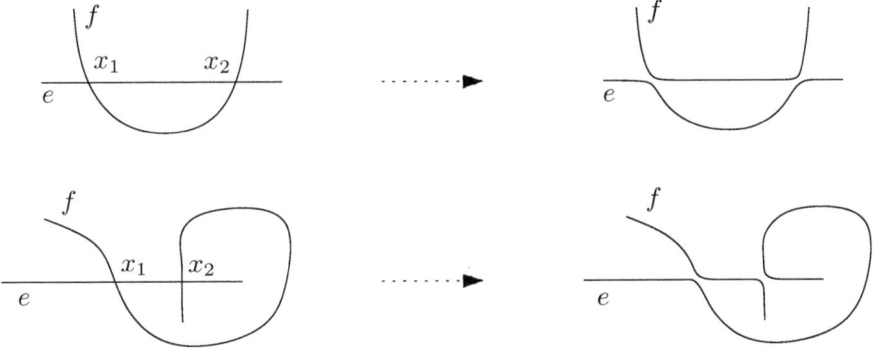

Figure 1. Swapping e, f between x_1, x_2 (two cases).

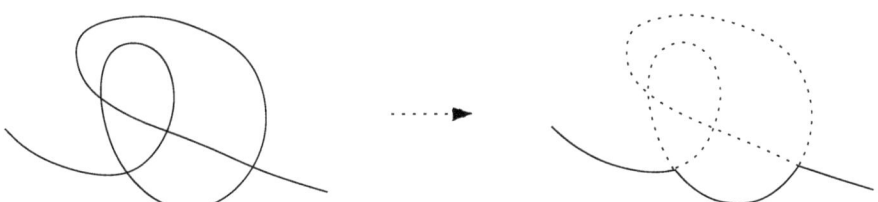

Figure 2. Eliminating selfintersections of an edge.

Thus, any two edges in D cross each other at most once.

It follows that there are at most $\binom{2k}{2} \leq 2k^2$ crossings in D.

The logarithmic improvement. Here we prove Theorem 1. Let G be a graph with pair-cr$(G) = k$. Let us consider a drawing D_0 of G witnessing pair-cr$(G) = k$. Let t be a suitable parameter to be fixed later (it will be of order $\log k$). Let us call an edge of G *good* if it crosses no edge in the drawing D_0, *light* if it crosses at least one and at most t edges in D_0, and *heavy* if it crosses more than t edges in D_0. Although we later redraw light and heavy edges several times, the notation "good", "light", or "heavy" is fixed for each edge of G by the above definition. Let l be the number of light edges and h the number of heavy edges.

Let D_1 be the subdrawing of D_0 formed by the good and light edges, and let D_{pl} be its planar subdrawing formed by the good edges only.

Consider a cell of D_{pl}. Suppose that some light edge in this cell crosses at least 2^t other light edges. Then we can decrease the number of crossings in D_1 without introducing any new crossing pair of edges, as can be seen from the following result of Schaefer and Štefankovič [2004] (implicitly contained in the proof of their Theorem 3.2): *Let D be a drawing of a graph G, and let e be an edge of G that crosses at most t other edges in D. Suppose that e has at least 2^t crossings in D. Then the edge e and the edges crossing it can be redrawn (within a small neighborhood of e) in such a way that the obtained drawing D' of G has fewer crossings than D and that there are no new crossing pairs of edges in D' (compared to D).*

Applying the result of Schaefer and Štefankovič finitely many times, we obtain a redrawing D_2 of D_1 with the same or smaller number of crossing pairs, such that each light edge is redrawn within the same face of D_{pl} and is involved in at most $2^t - 1$ crossings. Thus, there are at most $l \cdot (2^t - 1)/2$ crossings in D_2 (recall that l is the number of light edges).

Now, let D_3 be a redrawing of D_2 such that each light edge is redrawn within the same face of D_{pl} and that the number of crossings in D_3 is minimized. D_3 has at most as many crossings as D_2, i.e., at most $l \cdot (2^t - 1)/2$ crossings. Moreover, every two edges in D_3 cross each other at most once (otherwise we could argue analogously as in Figs. 1 and 2).

Finally, we add the heavy edges to the drawing D_3, in such a way that each heavy edge is drawn in the same face of D_{pl} as in D_0, the number of heavy-light[1] crossings is minimized, and subject to this, the number of heavy-heavy crossings is minimized. Let D_4 be the obtained drawing of G.

We claim that each heavy edge crosses any other edge at most once. To see this, first suppose that a heavy edge e crosses a light edge f at least twice, and let x_1 and x_2 be two crossings of e and f. Let z_e be the number of crossings of the portion of e between x_1 and x_2 with light edges, and similarly for z_f. If $z_f \leq z_e$, then e can be routed along f between x_1 and x_2, thereby decreasing

---

[1]A crossing is *heavy-light*, if it is a crossing of a heavy edge with a light edge. *Heavy-heavy* and *light-light* crossings are defined analogously.

the number of heavy-light crossings. See Figure 3. Possible selfintersections of e are eliminated as in Figure 2. If $z_f > z_e$, then the drawing D_3 did not have the

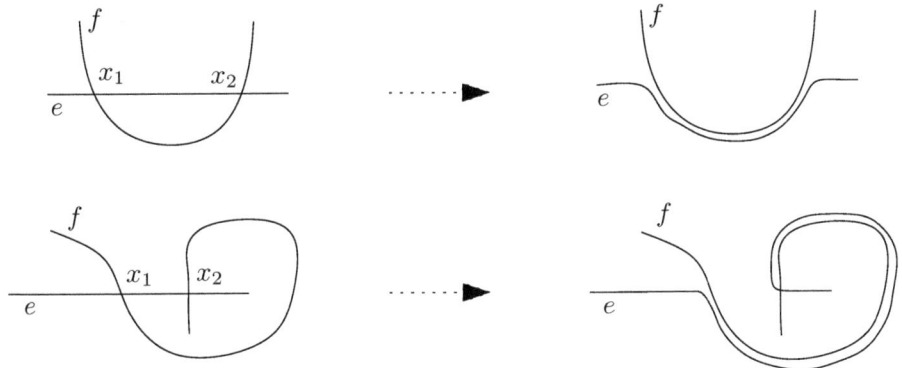

Figure 3. Rerouting e along f between x_1 and x_2 (two cases).

minimum number of crossings, as the number of crossings in it could be decreased by routing f along e. Again, possible selfintersections of f are eliminated as in Figure 2.

Similarly, suppose that two heavy edges e and f cross at least twice, and let x_1, x_2 be two of their common crossings. Then swapping the portions of e and f between x_1 and x_2 eliminates x_1 and x_2; see Figure 1. (As above, possible selfintersections of e or f are eliminated as in Figure 2.)

Thus, the heavy edges are involved in at most $\binom{h}{2} + h \cdot l \leq h(h+l) \leq h \cdot 2k$ crossings. The good edges are involved in no crossings and the number of light-light crossings is at most $l \cdot (2^t - 1)/2$. Thus, the total number of crossings in D_4 is at most $h \cdot 2k + l \cdot (2^t - 1)/2$. Using the obvious inequalities $l \leq 2k$ and $h \leq 2k/t$, this is at most $O(k^2/t + k2^t)$. Setting $t = \frac{1}{2}\log_2 k$, say, gives the claimed bound. The proof of Theorem 1 is complete.

3. Monotone Drawings

In this section we prove Theorem 2. Let G be a graph with pair-$\mathrm{cr}^{\mathrm{mon}}(G) = k$. Among all monotone drawings of G witnessing pair-$\mathrm{cr}^{\mathrm{mon}}(G) = k$, we choose a drawing D with the minimum number of crossings. We define *good, light,* and *heavy* edges in D in the same way as in the proof of Theorem 1 (now, the parameter t will be equal to $k^{1/3}$).

LEMMA 1. *Let e be a light edge in D. Then e intersects each edge at most $2t - 1$ times.*

PROOF. Consider an edge $f \in E(D), f \neq e$. Since D is monotone, each pair of consecutive common crossings of e, f determines a *lens* bounded by one of the edges e, f from above and by the other one from below. Let L be such a lens.

We claim that at least one edge intersecting e has an endpoint inside L. Suppose that this is not true. A *sling* in L is a continuous portion of an edge such that it is contained in L and its endpoints lie on e (see Figure 4). If there were some slings in L, we could reroute them along e (and outside L) in such a way that no new crossing pairs are introduced and the number of crossings is decreased (see Figure 4). Thus, there are no slings in L. It follows that rerouting

Figure 4. Three slings (bold) in a lens L determined by e, f and rerouting these slings along e.

f along e at the lens L (see Figure 5) decreases the number of crossings and introduces no new crossing pairs — a contradiction with the choice of D. Thus, there had to be an edge intersecting e and having an endpoint inside L.

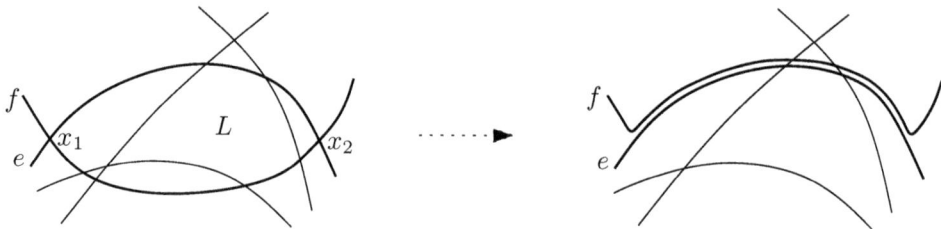

Figure 5. Rerouting f along e at the lens L.

Since at most t edges intersect e (e is light), it follows that there are at most $2(t-1)$ lenses determined by e, f. Thus, e, f cross each other at most $2t-1$ times. $\square$

There are k crossing pairs in D. By Lemma 1, each crossing pair involving at least one light edge has at most $2t-1$ common crossings. Thus, there are at most $k(2t-1)$ crossings involving at least one light edge.

We redraw the heavy edges so that there are no crossings with good edges, the number of heavy-light crossings is minimized, and subject to this, the number of heavy-heavy crossings is minimized.

The obtained drawing has at most $k(2t-1)$ crossings involving at least one light edge. Moreover, any two heavy edges cross at most once, for otherwise we could get a better drawing by swapping these two edges as in Figure 1 (top). Since there are at most $\lfloor 2k/t \rfloor$ heavy edges, the total number of crossings is at most $k(2t-1) + \binom{\lfloor 2k/t \rfloor}{2}$. Choosing $t = k^{1/3}$, this is at most $4k^{4/3}$. This finishes the proof of Theorem 2.

Acknowledgment

Daniel Král' and Ondřej Pangrác simplified the original proof of Theorem 1; Jiří Matoušek and Helena Nyklová wrote preliminary versions of parts of this paper. I thank them and other participants of a research seminar at the Charles University in Prague, where this paper originated.

References

[Kolman and Matoušek 2004] P. Kolman and J. Matoušek, "Crossing number, pair-crossing number, and expansion", *J. Combin. Theory Ser. B* **92**:1 (2004), 99–113.

[Kratochvíl and Matoušek 1994] J. Kratochvíl and J. Matoušek, "Intersection graphs of segments", *J. Combin. Theory Ser. B* **62**:2 (1994), 289–315.

[Pach and Tóth 2000] J. Pach and G. Tóth, "Which crossing number is it anyway?", *J. Combin. Theory Ser. B* **80**:2 (2000), 225–246.

[Schaefer and Štefankovič 2004] M. Schaefer and D. Štefankovič, "Decidability of string graphs", *J. Comput. System Sci.* **68**:2 (2004), 319–334.

PAVEL VALTR
DEPARTMENT OF APPLIED MATHEMATICS
and
INSTITUTE FOR THEORETICAL COMPUTER SCIENCE (ITI)
CHARLES UNIVERSITY
MALOSTRANSKÉ NÁM. 25, 118 00
PRAHA 1
CZECH REPUBLIC

Combinatorial and Computational Geometry
MSRI Publications
Volume **52**, 2005

Geometric Random Walks: A Survey

SANTOSH VEMPALA

ABSTRACT. The developing theory of geometric random walks is outlined
here. Three aspects — general methods for estimating convergence (the
"mixing" rate), isoperimetric inequalities in $\mathbb{R}^n$ and their intimate connection to random walks, and algorithms for fundamental problems (volume
computation and convex optimization) that are based on sampling by random walks — are discussed.

1. Introduction

A geometric random walk starts at some point in $\mathbb{R}^n$ and at each step, moves
to a "neighboring" point chosen according to some distribution that depends
only on the current point, e.g., a uniform random point within a fixed distance
δ. The sequence of points visited is a random walk. The distribution of the
current point, in particular, its convergence to a steady state (or stationary)
distribution, turns out to be a very interesting phenomenon. By choosing the
one-step distribution appropriately, one can ensure that the steady state distribution is, for example, the uniform distribution over a convex body, or indeed
any reasonable distribution in $\mathbb{R}^n$.

Geometric random walks are Markov chains, and the study of the existence
and uniqueness of and the convergence to a steady state distribution is a classical
field of mathematics. In the geometric setting, the dependence on the dimension
(called n in this survey) is of particular interest. Pólya proved that with probability 1, a random walk on an n-dimensional grid returns to its starting point
infinitely often for $n \leq 2$, but only a finite number of times for $n \geq 3$.

Random walks also provide a general approach to sampling a geometric distribution. To sample a given distribution, we set up a random walk whose steady
state is the desired distribution. A random (or nearly random) sample is obtained by taking sufficiently many steps of the walk. Basic problems such as
optimization and volume computation can be reduced to sampling. This con-

Supported by NSF award CCR-0307536 and a Sloan foundation fellowship.

nection, pioneered by the randomized polynomial-time algorithm of Dyer, Frieze and Kannan [1991] for computing the volume of a convex body, has lead to many new developments in recent decades.

In order for sampling by a random walk to be efficient, the distribution of the current point has to converge rapidly to its steady state. The first part of this survey (Section 3) deals with methods to analyze this convergence, and describes the most widely used method, namely, bounding the *conductance*, in detail. The next part of the survey is about applying this to geometric random walks and the issues that arise therein. Notably, there is an intimate connection with geometric isoperimetric inequalities. The classical isoperimetric inequality says that among all measurable sets of fixed volume, a ball of this volume is the one that minimizes the surface area. Here, one is considering all measurable sets. In contrast, we will encounter the following type of question: Given a convex set K, and a number t such that $0 < t < 1$, what subset S of volume $t \cdot \mathrm{vol}(K)$ has the smallest surface area inside K (i.e., not counting the boundary of S that is part of the boundary of K)? The inequalities that arise are interesting in their own right.

The last two sections describe polynomial-time algorithms for minimizing a quasi-convex function over a convex body and for computing the volume of a convex body. The random walk approach can be seen as an alternative to the ellipsoid method. The application to volume computation is rather remarkable in the light of results that no deterministic polynomial-time algorithm can approximate the volume to within an exponential (in n) factor. In Section 9, we briefly discuss the history of the problem and describe the latest algorithm.

Several topics related to this survey have been addressed in detail in the literature. For a general introduction to discrete random walks, the reader is referred to [Lovász 1996] or [Aldous and Fill $\geq$ 2005]. There is a survey by Kannan [1994] on applications of Markov chains in polynomial-time algorithms. For an in-depth account of the volume problem that includes all but the most recent improvements, there is a survey by Simonovits [2003] and an earlier article by Bollobás [1997].

Three walks. Before we introduce various concepts and tools, let us state precisely three different ways to walk randomly in a convex body K in $\mathbb{R}^n$. It might be useful to keep these examples in mind. Later, we will see generalizations of these walks.

The *Grid Walk* is restricted to a discrete subset of K, namely, all points in K whose coordinates are integer multiples of a fixed parameter δ. These points form a grid, and the neighbors of a grid point are the points reachable by changing any one coordinate by $\pm\delta$. Let $e_1, \ldots, e_n$ denote the coordinate vectors in $\mathbb{R}^n$; then the neighbors of a grid point x are $\{x \pm \delta e_i\}$. The grid walk tries to move to a random neighboring point.

Grid Walk (δ)

- Pick a grid point y uniformly at random from the neighbors of the current point x.
- If y is in K, go to y; else stay at x.

The *Ball Walk* tries to step to a random point within distance δ of the current point. Its state space is the entire set K.

Ball Walk (δ)

- Pick a uniform random point y from the ball of radius δ centered at the current point x.
- If y is in K, go to y; else stay at x.

Hit-and-run picks a random point along a random line through the current point. It does not need a "step-size" parameter. The state space is again all of K.

Hit-and-run

- Pick a uniform random line ℓ through the current point x.
- Go to a uniform random point on the chord $\ell \cap K$.

To implement the first step of hit-and-run, we can generate n independent random numbers, $u_1, \ldots, u_n$ each from the standard Normal distribution, and then the direction of the vector $u = (u_1, \ldots, u_n)$ is uniformly distributed. For the second step, using the membership oracle for K, we find an interval $[a, b]$ that contains the chord through x parallel to u so that $|a - b|$ is at most twice (say) the length of the chord (this can be done by a binary search with a logarithmic overhead). Then pick random points in $[a, b]$ till we find one in K.

For the first step of the ball walk, in addition to a random direction u, we generate a number r in the range $[0, \delta]$ with density $f(x)$ proportional to x^{n-1} and then $z = ru/|u|$ is uniformly distributed in a ball of radius δ.

Do these random walks converge to a steady state distribution? If so, what is it? How quickly do they converge? How does the rate of convergence depend on the convex body K?

These are some of the questions that we will address in analyzing the walks.

2. Basic Definitions

Markov chains. A *Markov chain* is defined using a σ-algebra $(K, \mathcal{A})$, where K is the state space and $\mathcal{A}$ is a set of subsets of K that is closed under complements and countable unions. For each element u of K, we have a probability measure P_u on $(K, \mathcal{A})$, i.e., each set $A \in \mathcal{A}$ has a probability $P_u(A)$. Informally, P_u is the distribution obtained on taking one step from u. The triple $(K, \mathcal{A}, \{P_u : u \in K\})$

along with a starting distribution Q_0 defines a Markov chain, i.e., a sequence of elements of K, $w_0, w_1, \ldots$, where w_0 is chosen from Q_0 and each subsequent w_i is chosen from $P_{w_{i-1}}$. Thus, the choice of w_{i+1} depends only on w_i and is independent of $w_0, \ldots, w_{i-1}$.

A distribution Q on $(K, \mathcal{A})$ is called *stationary* if one step from it gives the same distribution, i.e., for any $A \in \mathcal{A}$,

$$\int_A P_u(A) \, dQ(u) = Q(A).$$

A distribution Q is *atom-free* if there is no $x \in K$ with $Q(x) > 0$.

The *ergodic flow* of subset A with respect to the distribution Q is defined as

$$\Phi(A) = \int_A P_u(K \setminus A) \, dQ(u).$$

It is easy to verify that a distribution Q is stationary iff $\Phi(A) = \Phi(K \setminus A)$. The existence and uniqueness of the stationary distribution Q for general Markov chains is a rather technical issue that is not covered in this survey; see [Revuz 1975].[1] In all the chains we study in this survey, the stationary distribution will be given explicitly and can be easily verified. To avoid the issue of uniqueness of the stationary distribution, we only consider *lazy* Markov chains. In a lazy version of a given Markov chain, at each step, with probability $1/2$, we do nothing; with the rest we take a step according to the Markov chain. The next theorem is folklore and will also be implied by convergence theorems that we present later.

THEOREM 2.1. *If Q is stationary with respect to a lazy Markov chain then it is the unique stationary distribution for that Markov chain.*

For some additional properties of lazy Markov chains, see [Lovász and Simonovits 1993, Section 1]. We will hereforth assume that the distribution in the definition of Φ is the unique stationary distribution.

The *conductance* of a subset A is defined as

$$\phi(A) = \frac{\Phi(A)}{\min\{Q(A), Q(K \setminus A)\}}$$

and the conductance of the Markov chain is

$$\phi = \min_A \phi(A) = \min_{0 < Q(A) \leq \frac{1}{2}} \frac{\int_A P_u(K \setminus A) \, dQ(u)}{Q(A)}.$$

The *local* conductance of an element u is $\ell(u) = 1 - P_u(\{u\})$.

The following weaker notion of conductance will also be useful. For any $0 \leq s < \frac{1}{2}$, the *s-conductance* of a Markov chain is defined as

$$\phi_s = \min_{A : s < Q(A) \leq \frac{1}{2}} \frac{\Phi(A)}{Q(A) - s}.$$

---

[1] For Markov chains on discrete state spaces, the characterization is much simpler; see [Norris 1998], for example.

Comparing distributions. We will often have to compare two distributions P and Q (typically, the distribution of the current point and the stationary distribution). There are many reasonable ways to do this. Here are three that will come up.

(i) The *total variation distance* of P and Q is

$$\|P - Q\|_{tv} = \sup_{A \in \mathcal{A}} |P(A) - Q(A)|.$$

(ii) The L_2 *distance* of P with respect to Q is

$$\|P/Q\| = \int_K \frac{dP(u)}{dQ(u)} \, dP(u) = \int_K \left(\frac{dP(u)}{dQ(u)}\right)^2 \, dQ(u).$$

(iii) P is said to be M-*warm* with respect to Q if

$$M = \sup_{A \in \mathcal{A}} \frac{P(A)}{Q(A)}.$$

If Q_0 is $O(1)$-warm with respect to the stationary distribution Q for some Markov chain, we say that Q_0 is a *warm* start for Q.

Convexity. Convexity plays a key role in the convergence of geometric random walks. The following notation and concepts will be used.

The unit ball in $\mathbb{R}^n$ is B_n and its volume is $\mathrm{vol}(B_n)$. For two subsets A, B of $\mathbb{R}^n$, their Minkowski sum is

$$A + B = \{x + y : x \in A, \, y \in B\}.$$

The Brunn-Minkowski theorem says that if A, B and $A + B$ are measurable, then

$$\mathrm{vol}(A + B)^{1/n} \geq \mathrm{vol}(A)^{1/n} + \mathrm{vol}(B)^{1/n}. \tag{2-1}$$

Recall that a subset S of $\mathbb{R}^n$ is convex if for any two points $x, y \in S$, the interval $[x, y] \subseteq S$. A function $f : \mathbb{R}^n \to \mathbb{R}_+$ is said to be *logconcave* if for any two points $x, y \in \mathbb{R}^n$ and any $\lambda \in [0, 1]$,

$$f(\lambda x + (1 - \lambda)y) \geq f(x)^\lambda f(y)^{1-\lambda}.$$

The product and the minimum of two logconcave functions are both logconcave; the sum is not in general. The following fundamental properties, proved by Dinghas [1957], Leindler [1972] and Prékopa [1973; 1971], are often useful.

THEOREM 2.2. *All marginals as well as the distribution function of a logconcave function are logconcave. The convolution of two logconcave functions is logconcave.*

Logconcave functions have many properties that are reminiscent of convexity. For a logconcave density f, we denote the induced measure by π_f and its centroid by $z_f = \mathsf{E}_f(X)$. The second moment of f refers to $\mathsf{E}_f(|X - z_f|^2)$. The next three lemmas are chosen for illustration from [Lovász and Vempala 2003c]. The first one was proved earlier by Grünbaum [1960] for the special case of the uniform density over a convex body. We will later see a further refinement of this lemma that is useful for optimization.

LEMMA 2.3. *Let* $f : \mathbb{R}^n \to \mathbb{R}_+$ *be a logconcave density function, and let* H *be any halfspace containing its centroid. Then*

$$\int_H f(x)\,dx \geq \frac{1}{e}.$$

LEMMA 2.4. *If* X *is drawn from a logconcave distribution in* $\mathbb{R}^n$, *then for any integer* $k > 0$,

$$\mathsf{E}(|X|^k)^{1/k} \leq 2k\mathsf{E}(|X|).$$

Note that this can be viewed as a converse to Hölder's inequality which says that

$$\mathsf{E}(|X|^k)^{1/k} \geq \mathsf{E}(|X|).$$

LEMMA 2.5. *Let* $X \in \mathbb{R}^n$ *be a random point from a logconcave distribution with second moment* R^2. *Then* $\mathsf{P}(|X| > tR) < e^{-t+1}$.

A density function $f : \mathbb{R}^n \to \mathbb{R}_+$ is said to be *isotropic*, if its centroid is the origin, and its covariance matrix is the identity matrix. This latter condition can be expressed in terms of the coordinate functions as

$$\int_{\mathbb{R}^n} x_i x_j f(x)\,dx = \delta_{ij}$$

for all $1 \leq i, j \leq n$. This condition is equivalent to saying that for every vector $v \in \mathbb{R}^n$,

$$\int_{\mathbb{R}^n} (v^T x)^2 f(x)\,dx = |v|^2.$$

In terms of the associated random variable X, this means that

$$\mathsf{E}(X) = 0 \quad \text{and} \quad \mathsf{E}(XX^T) = I.$$

We say that f is *near-isotropic up to a factor of* C or C-*isotropic*, if

$$\frac{1}{C} \leq \int_{\mathbb{R}^n} (v^T x)^2\,d\pi_f(x) \leq C$$

for every unit vector v. The notions of "isotropic" and "near-isotropic" extend to nonnegative integrable functions f, in which case we mean that the density function $f / \int_{\mathbb{R}^n} f$ is isotropic. For any full-dimensional integrable function f with bounded second moment, there is an affine transformation of the space bringing it to isotropic position, and this transformation is unique up to an orthogonal transformation of the space. Indeed if f is not isotropic, we can make the centroid be the origin by a translation. Next, compute $A = \mathsf{E}(XX^T)$ for the associated random variable X. Now A must be positive semi-definite (since each XX^T is) and so we can write $A = BB^T$ for some matrix B. Then the transformation B^{-1} makes f isotropic.

It follows that for an isotropic distribution in $\mathbb{R}^n$, the second moment is

$$\mathsf{E}(|X|^2) = \sum_i \mathsf{E}(X_i^2) = n.$$

Further, Lemma 2.5 implies that for an isotropic logconcave distribution f,

$$\mathsf{P}(X > t\sqrt{n}) < e^{-t},$$

which means that most of f is contained in a ball of radius $O(\sqrt{n})$, and this is sometimes called its effective diameter.

Computational model. If the input to an algorithm is a convex body K in $\mathbb{R}^n$, we assume that it is given by a *membership oracle* which on input $x \in \mathbb{R}^n$ returns Yes if $x \in K$ and No otherwise. In addition we will have some bounds on K — typically, $B_n \subseteq K \subseteq RB_n$, i.e., K contains a unit ball around the origin and is contained in a ball of given radius. It is enough to have any point x in K and the guarantee that a ball of radius r around x is contained in K and one of radius R contains K (by translation and scaling this is equivalent to the previous condition). Sometimes, we will need a separation oracle for K, i.e., a procedure which either verifies that a given point x is in K or returns a hyperplane that separates x from K. The complexity of the algorithm will be measured mainly by the number of oracle queries, but we will also keep track of the number of arithmetic operations.

In the case of a logconcave density f, we have an oracle for f, i.e., for any point x it returns $Cf(x)$ where C is an unknown parameter independent of x. This is useful when we know a function proportional to the desired density, but not its integral, e.g., in the case of the uniform density over a bounded set, all we need is the indicator function of the support. In addition, we have a guarantee that the centroid of f satisfies $|z_f|^2 < Z$ and the eigenvalues of the covariance matrix of f are bounded from above and below by two given numbers. Again, the complexity is measured by the number of oracle calls. We will say that an algorithm is *efficient* if its complexity is polynomial in the relevant parameters.

To emphasize the essential dependence on the dimension we will sometimes use the $O^*(.)$ notation which suppresses logarithmic factors and also the dependence on error parameters. For example, $n \log n / \varepsilon = O^*(n)$.

Examples. For the ball walk in a convex body, the state space K is the convex body, and $\mathcal{A}$ is the set of all measurable subsets of K. Further,

$$P_u(\{u\}) = 1 - \frac{\text{vol}\,(K \cap (u + \delta B_n))}{\text{vol}(\delta B_n)}$$

and for any measurable subset A, such that $u \notin A$,

$$P_u(A) = \frac{\text{vol}\,(A \cap (u + \delta B_n))}{\text{vol}(\delta B_n)}.$$

If $u \in A$, then

$$P_u(A) = P_u(A \setminus \{u\}) + P_u(\{u\}).$$

It is straightforward to verify that the uniform distribution is stationary, i.e.,

$$Q(A) = \frac{\text{vol}(A)}{\text{vol}(K)}.$$

For hit-and-run, the one-step distribution for a step from $u \in K$ is given as follows. For any measurable subset A of K,

$$P_u(A) = \frac{2}{\text{vol}_{n-1}(\partial B_n)} \int_A \frac{dx}{\ell(u,x)|x - u|^{n-1}} \tag{2-2}$$

where $\ell(u, x)$ is the length of the chord in K through u and x. The uniform distribution is once again stationary. One way to see this is to note that the one-step distribution has a density function and the density of stepping from u to v is the same as that for stepping from v to u.

These walks can be modified to sample much more general distributions. Let $f : \mathbb{R}^n \to \mathbb{R}_+$ be a nonnegative integrable function. It defines a measure π_f (on measurable subsets of $\mathbb{R}^n$):

$$\pi_f(A) = \frac{\int_A f(x)\, dx}{\int_{\mathbb{R}^n} f(x)\, dx}.$$

The following extension of the ball walk, usually called the *ball walk with a Metropolis filter* has π_f as its stationary distribution (it is a simple exercise to prove, but quite nice that this works).

Ball walk with Metropolis filter (δ, f)

- Pick a uniformly distributed random point y in the ball of radius δ centered at the current point x.
- Move to y with probability $\min\left\{1, \frac{f(y)}{f(x)}\right\}$; stay at x with the remaining probability.

Hit-and-run can also be extended to sampling from such a general distribution π_f. For any line ℓ in $\mathbb{R}^n$, let $\pi_{\ell,f}$ be the *restriction* of π to ℓ, i.e.,

$$\pi_{\ell,f}(S) = \frac{\int_{p+tu \in S} f(p + tu)\, dt}{\int_\ell f(x)\, dx},$$

where p is any point on ℓ and u is a unit vector parallel to ℓ.

Hit-and-run (f)

- Pick a uniform random line ℓ through the current point x.
- Go to a random point y along ℓ chosen from the distribution $\pi_{\ell,f}$.

Once again, it is easy to verify that π_f is the stationary distribution for this walk. One way to carry out the second step is to use a binary search to find the point p on ℓ where the function is maximal, and the points a and b on both sides of p on ℓ where the value of the function is $\varepsilon f(p)$. We allow a relative error of ε, so the number of oracle calls is only $O(\log(1/\varepsilon))$. Then select a uniformly distributed random point y on the segment $[a, b]$, and independently a uniformly distributed random real number in the interval $[0, 1]$. Accept y if $f(y) > rf(p)$; else, reject y and repeat.

3. Convergence and Conductance

So far we have seen that random walks can be designed to approach any reasonable distribution in $\mathbb{R}^n$. For this to lead to an efficient sampling method, the convergence to the stationary distribution must be fast. This section is devoted to general methods for bounding the rate of convergence.

One way to define the *mixing rate* of a random walk is the number of steps required to reduce some measure of the distance of the current distribution to the stationary distribution by a factor of 2 (e.g., one of the distance measures from page 581). We will typically use the total variation distance. For a discrete random walk (i.e., the state space is a finite set), the mixing rate is character-ized by the *eigenvalues gap* of the transition matrix P whose ijth entry is the probability of stepping from i to j, conditioned on currently being at i. Let $\lambda_1 \geq \lambda_2 \ldots \geq \lambda_n$ be the eigenvalues of P. The top eigenvalue is 1 (by the defini-tion of stationarity) and let $\lambda = \max\{\lambda_2, |\lambda_n|\}$ (in the lazy version of any walk, all the λ_i are nonnegative and $\lambda = \lambda_2$). Then, for a random walk starting at the point x, with Q_t being the distribution after t steps, the following bound on the convergence can be derived (see [Lovász 1996], for example). For any point $y \in K$,

$$|Q_t(y) - Q(y)| \leq \sqrt{\frac{Q(y)}{Q(x)}} \lambda^t. \tag{3-1}$$

Estimating λ can be difficult or impractical even in the discrete setting (if, for example, the state space is too large to write down P explicitly).

Intuitively, a random walk will "mix" slowly if it has a bottleneck, i.e., a partition $S, K \setminus S$ of its state space, such that the probability of stepping from S to $K \setminus S$ (the ergodic flow out of S) is small compared to the measures of S and $K \setminus S$. Note that this ratio is precisely the conductance of S, $\phi(S)$. It takes about $1/\phi(S)$ steps in expectation to even go from one side to the other. As we will see in this section, the mixing rate is bounded from above by $2/\phi^2$. Thus, conductance captures the mixing rate upto a quadratic factor. This was first proved for discrete Markov chains by Jerrum and Sinclair [1989] who showed that conductance can be related to the eigenvalue gap of the transition matrix. A similar relationship for a related quantity called *expansion* was found by Alon

[1986] and by Dodziuk and Kendall [1986]. The inequality below is a discrete analogue of Cheeger's inequality in differential geometry.

THEOREM 3.1.
$$\frac{\phi^2}{2} \leq 1 - \lambda \leq 2\phi.$$

As a consequence of this and (3–1), we get that for a discrete random walk starting at x, and any point $y \in K$,

$$|Q_t(y) - Q(y)| \leq \sqrt{\frac{Q(y)}{Q(x)}} \left(1 - \frac{\phi^2}{2}\right)^t. \tag{3-2}$$

For the more general continuous setting, Lovász and Simonovits [1993] proved the connection between conductance and convergence. Their proof does not use eigenvalues. We will sketch it here since it is quite insightful, but does not seem to be well-known. It also applies to situations where the conductance can be bounded only for subsets of bounded size (i.e., the s-conductance, ϕ_s, can be bounded from below for some $s > 0$). We remind the reader that we have assumed that our Markov chains are lazy.

To show convergence, we need to prove that $|Q_t(A) - Q(A)|$ falls with t for every measurable subset A of K. However, this quantity might converge at different rates for different subsets. So we consider

$$\sup_{A:Q(A)=x} Q_t(A) - Q(A)$$

for each $x \in [0, 1]$. A bound for every x would imply what we want. To prove inductively that this quantity decreases with t, Lovász and Simonovits define the following formal upper bound. Let $\mathcal{G}_x$ be the set of functions defined as

$$\mathcal{G}_x = \left\{ g : K \to [0, 1] \ : \ \int_{u \in K} g(u) \, dQ(u) = x \right\}.$$

Using this, define

$$h_t(x) = \sup_{g \in \mathcal{G}_x} \int_{u \in K} g(u) \, (dQ_t(u) - dQ(u)) = \sup_{g \in \mathcal{G}_x} \int_{u \in K} g(u) \, dQ_t(u) - x,$$

It is clear that for A with $Q(A) = x$, $h_t(x) \geq Q_t(A) - Q(A)$ since the indicator function of A is in $\mathcal{G}_x$. The function $h_t(x)$ has the following properties.

LEMMA 3.2. *Let t be a positive integer.*

a. *The function h_t is concave.*
b. *If Q is atom-free, then $h_t(x) = \sup_{A:Q(A)=x} Q_t(A) - Q(A)$ and the supremum is achieved by some subset.*
c. *Let Q be atom-free and $t \geq 1$. For any $0 \leq x \leq 1$, let $y = \min\{x, 1 - x\}$. Then,*
$$h_t(x) \leq \tfrac{1}{2} h_{t-1}(x - 2\phi y) + \tfrac{1}{2} h_{t-1}(x + 2\phi y).$$

The first part of the lemma is easily verified. We sketch the second part: to maximize h_t, we should use a function g that puts high weight on points u with $dQ_t(u)/dQ(u)$ as high as possible. Let A be a subset with $Q(A) = x$, so that for any point y not in A, the value of $dQ_t(y)/dQ(y)$ is no more than the value for any point in A (i.e., A consists of the top x fraction of points according to $dQ_t(u)/dQ(u)$). Let g be the corresponding indicator function. These points give the maximum payoff per unit of weight, so it is optimal to put as much weight on them as possible. We mention in passing that the case when Q has atoms is a bit more complicated, namely we might need to include one atom fractionally (so that $Q(A) = x$). In the general case, $h_t(x)$ can be achieved by a function g that is $0 - 1$ valued everywhere except for at most one point.

The third part of the lemma, which is the key to convergence, is proved below.

PROOF OF LEMMA 3.2C. Assume that $0 \leq x \leq \frac{1}{2}$. The other range is proved in a similar way. We will construct two functions, g_1 and g_2, and use these to bound $h_t(x)$. Let A be a subset to be chosen later with $Q(A) = x$. Let

$$g_1(u) = \begin{cases} 2P_u(A) - 1 & \text{if } u \in A, \\ 0 & \text{if } u \notin A, \end{cases} \quad \text{and} \quad g_2(u) = \begin{cases} 1 & \text{if } u \in A, \\ 2P_u(A) & \text{if } u \notin A. \end{cases}$$

First, note that $\frac{1}{2}(g_1 + g_2)(u) = P_u(A)$ for all $u \in K$, which means that

$$\frac{1}{2}\int_{u \in K} g_1(u)\, dQ_{t-1}(u) + \frac{1}{2}\int_{u \in K} g_2(u)\, dQ_{t-1}(u) = \int_{u \in K} P_u(A)\, dQ_{t-1}(u)$$

$$= Q_t(A) \tag{3-3}$$

By the laziness of the walk ($P_u(A) \geq \frac{1}{2}$ iff $u \in A$), the range of the functions g_1 and g_2 is between 0 and 1 and if we let

$$x_1 = \int_{u \in K} g_1(u)\, dQ(u) \quad \text{and} \quad x_2 = \int_{u \in K} g_2(u)\, dQ(u),$$

then $g_1 \in \mathcal{G}_{x_1}$ and $g_2 \in \mathcal{G}_{x_2}$. Further,

$$\frac{1}{2}(x_1 + x_2) = \frac{1}{2}\int_{u \in K} g_1(u)\, dQ(u) + \frac{1}{2}\int_{u \in K} g_2(u)\, dQ(u)$$

$$= \int_{u \in K} P_u(A)\, dQ(u) = Q(A) = x$$

since Q is stationary. Next, since Q is atom-free, there is a subset $A \subseteq K$ that achieves $h_t(x)$. Using this and (3–3),

$$h_t(x) = Q_t(A) - Q(A)$$

$$= \frac{1}{2}\int_{u \in K} g_1(u)\, dQ_{t-1}(u) + \frac{1}{2}\int_{u \in K} g_2(u)\, dQ_{t-1}(u) - Q(A)$$

$$= \frac{1}{2}\int_{u \in K} g_1(u)\, (dQ_{t-1}(u) - dQ(u)) + \frac{1}{2}\int_{u \in K} g_2(u)\, (dQ_{t-1}(u) - dQ(u))$$

$$\leq \frac{1}{2}h_{t-1}(x_1) + \frac{1}{2}h_{t-1}(x_2).$$

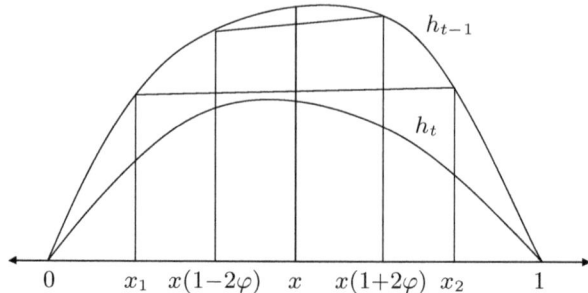

Figure 1. Bounding h_t.

We already know that $x_1 + x_2 = 2x$. In fact, x_1 and x_2 are separated from x.

$$
\begin{aligned}
x_1 &= \int_{u \in K} g_1(u) \, dQ(u) \\
&= 2 \int_{u \in A} P_u(A) \, dQ(u) - \int_{u \in A} dQ(u) \\
&= 2 \int_{u \in A} (1 - P_u(K \setminus A)) \, dQ(u) - x \\
&= x - 2 \int_{u \in A} P_u(K \setminus A) \, dQ(u) \\
&= x - 2\Phi(A) \\
&\leq x - 2\phi x \\
&= x(1 - 2\phi).
\end{aligned}
$$

(In the penultimate step, we used the fact that $x \leq \frac{1}{2}$.) Thus we have,

$$
x_1 \leq x(1 - 2\phi) \leq x \leq x(1 + 2\phi) \leq x_2.
$$

Since h_{t-1} is concave, the chord from x_1 to x_2 on h_{t-1} lies below the chord from $x(1 - 2\phi)$ to $x(1 + 2\phi)$ (see Figure 1). Therefore,

$$
h_t(x) \leq \tfrac{1}{2} h_{t-1}(x(1 - 2\phi)) + \tfrac{1}{2} h_{t-1}(x(1 + 2\phi))
$$

which is the conclusion of the lemma. ☐

In fact, a proof along the same lines implies the following generalization of part (c).

LEMMA 3.3. *Let Q be atom-free and $0 \leq s \leq 1$. For any $s \leq x \leq 1 - s$, let $y = \min\{x - s, 1 - x - s\}$. Then for any integer $t > 0$,*

$$
h_t(x) \leq \frac{1}{2} h_{t-1}(x - 2\phi_s y) + \frac{1}{2} h_{t-1}(x + 2\phi_s y).
$$

Given some information about Q_0, we can now bound the rate of convergence to the stationary distribution. We assume that Q is atom-free in the next theorem and its corollary. These results can be extended to the case when Q has atoms with slightly weaker bounds [Lovász and Simonovits 1993].

THEOREM 3.4. *Let $0 \leq s \leq 1$ and C_0 and C_1 be such that*

$$h_0(x) \leq C_0 + C_1 \min\{\sqrt{x-s}, \sqrt{1-x-s}\}.$$

Then

$$h_t(x) \leq C_0 + C_1 \min\{\sqrt{x-s}, \sqrt{1-x-s}\} \left(1 - \frac{\phi_s^2}{2}\right)^t.$$

PROOF. By induction on t. The inequality is true for $t = 0$ by the hypothesis. Now, suppose it holds for all values less than t. Assume $s = 0$ (for convenience) and without loss of generality that $x \leq 1/2$. From Lemma 3.3, we know that

$$h_t(x) \leq \frac{1}{2} h_{t-1}(x(1-2\phi)) + \frac{1}{2} h_{t-1}(x(1+2\phi))$$

$$\leq C_0 + \frac{1}{2} C_1 \left(\sqrt{x(1-2\phi)} + \sqrt{x(1+2\phi)}\right) \left(1 - \frac{\phi^2}{2}\right)^{t-1}$$

$$= C_0 + \frac{1}{2} C_1 \sqrt{x} \left(\sqrt{1-2\phi} + \sqrt{1+2\phi}\right) \left(1 - \frac{\phi^2}{2}\right)^{t-1}$$

$$\leq C_0 + C_1 \sqrt{x} \left(1 - \frac{\phi^2}{2}\right)^t.$$

Here we have used $\sqrt{1-2\phi} + \sqrt{1+2\phi} \leq 2(1 - \frac{\phi^2}{2})$. □

The following corollary, about convergence from various types of "good" starting distributions, gives concrete implications of the theorem.

COROLLARY 3.5. a. *Let $M = \sup_A Q_0(A)/Q(A)$. Then,*

$$\|Q_t - Q\|_{tv} \leq \sqrt{M} \left(1 - \frac{\phi^2}{2}\right)^t.$$

b. *Let $0 < s \leq \frac{1}{2}$ and $H_s = \sup\{|Q_0(A) - Q(A)| : Q(A) \leq s\}$. Then,*

$$\|Q_t - Q\|_{tv} \leq H_s + \frac{H_s}{s} \left(1 - \frac{\phi_s^2}{2}\right)^t.$$

c. *Let $M = \|Q_0/Q\|$. Then for any $\varepsilon > 0$,*

$$\|Q_t - Q\|_{tv} \leq \varepsilon + \sqrt{\frac{M}{\varepsilon}} \left(1 - \frac{\phi^2}{2}\right)^t.$$

PROOF. The first two parts are straightforward. For the third, observe that the L_2 norm,

$$\|Q_0/Q\| = \mathsf{E}_{Q_0}\left(\frac{dQ_0(x)}{dQ(x)}\right).$$

So, for $1 - \varepsilon$ of Q_0, $dQ_0(x)/dQ(x) \leq M/\varepsilon$. We can view the starting distribution as being generated as follows: with probability $1 - \varepsilon$ it is a distribution with

$\sup Q_0(A)/Q(A) \le M/\varepsilon$; with probability ε it is some other distribution. Now using part (a) implies part (c). $\qquad\square$

Conductance and s-conductance are not the only known ways to bound the rate of convergence. Lovász and Kannan [1999] have extended conductance to the notion of *blocking conductance* which is a certain type of average of the conductance over various subset sizes (see also [Kannan et al. 2004]). In some cases, it provides a sharper bound than conductance. Let $\phi(x)$ denote the minimum conductance over all subsets of measure x. Then one version of their main theorem is the following.

THEOREM 3.6. *Let π_0 be the measure of the starting point. Then, after*

$$ t > C \ln\left(\frac{1}{\varepsilon}\right) \int_{\pi_0}^{\frac{1}{2}} \frac{dx}{x\phi(x)^2} $$

steps, where C is an absolute constant, we have $\|Q_t - Q\|_{tv} \le \varepsilon$.

The theorem can be extended to continuous Markov chains. Another way to bound convergence which we do not describe here is via the *log-Sobolev* inequalities [Diaconis and Saloff-Coste 1996].

4. Isoperimetry

How to bound the conductance of a geometric random walk? To show that the conductance is large, we have to prove that for any subset $A \subset K$, the probability that a step goes out of A is large compared to $Q(A)$ and $Q(K \setminus A)$. To be concrete, consider the ball walk. For any particular subset S, the points that are likely to "cross over" to $K \setminus S$ are those that are "near" the boundary of S inside K. So, showing that $\phi(S)$ is large seems to be closely related to showing that there is a large volume of points near the boundary of S inside K. This section is devoted to inequalities which will have such implications and will play a crucial role in bounding the conductance.

To formulate an isoperimetric inequality for convex bodies, we consider a partition of a convex body K into three sets S_1, S_2, S_3 such that S_1 and S_2 are "far" from each other, and the inequality bounds the minimum possible volume of S_3 relative to the volumes of S_1 and S_2. We will consider different notions of distance between subsets. Perhaps the most basic is the Euclidean distance:

$$ d(S_1, S_2) = \min\{|u - v| : u \in S_1, v \in S_2\}. $$

Suppose $d(S_1, S_2)$ is large. Does this imply that the volume of $S_3 = K \setminus (S_1 \cup S_2)$ is large? The classic counterexample to such a theorem is a dumbbell — two large subsets separated by very little. Of course, this is not a convex set!

The next theorem, proved in [Dyer and Frieze 1991] (improving on a theorem in [Lovász and Simonovits 1992] by a factor of 2; see also [Lovász and Simonovits 1993]) asserts that the answer is yes.

THEOREM 4.1. *Let S_1, S_2, S_3 be a partition into measurable sets of a convex body K of diameter D. Then,*

$$\text{vol}(S_3) \geq \frac{2d(S_1, S_2)}{D} \min\{\text{vol}(S_1), \text{vol}(S_2)\}.$$

A limiting version of this inequality is the following: For any subset S of a convex body of diameter D,

$$\text{vol}_{n-1}(\partial S \cap K) \geq \frac{2}{D} \min\{\text{vol}(S), \text{vol}(K \setminus S)\}$$

which says that the surface area of S inside K is large compared to the volumes of S and $K \setminus S$. This is in direct analogy with the classical isoperimetric inequality, which says that the surface area to volume ratio of any measurable set is at least the ratio for a ball.

How does one prove such an inequality? In what generality does it hold? (i.e., for what measures besides the uniform measure on a convex set?) We will address these questions in this section. We first give an overview of known inequalities and then outline the proof technique.

Theorem 4.1 can be generalized to arbitrary logconcave measures. Its proof is very similar to that of 4.1 and we will presently give an outline.

THEOREM 4.2. *Let f be a logconcave function whose support has diameter D and let π_f be the induced measure. Then for any partition of $\mathbb{R}^n$ into measurable sets S_1, S_2, S_3,*

$$\pi_f(S_3) \geq \frac{2d(S_1, S_2)}{D} \min\{\pi_f(S_1), \pi_f(S_2)\}.$$

In terms of diameter, this inequality is the best possible, as shown by a cylinder. A more refined inequality is obtained in [Kannan et al. 1995; Lovász and Vempala 2003c] using the average distance of a point to the center of gravity (in place of diameter). It is possible for a convex body to have much larger diameter than average distance to its centroid (e.g., a cone). In such cases, the next theorem provides a better bound.

THEOREM 4.3. *Let f be a logconcave density in $\mathbb{R}^n$ and π_f be the corresponding measure. Let z_f be the centroid of f and define $M(f) = \mathsf{E}_f(|x - z_f|)$. Then, for any partition of $\mathbb{R}^n$ into measurable sets S_1, S_2, S_3,*

$$\pi_f(S_3) \geq \frac{\ln 2}{M(f)} d(S_1, S_2)\pi_f(S_1)\pi_f(S_2).$$

For an isotropic density, $M(f)^2 \leq \mathsf{E}_f(|x - z_f|^2) = n$ and so $M(f) \leq \sqrt{n}$. The diameter could be unbounded (e.g., an isotropic Gaussian).

A further refinement, based on isotropic position, has been conjectured in [Kannan et al. 1995]. Let λ be the largest eigenvalue of the inertia matrix of f, i.e.,

$$\lambda = \max_{v:|v|=1} \int_{\mathbb{R}^n} f(x)(v^T x)^2 \, dx. \qquad (4\text{--}1)$$

Then the conjecture says that there is an absolute constant c such that

$$\pi_f(S_3) \geq \frac{c}{\sqrt{\lambda}} d(S_1, S_2) \pi_f(S_1) \pi_f(S_2).$$

Euclidean distance and isoperimetric inequalities based on it are relevant for the analysis of "local" walks such as the ball walk. Hit-and-run, with its nonlocal moves, is connected with a different notion of distance.

The *cross-ratio distance* between two points u, v in a convex body K is computed as follows: Let p, q be the endpoints of the chord in K through u and v such that the points occur in the order p, u, v, q. Then

$$d_K(u, v) = \frac{|u - v||p - q|}{|p - u||v - q|} = (p : v : u : q).$$

where $(p : v : u : q)$ denotes the classical cross-ratio. We can now define the cross-ratio distance between two sets S_1, S_2 as

$$d_K(S_1, S_2) = \min\{d_K(u, v) : u \in S_1, v \in S_2\}.$$

The next theorem was proved in [Lovász 1999] for convex bodies and extended to logconcave densities in [Lovász and Vempala 2003d].

THEOREM 4.4. *Let f be a logconcave density in $\mathbb{R}^n$ whose support is a convex body K and let π_f be the induced measure. Then for any partition of $\mathbb{R}^n$ into measurable sets S_1, S_2, S_3,*

$$\pi_f(S_3) \geq d_K(S_1, S_2) \pi_f(S_1) \pi_f(S_2).$$

All the inequalities so far are based on defining the distance between S_1 and S_2 by the *minimum* over pairs of some notion of pairwise distance. It is reasonable to think that perhaps a much sharper inequality can be obtained by using some *average* distance between S_1 and S_2. Such an inequality was proved in [Lovász and Vempala 2004]. As we will see in Section 6, it leads to a radical improvement in the analysis of random walks.

THEOREM 4.5. *Let K be a convex body in $\mathbb{R}^n$. Let $f : K \to \mathbb{R}_+$ be a logconcave density with corresponding measure π_f and $h : K \to \mathbb{R}_+$, an arbitrary function. Let S_1, S_2, S_3 be any partition of K into measurable sets. Suppose that for any pair of points $u \in S_1$ and $v \in S_2$ and any point x on the chord of K through u and v,*

$$h(x) \leq \frac{1}{3} \min(1, d_K(u, v)).$$

Then

$$\pi_f(S_3) \geq \mathsf{E}_f(h(x)) \min\{\pi_f(S_1), \pi_f(S_2)\}.$$

The coefficient on the RHS has changed from a "minimum" to an "average". The weight $h(x)$ at a point x is restricted only by the minimum cross-ratio distance between pairs u, v from S_1, S_2 respectively, such that x lies on the line between them (previously it was the overall minimum). In general, it can be much higher than the minimum cross-ratio distance between S_1 and S_2.

The localization lemma. The proofs of these inequalities are based on an elegant idea: integral inequalities in $\mathbb{R}^n$ can be reduced to one-dimensional inequalities! Checking the latter can be tedious but is relatively easy. We illustrate the main idea by sketching the proof of Theorem 4.2.

For a proof of Theorem 4.2 by contradiction, let us assume the converse of its conclusion, i.e., for some partition S_1, S_2, S_3 of $\mathbb{R}^n$ and logconcave density f,

$$\int_{S_3} f(x) \, dx < C \int_{S_1} f(x) \, dx \quad \text{and} \quad \int_{S_3} f(x) \, dx < C \int_{S_2} f(x) \, dx$$

where $C = 2d(S_1, S_2)/D$. This can be reformulated as

$$\int_{\mathbb{R}^n} g(x) \, dx > 0 \quad \text{and} \quad \int_{\mathbb{R}^n} h(x) \, dx > 0 \tag{4-2}$$

where

$$g(x) = \begin{cases} Cf(x) & \text{if } x \in S_1, \\ 0 & \text{if } x \in S_2, \\ -f(x) & \text{if } x \in S_3. \end{cases} \quad \text{and} \quad h(x) = \begin{cases} 0 & \text{if } x \in S_1, \\ Cf(x) & \text{if } x \in S_2, \\ -f(x) & \text{if } x \in S_3. \end{cases}$$

These inequalities are for functions in $\mathbb{R}^n$. The main tool to deal with them is the *localization lemma* [Lovász and Simonovits 1993] (see also [Kannan et al. 1995] for extensions and applications).

LEMMA 4.6. *Let $g, h : \mathbb{R}^n \to \mathbb{R}$ be lower semi-continuous integrable functions such that*

$$\int_{\mathbb{R}^n} g(x) \, dx > 0 \quad \text{and} \quad \int_{\mathbb{R}^n} h(x) \, dx > 0.$$

Then there exist two points $a, b \in \mathbb{R}^n$ and a linear function $\ell : [0,1] \to \mathbb{R}_+$ such that

$$\int_0^1 \ell(t)^{n-1} g((1-t)a + tb) \, dt > 0 \quad \text{and} \quad \int_0^1 \ell(t)^{n-1} h((1-t)a + tb) \, dt > 0.$$

The points a, b represent an interval A and one may think of $l(t)^{n-1}dA$ as the cross-sectional area of an infinitesimal cone with base area dA. The lemma says that over this cone truncated at a and b, the integrals of g and h are positive. Also, without loss of generality, we can assume that a, b are in the union of the supports of g and h.

The main idea behind the lemma is the following. Let H be any halfspace such that

$$\int_H g(x) \, dx = \frac{1}{2} \int_{\mathbb{R}^n} g(x) \, dx.$$

Let us call this a bisecting halfspace. Now either

$$\int_H h(x)\,dx > 0 \quad \text{or} \quad \int_{R^n \setminus H} h(x)\,dx > 0.$$

Thus, either H or its complementary halfspace will have positive integrals for both g and h. Thus we have reduced the domains of the integrals from $\mathbb{R}^n$ to a halfspace. If we could repeat this, we might hope to reduce the dimensionality of the domain. But do there even exist bisecting halfspaces? In fact, they are aplenty: for any $(n-2)$-dimensional affine subspace, there is a bisecting halfspace with A contained in its bounding hyperplane. To see this, let H be halfspace containing A in its boundary. Rotating H about A we get a family of halfspaces with the same property. This family includes H', the complementary halfspace of H. Now the function $\int_H g - \int_{R^n \setminus H} g$ switches sign from H to H'. Since this is a continuous family, there must be a halfspace for which the function is zero, which is exactly what we want (this is sometimes called the "ham sandwich" theorem).

If we now take all $(n-2)$-dimensional affine subspaces given by some $x_i = r_1$, $x_j = r_2$ where r_1, r_2 are rational, then the intersection of all the corresponding bisecting halfspaces is a line (by choosing only rational values for x_i, we are considering a countable intersection). As long as we are left with a two or higher dimensional set, there is a point in its interior with at least two coordinates that are rational, say $x_1 = r_1$ and $x_2 = r_2$. But then there is a bisecting halfspace H that contains the affine subspace given by $x_1 = r_1, x_2 = r_2$ in its boundary, and so it properly partitions the current set. With some additional work, this leads to the existence of a concave function on an interval (in place of the linear function ℓ in the theorem) with positive integrals. Simplifying further from concave to linear takes quite a bit of work.

Going back to the proof sketch of Theorem 4.2, we can apply the lemma to get an interval $[a, b]$ and a linear function ℓ such that

$$\int_0^1 \ell(t)^{n-1} g((1-t)a + tb)\,dt > 0 \quad \text{and} \quad \int_0^1 \ell(t)^{n-1} h((1-t)a + tb)\,dt > 0.$$

$$(4\text{--}3)$$

(The functions g, h as we have defined them are not lower semi-continuous. However, this can be easily achieved by expanding S_1 and S_2 slightly so as to make them open sets, and making the support of f an open set. Since we are proving strict inequalities, we do not lose anything by these modifications).

Let us partition $[0, 1]$ into Z_1, Z_2, Z_3.

$$Z_i = \{t \in [0, 1] : (1-t)a + tb \in S_i\}.$$

Note that for any pair of points $u \in Z_1, v \in Z_2, |u - v| \geq d(S_1, S_2)/D$. We can rewrite (4–3) as

$$\int_{Z_3} \ell(t)^{n-1} f((1-t)a + tb) \, dt < C \int_{Z_1} \ell(t)^{n-1} f((1-t)a + tb) \, dt$$

and

$$\int_{Z_3} \ell(t)^{n-1} f((1-t)a + tb) \, dt < C \int_{Z_2} \ell(t)^{n-1} f((1-t)a + tb) \, dt.$$

The functions f and $\ell(.)^{n-1}$ are both logconcave, so $F(t) = \ell(t)^{n-1} f((1-t)a + tb)$ is also logconcave. We get,

$$\int_{Z_3} F(t) \, dt < C \min \left\{ \int_{Z_1} F(t) \, dt, \int_{Z_2} F(t) \, dt \right\}. \qquad (4\text{--}4)$$

Now consider what Theorem 4.2 asserts for the function $F(t)$ over the interval $[0, 1]$ and the partition Z_1, Z_2, Z_3:

$$\int_{Z_3} F(t) \, dt \geq 2d(Z_1, Z_2) \min \left\{ \int_{Z_1} F(t) \, dt, \int_{Z_2} F(t) \, dt \right\}. \qquad (4\text{--}5)$$

We have substituted 1 for the diameter of the interval $[0, 1]$. Also, $d(Z_1, Z_2) \geq d(S_1, S_2)/D = C/2$. Thus, Theorem 4.2 applied to the function $F(t)$ contradicts (4–4) and to prove the theorem in general, it suffices to prove it in the one-dimensional case.

In fact, it will be enough to prove (4–5) for the case when each Z_i is a single interval. Suppose we can do this. Then, for each maximal interval (c, d) contained in Z_3, the integral of F over Z_3 is at least C times the smaller of the integrals to its left $[0, c]$ and to its right $[d, 1]$ and so one of these intervals is "accounted" for. If all of Z_1 or all of Z_2 is accounted for, then we are done. If not, there is an unaccounted subset U that intersects both Z_1 and Z_2. But then, since Z_1 and Z_2 are separated by at least $d(S_1, S_2)/D$, there is an interval of Z_3 of length at least $d(S_1, S_2)/D$ between $U \cap Z_1$ and $U \cap Z_2$ which can account for more.

We are left with proving (4–5) when each Z_i is an interval. Without the factor of two, this is trivial by the logconcavity of F. To get C as claimed, one can reduce this to the case when $F(t) = e^{ct}$ and verify it for this function [Lovász and Simonovits 1993]. The main step is to show that there is a choice of c so that when the current $F(t)$ is replaced by e^{ct}, the LHS of (4–5) does not increase and the RHS does not decrease.

5. Mixing of the Ball Walk

With the isoperimetric inequalities at hand, we are now ready to prove bounds on the conductance and hence on the mixing time. In this section, we focus on the ball walk in a convex body K. Assume that K contains the unit ball.

A geometric random walk is said to be *rapidly mixing* if its conductance is bounded from below by an inverse polynomial in the dimension. By Corollary 3.5, this implies that the number of steps to halve the variation distance to stationary is a polynomial in the dimension. The conductance of the ball walk in a convex body K can be exponentially small. Consider, for example, starting at point x near the apex of a rotational cone in $\mathbb{R}^n$. Most points in a ball of radius δ around x will lie outside the cone (if x is sufficiently close to the apex) and so the local conductance is arbitrarily small. So, strictly speaking, the ball walk is not rapidly mixing.

There are two ways to get around this. For the purpose of sampling uniformly from K, one can expand K a little bit by considering $K' = K + \alpha B_n$, i.e., adding a ball of radius α around every point in K. Then for $\alpha > 2\delta\sqrt{n}$, it is not hard to see that $\ell(u)$ is at least $1/8$ for every point $u \in K'$. We can now consider the ball walk in K'. This fix comes at a price. First, we need a membership oracle for K'. This can be constructed as follows: given a point $x \in \mathbb{R}^n$, we find a point $y \in K$ such that $|x - y|$ is minimum. This is a convex program and can be solved using the ellipsoid algorithm [Grötschel et al. 1988] and the membership oracle for K, Second, we need to ensure that $\mathrm{vol}(K')$ is comparable to $\mathrm{vol}(K)$. Since K contains a unit ball, $K' \subseteq (1 + \alpha)K$ and so with $\alpha < 1/n$, we get that $\mathrm{vol}(K') < e\,\mathrm{vol}(K)$. Thus, we would need $\delta < 1/2n\sqrt{n}$.

Does large local conductance imply that the conductance is also large? We will prove that the answer is yes. The next lemma about one-step distributions of nearby points will be useful.

LEMMA 5.1. *Let u, v be such that $|u - v| \leq \frac{t\delta}{\sqrt{n}}$ and $\ell(u), \ell(v) \geq \ell$. Then,*

$$\|P_u - P_v\|_{tv} \leq 1 + t - \ell.$$

Roughly speaking, the lemma says that if two points with high local conductance are close in Euclidean distance, then their one-step distributions have a large overlap. Its proof follows from a computation of the overlap volume of the balls of radius δ around u and v.

We can now state and prove a bound on the conductance of the ball walk.

THEOREM 5.2. *Let K be a convex body of diameter D so that for every point u in K, the local conductance of the ball walk with δ steps is at least ℓ. Then,*

$$\phi \geq \frac{\ell^2 \delta}{16\sqrt{n}D}.$$

The structure of most proofs of conductance is similar and we will illustrate it by proving this theorem.

PROOF. Let $K = S_1 \cup S_2$ be a partition into measurable sets. We will prove that

$$\int_{S_1} P_x(S_2)\, dx \geq \frac{\ell^2 \delta}{16\sqrt{n}D} \min\{\mathrm{vol}(S_1), \mathrm{vol}(S_2)\} \tag{5-1}$$

Note that since the uniform distribution is stationary,

$$\int_{S_1} P_x(S_2)\, dx = \int_{S_2} P_x(S_1)\, dx.$$

Consider the points that are "deep" inside these sets, i.e. unlikely to jump out of the set (see Figure 2):

$$S_1' = \left\{ x \in S_1 : P_x(S_2) < \frac{\ell}{4} \right\}$$

and

$$S_2' = \left\{ x \in S_2 : P_x(S_1) < \frac{\ell}{4} \right\}.$$

Let S_3' be the rest i.e., $S_3' = K \setminus S_1' \setminus S_2'$.

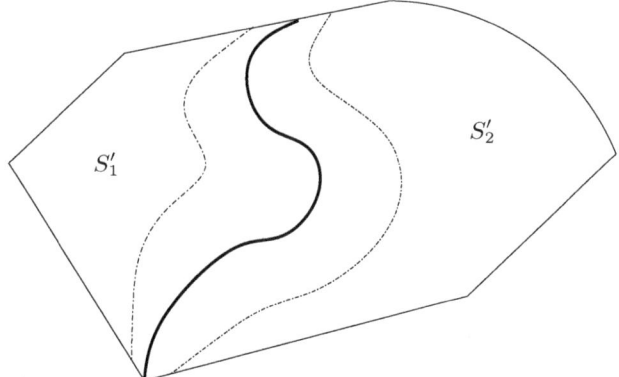

Figure 2. The conductance proof. The dark line is the boundary between S_1 and S_2.

Suppose $\mathrm{vol}(S_1') < \mathrm{vol}(S_1)/2$. Then

$$\int_{S_1} P_x(S_2)\, dx \geq \frac{\ell}{4}\mathrm{vol}(S_1 \setminus S_1') \geq \frac{\ell}{8}\mathrm{vol}(S_1)$$

which proves (5–1).

So we can assume that $\mathrm{vol}(S_1') \geq \mathrm{vol}(S_1)/2$ and similarly $\mathrm{vol}(S_2') \geq \mathrm{vol}(S_2)/2$. Now, for any $u \in S_1'$ and $v \in S_2'$,

$$\|P_u - P_v\|_{tv} \geq 1 - P_u(S_2) - P_v(S_1) > 1 - \frac{\ell}{2}.$$

Applying Lemma 5.1 with $t = \ell/2$, we get that

$$|u - v| \geq \frac{\ell\delta}{2\sqrt{n}}.$$

Thus $d(S_1, S_2) \geq \ell\delta/2\sqrt{n}$. Applying Theorem 4.1 to the partition S_1', S_2', S_3', we have

$$\operatorname{vol}(S_3') \geq \frac{\ell\delta}{\sqrt{n}D} \min\{\operatorname{vol}(S_1'), \operatorname{vol}(S_2')\}$$

$$\geq \frac{\ell\delta}{2\sqrt{n}D} \min\{\operatorname{vol}(S_1), \operatorname{vol}(S_2)\}$$

We can now prove (5–1) as follows:

$$\int_{S_1} P_x(S_2)\, dx = \frac{1}{2}\int_{S_1} P_x(S_2)\, dx + \frac{1}{2}\int_{S_2} P_x(S_1)\, dx \geq \tfrac{1}{2}\operatorname{vol}(S_3')\frac{\ell}{4}$$

$$\geq \frac{\ell^2\delta}{16\sqrt{n}D} \min\{\operatorname{vol}(S_1), \operatorname{vol}(S_2)\}. \qquad \square$$

As observed earlier, by going to $K' = K + (1/n)B_n$ and using $\delta = 1/2n\sqrt{n}$, we have $\ell \geq 1/8$. Thus, for the ball walk in K', $\phi = \Omega(1/n^2 D)$. Using Corollary 3.5, the mixing rate is $O(n^4 D^2)$.

We mentioned earlier that there are two ways to get around the fact that the ball walk can have very small local conductance. The second, which we describe next, is perhaps a bit cleaner and also achieves a better bound on the mixing rate. It is based on the observation that only a small fraction of points can have small local conductance. Define the points of high local conductance as

$$K_\delta = \left\{ u \in K : \ell(u) \geq \frac{3}{4} \right\}$$

LEMMA 5.3. *Suppose that K is a convex body containing a unit ball in $\mathbb{R}^n$.*

a. K_δ *is a convex set.*
b. $\operatorname{vol}(K_\delta) \geq (1 - 2\delta\sqrt{n})\operatorname{vol}(K)$.

The first part follows from the Brunn-Minkowski inequality (2–1). The second is proved by estimating the average local conductance [Kannan et al. 1997] and has the following implication: if we set $\delta \leq \varepsilon/2\sqrt{n}$, we get that at least $(1 - \varepsilon)$ fraction of points in K have large local conductance. Using this, we can prove the following theorem.

THEOREM 5.4. *For any $0 \leq s \leq 1$, we can choose the step-size δ for the ball walk in a convex body K of diameter D so that*

$$\phi_s \geq \frac{s}{200nD}.$$

PROOF. The proof is quite similar to that of Theorem 5.2. Let S_1, S_2 be a partition of K. First, since we are proving a bound on the s-conductance, we can assume that $\operatorname{vol}(S_1), \operatorname{vol}(S_2) \geq s\operatorname{vol}(K)$. Next, we choose $\delta = s/4\sqrt{n}$ so that by Lemma 5.3,

$$\operatorname{vol}(K_\delta) \geq (1 - \tfrac{1}{2}s)\operatorname{vol}(K).$$

So only an $s/2$ fraction of K has small local conductance and we will be able to ignore it. Define

$$S_1' = \left\{ x \in S_1 \cap K_\delta : P_x(S_2) < \tfrac{3}{16} \right\},$$
$$S_2' = \left\{ x \in S_2 \cap K_\delta : P_x(S_1) < \tfrac{3}{16} \right\}.$$

As in the proof of Theorem 5.2, these points are "deep" in S_1 and S_2 respectively and they are also restricted to be in K_δ. Recall that the local conductance of every point in K_δ is at least $3/4$. We can assume that $\text{vol}(S_1') \geq \text{vol}(S_1)/3$. Otherwise,

$$\int_{S_1} P_x(S_2)\, dx \geq \left(\tfrac{2}{3}\text{vol}(S_1) - \tfrac{1}{2}\text{svol}(K) \right) \tfrac{3}{16} \geq \tfrac{1}{32}\text{vol}(S_1).$$

which implies the theorem. Similarly, we can assume that $\text{vol}(S_2') \geq \text{vol}(S_2)/3$.
For any $u \in S_1'$ and $v \in S_2'$,

$$\| P_u - P_v \|_{tv} \geq 1 - P_u(S_2) - P_v(S_1) > 1 - \tfrac{3}{8}.$$

Applying Lemma 5.1 with $t = 3/8$, we get that

$$|u - v| \geq \frac{3\delta}{8\sqrt{n}}.$$

Thus $d(S_1, S_2) \geq 3\delta/8\sqrt{n}$. Let $S_3' = K_\delta \setminus (S_1' \cup S_2')$ Applying Theorem 4.1 to the partition S_1', S_2', S_3' of K_δ, we have

$$\text{vol}(S_3') \geq \frac{3\delta}{4\sqrt{n}D} \min\{\text{vol}(S_1'), \text{vol}(S_2')\}$$

$$\geq \frac{s}{16nD} \min\{\text{vol}(S_1), \text{vol}(S_2)\}$$

The theorem follows:

$$\int_{S_1} P_x(S_2)\, dx \geq \tfrac{1}{2}\text{vol}(S_3')\tfrac{3}{16} > \frac{s}{200nD} \min\{\text{vol}(S_1), \text{vol}(S_2)\}. \qquad \square$$

Using Corollary 3.5(b), this implies that from an M-warm start, the variation distance of Q_t and Q is smaller than ε after

$$t \geq C\frac{M^2}{\varepsilon^2}n^2 D^2 \ln\left(\frac{2M}{\varepsilon}\right) \tag{5-2}$$

steps, for some absolute constant C.

There is another way to use Lemma 5.3. In [Kannan et al. 1997], the following modification of the ball walk, called the *speedy* walk, is described. At a point x, the speedy walk picks a point uniformly from $K \cap x + \delta B_n$. Thus, the local conductance of every point is 1. However, there are two complications with this. First, the stationary distribution is not uniform, but proportional to $\ell(u)$. Second, each step seems unreasonable — we could make $\delta > D$ and then we would only need one step to get a random point in K. We can take care of the first problem with a rejection step at the end (and using Lemma 5.3). The root of the second problem is the question: how do we implement one step of the speedy walk? The only general way is to get random points from the ball around the current point till one of them is also in K. This process is the ball walk and it requires $1/\ell(u)$ attempts in expectation at a point u. However, if we count only the proper steps, i.e., ones that move away from the current point, then it is possible to show that the mixing rate of the walk is in fact $O(n^2D^2)$ from *any* starting point [Lovász and Kannan 1999]. Again, the proof is based on an isoperimetric inequality which is slightly sharper than Theorem 4.2. For this bound to be useful, we also need to bound the total number of improper or "wasted" steps. If we start at a random point, then this is the number of proper steps times $\mathsf{E}(1/\ell(u))$, which can be unbounded. But, if we allow a small overall probability of failure, then with the remaining probability, the expected number of wasted steps is bounded by $O(n^2D^2)$ as well.

The bound of $O(n^2D^2)$ on the mixing rate is the best possible in terms of the diameter, as shown by a cylinder. However, if the convex body is isotropic, then the isoperimetry conjecture (4–1) implies a mixing rate of $O(n^2)$.

For the rest of this section, we will discuss how these methods can be extended to sampling more general distributions. We saw already that the ball walk can be used along with a Metropolis filter to sample arbitrary density functions. When is this method efficient? In [Applegate and Kannan 1990] and [Frieze et al. 1994] respectively, it is shown that the ball walk and the lattice walk are rapidly mixing from a warm start, provided that the density is logconcave and it does not vary much locally, i.e., its Lipschitz constant is bounded. In [Lovász and Vempala 2003c], the assumptions on smoothness are eliminated, and it is shown that the ball walk is rapidly mixing from a warm start for any logconcave function in $\mathbb{R}^n$. Moreover, the mixing rate is $O(n^2D^2)$ (ignoring the dependence on the start), which matches the case of the uniform density on a convex body. Various properties of logconcave functions are developed in [Lovász and Vempala 2003c] with an eye to the proof. In particular, a smoother version of any given logconcave density is defined and used to prove an analogue of Lemma 5.3. For a logconcave density f in $\mathbb{R}^n$, the smoother version is defined as

$$\hat{f}(x) = \min_{C} \frac{1}{\mathrm{vol}(C)} \int_C f(x+u)\,du,$$

where C ranges over all convex subsets of the ball $x + rB_n$ with $\mathrm{vol}(C) = \mathrm{vol}(B_n)/16$. This function is logconcave and bounded from above by f everywhere (using Theorem 2.2). Moreover, for δ small enough, its integral is close to the integral of f. We get a lemma very similar to Lemma 5.3. The function $\hat{f}$ can be thought of as a generalization of K_δ.

LEMMA 5.5. *Let f be any logconcave density in $\mathbb{R}^n$. Then*

(i) *The function $\hat{f}$ is logconcave.*
(ii) *If f is isotropic, then $\int_{\mathbb{R}^n} \hat{f}(x)\,dx \geq 1 - 64\delta^{1/2}n^{1/4}$.*

Using this along with some technical tools, it can be shown that ϕ_s is large. Perhaps the main contribution of [Lovász and Vempala 2003c] is to move the smoothness conditions from requirements on the input (i.e., the algorithm) to tools for the proof.

In summary, analyzing the ball walk has led to many interesting developments: isoperimetric inequalities, more general methods of proving convergence (ϕ_s) and many tricks for sampling to get around the fact that it is not rapidly mixing from general starting points (or distributions). The analysis of the speedy walk shows that most points are good starting points. However, it is an open question as to whether the ball walk is rapidly mixing from a pre-determined starting point, e.g., the centroid.

6. Mixing of Hit-and-Run

Hit-and-run, introduced in [Smith 1984], offers the attractive possibility of long steps. There is some evidence that it is fast in practice [Berbee et al. 1987; Zabinsky et al. 1993].

Warm start. Lovász [1999] showed that hit-and-run mixes rapidly from a warm start in a convex body K. If we start with an M-warm distribution, then in

$$O\left(\frac{M^2}{\varepsilon^2}n^2D^2 \ln\left(\frac{M}{\varepsilon}\right)\right)$$

steps, the distance between the current distribution and the stationary is at most ε. This is essentially the same bound as for the ball walk, and so hit-and-run is no worse. The proof is based on cross-ratio isoperimetry (Theorem 4.5) for convex bodies and a new lemma about the overlap of one-step distributions. For $x \in K$, let y be a random step from x. Then the step-size $F(x)$ at x is defined as

$$\mathsf{P}\left(|x - y| \le F(x)\right) = \tfrac{1}{8}.$$

The lemma below asserts that if u, v are close in Euclidean distance and cross-ratio distance then their one-step distributions overlap substantially. This is analogous to Lemma 5.1 for the ball walk.

LEMMA 6.1. *Let $u, v \in K$. Suppose that*

$$d_K(u, v) < \frac{1}{8} \ and \ |u - v| < \frac{2}{\sqrt{n}}\max\{F(u), F(v)\}.$$

Then

$$\|P_u - P_v\|_{tv} < 1 - \frac{1}{500}.$$

Hit-and-run generalizes naturally to sampling arbitrary functions. The isoperimetry, the one-step lemma and the bound on ϕ_s were all extended to arbitrary logconcave densities in [Lovász and Vempala 2003d]. Thus, hit-and-run is rapidly mixing for any logconcave density from a warm start. While the analysis is along the lines of that in [Lovász 1999] and uses the tools developed in [Lovász and Vempala 2003c], it has to overcome substantial additional difficulties.

So hit-and-run is at least as fast as the ball walk. But is it faster? Can it get stuck in corners (points of small local conductance) like the ball walk?

Any start. Let us revisit the bad example for the ball walk: starting near the apex of a rotational cone. If we start hit-and-run at any interior point, then it exhibits a small, but inverse polynomial, drift towards the base of the cone. Thus, although the initial steps are tiny, they rapidly get larger and the current point moves away from the apex. This example shows two things. First, the "step-size" of hit-and-run can be arbitrarily small (near the apex), but hit-and-run manages to escape from such regions. This phenomenon is in fact general as shown by the following theorem, proved recently in [Lovász and Vempala 2004].

THEOREM 6.2. *The conductance of hit-and-run in a convex body of diameter D is $\Omega(1/nD)$.*

Unlike the ball walk, we can bound the conductance of hit-and-run (for arbitrarily small subsets). From this we get a bound on mixing time.

THEOREM 6.3. *Let K be a convex body that contains a unit ball and has centroid z_K. Suppose that $\mathsf{E}_K(|x - z_K|^2) \le R^2$ and $\|Q_0/Q\| \le M$. Then after*

$$t \ge Cn^2 R^2 \ln^3 \frac{M}{\varepsilon},$$

steps, where C is an absolute constant, we have $\|Q_t - Q\|_{tv} \le \varepsilon$.

The theorem improves on the bound for the ball walk (5–2) by reducing the dependence on M and ε from polynomial (which is unavoidable for the ball walk) to logarithmic, while maintaining the (optimal) dependence on R and n. For a body in near-isotropic position, $R = O(\sqrt{n})$ and so the mixing time is $O^*(n^3)$. One also gets a polynomial bound starting from *any single* interior point. If x is at distance d from the boundary, then the distribution obtained after one step from x has $\|Q_1/Q\| \le (n/d)^n$ and so applying the above theorem, the mixing time is $O(n^4 \ln^3(n/d\varepsilon))$.

The main tool in the proof is a new isoperimetric inequality based on "average" distance (Theorem 4.5). The proof of conductance is on the same lines as shown for Theorem 5.2 in the previous section. It uses Lemma 6.1 for comparing one-step distributions.

Theorems 6.2 and 6.3 have been extended in [Lovász and Vempala 2004] to the case of sampling an exponential density function over a convex body, i.e., $f(x)$ is restricted to a convex body and is proportional to $e^{a^T x}$ for some fixed vector a. It remains open to determine if hit-and-run mixes rapidly from any starting point for arbitrary logconcave functions.

As in the ball walk analysis, it is not known (even in the convex body case) if starting at the centroid is as good as a warm start. Also, while the theorem is the best possible in terms of R, it is conceivable that for an isotropic body the mixing rate is $O(n^2)$.

7. Efficient Sampling

Let f be a density in $\mathbb{R}^n$ with corresponding measure π_f. Sampling f, i.e., generating independent random points distributed according to π_f is a basic algorithmic problem with many applications. We have seen in previous sections that if f is logconcave there are natural random walks in $\mathbb{R}^n$ that will converge to π_f. Does this yield an efficient sampling algorithm?

Rounding. Take the case when f is uniform over a convex body K. The convergence depends on the diameter D of K (or the second moment). So the resulting algorithm to get a random point would take $\text{poly}(n, D)$ steps. However, the input to the algorithm is only D and an oracle. So we would like an algorithm whose dependence on D is only logarithmic. How can this be done? The ellipsoid algorithm can be used to find a transformation that achieves $D = O(n^{1.5})$ in $\text{poly}(n, \log D)$ steps.

Isotropic position provides a better solution. For a convex body in isotropic position $D \leq n$. For an isotropic logconcave distribution, $(1 - \varepsilon)$ of its measure lies in a ball of radius $\sqrt{n} \ln(1/\varepsilon)$. But how to make f isotropic? One way is by sampling. We get m random points from f and compute an affine tranformation that makes this set of points isotropic. We then apply this transformation to f. It is shown in [Rudelson 1999], that the resulting distribution is near-isotropic with $m = O(n \log^2 n)$ points for convex bodies and $m = O(n \log^3 n)$ for logconcave densities [Lovász and Vempala 2003c] with large probability.

Although this sounds cyclic (we need samples to make the sampling efficient) one can bootstrap and make larger and larger subsets of f isotropic. For a convex body K such an algorithm was given in [Kannan et al. 1997]. Its complexity is $O^*(n^5)$. This has been improved to $O^*(n^4)$ in [Lovász and Vempala 2003b]. The basic approach in [Kannan et al. 1997] is to define a series of bodies, $K_i = K \cap 2^{i/n} B_n$. Then $K_0 = B_n$ is isotropic upto a radial scaling. Given that K_i is 2-isotropic, K_{i+1} will be 6-isotropic and so we can sample efficiently from it. We use these samples to compute a transformation that makes K_{i+1} 2-isotropic and continue. The number of samples required in each phase is $O^*(n)$ and the total number of phases is $O(n \log D)$. Since each sample is drawn from a near-isotropic convex body, the sample complexity is $O^*(n^3)$ on average ($O^*(n^4)$ for the first point and $O^*(n^3)$ for subsequent points since we have a warm start). This gives an overall complexity of $O^*(n^5)$. The improvement to $O^*(n^4)$ uses ideas from the latest volume algorithm [Lovász and Vempala 2003b], including sampling from an exponential density and the pencil construction (see Section 9).

A similar method also works for making a logconcave density f isotropic [Lovász and Vempala 2003c]. We consider a series of level sets

$$L_i = \left\{ x \in \mathbb{R}^n : f(x) \geq \frac{M_f}{2^{(1+1/n)^i}} \right\}$$

where M_f is the maximum value of f. In phase i, we make the restriction of f to L_i isotropic. The complexity of this algorithm is $O^*(n^5)$. It is an open question to reduce this to $O^*(n^4)$.

Independence. The second important issue to be addressed is that of independence. If we examine the current point every m steps for some m, then are these points independent? Unfortunately, they might not be independent even if m is as large as the mixing time. Another problem is that the distribution might not be exactly π_f. The latter problem is easier to deal with. Suppose that x is from some distribution π so that $\|\pi - \pi_f\|_{tv} \leq \varepsilon$. Typically this affects the algorithm using the samples by some small function of ε. There is a general way to handle this (sometimes called *divine intervention*). We can pretend that x is drawn from π_f with probability $1 - \varepsilon$ and from some other distribution with probability at most ε. If we draw k samples, then the probability of success (i.e., each sample is drawn from the desired distribution) is at least $1 - k\varepsilon$.

Although points spaced apart by m steps might not be independent, they are "nearly" independent in the following sense. Two random variables X, Y will be called μ-*independent* $(0 < \mu < 1)$ if for any two sets A, B in their ranges,

$$\left| P(X \in A,\ Y \in B) - P(X \in A)P(Y \in B) \right| \leq \mu.$$

The next lemma summarizes some properties of μ-independence.

LEMMA 7.1. (i) *Let X and Y be μ-independent, and f, g be two measurable functions. Then $f(X)$ and $g(Y)$ are also μ-independent.*

(ii) *Let X, Y be μ-independent random variables such that $0 \leq X \leq a$ and $0 \leq Y \leq b$. Then*

$$\left| E(XY) - E(X)E(Y) \right| \leq \mu ab.$$

(iii) *Let $X_0, X_1, \ldots$, be a Markov chain, and assume that for some $i > 0$, X_{i+1} is μ-independent from X_i. Then X_{i+1} is μ-independent from $(X_0, \ldots, X_i)$.*

The guarantee that π is close to π_f will imply the following.

LEMMA 7.2. *Let Q be the stationary distribution of a Markov chain and t be large enough so that for any starting distribution Q_0 with $\|Q_0/Q\| \leq 4M$ we have $\|Q_t - Q\|_{tv} \leq \varepsilon$. Let X be a random point from a starting distribution Q_0 such that $\|Q_0/Q\| \leq M$. Then the point Y obtained by taking t steps of the chain starting at X is 2ε-independent from X.*

PROOF. Let $A, B \subseteq \mathbb{R}^n$; we claim that

$$\left| \mathsf{P}(X \in A, Y \in B) - \mathsf{P}(X \in A)\mathsf{P}(Y \in B) \right|$$
$$= \mathsf{P}(X \in A)\left| \mathsf{P}(Y \in B | X \in A) - \mathsf{P}(Y \in B) \right|$$
$$\leq 2\varepsilon.$$

Since

$$\left| \mathsf{P}(X \in A, Y \in B) - \mathsf{P}(X \in A)\mathsf{P}(Y \in B) \right|$$
$$= \left| \mathsf{P}(X \in \bar{A}, Y \in B) - \mathsf{P}(X \in \bar{A})\mathsf{P}(Y \in B) \right|$$

we may assume that $Q_0(A) \geq 1/2$. Let Q_0' be the restriction of Q_0 to A, scaled to a probability measure. Then $Q_0' \leq 2Q_0$ and so $\|Q_0'/Q\| \leq 4\|Q_0/Q\| \leq 4M$. Imagine running the Markov chain with starting distribution Q_0'. Then, by the assumption on t,

$$\left| \mathsf{P}(Y \in B | X \in A) - \mathsf{P}(Y \in B) \right| = \|Q_t'(B) - Q_t(B)\|_{tv}$$
$$\leq \|Q_t'(B) - Q(B)\|_{tv} + \|Q_t(B) - Q(B)\|_{tv}$$
$$\leq 2\varepsilon,$$

and so the claim holds.   □

Various versions of this lemma, adapted to the mixing guarantee at hand, have been used in the literature. See [Lovász and Simonovits 1993; Kannan et al. 1997; Lovász and Vempala 2003b] for developments along this line.

We conclude this section with an effective theorem from [Lovász and Vempala 2003a] for sampling from an arbitrary logconcave density.

THEOREM 7.3. *If f is a near-isotropic logconcave density function, then it can be approximately sampled in time $O^*(n^4)$ and in amortized time $O^*(n^3)$ if n or more nearly independent sample points are needed; any logconcave function can be brought into near-isotropic position in time $O^*(n^5)$.*

8. Application I: Convex Optimization

Let $S \subset \mathbb{R}^n$, and $f : S \to \mathbb{R}$ be a real-valued function. Optimization is the following basic problem: $\min f(x)$ s.t. $x \in S$, that is, find a point $x \in S$ which minimizes by f. We denote by x^* a solution for the problem. When the set S and the function f are convex[2], we obtain a class of problems which are solvable in $\text{poly}(n, \log(1/\varepsilon))$ time where ε defines an optimality criterion. If x is the point found, then $|x - x^*| \leq \varepsilon$.

The problem of minimizing a convex function over a convex set in $\mathbb{R}^n$ is a common generalization of well-known geometric optimization problems such as linear

---

[2]In fact, it is enough for f to be quasi-convex.

programming as well as a variety of combinatorial optimization problems including matchings, flows and matroid intersection, all of which have polynomial-time
algorithms [Grötschel et al. 1988]. It is shown in [Grötschel et al. 1988] that the
ellipsoid method [Judin and Nemirovskiĭ 1976; Hačijan 1979] solves this problem
in polynomial time when K is given by a separation oracle. A different, more efficient algorithm is given in [Vaidya 1996]. Here, we discuss the recent algorithm
of [Bertsimas and Vempala 2004] which is based on random sampling.

Note that minimizing a quasi-convex function is easily reduced to the feasi-
bility problem: to minimize a quasi-convex function $f(x)$, we simply add the
constraint $f(x) \leq t$ and search (in a binary fashion) for the optimal t.

In the description below, we assume that the convex set K is contained in the
axis-aligned cube of width R centered at the origin; further if K is nonempty
then it contains a cube of width r. It is easy to show that any algorithm with
this input specification needs to make at least $n \log(R/r)$ oracle queries. The
parameter L is equal to $\log \frac{R}{r}$.

Algorithm.

Input: A separation oracle for a convex set K and L.

Output: A point in K or a guarantee that K is empty.

1. Let P be the axis-aligned cube of side length R and center $z = 0$.
2. Check if z is in K. If so, report z and stop. If not, set

$$H = \{x \mid a^T x \leq a^T z\}.$$

   where $a^T x \leq b$ is the halfspace containing K reported by the oracle.
3. Set $P = P \cap H$. Pick m random points $y^1, y^2, \ldots, y^m$ from P. Set z to
   be their average.
4. Repeat steps 2 and 3 at most $2nL$ times. Report K is empty.

The number of samples required in each iteration, m, is $O(n)$. Roughly speaking,
the algorithm is computing an approximate centroid in each iteration. The idea
of an algorithm based on computing the *exact* centroid was suggested in [Levin
1965]. Indeed, if we could compute the centroid in each iteration, then by Lemma
2.3, the volume of P falls by a constant factor $(1 - 1/e)$ in each iteration. But,
finding the centroid, is #P-hard, i.e., computationally intractable.

The idea behind the algorithm is that an approximate centroid can be computed using $O(n)$ random points and the volume of P is likely to drop by a
constant factor in each iteration with this choice of z. This is formalized in the
next lemma. Although we need it only for convex bodies, it holds for arbitrary
logconcave densities.

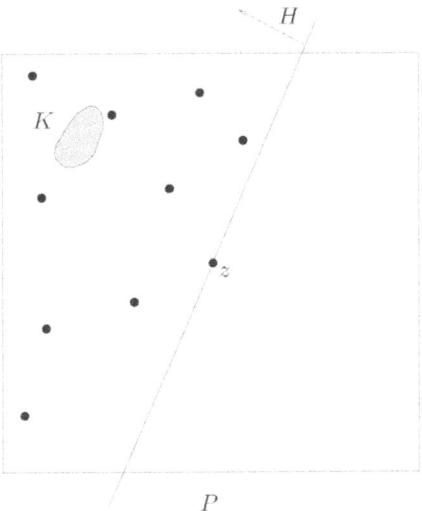

Figure 3. An illustration of the algorithm.

LEMMA 8.1. *Let g be a logconcave density in $\mathbb{R}^n$ and z be the average of m random points from π_g. If H is a halfspace containing z,*

$$\mathsf{E}\left(\pi_g(H)\right) \geq \left(\frac{1}{e} - \sqrt{\frac{n}{m}}\right).$$

PROOF. First observe that we can assume g is in isotropic position. This is because a linear transformation A affects the volume of a set S as $\mathrm{vol}(AS) = |\det(A)|\mathrm{vol}(S)$ and so the ratio of the volumes of two subsets is unchanged by the transformation. Applying this to all the level sets of g, we get that the ratio of the measures of two subsets is unchanged.

Since $z = \frac{1}{m}\sum_{i=1}^{m} y^i$,

$$\mathsf{E}\left(|z|^2\right) = \frac{1}{m^2}\sum_{i=1}^{m}\mathsf{E}\left(|y^i|^2\right) = \frac{1}{m}\mathsf{E}\left(|y^i|^2\right) = \frac{1}{m}\sum_{j=1}^{n}\mathsf{E}\left((y_j^i)^2\right) = \frac{n}{m},$$

where the first equality follows from the independence between y^i's, and equalities of the second line follow from the isotropic position. Let h be a unit vector normal to H. We can assume that $h = e_1 = (1, 0, \ldots, 0)$.

Next, let f be the marginal of g along h, i.e.,

$$f(y) = \int_{x:x_1=y} g(x)\,dx. \tag{8-1}$$

It is easy to see that f is isotropic. The next lemma (from [Lovász and Vempala 2003c]; see [Bertsimas and Vempala 2004] for the case of f arising from convex bodies) states that its maximum must be bounded.

LEMMA 8.2. *Let $f : \mathbb{R} \to R_+$ be an isotropic logconcave density function. Then,*

$$\max_y f(y) < 1.$$

Using Lemma 2.3,

$$\int_{z_1}^{\infty} f(y)\,dy = \int_0^{\infty} f(y)\,dy - \int_0^{z_1} f(y)\,dy$$

$$\geq \frac{1}{e} - |z_1| \max_y f(y)$$

$$\geq \frac{1}{e} - |z|.$$

The lemma follows from the bound on $\mathsf{E}(|z|)$.   □

The guarantee on the algorithm follows immediately. This optimal guarantee is also obtained in [Vaidya 1996]; the ellipsoid algorithm needs $O(n^2 L)$ oracle calls.

THEOREM 8.3. *With high probability, the algorithm works correctly using at most $2nL$ oracle calls (and iterations).*

The algorithm can also be modified for optimization given a membership oracle only and a point in K. It has a similar flavor: get random points from K; restrict K using the function value at the average of the random points; repeat. The oracle complexity turns out to be $O(n^5 L)$ which is an improvement on previous methods. This has been improved for linear objective functions using a variant of simulated annealing [Kalai and Vempala 2005].

9. Application II: Volume Computation

Finally, we come to perhaps the most important application and the principal motivation behind many developments in the theory of random walks: the problem of computing the volume of a convex body.

Let K be a convex body in $\mathbb{R}^n$ of diameter D such that $B_n \subset K$. The next theorem from [Bárány and Füredi 1987], improving on [Elekes 1986], essentially says that a deterministic algorithm cannot estimate the volume efficiently.

THEOREM 9.1. *For every deterministic algorithm that runs in time $O(n^a)$ and outputs two numbers A and B such that $A \leq \mathrm{vol}(K) \leq B$ for any convex body K, there is some convex body for which the ratio B/A is at least*

$$\left(\frac{cn}{a \log n} \right)^n$$

where c is an absolute constant.

So, in polynomial time, the best possible approximation is exponential in n and to get a factor 2 approximation (say), one needs exponential time. The basic idea of the proof is simple. Consider an oracle that answers "yes" for any point

Reference	Complexity	New ingredient(s)
[Dyer et al. 1991]	n^{23}	Everything
[Lovász and Simonovits 1990]	n^{16}	Localization lemma
[Applegate and Kannan 1990]	n^{10}	Logconcave sampling
[Lovász 1990]	n^{10}	Ball walk
[Dyer and Frieze 1991]	n^8	Better error analysis
[Lovász and Simonovits 1993]	n^7	Many improvements
[Kannan et al. 1997]	n^5	Isotropy, speedy walk
[Lovász and Vempala 2003b]	n^4	Annealing, hit-and-run

Table 1. Complexity comparison

in a unit ball and "no" to any point outside. After m "yes" answers, the convex body K could be anything between the ball and the convex hull of the m query points. The ratio of these volumes is exponential in n.

Given this lower bound, the following result of Dyer, Frieze and Kannan [Dyer et al. 1991] is quite remarkable.

THEOREM 9.2. *For any convex body K and any $0 \le \varepsilon, \delta \le 1$, there is a randomized algorithm which computes an estimate V such that with probability at least $1 - \delta$, we have $(1 - \varepsilon)\mathrm{vol}(K) \le V \le (1 + \varepsilon)\mathrm{vol}(K)$, and the number of oracle calls is $\mathrm{poly}(n, 1/\varepsilon, \log(1/\delta))$.*

Using randomness, we can go from an exponential approximation to an arbitrarily small one!

The main tool used in [Dyer et al. 1991] is sampling by a random walk. They actually used the grid walk and showed that by "fixing up" K a bit without changing its volume by much, the grid walk can sample nearly random points in polynomial time. Even though the walk is discrete, its analysis relies on a continuous isoperimetric inequality, quite similar to the one used for the analysis of the ball walk. The original algorithm of Dyer, Frieze and Kannan had complexity $O^*(n^{23})$. In the years since, there have been many interesting improvements. These are summarized in Table 9. In this section, we describe the latest algorithm from [Lovász and Vempala 2003b] whose complexity is $O^*(n^4)$.

Let us first review the common structure of previous volume algorithms. Assume that the diameter D of K is $\mathrm{poly}(n)$. All these algorithms reduce volume computation to sampling from a convex body, using the "Multi-Phase Monte-Carlo" technique. They construct a sequence of convex bodies $K_0 \subseteq K_1 \subseteq \cdots \subseteq K_m = K$, where $K_0 = B_n$ or some body whose volume is easily computed. They estimate the ratios $\mathrm{vol}(K_{i-1})/\mathrm{vol}(K_i)$ by generating sufficiently many independent (nearly) uniformly distributed random points in K_i and counting the fraction that lie in K_{i-1}. The product of these estimates is an estimate of $\mathrm{vol}(K_0)/\mathrm{vol}(K)$.

In order to get a sufficiently good estimate for the ratio $\mathrm{vol}(K_{i-1})/\mathrm{vol}(K_i)$, one needs about $m\mathrm{vol}(K_i)/\mathrm{vol}(K_{i-1})$ random points. So we would like to have the ratios $\mathrm{vol}(K_i)/\mathrm{vol}(K_{i-1})$ be small. But, the ratio of $\mathrm{vol}(K)$ and $\mathrm{vol}(K_0)$ could be $n^{\Omega(n)}$ and so m has to be $\Omega(n)$ just to keep the ratios $\mathrm{vol}(K_i)/\mathrm{vol}(K_{i-1})$ polynomial. The best choice is to keep these ratios bounded; this can be achieved, for instance, if $K_0 = B_n$ and $K_i = K \cap (2^{i/n} B_n)$ for $i = 1, 2, \ldots, m = \Theta(n \log n)$. Thus, the total number of random points used is $O(m^2) = O^*(n^2)$. Since $\mathrm{vol}(K_i) \leq 2\mathrm{vol}(K_{i-1})$ for this sequence, a random point in K_{i-1} provides a warm start for sampling from K_i. So each sample takes $O^*(n^3)$ steps to generate, giving an $O^*(n^5)$ algorithm. In [Applegate and Kannan 1990; Lovász and Simonovits 1993], sampling uniformly from K_i was replaced by sampling from a smooth logconcave function to avoid bad local conductance and related issues.

The number of phases, m, enters the running time as its square and one would like to make it as small as possible. But, due to the reasons described above, $m = \Theta(n \log n)$ is optimal for this type of algorithm and reducing m any further (i.e., $o(n)$) seems to be impossible for this type of method.

The algorithm in [Lovász and Vempala 2003b] can be viewed as a variation of *simulated annealing*. Introduced in [Kirkpatrick et al. 1983], simulated annealing is a general-purpose randomized search method for optimization. It walks randomly in the space of possible solutions, gradually adjusting a parameter called "temperature". At high temperature, the random walk converges to the uniform distribution over the whole space; as the temperature drops, the stationary distribution becomes more and more biased towards the optimal solutions.

Instead of a sequence of bodies, the algorithm in [Lovász and Vempala 2003b] constructs a sequence of functions $f_0 \leq f_1 \leq \ldots \leq f_m$ that "connect" a function f_0 whose integral is easy to find to the characteristic function f_m of K. The ratios $(\int f_{i-1})/(\int f_i)$ can be estimated by sampling from the distribution whose density function is proportional to f_i, and averaging the function f_{i-1}/f_i over the sample points. Previous algorithms can be viewed as the special case where the functions f_i are characteristic functions of the convex bodies K_i. By choosing a different set of f_i, the algorithm uses only $m = O^*(\sqrt{n})$ phases, and $O^*(\sqrt{n})$ sample points in each phase. In fact, it uses exponential functions of the form $f(x) = e^{-a^T x/T}$ restricted to some convex body. The temperature T will start out at a small value and increase gradually. This is the reverse of what happens in simulated annealing.

Besides annealing, the algorithm uses a pre-processing step called the *pencil* construction. We describe it next.

Let K be the given body in $\mathbb{R}^n$ and $\varepsilon > 0$. Let C denote the cone in $\mathbb{R}^{n+1}$ defined by

$$C = \left\{ x \in \mathbb{R}^{n+1} : x_0 \geq 0, \sum_{i=1}^{n} x_i^2 \leq x_0^2 \right\}$$

where $x = (x_0, x_1 \ldots, x_n)^\mathsf{T}$. We define a new convex body $K' \in \mathbb{R}^{n+1}$ as follows:

$$K' = \big([0, 2D] \times K\big) \cap C.$$

In words, K' is a sharpened $(n + 1)$-dimensional "pencil" whose cross-section is K and its tip is at the origin. See Figure 4. The idea of the algorithm is to start

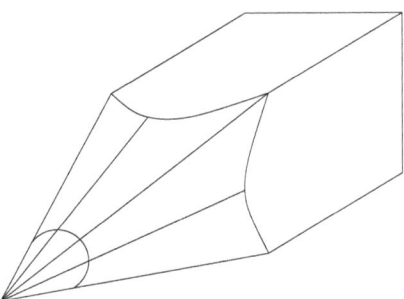

Figure 4. The pencil construction when K is a square.

with a function that is concentrated near the tip of the pencil, and is thus quite close to an exponential over a cone, and gradually move to a nearly constant function over the whole pencil, which would give us the volume of the pencil. The integral of the starting function is easily estimated. The sharpening takes less than half of the volume of the pencil away. Hence, if we know the volume of K', it is easy to estimate the volume of K by generating $O(1/\varepsilon^2)$ sample points from the uniform distribution on $[0, 2D] \times K$ and then counting how many of them fall into K'.

We describe the annealing part of the algorithm in a bit more detail. For each real number $a > 0$, let

$$Z(a) = \int_{K'} e^{-a x_0} \, dx$$

where x_0 is the first coordinate of x. For $a \leq \varepsilon^2/D$, an easy computation shows that

$$(1 - \varepsilon)\text{vol}(K') \leq Z(a) \leq \text{vol}(K).$$

On the other hand, for $a \geq 2n$ the value of $Z(a)$ is essentially the same as the integral over the whole cone which is easy to compute. So, if we select a sequence $a_0 > a_1 > \ldots > a_m$ for which $a_0 \geq 2n$ and $a_m \leq \varepsilon^2/D$, then we can estimate $\text{vol}(K')$ by

$$Z(a_m) = Z(a_0) \prod_{i=0}^{m-1} \frac{Z(a_{i+1})}{Z(a_i)}.$$

The algorithm estimates each ratio $R_i = Z(a_{i+1})/Z(a_i)$ as follows. Let μ_i be the probability distribution over K' with density proportional to $e^{-a_i x_0}$, i.e., for $x \in K'$,

$$\frac{d\mu_i(x)}{dx} = \frac{e^{-a_i x_0}}{Z(a_i)}.$$

To estimate the ratio R_i, the algorithm draws random samples $X^1, \ldots, X^k$ from μ_i, and computes

$$W_i = \frac{1}{m} \sum_{j=1}^{m} e^{(a_i - a_{i+1})(X^j)_0}.$$

It is easy to see that $\mathsf{E}(W_i) = R_i$. The main lemma in the analysis is that the second moment of W_i is small.

LEMMA 9.3. *Let X be a random sample from $d\mu_i$ and*

$$Y = e^{(a_i - a_{i+1})X_0}.$$

Then,

$$\frac{\mathsf{E}(Y^2)}{\mathsf{E}(Y)^2} \leq \frac{a_{i+1}^2}{a_i(2a_{i+1} - a_i)}.$$

From the lemma, it follows that with

$$a_{i+1} = a_i \left(1 - \frac{1}{\sqrt{n}}\right)$$

we get that $\mathsf{E}(Y^2)/\mathsf{E}(Y)^2$ is bounded by a constant. So with k samples,

$$\frac{\mathsf{E}(W_i^2)}{\mathsf{E}(W_i)^2} \leq \left(1 + \frac{O(1)}{k}\right).$$

Hence, the standard deviation of the estimate $Z = W_1 W_2 \ldots W_m$ is at most ε times $\mathsf{E}(W_1 W_2 \ldots W_m) = \mathrm{vol}(K)$, for $k = O(m/\varepsilon^2)$. Further, the number of phases needed to go from $a_0 = 2n$ to $a_m \leq \varepsilon^2/D$ is only $\sqrt{n} \log(D/\varepsilon^2)$. So the total number of sample points needed is only $O^*(n)$ (it would be interesting to show that this is a lower bound for any algorithm that uses a blackbox sampler).

As mentioned earlier, the samples obtained are not truly independent. This introduces technical complications. In previous algorithms, the random variable estimating the ratio was bounded (between 1 and 2) and so we could directly use Lemma 7.1. For the new algorithm, the individual ratios being estimated could be unbounded. To handle this further properties of μ-independence are developed in [Lovász and Vempala 2003b]. We do not go into the details here.

How fast can we sample from μ_i? Sampling from μ_0 is easy. But each μ_{i-1} no longer provides a warm start for μ_i, i.e., $d\mu_{i-1}(x)/d\mu_i(x)$ could be unbounded. However, as the next lemma asserts, the L_2 distance is bounded.

LEMMA 9.4. $\|\mu_{i-1}/\mu_i\| < 8$.

The proof of this lemma and that of Lemma 9.3 are both based on the following property of logconcave functions proved in [Lovász and Vempala 2003b].

LEMMA 9.5. *For $a > 0$, any convex body K and logconcave function f in $\mathbb{R}^n$, the function*

$$Z(a) = a^n \int_K f(ax) \, dx$$

is logconcave.

Finally, hit-and-run only needs bounded L_2 norm to sample efficiently, and by the version of Theorem 6.3 for the exponential density, we get each sample in $O^*(n^3)$ time. Along with the bound on the number of samples, this gives the complexity bound of $O * (n^4)$ for the volume algorithm.

It is apparent that any improvement in the mixing rate of random walks will directly affect the complexity of volume computation. Such improvements seem to consistently yield interesting new mathematics as well.

References

[Aldous and Fill ≥ 2005] D. Aldous and J. Fill, "Reversible Markov chains and random walks on graphs". Available at http://stat-www.berkeley.edu/users/aldous/RWG/book.html.

[Alon 1986] N. Alon, "Eigenvalues and expanders", *Combinatorica* **6**:2 (1986), 83–96.

[Applegate and Kannan 1990] D. Applegate and R. Kannan, "Sampling and integration of near log-concave functions", pp. 156–163 in *Proc. 23th ACM STOC*, 1990.

[Bárány and Füredi 1987] I. Bárány and Z. Füredi, "Computing the volume is difficult", *Discrete Comput. Geom.* **2**:4 (1987), 319–326.

[Berbee et al. 1987] H. C. P. Berbee, C. G. E. Boender, A. H. G. Rinnooy Kan, C. L. Scheffer, R. L. Smith, and J. Telgen, "Hit-and-run algorithms for the identification of nonredundant linear inequalities", *Math. Programming* **37**:2 (1987), 184–207.

[Bertsimas and Vempala 2004] D. Bertsimas and S. Vempala, "Solving convex programs by random walks", *J. Assoc. Comput. Mach.* **51**:4 (2004), 540–556.

[Bollobás 1997] B. Bollobás, "Volume estimates and rapid mixing", pp. 151–182 in *Flavors of geometry*, edited by S. Levy, Math. Sci. Res. Inst. Publ. **31**, Cambridge Univ. Press, Cambridge, 1997.

[Diaconis and Saloff-Coste 1996] P. Diaconis and L. Saloff-Coste, "Logarithmic Sobolev inequalities for finite Markov chains", *Ann. Appl. Probab.* **6**:3 (1996), 695–750.

[Dinghas 1957] A. Dinghas, "Über eine Klasse superadditiver Mengenfunktionale von Brunn-Minkowski-Lusternikschem Typus", *Math. Z.* **68** (1957), 111–125.

[Dodziuk and Kendall 1986] J. Dodziuk and W. S. Kendall, "Combinatorial Laplacians and isoperimetric inequality", pp. 68–74 in *From local times to global geometry, control and physics* (Coventry, 1984/85), edited by K. D. Ellworthy, Pitman Res. Notes Math. Ser. **150**, Longman Sci. Tech., Harlow, 1986.

[Dyer and Frieze 1991] M. Dyer and A. Frieze, "Computing the volume of convex bodies: a case where randomness provably helps", pp. 123–169 in *Probabilistic combinatorics and its applications* (San Francisco, CA, 1991), edited by B. Bollobás, Proc. Sympos. Appl. Math. **44**, Amer. Math. Soc., Providence, RI, 1991.

[Dyer et al. 1991] M. Dyer, A. Frieze, and R. Kannan, "A random polynomial-time algorithm for approximating the volume of convex bodies", *J. Assoc. Comput. Mach.* **38**:1 (1991), 1–17.

[Elekes 1986] G. Elekes, "A geometric inequality and the complexity of computing volume", *Discrete Comput. Geom.* **1**:4 (1986), 289–292.

[Frieze et al. 1994] A. Frieze, R. Kannan, and N. Polson, "Sampling from log-concave distributions", *Ann. Appl. Probab.* **4**:3 (1994), 812–837. Correction in **4** (1994), 1255.

[Grötschel et al. 1988] M. Grötschel, L. Lovász, and A. Schrijver, *Geometric algorithms and combinatorial optimization*, Algorithms and Combinatorics **2**, Springer, Berlin, 1988.

[Grünbaum 1960] B. Grünbaum, "Partitions of mass-distributions and of convex bodies by hyperplanes.", *Pacific J. Math.* **10** (1960), 1257–1261.

[Hačijan 1979] L. G. a. Hačijan, "A polynomial algorithm in linear programming", *Dokl. Akad. Nauk SSSR* **244**:5 (1979), 1093–1096.

[Jerrum and Sinclair 1989] M. Jerrum and A. Sinclair, "Approximating the permanent", *SIAM J. Comput.* **18**:6 (1989), 1149–1178.

[Judin and Nemirovskiĭ 1976] D. B. Judin and A. S. Nemirovskiĭ, "Estimation of the informational complexity of mathematical programming problems", *Èkonom. i Mat. Metody* **12**:1 (1976), 128–142. Translated in *Matekon* **13**:2 (1976), 3–45.

[Kalai and Vempala 2005] A. Kalai and S. Vempala, "Simulated annealing for convex optimization", Preprint, 2005. Available at http://www-math.mit.edu/~vempala/papers/anneal.ps.

[Kannan 1994] R. Kannan, "Markov chains and polynomial time algorithms", pp. 656–671 in *35th Annual Symposium on Foundations of Computer Science* (Santa Fe, NM, 1994), IEEE Comput. Soc. Press, Los Alamitos, CA, 1994.

[Kannan et al. 1995] R. Kannan, L. Lovász, and M. Simonovits, "Isoperimetric problems for convex bodies and a localization lemma", *Discrete Comput. Geom.* **13**:3-4 (1995), 541–559.

[Kannan et al. 1997] R. Kannan, L. Lovász, and M. Simonovits, "Random walks and an $O^*(n^5)$ volume algorithm for convex bodies", *Random Structures Algorithms* **11**:1 (1997), 1–50.

[Kannan et al. 2004] R. Kannan, L. Lovász, and R. Montenegro, "Blocking conductance and mixing in random walks", 2004.

[Kirkpatrick et al. 1983] S. Kirkpatrick, C. D. Gelatt, Jr., and M. P. Vecchi, "Optimization by simulated annealing", *Science* **220**:4598 (1983), 671–680.

[Leindler 1972] L. Leindler, "On a certain converse of Hölder's inequality", pp. 182–184 in *Linear operators and approximation* (Oberwolfach, 1971), Internat. Ser. Numer. Math. **20**, Birkhäuser, Basel, 1972.

[Levin 1965] A. J. Levin, "An algorithm for minimizing convex functions", *Dokl. Akad. Nauk SSSR* **160** (1965), 1244–1247.

[Lovász 1990] L. Lovász, "How to compute the volume?", pp. 138–151 in *Jber. d. Dt. Math.-Verein, Jubiläumstagung 1990*, B. G. Teubner, Stuttgart, 1990.

[Lovász 1996] L. Lovász, "Random walks on graphs: a survey", pp. 353–397 in *Combinatorics: Paul Erdős is eighty* (Keszthely, Hungary, 1993), vol. 2, edited by D. Miklós et al., Bolyai Soc. Math. Stud. **2**, János Bolyai Math. Soc., Budapest, 1996.

[Lovász 1999] L. Lovász, "Hit-and-run mixes fast", *Math. Program.* **86**:3, Ser. A (1999), 443–461.

[Lovász and Kannan 1999] L. Lovász and R. Kannan, "Faster mixing via average conductance", pp. 282–287 in *Annual ACM Symposium on Theory of Computing* (Atlanta, 1999), ACM, New York, 1999.

[Lovász and Simonovits 1990] L. Lovász and M. Simonovits, "The mixing rate of Markov chains, an isoperimetric inequality, and computing the volume", pp. 346–354 in *31st Annual Symposium on Foundations of Computer Science* (St. Louis, MO, 1990), IEEE Comput. Soc. Press, Los Alamitos, CA, 1990.

[Lovász and Simonovits 1992] L. Lovász and M. Simonovits, "On the randomized complexity of volume and diameter", pp. 482–491 in *33st Annual Symposium on Foundations of Computer Science* (Pittsburg, 1992), IEEE Comput. Soc. Press, Los Alamitos, CA, 1992.

[Lovász and Simonovits 1993] L. Lovász and M. Simonovits, "Random walks in a convex body and an improved volume algorithm", *Random Structures Algorithms* 4:4 (1993), 359–412.

[Lovász and Vempala 2003a] L. Lovász and S. Vempala, "Logconcave functions: Geometry and efficient sampling algorithms", pp. 640–649 in *44th Annual Symposium on Foundations of Computer Science* (Cambridge, MA, 2003), IEEE Comput. Soc. Press, Los Alamitos, CA, 2003. Available at http://www-math.mit.edu/~vempala/papers/logcon.ps.

[Lovász and Vempala 2003b] L. Lovász and S. Vempala, "Simulated annealing in convex bodies and an $O^*(n^4)$ volume algorithm", pp. 650–659 in *44th Annual Symposium on Foundations of Computer Science* (Cambridge, MA, 2003), IEEE Comput. Soc. Press, Los Alamitos, CA, 2003. Available at http://www-math.mit.edu/~vempala/papers/vol4.ps. Also to appear in *JCSS*.

[Lovász and Vempala 2003c] L. Lovász and S. Vempala, "The geometry of logconcave functions and an $O^*(n^3)$ sampling algorithm", Technical report MSR-TR-2003-04, 2003. Available at http://www-math.mit.edu/~vempala/papers/logcon-ball.ps.

[Lovász and Vempala 2003d] L. Lovász and S. Vempala, "Hit-and-run is fast and fun", Technical report MSR-TR-2003-05, 2003. Available at http://www-math.mit.edu/~vempala/papers/logcon-hitrun.ps.

[Lovász and Vempala 2004] L. Lovász and S. Vempala, "Hit-and-run from a corner", in *36th Annual Symposium on the Theory of Computing* (Chicago, 2004), ACM, New York, 2004. Available at http://www-math.mit.edu/~vempala/papers/start.ps.

[Norris 1998] J. R. Norris, *Markov chains*, Cambridge University Press, Cambridge, 1998.

[Prékopa 1971] A. Prékopa, "Logarithmic concave measures with application to stochastic programming", *Acta Sci. Math. (Szeged)* **32** (1971), 301–316.

[Prékopa 1973] A. Prékopa, "On logarithmic concave measures and functions", *Acta Sci. Math. (Szeged)* **34** (1973), 335–343.

[Revuz 1975] D. Revuz, *Markov chains*, North-Holland, Amsterdam, 1975.

[Rudelson 1999] M. Rudelson, "Random vectors in the isotropic position", *J. Funct. Anal.* **164**:1 (1999), 60–72.

[Simonovits 2003] M. Simonovits, "How to compute the volume in high dimension?", *Math. Program.* **97**:1-2, Ser. B (2003), 337–374.

[Smith 1984] R. L. Smith, "Efficient Monte Carlo procedures for generating points uniformly distributed over bounded regions", *Oper. Res.* **32**:6 (1984), 1296–1308.

[Vaidya 1996] P. M. Vaidya, "A new algorithm for minimizing convex functions over convex sets", *Math. Programming Ser. A* **73**:3 (1996), 291–341.

[Zabinsky et al. 1993] Z. B. Zabinsky, R. L. Smith, J. F. McDonald, H. E. Romeijn, and D. E. Kaufman, "Improving hit-and-run for global optimization", *J. Global Optim.* **3**:2 (1993), 171–192.

SANTOSH VEMPALA
DEPARTMENT OF MATHEMATICS
MASSACHUSETTS INSTITUTE OF TECHNOLOGY
CAMBRIDGE, MA 02139
vempala@mit.edu

For EU product safety concerns, contact us at Calle de José Abascal, 56–1°, 28003 Madrid, Spain or eugpsr@cambridge.org.

www.ingramcontent.com/pod-product-compliance
Ingram Content Group UK Ltd.
Pitfield, Milton Keynes, MK11 3LW, UK
UKHW020455240426
470322UK00016B/370